Developing Digital RF Memories and Transceiver Technologies for Electromagnetic Warfare

For a complete listing of titles in the
Artech House Electronic Warfare Series,
turn to the back of this book.

Developing Digital RF Memories and Transceiver Technologies for Electromagnetic Warfare

Phillip E. Pace

ARTECH HOUSE

BOSTON | LONDON
artechhouse.com

Library of Congress Cataloging-in-Publication Data
A catalog record for this book is available from the U.S. Library of Congress.

British Library Cataloguing in Publication Data
A catalogue record for this book is available from the British Library.

Cover design by Charlene Stevens

ISBN 13: 978-1-63081-697-1

For software accompanying this book go to:
https://us.artechhouse.com/Assets/downloads/pace_697.zip

© 2022 ARTECH HOUSE
685 Canton Street
Norwood, MA 02062

All rights reserved. Printed and bound in the United States of America. No part of this book may be reproduced or utilized in any form or by any means, electronic or mechanical, including photocopying, recording, or by any information storage and retrieval system, without permission in writing from the publisher.

All terms mentioned in this book that are known to be trademarks or service marks have been appropriately capitalized. Artech House cannot attest to the accuracy of this information. Use of a term in this book should not be regarded as affecting the validity of any trademark or service mark.

10 9 8 7 6 5 4 3 2 1

Contents

PREFACE   xvii

I Spectrum Sensing and Practical Considerations   1

Chapter 1 **Electromagnetic Spectrum Dominance**   3
    1.1   Introduction   3
    1.2   The Electromagnetic Spectrum Domain   4
    1.3   The Developing Role of Digital RF Memories   5
    1.4   Electromagnetic Warfare   7
    1.5   Airborne Assets and Unmanned Systems   9
    1.6   Electromagnetic Maneuver Warfare   10
        1.6.1   Cognitive Electromagnetic Warfare   11
        1.6.2   Open System Standards   13
    1.7   Commercial Technologies Driving A Difference   14
        1.7.1   DARPA's Mosaic and CONCERTO   15
    1.8   The Multifunction Digital RF Memory   17
        1.8.1   Fusing Enabling Technologies   18
        1.8.2   Network-Enabled   19
        1.8.3   Uses of a DRFM System   19
    1.9   Electronically Scanned Antennas   20
        1.9.1   Passive-Element Electronically Scanned Arrays   20
        1.9.2   Active-Element Electronically-Scanned Array   21
    1.10 Quantifying DRFM Spectrum Dominance   23
        1.10.1 DRFM Operational Tempo   27
        1.10.2 Electromagnetic Maneuver Agility   28
        1.10.3 A DRFM Measure of Effectiveness   29
    1.11 DRFM Signal Processing Techniques   30
        1.11.1 Artificial Intelligence and Machine Learning   30

	1.12	Summary	39
		References	41
Chapter 2		**Digital RF Memory Receiver Architectures**	47
	2.1	DRFM Architectures	47
		2.1.1 Addressing the Requirements	47
	2.2	Superheterodyne Kernels	48
	2.3	Channelized Kernels	50
	2.4	Phase Sampling Architectures	52
		2.4.1 Quantization Noise	57
	2.5	Use of DSP and FPGA Technology	61
	2.6	Global Positioning System Receivers	63
		2.6.1 An Important Unmanned Vehicle Companion	63
		2.6.2 INS/GPS Integration	63
		2.6.3 GPS Antennas	64
		2.6.4 GPS Receivers	67
		2.6.5 Codes Used by the GPS System	68
		2.6.6 Robust Receiver Design with AI and Machine Learning	71
	2.7	Summary	77
		References	77
Chapter 3		**Designing DRFM AESA Antennas**	81
	3.1	Introduction	81
	3.2	Beamforming Concepts	81
		3.2.1 Analog Beamforming	82
		3.2.2 Digital Beamforming	83
		3.2.3 Hybrid Beamforming	84
	3.3	Active-Element Electronically Scanned Arrays	88
		3.3.1 Example: Multifunction AESAs for Fire Control Radar	93
		3.3.2 Dual-Polarization Panels	97
		3.3.3 Planar-Fed Folded Notch	98
		3.3.4 Broadband Techniques	102
		3.3.5 Conformal Systems with Metamaterials	103
		3.3.6 Fractal Antennas	104
		3.3.7 AESA Performance Comparison	107
	3.4	Transmit / Receive Technologies	107
		3.4.1 Schematic of a T/R Module	109
		3.4.2 SiGe BiCMOS Contributions	110
		3.4.3 GaN Contributions	113
		3.4.4 Alternating Transmit/Receive	116
		3.4.5 Zero-IF Example	119
		3.4.6 Geometry Constraints	120
	3.5	Predicting Impact of Beasmforming on Rx Sensitivity	121

	3.6	AESA Noise Figure and Dynamic Range	123
		3.6.1 Noise Figure	123
	3.7	A Look to the Future	127
	3.8	Summary	132
		References	132
Chapter 4		**Choosing the Correct Wideband Receiver**	137
	4.1	AESA Antenna-to-Receiver Interface	137
		4.1.1 RF CMOS Technology	139
	4.2	Analog-to-Information Sampling	141
	4.3	Compressive Sensing Receivers	143
		4.3.1 Nyquist Folding Analog-to-Information Receiver	144
		4.3.2 The Modulated Wideband Converter	149
	4.4	Blind Compressive Sensing Without A Priori Basis Matrix Knowledge	153
	4.5	Bandpass Sampling Channelizer	154
	4.6	Polyphase Analysis Synthesis Channelizers	159
		4.6.1 Nonmaximally Decimated Analysis/Synthesis Filtering	162
		4.6.2 Another Form Using the Noble Identity	167
		4.6.3 Synthesis Filter Bank	169
		4.6.4 Split N-Point FFT	170
		4.6.5 Scalable Graphical Processing Unit Architecture	172
	4.7	Digital Receiver / Exciters	178
	4.8	Summary	182
		References	184
Chapter 5		**Transceiver Design and Practical Considerations**	189
	5.1	Mathematical Models of the Transceiver Process	189
		5.1.1 Quadrature Conversion Process	190
		5.1.2 Single Sideband Output Modulation	193
		5.1.3 Hilbert Transform	195
	5.2	Receive Process	196
		5.2.1 Sampling Operations	196
		5.2.2 Sample Time Uncertainty	199
		5.2.3 Digital Sampling Clock Accuracy	200
		5.2.4 Sampling Circuit Designs	203
	5.3	Basic Definitions	208
	5.4	Analog-to-Digital Conversion	209
		5.4.1 Transfer Functions to Determine Linearity	212
		5.4.2 Operational Amplifiers	216
		5.4.3 Comparator Circuits	218
	5.5	ADC Circuit Concepts	220
		5.5.1 Flash ADCs	220

	5.6	ADC Noise Floor with Windowing	228
		5.6.1 Quantization Error	234
		5.6.2 Signal-to-Noise Relationships	235
	5.7	FFT Spectrum Analysis	236
	5.8	Digital-to-Analog Conversion	239
	5.9	Transfer Functions to Determine Linearity	240
		5.9.1 Digital-to-Analog Conversion Models	245
	5.10	Technical Issues and Limitations	246
		5.10.1 DRFM Problems	248
		5.10.2 Spurious Signals and Their Dependence on Bit Resolution	252
		5.10.3 2-Bit DRFM Example	253
		5.10.4 Reducing Harmonic Spurs	256
		5.10.5 Mixers and the RF Up / Downconversion Process	258
	5.11	Phase Angle Sampling System: Quantization Noise	262
		5.11.1 Comparison Between Amplitude and Phase Sampling	263
	5.12	Digitization Figures of Merit	264
		5.12.1 DRFM ADC Requirements	266
		5.12.2 Direct Sampling Signal-to-Noise Considerations	267
	5.13	DRFM Transceiver Design Challenges	270
		5.13.1 Electronic Support	270
	5.14	Summary	272
		References	273
Chapter 6		**High-Performance Transceiver Technologies**	277
	6.1	Comparator Designs	277
		6.1.1 Two-Stage Open-Loop Design	278
		6.1.2 Conventional Double-Tail Dynamic Design	278
		6.1.3 New Double Tail with Clock Gating	279
		6.1.4 Low Power Ultrahigh-Speed Design	281
	6.2	Flash ADC Examples	283
		6.2.1 SiGe BiCMOS	287
	6.3	Time-Interleaving Flash Techniques	296
		6.3.1 Time Interleaved Flash Examples	298
	6.4	Time-Interleaving Pipeline Techniques	308
		6.4.1 Time-Interleaved Pipeline Examples	309
	6.5	TI-Successive Approximation Register Techniques	316
		6.5.1 Time-Interleaved SAR Examples	319
	6.6	Figure of Merit Plots for ADCs	323
	6.7	FinFET Transistor Technologies	328
	6.8	Embedded Dual-Port Memory	330
		6.8.1 FinFET Technology for SRAM	331

	6.8.2	MOSFET Technology	333
	6.8.3	NEMFET Technology	334
6.9	Practical Considerations for DACs		339
	6.9.1	Time-Interleaved ΣΔ DACs	339
6.10	Machine Learning Calibration Technology		347
6.11	Direct Digital Synthesis		351
	6.11.1	Early DDS Architectures	353
	6.11.2	ROM-Based DDS Architecture	355
	6.11.3	CORDIC DDS Architecture	356
	6.11.4	A Comparative Study	358
6.12	DRFM Oscillators and Phase Noise		360
	6.12.1	Oscillator Phase Noise Limitations	360
	6.12.2	Oscillator Choices	361
	6.12.3	Definition of Phase Noise	362
	6.12.4	Estimation of the Composite Power Law Parameters	365
6.13	Summary		366
	References		367

Chapter 7 Microwave-Photonic Transceiver Technologies — 375

7.1	Photonic Receiver Digital Antenna Components		376
7.2	Lasers		378
	7.2.1	Active Mode-Locking Mechanism	380
	7.2.2	Mode-Locked Fiber Lasers	381
	7.2.3	Sigma Mode-Locked Fiber Laser	384
7.3	Detectors		389
	7.3.1	Detector Physics and Detection Mechanism	390
	7.3.2	Detector Figures of Merit	392
7.4	Optical Link and Components		394
7.5	Photonic Local Oscillator		396
7.6	Electro-Optical Modulators		397
	7.6.1	Electro-Optical Sampling Error	402
	7.6.2	Maximum Amplitude Jitter	403
	7.6.3	Modulator Bias Controller	403
7.7	Signal Processing		406
7.8	Photonic RF Memory		407
7.9	Designing Microwave-Photonic Antennas		411
	7.9.1	Microwave-Photonic Antenna Modules	412
	7.9.2	Microwave-Photonic Antenna Arrays	416
	7.9.3	Microwave-Photonic Beamforming Phased Arrays	420
7.10	Photonic Analog-to-Digital Converters		425
	7.10.1	Jitter - the Limiting Factor	425
	7.10.2	The Quantum Limit	426

7.10.3	The Taylor Scheme: How It All Began	427
7.10.4	Limitations of the Taylor Scheme	430
7.10.5	Optical (Phase-Encoded) Sampling	430
7.11	High-Resolution Encoding Process for Photonic ADCs	432
7.11.1	Robust Symmetrical Number System	433
7.11.2	Lessons from a Photonic RSNS ADC Prototype	434
7.12	Configuring Photonic Compressive Sampling Systems	436
7.12.1	Multicoset Wideband Compressive Sensing Concepts	437
7.12.2	Creating a Wideband Compressive Nonuniform Sampler	438
7.13	Designing a Photonic Nyquist Folding Receiver	441
7.13.1	Understanding the Double-Modulator Design Concept	444
7.13.2	pNYFR T&E Simulation Results	449
7.13.3	Receiver Hardware Prototype Results	452
7.13.4	Single-Tone Test Results	455
7.14	Design of Photonics Compressive Sampling Systems	458
7.14.1	Single-Channel CS With Double-Parallel MZI	461
7.14.2	Multichannel Photonics-Assisted CS System	463
7.14.3	Extending the Bandwidth of the System	465
7.15	Wideband Spectrum Sensing and Analysis	467
7.15.1	Channelizer Using Fiber Bragg Gratings	469
7.15.2	Spectrum Scanning Approach	472
7.15.3	Direct Conversion	474
7.15.4	Cryogenic Spatial-Spectral Crystal Photonics	477
7.16	Summary	485
	References	485

II Modern ES and EA Techniques Using Deep Learning — 495

Chapter 8 Modern Spectral Sensing and Detection — 497

8.1	Persistent Spectrum Sensing	498
8.2	Centralized or Decentralized EMS Sensing	500
8.2.1	Sequential Spectrum Sensing Methods	503
8.2.2	Quickest Spectrum Detection Methods	504
8.2.3	Single DRFM: Spectrum Sensing and Access	505
8.2.4	Multi-DRFM: Spectrum Sensing and Access	506
8.2.5	Bandit Problems	507
8.3	Detection Methods: Time-Frequency	509
8.3.1	Cohen's Time-Frequency Distributions	511
8.3.2	Choi-Williams Distribution	511
8.3.3	The Wigner-Ville Distribution	514
8.3.4	Wigner-Ville Distribution Analysis	515
8.3.5	Continuous WVD	515

		8.3.6	Example Calculation: Real Input Signal	518
		8.3.7	Example Calculation: Complex Input Signal	521
		8.3.8	Two-Tone Input Signal Results	525
	8.4	Wavelet Decomposition Using Quadrature Mirrors		528
		8.4.1	Wavelets and the Wavelet Transform	528
		8.4.2	Wavelet Filters	530
		8.4.3	Sinc and Modified Sinc Filter	531
	8.5	QMFB Tree Receiver		533
		8.5.1	Example Calculations	534
	8.6	Cyclostationary Spectral Analysis		544
		8.6.1	Cyclic Autocorrelation	544
		8.6.2	Spectral Correlation Density	545
		8.6.3	Spectral Correlation Density Estimation	547
	8.7	Discrete Time Cyclostationary Algorithms		550
		8.7.1	The Time-Smoothing FFT Accumulation Method	550
		8.7.2	Direct Frequency-Smoothing Method	553
		8.7.3	Test Signals	554
	8.8	Atomic Decomposition		558
	8.9	Summary		559
		References		560
Chapter 9		**Machine Learning in Electromagnetic Warfare**		**563**
	9.1	Modern AI Concepts for Distributed Sensing		564
		9.1.1	Reasons to Use Neural Networks/AI in Electromagnetic Warfare	565
		9.1.2	Automation and the Human-Computer Interface	566
		9.1.3	Autonomous Modulation Classification	568
	9.2	Nonlinear Classification Networks		569
		9.2.1	Single Perceptron Networks	569
		9.2.2	Multilayer Perceptron Networks	573
		9.2.3	Modified Feature Extraction Signal Processing	577
		9.2.4	Low-Pass Filtering for Cropping Consistency	577
		9.2.5	Calculating the Marginal Frequency Distribution	581
		9.2.6	Wiener Filtering	583
	9.3	Principal Components Analysis		586
		9.3.1	Classification Using Modified Feature Extraction	589
		9.3.2	Extended Database	589
		9.3.3	Signals Used for TestSNR and TestMod	590
		9.3.4	Optimizing the Feature Extraction and Classification Network	591
	9.4	Artificial Intelligence		595
		9.4.1	Machine Learning	595

	9.4.2	Data Parallelism	596
	9.4.3	Model Parallelism	597
	9.4.4	Multi-Task Learning	599
9.5	Deep Learning Architectures and EMW Applications		601
9.6	Autoencoder		603
	9.6.1	Regularization Function	606
	9.6.2	Variational Auto Encoder	607
9.7	Convolutional Neural Network		608
	9.7.1	Unmanned System Example	609
	9.7.2	Modulation Recognition Example	613
	9.7.3	Antenna Scan Modulation Example	614
	9.7.4	Determining the Radar Pulse Repetition Interval Example	616
	9.7.5	Recognition of LPI Radar Waveforms Example	619
	9.7.6	Electromagnetic Warfare Example	622
9.8	Recurrent Neural Networks		623
	9.8.1	Loss Function	625
	9.8.2	Disadvantages of RNN	626
	9.8.3	Naval Ship Class Example	629
9.9	Long-, Short-Term Memory		630
	9.9.1	LSTM Architecture	631
	9.9.2	Recognition of Unknown Radar Emitters Example	633
	9.9.3	InSAR Sentinel-1 Example	642
9.10	Gated Recurrent Units		645
	9.10.1	Original Concept	645
9.11	Boltzmann Machine		648
9.12	Restricted and Deep Boltzmann Machines		649
9.13	Deep Belief Networks		651
9.14	Generative Adversarial Network		652
	9.14.1	SAR ATR Example	653
9.15	Transfer Learning		654
	9.15.1	The Problem	655
	9.15.2	Framework for the ATLA Method	656
	9.15.3	Losses for Target, Classification, and GAN	658
	9.15.4	Experiments and Data	658
	9.15.5	Results	659
	9.15.6	ATLA Summary	660
	9.15.7	Recognizing Unknown Radar Waveforms	660
9.16	Addressing the Security in DRFM Networks		660
	9.16.1	Federated Learning	661
	9.16.2	How Federated Learning Works	662
	9.16.3	Unmanned Vehicle Considerations	664

		9.16.4 Federated Transfer Learning	665
	9.17	Neuromorphic Computing for DRFMs	667
		9.17.1 Spiking Neural Networks	668
	9.18	Wireless Comms and Networking with UAVs	669
	9.19	Summary Comments	673
		References	674
Chapter 10	**Electronic Attack Using Deep Learning**		**681**
	10.1	Spectrum Dominance with UAVs and DRFMs	681
		10.1.1 Autonomous UAV Navigation Using Deep Reinforcement Learning	683
		10.1.2 Autonomous UAV Navigation Using Deep Transfer Learning	689
	10.2	Joint Electromagnetic Spectrum Operations	691
	10.3	Types of Electronic Attack	692
		10.3.1 Airborne EA: Suppression Techniques	693
		10.3.2 Surface EA: Defeating Anti-Ship Missiles	694
		10.3.3 Offensive EA	696
		10.3.4 Defensive EA	697
	10.4	Range-Doppler Imaging Emitter: Signal Processing	697
		10.4.1 Pulse Doppler Signal Processing	697
		10.4.2 Synthetic Aperture Signal Processing	698
		10.4.3 Inverse Synthetic Aperture Imaging	699
	10.5	Obscuration-EA and Deep Learning	703
		10.5.1 Obscuration Analysis	704
		10.5.2 The Beacon Equation	705
		10.5.3 Offensive Engagement Algorithms:	708
		10.5.4 Cognitive Deep Learning Control	712
		10.5.5 Practical Considerations	714
		10.5.6 Deep Reinforcement Learning for Unmanned Combat Systems	715
		10.5.7 Selective-Reactive EA	718
		10.5.8 DRFM Subsumptive AI Rules	719
		10.5.9 Smart Interference Using Generative Adversarial Nets	720
	10.6	Deception-EA Against Synthetic Imaging Apertures	724
		10.6.1 Generating Noise Patches in SAR Images	724
	10.7	Target EA and Deep Learning Algorithms	732
		10.7.1 Characteristics of Target Jamming	732
		10.7.2 Transponder False Targeting	732
		10.7.3 Repeater False Targeting	739
		10.7.4 EA: Range Gate Pull-Off	741
		10.7.5 EA: Velocity Gate Pull-Off	750

xiv   Contents

10.7.6 Coordinated RGPO-VGPO	753
10.7.7 SAR Active Decoy EA Technique	757
10.8 Countertargeting the Imaging Sensor	760
10.8.1 Confusing the Adversary's Targeting Process	762
10.9 Digital Synthesis for Structured False Targets	763
10.9.1 Early Analog Technology	763
10.9.2 Embedded Digital Integration	764
10.10 Programmable Digital Image Synthesizer	766
10.10.1 Warfighter Payoffs	767
10.10.2 DIS Complex Range Bin Processing	769
10.10.3 Bit-Level Simulation of the DIS	772
10.10.4 LFM Chirp Pulse Model	772
10.10.5 Deriving Digital Image Synthesizer Coefficients for the Test Target	776
10.10.6 ISAR Image Compression for Testing DIS Architecture	782
10.10.7 Pipeline Architecture: Simulation Results	784
10.10.8 False Target in ISAR	787
10.10.9 Overflow is the Biggest Problem	788
10.11 AI and DRFM DIS CONOPS	791
10.11.1 Experimental Determination of False Target: Phase and Gain Coefficients	791
10.12 Deriving the Sea Clutter Coefficients	792
10.12.1 NRL Normalized RCS Model	794
10.12.2 Clutter Parameter Comparison	796
10.13 Phase and Gain Coefficients for Sea Clutter	800
10.13.1 Phase Coefficient for Sea Clutter	804
10.13.2 Gain Coefficient for Sea Clutter	805
10.13.3 Summary	806
10.14 Concluding Remarks	807
References	811
Chapter 11 Counter-DRFM Methods	819
11.1 Pulse Diversity Techniques	819
11.1.1 Perturbing the Phase of the LFM Chirp	820
11.1.2 Orthogonal Coding	825
11.2 Signal Processing Techniques	831
11.2.1 Polarization Discrimination Methods	831
11.2.2 Statistical Counter-DRFM Theory Using Estimation and Neyman-Pearson Receivers	835
11.2.3 Detecting DRFM Decoys Using Neyman-Pearson Receivers	842

11.2.4	Frequency Diversity Characteristics	847
11.2.5	Counter-DRFM Pull-Off Using LSTM	850
11.2.6	Countering the Coordinated RGPO-VGPO	858
11.2.7	Statistical Signal Processors	865
11.2.8	Use of Phase Noise Measurements	867
11.2.9	Converse Beam Cross Sliding Spotlight SAR	868
11.3	Concluding Remarks	874
	References	876

About the Author   **879**

Index   **881**

PREFACE

INTRODUCTION

This book has grown out of teaching and research in the field of microwave-photonic transceiver design for electromagnetic warfare at the U. S. Naval Postgraduate School. Also, while as a technical editor for the *IEEE Trans. on Aerospace and Electronic Systems* from 2013 – 2020, it has become evident that RF transceivers for digital radio frequency (RF) memories (DRFM) are seeing an unprecedented level of growth in performance and capability. Along with the explosion in unmanned air, land and sea vehicles, these on-board microdata centers are pushing the bounds of size, weight, and power consumption while their size continues to shrink. The on-board processors must handle the massive amounts of data from the sensors to avoid collision and perform the surveillance and reconnaissance mission – transforming the raw sensor data into the required decision-level knowledge, all while, for example, navigating an ingress for suppression. The biggest news item nevertheless, is *artificial intelligence* (AI), *autonomous weapons*, and *machine learning* (ML), grabbing the headlines and they will continue to do so into the foreseeable future. Being able to easily handle the complex big data problems as a response management server and friendly copilot system deserves special attention! However, the topic of autonomous weapons, with its many ethical implications, demands some formal inquiry and doctrine especially for unmanned airborne systems (UAS). With a network of UAS, AI, and highly capable DRFM transceivers, a flexible design can be realized for any desired mission.

OVERVIEW OF THE BOOK

The book is divided into two parts. The main objective in Part I is the concept of embedded transceiver design and architecture development for electromagnetic spectrum dominance and electromagnetic maneuver warfare. Here we examine the design by bringing together a host of techniques and solutions to provide a comprehensive treatment of the technologies used in modern DRFM transceivers. Recent antenna and microwave-photonic receiver technology is also discussed for wideband spectrum sensing.

At the heart of next generation unmanned vehicles and sensor requirements, the AI-centric systems and electronic support (ES) processing, Part II introduces the concepts and techniques of modern spectral sensing using AI and ML for emitter modulation classification and electronic attack (EA) for counter-targeting to interrupt the kill web. Special emphasis is placed on EA techniques creating false targets against high range resolution profiling radar systems such as synthetic aperture radar (SAR) and inverse SAR (ISAR). A final chapter is included on recent counter-DRFM techniques discussing the methods used to defeat the DRFM.

A key feature of this book is the online resources that are available including MATLAB® software that is present on the Artech House website:

https://us.artechhouse.com/Assets/downloads/pace_697.zip.

These resources can be used to reproduce many of the results in the book and provide exercises to enhance the learning of the material. The text contains sufficient mathematical detail to enable the average undergraduate electrical engineering student and physics student to follow, without too much difficulty. A certain amount of analytical detail, rigor, and thoroughness allows many of the topics to be investigated further with the aid of many references. A brief overview of each chapter is given below.

PART I:
SPECTRUM SENSING, ARCHITECTURES AND PRACTICAL CONSIDERATIONS

To begin Part I, Chapter 1 introduces the concept of electromagnetic spectrum dominance and the important role that network-enabled DRFMs have in maintaining in sampling this domain. The roles of unmanned systems and electromagnetic maneuver warfare are emphasized. The DRFM architecture and both the active and passive electronically scanned antenna are introduced. To quantify the DRFM's spectrum dominance and agility, the measures of effectiveness (MOE) in the observe, orient, decide, action (OODA) loop are examined.

In Chapter 2, an overview of the DRFM kernels is given. These include the amplitude analyzing architectures, the phase sampling kernels, and the channelized phase sampling kernels. Also examined, are the Global Positioning System and the inertial navigation system (INS).

In Chapter 3, the recent progress in active electronically scanned antennas (AESAs) that are compatible with DRFMs are presented. Analog, digital, and hybrid beamforming concepts are introduced. Transmit receive module schematics are also examined and the sensitivity is examined as well.

In Chapter 4, the concept of choosing the correct wideband spectrum sensing receiver is presented. Designing the AESA antenna-to-receiver interface is discussed first. The concept of configuring wideband, software defined DRFM functions is discussed. Several wideband spectrum sensing architectures are then examined. These include the compressive sensing receivers such as the analog-to-information, Nyquist folding receiver (NYFR), the modulated wideband converter, and the blind compressive sensing receiver. The impact of the bandpass sampling channelizer is also presented along with the polyphase analysis/synthesis channelizers.

In Chapter 5, the details of transceiver design are investigated starting with the mathematical model of the transceiver process. This includes the receiver's process (sampling operations, time uncertainty, clock accuracy, and sampling circuit designs), and the analog-to-digital converter (ADC) transfer function and circuit concepts. For example, a flash converter is used to demonstrate the linearity, quantization error, and noise floor. A mathematical model for the digital-to-analog converter (DAC) is also shown along with an explanation of the spurious signal technical issues and limitations. Finally, digitization figures of merit and the direct sampling signal-to-noise relationships are derived and the DRFM transceiver design challenges are summarized.

In Chapter 6, the transceiver component technologies and architectures are summarized. A look at the different high-speed comparator designs is presented first. Then the flash ADC examples in SiGe BiCMOS technology are studied. Time-interleaved techniques are then explored followed by the pipelined approach and a successive approximation. Figure-of-merit plots for the ADC are given for the important technologies. FinFET technologies are also explored. An important requirement of the DRFM transceiver is the embedded, dual-port memory and several technologies and architectures are shown. Practical considerations are then presented for the output DAC including ML calibration technology. Direct digital synthesis (DDS) for waveform generation and storage is also discussed, including the ROM-based DDS and the co-ordinate rotation digital computer (CORDIC) is also presented. DRFM oscillators and their phase noise and characteristics are also discussed.

In Chapter 7, the microwave-photonic transceiver technologies are introduced beginning with the digital antenna components. These include the passive and active mode-locked laser (for sampling the antenna voltage), optical fiber links, and the electro-optical Mach Zehnder modulators (MZMs) for direct inception of the antenna voltage into the optical domain are presented. Wideband component technologies are also discussed including photonic memories, detectors, and antenna systems. Wideband sampling of RF signals using lasers, and MZMs for photonic ADC and direct digitization approaches are then introduced followed by a discussion on compressive sampling techniques and the wideband spectrum sensing techniques using cryogenic crystals.

PART II:
MODERN ES AND EA TECHNIQUES USING DEEP LEARNING METHODS

Chapter 8 begins Part II with the discussion of modern ES and EA techniques using deep learning methods with an examination of persistent spectrum sensing and signal detection methods starting with an examination of the distributed spectrum sensing and persistent sensing, defining the spatial diversity order D. Both centralized and decentralized electromagnetic spectrum sensing techniques are discussed and their relationship to the health or freshness of the state of the spectrum.

Sequential spectrum sensing methods and quickest detection problem are presented and include single DRFM and multiple DRFM spectrum sensing and access techniques resulting in the value of the policy known as the Bellman optimality equation from reinforcement learning. Reinforcement learning is an area of machine learning concerned with how intelligent agents must take actions in a mission in order to maximize the notion of cumulative reward and works on the principle of feedback and improvement. Bandit problems are also discussed (taken from the gambling machine). For example, the arms of the multiarmed bandit formulation correspond to different frequency bands and choosing an arm to play corresponds to sensing or accessing a particular frequency band.

Detection methods are then discussed starting with Cohen's time frequency distributions. These include Choi-Williams distribution, a derivative of this, the Wigner-Ville distribution and examples of these calculations for both real input signals and complex input signals are given. Time-frequency decomposition through quadrature mirror filtering is also discussed and a modified sink filter is derived. Cyclostationary spectral estimation techniques are presented starting with the cyclic autocorrelation followed by the spectral correlation density function. Estimation of the spectral correlation density is shown using two techniques. These include the discrete time, time-smoothing fast Fourier transform (FFT) accumulation method and the direct frequency-smoothing algorithm. Examples are presented for both techniques. Finally, the atomic decomposition of signals formulation is given.

Chapter 9 covers further, the subject of artificial intelligence, machine learning, and deep learning, beginning with a quote from the late Dr. Richard Hamming, computer science professor at the U.S. Naval Postgraduate School. He had investigated artificial intelligence at its beginning while he was at Bell Laboratories, posing the famous question ... "can machines think?" The modern artificial intelligence concepts for distributed sensing to those in automation in the human-computer interface and nonlinear classification networks to multi-layer perceptron (MLP) networks are covered in a spectrum sensing framework. To use a MLP network, a modified feature extraction signal processing algorithm is demonstrated as a feature vector formation process as an example input to the MLP modulation classifier neural network. Machine learning, deep learning architectures are then discussed including autoencoders, convolutional neural networks deep belief networks and recurrent neural networks. Long, short-term memory and hardware considerations for gated recurrent units are also covered, generative of serial networks are also emphasized. Transfer learning and the problem framework free TLA method is explored. Neuromorphic computing for digital memories including spiking neural networks are also discussed. Wireless communications and networking with unmanned aerial vehicles is discussed as well.

In Chapter 10, electronic attack (EA) using deep learning is discussed beginning with spectrum dominance using UAVs. Autonomous UAV navigation using deep reinforcement learning is also presented. Joint electromagnetic spectrum operations is presented along with airborne and surface EA. Both offensive and defensive types of EA are examined. Obscuration EA and deep learning from unmanned combat systems are covered including the beacon equation, offensive engagement algorithms, cognitive control, and deep reinforcement learning. Selective-reactive EA in smart interference using generative adversarial networks is also covered. Range Doppler imaging signal processing is reviewed including synthetic aperture signal processing. Deception EA against SAR including generating noise patches in SAR images, and false targeting EA using deep learning algorithms are presented. The characteristics of target jamming and transponder jamming are examined and compared. Repeater techniques include range gate pull-off, velocity gate pull-off in a coordinated range, and velocity gate pull-off. SAR active decoy EA techniques are discussed as well.

Countertargeting techniques against the imaging radar to confuse the adversaries targeting process is discussed. Digital image synthesis for creating structured false target images against the synthetic aperture is also presented. Programmable, digital image synthesizer (DIS) technology that provides significant warfighter payoffs is presented. Deriving the image synthesizer coefficients for the false targets are shown and the pipeline architecture simulation results are also shown. Digital overflow effects are also quantified. Artificial intelligence and the DIS concept of operations is discussed. Generation of sea clutter to be included with the false (ship) targets for the appearance of realism is presented. The normalized phase and gain coefficients for sea clutter are derived. The chapter concludes with an appendix that describes the noise obscuration techniques and the DRFM false targeting techniques.

In Chapter 11, *counter*-DRFM techniques are examined and are divided into *pulse diversity* methods and *signal processing* methods. Pulse diversity techniques involve perturbing the phase of the waveform and orthogonal coding of the waveform on each successive pulse. The signal processing techniques involve several methods such as polarization discrimination, statistical detection theory using estimation, and the Neyman-Pearson receiver, frequency diversity characteristics, and defeating coordinated pull-off using LSTM deep learning methods. Countering the coordinated RGVGPO using statistical signal processors and by using phase noise measurements is also examined. Finally, the use of converse beam cross-sliding spotlight SAR techniques to avoid DRFM EA is presented.

ACKNOWLEDGMENTS

This book would not have been possible without the help, encouragement, and support received during its preparation. First, I am eternally grateful to GOD ALMIGHTY for without His graces and blessings, this work would not have been possible.

Immeasurable appreciation and deepest gratitude for the help and support are extended to the following persons who in one way or another have contributed in making this work possible. I would like to thank my wonderful family most importantly my wife, Ann, for her patience and endurance. I could not of completed this enormous task without her love, support, patience, sacrifice, and understanding. I would also like to thank my children Amanda, Zachary, and Molly for being such a blessing to us.

Many thanks go to Prof. Douglas J. Fouts, Chair, Dept. of Electrical and Computer Engineering, Naval Postgraduate School, for his many hours of hard work on our collaborative research efforts over the many years. Also thanks Prof. David C. Jenn for his many inspiring collaborations over the past 30 years and recently Ric Romero. As well to Prof. David Garren and to Prof. Nadav Levanon, Tel Aviv University for many insightful discussions over the years. The author is very grateful for having such a good friends. I am especially appreciate of the relationship I had with Bob Surratt who inspired me to go further than I thought possible. I am also very appreciative of the many, numerous graduate students and post-doc's that helped in this work including Stig Ekstorm, Chris Karow, Ken Hollinger, Dr. Thor Martinsen, Dr. Brandon Hamschin, Pak Ang, Richard Schroyer, Han Wei Lim, Marcelo Magalhaes, Capt. Richard Shmel, USMC, Maximilian Hainz, Dr. Sebastian Teich, Carsten Sewing, Robert Humeur, Owen Brooks, Sascha Pauly, Anthony Queck, Dr. James Calusdian and Admiral Jim Butler, USN. A big thanks also goes to David Boyle and Catherine Boyle who were tremendous help, correcting the many mistakes in the first draft of the manuscript. The many hours they put in made all the difference.

This book would not have been possible without the wonderful and extremely professional staff at Horizon House Artech Publishing. Especially David Michelson, who offered me much encouragement especially when at times, I was feeling overwhelmed, and Judi Stone, who helped me manage the copyedits which improved the quality of this manuscript considerably, and Kati Klotz who provided the marketing assistance.

FINAL MESSAGE

Every attempt has been made to ensure the accuracy of all materials in this book including the MATLAB programs contained on the website. I would, however, appreciate readers bringing to my attention any errors that may appear.

Phillip E. Pace, IEEE Fellow

Part I

Spectrum Sensing and Practical Considerations

Chapter 1

Electromagnetic Spectrum Dominance

1.1 INTRODUCTION

There is a huge demand for sensors onboard unmanned vehicles. The network-enabled digital RF (radio frequency) memory (DRFM) represents *the eyes and the ears* of U.S. and allied military forces, and are becoming even more important in the future. Top commanders rely on unmanned vehicles to get them the important information they need, when they need it, with zero (or little) time lag involved. As unmanned, miniature air launched vehicles are shrinking in size, this creates an unprecedented demand on the DRFM design and the embedded computing size, weight, power consumption, and cost (SWaP-C) as more is being asked of the DRFM but doing it with less of everything. Missions today require the DRFM to perform a large sensor fusion task, tying together electro-optical, infrared, and RF for C4ISR (command, control communications computers intelligence, surveillance, and reconnaissance), and performing signal intelligence and electromagnetic warfare missions including electronic attack. All of this while performing persistent surveillance and a huge amount of data mining.

This chapter begins with a discussion of the *electromagnetic spectrum* as a domain and the role of the DRFM in electromagnetic warfare. Airborne, surface, and sub-surface, unmanned systems that can participate in spectrum warfare are discussed next. Knowing its location and its communication reach are both critical for the unmanned system. With the growth of the commercial networking market, the significance of networking and the use of *open system standards* is presented. A brief overview of the DARPA's Mosaic program and the CONCERTO program is also presented. It is also important for the unmanned system/DRFM to know its location. The global positioning system receivers, used mainly for determining location are then examined, including some of the antennas that can be used onboard.

The DRFM plays a critical role in spectrum dominance. The key to quantifying the DRFM's spectrum dominance is also examined and a model for electromagnetic *maneuvering agility* is also proposed as a measure of effectiveness. An overview of the DRFMs role in autonomous detection and classification of RF emitter modulations is emphasized, and the response management of the embedded DRFM using artificial intelligence techniques that can be used to form an autonomous electronic attack are discussed. These include machine learning and deep learning methods for counter-targeting. Counter-DRFM techniques are also emphasized and include pulse diversity and signal processing methods.

1.2 THE ELECTROMAGNETIC SPECTRUM DOMAIN

The entire electromagnetic spectrum (EMS) extends from the lowest to the highest frequency (longest to shortest wavelength, respectively) and includes all radio waves. In addition, the EMS consists of many subranges, commonly referred to as portions. Portion names include commercial radio and television, microwaves, radar, infrared radiation, visible light, ultraviolet radiation, X-rays, and gamma rays. These various portions bear different names based on differences in behavior in the emission, transmission, and absorption of the corresponding waves and also based on their different practical applications. There are no precise accepted boundaries between any of these contiguous portions, so the ranges tend to overlap.

The spectrum outline is shown in Figure 1.1. The electromagnetic waves travel at the speed of light in a vacuum. That is, the entire EMS propagates at the speed of light (in a vacuum). Although all electromagnetic waves travel at the speed of light in a vacuum, they do so at a wide range of frequencies, wavelengths, and photon energies. Consider the spectrum interval conversion law for energy $E = h\nu$ using Max Planck's constant[1] $h = 6.626 \times 10^{-34}$ W s^2 where ν is the frequency of the radio frequency radiation. Note that as the *wavelength decreases*, the *energy increases*. This is an important consideration in understanding maneuvering within the EMS. As the energy increases, the *attenuation* in the atmosphere also increases due to absorption mechanisms. Thus, the operating frequency (or carrier frequency) of the system is a fundamental consideration to understanding the maneuver within the EMS. For example, multidomain weapon and communication systems operate (or move) at the speed of light with C2 command operations moving at the speed of sound (e.g., voice commands of a secure channel). Maneuver in the other domains varies greatly, but typically at a much slower speed than the EMS (for example, ground vehicles, surface vessels, and aircraft).

1 Max Planck developed the law that expresses thermal radiation as a function of temperature and wavelength for *all* wavelengths.

1.3 THE DEVELOPING ROLE OF DIGITAL RF MEMORIES

DRFM development started in the early 1970s. The basic principle of the DRFM was put in British patent 1605203 by Chris Haynes, EMI Electronics, Ltd., 1974. Another important digital storage system for high frequency signals was also put forth in U.S. patent 3947827 by the Wittaker Corporation, March 1976 [1]. These devices were developed to digitize the EMS and used high-speed sampling and fast digital memory for storing and replicating radiating signals. They provide the ability to capture radiated emissions and generate precise, coherent replicas, making them important in applications such as signal jamming, deception of covert communications, signal intelligence (SIGINT) operations, decoys, radar transmitters, simulations, and test equipment. Their most encompassing role, however, comes from their use across all aspects of electronic warfare and for our purposes in this book, we will call this *electromagnetic warfare* (EMW), which is where this book is focused.

For most of its history, EMW concepts have been treated as something special, an activity that was never very well integrated into military operations, because it was always considered an add-on, after the fact. In the past, EMW operators with this special knowledge were organized into special units. From an equipment standpoint, military leaders tended to turn EMW funding on and off, depending on whether or not we were at war (or near to war, in the case of the Cold War). We relied on weak organizing principles for how we would use, manage, and control the EM domain. From the mid- to late 1990s we tended to organize EM activities based on equipment (material), and we trained personnel accordingly as radio operators, radar operators, and EMW operators [2].

As these radios, radars, and EMW systems became more automated, the need for dedicated equipment operators diminished, as did our concept of tactical maneuvering within the EM spectrum (EMS). That is, now the radar systems were responding to electronic attacks automatically and radios were finding open, white spaces in the spectrum to operate, using a more cognitive approach to communication. Consequently, it was the weapons systems doing most of the maneuvering in the EMS, instead of operators twisting knobs and turning dials. As the equipment operators began to drop away, a lot of EMS knowledge, understanding, and awareness were seemingly lost, and the operational notion of EM maneuvering worsened.

6 ELECTROMAGNETIC SPECTRUM DOMINANCE

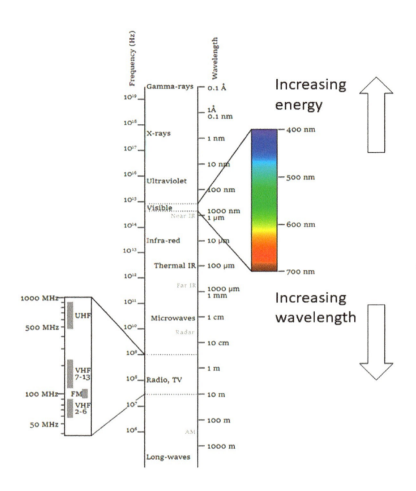

Figure 1.1 Electromagnetic spectrum showing the different bands from long-waves to gamma rays. The DRFM can be configured to work at radio frequencies up through millimeter waves depending on the technology being used in the receiver to sample the EMS.

A photo of high-speed data acquisition early prototype DRFM developed by the Mayo clinic is shown in Figure 1.2. It used an 8 metal layer board with impedance wirewrap interconnects technology and a 4-bit analog-to-digital converter (ADC), digital-to-analog converter (DAC). It also had a 125-MHz motherboard clock rate and a 1-GHz daughter board with Fairchild emitter coupled logic components. Compare this today to just the integration shown in Figure 1.3.

1.4 Electromagnetic Warfare 7

Figure 1.2 Photo of an early high-speed data acquisition prototype DRFM developed for the Mayo Clinic.

Figure 1.3 Photo of a more recent 4-bit DRFM (courtesy of Mercury Defense Electronics).

As technology has progressed, data acquisition and response time has become exponentially faster and the capabilities have also grown tremendously. Figure 1.3 shows an example of a 19" airborne self-contained rack-mountable unit also with 4-bits and 800-MHz bandwidth.

1.4 ELECTROMAGNETIC WARFARE

EMW is the use or manipulation of the EMS spectrum in warfare from the air, land, sea, and space. That is, EMW includes any military action involving the use of EM and directed energy to control the EMS or attack the enemy. It involves the use of

EM energy, directed energy, or anti-radiation weapons (such as antiradiation missiles) for use against an adversary. EMW comprises three main categories: electronic attack (EA), electronic protection (EP), and electronic support (ES) [3][2]. EMW typically uses the radio and microwave frequencies shown in Figure 1.1 for communications, radar systems, and satellites. Certain EMW solutions also leverage electro-optical/infrared sensors for intelligence and enemy targeting. Whether the EMW use is target detection, denial of detection, deception, destruction, or protection, a network-enabled digital radio frequency memory plays a critical role. Figure 1.4 shows the breakout of all the subdivisions of the DRFM's electromagnetic actions for both offensive EA (left) and defensive EP (right) activities. In the middle is the receive DRFM activity (ES). Each division of EMW has a unique role in supporting unified land operations. EA is the division of electromagnetic warfare EMW involving the use of electromagnetic energy, directed energy, or antiradiation weapons to attack personnel, facilities, or equipment with the intent of degrading, neutralizing, or destroying enemy combat capability and is considered a form of fires [5]. Electronic protection or EP is the division of electromagnetic warfare involving actions taken to protect personnel, facilities, and equipment from any effects of friendly or enemy use of the electromagnetic spectrum that degrade, neutralize, or destroy friendly combat capability [5]. Electromagnetic warfare support or ES is the division of EMW involving actions tasked by, or under direct control of, an operational commander *to search for, intercept, identify, and locate or localize sources of intentional and unintentional radiated electromagnetic energy* for the purpose of immediate threat recognition, targeting, planning, and conduct of future operations [5]. The emphasis here is to bring attention to the reader, the fact that these functions are *exactly* the function of the DRFM. In fact, in addition to the ES receiving action, the DRFM also takes on most of the functions under EA. More specifically, the DRFM can degrade, neutralize, and perform deception of emitters as well as share this information across his network of networks.

2   These are not just new definitions to replace the outdated and obsolete terms of ECM, ECCM, and ESM, respectively. They are new and expanded definitions [4].

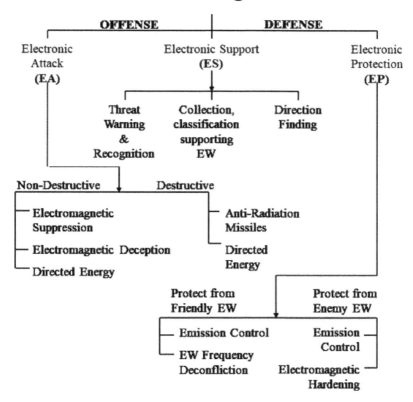

Figure 1.4 DRFM electromagnetic actions showing the offensive (EA), defensive (EP), and receive activity (ES).

1.5 AIRBORNE ASSETS AND UNMANNED SYSTEMS

There has been a maturing of EMW from a single dedicated receiver capability (stovepipe) to a multifunction capability focus with airborne EMW expanding as the technological target. Airborne EMW is heading in different directions as the U.S. Air Force, Navy, Army, and Marine Corps each stake out their own approaches, equipment requirements, tactics, techniques and procedures, and concepts of operation to deal with the rapidly evolving technologies flooding the 21st century world [6]. The Air Force has abandoned large-scale dedicated manned EMW aircraft for the EC-37B

Compass Call and the F-35A Raptor joint strike fighter, while the Navy has backed the Boeing EA-18G Growler shown in Figure 1.5 and the carrier-based F-35C [7].

Figure 1.5 EA-18G Growler showing it carrying three AN/ALQ-99 jammer pods. Note it can also carry several HARMs (high-speed anti-radiation missiles).

The Marines have chose aircraft such as the F-35B with cyber and electronic warfare capabilities (e.g., AN/ASQ-239, Intrepid Tiger II EMW payload systems). Also included are the AV-8B jump jets, MV-22 Osprey, and the UH-1Y transport helicopters. The Army is building a Multi-Function Electronic Warfare (MFEW) family of systems. Each of these platforms carry a DRFM as the central mission warfare component for receiving (ES) and transmitting (EA) warfare waveforms. The air component of the Army's MFEW family of systems uses an open architecture Silent CROW system that is designed to be easily configured for a variety of airborne and ground platforms, including a wing-mounted pod for the MQ-1C Gray Eagle UAV. The Silent CROW should enable U.S. soldiers to disrupt, deny, degrade, deceive, and destroy adversaries' electronic systems through ES, EA, and cyber techniques [7].

1.6 ELECTROMAGNETIC MANEUVER WARFARE

Electromagnetic Maneuver Warfare (EMMW) is where EA is not just an enabler dedicated to a special aircraft (e.g., EA-18G Growler[3]) but is the cornerstone of a network encompassing the entire force from unmanned, miniature air-launched vehicles to surface forces and submarines coordinated by an EM battle management system where all of these individual platforms will collect data on enemy signals to inform the network while dialing up and down their own emissions to deceive or attack the adversary [8]. It is about transmitting as little as possible in order to maximize the amount of deception towards the adversary. Note that this is contrary to a cell phone that is transmitting and receiving all of the time.

---

3   The role of the EA-18G is to be more robust and stand in the threat ring and fight-contrary to the Prowler. As A2/AD systems become increasingly more sophisticated with longer range, the EA-18G, is required to break kill-chain connecting adversary sensors to the C2 and their weapons.

The key, however, is receiving the best information more rapidly than the adversary and being able to use it more quickly than they do. Col. John Boyd formed his energy-maneuverability theory in 1966, relating the available energy in a fighter aircraft to life or "information equals life;" the concept that the available energy in a fighter aircraft provides a method to understand the relationship between its altitude and its kinetic energy, or the position and velocity of the aircraft as a means to define the aircraft's maneuverability [9]. Boyd's claim was that within 40s, from an initial position of disadvantage, he could defeat any opposing pilot via the use of his energy-maneuverability principles in mock combat. Spectrum warfare is just as important as any other traditional domain of war, Greenert insists.

> "This environment is so fundamental to naval operations - and so critical to our national interests - that we must treat it on par with our traditional domains of land, sea, air, and space," Greenert says. "In fact, future conflicts will not be won simply by using the electromagnetic spectrum and cyberspace; they will be won *within* the electromagnetic spectrum and cyberspace. This will require changes to our operating concepts, military systems, and - most importantly - a new way of thinking in our Navy" [10].

This is important because it will require multifunction and multimission systems for warfighters to operate safely in harm's way by moving away from a single box aboard a single platform to a more holistic mission look incorporating both offense and defense to get the job done. Another factor is that the signal structure you see will only be there momentarily. That is, the adversary using cognitive sensor waveforms adapts quickly and you have to be able to adapt quicker [10]. With the use of stand-in unmanned systems using onboard DRFMs, this is an advantage you can achieve using the power of EMMW. Below we show that EMMW can be quantified along with these types of principles. We will also show that we can form a measure of effectiveness for EMMW that can quite readily be used to predict the outcome we are looking for.

1.6.1 Cognitive Electromagnetic Warfare

The concept of cognitive EMW is not new. In fact, any signal processing algorithm with a feedback loop to incorporate an intelligent decision or calculated update is a cognitive process. However, advances in machine learning combined with software defined radios (SDRs) now offer a renewed opportunity to adapt and modify spectrum sensing and a multifaceted reactive interference within a highly contested and unique EMS environment. Cognitive SDRs have made a significant difference in the manner in which the EM spectrum can be accessed [11, 12]. A SDR receiver system digitizes the RF signals at the output of the antenna/low noise amplifier directly (direct sampling) and does not downconvert the signal to a low frequency. Although not a new concept, the rapidly evolving ADC and DAC capabilities using semiconductors, photonics and superconductivity, have revolutionized transceivers that can almost instantaneously receive, process and transmit a signal. There are currently numerous SDRs available on the market starting at $100. Cognitive SDRs can then receive a bandwidth of signals and intelligently (using e.g., neural networks or some nonlinear algorithm) determine their composition and which channels or frequencies are in use and which are not. It can then instantaneously make a decision whether to keep looking for an open white space in the spectrum or move into the open space it has found, thereby minimizing interference to other users. DRFM cognitive EMW processing aims to sense, characterize, and exploit the EMS by networking other companion DRFMs with their own machine learning artificial intelligence toolkits. Using cognitive EMW tools, DRFMs on stand-in unmanned aerial vehicles (UAVs), for example, can observe the EMS threat, orient for the current state of the spectrum, send the information to multi-domain warfighters and the C2 and rapidly consolidate assets needed to neutralize the threat [13–15].

Using a cognitive approach, the interception and classification of radar modulations and communication modulations has become a more refined process as discussed in [16]. For example, the determination of the antenna scan modulation using cognitive signal processing approach is discussed in [17]. Cognitive electronic attack techniques are presented in [18]. Finding holes in the EMS or spectrum hole detection is typically quantified by specifying probabilities of false alarm P_{FA} and probability of missed detection P_{MD}, the minimum detectable signal level, and/or the minimum signal-to-noise ratio (SNR) for sensing [19]. While most of the efforts in analysis of detection have focused on the Neyman-Pearson detectors, it is well know that *sequential detection* provides substantially faster operation on average. Sequential detection is in contrast to a Neyman-Pearson test where the logarithm of the maximum likelihood ratio (MLR) is compared to a single threshold Λ_0, at a predefined and fixed observation interval. The sequential test of A. Wald compares the MLR to two thresholds Λ_1 and Λ_2 sequentially, until certain conditions are not satisfied. In both cases parameters of the test are calculated based on the required P_{FA} and P_{MD}.

A spectrum sensing policy employing a *recency-based exploration* approach has also been explored for cognitive EMS sensing. In this formulation, the problem is to find a spectrum sensing policy for a multiband dynamic spectrum access as a stochastic, restless, multiarmed bandit problem with stationary unknown reward distributions [20]. In cognitive radio networks, the multiarmed bandit problem arises when deciding where in the radio spectrum to look for idle frequencies that could be efficiently exploited for data transmission [21]. Two models for the dynamics of the EMS frequency bands have been explored: (1) the independent model in which the state of the band evolves randomly independently from the past, and (2) the Gilbert–Elliot model, where the states evolve according to a two-state Markov chain. It is shown that, in these conditions, the proposed sensing policy attains asymptotically logarithmic weak regret. The policy examined in [20] represents an *index policy*, in which the index of a frequency band comprises a sample mean term and a recency-based exploration bonus term. The sample mean promotes spectrum exploitation, whereas the exploration bonus encourages further exploration for idle bands providing high data rates. The recency-based approach readily allows constructing the exploration bonus such that it will grow the time interval between consecutive sensing time instants of a suboptimal band exponentially, which then leads to logarithmically increasing weak regret.

1.6.2 Open System Standards

Embedded computing standardization is of significant importance in making EMW ubiquitous on manned and unmanned aircraft as is the growing case of data processors like field programmable gate arrays (FPGAs) and Nvidia graphics processing units, (GPUs) that are trained to look for specific targets based on existing databases, all primary targets for DRFM integration.

Three open-systems electronics standards for military embedded computing systems have gained traction and support from U.S. defense acquisition programs. [22]. These standards are the:

- Hardware Open Systems Technologies (HOST);
- Command, Control, Communications, Computers, Intelligence, Surveillance and Reconnaissance (C4ISR) Electronic Warfare Modular Open Suite of Standards (CMOSS);
- Sensor Open System Architecture Standard (SOSA) .

The HOST program seeks to decompose military embedded systems into functional blocks to ease system design and reuse of embedded computing hardware and software. It's intended to open systems for several different vendors and avoid locking single vendors into large system designs.

The CMOSS program is intended to move the embedded industry away from costly, complex, proprietary solutions and toward readily available, cost-effective, and open-architecture commercial off-the-shelf (COTS) technologies [23].

The SOSA standard seeks to drive down the cost and complexity of aerospace and defense electronics components and systems by adopting and then enforcing the use of widely accepted open-systems standards to promote component interoperability, rapid technology insertion and upgrades, reuse, and critical mass in the supplier market.

It appears that SOSA may win out as it seeks to take the best of several influential industry standards, such as OpenVPX. It also seeks to narrow down OpenVPX standards to a manageable level while reducing the OpenVPX slot profiles from 37 to three; the 3U switch profiles from 24 to three; and the 6U slot profiles from 19 to three. In addition, SOSA also seeks to settle on 12-volt power to make it less likely that systems designers will need custom-designed power conditioning and control, as well as offer certified solutions for software packages such as security and system management [22].

Open-systems industry standards for embedded computing are now locked into transceivers and DRFM designs and making for wideband, agile spectrum dominators. In addition, open-systems standards policy has made small-form-factor embedded computing for distributed DRFMs a reality and the growth of airborne EMW possible along with continued advancement in performance of ADCs and DACs. Open-system standards like SOSA help to ensure components share a common platform and can interchange information across services. This will then lead to the DRFM having assured connectivity and access in all operating environments, enabling its ability for persistent network awareness and control having bandwidth efficient communications with other DRFMs on the network [23].

The DRFM is now part of a distributed sensor network for battle-space awareness with optimal netting and synchronization providing the C2 commander with high-definition battle-space displays and visualization capabilities for improved decision making. The DRFM should now have the software to dynamically handle large amounts of data and the signal processing tools for processing the data for all the multimissions assigned.

1.7 COMMERCIAL TECHNOLOGIES DRIVING A DIFFERENCE

The DRFM integration of cloud native technology systems such as the NVIDIA's A100 Tensor Core GPU delivers unprecedented acceleration at every scale for AI, data analysis, and all EMW applications such as modulation classification repeater EA and taking on the toughest computing assignments in the harshest environments. For the past three years Nvidia has been making graphics chips that feature extra cores, known as tensor cores. These units can be found in desktop PCs, laptops, workstations, and data centers around the world. As the engine of the NVIDIA data center platform, the A100 can efficiently scale up to thousands of GPUs or, using new Multi-Instance GPU (MIG) technology, can be partitioned into seven isolated GPU instances to accelerate workloads of all sizes[4]. The NVIDIA Ampere GPU architecture retains and extends the same CUDA programming model provided by previous NVIDIA GPU architectures such as Turing and Volta, and applications that follow the best practices for those architectures should typically see speedups on the NVIDIA A100 GPU without any code changes.

4   Note: MIG is supported only on Linux operating system distributions supported by CUDA 11/R450 or higher.

Xilinx Zynq Ultrascale+ RFSoC (radio frequency system on chip) has become a significant game changer due to the multi-gigasample RF data converter and soft-decision, forward error correction capabilities that are available. This Xlinx component also has one of the densest FPGA and DSP boards available on the commercial market.

Boards such as Abaco's 3U VPX VP430 that implement this technology are able to support MIMO (multiple input multiple output) beamforming, sensor processing, and high-speed radar signal processing. The alignment of CMOSS- and SOSA-aligned solutions for embedded computing is becoming a game changer [24]. Other supporting technology like the SBC3511 offers memory resources including 32 gigabytes of high-speed DDR4 SDRAM and as much as 256 gigabytes of NAND Flash (NVMe solid-state drive) with a range of I/O options [24].

Mercury Defense Systems has also introduced the EnsembleSeries HDS6605, powerful general-purpose 6U Open VPX embedded computing blade server with hardware-enabled support *for AI applications*. Here each 6U Open VPX blade provides as many as 22 cores from one 1.9-GHz processor, delivering 2.6 tera-FLOPS of processing power.

1.7.1 DARPA's Mosaic and CONCERTO

Below, DARPA's Mosaic and CONCERTO are explained to emphasize the importance of the network-enabled DRFM and the critical role it plays in their mission. The goal of the new DARPA-led effort, properly named "Mosaic Warfare," is to push beyond the monolithic weapons systems the Department of Defense (DOD) currently relies on today and to support the transition to a new force structure that incorporates a greater number of smaller, unmanned, autonomous, and networked sensor and weapons platforms that represent the "pointy end of the stick" [25].

The basic idea is to create a force with greater mass by *networking* these smaller, cheaper, autonomous tiles with each other and with the legacy weapons platforms that will launch them, fight alongside them, and sometimes recover them. The result should be more sensors and more effectors for an adversary to contend with. As discussed in this book, the DRFM plays a key role in the Mosaic structure as an integral, network-enabled one (or more) of the tiles.

In Mosaic Warfare the current monolithic force structure of large manned, unmanned platforms (transport aircraft and bombers, surface combatants, submarines, tanks, and APCs) carry hundreds, if not thousands, of unmanned autonomous platforms (unmanned air, ground, surface and subsurface vehicles) to the edge of the A2/AD threat environment and then launch them to go deep into the threat environment. The unmanned platforms will be collectively equipped with DRFMs and other IR, UV sensors, datalinked together with the C2 and the nonkinetic effectors and hard-kill munitions to form sensor-to-shooter networks between them – and with the larger monolithic weapons platforms. If some of these unmanned tiles are defeated, the remaining tiles will constantly re-form, re-assign, and adapt in real-time to complete the mission, even if the mission focus changes. This requires the DRFM to have several autonomous signal processing functions as discussed below. In addition, it *must be* network-enabled to participate in the EM spectrum multidomain operations as directed by the C2. Measuring the agility of Mosaic capability is discussed below. The construction of the Mosaic tile approach to war fighting has been presented in three categories: (a) planning and composition, (b) interoperability, and (c) execution. The programs that have been formed include the Adapting Cross-Domain Kill-Webs (ACK), which will help users with selecting sensors, effectors, and support elements across military domains to form and adapt kill webs to deliver desired effects on targets [25].

Another program is the Complex Adaptive System Composition and Design Environment (CASCADE), which assists military decision-makers with rapidly identifying and selecting options for tasking and retasking assets, within and across organizational boundaries to contend with both dynamic and unexpected contingencies that might occur. Within the interoperability category they have formed the Dynamic Network Adaptation for Mission Optimization or DyNAMO. These architectures are ad hoc networks of networks that can be formed on demand. In fact, DyNAMO can be seen as an ACK service that serves to figure out how to stitch together the physical tactical networks and route data over them [25]. To put the different systems and capabilities together, the System of Systems Technology Integration Tool Chain for Heterogeneous Electronic Systems (STITCHES) has been developed. STITCHES is a software tool that allows warfighters to automatically connect multiple, diverse, federated systems and capabilities, translating between different standards and software where interoperability doesn't already exist. In summary, Mosaic warfare is an infrastructure to enable the warfighter to build efficient architectures for mission execution. What is important to note is that the network-enabled DRFM with its distributed multifunction capabilities makes up a central building block of this infrastructure as will be shown.

Persistent intelligence, surveillance, and reconnaissance (ISR) is a collection strategy that emphasizes the ability of collection systems to linger on demand in an area to detect locate, characterize, identify, track, target, and possibly provide battle damage assessment and re-targeting in near- or real-time [26]. DARPA's CONverged Collaborative Elements for RF Task Operations (CONCERTO), focuses on supporting communications, radar, and EMW systems with a flexible RF architecture that uses shared common hardware, enabling multifunction systems that meet the low-SWaP requirements of compact UAS [27, 28]. The converged systems will be able to efficiently switch between intelligence, surveillance, and reconnaissance, C2, networking, and combat operations support missions without physical payload changes [29].

This use of the EM spectrum is to predict an adversary's activities, and formulate and execute preemptive activities to deter anticipated adversary courses of action. This is an active networks versus active countermeasures competition [30]. This platform sensor capability is also made possible by the advances in renewable energy, scalable open architectures, and software-defined signal processing.

As the ability to hit a target is no longer an issue, the issues now are (1) what effect do you want to achieve, and (2) what is the required balance of human processing versus machine processing and (3) how fast can the huge volume of (dissimilar) sensor data be overlaid and fused together to provide one comprehensive picture.

Deep learning and machine learning are discussed in Chapter 9 and, while also relating to many areas of EMW, they are particularly useful in the above problem of non-similar sensor's image registration, the efforts of selective reactive EA during suppression and ES problem situation for classifying adversary emitter modulations. Even further, another key development is the ES/EA that is able to detect and characterize previously unknown sensor emissions (never before intercepted), adapt to them, create effects that confuse or deceive rather than just overwhelm them and even predict the sensor platform's reaction [30]. Being able to maneuver in the electromagnetic spectrum is a fundamental tenet of all operations. For a distributed network of DRFM sensors it becomes important not to rely on monolithic single points of failure. With the recent Mosaic Warfare, concept being developed by DARPA, the warfighter will be linked to warfighting assets such as missile batteries, tanks, planes, ships, and so on. This new design concept will confuse enemies by presenting a highly adaptable *mosaic web* of sensors, shooters, and decision-makers enabled by advanced computing and enabled by AI and machine learning at the fastest speeds possible. AI is used to optimize the warfighter communications in controlling how data and bandwidth are used most effectively. As discussed in further context below, humans are better at making high-level decisions, while AI-powered machines can process complicated things at great speed.

That network—named for the adaptable, piecemeal art form—should be able to assemble and disassemble itself into infinite new combinations on the fly. For example, in a mosaic warfare ground scenario, AI might suggest sending an unmanned aerial vehicle or ground robot ahead of the main ground battle force. That unmanned system might spot an enemy tank and pass the coordinates back, which are then relayed to a non-line-of-sight strike system in the rear that, in turn, launches its munitions and takes out the target.

1.8 THE MULTIFUNCTION DIGITAL RF MEMORY

The multifunction capability of the DRFM is critical for supporting operational EMMW. The DRFM device today is network-enabled consisting of system-of-systems that are able to collect, store, and manipulate radiated electromagnetic signals while coherently replaying selected collections with a minimal amount of delay [31]. This capability makes the DRFM, shown in Figure 1.6, the single most important device technology in the electromagnetic operational environment. The DRFM can be broken down into three main subsystems: (a) RF to digital conversion (SDR); (b) the digital signal processing; and (c) the modular DRFM technique generation and retransmit

18 ELECTROMAGNETIC SPECTRUM DOMINANCE

section. Today, the DRFM device has positioned itself to play an ever-increasingly critical role for EM sensing and dominant control of the EMS.

1.8.1 Fusing Enabling Technologies

The DRFM shown in Figure 1.6 shows the trend to minimize the number of antenna apertures on board the sensing platform. The phased array aperture in the front is made up of several subarrays and functions as a single multifunction antenna (EA, ES, communications, passive target detection). Integration of a GPS, INS, and separate communications antenna is also shown in Figure 1.6.

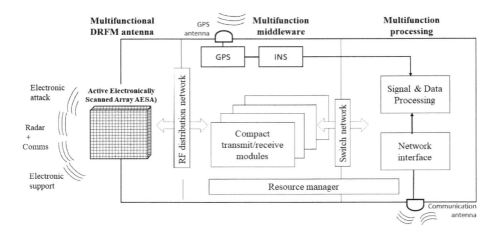

Figure 1.6 Block diagram of a multifunction DRFM.

The phased array antenna has an RF distribution network that interfaces with the compact transmit receive modules that are the interface to the signal and data processing functions and network processing at the backend. All of the applications are controlled by the resource manager's executive routines that manage the computer memory, clock distribution network, and frequency allocations being used. However, with the DRFM taking part in a distributed, decentralized sensing architecture, the frequency allocations may come from a central C2 over VITA-49 [12]. Control of the onboard frequencies being used at one time (frequency management) is then a major deconfliction task due to all the frequencies being used at once by the platform. These include for example, a global positioning satellite (GPS) system for determining ownship latitude, longitude and altitude (LLA), an inertial navigation system (INS) for determining the attitude of the aircraft platform including velocity, platform pitch, roll, and yaw, the ES/EA signals from the front antenna, the communications signals, and other onboard boutique RF sensors/transmitters in use.

1.8.2 Network-Enabled

A problem with wireless networks involves the network outages due to brief losses of the wireless links outages. Currently, DRFM networking modems send the network control information and the data using the same wireless link, which causes a network failure when that wireless link degrades. The loss of network connectivity that can take more than several minutes to recover once the wireless link is re-established. Network enabled DRFMs, however, require consistent network reliability and can achieve this by separating control information and communications data in *separate wireless links* to create a protected control channel. In this manner, the radio technology can maintain network reliability through periods of frequent signal degradation that routinely occur during military operations. Consequently, the DRFM can maintain network reliability even when the data channel might be lost.

1.8.3 Uses of a DRFM System

Electromagnetic warfare is undergoing a renaissance. Disruptive technology is accelerating at a rapid pace transforming both commercial and military systems alike. The use of the EMS makes for a very congested and well-contested EM maneuver space. The technology and EMW technique trends are driving up the tempo of EM maneuvering within the EMS, as new technologies enable the development of radars, radios and EMW systems that are increasingly faster and smarter.

Today's airborne platforms are confronted with tough requirements for reduced volume, prime power, weight, through-life cost, and numbers of antenna apertures. In airborne platforms and especially in unmanned aircraft vehicles with their hard space and weight constraints, multifunction DRFM sensors, that enable the merging of functionalities; that is, not only radar air-to-air and air-to-ground functions, but also communication (e.g., data links for missile guidance and/or to ground stations), ES for radar and missile warning, EA for jamming as well as EA against ground radars, are of extremely high interest.

There are four steps a DRFM takes to attack an adversary:

- Evaluate the electronic environment.
- Select a electronic attack tactic.
- Deploy the tactic, that includes clutter enhancement (blip enhancement) deception jamming, and noise jamming. The DRFM can act as a *repeater* (original signal is amplified, modulated, and reradiated) or a *transponder* (DRFM intercepts the signal and a new signal radiated at regular intervals).
- Evaluate the results.

In radar EMW the DRFM is known as a jammer or repeater and can take part in both electronic protection or electronic attack [32]. This is also due in part to the revolution in machine learning and AI necessary to process the vast amount of data that can now be digitized and collected [33]. Convergence of AI, machine learning technology, and *deep learning* are starting to work together to achieve better results [34].

1.9 ELECTRONICALLY SCANNED ANTENNAS

An electronically scanned antenna (ESA) system differs from a mechanically scanned antenna primarily through the use of multiple passive phase shifters located on a fixed aperture. The waveform transmitted is subdivided to the many individual radiating elements [35]. Controlled variation of the transmitted waveform phase at each element electronically steers the beam boresight. The inverse process is used upon receive. Both passive-element and active-element electronically scanned antennas can be used. Both active and passive ESA support multimission/multimode DRFM operation and performance improvements over mechanically scanned antenna systems.

Passive-element, electronically scanned array (PESA) antennas have been realized using ferrite- or semiconductor-phase shifters while active-element array technology is based solely on solid-state technology. Comparing the roles of the phase shifters, it is found that they have widely different requirements as discussed below.

Active-element, electronically scanned antennas are distinguished by the use of transmit/receive modules and include radio frequency amplification (active) in addition to the phase shift function. These are distributed over the aperture and result in the elimination of the single-point, high-power transmitter and its supporting hardware [36].

1.9.1 Passive-Element Electronically Scanned Arrays

A block diagram of a PESA is shown in Figure 1.7 [37].

1.9 Electronically Scanned Antennas 21

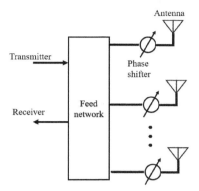

Figure 1.7 Block diagram of passive-element electronically scanning array (PESA) antenna.

In the passive array, the phase shifters are situated directly behind the radiator elements and carry the full transmit power and the raw received signals. Consequently, the phase shifter losses increase the antenna losses and reduce the system efficiency and also increase the noise figure. Another consequence of this is that the passive array employs only the absolute minimum number of bits in the monolithic micro-strip phase shifter circuit in order to keep the losses at a minimum (1 or 2 dB). The monolithic microstrip circuit (MIC) technology uses discrete diode chips for the necessary power handling capability. This results in large circuits with only a moderate bandwidth (10%).

The example in [37] uses integrated microstrip patch radiator/microstrip phase shifter construction. The planar phase shifters and bias patches on the back of the radiator substrate are limited in circuit area by the half-wavelength radiator patch in two dimensions. This limitation, in addition to low loss and cost, resulted in a phase shifter design with only 2 bits of resolution. Each phase shifter uses reflection design with three beam-lead PIN diodes exhibiting 1.1 dB loss averaged over the four phase states and over the 1-GHz bandwidth. Power handling capability was 5 W peak with only a 10 V reverse voltage and a 10 mA forward current per diode with only one diode being forward biased. The full array consists of 7,000 elements organized in 8 × 8 element modules. The beam pattern also had relatively high sidelobes due to the 2-bit phase shifter design.

1.9.2 Active-Element Electronically-Scanned Array

The use of microwave solid-state technology for T/R modules was first explored in the 1960s and the first deployed AESA T/R modules were the Raytheon PAVE PAWS radar modules developed in the 1970s [38]. An example of an active element electronically scanned antenna or *active electronically scanned array* (AESA) is shown in Figure 1.8 including a traditional T/R module. The AESA exhibits significant improvements in performance and reliability relative to both passive ESA and mechanically scanned systems at an affordable price. Some of the attributes of AESAs include instant repositioning of the antenna transmit power out and noise figure is set at the aperture.

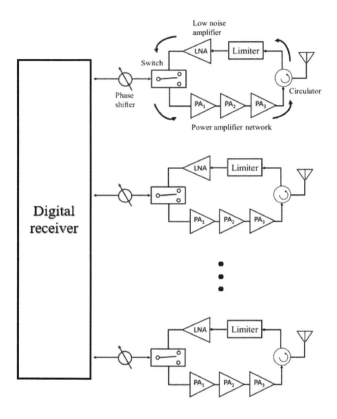

Figure 1.8 Schematic diagram of an active-element, electronically scanned array (AESA) antenna that uses T/R modules at each element.

The instantaneous scanning, beam agility characteristics provide for interlaced modes of operation, variable dwell-times, an increase in target update rates, and the ability to track multiple targets simultaneously using different frequencies and polarizations for each target. Independent control of the output power, the polarization phase, and receive gain of each TR module in the array allows almost unlimited flexibility to choose beam shapes optimized for the mission objectives. Independent control of output power and receive gain of each module also allows the transmit and receive apertures to be weighted and optimized separately for best performance in a given mission situation. The AESA has wider RF bandwidth potential and provides additional versatility for passive detection/tracking directive jamming and other electromagnetic warfare techniques [35].

The T/R module has a network of power amplifiers for transmit and a low noise amplifier (LNA) for receive. The T/R module is designed to support a transceiver type operation. A circulator or duplexer is used at the antenna to separate the transmit and receive analog radiation. A switch is used to flip between the transmit and receive channels in order to control the signal going to the attenuator and phase shifter. The switch can also be used to open-circuit the module in order to prevent any leakage from going out the antenna. Additional switches are required for modules that can transmit and receive in either horizontal or vertical polarization. A coordinated (incremental) phase shift sent to the matrix of modules is used to construct a beam and shift its angle within the field of view. By using more modules, the beamwidth can be narrowed and the sidelobes lowered. The phase shifter control voltage is changed each time the beam position is changed or the frequency changes.

Upon receive, the RF radiation is received at the antenna and is input to a circulator or (duplexer). The output of the circulator is first sent to a protection circuit or limiter before being sent to a LNA having a certain amount of gain. The LNA output is then sent through a switch to a phase shifter that is controlled to do the (analog) beam forming to steer the beam. The LNA sets the noise figure performance. The noise figure represents the amount of noise in the output that is above and beyond the thermal noise level.

The phase shifters used in an active array are realized as MMICs to integrate the high numbers of bits in a small, low-cost chip performing over considerable bandwidth (e.g., 30%-40%). The high insertion loss can be accepted since the circuits handle only small-signal power levels and reamplification is inexpensive. Unfortunately, the phase-to-amplitude conversion effect produces amplitude errors over the array. These depend on the phase distribution that can lead to high grating lobes and average sidelobes. Therefore, the insertion loss variation must be kept to a minimum and additional methods have to be taken to reduce the array degradation [37].

To summarize, the advantages of an AESA include a high degree of resistance to EA, the ability to enable low probability of intercept radar waveforms, high reliability due to the many T/R modules on the array, and an instantaneously agile multi-beam capability. Further, the very low radar cross-section and its ability to support multiple modes simultaneously make the AESA a significant component of the DRFM design as discussed in Chapter 2.

1.10 QUANTIFYING DRFM SPECTRUM DOMINANCE

The DRFM's coherent ability to collect wideband, radiated signals and replay them in near real time enables the ability for fast spectrum maneuvering. As an extremely agile component operating either autonomously or with a human-in-the-loop, the DRFM is changing the nature of spectrum domination. Integrated on the mission information grid with the Command and Control Warfare (C2W) officer, the sensing portion or receiver ES and active attack or EA activities can be interconnected as shown in Figure 1.9. As first conceptualized by the late Adm Arthur Cebrowski, often known as the father of network-centric warfare, many questions come up as the DRFM systems are integrated together.

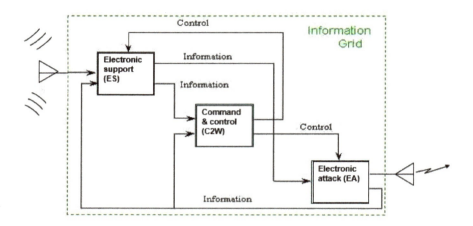

Figure 1.9 Network-enabled connection on the information grid that interconnects the receiver portion, the electronic attack portion, and the command and control section.

Questions to be asked include: What DRFM platform configurations can participate? How are they networked? What is the impact of these configurations on the network? What is the impact of the network on these systems? To measure and quantify the DRFM spectral maneuvering agility, it is first instructive to consider the network-enabled DRFM as part of an observe, orient, decide and act (OODA), loop as shown in Figure 1.10. As shown below, the EM agility is achieved through a technologically enabled OODA-loop and is the overarching key to dominating the spectrum maneuver and maintaining agility in the EM domain.

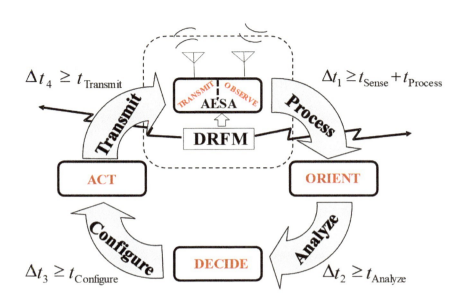

Figure 1.10 The DRFM as a network-enabled component within an observe, orient, decide and act (OODA) loop.

As a multifunction system-of-systems on a network, the DRFM can operate with several simultaneous RF functions, enabling it to maneuver about the EM spectrum and exploit highly defended or difficult spectral terrain. The DRFM functions can be broken down in terms of the loop constructs with the overall goal of electromagnetic spectrum domination operation.

Observation
In the observation phase the DRFM receives the spectrum density information to feed multiple DRFMs on the network, as well as to react to the incoming spectrum energy *autonomously* or by command and control (C2). A network of DRFMs in a cooperative spectrum sensing activities adopt cooperative spectrum sensing (CSS) to overcome these problems. Cooperation among the sensing users can be centralized or distributed. In distributed CSS, individual users share sensing information with each other without any central control station, while centralized CSS rely on the central control station to collect sensing reports of the individual users for an appropriate final decision. The combination schemes at the fusion center (FC) are categorized into soft and hard combination schemes. Hard fusion schemes (HFS), such as logical AND, logical OR, and voting allow the users to take local decisions and send hard binary information to the FC. The main goal here is to enable a common operational picture for situation awareness taking in all electromagnetic signatures.

For receiving, a high third-order-intercept (TOI) of approximately +15 dBm is necessary as well as low noise figure (below 3 dB). The TOI is required for the front-end in order to be resistant against parasitic RF signals and EA. A low noise figure, is required to be able to evaluate weak RF signals and this parameter also determines the range observable by the DRFM. This can be shown be examining the *sensitivity* as

$$\delta_I = kT_oB_I(NF)SNR_{Ii} \tag{1.1}$$

where k is Boltzmann's constant, T_o is the ambient temperature, B_I is the input bandwidth to the receiver, NF is the noise factor, and SNR_{Ii} is the input required signal-to-noise ratio of the intercepting receiver. Linear power amplification is also required for applications such as high-speed datalinks however, log-amplifiers are used for receivers to ensure that an unsaturated output is typically available for input to the video bandwidth section of the receiver.

Orientation
The orientation phase integrates the sensor data and creates a single, fused, real- or near-real-time *state of spectrum* (SOS). The orientation phase is also where cognitive assignments are determined. The automatic detection of available channels within the wireless spectrum frequencies are identified and directed to the emitters required to transmit. Another goal of the orientation phase is tracking multiple frequency contacts. For this case the DRFM chooses the proper gates to associate the spectrum observations with the frequency track files established. In addition, it updates the parameter changes for the emitters with confirmed track files. It also determines the kinematic states of the high-priority spectrum contacts that have been reported over time and is often done from a set of noisy or partial observations.

The multiple contract tracking in time, space, and frequency is similar to tracking a maneuvering aircraft in north, east, and down (NED) coordinates of a physical platform. In orientation phase, the derivation of *spectrum access policies* is considered. Here the confronting question is answered – are we approaching our spectral density limitations? Spectrum access policies being derived include adversary communication signaling and protocols and RF low probability of intercept modulation schemes. A significant capability in the orientation phase is *passive target detection*. Otherwise known as passive coherent location (PCL), the search, tracking, and imaging of targets with unattended self-configurable algorithms is accomplished by processing the target's reflection off a convenient broadcast tower and coherently demodulating the return from the target using a direct waveform from the source radiation. The advantage of this radar technique is the covertness in that no transmitting energy is required.

Decision Making
Decision making can take on two dynamic timelines. For automated decision support to C2 when the inputs are subjectively complex, the concept of machine learning

is included such that the automated analysis of the operational pictures triggers a reconfiguration event that in turn enables action in the form of a maneuver within the electromagnetic operational environment. The analysis and triggering of the EMS reconfiguration is done autonomously and quickly. Machine learning is also used for dynamically and autonomously monitoring frequencies and frequency assignments within the spectrum and detecting available channels in order to avoid congestion in order to maximize the number of concurrent users supportable within the network.

When the C2 commander has to decide the actions (or reactions), the decision making is done under pressure and the process is closer to "recognition" rather than rational analysis of alternatives and the decision speed is relatively slow by comparision. Here, the decision maker has a number of potential courses of action, each appropriate in one type of situation or another. Decision making theory can be simple even if the decisions themselves and the inputs are subjectively complex.

Actions

Actions that affect spectrum dominance include using the spectral configuration and applying integrated fires for kinetic kills and non-kinetic kills without any unintentional effects! Autonomous detection classification of LPI radar communications constitutes another action that can be taken to affect spectrum dominance. This includes cancellation of own-platform signatures and jammers and other interference. Actions to maintain spectrum dominance can also be passive in nature such as the passive detection, and tracking of targets using *emitters of opportunity* for example, television towers, cell towers, radio station towers using a technique known as passive coherent location (PCL). The tower emissions can provide the energy to be used by the DRFM for targeting information to aid in anti-radiation missile targeting and strike warfare.

1.10.1 DRFM Operational Tempo

Upon reception by the DRFM intercept receiver in the "observe" phase, the digitized spectrum is sent to the C2, ($\Delta t_1 \geq t_{Sense} + t_{Process}$) where t_{Sense} is the time required to intercept the spectrum and $t_{Process}$ is the time required to digitize and process the signals. The resulting sensor data is subsequently processed and a complete EM operational environment snapshot is built in the "orient" phase. From here, analysis is performed, ($\Delta t_2 \geq t_{Analyze}$), (either autonomously or by the C2) in order to "decide" what portion of the spectrum can best be utilized or reconfigured. Once this cognitive configuration has occurred ($\Delta t_3 \geq t_{Configure}$) the DRFM (C2) arrives at the "act" phase of its cycle where it is able to transmit the EM spectrum activity, where ($\Delta t_4 \geq t_{Transmit}$) represents the time from action to observation. This variable is limited by the time needed for a transmission to take place in the newly configured EM operational picture. Letting t_{OODA} represent the total time it takes to complete one cycle of the OODA loop (i.e., the period), we have

$$t_{OODA} \geq \sum_{i=1}^{4} \Delta t_i = t_{Sense} + t_{Process} + t_{Analyze} + t_{Configure} + t_{Transmit} \quad (1.2)$$

In other words, the frequency (tempo) of the EM maneuver OODA is

$$Tempo_{OODA} = 1/t_{OODA} =$$

$$\frac{1}{t_{Sense} + t_{Process} + t_{Analyze} + t_{Configure} + t_{Transmit}} \quad (1.3)$$

This measure can in turn be used to derive a few key DRFM insights. For example, if we let $t_{Maneuver} = t_{Configure} + t_{Transmit}$, the maneuver time (after some algebra) can be shown to be [8]

$$t_{Maneuver} = \frac{1}{t_{Maneuver}} - t_{Sense} + t_{Process} + t_{Analyze} \quad (1.4)$$

As expected, the time it takes to maneuver within the EM spectrum is related to the time required to sense, process, and analyze the spectrum received and the tempo of our OODA loop.

1.10.2 Electromagnetic Maneuver Agility

Maneuvering within a contested spectrum is a contest that depends upon the tempo, accuracy, and interaction of both parties' OODA loops. For simplification, let's first consider a worst-case scenario in which the enemy's sensing, processing, and analysis capabilities always result in a correct maneuver and their attacks are therefore restricted only by their tempo. Let $Tempo_{Enemy}$ represent the tempo of the enemy's OODA loop. Additionally, let ρ represent the *resistance measure* or number of enemy attacks or impediments encountered during a given unit of time t. Further, let μ represent our maneuver measure or the number of our reconfiguration events that occur during the same time t. With these considerations, we can compare OODA-loop tempos along with the *maneuver versus attack ratio*: μ/ρ [8].

Case I: $Tempo_{OODA} < Tempo_{Enemy}$
Then $\frac{\mu}{\rho} < 1$

In this case we are incapable of keeping up with enemy EM activities. The number of enemy attacks exceeds the number of our maneuver responses. *Consequently, our EM operations are degraded or destroyed.*

Case II: $Tempo_{OODA} \geq Tempo_{Enemy}$

IIa: Sensing, processing or analysis function are insufficient or incorrect.
Then $\frac{\mu}{\rho} \leq 1$

Despite superior tempo, we are incapable of making proper EMOE maneuver decisions. Enemy attacks exceed the number of correct maneuver responses we make. *For this case, our EM operations are degraded or destroyed.*

IIb: Sensing, processing or analysis function are correct and our maneuver actions are effective.b
Then $\frac{\mu}{\rho} = 1$

Equilibrium is achieved and we are able to match the adversary move-for-move in the EMOE. *EM operations are sustained.*

IIc: Sensing, processing, and analysis are correct and our system is designed to maneuver in the absence of enemy action (i.e., preemptive maneuver).
or
Sensing, processing or analysis functions are insufficient or incorrect and the DRFM conducts unnecessary or incorrect maneuvers within the EMOE.
Then $\frac{\mu}{\rho} > 1$

The DRFM either purposely operates using an excessive maneuver in order to preempt an electromagnetic attack (e.g., frequency hopping spread spectrum DRFM) or the DRFM incorrectly engages in maneuver beyond what the enemy actions dictate. *In this case, EM operations are sustained or possibly degraded.*

1.10.3 A DRFM Measure of Effectiveness

In summary, we show that the ratio μ/ρ can be a DRFM *measure of EM effectiveness* (MOE). If this quantity $\mu/\rho < 1$, it means that attacks are taking place in the EM spectrum that are going unanswered. This in turn leads to degradation or total loss of EM spectrum capability at this particular node. A $\mu/\rho = 1$, provided that accurate sensing, processing and analysis occurs, means that enemy actions within the EM spectrum are being answered with effective defensive maneuvers. Perhaps surprisingly, a $\mu/\rho > 1$ isn't necessarily desirable. In the case of preemptive maneuvering, such as frequency hopping spread spectrum communications, the elevated level or maneuver is part of the DRFM's operating parameters and therefore does not adversely affect performance. However, in other circumstances, an excess level of maneuver can lead to configuration associated time delays that in turn result in degraded performance. In the extreme case, a large μ/ρ number can be indicative of a situation known as *thrashing* in which the DRFM is in a constant state of reconfiguration and the usable time within the EMOE along with the information throughput decreases to zero.

The DRFM (and network) technologies are the key to dominating the spectrum maneuver and maintaining the agility in the EM domain. The test of EM spectrum agility is the ability to move quickly and change directions (e.g., polarization) while maintaining EM dominance towards the mission objective. The OODA allows a C2 commander the capability to quantify his or her ability to deny spectral information to the enemy and so disrupt their C2 capabilities. At the same time safeguards are exercised to protect friendly spectrum activities (EP) against retaliation, EM attacks, and jamming.

Of course the outcome of an EM spectrum engagement doesn't just depend upon the equipment technology, capabilities, and tempos of both player's OODA loops. What we observe (sense) depends on the adversary's action (transmission). How we orient the spectrum (analyze) takes into account their previous observation (sensing). What we decide (configure) must consider the adversary's previous orientation. How we act (transmit) will depend upon the adversary's previous decision (configuration). That is, each individual phase of our OODA loop is interconnected with each corresponding phase of the adverary's "OODA loop" and vice versa. Continuous analysis of each move and counter-move is critical. Given the time constraints that the DRFM node faces in the EM operational environment, the analysis most often must take place autonomously, and if not, the C2 must consider the time constraint even more carefully.

1.11 DRFM SIGNAL PROCESSING TECHNIQUES

Small form factor transceivers, with high capacity FPGAs, GPUs, and multi-core central processing units (CPUs) are being targeted for embedded DRFM micro computers. These systems are being used onboard high endurance forward sensing, unmanned aerial vehicles and unmanned surface robot warships to increase the distribution of the spectrum sensing and power projection capabilities of the warfighter. For computational capabilities beyond the local DRFM sensor, the data is networked back to manned surface combatants that can process and act on the data. Signal processing techniques onboard the DRFM sensor today are briefly introduced below and discussed further in later chapters.

1.11.1 Artificial Intelligence and Machine Learning

The concept of artificial intelligence (AI) was first proposed by *John McCarthy* (1927 - 2011) while working as a mathematics and computer scientist professor at Dartmouth College. It was there where he organized the now-famous Dartmouth Conference on AI along with *Claude Shannon* in the summer of 1956 [39].

The development of AI technology has had a profound impact on military development trends, leading to major changes. AI technology has entered a new period of high-speed growth and is recognized as the most likely disruptive technology to change the world in the future. AI is making great strides toward a long-term goal of superhuman intelligent machines. Whether in the field of bioimaging, avionics, mental health, or health care in general, AI is finding its way into all walks of life.

At the rise of technology innovation, people have been wary, from the weavers throwing their shoes at the mechanical looms at the beginning of the industrial era to today's fear of killer robots that reject humans and begin to reproduce themselves. If AI succeeds in its present endeavor, that could well be catastrophic for the human race [40]. The reason is that the standard model of AI requires machines to pursue a fixed objective specified by humans.

As the 18th century Scottish philosopher David Hume wrote in his 1738 classic *A Treatise of Human Nature,* the AI program, as a result of the consequence of human intelligence, will develop the right goals on its own, presenting his position as an orthogonality thesis:

> .... Intelligence and final goals are orthogonal: more or less any level of intelligence could in principle be combined with more or less any final goal.

The Orthogonality Thesis asserted by Hume is that there can exist arbitrarily intelligent agents pursuing any kind of goal. The strong form of the Orthogonality Thesis says that there's no extra difficulty in creating an intelligent agent to pursue a goal, above and beyond the computational tractability of that goal. That is, the Orthogonality Thesis is a statement about computer science, and is an assertion about the logical *design space* of possible cognitive agents. But no matter how much technology we put into a machine we wonder, *"Can a machine ever think* [41]?"

I had the distinct privilege to know Richard W. Hamming while I was at the U. S. Naval Postgraduate School (NPS) as his office was across the hall from mine when computer science was colocated with electrical and computer engineering in the mid-1990s. He was a tall, sometimes loud but always gentle man. As I remember when we went out for lunch one afternoon, he was complaining to Prof. Herschel Loomis, in a jovial sort of manner, how everyone was developing Hamming-this and Hamming-that but he never knew anything about any of it! He had everyone in stitches.

Anyway, the reason I bring this up, is that he taught a course at NPS, fondly called "Hamming-on-Hamming" that was all about his life stories when he was a researcher at Bell Labs and learning to learn. The classroom was always packed and his notes were finally turned into a textbook just before his death in 1998 [41]. He was always interested in the question about AI and if machines could think, and dedicates three chapters to the subject.

Another question is how much of the decision making is left to the AI embedded into the platform (e.g., for integrated fires?). AI has the potential to play a significant role in several domains of DRFM processing. AI can exploit big data analytics to improve the EMS analysis, prediction of white spaces, as well as adversary social behaviors. The DRFM wireless network has to analyze a large amount of processed SoS data generated by other network-enabled DRFMs while also reducing the intercepted EMS to derive its own EMS to share.

While each DRFM on the network is an open-system architecture using embedded computing with processed EMS data sent to decision makers over a secure network, AI acts as an enabler for the DRFM's position within the cognitive, self-organizing network. There are four trends in embedded computing that are important for AI. Those include SWaP, performance, open systems architectures, and interoperability (e.g., SOSA), and increased features and applications. Boards are now being liquid cooled to take heat away from the FPGAs and GPUs. A rugged design is also important. For example, Mercury's Ensemble LDS3517 Xeon D embedded server with support for FPGAs and switch mezzanine card XMC modules puts together a server-class Intel processor with the Xilinx UltraScale family of FPGAs in the 3U Open VPX (rugged) form factor using VITA-48. Another example is Aitech of Chatsworth CA's C877 SBC, which is also able to integrate the Xilinx Zynq UltraScale+ FPGA. The C877 has a high bandwidth bus architecture and a variety of onboard I/O options for AI.

In order to handle the extreme amounts of data and contacts, data caching is required (e.g., at a base station). Caching the data can alleviate the DRFM network congestion by avoiding data access from the core network because they can share data with each other via a wireless link [42]. The importance of AI in data caching is illustrated in Figure 1.11. AI can enable intelligent resource management in contrast to traditional optimization techniques. An overview of the traditional caching approaches including the structures, policies, performance metrics, and challenges is provided in [42]. AI can be partitioned into two categories. The first type is basic in nature in that it enables the DRFM to respond to the EMS environment in a deterministic way. It enables the network to configure itself based on the performance indicators that are measured. The second type is based on the complete capability of DRFM to interact with the EMS, which enables the system to make decisions even in the case of an unknown environment. AI can make nearly instantaneous decisions using huge amounts of information because it is able to conduct the learning process on a high-dimensional input dataset.

1.11.1.1 Neural Networks

Neural networks use a set of processing elements (or nodes) loosely analogous to *neurons* in the brain (hence the name, neural networks). One example is illustrated in Figure 1.12.

1.11 DRFM Signal Processing Techniques   33

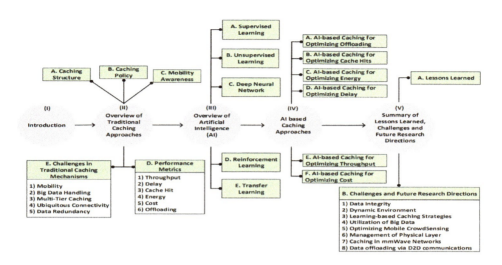

Figure 1.11 Survey of AI techniques used in data caching for wireless networking. (From [42].)

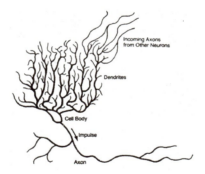

Figure 1.12 Schematic illustration of a neuron in the brain and its different parts.

These nodes are interconnected in a larger network by the axons and dendrites from many other neurons within the brain. As the network is exposed to data or images it trains autonomously to identify the patterns in data as it is repeatedly exposed to the data. The neural network example in Figure 1.13 behaves in a similar manner. By applying data and training the weights (circles) the network can begin to learn patterns and identify inputs.

34 ELECTROMAGNETIC SPECTRUM DOMINANCE

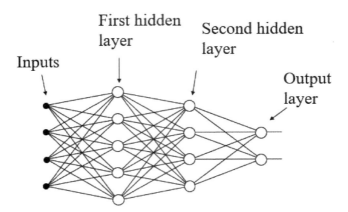

Figure 1.13 Architecture of a multilayer perceptron neural network with an input layer, an output layer, and two-hidden layers.

In a sense, the network learns from experience just as people do. The circles represent the nodes (or cells in Figure 1.12). One particularly useful model of a neuron that is used is the Rosenblatt perceptron as shown in Figure 1.14. Here X_n represent the n−dimensional input vector and W_n the n−dimensional weighting vector. The neuron forms a weighted sum of n−components of the input vector and adds a bias value $W_0 = \theta$. The result of the summation is passed through a nonlinearity.

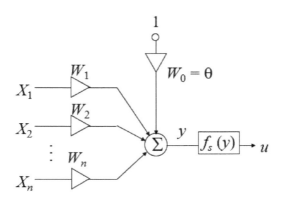

Figure 1.14 Block diagram of a Rosenblatt perceptron.

This distinguishes neural networks from traditional computing programs that simply follow instructions in a fixed sequential order. With I input nodes and H hidden layers, the transfer function of this multilayer perceptron network is [43]

$$y_k(l) = f_s \left[\sum_{h=1}^{H} w_{kh} f_s \left(\sum_{i=1}^{i} w_{hi} x_i(l) \right) \right] \quad (1.5)$$

where y_k is the output, x_i is the input, l is the sample number, i is the input node index, h is the number of hidden layers index, and k is the output node index. Here w_{kh} and w_{hi} represent the weight value from neuron h to k and from neuron i to h, respectively, and f_s represents the sigmoid activation function. All weight values in the network are determined at the same time in a global nonlinear training strategy involving, for example, supervised learning. The sigmoid nonlinearity f_s can be expressed as

$$f_s(y) = \left(1 + e^{-\beta y}\right)^{-1} \quad (1.6)$$

where β is the gain of the sigmoid. Note the important property that it is continuous, differentiable, and varies monotonically from 0 to 10 as y varies from $-\infty$ to ∞.

Going by many names such as connectionist models, parallel distributed processing models, and neuromorphic systems, these architectures are specified by three things:

1. Network topology: How many nodes are in the input layer? How many nodes are in the output layer? How many hidden nodes are there? How are the nodes connected?

2. Node characteristics: What is the transfer function of the nodes? That is, when the weighted inputs are summed, what form does the nonlinearity $f_s(y)$ take on?

3. Training methods: What type of training is used to update the weights in the hidden layers and output layers? This includes back propagation, gradient search, and temporal difference as examples. Other training concepts include *pruning* or eliminating the weights that disturb the solution the least, and *complexity regularization* discouraging the learning algorithm from seeking solutions that are too complex (brain damage!).

In EMMW, AI is an effective technique that can be used for data mining to learn the adversary's method of operations. Its main goal is to explore the relationship between the input EMS data, the output contacts, and the adversary spectrum activity. This enables the DRFM to "auto-process" patterns of EMS input frequency data for making instantaneous decisions. For example, AI-based decision-making caches and content caching in wireless networks can be performed based on a learning process. That is, by using AI the potential to accurately predict the "data popularity profile" can be learned by utilizing an adversary's social behaviors, mobility patterns, and their localization.

The raw dataset is provided to the training process for decision making and provides accurate prediction. The approaches for *intelligent learning* can be classified into supervised learning, unsupervised learning, reinforcement learning (RL), and transfer learning (TL).

In *supervised learning*, a labeled dataset is provided that contains inputs and known outputs for building or training a model based on the relations included in the dataset. Then, a new dataset is collected and given to the learned model. This enables an algorithm to perform optimal predictions.

In *unsupervised learning*, a set of unlabeled input dataset is provided to the algorithm. The objective is to determine patterns and learn by clustering data into multiple groups based on the similarity. Unsupervised learning is most widely utilized in data collection and clustering problems.

Deep neural networks (DNNs) are used to perform complex computations and handle high-dimensional input data via multiple hidden layers. DNN transforms the input into an output that has either a linear or nonlinear relationship. Each hidden layer's neurons are based on the output of the previous layer's feature-training. A DNN is a feed-forward network where data flow occurs from input to output without complex feedback loops.

Transfer learning is a method of learning from one problem and applying the gained knowledge or information to another related problem. Transfer learning helps to enhance the learning process by transferring knowledge between source and target domains and is beneficial when the training dataset is insufficient to train the network from scratch. Transfer learning can be devised in regression, clustering, and classification problems. Despite the benefit, this approach may show low performance if the relationship between source and target domains is inadequate.

Transfer learning differs from traditional machine learning in that it is the use of pretrained models that have been used for another task to jump-start the development process on the new task or problem. DRFM transfer learning involves the spectrum domain and a particular task such as modulation classification.

As an example, consider the spectral domain $S(\omega)$ consisting of a feature space $FS(\omega)$ and a marginal probability distribution $P(F)P(F)$ over the feature space, where $F = \omega_1 \ldots \omega_n \in F(\omega)$ and also $F = \omega_1 \ldots \omega_n \in FS(\omega)$. For modulation classification (one key task for the DRFM), with a bag-of-frequencies representation, $FS(\omega)$ is the space of all frequencies within the intermediate bandwidth, $\omega_i \omega_i$ is the iith term vector corresponding to some frequency and $FS(\omega)$ is the sample of frequencies used for training.[1]

The DRFM's *reinforcement learning* is a machine learning algorithm hosted within the permanent memory, concerned with how software agents need to take actions in the spectrum in order to maximize the desired SOS with some notion of cumulative reward. Reinforcement learning is one of three basic machine learning algorithms that reside with the supervised learning and unsupervised learning used in modulation classification (See Chapter 9).

1.11 DRFM Signal Processing Techniques 37

DRFM functions where AI can make a significant difference include:

1. Deriving the state of the spectrum (SOS): Can only be accomplished by a machine learning AI algorithm on a computer due to the rapidly changing EMS radiation. A DRFM application will experience channel selection collision when other spectrum users compete for the same channel. Thus, the decisions of spectrum users are coupled. Such couplings cannot be easily analyzed by traditional game-theoretical approaches because the perfect information of other game players as well as the channel state information is unknown [44].

2. Situation assessment, surveillance, and geolocation of targets [45].

3. Autonomous recognition of emitter modulations: Machine learning through the use of convolutional neural networks (CNN), clustering support vector machines, and other nonlinear processing tools, can process the EMS quickly, passing the detects and classifying the known and unknown emitters out to the C2 and others on the network.

Today there is considerable disagreement across industry on when fully autonomous systems will be deployed by the military. All agree such systems will require at least a low level of AI. The most oft-cited problems are legal, moral, and ethical involving a robotic platform capable of self-guidance and action on the battlefield. But in fact, AI will touch nearly every area of technology. An outline of the AI algorithms is shown in Figure 1.15. The organization is broken out by the type of learning used. This includes supervised learning, unsupervised learning, deep neural networks, reinforcement learning, and transfer learning.

With the data analysis using AI (e.g., processing thousands of satellite images) better protection for the warfighter can be accomplished; for example, identifying wounded personnel from the battlefield, locating mines within mine fields while maintaining a strong cyber defense online eliminating any data poisoning inserts. With the evolution of autonomous systems on the battlespace, from UAVs to RF weapons and robots, with each generation relying more and more on AI capabilities, the DRFM remains a critical center piece for all cyber and electronic warfare activities.

With the DRFM as a smart digital transceiver, infusion of technologies such as *superconductivity electronics* (SCE) will improve sensitivies far below -100 dBm [46, 47]. In addition, the inclusion of *photonics* and integrated optics will improve microwave receiver bandwidths and enable direct digitization by efficient coupling of the antenna signals directly into the optical domain for high-speed sampling and signal processing [48]. In addition, new antenna technologies such as AESA and high power amplifier material such as *Galium Nitride* (GaN) technologies are making a significant impact in DRFM transducer reliability [49].

For the DRFM's self-healing stage where the transmit waveforms are reconfigured to set up the spectrum in the desired plan according to the time-frequency-space

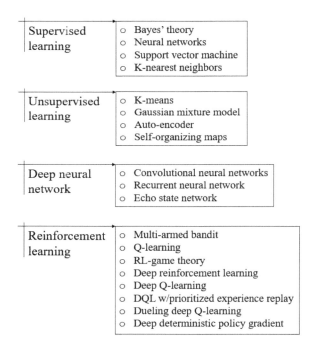

Figure 1.15 Breakout of the artificial intelligence architectures by method of learning.

constellation required. Self-healing algorithms include the use of *logistic regression, support vector machines*, and *hidden Markov models*.

Logistic regression is a type of statistical analysis used for predictive analytics and modeling, and easily extends to our DRFM applications in machine learning. It is an extension of linear regression, which is a standard algorithm that is not as complicated/accurate as other models. Yet, it is the building blocks for more complicated models. In the logistic regression approach, the dependent variable is finite (or categorical). That is, it is either A or B (binary regression) or a range of finite options A, B, C, or D (multinomial regression). The DRFM uses this algorithm approach to understand the relationship between the dependent variable and one or more independent variables by estimating probabilities using a logistic regression equation.

This helps the DRFM predict the likelihood of an event happening (e.g., presence of an emitter frequency) or a choice being made. For example, you may want to know the likelihood of a frequency being jammed — or not (dependent variable). Your analysis can look at known characteristics of spectral activity, such as frequency activity around this particular band, repeat visits by an emitter to a particular frequency, and behavior on your emitter frequency of interest (independent variables). Logistic regression models help you determine a probability of what type of emitters are likely to be present — or not. As a result, you can make better decisions about the reconfiguration of the spectrum and the DRFM activity or make decisions about the SOS itself.

1.12 SUMMARY

The DRFM is a software-defined transceiver and if it is on a mobile platform (e.g., miniature air-launched, surface-launched decoy) is able to adapt time, frequency, and space. The concept of *maneuver agility* or electromagnetic agility is able to quantify the DRFM's ability to maneuver around highly defended (or difficult) EM terrain. As a software-defined transceiver (SDT) that is able to cognitively adapt to changes in the *state of the spectrum* it uses a self-organizing network for self-configuration and self-optimization. The DRFM actions within the EMS must adapt quickly to the changes that are taking place continuously. These actions that are adapting, reacting – including the preemptive actions – can be modeled within a time-adaptability cycle with a specific time constant. This leads to the spectrum adaptability cycle. The adaptability cycle gives the DRFM a competitive advantage.

The DRFM uses statistical concepts as part of the machine learning to enable its onboard computer to "learn" without explicit programming constructs. The DRFM *transfer learning* is also a machine learning method where a spectrum model or activity developed for a task (e.g., electronic attack) is reused as the starting point for a model on a second task. The use of AI machine learning and deep learning algorithms in order to *maintain spectrum maneuver agility* (Chapter 10) forcing the enemy into defensive positions towards unfavorable and exploitable EM terrain.

These defensive spectrum operations (a) defeat an adversary attack, (b) buy time, (c) economize transmitter and system power on the UAS, and (d) develop a SOS condition that is conducive for offensive operations such as electronic attack. These defensive operations alone cannot achieve the desired SOS domain situation. The purpose of forcing the enemy into this position is to create circumstances for a counter-EM offensive that allows the DRFM regain the spectral advantage. Additional reasons for conducting defensive operations include:

- Retaining decisive bandwidth or denying a section of bandwidth to the adversary;
- Fixing the adversary spectral position as a preliminary step to offensive operations;
- Surprise maneuver by the adversary;
- Increasing the adversary's exposure by forcing them to increase his transmit power and/or concentrate their spectral activity into a particular bandwidth.

While an "offense" (e.g., electronic attack) is the most *decisive* type of combat operation, a "defense" is the *stronger* type. The inherent strengths of the DRFM defense include: (a) the ability to occupy the spectral positions *before* an attack and (b) use the available time *to prepare* defensive protection measures. The DRFM's ultimate goal is to *prevent* an adversary's spectral maneuvers through a combination of transmitted tones, noise waveforms and other modulations including coordinated, coherent attacks from multiple, networked DRFM platforms.

Generating and participating in *coordinated effects* is especially critical to a DRFM conducting the defense of a particular bandwidth against an adversary with a significant advantage in transmit power and the ability to select the frequency and time of the attack. The adversary can amass their transmitter power at a specific frequency or bandwidth, thus dramatically influencing the jam-to-signal ratio of forces at that frequency. An adversary 3-to-1 advantage in overall combat power can easily turn into a local 6-to-1 (or higher) ratio. Special limiters must be built at the front-end to prevent this type of saturation and a good amount of forward thinking must go into the planning of each deployed transceiver design.

Unmanned systems together with AI now offers the capability to deploy diverse, persistent and robust (and energy efficient) DRFM sensor networks to play a critical role in (a) understanding and awareness of the EMS environment, (b) EMS agility, decentralized operations and cooperative deception, and (c) full spectrum access and EMS maneuverability.

It is the unique situational awareness in support of operational planning, as a result of distributed sensor coordination and collaboration, dynamic sensor tasking, and management that also enables the *unparalleled collection and exploitation capabilities* offered by the DRFM. No better global picture can be offered than that of the sensor data transfer and fusion into a common, collaborative picture.

REFERENCES

[1] S. J. Roome, "Digital radio frequency memory," *Electronics Communication Engineering Journal*, vol. 2, no. 4, pp. 147–153, 1990.

[2] H. Pace and P. E. Pace, "Frequency management for the 21st century," in *Journal of Electronic Defense*, vol. 21, no. 1, 1998, pp. 21–25.

[3] D. C. Schleher, *Electronic Warfare in the Information Age*. Norwood, MA: Artech House, 1999.

[4] Headquarters, Department of the Army, "Electronic warfare techniques," vol. ATP 3-12.3, no. 1, 2019.

[5] Chairman, Joint Chiefs of Staff, "Information operations," in *Joint Publication 3-13*, vol. JP 3-13, 2012.

[6] J. R. Wilson, "Today's battle for the electromagnetic spectrum," in *Military and Aerospace Electronics Magazine*, vol. 34, no. 2, 2020.

[7] ——, "Enabling technologies for airborne electronic warfare," in *Military and Aerospace Electronics Magazine*, vol. 31, no. 2, 2020, pp. 12 –21.

[8] C. Martinsen, P. E. Pace, and E. L. Fisher, "Maneuver warfare in the electromagnetic battlespace," in *Journal of Electronic Defense*, vol. 37, no. 10, 2014, pp. 30–44.

[9] T. Schuck. (2017) OODA Loop 2.0: Information, Not Agility, Is Life. [Online]. Available: https://breakingdefense.com/2017/05/ooda-loop-2-0-information-not-agility-is-life/

[10] J. R. Wilson, "Today's battle for the electromagnetic spectrum," in *Military and Aerospace Electronics Magazine*, vol. 30, no. 8, 2016.

[11] T. Hentschel, *Sample Rate Conversion in Software Configurable Radios*. Norwood, MA: Artech House, 2002.

[12] J. Lunden, V. Koivunen, and H. V. Poor, "Spectrum exploration and exploitation for cognitive radio: Recent advances," *IEEE Signal Processing Magazine*, vol. 32, no. 3, pp. 123–140, 2015.

REFERENCES

[13] A. Orduyilmaz, M. Ispir, M. Serin, and M. Efe, "Ultra wideband spectrum sensing for cognitive electronic warfare applications," in *2019 IEEE Radar Conference (RadarConf)*, 2019, pp. 1–6.

[14] S. You, M. Diao, and L. Gao, "Deep reinforcement learning for target searching in cognitive electronic warfare," *IEEE Access*, vol. 7, pp. 37 432–37 447, 2019.

[15] Ryno Strauss Verster and Amit Kumar Mishra, "Selective spectrum sensing: A new scheme for efficient spectrum sensing for ew and cognitive radio applications," in *2014 IEEE International Conference on Electronics, Computing and Communication Technologies (CONECCT)*, 2014, pp. 1–6.

[16] Q. Guo, X. Zhang, and Z. Li, "A novel sorting method of radar signals based on support vector clustering and delaminating coupling," in *2006 5th IEEE International Conference on Cognitive Informatics*, vol. 2, 2006, pp. 839–844.

[17] S. Ayazgok, C. Erdem, M. T. Ozturk, A. Orduyilmaz, and M. Serin, "Automatic antenna scan type classification for next-generation electronic warfare receivers," *IET Radar, Sonar Navigation*, vol. 12, no. 4, pp. 466–474, 2018.

[18] B. Zhang and W. Zhu, "Research on decision-making system of cognitive jamming against multifunctional radar," in *2019 IEEE International Conference on Signal Processing, Communications and Computing (ICSPCC)*, 2019, pp. 1–6.

[19] O. F. Rodriguez, V. Kontorovich, and S. Primak, "Characteristics of sequential detection in cognitive radio networks," in *13th International Conference on Advanced Communication Technology (ICACT2011)*, 2011, pp. 307–312.

[20] J. Oksanen and V. Koivunen, "An order optimal policy for exploiting idle spectrum in cognitive radio networks," *IEEE Transactions on Signal Processing*, vol. 63, no. 5, pp. 1214–1227, 2015.

[21] S. Bagheri and A. Scaglione, "The restless multi-armed bandit formulation of the cognitive compressive sensing problem," *IEEE Transactions on Signal Processing*, vol. 63, no. 5, pp. 1183–1198, 2015.

[22] J. Keller, "Open-systems electronics standards for military embedded computing gaining money and traction," in *Military and Aerospace Electronics Magazine*, vol. 33, no. 2, 2019.

[23] B. Manz, "For multifunction systems, the future is open," in *Journal of Electromagnetic Dominance*, vol. 43, no. 3, 2020.

[24] J. Whitney, "The shrinking world of small-form-factor embedded computing," in *Military and Aerospace Electronics*, vol. 31, no. 3, 2020.

[25] J. Haystead, "DARPA's mosaic warfare," in *Journal of Electronic Defense*, vol. 43, no. 2, 2020, pp. 20 – 25.

[26] J. R. Wilson, "What is global persistent surveillance," in *Military and Aerospace Electronics Magazine*, vol. 31, no. 2, 2017, pp. 8 –19.

[27] Rajesh. (April 22, 2021) DARPA CONCERTO developing single reconfigurable UAV payload for communications, radar and electronic warfare functions to enhance UAS adaptability and mission efficiency. [Online]. Available: https://idstch.com/technology/electronics/darpa-concerto-developing-single-reconfigurable-uav-payload-for\ \-communications-radar-and\ \electronic-warfare-functions-to-enhance-uas\ \-adaptability-and-mission-efficiency/

[28] J. R. Wilson. (Feb. 8, 2020) Enabling technologies for airborne electronic warfare. [Online]. Available: https://www.militaryaerospace.com/sensors/article/14168864/airborne-electronic-warfare

[29] J. Keller. (June 11, 2019) DARPA hires three companies for blended RF system to reduce size and ease technology insertion aboard UAVs. [Online]. Available: https://www.militaryaerospace.com/rf-analog/article/14034822/blended-rf-system-uavs-technology-insertion#:~:text=DARPA%20hires%20three%20companies%20for%20blended%20RF%20system,into%20one%20integrated%20RF%20system%20for%20medium-sized%20UAVs.

[30] B. Clark and M. Gunzinger, "Winning the airwaves," in *Center for Strategic and Budgetary Assessments*, vol. 1, no. 1, 2015, pp. 1–49.

[31] D. Herskovitz, "A sampling of digital RF memories," in *Journal of Electronic Defense*, March 1998, pp. 51–54.

[32] M. A. Govoni and H. Li, "Complex, aperiodic random signal modulation on pulse-LFM chirp radar waveform," in *Radar Sensor Technology XIV*, K. I. Ranney and A. W. Doerry, Eds., vol. 7669, International Society for Optics and Photonics.   SPIE, 2010, pp. 257–267.

[33] I. Arel, D. C. Rose, and T. P. Karnowski, "Deep machine learning - a new frontier in artificial intelligence research [research frontier]," *IEEE Computational Intelligence Magazine*, vol. 5, no. 4, pp. 13–18, 2010.

[34] J. Hou and Y. Wen, "Prediction of learners' academic performance using factorization machine and decision tree," in *2019 International Conference on Internet of Things (iThings) and IEEE Green Computing and Communications (GreenCom) and IEEE Cyber, Physical and Social Computing (CPSCom) and IEEE Smart Data (SmartData)*, 2019, pp. 1–8.

REFERENCES

[35] W. P. Hull and R. D. Nordmeyer, "Active-element, phased-array radar: affordable performance for the 1990s," in *NTC '91 - National Telesystems Conference Proceedings*, 1991, pp. 193–197.

[36] M. Brandfass, M. Boeck, and R. Bil, "Multifunctional aesa technology trends - a radar system aspects view," in *2019 IEEE International Symposium on Phased Array System Technology (PAST)*, 2019, pp. 8138–8143.

[37] H. . Feldle and K. Solbach, "Passive and active phased arrays using solid state technologies," in *IEE Colloquium on Phased Arrays*, 1991, pp. 3/1–3/4.

[38] R. Sturdivant and M. Harris, *Transmit Receive Modules for Radar and Communication Systems*. Norwood, MA: Artech House, 2016.

[39] P. Hyman, "John McCarthy, 1927 - 2011," *Communications, ACM*, vol. 55, no. 1, p. 29, 2012.

[40] S. Russell, "It's not too soon to be wary of ai: We need to act now to protect humanity from future superintelligent machines," *IEEE Spectrum*, vol. 56, no. 10, pp. 46–51, 2019.

[41] R. W. Hamming, *The Art of Doing Science and Engineering, Learning to Learn*. Gordon and Breach Science Publishers, 1997.

[42] M. Sheraz, M. Ahmed, X. Hou, Y. Li, D. Jin, and Z. Han, "Artificial intelligence for wireless caching: Schemes, performance, and challenges," *IEEE Communications Surveys Tutorials*, pp. 1–1, 2020.

[43] P. E. Pace, *Detecting and Classifying Low Probability of Intercept Radar*. Norwood, MA: Artech House, 2009.

[44] P. Zhu, J. Li, D. Wang, and X. You, "Machine-learning-based opportunistic spectrum access in cognitive radio networks," *IEEE Wireless Communications*, vol. 27, no. 1, pp. 38–44, 2020.

[45] F. Amigoni, J. Banfi, and N. Basilico, "Multirobot exploration of communication-restricted environments: A survey," *IEEE Intelligent Systems*, vol. 32, no. 6, pp. 48–57, 2017.

[46] O. A. Mukhanov, D. Gupta, A. M. Kadin, and V. K. Semenov, "Superconductor analog-to-digital converters," *Proceedings of the IEEE*, vol. 92, no. 10, pp. 1564–1584, 2004.

[47] I. V. Vernik, D. E. Kirichenko, T. V. Filippov, A. Talalaevskii, A. Sahu, A. Inamdar, A. F. Kirichenko, D. Gupta, and O. A. Mukhanov, "Superconducting high-resolution low-pass analog-to-digital converters," *IEEE Transactions on Applied Superconductivity*, vol. 17, no. 2, pp. 442–445, 2007.

[48] P. R. Herczfeld and A. S. Daryoush, "Integrated Optic Components for An Optically Controlled Phased Array Antenna–System Considerations," in *Integrated Optical Circuit Engineering IV*, M. A. Mentzer and S. Sriram, Eds., vol. 0704, International Society for Optics and Photonics. SPIE, 1987, pp. 174–182.

[49] G. D'Amato, G. Piccinni, G. Avitabile, G. Coviello, and C. Talarico, "An integrated phase shifting frequency synthesizer for active electronically scanned arrays," in *2018 IEEE 21st International Symposium on Design and Diagnostics of Electronic Circuits Systems (DDECS)*, 2018, pp. 91–94.

Chapter 2

Digital RF Memory Receiver Architectures

2.1 DRFM ARCHITECTURES

There are as many DRFM transceiver architectures as there are applications for which they are used. Most of these are adaptations of the basic kernel structure suited for the purpose at hand. The basic kernels are the subject of this section. In addition, the GPS receiver is examined.

2.1.1 Addressing the Requirements

Depending on the mission at hand that the DRFM is supporting and what the end objectives are, DRFM configuration can be changed accordingly. The ES function of the DRFM is to receive, memorize, and possibly retransmit an adversary's continuous wave (CW), low probability of intercept (LPI) signal, pulse, or frequency modulated, and/or phase modulated radar signal as directed by the C2 or mode executive communication signals, which are also included here.

For example, if pulsed signals are of interest, the receiver will measure the frequency, pulse width (PW), pulse repetition interval (PRI), PRI type, stagger, stagger level, jitter, jitter rate, jitter waveform, power, scan patterns, frequency modulation on pulse (FMOP), and phase modulation on pulse (PMOP) of incoming radar signals. These measurements can be made very accurately.

After the measurements are made, the results may be compared with the emitter parameters stored in an internal memory database. After an emitter is identified against a preprogrammed emitter library, the emitter logic may assign a priority level to the measured RF signal to enable the correct EA waveform response management. Each DRFM technique generator has the capability to respond to many simultaneous radar signals depending on the architecture and whether it uses a wideband or narrowband approach.

2.2 SUPERHETERODYNE KERNELS

A block diagram of a single sideband DRFM kernel is shown in Figure 2.1. There are many issues in the development of a DRFM kernel and an overview is given here with the details left to Chapter 4, showing how to pick out the correct wideband spectrum sensing receiver, and to Chapter 5, where practical considerations are addressed for the complete transceiver design. These include the digitization or ADC issues and the clocking system that drives them. The quality of the sampling clock will affect the performance and stability of the the ADC. The ADC determines the instantaneous bandwidth coverage of any intercepted radiation and the frequency bandwidth over which frequency hopping will occur [1]. A bandlimiting preshaping filter should be included to limit the frequency range into the ADC while also minimizing the total harmonic distortion.

The digital signal processors (DSPs) and field programmable gate arrays (FPGAs) must also be chosen carefully to give the custom I/O resources the data path necessary to maintain signal integrity with fast transfer speeds. The block random access memory (RAM) resources must also be such that the I/O are symmetrical and independent. The DSP devices also play the important role of modulation and even interface to the application controller through an integrated PCI protocol communication port.

The digital-to-analog converters (DACs) are also required at the output as shown in the block diagram in Figure 2.1. Figure 2.1 shows a single-sideband DRFM kernel. The DAC shares the same clock as the ADC to keep the coherence of the reproduce signal with the original signal intercepted. The circuit design around the DAC should follow the similar technology rules as the ADC and meet the requirement for the model in driving ability and signal standards.

A printed circuit board, a brass board, or a monolithic microwave integrated circuit (MMIC) can serve as the integration platform depending on the sophistication of the design, the interfaces that are required to the ADC and FPGA, and DSP. Grounding is also an important consideration. The kernel consists of a receive antenna with a bandpass pre-select filter used to select and pass only the signals of interest [1]. A local oscillator (LO) is used to tune the DRFM to intercept the desired signal in the downconversion process. A LPF removes the components above Nyquist (antialiasing) along with unwanted mixer products.

At the output of the LPF, the signal is digitized by an ADC with resolution typically on the order of 2–10 bits depending on the DRFM throughput. The higher the resolution, the slower the conversion process. After digitization, the samples are strobed into memory. High-speed dual-ported memory is often used so that the stored digital signal can be recorded and replayed simultaneously through memory control.

With the use of multiported memory, recording and multiple replays can occur simultaneously. The retrieved digital signal is strobed from memory to a DAC in order to reconstitute the signal back into an analog waveform. After lowpass filtering, the

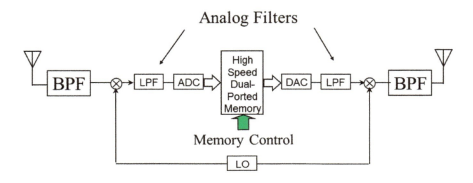

Figure 2.1 Single sideband digital RF memory architecture. Only the magnitude is digitized.

baseband signal is mixed with the LO in order to reconstruct the coherent waveform for retransmission on the carrier [2]. A BPF at the output serves to transmit only the desired frequencies of interest.

Figure 2.2 shows a block diagram of a double sideband DRFM [3]. This architecture is similar to Fig. 2.1 except that the phase of the DRFM signal is retained throughout the process using both an in-phase and quadrature channel. These signals are produced by the quadrature IF modulator at the input. The quadrature IF modulator also downconverts the input RF signal. Also shown in this particular configuration is the capability to retrieve the stored digital signal for further more complicated signal processing using FPGAs and DSPs.

This additional processing power can be used to create a variety of complex waveforms. For example, digital image synthesis for inverse synthetic aperture (ISAR) counter-targeting applications requires a level of signal processing that cannot be accomplished with simple memory recall and bit manipulation. The synthesis of the image requires a special-purpose processor that is able to generate the complex, coherent signal necessary (e.g., for counter-surveillance, counter-targeting).

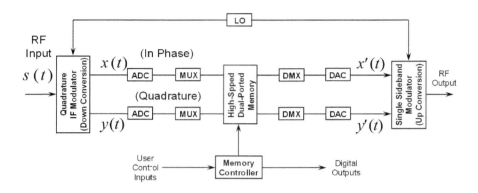

Figure 2.2 Double sideband digital RF memory architecture.

The superheterodyne configurations shown in Figures 2.1 and 2.2 give good rejection of spurious signals while retaining all the conventional advantages of the superheterodyne architecture [4]. It has been shown that, while the frequency conversion process of a fixed-tuned superheterodyne receiver displays maximum sensitivity to a particular signal frequency, the receiver also has some sensitivity to all input frequencies ω_{in} that satisfy the intermediate frequency relationship $\omega_{IF} = m\omega_{in} - n\omega_{LO}$ where m and n are integers and ω_{LO} is the local oscillator frequency [4]. Thus, with the local oscillator frequency fixed at ω_{LO} radians per second, there is an output at the IF for each signal frequency that satisfies the relationship above. These input frequencies are the well-known spurious responses. One measure of the receiver sensitivity to these undesired signals is the spurious-response rejection. This is defined as the ratio of the amplitudes of the undesired signal and the desired signal that produce the same magnitude signal at the converter output. In general, this ratio is a function of the local oscillator power, the particular spurious-response frequency, and the reference power of the desired signal.

2.3 CHANNELIZED KERNELS

SDR concept has been introduced as a means to reduce or eliminate the problems with downconversion and upconversion by placing the ADC and DAC of the DRFM as close to the antenna as possible so that most DRFM functions can be performed in the digital domain. Such an approach gives SDR systems the flexibility to be ported to different missions and standards and the ability to serve multiple goals simultaneously by software modification. However, SDR systems require powerful DSPs for performing functions such as channelization digitally, and they also require functions that are not required in conventional RF transceivers, such as sample rate conversion (SRC). Channelized receivers are widely used in modern wideband software-defined DRFMs. The channelized receiver architecture can achieve wideband intercepts by splitting the frequency spectrum into multiple parts and processing them in parallel as shown in Figure 2.3.

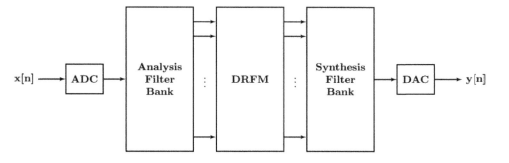

Figure 2.3 Block diagram of a channelized receiver used within a DRFM.

The channelized receiver achieves broad instantaneous frequency coverage and can slow down the data rate to realize the most complicated amplitude and frequency modulation schemes. That is, instead of processing the entire frequency band of interest, the frequency spectrum is split into multiple bands, each processed in parallel, resulting in a lower overall noise power.

The channelized receiver approach also helps to resolve multiple frequencies if channel bandwidths are small enough, thereby improving the instantaneous frequency measurement and direction of arrival estimation performance using only a single target frequency [5]. Using this approach a DRFM can also employ a channelized transmitter to fulfill simultaneous transmitting of multichannel signals. Consequently, channelized DRFMs are being used to enable ES and most recently EA.

The DRFM performs major functions such as false target generation, frequency modulation, and amplitude modulation. The output signal then goes to the synthesis bank, which reconstructs the wideband signal using the narrowband signals from the DRFM kernel. The architecture contains parallel channels of downsamplers, filters, exponential multipliers, and depending on the particular arrangement, increasingly more efficient architectures can be formed [6]. For the DRFM attack system, the analysis and synthesis reside at the intercept (ES) and jammer (EA) side, and for the communication system, the coupled filter bank pair resides at both the modulator side and at the demodulator side of the communication link. An example is shown in Figure 2.4.

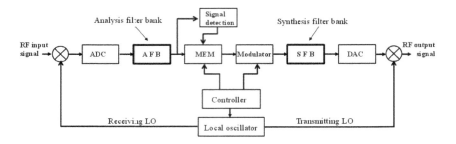

Figure 2.4 DRFM using an analysis filter bank at the front end and a synthesis filter bank at the output for EA.

2.4 PHASE SAMPLING ARCHITECTURES

Phase sampling the received signal represents an attractive alternative for a DRFM architecture since the intercepted signal information is mainly carried in the phase modulation. That is, no intentional AM is present during, for example, a radar pulse since the maximum transmitted power is needed to minimize the effect of amplitude variations due to multipath propagation. In a phase-sampled DRFM, the amplitude information is discarded and the instantaneous phase is sampled and quantized. Figure 2.5 shows a block diagram of a phase-sampled DRFM that uses both a coarse and fine LO to downconvert the intercepted signal.

The limiter output feeds a pair of mixers: one fed directly from the fine LO, the other from the fine LO through a 90° phase shift. The filtered mixer outputs now contain in-phase (I) and quadrature (Q) of the beat signal between the IF signal and the fine LO. By direct comparison between the amplitudes of the I and Q signals it is possible to determine an instantaneous phase measurement (with respect to the LO). At a low bit resolution, a small set of comparators amplitude analyze the signs and relative magnitudes of the I/Q components, providing the digital information required to identify the phase of the signal over a wide dynamic range. At higher bit resolutions, where the number of analog comparators is greater than a few, it is convenient to use a pair of flash ADCs and perform the Cartesian-to-phase operation digitally.

2.4 Phase Sampling Architectures

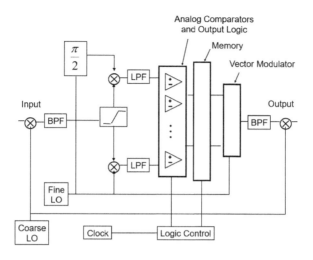

Figure 2.5 Block diagram of a phase-sampling DRFM.

The block diagram of a digital I/Q quadrature demodulator is shown in Figure 2.6. Special integrated designs have been specifically developed for various bands.

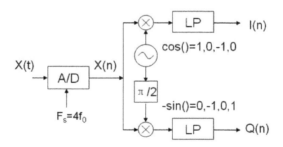

Figure 2.6 Block diagram digital quadrature demodulator.

An E-band (76-88 GHz) I/Q demodulator in a 90-nm complementary metal-oxide semiconductor (CMOS) package was developed with a −3-dB conversion gain and 26.6-mW power consumption requiring only 4-dBm LO power [7]. A D-band (110-170 GHz) monolithic microwave integrated quadrature modulator and demodulator circuit were developed in 250-nm indium-phosphide InP using double heterojunction bipolar transistor technology [8]. The mixers require an external LO signal and can be used as a direct carrier quadrature modulator and demodulator.

The upconverter has a single-sideband (SSB) conversion gain of 6 dB with image and LO suppression of 32 and 27 dBc, respectively, and can provide a maximum output

RF power of 2.5 dBm, a third-order output intercept point of 4 dBm, and consumes 78-mW dc power. A key advantage of direct I/Q phase modulation of a LO using for example, SSB, is that there is no unwanted sideband involved in the conversion back to IF. It is therefore possible to operate the ADC and DAC signals down to dc.

The coarse LO and filtering provide the IF signal carrying the amplitude and phase information of the input. After the coarse LO, the signal is limited in order that the phase can be sampled and quantized by the mixed signal comparators[1]. The fine LO provide the I and Q components. The reconstructed I and Q signals are put up on the carrier by the vector modulator and then passband filtered before transmission [9].

The advantages of a phase-sampled DRFM include performance similar to DSB DRFM but requiring storage for only *half* the number of bits. Also, the harmonic suppression is improved over the amplitude DRFM (worst-case harmonic for the amplitude DRFM is the third while for the phase it is the seventh). Most importantly, the phase-sampling DRFM has no AGC requirement and performance is independent of the signal amplitude and therefore has less dynamic range problems. Also, the user can modulate the phase of the output signal directly.

Phase sampled DRFMs do have their disadvantages, however. Due to the digital reconstruction of the stored signals (e.g., addition of signals) the processing becomes more difficult. Also, when more than one signal is simultaneously received, small signals are suppressed by up to 6 dB and spurious products will be generated. In fact, bit for bit, an amplitude DRFM gives better spurious performance. Since the amplitude information is not used, an additional subsystem is required to ensure that the recreated signal is transmitted at the appropriate level. Figure 2.7 shows a comparison between a 3-bit phase sampling and a 3-bit amplitude sampling DRFM.

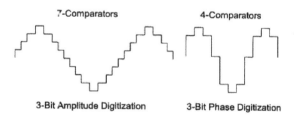

Figure 2.7 Comparison of phase and amplitude digitizing for 3 bits. On the left, seven comparators are used to achieve eight levels for a sinewave reconstruction (level sensitive). On the right, only four comparators are required to achieve a five level sinewave reconstruction (level insensitive). (©IEEE, reprinted with permission from [10].)

1 The comparator consists of an analog (differential) amplifier network to amplify the difference between two inputs and a digital latch that may be clocked to provide the output logic. Comparator designs are discussed in Chapter 6.

2.4 Phase Sampling Architectures

The 3-bit amplitude sampling DRFM is level sensitive and reconstructs the sinewave using eight levels and requires seven comparators. The phase sampling reconstructs the sinewave using five levels and requires only four comparators and is level insensitive [10].

In phase sampling, phase comparisons are made between the input signal and either a 180° shifted version of the signal or the signal ground. Figure 2.8 shows the phase sampling necessary for an 8-state DRFM.

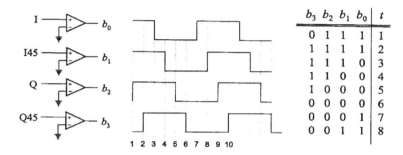

Figure 2.8 Phase sampling states necessary for an 8-state $n = 3$-bit DRFM (Adapted from [10].)

The zones within each of the circles represent the phase of the input signal. Four comparators are used to make the comparisons. For example, a high output from a comparator puts the signal in zone A and a low output puts the signal in zone B. Note that the phase sampling for the 8-state DRFM requires not only the I and Q but also a 45° shifted version of both the I and the Q. The resulting code for the 8-state shown in the figure is called the Johnson code [13]. Generally to get a $2n$ phase state DRFM requires n comparators resulting in an n-bit Johnson code.

Figure 2.9 shows the angular resolution as a function of the number of phase quantization bits used. Also shown for comparison is the discrete amplitude states as a function of the number of bits used.

56   DIGITAL RF MEMORY RECEIVER ARCHITECTURES

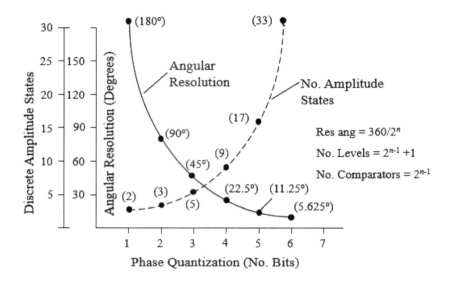

Figure 2.9 Phase quantization level versus angular resolution and number of amplitude states. (©IEEE, reprinted with permission from [10].)

For the $n = 3$-bit phase system there are five amplitude states (four comparators) and eight phase states with 45° angular resolution. In the amplitude DRFM the resolution is specified in amplitude. There are eight amplitude states required for the 3-bit amplitude system instead of five as in the phase sampled approach. This helps in reducing the resolution requirements of the comparator to only 2 bits for a 3-bit phase sampled system. The summer only requires a 4-bit DAC. This hardware trade-off is more advantageous to implement in GaAs since greater bit accuracy can be implemented in a GaAs DAC than a GaAs comparator or ADC.

The advantage of the phase approach over the amplitude approach diminishes as the number of bits increases. By using a multibit differential summing amplifier both the I and the Q signals can be reconstructed. Figure 2.10 shows the reconstruction of I and Q signals for the 3-bit system.

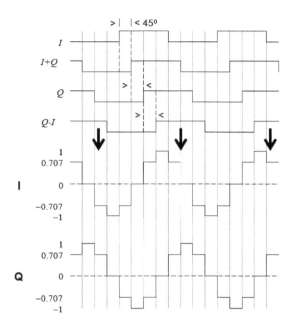

Figure 2.10 Vector reconstruction of *I* and *Q*.

The four digital inputs are phase-shifted in 45° increments. As mentioned previously, the user can directly modulate the phase by appropriate rotations of the *I*, *Q*, *I*45, and *Q*45 inputs (e.g., using a barrel shifter) [9].

2.4.1 Quantization Noise

The quantization noise performance of a phase-sampled system is very dependent on the nature of the input signal.

For sinusoidal signals the quantization noise appears as spurious signals at harmonic frequencies aliased into the system bandwidth. In an ideal n–bit system, the spurious signals occur at $\omega = (1 \pm m2^n)\omega_c$ where m is any integer $\neq 0$. The amplitude of the spectral components is

$$F = (1 \pm m2^n)\omega_c \left(\frac{\sin\left[\frac{(1 \pm m2^n)\pi}{2^n}\right]}{\left[\frac{(1 \pm m2^n)\pi}{2^n}\right]} \right) \quad (2.1)$$

The peak spurious signal for a range of phase quantization schemes is shown in Figure 2.11. Typically the primary source of spurious signals is likely to be the simultaneous reception of multiple signals.

58   DIGITAL RF MEMORY RECEIVER ARCHITECTURES

n-bits	H_{max} (No. of the largest harmonic present)	$F(\omega_c)$ (amplitude of the fundamental)	$F(H_{max}\,\omega_c)$ (amplitude of largest harmonic)	dB
1	−1	0.6366	0.6366	0
2	−3	0.9003	0.3001	−9.5
3	−7	0.9745	0.1392	−16.9
4	−15	0.9936	0.0662	−23.5
5	−31	0.9984	0.0322	−29.8
6	−63	0.9996	0.0159	−36.0
7	−127	0.9999	0.0079	−42.1
8	−255	1.0000	0.0039	−48.1

Figure 2.11   Peak spurious signal for a range of phase quantization schemes.

A comparison of the S/N and the % power in the spurious signals for *both* the amplitude sampling and the phase sampling DRFM architectures is shown in Figure 2.12. One can notice by examination that there are advantages to an amplitude quantization scheme. For example, for the 4-bit comparison the S/N is almost 7 dB higher for the amplitude quantization scheme and the % power in the spurious signals is considerably lower. However, there are advantages to the phase quantization scheme for $n < 5$-bits. For example, when false Doppler is being created it is much easier to rotate the phase (e.g., with an accumulator) than trying to do it with the amplitude. This is discussed more in Chapter 10.

No. of Bits	No. of Levels	S/N in dB AMP Quant.	% Powers in Spurs AMP Quant.	S/N in dB Phase Quant.	% Powers in Spurs Phase Quant.
1	2	6.3	18.9		
2	4	13.1	4.47	6.3	18.9
3	8	19.5	1.11	12.8	5.04
4	16	25.6	0.275	18.9	1.28
5	32	31.7	0.0684	24.9	0.321
6	64	37.7	0.0170	31.0	0.0803
7	128	43.8	0.00421	37.0	0.0201
8	256	49.8	0.00104	43.0	0.00502

Figure 2.12   Summary of S/N (in dB) and % power in amplitude quantization spurs and phase quantization spurs.

2.4 Phase Sampling Architectures

A 3.4-GSamp/s 3-bit phase sampling architecture is shown in Figure 2.13. The architecture uses a flash ADC and the I & Q are being amplified by a preamplifier prior to the comparator arrangement. Also note the use of single-ended to differential comparator structure.

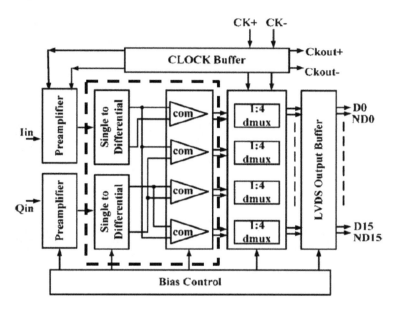

Figure 2.13 The ADC section for the 3.4-GSamp/s 3-bit phase sampling DRFM (©IEEE, reprinted with permission from [11].)

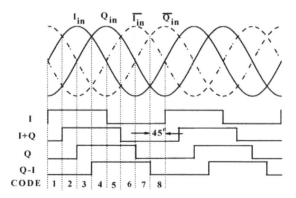

Figure 2.14 The ADC input and output timing diagram for the 3.4-GSamp/s 3-bit phase sampling DRFM. (©IEEE, reprinted with permission from [11].)

A detailed block diagram of the 3-bit phase digitizing DAC is shown in Figure 2.15 including low voltage differential signaling (LVDS) input buffer, multiplexer components, switch amplifier, switch stages, and current steering parts.

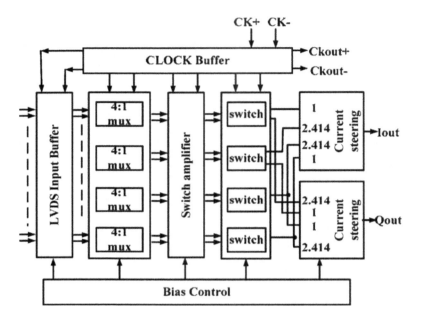

Figure 2.15 The DAC section for the 3.4-GSamp/s 3-bit phase sampling DRFM (©IEEE, reprinted with permission from [11].)

The LVDS was standardized in 1995 as a physical layer specification for serial interfaces. LVDS is activated by the 3.5-mA constant current, and transmits high-speed differential signal data with a very low voltage swing of 350 mV terminated with a 100Ω load. The data transmission speed is designated by the standard as 655M bits/s at the maximum. However, this is not the limit. Semiconductor manufacturers incorporating their own technology have achieved higher data transmission speeds of up to 3G bits/s. A specific example of LVDS, differential signaling is shown in Figure 2.16.

2.5 Use of DSP and FPGA Technology

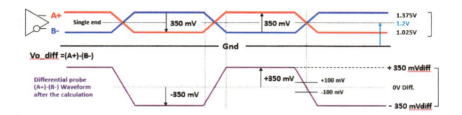

Figure 2.16 LVDS low voltage differential signaling example.

The input output timing diagram for the 3-bit phase representation is shown in Figure 2.17.

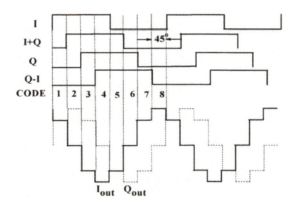

Figure 2.17 Regeneration code for the 3.4-GSamp/s 3-bit phase-sampling DRFM. (©IEEE, reprinted with permission from [11].)

2.5 USE OF DSP AND FPGA TECHNOLOGY

More frequently now, the DRFM hardware architecture is software-defined and incorporates both DSPs FPGAs. The concept presented in [12] takes advantage of the high-speed data processing capability of the FPGA and the powerful computing and control capability of the DSP. The unique aspect developed here uses an extensible interface 4.0 (AX14) bus to efficiently assemble each module together. By using the same interface for all modules, the flexibility of the system is enhanced. Figure 2.18 shows the architecture taking advantage of a AX14 Aurora bus interface and rapid IO between the signal processing subsystem and the acquisition and modulation subsystem.

In Figure 2.18, the hardware requirement for a general architecture based on the FPGA and DSP is analyzed. Due to a reasonable division of the functionality of the modules, the overall power consumption is very low while at the same time, online-updating and dynamic loading of the FPGA program is easily achieved through a serial peripheral interface (SPI) for EA operations and debugging [13].

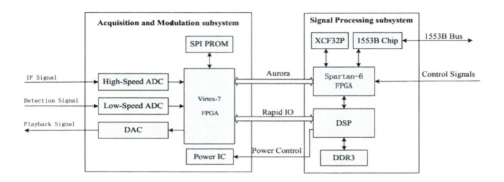

Figure 2.18 Block diagram of an extensible architecture utilizing both DSPs and FPGAs with a special controlling board for low power consumption. (©IEEE, reprinted with permission from [14]).

As shown in Figure 2.18, the contribution lies in the communicating and controlling board and the sampling and modulating board. The controlling board is used to communicate with peripheral devices and undertake certain parameter calculations. The sampling board is used for data acquisition, modulation, and playback [13]. The input signal is sampled by the ADC and is transmitted to the FPGA on the sampling and modulating board. The signal is then sorted and the results are transmitted to the DSP through a high-speed Serial Rapid I/O (SRIO) interface.

When the quantity of data is too large, an intermediate set of the data can be cached in the QDRII around the FPGA. Next, according to the instructions issued by DSP, the digital signal stored in the FPGA (on the sampling and modulating board) will be modulated. Finally, the modulated signals are converted into analog signals by the DAC. The DRFM system achieves communication with all peripheral devices through a 1553B bus. In addition, to make the system flexible and versatile, a PROM with SPI is added. The FPGA on the communicating and controlling board can also send updating programs to the FPGA on the sampling and modulating board via a LVDS interface between the boards, achieving online-update of the FPGA on the sampling and modulating board.

The system is based on the design of modularization, and different parts are powered respectively. Consequently, the FPGA on the communicating and controlling board can switch the power chips on the sampling and modulating board [14]. If there is no need to sample and modulate the input signal, the chips on the sampling and modulating board can be powered off in order to reduce the power consumption – an important advantage when used on an unmanned air vehicle. In summary, the hardware requirement for a general-purpose DRFM system is analyzed and the hardware architecture based on the FPGA and DSP techniques is examined. Due to a reasonable division of the functions, the overall power consumption can be kept very low while at the same time, online-updating and dynamic loading of the FPGA program can be easily achieved through SPI for operation and debugging. This framework has been actually applied to a radar deceptive jammer in [14] with good success.

2.6 GLOBAL POSITIONING SYSTEM RECEIVERS

The Global Positioning System (GPS) was conceived in 1960 under the auspices of the U.S. Air Force, but in 1974 the other branches of the U.S. military joined the effort. The first satellites were launched into space in 1978. The System was declared fully operational in April 1995. Other systems include the European led GALILEO, Russian led GLONASS, Chinese led BeiDou, Indian led INRSS and the Japanese led QZSS.

2.6.1 An Important Unmanned Vehicle Companion

The Global Positioning System consists of 24 satellites that circle the globe once every 12 hours to provide worldwide position, time, and velocity information. GPS makes it possible to precisely identify locations on the earth by measuring distance from the satellites. GPS allows you to record or create locations from places on the earth and help you navigate to and from those places. For a DRFM in an unmanned vehicle, it is critical to have a hardened GPS receiver on board to locate itself for targeting and communicating its position location.

2.6.2 INS/GPS Integration

The use of the INS/GPS information can enable real-time relative targeting methods and increased accuracy during passive location, geo-targeting solutions. In addition, airborne system requirements also necessitate the use of detect and avoid signal processing using, the navigational elements (GPS, INS, etc.) that are shown. The multifunctional diversities discussed above require new operational concepts, systems, and components. RF technology has evolved over the last 30 years that address the stringent RF requirements for the year 2030 and beyond. This includes phased array electronically scanned antennas as discussed below.

2.6.3 GPS Antennas

The small unmanned aerial system (SUAS) is an enabling technology with the potential to impact a wide range of industries. Central to the design and successful employment of most SUAS is the GPS. However, GPS has a known susceptibility to jamming, both from intentional and unintentional sources. This jamming can be mitigated using antenna design, although special care must be given to balance the robustness of the antijam capability with the practicality of the design on a size and power constrained SUAS footprint.

An ideal GPS receive antenna must provide good multipath and interference rejection. Perfect right-hand circulatory polarized (RHCP) (axial ratio of 0 dB), can provide multipath mitigation as polarization will change direction upon reflection (RHCP will become left-hand circulatory polarized (LHCP), and thus get naturally rejected by the antenna). Dualpolarized (LHCP and RHCP) or combinations of linearly polarized antenna solutions can provide polarization diversity, allowing for additional rejection through receiver processing techniques [15]. Uniform gain in positive elevations and no radiation response in negative elevations provide for mitigation of reflected signals below the receive antenna.

Several antenna designs have been examined to address jamming mitigation for GPS. Small unmanned systems such as the RQ-11B Raven require autonomous and continuous protection against EA since they are employed in scenarios where it is either impossible or undesirable to fly manned aircraft or larger UAS.

GPS operates using the principle of trilateration, wherein range measurements calculated using RHCP RF signals transmitted from at least four Earth-orbiting satellites are used to determine the longitude, latitude, and altitude of a receiver. Two levels of GPS service are possible: Standard Positioning Service (SPS) and Precise Positioning Service (PPS), where PPS is restricted for military use only. Currently SPS signals are transmitted on the L1 frequency at 1575.42 MHz, while PPS signals are on the L1 frequency and an additional L2 frequency at 1227.6 MHz. The minimum RF ground signal levels are -158.5 dBW for L1 SPS, -161.5 dBW for L1 PPS, and -164.5 dBW for L2 PPS.

A multiband segmented loop antenna for unmanned aerial vehicle applications proposed for GPS in [16] is shown in Figure 2.19. The antenna is composed of a segmented loop including eight segments, a patch element, and a shorting strip and shown in Figure 2.22. The antenna operates with an omnidirectional radiation at 956 MHz because the eight segments are electrically connected with seven capacitive reactances. Due to the addition of both the patch element and shorting strip, the impedance matching characteristic is improved. The antenna design has the multiband performance covering GPS L1, GSM1800, GSM1900, UMTS, LTE2300, LTE2500, and 2400 MHz ISM bands antennas were designed in [15] and in [17]. The dual-band RF receiver front-end architecture is designed in [18].

2.6 Global Positioning System Receivers 65

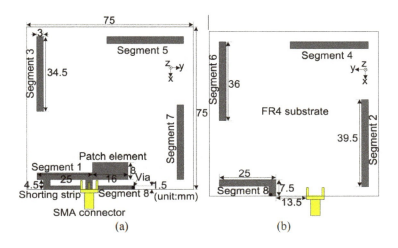

Figure 2.19 Multiband antenna. (©IEEE, reprinted with permission from [16].)

The integration of GPS and VHF/UHF antennas occupying the same physical volume conformal to the tail of a small UAV is examined in [17]. Figure 2.20 shows a GPS antenna covering both the L1 and the L2 bands.

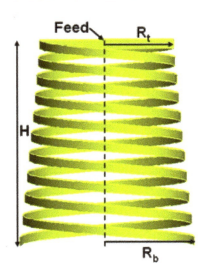

Figure 2.20 GPS antenna with 1:5 turn bifilar helix antenna. (©2009 VDE Verlag GmbH, reprinted with permission from [17].)

Here $H = 46.11$ cm, $R_b = 2.26$ cm and $R_t = 1.74$ cm. The VHF/UHF antenna can operate from 30 MHz up to 300 MHz. It is integrated with the GPS antenna. The GPS antenna and VHF/UHF antennas are conformal to the tail of the Dakota UAV as shown in Figure 2.21. The GPS antenna fits completely within the VHF/UHF antenna. At 30 MHz the height of the tail is $\lambda/17$. Electrically small antennas have poor input impedance. However, the input impedance can be matched with a passive matching network as long as the Q is not too high.

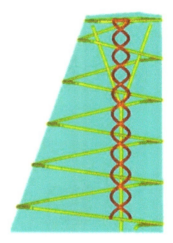

Figure 2.21 Integration of VHF/UHF (yellow) and GPS (red) antennas conformal to tail of Dakota UAV. (©2009 VDE Verlag GmbH reprinted with permission from [17].)

Other types of GPS antennas include patch antennas, which are resonant devices consisting of a flat plate attached to a metal base plate or ground plane with a dielectric in between. Quadrifilar helix antennas are resonant devices consisting of four twisted filars shaped in eggbeater patterns. Choke ring antennas have also been used and typically consist of 3 to 5 concentric ring structures surrounding an antenna element. Smart antennas are multielement antenna arrays coupled with antenna electronics that shape a receiver radiation pattern in response to a signal environment.

Adaptive antennas cross correlate each antenna input to generate a set of complex (phase and amplitude) weights for each antenna element. The weights are calculated in such a way as to provide a desired effect, either in canceling signals from inputs that contain interference (null steering) or magnifying signals that contain the desired content (beamforming). A controlled reception pattern antenna is one where only null steering is used. Synthetic arrays utilize a moving antenna and signal processing techniques to simulate a larger antenna aperture. The large synthetic aperture allows for the creation of narrow beams that can be steered toward a desired signal or away from

2.6 Global Positioning System Receivers

interference. In this way, moving antennas can be used to synthesize the functionality of adaptive antenna arrays without having the footprint of a large physical array.

2.6.4 GPS Receivers

A GPS receiver measures transmitted radio waves from multiple GPS satellites to determine the different propagation times associated with the satellites. The GPS receivers (and their associated digital processor) must determine their own position in 3-D. Synchronization of the clocks on the GPS satellites must be maintained in order to relay accurate information to the receivers. Precise synchronization is achieved by employing atomic clocks on each GPS satellite. The atomic clocks are constantly updated by a designated master control station (MCS). This is also important for UAS requiring synchronization between themselves for distributed EA techniques.

In order for the GPS receiver to calculate its position and speed precisely, the satellite must relay its orbital parameter information to the GPS receiver. These parameters are transmitted to the satellites from the MCS and then relayed to the GPS receivers. The GPS receiver measures the transmitted L-band RF signal from multiple satellites to determine the different propagation times. The satellites relay their orbital parameter information to the GPS receiver in order that it may resolve its position in 3-D from its time-offset from the respective satellite.

A functional block diagram of a generic GPS receiver is shown in Figure 2.22.

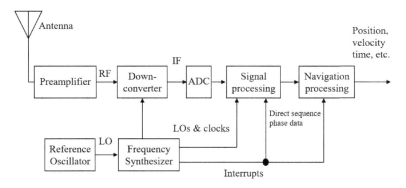

Figure 2.22 Generic GPS receiver functional block diagram.

The received signal is buried in the noise from the antenna through the ADC until it is demodulated using the direct sequence phase data in the signal processing. After this, the data is available for navigation processing. The preamplifier consists of burnout protection, filtering, and a LNA. A downconverter to an intermediate frequency (IF) section, and a ADC then provide the digitized data. The reference LO provides the time and frequency reference for the receiver.

The direct sequence made up of pseudorandom phase data within each subcode or chip is also made available for compression of the GPS signal to pull it out of the noise. By using different PR spreading codes with low cross-correlation the satellites are able to share the same band a technique known as code division multiple access (CDMA) [19]. Correlation of the received signal with the locally generated codes is done using search and acquisition processing to acquire the correct PR code. Tracking loops also included to maintain coherence [19].

2.6.5 Codes Used by the GPS System

The GPS received signal can be represented as

$$s(t) = A(t)C(t)D(t)cos[(\omega_0 + \Delta\omega)t + \phi_0] \quad (2.2)$$

where $A(t)$ is the signal amplitude that models the gains/losses on-board the satellite, $C(t)$ is the biphase PR code and is a direct sequence spread spectrum modulation (± 1). $D(t)$ is the biphase data modulation that carries the navigation code containing the orbital parameters and other information needed by the receiver, ω_0 is the carrier frequency (link L_1 or L_2), $\Delta\omega$ is the frequency offset (Doppler), and ϕ_0 is the nominal but ambiguous carrier phase.

A. Biphase PR Code $C(t)$

The first code is the C-code and is the direct sequence, spread spectrum pseudo random binary code. The C-code or $C(t)$ can be either the course acquisition, C/A-code or precision P-code. Each code consists of a stream of binary digits, zeros and ones, known as bits or chips. The C/A-code is generated by a product of equal period 1023-bit PR codes using 10-stage maximal lengthlinear shift registers and has a period of 1 ms. It is transmitted at 1.023 Mega-chips per second (Mcps). That is, the chipping rate is 1.023 Mbps and contains 1023 binary digits [20]. The P-Code (or Precision-code) is 6.1871×10^{12} bits long with a chipping rate of 10.23 Mbps that repeats itself every week.

B. Carrier Signals $L1, L2, L5$

The second code used is the link code. The above codes are modulated on RF links or carrier signals $L1, L2$ and recently $L5$. Link 1 $L1 = 1575.42$ MHz ($= 154 \times 10.23$ MHz) and is used to modulate the C/A code (in-phase) and has a bandwidth of $B = 1.023$ MHz. And on the quadrature channel, the P-code is transmitted with a $B = 10.23$ MHz bandwidth. For the Link 2, $L2 = 1227.6$ MHz $= 120 \times 10.23$ MHz, either the P-code or the C/A-code can be transmitted (one or the other).

2.6 Global Positioning System Receivers

The modulated PR codes are used by the receiver to compute a relative velocity between the satellite and receiver. The relative velocity is calculated by determining the Doppler shift of the incoming carrier frequency. The Doppler frequency is detected within the receiver using a phase locked loop (PLL). Four such measurements are used to calculate the receiver velocity. The reason two carriers are employed instead of just one carrier is to compensate for the effects of ionospheric refraction. Since ionospheric refraction is inversely proportional to the square of the frequency of propagation, the difference in propagation time between two different carriers can be used to remove the effects of refraction. Although using both the L1 and L2 carriers greatly improves the positional accuracy of the GPS unit by correcting for refraction, two carriers are not always required.

C. Biphase Navigation Code $D(t)$

The third code used in GPS is the navigation data code or D-code. The GPS signal carries the data in the form of a NAV-message from the satellite that the user receiver needs to solve for position, velocity, and time. The navigation code has a low frequency and is modulated onto the GPS carriers. The navigation code $D(t) = 50$ bps (50 Hz) as ± 1 and has a 30 s frame period, carries the satellite data that the receiver needs to solve for position, velocity and time. The binary data signals are formed by modulo 2 addition of the P-, or C/A-code to the binary data modulation $D(t)$. The characteristics of the navigation data transferred down include:

- Precise satellite position
- Precise satellite time
- P(Y) code acquisition from C/A code
- Selection of best set of satellites
- GPS time to Universal Coordinated Time (UTC) conversion data
- Ionospheric corrections
- Quality of satellite signals/data

The entire navigation message contains 25 frames. Each frame is 1500 bits long and is divided into five subframes. Each subframe contains 10 words and each word is comprised of 30 bits. Therefore, the entire Navigation message contains 37,500 bits. At a rate of 50 bps takes 12.5 minutes to broadcast and to receive. The D code signal is transmitted at 50 bps with a 30 s frame period. The resulting binary modulating signal (D code with P or C/A code) is formed by a modulo 2 addition of the P or C/A code to the binary data signal. Modulo 2 addition of 0, 1 is equivalent to multiplication of +1 and -1, respectively.

2.6.5.1 Example Calculation: L1 Received Power

In this example, we show the calculation of the received power using the Link 1. Consider the satellite transmitted input power to satellite transmit antenna is $P_t = 21.88W = 13.4dBW$. Also consider the range from satellite to weapon receiver is $R = 25,150$km. The transmit antenna gain (RHCP) is considered $G_{ta} = 21.88 = 13.4$dBi(minimum). In addition, consider that the atmospheric losses are $L_{atmos} = 1.58 = 2$dB. For Link 1, $\lambda_{L1} = 3 \times 10^8/1.575 \times 10^9 = 0.19$m. We can also consider that the receive antenna gain $G_{ra} = 1 = 0$dBi.

The minimum signal power density incident on the receiver's antenna can be calculated as

$$P_d = \frac{P_t}{4\pi R^2} \frac{G_{ta}}{L_{atmos}}$$

Substituting into the equation gives $P_d = 3.81 \times 10^{-14} W/m^2$. Calculation of the effective receive antenna aperture area results in

$$A_{er} = \frac{\lambda_{L1}^2 G_{ra}}{4\pi} = 2.87 \times 10^{-3} m^2$$

Giving a minimum received signal power for the L1 signal as

$$P_r = P_d A_{er} = (3.81 \times 10^{-14} W/m^2)(2.87 \times 10^{-3} m^2)$$
$$P_r = 1.09 \times 10^{-16} W$$
$$P_r = -160 dBW$$

The thermal noise in the receiver within the spread spectrum bandwidth is $N_{ss} = kT_0 B_{ss} F$. with the B_{ss} being the spread spectrum bandwidth of coded signal. For example, $B_{ss} = 1.023$ MHz for the C/A coded signal. With $F = 2.51$, $N_{ss} = -140$ dBW (20 dB above the GPS received signal power).

2.6.5.2 Correlation Processing

To provide the navigation information, the signal must be detected and account for this -20 dB deficit in signal power, correlation processing is used to provide the bandwidth compression. For correlation analysis, we used the well-known Pearson correlation

$$r_{x,y} = \frac{\sum_{i=1}^n (x_i - \bar{x})(y_i - \bar{y})}{\sqrt{\sum_{i=1}^n (x_i - \bar{x})^2}\sqrt{\sum_{i=1}^n (y_i - \bar{y})^2}} \quad (2.3)$$

where n is the sample size, x_i, y_i are the individual sample points indexed with i, and $\bar{x}$ and $\bar{y}$ are the sample means. The spread spectrum signal processing gain can then be

2.6 Global Positioning System Receivers

expressed as

$$G_{ss} = \frac{\text{Spread spectrum BW}(B_{ss})}{\text{Baseband signal bandwidth}(B_{bb})} \tag{2.4}$$

For example, for the L1 C/A code, the processing gain is

$$G_{ss} = \frac{1.023 \times 10^6 \text{ Hz}}{50 \text{ Hz}} = 2.05 \times 10^4$$

or $G_{ss} = 43$ dB. With 43 dB processing gain, the receiver is able to pull the signal out of the noise.

2.6.5.3 Frequency Domain Description

The spectral density of the modulated signal is the spectral density of the PR code centered at $\pm(\omega_0 + \Delta\omega)$. At baseband the spectral density is

$$S_s(\omega) = \frac{A^2 T_c}{2} \frac{\sin^2(\omega T_c/2)}{(\omega T_c/2)^2} \tag{2.5}$$

where T_c is the PR code chip width or the inverse of the PR code chipping rate. The signal power in a 2B-Hz two-sided bandwidth is given by

$$P_s = \frac{1}{2\pi} \int_{-2\pi B}^{2\pi B} S_s(\omega) d\omega. \tag{2.6}$$

When $B = \infty$, $P_s = A^2/2$.

2.6.5.4 Modernization

Because of the extra demands, new signals have been added to GPS and new systems have been designed. Consequently, the International Civil Aviation Organization coined the term GNSS to refer to the global collection of navigation satellite systems (GNSS). In more recent times, however, GNSS has come also to be used as a generic moniker for any single global satellite navigation system [19]. Other examples include new civilian signals being introduced at L2 and an L5 carrier (1,176.45 MHz) and a new civilian signal, in addition to the traditional C/A signal, is also being planned at L1.

2.6.6 Robust Receiver Design with AI and Machine Learning

GPS signals are becoming essential for the UAS navigation systems however, from a security perspective, the GPS signals are vulnerable to EA and spoofing attacks. The susceptibility of global navigation satellite systems (GNSS) signals to spoofing leads them vulnerable to potentially effective threat deception measures (e.g., false location information) and disruption of DRFM timing synchronization modules thereby disrupting any UAS to UAS synchronization signals and distributed EA [21].

2.6.6.1 Interference EA

EA can come in the form of wideband jamming where the jammer bandwidth is matched to the spread spectrum bandwidth of the receiver B_{ss}. In the presence of interference the power at the GPS receiver from the jammer can be expressed as

$$P_{gj} = \frac{P_j G_j}{4\pi R_j^2} \left(\frac{\lambda_{L1}^2 G_{raj}}{4\pi L_r} \right) \left(\frac{B_{bb}}{2B_{ss}} \right) \tag{2.7}$$

where P_j is jammer transmitter power (W), G_j is the jammer antenna gain, G_{raj} is the gain of the GPS receiver in direction of jammer (at or below horizon), R_j is the range of the GPS receiver and L_r is the same loss as the satellite signal traversing the receiver (2 dB). The correlation process described above compresses the navigation signal into the receiver's base bandwidth, however, it spreads the jamming interference out in frequency. If the noise jamming waveform was initially matched to the spread spectrum bandwidth B_{ss} of the GPS signal, the output spectrum will be doubled in width at output of the correlation process $2(B_{ss})$.

Effective suppression of the GPS receiver's ability will set up a *keep-out-zone* where in no GPS signals will be received. The keep-out-zone or denial zone is where the EA power exceeds the signal power. When the weapon gets within this zone, its receiver cannot track the GPS signals and the receiver can no longer navigate. Actually, the signal level needs to be somewhat greater than the interference for acquisition and tracking. Small, GPS jammers can be fielded in large numbers to create GPS denial zones covering tens of thousands of square kilometers.

Other more sophisticated interference techniques include CW tone jammers, and are more effective since higher power densities can be created at the receiver. However, these can be easily located and filtered out [22]. Wideband pulse jammers can deliver high peak power at low duty cycles to damage and saturate the front-end of the receiver. These are a minimal threat since microwave limiters used in the front-end prevent amplifiers from being driven into saturation.

2.6.6.2 Spoofing Deception EA

Spoofing involves broadcasting counterfeit look-alike GNSS signals to mislead the DRFM transceiver by maximizing the time error induced while minimizing the probability of being detected. Several spoofing strategies and techniques are presented in

[21]. Spoofing attacks can be divided into (a) meaconing: in where the authentic satellite signals are recorded and replayed at a later time, and (b) data level spoofing: where incorrect ephemeris data is transmitted [23].

For a spoofer that uses a *following strategy*, the first step is to capture the transmitted waveform (for example, a direct sequence phase modulation). The examination of the captured target signal's timing characteristics is important as the DRFM jammer waveform must satisfy certain constraints due to the finite propagation and processing time of the signals. In general, to meet the timing requirement, the jammer must lie anywhere on or within the ellipse shown in Fig. 2.23.

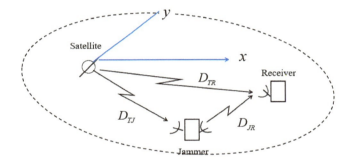

Figure 2.23 Navigation satellite based jamming scenario.

The equation for this ellipse where the satellite transmitter and the GPS receiver are the foci along the major axis can be found from

$$\frac{D_{TJ} + D_{JR}}{c} + T_J \leq \frac{D_{TR}}{c} + \gamma T_d \qquad (2.8)$$

where the distances are as shown in Fig. 2.23. The variable T_J is the processing time at the jammer, T_d is the dwell time of the target signal, γ is the fraction of the dwell that must be jammed to be effective, and c is the speed of propagation. This equation can be manipulated to the form of an ellipse and the differences in the propagation times can be plotted and analyzed [24].

The spoofer aims to manipulate the GPS time in a variety of ways ranging from gradual attacks, such as data-level spoofing to sudden jumps caused by meaconing. For the spoofing EA, the receiver locks on to the deception signal thereby denying valid measurements and it can be potentially devastating.

Recently, a mass-spoofing incident that occurred near the Russian port of Novorossiysk in the Black Sea has gained worldwide attention. During this attack by unknown sources, the GPS of more than 20 ships reported the same wrong location that is around 25 nm off [25].

2.6.6.3 Protection from the Spoofing EA

Significant progress has been made in boosting the robustness of GPS receivers against jammers and spoofing devices. This includes the antenna designs discussed above and also includes AI and machine learning. A redrawing of a robust GPS receiver incorporating functions that actively protect against a spoofer are identified with a star (*), is shown in Figure 2.24.

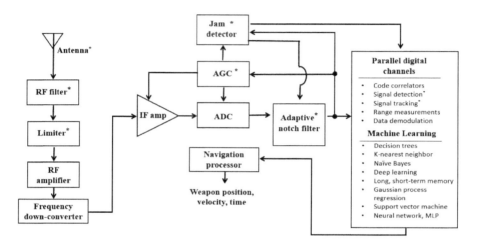

Figure 2.24 Robust GPS receiver with electronic protection (EP measures within the receiver indicated by *. (From [23, 26–28].)

Significant work has been done using signal processing to detect the spoofer. Detection techniques that range from those that analyze the physical characteristics of the signal, such as the carrier to noise ratio (C/N_0) monitoring and AGC monitoring, to more sophisticated detection techniques based on multihypothesis Bayesian classifiers that monitor, the received power and the distortion of the complex correlation function have been reported [25].

In [26] a method based on post-correlation deduction is put forward to detect the spoofing. In this method, the tracked GNSS signal subtracted from the correlation results, and the residual signal captured or tracked once again, by testing whether there is more than one GNSS signal of the same satellite we can determine whether there are spoofing attacks. Although this method is simple, high detection accuracy and no restriction on the type of spoofing attacks, it can be also carried out in any stage of the spoofing attacks.

A technique for the joint estimation of spoofer location (LS) and GPS time using a multireceiver direct time estimation (MRDTE) algorithm was developed in [25]. Utilization of the geometry and known positions of multiple, static GPS receivers that are distributed within the area of regard are used. The direct time estimation computes the most likely clock parameters by evaluating a range of multipeak vector correlations, each of which is constructed via different pregenerated clock candidates. Next, they compare the time-delayed similarity in the identified peaks across the receivers to detect and distinguish the spoofing signals. They can then localize the spoofer and estimate the GPS time using a joint particle and Kalman filter. The probability of spoofing detection and false alarm using Neyman Pearson decision rule are also calculated in [25].

There is growing interest in the use of machine learning changing the ways that navigation problems are prevented and resolved, and it is taking on a significant role in advancing positioning, navigation and timing technologies for the future [29]. In addition to solving the normal navigation issues such as signal acquisition (MLP, CNN), signal detection, and classification (logistic regression (LR), support vector machines (SVMs), CNN, Recurrent neural networks (RNN)) and are playing a significant part in detecting and mitigating spoofing and jammer attacks [29].

Advanced forms of GNSS spoofing have been conceived, for example, a spoofing method called "in the wild," which is an actual malicious spoofing attack is described in [28]. Receivers employing receiver autonomous integrity monitoring (RAIM) at the pseudorange level already have a rudimentary defense against spoofing. An inconsistent set of five or more pseudoranges would allow the receiver to detect an unsophisticated spoofer that broad casts one or more false signals with no attempt to achieve a believable consistency [28]. ML has been implemented to detect and defend against this type of spoofing; for example, one- and two-hidden-layer neural networks with various numbers of hidden neurons were implemented to detect GPS spoofing signals by making use of different features, such as pseudo-range, Doppler shift and SNR, to perform the classification of GPS signals. Other techniques using MLP, LSTM SVM with PCA and SVM with CNN have been used and are discussed in [29].

GNSS and inertial navigation systems (INS) integration has also benefited from ML using a CNN-based Kalman filter. The ML is integrated with 3D model-assisted GNSS to achieve high accuracy in the urban environments [29]. Satellite selection can also benefit under attack with the use of ML. Figure 2.25(a) shows a breakdown summary of the ML algorithms utilized in GNSS and Figure 2.25(b) shows the distribution of the use of deep learning models in the GNSS studies examined in [29].

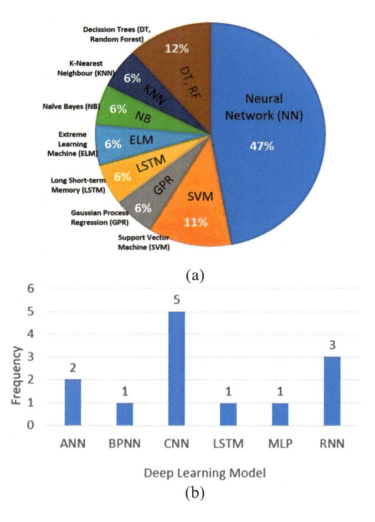

Figure 2.25 Machine learning in GNSS showing (a) a summary of the ML algorithms mostly utilized in GNSS, and, (b) distribution of the use of deep learning models in the GNSS studies examined in [29]. (©IEEE, reprinted with permission from [29])

2.7 SUMMARY

This chapter has given an introduction to the different architectures used for DRFM transceivers and included the amplitude sampling, phase sampling, and channelized kernels. Also included was an overview of the GPS receiver and the AI and machine learning techniques that are being used to improve the GNSS performance and its robustness and resistance to EA.

REFERENCES

[1] Zongbo Wang, Meiguo Gao, Yunjie Li, Haiqing Jiang, and Sunguo Ying, "The hardware platform design for DRFM system," in *2008 9th International Conference on Signal Processing*, 2008, pp. 426–430.

[2] M. Brandfass, M. Boeck, and R. Bil, "Multifunctional AESA technology trends - a radar system aspects view," in *2019 IEEE International Symposium on Phased Array System Technology (PAST)*, 2019, pp. 8138–8143.

[3] P. E. Pace, "Advanced techniques for digital receivers," in *Artech House*, 2000.

[4] R. Nitzberg, "Spurious frequency rejection," *IEEE Transactions on Electromagnetic Compatibility*, vol. 6, no. 1, pp. 33–36, 1964.

[5] O. Ozdil, M. Ispir, I. E. Ortatatli, and A. Yildirim, "Channelized DRFM for wideband signals," in *IET International Radar Conference 2015*, 2015, pp. 1–5.

[6] W. Liu, J. Meng, and L. Zhou, "Impact analysis of DRFM-based active jamming to radar detection efficiency," *The Journal of Engineering*, vol. 2019, no. 20, pp. 6856–6858, 2019.

[7] Y. Chou, Y. Lin, and H. Wang, "A high image rejection e-band sub-harmonic IQ demodulator with low power consumption in 90-nm CMOS process," in *2016 11th European Microwave Integrated Circuits Conference (EuMIC)*, 2016, pp. 488–491.

[8] S. Carpenter, M. Abbasi, and H. Zirath, "Fully integrated D-band direct carrier quadrature (I/Q) modulator and demodulator circuits in InP DHBT technology," *IEEE Transactions on Microwave Theory and Techniques*, vol. 63, no. 5, pp. 1666–1675, 2015.

[9] P. C. J. Pring, G. E. James, D. Hayes, and M. P. White, "The phase performance of digital radio frequency memories (DRFMs)," in *1994 Second International Conference on Advanced A-D and D-A Conversion Techniques and their Applications*, 1994, pp. 18–23.

REFERENCES

[10] T. T. Vu and J. M. Hattis, "A GaAs phase digitizing and summing system for microwave signal storage," *IEEE Journal of Solid-State Circuits*, vol. 24, no. 1, pp. 104–117, 1989.

[11] Min Zhang, Youtao Zhang, Xiaopeng Li, Ao Liu, and Feng Qian, "3.4GS/s 3 bit phase digitizing ADC and DAC for DRFM," in *2009 IEEE 8th International Conference on ASIC*, 2009, pp. 226–229.

[12] L. Yin, M. Xie, H. Li, C. Wang, and X. Fu, "Design and implementation of the digital radio frequency memory system based on advanced extensible interface 4.0," in *2016 9th International Congress on Image and Signal Processing, BioMedical Engineering and Informatics (CISP-BMEI)*, 2016, pp. 1307–1311.

[13] M. Xie, J. Huang, Y. Jiang, and X. Fu, "Design and realization of DRFM system based on FPGA and DSP," in *IET International Radar Conference 2015*, 2015, pp. 1–6.

[14] ——, "Design and realization of DRFM system based on FPGA and DSP," in *IET International Radar Conference 2015*, 2015, pp. 1–6.

[15] J. Patton and A. I. Zaghloul, "GPS antennas for small unmanned aerial systems (suas)," in *2016 URSI Asia-Pacific Radio Science Conference (URSI AP-RASC)*, 2016, pp. 181–184.

[16] D. Kang and J. Choi, "Design of multiband segmented loop antenna for unmanned aerial vehicle applications," in *2017 11th European Conference on Antennas and Propagation (EUCAP)*, 2017, pp. 962–964.

[17] B. T. Strojny and R. G. Rojas, "Integration of conformal GPS and VHF/UHF communication antennas for small UAV applications," in *2009 3rd European Conference on Antennas and Propagation*, 2009, pp. 2488–2492.

[18] R. L. La Valle, J. G. García, and P. A. Roncagliolo, "A dual-band RF front-end architecture for accurate and reliable GPS receivers," in *2018 IEEE/MTT-S International Microwave Symposium - IMS*, 2018, pp. 995–998.

[19] M. Braasch and A. Dempster, "Tutorial: GPS receiver architectures, front-end and baseband signal processing," *IEEE Aerospace and Electronic Systems Magazine*, vol. 34, no. 2, pp. 20–37, 2019.

[20] E. Elezi, G. Çankaya, A. Boyacı, and S. Yarkan, "The effect of electronic jammers on GPS signals," in *2019 16th International Multi-Conference on Systems, Signals Devices (SSD)*, 2019, pp. 652–656.

[21] P. Bethi, S. Pathipati, and A. P, "Stealthy GPS spoofing: Spoofer systems, spoofing techniques and strategies," in *2020 IEEE 17th India Council International Conference (INDICON)*, 2020, pp. 1–7.

[22] E. Elezi, G. Çankaya, A. Boyacı, and S. Yarkan, "A detection and identification method based on signal power for different types of electronic jamming attacks on GPS signals," in *2019 IEEE 30th Annual International Symposium on Personal, Indoor and Mobile Radio Communications (PIMRC)*, 2019, pp. 1–5.

[23] S. Bhamidipati and G. X. Gao, "GPS multireceiver joint direct time estimation and spoofer localization," *IEEE Transactions on Aerospace and Electronic Systems*, vol. 55, no. 4, pp. 1907–1919, 2019.

[24] R. Poisel, *Modern Communications Jamming Principles and Techniques*. Norwood, MA: Artech House, 2004.

[25] M. Jones, "Spoofing in the black sea: What really happened?" *GPS World*, Oct. 2017. [Online]. Available: https://www.gpsworld.com/spoofing-in-the-black-sea-what-really-happened/

[26] L. Cheng, S. Zhou, and J. Zheng, "Research on post-correlation deduction method for spoofing detection," in *2019 IEEE International Conference on Signal, Information and Data Processing (ICSIDP)*, 2019, pp. 1–7.

[27] Q. Wang, Z. Lu, M. Gao, and G. Qu, "Edge computing based GPS spoofing detection methods," in *2018 IEEE 23rd International Conference on Digital Signal Processing (DSP)*, 2018, pp. 1–5.

[28] M. L. Psiaki and T. E. Humphreys, "GNSS spoofing and detection," *Proceedings of the IEEE*, vol. 104, no. 6, pp. 1258–1270, 2016.

[29] A. Siemuri, H. Kuusniemi, M. S. Elmusrati, P. Välisuo, and A. Shamsuzzoha, "Machine learning utilization in GNSS—use cases, challenges and future applications," in *2021 International Conference on Localization and GNSS (ICL-GNSS)*, 2021, pp. 1–6.

Chapter 3

Designing DRFM AESA Antennas

3.1 INTRODUCTION

Digital beamforming using AESA antennas are now commonplace among modern multifunction, software-defined DRFM and transceiver designs [1]. The multi-function capabilities such as passive target detection, EMS sensing and EMMW, communications, and resource management of multiple input, multiple-output waveforms must be handled efficiently. This requires software-defined solutions for the digital beamforming for the different DRFM functions over a large frequency range (e.g., 5 GHz-30 GHz) resulting in unprecedented capabilities [2].

A new generation of RF sensor modules for multifunction AESA systems provide a combination of different operating modes for the DRFM (e.g., EMW and communications/datalink within the same antenna front-end). Typical operating frequencies covering C-Band, X-Band, and Ku-Band will require a bandwidth of > 10 GHz [3]. The input and output radiation is through a matrix of *transmit and receive elements* (T/R) that are shared among the embedded signal processing applications. Signal processing applications can consist of EA waveforms, communications waveforms, and ES activities that sense the electromagnetic environment. The antennas and T/R modules that can be used for DRFM designs are discussed in this chapter.

3.2 BEAMFORMING CONCEPTS

The beamforming concept starts with an array of antenna elements that convert the incoming RF signals into EM electrical signals are typically reciprocal. Each antenna has a radiation pattern that depends on how it radiates the electrical currents that are distributed on the antenna elements. The radiation pattern defined by the directivity (gain) indicates the direction of most of the energy (mainlobe) and the sidelobes contain all of the energy not in the mainlobe.

The beamformers are used to sum together the antenna elements, making the combined aperture beam directive. They control the radiation pattern through the constructive and destructive superposition of the antenna signals from the different elements. Passive beamformers use passive phase shifter components such as transmission lines that point the beam to a certain direction. The active beamforming uses active phase shifters at each antenna element to change the relative phase among the elements. Because they are active, the beam can be instantaneously and dynamically steered. Electronically steerable antennas can adopt one of these approaches to beamforming: analog, digital, and hybrid.

3.2.1 Analog Beamforming

The original analog beamforming antenna systems used fixed delays created by phase shifters at each antenna element to create a static beam pattern for a single, specific frequency. The switches were added to select among several fixed phase shifters creating a set of predesigned antenna patterns [4]. Adjustable phases shifters that could be actively controlled then came along. This gave rise to the AESA antennas.

Although many variations have been investigated, the digital signals are typically generated at baseband using a DAC converted to RF using a frequency conversion and split to feed the T/R modules with phase shifters at each element. The reciprocal path for the received signal is also followed. That is, after downconversion, the RF is digitized with an ADC. The T/R module contains the phase shifter, amplitude control and power amplifier, and low noise amplifier. This configuration is typically referred to as an analog beamforming network.

A schematic of an analog beamforming approach is shown in Figure 3.1. This results in reduced component cost particularly at millimeter wave frequencies where the small size of the phase shifters allow better integration. Unfortunately the precision and noise figure degradation due to these types of phase shifters are challenges for this technique. As well, the phase shifters and beam forming networks must be designed for the particular frequency of operation [5].

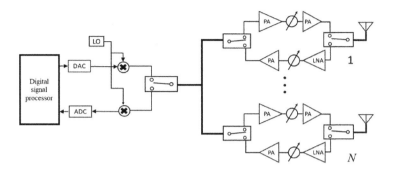

Figure 3.1 Block diagram of an analog beamforming network. (After [4].)

3.2.2 Digital Beamforming

The second technique is a fully digital beamforming approach that offers a higher degree of control and is considered the most flexible beamforming approach and a superior method for multi-beam applications. As the cost, size, and power consumption of digital receivers/exciters and processors go down, the use of element-level digital beamforming becomes a realistic approach to implementing large phased arrays. There are several key features to the element-level digital beamforming approach as shown in Figure 3.2. First, the total information available at the aperture is *preserved*, represented by the N individual element signals, in contrast to an analog beamformer, which produces the weighted sum of these signals reducing the signal dimensionality from N to 1. The DACs and the ADCs are directly connected to each T/R module (elemental digital beamforming) and digital baseband processing is then capable multiple spatial streams simultaneously [6].

This is particularly beneficial for the DRFM participating in a MIMO architecture where a single antenna array can dynamically create the MIMO data streams and beams optimized in real time for the user and load requirements [4]. Once the element receive signals have been digitized, they can be manipulated indefinitely because a digital representation of the signal is used rather than the real received signal power and any number of beams can be formed. Consequently, the signal can be processed using multiple *hypothesis testing* without degrading signal quality [6]. In DBF, the phase shifting operations of the array are implemented digitally, which allows for arbitrary time delays at each element.

84   DESIGNING DRFM AESA ANTENNAS

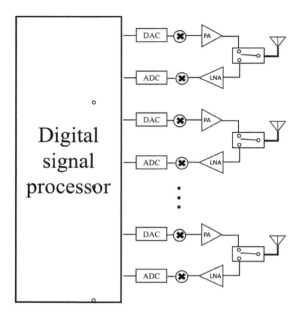

Figure 3.2 Fully digital array of T/R modules as a beamforming input.

T/R module development for phased arrays began in 1964 with the application of silicon technology to the Molecular Electronics for Radar Applications (MERA) Program [7]. The demands for increased radar performance and the requirement to cope with an ever-increasing threat environment have driven the development of active phased arrays for ground-based, naval, and airborne radar systems. For airborne systems, advantages come from the beam pointing ability of electronic scanning, which allows flexible and efficient scan volume coverage and interleaved mode operation (e.g., simultaneous search, multiple track and terrain avoidance) along with dynamic beam pattern shaping for optimization in each operational mode [8]. Many variations of the T/R module and the digital beamforming architecture have been developed as discussed below.

3.2.3 Hybrid Beamforming

The third technique is a hybrid technique that involves the best of both alternatives above and is shown in Figure 3.3. In analog baseband beamforming, the beamforming occurs in baseband after downconversion and before a conversion, enabling the use of higher-precision phase shifters. However, the size of the phase shifters and the complexity of the beamforming network mixers in each RF chain and the network of baseband splitters and combiners are challenges. In addition, as the data conversion requirements of elemental digital beamforming systems are becoming the bottleneck for provisioning digital beamforming solutions, cost and power considerations (especially for mmWave antenna arrays) have led to an interim approach using hybrid techniques. There are various methods for hybrid topologies for multifunction [9] and MIMO [10] applications. Architectures for massive MIMO OFDM [11] including minimizing the number of phase shifters [12] have also been investigated.

With this approach, the processing load on the digital electronics the power consumption is less than with elemental digital beamforming. The hybrid approach can be designed to allow for multiple spatial streams; however, it is less flexible than elemental digital beamforming since the number of beams is limited to the number of hybrid subsets, as shown in Figure 3.3.

In summary, with digital techniques, the beamforming is performed digitally at baseband requiring one beamformer and an RF switch at each antenna element. This offers a high degree of control and is considered the most flexible beamforming approach and superior to analog beamforming for receive and transmit of wide being singles and, more importantly, for multibeam applications the digital implementation has *greater reconfigurability*. The digital electronics, however, must be capable of processing Gbps of throughput in real time while minimizing latency. As well, more intelligence is required and its application is energy intensive with a significant complexity in the physical layout using SDR FPGAs, DSPs, and GPUs.

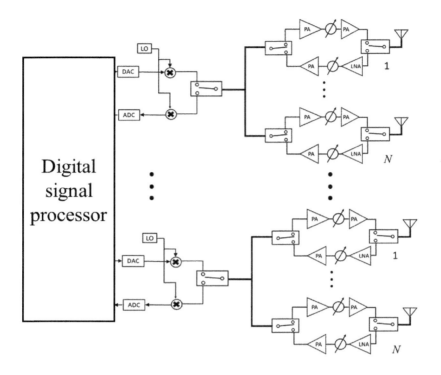

Figure 3.3 Hybrid (analog/digital) beamforming approach.

Thermal management and power management are also significant. Fully integrated ASICs with multichannel wide bandwidth data conversion and efficient FinFET CMOS DSP is also needed to meet the requirements of the beamforming. (FinFET technology and others are discussed in later in Chapter 6). While analog beamforming is less complicated and more power efficient than the comparable hybrid or digital approach, the size and cost of discrete analog components and interconnects are significant complexity issues as the antenna array grows beyond a small size [4]. For antenna arrays greater than 4×4 or 8×8, hybrid beamforming or digital beamforming offer size, weight, and cost advantages for the DRFM designer.

3.2.3.1 Advantages of Digital Beamforming

The advantages of the digital beamforming approach include:

- *Wideband signal reception and transmission*: Requires high-precision phase shifters and delay compensation so the array can operate over a large signal bandwidth without beam squint.
- *Larger antennas can be built*: Beamformer needs to be modular to enable simple scaling. To reduce beam squint, large antennas need to correct for the delays from scanning and system routing (a challenge for larger antennas).
- *Larger number of beams*: MIMO with multibeam capability is the most effective way to increase channel capacity. Digital beamforming supports large numbers of beams using the entire antenna aperture (same directivity and gain for each antenna).
- *Fast instantaneous beam steering*: Nanosecond switch times provide for fast target acquisition and tracking in dynamic clutter environments.
- *Flexible reconfiguration*: Supports online calibration, configuring dynamic sub-arrays, and synchronization.
- *Precise beamforming and nulling of interference sources*: Precise control of phase and gain enables fine control of the radiation patter, sidelobe level, null depth, and null positioning to suppress unwanted directional interference maintaining a high signal-to-noise ratio.
- *Conformal structures*: The ability of digital beamforming to calibrate and compensate for delay allows decoupling the antenna geometry making conformal antennas feasible.

Digital beamforming application advantages include: [13]

- *Passive Radar*: Simultaneous reception of both polarizations; full reception of backscatter independent of the target's orientation;
- *Data Link*: When using circular polarization, link is independent of the platform orientation and the volumetric clutter (e.g., rain clutter) attenuation is reduced;
- *Electronic Support*: By receiving both polarizations, polarization loss is minimized;
- *Electronic Attack*: Cross-polarization techniques against monopulse radars are enabled.

3.3 ACTIVE-ELEMENT ELECTRONICALLY SCANNED ARRAYS

Consider the linear antenna array with N equally spaced isotropic radiating elements placed along the x-axis of a spherical coordinate system, as shown in Figure 3.4.

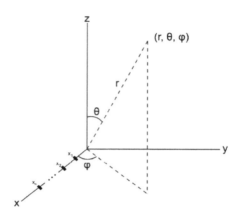

Figure 3.4 Linear antenna array placed along the x–axis.

The radiation pattern $F_{array}(\theta, \phi)$ of the linear antenna array with N elements can be approximated by multiplying the array factor $AF_{array}(\theta, \phi)$ by the element radiation pattern $F_{element}(\theta, \phi)$ as

$$AF(\theta, \phi) = AF_{array}(\theta, \phi) F_{array}(\theta, \phi) \tag{3.1}$$

and assumes the number of radiation elements is fairly large. Otherwise, the single radiation pattern holding for all elements doesn't hold. This equation must also be modified to take into account mutual coupling and leakage losses in the elements, which are also not considered in this equation. The array factor depends on the wavelength of the incoming, outgoing waveform λ, the incoming angle of arrival θ, the distance between the elements d, and the number of elements N as

$$AF(\theta, \phi) = \sum_{i=1}^{N} a_n e^{j(i)\psi} \tag{3.2}$$

where

$$\psi = kd \sin\theta \sin\phi + \Delta\varphi$$

and the element weights are applied per element by the factor a_n and $k = 2\pi/\lambda$. The angle $\Delta\varphi$ is a also phase value that is calculated with a beamsteering equation to guide the beam to a specific angle with an accuracy that depends on the resolution of the phase shifters used in the digital beamforming. The phase shift required for a specific angle $\sin\theta$ is

$$\Delta\varphi = \frac{2\pi d \sin\theta}{\lambda}$$

or expressed as a frequency independent term using time delays (instead of frequency offsets)

$$\Delta t = \frac{d \sin\theta}{c}$$

where c is the speed of light typically taken to be $c = 3 \times 10^8$ m/s. Consequently, the frequency dependency is eliminated when the *beamforming network is set up with delay lines instead of phase shifters*. The array factor can be written as

$$AF(\theta,\phi) = \sum_{i=1}^{N} e^{j(i)[kd\sin\theta\sin\phi+\Delta\varphi]} \qquad (3.3)$$

The $\Delta\varphi$ term represents a linear progressive phase that is added to each element (rad).

If the reference point is moved to the physical center of the array and the AF is normalized, then it is possible to show that the array factor can be written as [1]

$$AF = \frac{1}{N}\left[\frac{\sin N\psi/2}{\sin \psi/2}\right] \qquad (3.4)$$

The 3-dB beamwidth in either azimuth θ or elevation ϕ can be approximated as

$$BW_{3dB} = \frac{(50.76)\lambda b}{\sin\phi_0 Nd}\text{deg} \qquad (3.5)$$

where b is a beam broadening factor and ϕ_0 is the scan angle (where 90° is broadside and is perpendicular to the linear array). The value of the phase shift that is provided by analog phase shifters is directly affected by the noise on the control line and this is the reason digital phase shifters are used to provide the phase shift (ψ) [1].

The typical implementation of an AESA uses patch antenna elements configured in equally spaced rows and columns for example, with a 4 × 4 design, 16 elements in a square pattern on a panel. Antenna arrays built from multiple 4 × 4 cells can grow as large as 100,000 elements or more in ground-based systems. There are design trade-offs with the size of the array versus the power of each radiating element. The number of elements summed into the beam pattern determines the directivity of the beam and the effective radiated power as well as other parameters. Antenna performance can be predicted for important figures of merit that include antenna gain, effective isotropic radiated power (EIRP), the EIRP Density (EIRPD), and G_t. The antenna gain and effective EIRP are directly proportional to the number of elements in the array.

Achieving low EIRP and EIRPD can lead to the arrays often used in low probability of intercept radar.

The antenna gain of the antenna is the radiation intensity in the desired direction divided by the radiation intensity of an isotropic antenna (from all angles). The antenna gain can be calculated as

$$G_t = \frac{4\pi}{\theta_{AZ}\theta_{EL}} \tag{3.6}$$

where θ_{AZ} and θ_{EL} are the 3-dB azimuth and elevation beamwidth of the antenna's mainlobe in radians. With N number of elements in an AESA configuration, the gain of the antenna (in dB) can be calculated as

$$G_N = 10\log(N) + G_t \quad \text{dB} \tag{3.7}$$

The EIRP is a function of the transmitted power as

$$EIRP = P_t G_N \quad \text{W} \tag{3.8}$$

and the EIRPD is

$$EIRPD = \frac{P_t G_N}{BW} \quad \text{W/Hz} \tag{3.9}$$

where BW is the bandwidth of the radiation. In decibels, the EIRP and EIRPD are

$$EIRP_{dB} = 10\log(P_t) + 10\log(G_N) \quad \text{dBW} \tag{3.10}$$

and the EIRPD is

$$EIRPD_{dB} = \log(P_t) + 10\log(G_N) - 10\log(BW) \quad \text{dBW-MHz} \tag{3.11}$$

When working with antennas and power, transmitter distance, and propagation loss, its important to know how to work in dB. An excellent choice for a tutorial on this subject is the airborne radar book by Stimson [14].

The 3-dB azimuth and elevation beamwidth θ_{AZ} and θ_{EL} can also be approximated for a linear array in azimuth and elevation, respectively, as

$$\theta_{AZ} = \frac{\lambda}{L_a}$$

and

$$\theta_{EL} = \frac{\lambda}{L_e}$$

3.3 Active-Element Electronically Scanned Arrays

where L_a and L_a are the effective capture area Az of the antenna in the azimuth and elevation dimension, respectively. Consequently, the gain of the antenna can be shown to be [15]

$$G_t = \frac{4\pi A_e}{\lambda^2} \qquad (3.12)$$

where A_e is the effective capture area of the antenna[1]. The effective area of an antenna is a function of the aperture's true area A and the aperture efficiency η as

$$A_e = \eta A = \frac{\lambda^2 G_t}{4\pi} \qquad (3.13)$$

The aperture efficiency η is a function of illumination efficiency (ratio of directivity of antenna to directivity of uniformly illuminated antenna with the same aperture size), phase error loss or loss due to the aperture not being a uniform phase surface, spillover loss (for reflector antennas) mismatch loss derived from the reflection at feed port due to impedance mismatch, and RF losses between antenna and antenna feed port or measurement point [15].

The antenna beamwidth and near sidelobe levels are a function of wavelength, aperture size, and tapering of illumination on the aperture. Imperfections in the antenna aperture and blockage of radiation caused by radome imperfections can cause distant sidelobes. The antenna's integrated sidelobe (ISL) is the defining property of overall sidelobe characteristics of an antenna. ISL is defined as

$$ISL_{hh,vv} = 10\log_{10}\left[\frac{\int_{Sidelobes}|f_{hh,vv}(\theta,\phi)|^2 \sin(\theta)d\theta d\phi}{\int_{Mainlobes}|f_{hh,vv}(\theta,\phi)|^2 \sin(\theta)d\theta d\phi}\right] \qquad (3.14)$$

where $f_{hh,vv}$ are the horizontally and vertically copolar antenna patterns (amplitude and phase). The desired integrated ISL for the antenna is < -65 dB for detecting intrinsic reflectivity of clouds and precipitation. Amplitude tapering of elements is used for the desired sidelobe level. Traditionally, tapering of transmit elements is avoided for maximizing transmit power except in the case of an LPI emitter. A circular array uniform weighting of transmit elements produces -17 dB PSL with directivity of 36.73 dB. The remaining 48 dB sidelobe level reduction necessary for realizing a 65 dB two-way ISL requires a bit wider receive beam. The combination of narrower transmit beam and wider receive beam only produces an ISL of -45 dB, 20 dB short of the intended goal.

The Taylor weighting is often applied for reducing the sidelobes of an antenna [16]. Figure 3.5(a) shows the Taylor tapering of an array and the two-way antenna pattern. The ISL is -49.54 dB[17]. Directivity of antenna pattern with 25-dB Taylor tapering is 35.43 dB. The directivity for a 25-dB Taylor tapering is 0.4 dB lower than an array with uniform tapering.

1 Note for a unit gain antenna, $G_t = 1$ and $A_e = \lambda^2/(4\pi^2)$.

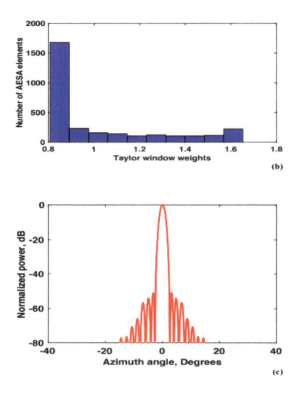

Figure 3.5 Distribution of Taylor amplitude weights on AESA shown as a histogram and the two-way radiation pattern of the circular aperture for 25-dB Taylor taper of transmit and receive aperture.

For a specified high-power amplifier (HPA) maximum power is transmitted when there is no tapering of transmit elements. As alluded to earlier, a circular array uniform weighting of transmit elements produces -17 dB PSL. Since the transmit AESA aperture is tapered for -25-dB PSL, the peak power from high-power amplifier should be increased proportionately to maintain the same transmit power as in the case of the uniform weighting of transmit elements. The gradient in HPA peak power varies only by a factor of two, as shown in Figure 3.5(b). There, are however, limitations when controlling the AESA sidelobes with monopulse beamforming [18]. Here the uniform, Taylor, split-Taylor, and Bayliss amplitude weightings must be evaluated from a system-level perspective, when balancing the need for sidelobe control versus size, weight, power, and cost (SWAP-C), all factors that are important in DRFMs in UAVs.

3.3.1 Example: Multifunction AESAs for Fire Control Radar

AESAs are now the mainstream for use on multimode fire-control radar systems in fighter aircraft. Figure 3.6(a) shows a photo of the AN/APG-83 AESA for use on the F-16 Fighting Falcon. The AESA here contains approximately 1,020 elements. In Figure 3.6(b), a picture of the F-22 Raptor AESA used on the AN/APG-77 radar is shown with approximately 2,000 elements and in Figure 3.6(c) with approximately 1,100 elements, the AESA for the AN/APG-79 radar used on the Super Hornet is shown [19]. Note the angle that the antenna face makes upward with the aircraft center line. This is done to reduce the RCS of the antenna. In addition, as with phased arrays, when the beam is electronically steered off to the side, its gain decreases (due to decrease in projected aperture area) and beam width increases (due to decreased projected aperture width). These effects impact range and resolution.

Studies also indicate that for fighter applications the critical fit parameter for the passive ESA is the volume available for the transmitter. Depending on the design choices, for example, the transmitted duty cycle, referring to (3.20), a 6- to 9-dB higher FOM is achieved for the active ESA (AESA) compared to the passive ESA and the size and weight of the passive system components required to recover this performance deficit exceed the allowable aircraft limits [20].

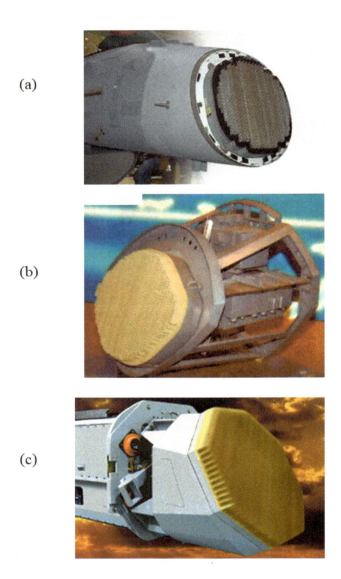

Figure 3.6 Photograph of the Actively Electronically Scanned Arrays (AESAs) for the (a) F-16, (b) F-22 Raptor, and (c) F/A-18 Super Hornet.

A summary was published by Air Power Australia that compared the power aperture for the major fighter aircraft AESAs. This is shown in Figure 3.7. A plot of the detection range (nmi) as a function of target RCS is shown in Figure 3.8. X-band

3.3 Active-Element Electronically Scanned Arrays

fighter radars with peak power ratings well above 20 kW have the potential to render all but top end stealth technology ineffective.

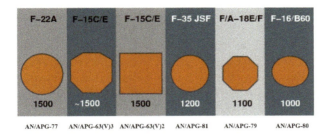

Figure 3.7 Comparison of the AESA T/R module counts for the radar systems used on the world's major multimode fire-control radar systems. (From [21].)

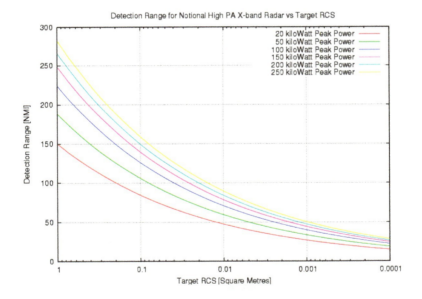

Figure 3.8 Comparison of the AESA detection range as a function of target RCS for different power aperture products (transmitted peak power (dB) + antenna gain (dB)). (From [21].)

The effects of array weighting on the detection capability of fire-control radar using medium PRF is examined in [22]. They consider the presence of ground clutter and examine the detectability as a function of the transmitting and receiving array weighting functions. They determine that the best target detectability is achieved using those weighting functions on transmit and receive which result in the lowest average sidelobe levels, but that the margins between the more highly tapered weighting functions were small [22]. Furthermore, they conclude that target detectability degrades as the proportion of failed elements increases. A failure of 5% of the elements gave modest, though meaningful, degradations in target detectability and would therefore form a suitable upper limit.

Phase comparison monopulse works by comparing the phase of the output voltages of two adjacent subapertures and relating that phase to the component of angle of arrival in the direction normal to the common edge of the subapertures. If the two subaperture output voltages are z_1 and z_2, then sum and difference voltages are formed as

$$\Sigma = z_1 + z_2 \quad \text{and} \quad \Delta = z_1 - z_2 \tag{3.15}$$

The monopulse measurement in one dimension is then [18]

$$\hat{m} = \tan^{-1}\frac{j\Delta}{\Sigma} \tag{3.16}$$

In [18], four different aperture weighting functions were compared for use in a phase comparison monopulse system: Taylor, split Taylor, Bayliss, and uniform. They show that a low sidelobe monopulse system requires the separation of the aperture weighting functions into the delta azimuth, delta elevation, and Rx sum RF manifolds and that this increases board complexity, power requirement, and cost of the AESA.

Below we detail several approaches to constructing the AESA. A block diagram of a RF front end is shown in Figure 3.9.

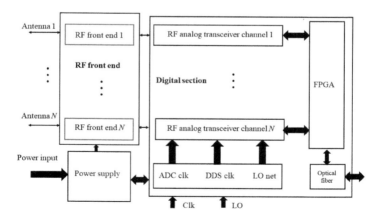

Figure 3.9 Block diagram of the RF and digital components on an AESA. (©IEEE, reprinted with permission from [23].)

3.3 Active-Element Electronically Scanned Arrays 97

We discuss the various technologies and design methods for construction of the antenna. A major capability of the antenna is the control of the transmit waveform amplitude in time (for example, to keep the target return SNR within the receiver at a constant as a function of the range-to-target). In addition, the ability to optimize the polarization on both transmit and receive can decrease the range resolution below the $\Delta R = ct_b/2$ normally obtained [24].

A number of different approaches for the design of the AESA antenna systems have been proposed and are the subject of this section. Then the technology of the transmit and receive modules is discussed.

3.3.2 Dual-Polarization Panels

The development of a third generation S-band, 64-element dual-polarization phased array panel with 6W radiated power per element per polarization is described in [25]. The array has 76 panels forming a 4-meter diameter, 4,864 element, S-band AESA. The work was done by MIT LL in cooperation with MACOM under the NOAA contract for MPAR (multifunction phased array) with its program goal to use AESAs for the dual purpose of aircraft surveillance and weather forecasting. Illustrated in Figure 3.10 are the T/R modules being air cooled along with the aperture PCB, backplane, and the panel structure. Also illustrated in the lower right corner is a photograph of the complete MPAR panel assembly. Figure 3.11 shows a table of the near-field results.

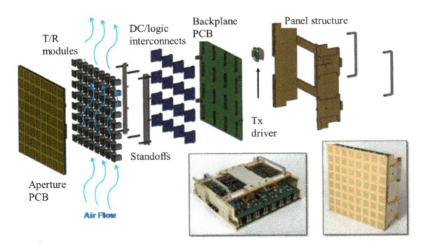

Figure 3.10 MPAR panel assembly diagram showing a photograph of complete panel assembly at lower right (©IEEE, reprinted with permission (©IEEE, reprinted with permission from [25].)

Antenna Metric	Goal	Measured
EIRP	85.4 dBW	85.3 dBW TxV 84.8 dBW TxH
Broadside Directivity	41.1 dB (Rx) 42.3 dB (Tx)	40.8 dB (Rx) 41.9 dB (Tx)
Relative EIRP, Gain between V& H Beams	< 0.1 dB delta between V and H	0.5 dB delta between V and H
Beamwidth	1.8° (Rx), 1.4° (Tx)	1.7° (Rx), 1.3° (Tx)
Mean Squared Sidelobe Level (MSSL)	< -50 dB	-53.9 dB (RxV) -53.1 dB (RxH) -50.4 dB (TxV) -49.3 dB (TxH)
Beampoint Error	< 0.05°	< 0.04°
Cross Pol Isolation	> 35dB	> 35 dB (Rx) > 40 dB (Tx)

Figure 3.11 MPAR ATD near field test results [25].

Near-field testing occurred in two phases at the MIT LL RF test facility and is described in detail in [25]. Near-field testing of the ATD demonstrated 85-dBW EIRP, with less than 0.5-dB delta EIRP between vertical and horizontal polarization, 40-dB broadside directivity.

3.3.3 Planar-Fed Folded Notch

A new form of a wideband, wide-scan notch planar-fed folded notch (PFFN) array antenna is shown in Figure 3.12.

3.3 Active-Element Electronically Scanned Arrays

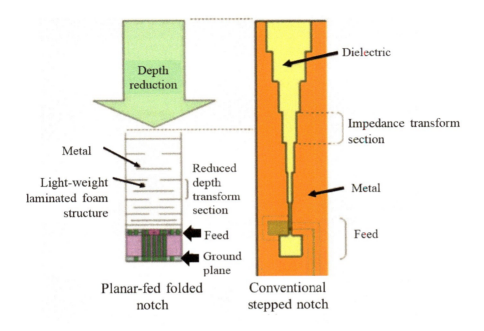

Figure 3.12 Wideband array showing (a) the planar-fed folded notch and (b) the conventional stepped notch array radiators. The folded transformer in the folded notch significantly reduces the required element height (©IEEE, reprinted with permission from [26].)

The PFFN array/radiator design incorporates a folded virtual slotline structure and a single multilayer PCB stripline-to-slotline feed/balun that is parallel to the array face. The architecture is a simple multilayered structure with metal patches arranged in a manner that creates virtual slotlines. The PFFN can be designed to cover an ultrawide bandwidth and a wide scan volume with a shorter physical depth than a comparable PFSN array. Its planar laminated structure also lends itself to integration with existing planar tile-based AESA technology.

The PFFN radiator is a novel physical implementation of a stepped impedance transformer that operates similarly to a traditional stepped notch radiator, but offers some significant advantages. The taper, number of sections, and overall height depends on the required bandwidth. PFFN designs realize stepped impedances via use of virtual slotlines to achieve a similar wideband impedance transformation. This implementation results in the potential for a lighter and significantly shorter element. The opportunity for reduction in size and/or weight ultimately depends upon the frequency band and bandwidth of the design.

Figure 3.12 shows how the arrangement of metal plates can be extended to realize dual-linear polarization. This dual-polarized PFFN element can be integrated with associated RF electronics (T/R modules, phase shifters, etc.) in an AESA to provide full polarimetric operation. A 1,024 element dual-polarized PFFN AESA was constructed using the 10:1 bandwidth element. Scanned beam pattern mosaics for this PFFN AESA were measured at various frequencies across the wide operational bandwidth (0.5 – 5.0 GHz). Figure 3.13 shows representative measured scanned beam patterns out to ±60 degs of scan in Az and ±40 degs of scan in El at 2.45 GHz.

Northrop Grumman has developed and demonstrated numerous wideband PFFN AESAs and arrays from VHF through X-band with array fractional bandwidths varying from $\sim 3:1$ to $> 10:1$. Most of these designs support wide-angle 2D electronic scanning (approximately ±60 degrees) over these wide design bandwidths [26].

For DRFM multifunction antennas the issues of intelligent power management, packaging, and cooling must also be addressed [27]. The use of metamaterials can be used to suppress the effects of the ground plane and reduce the effects of blind angles. Here the tile antenna is obtained with a multilayer structure where the radiating elements are connected to 3D RF modules, RF combiners, and dividers with the RF transiting along a cooling plate. The use of GaN technology amplifiers exhibit high power capacity, robustness, and good linearity over a wide frequency band. Also increasingly popular are the use of switches with power MEMS in place of circulators for wideband operation and the use of SiGe technology for reduced-size chips operating the amplitude and phase control [27]. Using electromagnetic simulation (such as CST), the capability to compute and accurately simulate the performances of large finite array antennas (e.g., 1000 radiating elements) is possible within a few hours.

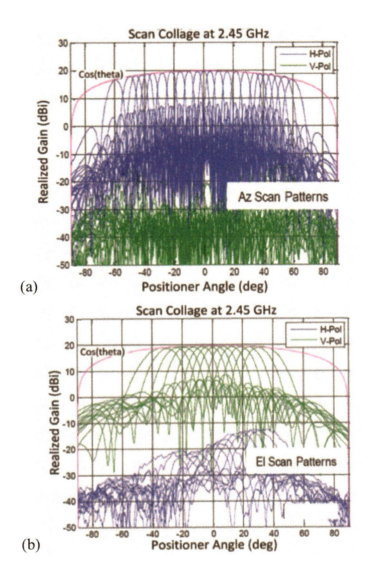

Figure 3.13 A 1,024-element PFFN wideband array (64" × 16") showing collages of (a) measured azimuth and (b) measured elevation scanned patterns at 2.45 GHz. (©IEEE, reprinted with permission from [26].)

3.3.4 Broadband Techniques

The wide operational bandwidth targeted with multifunction AESA comes with a number of technical challenges that make multifunction AESA one of the most intensive research topics in today's DRFM industry. To reach a sufficient technology maturity the large required operational bandwidth has to be resolved. Dual-polarized low-loss apertures with array grids designed for a grating lobe and scan blindness free ±60° conical field of view (FoV) at the highest operational frequency must be considered.

Today the AESA must also provide an adaptable RF resource that can be reconfigured to give the antenna real estate necessary for any particular DRFM mode. The DRFM system resource manager assigns the aperture or a part of it to a dedicated task depending on its priority.

The *spectrum management* of the broadband AESA represents another challenge toward multifunction front ends. Since the RF chain is designed to support a wide bandwidth, this makes the receive front-end circuitry more sensitive to jammer signals than a narrowband AESA. In transmit the 2nd and 3rd harmonics of the operation frequency are typically within the bandwidth of the front end and techniques enabling a high-power amplifier that minimizes the harmonics generated is needed.

Another challenge of multifunction AESA is the higher density of the array lattice, thus limiting the T/R layout space. For example, an array designed for a ±60° FoV with no grating lobes up to 15.5 GHz results in a unit cell about 10 mm in both the azimuth and elevation planes. Also, a third order intercept (TOI) of approximately +15 dBm and a low noise figure (NF) < 6 dB is necessary for the ES mission as well as the transmit task.

In addition to covering a wide bandwidth (e.g., 20 GHz), the antenna should be polarization-agile between at least two orthogonal linear polarizations: H-(horizontal) and V(vertical) or even fully reconfigurable between H-/V/RHCP/LHCP polarizations (R/LHCP) [28]. Polarization agility enables a maximum front-end flexibility and covers the technical requirements of the different DRFM tasks. It also requires a low-loss polarization switching in order to preserve the performance of the different operational modes in terms of noise figure and radiated power. In addition, a good cross-polarization rejection better than 20 dB and a grating lobe-free electrical FoV of ±60° void of scan blindness up to at least $0.75B_l$ are necessary.

With the criteria above, Figure 3.14 shows a matrix of candidate radiator technologies for broadband polarization-agile multifunction apertures [28].

3.3 Active-Element Electronically Scanned Arrays

Radiator	Photograph	Advantages	Disadvantages
Vivaldi / Tapered Notch	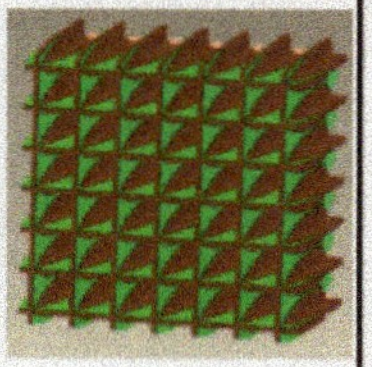	Low Risk Design	Scan-Blindness Cross-Polarisation >-20dB Complex Assembly (electrical connections between orthogonal substrates) Large Casing Depth
Bunny Ear		Cross-Polarisation <-20dB Small Casing Depth No Scan Blindness	Complex Mechanical Assembly
Capacitive Highly Coupled Dipole (C-HCD)	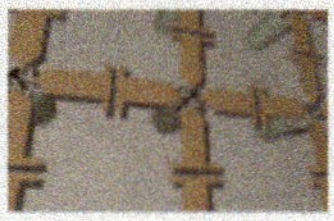	Low Profile Low Fab. Costs & Integration Excellent Tolerance Repeatability No Scan Blindness Cross-Polarisation <-20dB	Design Effort
Body of Revolution (BOR)		Low Risk Design No Scan Blindness Cross-Polarisation <-20dB	Bulky Construct Mass Casing Depth

Figure 3.14 Matrix of candidate radiator technologies for broadband polarization-agile multifunction apertures. These include a Vivaldi aperture, a bunny ear, a capacitive highly coupled dipole and the body of revolution design (©2016 EuMA, reprinted with permission from [28].)

3.3.5 Conformal Systems with Metamaterials

Multifunction conformal systems-of-sensors are built around several multifunction apertures distributed on the platform. These apertures are connected to modular and reconfigurable multifunction RF devices and DRFM signal and data processing units. These structures give access to all major operational RF functions required in a modern aircraft or unmanned air vehicles [29]. AESA systems that use tile architectures are well suited since these antennas must be thin to be structurally integrated within the platform for issues of available space and aerodynamics. They also need to be low observable. Tiles consist of low-profile, multilayer, highly efficient radiating elements, and array antennas being used to improve a new generation of active wideband systems. One example is shown in Figure 3.15.

Figure 3.15 Wideband self-complementary conformal antenna.

This antenna makes use of a dual-polarized self-complementary radiating structure embedded in a multilayer dielectric structure widening the angular scanning range and the frequency bandwidth of the impedance adaptation [29]. This particular structure exhibits a typical 3:1 frequency bandwidth while being compact (low profile), potentially conformable to the carrier structure. A major technical issue is the design of the dual-polarized feeding baluns in the multilayer technology, compatible with array cell size that is optimized for operation at the upper frequencies. Use of metamaterial structures are helpful to suppress the effects of the ground plane in wideband printed arrays. They also reduce the effects of the blind angles phenomenon common in printed arrays.

3.3.6 Fractal Antennas

Metamaterial fractal antennas have also been studied for AESA systems [30]. The word *fractal* is taken from the Latin dictionary meaning broken into parts or fragments. Fractals are self-similarity patterns that are inspired from phenomena of nature such as coastlines, mountains, and clouds and can be applied for antenna designs to achieve goals of multiband, wideband, and size miniaturization, making them ideal for DRFM antenna architectures. Many fractal shapes such as Hilbert curve, Sierpinski gasket, Sierpinski carpet, and Koch snowflake have been adopted for designing antennas having significantly different geometries [31].

An H-shaped fractal used to design a multiband planar antenna is shown in Figure 3.16. In addition to the antenna technology, various new physical architectures have been proposed. One technique reported to reduce the shape of the antenna and to achieve multi-band performance is by using fractal geometry in the patch antennas

[31]. This figure shows the different stages of the H-shaped fractal. The geometry of Stage 7 was used in the proposed antenna.

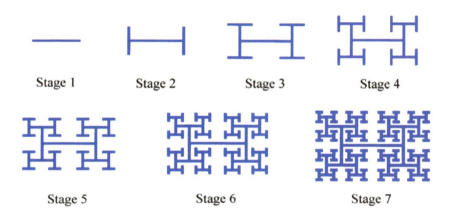

Figure 3.16 Different stages of the H-shaped fractal. The geometry of Stage 7 is used in the proposed antenna. (©IEEE, reprinted with permission from [31].)

The length of strips in each stage can be determined by

$$L(i+1) = \delta L(i) \qquad (3.17)$$

where $i = 1, 2, \ldots, 6$. The width of all metal strips is identical as 3.0 mm. The geometry of the H-fractal is symmetric to the 50-Ω microstrip-fed line with the corresponding width of W0. The size of the ground plane is $G_p = L_g \times W_g$. Other dimensions of the antenna are listed in Figure 3.17.

Parameter	H	Wg	W0	W1–W7	Lg	L0
Size (mm)	1.6	20.0	1.69	3.0	120.0	43.71
Parameter	L1	L2	L3	L4	L5	L6	L7
Size (mm)	53.02	36.96	25.77	17.96	12.52	8.72	6.08

Figure 3.17 Dimensions of the optimized antenna. (From [31].)

The following equations can be used for determining the range of the scale factor δ in the y-direction to avoid overlapping as a function of the stage, N

$$1 - \sum_{s=3}^{N} \delta^{s-1} > \frac{W0}{L1} \qquad (3.18)$$

for s odd and

$$1 - \sum_{s=4}^{N} \delta^{s-2} > \frac{2W0}{L1} \qquad (3.19)$$

for s odd. Here δ is a real number $0 < \delta < 1$ and is $\delta = 0.7259$ for the antenna 3-D pattern results shown in Figure 3.18.

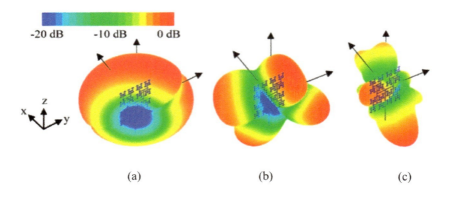

Figure 3.18 Simulated 3-D radiation patterns (normalized to 0 dB) of the proposed H-fractal antenna at frequencies of (a) 0.36,(b) 1.32 and (c) 5.50 GHz. (©IEEE, reprinted with permission from [31].)

Some are Hilbert curve, Sierpinski carpet, Sierpinski gasket, Giusepe piano, Koch snowflake, and Minkowski loop. A triple band hybrid fractal boundary antenna was recently reported that operates at multiple bands and was constructed using a combination of 3-D and screen printing on a package. The design uses a cantor fractal approach that enables it to operate at three communication frequencies, GSM 900, GSM1800, and 3G at 2,100 MHz [32]. More recently, a dual-band crown fractal antenna was proposed. Both simulation and experimental results of the proposed antenna constructed in FR-4 substrate were observed to verify the dual-band operation [33]. Fractal geometries have two main properties. These are *self-similarity* and *space filling*. These geometries were first proposed by Mandelbrot.

3.3.7 AESA Performance Comparison

The antenna balun or manifold, transports the received data that is collected by the antenna The analog signals are sampled and quantized using a digitization process. Typically done with an ADC, the digital signals are strobed into memory for storage and manipulation. The ADC is the most important component in the receiver as it (a) is the interface between the analog world and the digital world, (b) determines the dynamic range of processed analog signals, and (c) determines the resolution (bits) of the digitized signals. In multifunction apertures, the receive antenna function is typically a broadband, active phased array with a large RF bandwidth. Currently, the AESA is a common component of modern RF systems. Due to the extra complexity and cost of hardware and software, it has traditionally played a limited role in DRFM design. This was also due in part to the efficiency of the broadband components required [28].

A figure of merit for AESA performance comparison was presented in [20]. The figure of merit of FOM is for noise-limited, high-pulse repetition frequency (PRF) search operation and is defined as

$$FOM = \frac{1}{L_{TXW}L_{TXL}} \left(\frac{A}{L_{RXW}L_{RXL}} \right) \left(\frac{D_{TX}P_P}{L_{E/RGS}F} \right) \quad (3.20)$$

where L_{TXW} is the transmit pattern taper loss, L_{TXL} is the transmit dissipative loss and includes power loss for a variable taper if used, and A is the receiving aperture area. As well, L_{RXW} is the receive pattern taper loss, L_{RXL} is the receiver and dissipative loss and includes power loss for a variable taper if used. Here D_{TX} is the transmit duty cycle, P_P is the peak transmit power generated, F is the system noise factor at the receive module input, and $L_{E/RGS}$ is the eclipsing loss and range gate straddle loss.

3.4 TRANSMIT / RECEIVE TECHNOLOGIES

The solid-state T/R module technology has progressed significantly since it was first investigated in the 1960s. During the early development, designers were limited by the availability of silicon microwave transistors [21]. Silicon transistor technology required power to be generated at lower frequencies as band with frequency multiplication to reach operating frequencies at X band. This resulted in low DC to RF efficiencies. Advances in GaAsFETS in the mid-70s provided a technical base that supported the implementation of the AES antenna to generate waveforms at higher frequencies. The higher-density T/R module antenna has two main advantages [20]: it allows a modular approach where more power modules can be used depending on the resize, and it allows redundant power supply modules to be included in the system as required to achieve the desired mission reliability. The availability is also enhanced due to no moving parts, the solid-state reliability, and it is affordable. These active element electronically scanned phased array systems offer unique solutions for future DRFM. Other advantages include greater than 20% bandwidth and a graceful degradation when T/R elements fail.

The use of microwave solid-state technology for T/R modules was first explored in the 1960s and the first deployed AESA T/R modules were the Raytheon PAVE PAWS radar modules developed in the 1970s [1]. Up until just recently, the transmit, receive, and control functions of the T/R modules were realized with single-function GaAs and Si-MMICs [13]. The problem was that to provide minimal grating lobes, the modules must fit within the half-wavelength grid demand. For example, the final module width for a 9-GHz antenna is limited to 15 mm (maximum).

However with UAVs, these will have to be integrated into even smaller packages. In [13], they demonstrate a single T/R channel that forms part of a 3D T/R active zone embedded in an RF multilayer board with the radiating element on its top layer. Using their multifunction MMICs they can enable very compact module designs that offer separate T/R channels for the two polarizations within one grid-compatible 9-GHz module and reasonable wideband performance in housing of about 10 mm.

Today the important technologies being investigated for multifunction AESA antennas include BiCMOS, SiGe, and GaN [1]. Most of the AESA models currently fielded are based on gallium arsenide (GaAs)-based T/R modules. However, the emerging standard now for these models is the use of gallium nitride (GaN) technology for the output power amplifier. GaN is more efficient than the roughly 40% to 43% percent signal generation/power consumed rating of GaAs modules, and also generates less heat and extends the range of the emitter. Typical operating frequencies covering C-Band, X-Band, and Ku-Band will require a bandwidth of >10 GHz [3]. The T/R modules have to match with geometry constraints and a major challenge for multifunction RF sensor modules is the half-wavelength grid demand that is derived from the highest frequency with a grating-lobe-free FOV. Reducing the size of the total MMIC chip area can be realized in SiGe and GaN technology [3]. By using GaN along with BiCMOS SiGe technology, densely filled two-chip transmit-receive elements can be constructed to cover frequency ranges from C-band to Ku-band, opening up the further operational possibilities for next generation digital front-end AESA antenna systems.

During the last decades GaAs technology built the baseline for the active RF components within our X- or C-Band Tx/Rx modules. But in the last years SiGe and GaN components have increasingly replaced GaAs due to the superior properties of these technologies with regard to integration level (SiGe), power capability (GaN), and bandwidth (both). For a multifunctional wideband application these technologies display the key enabler to master the design challenges.

3.4.1 Schematic of a T/R Module

To analyze the input power and output power of a traditional T/R module, a block diagram of the architecture is shown in Figure 3.19.

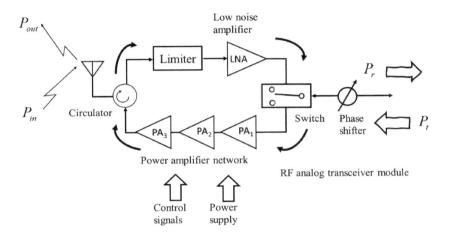

Figure 3.19 Schematic diagram of a T/R module for use in a multifunction AESA.

The T/R module has a network of power amplifiers for transmit and a LNA for receive. The T/R module is designed to support a transceiver type operation. A circulator or duplexer is used at the antenna to separate the transmit and receive analog radiation. A switch is used to flip between the transmit and receive channels in order to control the signal going to the attenuator and phase shifter. The switch can also be used to open circuit the module in order to prevent any leakage from going out the antenna. Additional switches are required for modules that can transmit and receive in either horizontal or vertical polarization. A coordinated (incremental) phase shift sent to the matrix of modules is used to construct a beam and shift its angle within the field of view. By using more modules, the beamwidth can be narrowed and the sidelobes lowered. The phase shifter control voltage is changed each time the beam position is changed or the frequency changes.

Upon receive, the RF radiation is received at the antenna and is input to a circulator or (duplexer) having an insertion loss of L_{inC}. The output of the circulator is first sent to a protection circuit or limiter (insertion loss L_{inL}) before being sent to a LNA having a gain of G_{LNA}. The LNA output is then sent through a switch (with insertion loss L_{inS}) to a phase shifter (L_{inPS}) that is controlled to do the (analog) beamforming to steer the beam as shown in (3.3). The LNA sets the noise figure performance. The noise figure represents the amount of noise in the output that is above and beyond the thermal noise level. The receive power P_r is then calculated (in dB) as

$$P_r = P_{in} - L_{inC} - L_{inL} + G_{LNA} - L_{inS} - L_{inPS} - L_{inAT} \qquad (3.21)$$

where P_{in} is the input radiation power to the module and L_{inAT} represents the loss in the signal power due to the attenuation by the attenuator. The attenuator is a passive resistive element in which the attenuation of the signal can be controlled. The attenuation is used to reduce the sidelobes of the antenna through weighting. The weighting typically takes the form of a Hamming or Taylor weighting; however, other weighting formulas can be used. If the weighting is not applied, from (3.3), the sidelobe level is monotonically decreasing with the lowest sidelobe being 13 dB down from the mainlobe.

The transmitted signal is P_{out} and depends upon the power amplifier network configuration when transmitting. The output power can be approximated as

$$P_{out} = P_t - L_{inPS} - L_{inS} + G_{PA1} + G_{PA2} - L_{inC} \qquad (3.22)$$

where G_{PA1} and G_{PA2} represent the gain values (in dB) of the power amplifiers in the transmit path. The amplifiers in the chain provide a linear signal amplification and is usually divided up into several stages. In Figure 3.19, the PA_1 is typically called the predriver amplifier, PA_2 is called the driver-amplifier and PA_3 is called the high-power amplifier. Splitting the amplification up into several stages allows the required gain to be spread across the amplifier chain. This is done to maintain linearity and prevent saturation in any one stage. The capabilities of a matrix configuration of T/R modules include the ability to do subarray beamforming and forming multiple simultaneous antenna beams very efficiently. In addition, the array factor degrades gracefully as the T/R modules begin to fail.

3.4.2 SiGe BiCMOS Contributions

SiGe BiCMOS provides the ideal combination of high-performance heterojunction bipolar transistors (HBTs) together with the high-integration capability of CMOS transistors, making this technology a real asset for phased-array applications. The beamforming functionality (i.e., phase and amplitude setting of the RF signal) together with the control functions (e.g., shift register, serial-parallel conversion, logic operations, ADCs and DACs etc.) can be integrated into one SiGe BiCMOS chip. Another possible increase in integration level is the combination of multiple channels on one chip.

In order to achieve an accurate beam control over the large bandwidth the usual attenuator and phase shifter architecture was replaced by an innovative vector modulator (VM) architecture. An architecture used to implement a VM for broadband analog beam control is shown in Figure 3.20(a) [28].

3.4 Transmit / Receive Technologies   111

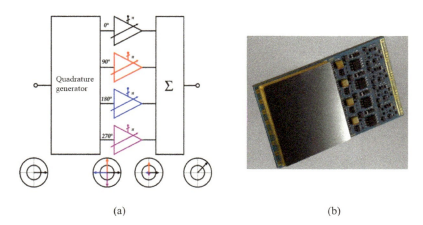

Figure 3.20 Vector modulator for analog beam control showing (a) generic block diagram and (b) photograph of a transmit/receive quad-pack with the lid put onto the frame (©IEEE, reprinted with permission from [3]).

Implemented in SiGe technology the chip area required for one vector modulator arrangement is very compact and two such structures can be implemented within one single chip: one VM designed and fitted to the Rx chain requirements and a second one more appropriately designed for the Tx requirements of the module [3]. By this Tx and Rx separation, noise figure and linearity performance in the receive mode and maximum available output power in the transmit mode could be optimized independently. An arrangement presented in [3] together with the whole functionality of a former mixed-signal Si ASIC was integrated in a chip size of about 5 mm x 5 mm.

Four identical RF channels are arranged side by side, separated by the inner walls of a frame with four compartments as shown in the photo in Figure 3.20(b). For the DC and control part such channel separation was not an issue due to the absence of open RF lines and circuitry. Figure 3.21 show the measured Rx phase and gain for 64 vector states that were precalculated using the nominal theoretical values. Both sets of curves are normalized to a reference state [23].

112 DESIGNING DRFM AESA ANTENNAS

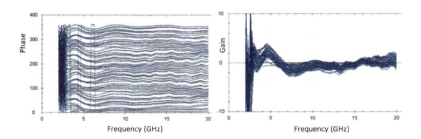

Figure 3.21 Rx phase (left) and corresponding gain variation (right) versus frequency for 64 nominal values (©IEEE, reprinted with permission from [3].)

Figure 3.22 shows the Rx noise figure at different frequencies and is displayed versus the corresponding Rx gain that was swept by vector settings.

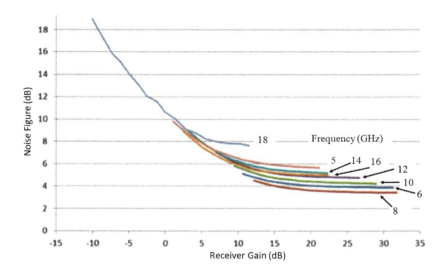

Figure 3.22 Rx noise figure (dB) versus receiver gain (dB) as a function of frequency (in GHz) (©IEEE, reprinted with permission from [3].)

3.4.3 GaN Contributions

Due to its specific properties GAN technology features several options to organize the front end of a transmit and receive module that are much more efficient with regard to insertion loss and bandwidth and also to the number of components and costs. Due to its high bandgap much greater than double gallium arsenide and consequentially a higher breakdown voltage operation is enabled. Further, the larger thermal conductivity results in lower temperature rise due to self-heating. Consequently gallium nitride devices have a higher power density five times higher than gallium arsenide. It's improved thermal conductivity as well as the higher power per unit width translates into smaller devices that are not only easier to fabricate but also offer much higher impedance. This enables lower power recombination losses in multitransistor amplifiers and lower circuit complexity and significantly eases the design of efficient wideband components.

Gallium nitride technology is an option not only for power amplifiers but also for low noise amplifiers. As a result many applications no longer require a limiter structure in front of the LNA. Not only is the performance enhanced by avoiding the limiters loss of up to 1-dB but limiter architectures and structures are often bulky and implicate elaborate assemblies with tiny diodes bringing their cost into significant ranges.

A further option with GaN is to replace the circulator by a GaN SPDT-switch. By doing this, not only the required area in the frontend of the module, but also the complexity (and cost) of both the LTCC-substrate and the assembly can be significantly reduced. A further advantage is the capability to realize wideband switches in GaN (e.g., from 6-18 GHz.) Hermetically sealed low temperature co-fired ceramics (LTCC) technology can also be used allowing, for extremely compact packing of the connecting lines using a multilayer architecture [3].

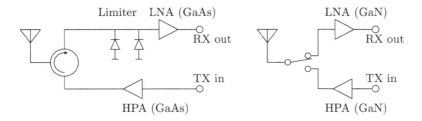

Figure 3.23 Rx noise figure (dB) versus receiver gain (dB) as a function of frequency (in GHz) (©IEEE, reprinted with permission from [3].)

For the GaAs T/R module front end, the receive path consists of a (more or less customized) limiter and 1 or 2 GaAs-LNAs. The technology of the GaN-based T/R module allow all the control, supply, and RF lines to be realized within a six-layer LTCC (low temperature cofired ceramics) substrate. A hybrid limiter circuit protects the sensitive LNA against high RF power levels, which could occur in the case of high antenna reflection or could be injected from external sources.

A comparison of the T/R module front end circuit design using both GaAs (left) and optimized GaN design are shown in Figure 3.23. To deal efficiently with both possibilities the limiter typically is switched active during Tx while in Rx the self-limiting effect of the limiter diodes is used. A three-port ferrite circulator routes the signal from the transmit path to the antenna and respectively, from the antenna to the receive path and also serves as an isolator in both directions. Test results for the GaN T/R module showing the Tx output power (dBm) are shown in Figure 3.24.

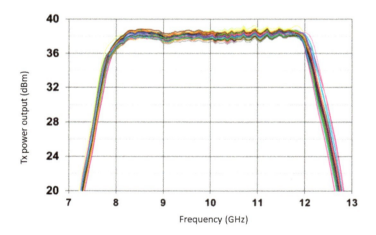

Figure 3.24 Transmit power from a 40 GaN T/R modules as a function of frequency (GHz) (©2015 EuMA, reprinted with permission from [34].)

The Rx gain and noise figure for the 40 GaN T/R modules is shown in Figure 3.25. The GaN technology is a much more efficient with regard to insertion loss and bandwidth and also to the number of components and cost. The most promising difference is replacement of the GaAs circulator by a GaN SPDT switch. This reduces the area (and cost) in the front-end LTCC substrate. A further advantage is the capability to realize wideband switches in GaN from 6 -18 GHz.

3.4 Transmit / Receive Technologies 115

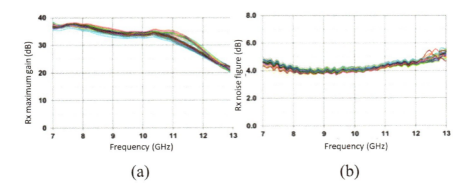

Figure 3.25 (a) Rx max gain (dB) and (b) Rx noise figure (dB) for the 40 GaN T/R modules as a function of frequency (GHz) (©2015 EuMA, reprinted with permission from [34].)

As well, a major difference is eliminating the requirement for a limiter in the GaN design. In the GaAs circuit, the nonreflective limiter absorbs the reflected Tx power at the Rx port of the circulator. Regarding Rx operation, the power reflected at the limiter LNA-input is absorbed by a Wilkinson divider or a switched load at the Tx-port of the circulator. With a switch, the respective power will be directly reflected, which may result in much more sensitive load-pull behavior. For very bad or total reflection even destructive effects for the HPA are probable. Such GaN technologies also operate at high voltage and high currents levels, which can lead to high densities of heat generation. With power densities in the range of several watts per millimeter in transistors, self-heating effects are critical and could impact the device reliability.

A Xilinx FPGA-based T/R module controller (TRMC) interface with a high level controller using a low voltage differential signaling (LVDS) scheme is described in [35]. The TRMC which controls the T/R module is internally divided into three major subblocks: namely the universal asynchronous receiver/transmitter (UART) logic, decoding logic and control logic. The architecture was developed in firmware using VHDL (very large scale integrated circuits hardware description language). The controller also provides real-time status of critical components of the T/R module, such as DC power supply temperature, forward power monitoring, and reverse power monitoring. The control logic commands the proper timing sequences required for all the data routing. In the proposed design, datas are multiplexed to achieve the switching time of < 50 ns, which is best suited for phased array and EMMW applications [35]. The Xilinx FPGA also achieved a switching time of <50 ns.

3.4.4 Alternating Transmit/Receive

MIT Lincoln Laboratory (MIT LL), in an effort sponsored by the National Center of Atmospheric Research (NCAR), engaged in the development of a C-Band active element phased array front-end hardware for polarimetric radar applications and is described in [36]. This was designed for use in Hurricane Hunter–like missions on a C-130, as shown in Figure 3.26. The phased array aperture design is summarized in Figure 3.26. Each AESA is comprised of a 4 × 4 array of tiles or panels with 252 each, resulting in an element count of about 4000 per aperture. The T/R module was implemented in a four-pack configuration. The AESA was designed to support alternating transmit simultaneous receive (ATSR) polarimetric operation with good co-pol to cross-pol isolation [36]. A table of the parameters for the NCAR AESA is also given in Figure 3.27 and shows the transmit and receive array elements and panel designs. A schematic diagram of the single alternating-transmit, simultaneous-receive T/R module is shown in Figure 3.28 for both the transmit V-pol and receive modes.

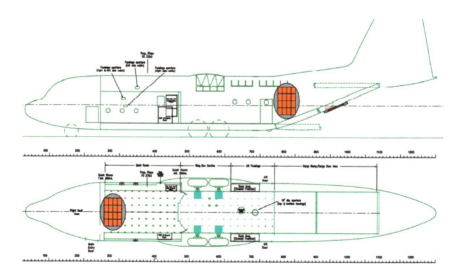

Figure 3.26 NCAR phased array radar (PAR) installation concept on C-130 with four independent aperture faces located on fuselage sides, top, and aft. (©IEEE, reprinted with permission from [36].)

3.4 Transmit / Receive Technologies 117

Parameter	Requirement	Design	Units
Wavelength	5.0		cm
Peak Tx Po at Antenna	10.7	10.3	kW
Pulse Width	33.0	33.3	μs
Rx Bandwidth	4.0	4.0	MHz
T*BW	133.0	133.2	
Receiver NF	5.0	5.0	dB
Receive Noise Floor	-104.0	-103.0	dBm
Antenna Gain (Tx, Rx)	41.7, 38.7	41.6, 39.4	dB
Tx Beamwidths (El, Az)		1.3, 1.7	°
Rx Beamwidths (El, Az): Taylor, SLL = 25dB, Nbar = 5	1.6, 1.9	1.6, 2	°
Effective 1-Way Beamwidths (El, Az)		1.4, 1.8	°
EIRP @ 10% duty	102.0	101.5	dB
FOM (tBW Compensated)	125.0	126.2	dB
Along Track Resolution		130.0	m
Array Elements, Total		4032	
Array Elements (X, Y)		56, 72	
Array Size (w, h)	1.5, 1.9	1.49, 1.89	m
Panel Element Count (X, Y)		14, 18	( )
Panel Size (w, h)		0.37, 0.47	m
Element Spacing (w, h)		2.67, 2.62	cm

Figure 3.27 Table of parameters for the NCAR AESA. (©IEEE, reprinted with permission from [36].)

The advantages of ESA (compared to mechanically steered ones) have long been recognized. The most important one is the ability for rapid (quasi-instantaneous) beam-steering that allows different Tx and Rx beam shapes, decoupling of search and track modes, and simultaneous modes (air-to-air & air-to-ground, SAR & GMTI) [34]. For a search mode, the scan pattern can take place using freely selectable sequences of beam positions, making own-ship detectability considerably more difficult. Within the operational context AESA-inherent outstanding capabilities compared to passive ESAs (i.e., w/o active T/R modules at each radiating element) are a dramatically increased power efficiency by RF generation nearby the radiating element, adaptive beamforming to suppress jammers, STAP to suppress clutter for more efficient GMTI, improved system availability due to graceful degradation of array performance resulting in a low life cycle cost, and LPI/stealthiness due to spatial power management, low side lobes, and low RCS of the array [34].

118 DESIGNING DRFM AESA ANTENNAS

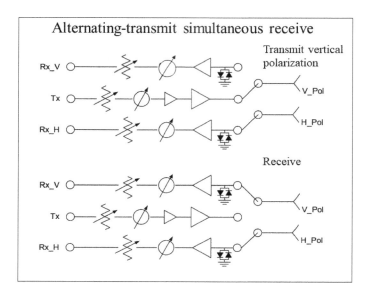

Figure 3.28 Block diagram of single alternating-transmit simultaneous-receive T/R modules for the the NCAR shown in transmit V-pol and receive modes (©IEEE, reprinted with permission from [36].)

The design of an AESA using the tile architecture is shown in Figure 3.9. This module consists of two parts: the RS front end and the digital part where there are 16 TR channel in the module. The digital part includes the FPGA, a highly integrated transceiver chip with a *digital predistortion* (DPD) function. The DPD feature enables use of high-efficiency PAs, which is beneficial for multifunction AESAs. The transceiver chip also integrates a duel a ADC and DAC for each RF analog transceiver channels. Due to digitization at the antenna, parallel adaptive beamforming is possible so that instantaneous angular area surveillance can also be performed without electronic antenna scanning.

In a traditional T/R module design, the transmit mode of radar, a pulsed RF signal (exciter output) is applied to the module from the array manifold. This signal is phase shifted in a digital phase shifter, amplified by amplifiers in the transmit chain of the T/R module and routed through the duplexer to the radiating element of the array. In the receive mode, return echo signals are routed through duplexer, receiver protector, and the low noise amplifier, which establishes the system noise figure. The amplified signal is amplitude adjusted by the digital attenuator and phase shifted by digital phase shifter and routed to the array manifold. The amplitude weighting (through digitally

3.4 Transmit / Receive Technologies   119

controlled attenuator) on the receive mode is used for synthesizing the low side lobe receive pattern of the array.

The phase shifter in both transmit and receive modes is used for electronically steering the array. For turning off the receive amplifier output during transmit mode and turning off the transmit amplifier input during receive mode, a SPDT T/R switch is employed in cascade with phase shifter. The radar dead time is utilized for changing the phase values and for switching the channel select T/R switches. A control logic circuitry serves to interface the module to the array controller, providing beamsteering, side lobe level control, and timing information needed by the module. Total time taken for switching of a 6-bit phase value for example, for the T/R module is $\approx 2 \: \mu s$.

A diagram of a traditional T/R module is shown in Figure 3.29. The T/R module is designed to have two transmitting parts, two receiving parts, and a common part. Equalizers are placed on the common RF path of each channel for the uniform gain with low ripple, and two SPTT switches are used to select the transmitting and receiving status. The microwave control circuits used are a series-to-parallel converter ASIC with transmit/receive mode converter and phase/magnitude setting.

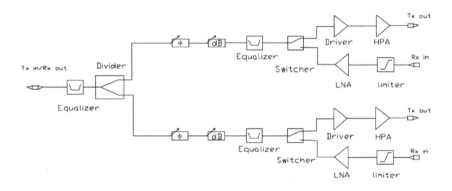

Figure 3.29 Schematic diagram of a T/R module for use in the multifunction AESA. (©IEEE, reprinted with permission from [37].)

LTCC is used as the base of the module so that RF, digital, and power supply lines can be incorporated into the same substrate in a miniature package for tight antenna spacing [37].

3.4.5   Zero-IF Example

A block diagram of the RF front end of a compact S-band T/R module for massive MIMO is shown in Figure 3.30. A circulator receives the wideband antenna input from the EM environment and sends it to a LNA. The circulator also transmits the output signal out to the antenna, after being processed by two power amplifiers (instead of three). A transceiver high-speed switch is used to control the input to the circulator. The multichannel baseband signals are decimated and filtered in an FPGA.

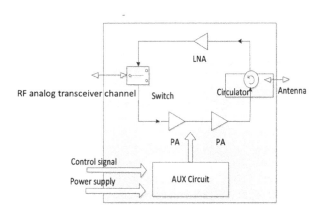

Figure 3.30 Block diagram of a transmit/receive module. (After [23].)

In this example, the system configuration is designed at S-band with zero IF (0 IF) for use in a massive MIMO system. Consequently, the size of the digital part is much smaller than the traditional approach where the IF band starts considerably above zero. In a large-scale system with many T/R elements, the most serious effect is the cumulative effect of the zero IF architecture [23]. The zero IF architecture leads to a large LO leakage and image frequencies.

In the transmit path the power amplifier chain must work in the linear region. To do this a digital predistortion (DPD) algorithm is used. The predistortion algorithm is a method to improve the linearity of RF amplifiers and compensate for any nonlinearities. The linearity requirement is so that they can accurately reproduce the signal present at their input. Each T/R module also integrates a quadrature error correction (QEC) algorithm to calibrate the effects to a low level.

3.4.6 Geometry Constraints

The T/R modules also have to match with *geometry constraints*. A major challenge for these future multifunction RF sensor modules is given by the half-wavelength grid demand derived from the highest frequency with accordant need for grating lobe-free field of view. One key to overcome this geometry demand can be the reduction of the total MMIC chip area with a high-level integration of single chip based RF functions into new multifunctional MMICs realized in SiGe- and GaN-technology.

There are also two other conflicting requirements. On one hand the extremely wide bandwidth that has to be supported by the front end (i.e., Tx/Rx modules

(TRMs) and antenna elements) limits the peak performance and efficiency of the AESA, especially in transmit. On the other hand the spectrum management of such a broadband AESA represents another major challenge toward multifunction front ends. The RF chain has to be designed to support the full bandwidth (e.g., 6–18 GHz), and this in consequence makes the Rx front-end circuitry more sensitive to RF signals or jamming than in a narrowband AESA. In transmit for the lower signal frequencies the 2nd and even the 3rd harmonic of the operation frequency are within the absolute bandwidth of the front end, which makes accordant filtering complex as it has to be switchable or adaptive.

Another challenge of a multifunction array is the higher density of the array lattice and thus the limited space available for the Tx/Rx Modules. An array designed for a ±60° grating lobe free FoV up to 15.5 GHz involves a unit cell of about 10 mm in both the azimuth and elevation plane. This area of 100 mm² is smaller than one half unit cell of an X-band array with the same FoV requirement. The space available for integration of the Tx/Rx module is reduced accordingly and the smaller array grid implies a larger number of Tx/Rx channels necessary to populate a given aperture [1].

3.5 PREDICTING IMPACT OF BEASMFORMING ON RX SENSITIVITY

In this section we examine the prediction of the analog and hybrid beamforming AESA's signal and noise on the receive system sensitivity, as well as the impacts of transmit and receive AESA directivity and beam pointing angle on the received interference. The parameters involved in the calculation include the transmit power, the transmit antenna's directivity, the receive antenna's directivity, the coupling between transmit and receive antennas, the Rx noise figure, and the receive bandwidth. We now include an additional factor in the expression for sensitivity to take into account the coupled power from the transmitter to the receiver, P_{coup} or [38]

$$\delta'_I = \{kT_oB_I(NF) + P_{conp}\}SNR_{Ii} \tag{3.23}$$

The coupled power into the receiver from the transmitter can be written as

$$P_{coup} = P_e N_{TE} I_{coup} \tag{3.24}$$

where P_e is the transmitted power per AESA antenna element, N_{TE} is the number of transmit elements, D_t is the transmit antenna directivity, D_r is the receive antenna directivity, and κ is the *coupling coefficient*. Here I_{coup} is the isolation term

$$I_{coup} = D_t D_r \kappa \tag{3.25}$$

Consider Figure 3.31 where there are three sources of noise within the transmit AESA: the HPA (elemental or unfocused), the array driver (subarray), and the exciter (focused). All of these sources are analyzed separately and the total impact on the receive sensitivity is determined from the component parts.

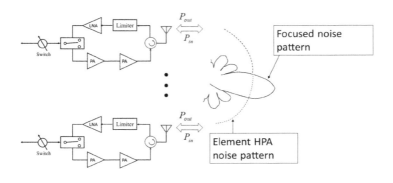

Figure 3.31 Transmitted RF noise as a summation of noise from each source within the transmit array. Each noise source contributes over the spatial region of interest. (After [38].)

The radiated noise analysis for each noise source begins with the first amplifier after an RF split [38]. The noise power per radiating element, P_m, for the mth split is computed from 3.25 as

$$P_m = kT_o(NF)_m B_I G_m \qquad (3.26)$$

and is the transmitted noise power per element in computing 3.25 for each level of splitting. The noise factor calculation for the array driver $(NF)_m$ begins with the array driver and ends at the stage prior to the HPA. The cumulative gain, G_m, of the array driver begins with the array driver and continues through the entire RF chain including the radiator ohmic loss. The radome ohmic losses should also be included. The radiator and radome mismatch are typically captured in the coupling term κ. The same approach is used for computing noise figure and cumulative gain for the remaining levels (HPA and exciter).

Since the transmit and receive antennas are typically mounted on an approximately flat ground plane, the radiation between them is in the endfire direction, where we approximate the radiator scan loss to be ≈ 6 dB. The coupling coefficient can be determined with standard propagation losses and absorber loss models in a modified Friss equation, (or a more rigorous method can be used such as FEKO).

Each noise contributor in Figure 3.31 is correlated over a specific region of the transmitting aperture so that the noise radiates with directive pattern that depends on the location of the noise source in the array and consequently, it is necessary to compute P_{coup} for each noise source by determining P_m, the transmit noise source directivities, and the coupling terms for each level. Note the receive array directivity does not depend on the transmit noise source. The total power P_{coup} is the sum of the coupled power from each noise source.

The receive antenna is typically the limiting technology in terms of bandwidth capability. That is, the bandwidth capabilities for the post-processing and digitization components available typically far exceed the antenna technology. The radiated signals the DRFM receive antenna collects depend on the antenna's beamwidth (degrees) and the bandwidth (Hz) and where the antenna bandwidth is positioned within the EM spectrum. They must also have the size, weight, and power characteristics to adapt to the various types of platforms such as unmanned aircraft and miniature air-launched decoys.

3.6 AESA NOISE FIGURE AND DYNAMIC RANGE

Now we can consider the analog, hybrid AESA as a matrix of these individual modules connected by a multiport lossy combiner. The combiner has a loss L ($L > 1$) that receives n transmit/receive elements, as shown in Figure 3.32. Below we calculate the noise figure followed by an example and the analysis of the dynamic range.

3.6.1 Noise Figure

The noise temperature contributed in passing through the combiner circuit element with loss L can be referred to the input of that element as

$$T_{eL} = (L-1)T_p \qquad (3.27)$$

and is the equivalent input noise temperature where T_p is the physical temperature of the lossy combiner. Thus, the temperature at the output of the combiner is

$$T_0 = T_{el}/L = T_p\left(1 - \frac{1}{L}\right) \qquad (3.28)$$

and the combiner output noise power is then

$$N_{0L} = kT_{el}B/(L) = kT_p(1 - 1/L)B \qquad (3.29)$$

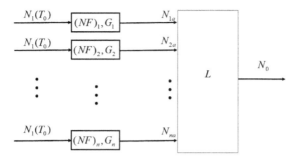

Figure 3.32 Analog hybrid AESA model as an active two-port network.

Now consider the connection of n two-port active transmit/receive elements at the input. If the loss of the multiport passive network is L, the bandwidth is B, the noise figure of the ith active network is NF_i then the equivalent noise temperature of the ith active network is

$$T_{ea_i} = (NF_i - 1)T_p \qquad (3.30)$$

The noise power at the ith active network at its output is

$$N_{ai} = kT_p BG_i + kT_{ea_i} BG_i \qquad (3.31)$$

The noise power of the ith active network at the passive combiner output is

$$N_{0ai} = (kT_p BG_i + kT_{ea_i} BG_i)/nL \qquad (3.32)$$

Now from (3.29) and (3.32) the composite network output noise power is the sum

$$N_0 = N_{0L} + \sum_{i=1}^{n} N_{0ai} = \sum_{i=1}^{n} (kT_p BG_i + kT_{ea_i} BG_i)/nL + kT_p(1 - 1/L)B \qquad (3.33)$$

$$= \left[\frac{kB}{nL} \sum_{i=1}^{n} ((T_p T_{ea_i})G_i) \right] + kT_p(1 - 1/L)B$$

and the total noise figure of the composite network NF is

$$NF = \frac{N_0}{kT_p BG} = N_0 / \left[kT_p B \sum_{i=1}^{n} G_i/nL \right] \qquad (3.34)$$

3.6 AESA Noise Figure and Dynamic Range

$$= \left[\frac{kB}{nL}\sum_{i=1}^{n}((T_p T_{ea_i})G_i\right] + kT_p(1-1/L)B/\left[kT_pB\sum_{i=1}^{n}G_i/nL\right]$$

If the noise figure and gain of the T/R modules are all the same; that is, $G_i = G$, $NF_i = NF_1$, then the composite total noise figure (3.34) becomes

$$NF = 1 + T_{ea}/T_p = (L-1)/G = NF_1 + (L-1)/G \tag{3.35}$$

To calculate the dynamic range of the AESA with the combiner shown in Figure 3.32 we consider each T/R module to have the same noise figure NF_1 and gain G. The sum of output noise power from (3.33) becomes

$$N_0 = kT_pB(NF_1 G/L + 1 - 1/L) \tag{3.36}$$

Substitution of the total noise figure of the composite network, above becomes,

$$N_0 = kT_p GBNF_1 G/L \tag{3.37}$$

3.6.1.1 Example

As an example consider the bandwidth of the receiver to be 1 MHz, the noise figure of each T/R module to be 2 dB, and the gain of each T/R module to be 20 dB. With the loss of the combiner being 1.2 dB, the output noise power of the composite network is $N_0 = -114 + +20 - 1.2 = -93.21$ dBm. As long as the gain of the active network is higher than the loss of the passive combiner, replacing the total noise figure of the combiner with the noise figure of the active network, this error can be neglected. If the input signal level to each T/R module is -100 dBm, the output signal level of the composite network is

$$S_0 = -100 + 20 - 1.2 + 12 = -69.2 \text{ dBm} \tag{3.38}$$

3.6.1.2 Dynamic Range

The above example reveals that the input dynamic range of the signal relative to the noise floor (disregarding the antenna noise) is

$$DR = -100 - (-114) = 14 \text{dB} \tag{3.39}$$

After the signal passes through this network, the output DR of the signal relative to noise is

$$DR_s = -69.2 - (-93.21) = 24.01 \text{dB} \tag{3.40}$$

The main issues with building the array of AESAs is the signal integrity or quality of the electrical signals as they go through the analog and digital portions of the system. The signal integrity revolves around the chip level interconnections as well as the package interconnects. Transmission-line interconnects and coupling is also a concern in addition to module resonances. The resonances arise from having electrical conductors enclosed in a volume to create a cavity with the cavity having a resonant frequency (Hz) that depends on the cavity size. For an excellent treatment of the signal integrity issues in T/R module construction the reader is referred to [1].

In summary, the next generation in T/R modules consists of single-chip T/R modules (in SiGe and Si-CMOS) leading to wafer-scale phased arrays. A block diagram of a digital beamforming AESA T/R frontend is shown in Figure 3.33 and represents a departure from the traditional approaches discussed. Since the phase shift required for the beamforming is in the digital domain, the phase shift is realized using a digital word. That is, the phase shifter is not part of the electronics in the T/R module. One important parameter is the data rate out of the ADC as it impacts the digital interface requirements. The power consumption is also a significant issue [1].

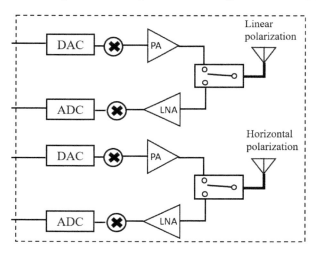

Figure 3.33 Modern approach to T/R module architecture.

A recent fully digital AESA with beamforming in GaN is shown in Figure 3.34. Due to a significant investment in GaN, metallic, electronic, and hybrid printing, metamaterials, nanomaterials, and artificial intelligence methods have been explored [39].

The more recent digital approach uses DACs and ADCs directly connected to each T/R module. This method, known as elemental digital beamforming, enables the beamforming algorithms and digital-based beam processing to be entirely implemented with robust digital hardware illuminating the need for sensitive analog face shifters in the RF domain. Digital processing is capable of creating multiple spatial streams simultaneously, such that a single antenna array can dynamically create MIMO data streams and beams optimized in real time for the user and load requirements.

The biggest challenge for digital beamforming is power consumption, whereas analog beamforming requires low power DC. However, since each analog beamforming only supports a single beam and a digital beamforming system enables multiple concurrent beams, digital beamforming is favored in higher density environments demanding low latency and uncongested performance. Digital beamforming is also attractive for infrastructure networks supporting DRFMs. When digital beamforming at millimeter-wave frequencies often a hybrid beamforming approach is necessary, which combines the analog beamforming phase-shifting topology with the RF shifter attenuators low noise amplifiers in power amplifiers switches and circulators still being used. Each separate antenna element is driven by data converters with some level of digital precoding. For a good review of MMW beamforming see [4].

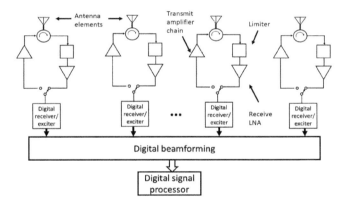

Figure 3.34 Fully digital array of T/R modules as a beamforming input.

3.7 A LOOK TO THE FUTURE

Recent advances in 3D printing promise to enable complex RF structures even more than what have already been realized. The characterization of the materials used in 3D printing processes has been shown to be critical in designing and accurately predicting the performance of these structures [40]. Understanding the RF properties of the materials through EM characterization has led to the development of novel structures that could not be realized with traditional manufacturing techniques. 3D printing has also allowed manufacturers to produce traditional antenna shapes with less weight and at a lower cost.

SWISSto12 in Switzerland has developed waveguides, filters, beamforming networks, antenna feed chains, and array antennas using 3D printing based on polymer materials that are metal plated or on metallic materials (such as aluminum or titanium) combined with advanced surface treatments and surface plating [40]. Optisys, another company that focuses on lightweight antenna using metal 3D printing, uses a powder bed fusion process, where thin layers of powder are welded into solid metal by a high-power laser that builds one small layer at a time. A Ka-band 16-element tracking array is shown in Figure 3.35.

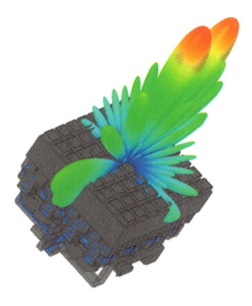

Figure 3.35 Optisys Ka-band 64-element tracking array with signal pattern (©IEEE, reprinted with permission from [40].)

Signal patterns show the sum and delta elevation performance with measured and simulated performance overlaid. This part includes a 16-element array of horns that are circularly polarized, with a waveguide combiner network on both left- and right-hand circular polarizations. Also the right-hand combiner network feeds into a dual-axis monopulse comparator [40]. All of this is implemented in a single part that weighs less than 2 oz and fits in the palm of the hand.

MITRE also has developed a wideband phased array concept that has a complex metamaterial design based on a PCB design that resembles an egg-crate construction with contiguous electrical connection interdigitated fingers embedded within an orthogonal board interface as shown in Figure 3.36.

3.7 A Look to the Future   129

Figure 3.36 A small test coupon of MITRE's biaxial metamaterial created with a Voxel8 multimaterial 3D printer (©IEEE, reprinted with permission from [41].)

Realized with metamaterials, by arranging naturally occurring materials in a specific pattern to produce an EM response not found in nature, periodic structures can be created at scales that are smaller than the wavelengths of the phenomena that they influence and can create materials with *negative indexes* that control the EM energy in ways that cannot be done with natural materials [41]. The periodic structures created by arranging naturally occurring materials in a specfic pattern, are shown in Figure 3.37. The EM energy response can be steered instantly for searching a volume of space. This is important because in traditional AESA antennas, phase shifters embedded in the control circuitry are used to steer the beam direction. Metamaterial-based AESAs can steer the beam without phase shifters, which reduces system complexity, eliminates a source of power loss, and simplifies waste-heat dissipation.

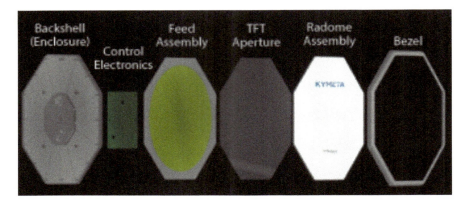

Figure 3.37 Periodic structures created by arranging naturally occurring materials in a specific pattern to produce an EM response that is not found in nature. Kymeta antenna construction (©IEEE, reprinted with permission from [40].)

Kymeta mTenna™ technology is manufactured using a different process, components than both traditional antennas and phased array antennas. The metamaterial in mTenna technology is a metasurface in a glass structure (glass-on-glass structure manufactured as LCD flat screen televisions), making it low-cost. They use the thin film transistor, liquid crystal as a tunable dielectric. Instead of reflecting microwaves like a traditional dish antenna or creating thousands of separate signals like a phased array, Kymeta uses a thin structure with tunable metamaterial elements to create a holographic beam that can transmit and receive satellite signals.

Gapwaves AB was founded with the aim of enabling wireless communication through the gap waveguide technology that provides a unique packaging for mmWave and terahertz circuits and components. The technology has advantages compared to existing transmission line and waveguide technology based on an artificial magnetic conductor that enables contactless propagation of electromagnetic waves, significantly reducing transmission losses as illustrated in Figure 3.38 [42].

3.7 A Look to the Future

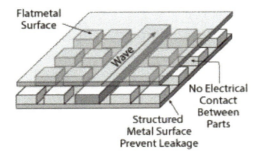

Figure 3.38 Gap waveguide structure (©IEEE, reprinted with permission from [42].)

The GAP waveguide is built up of two parts: a structured metal surface and a flat metal surface placed close to one another, allowing for an air gap between the two part as shown in Figure 3.39.

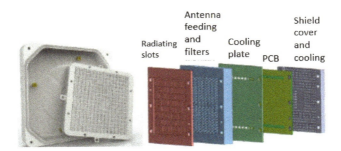

Figure 3.39 Active antenna system construction using gap waveguide technology (©IEEE, reprinted with permission from [42].)

The structured surface is characterized by pins forming a barrier, preventing the electromagnetic waves from propagating in undesired directions. In this way, the pins replace the walls in traditional rectangular waveguides without requiring perfect metallic contact. The waves are guided by ridges or grooves within the pin structure and propagate in air, resulting in low power losses. Antennas based on gap technology have more than 10x lower losses than microstrip lines, more than 3x lower losses than substrate integrated waveguide (SIW), and approximately the same losses as rectangular waveguides [40].

Finally, the concept of a future AESA composed of a panel-type module is shown in Figure 3.40. The fully digital transmitting and receiving module is located beneath the antenna elements. The digital signals flow between the fully digital T/R module and the signal processing unit. To realize the panel-type module-based AESA, it is important to realize the fully digital T/R circuit [43]. These digital technologies are discussed in the next chapter including figures of merit (FOM) that aid in comparison of the different technologies and architectures available.

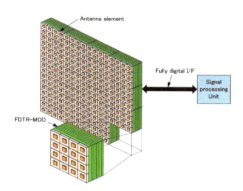

Figure 3.40 Concept of a future AESA using a full digital transmit receive module showing the intermediate frequency (IF) interface to the signal processing unit (©2020 EuCAP, reprinted with permission from [43].)

3.8 SUMMARY

In summary, AESA antenna architectures have come a long way. Now with the inclusion of FPGAs and signal processing in the elements, along with new amplifier materials such as SiGe and GaN and other direct bandgap semiconductors, DRFM transceivers have a variety of reliable, wideband antenna technologies to choose from. The advantages of AESAs have long been recognized with the key enabling T/R technologies giving both performance and affordability with reliability getting better all the time.

REFERENCES

[1] R. Sturdivant and M. Harris, *Transmit Receive Modules for Radar and Communication Systems*. Norwood, MA: Artech House, 2016.

[2] G. D'Amato, G. Piccinni, G. Avitabile, G. Coviello, and C. Talarico, "An integrated phase shifting frequency synthesizer for active electronically scanned

arrays," in *2018 IEEE 21st International Symposium on Design and Diagnostics of Electronic Circuits Systems (DDECS)*, 2018, pp. 91–94.

[3] R. Rieger, A. Klaaßen, P. Schuh, and M. Oppermann, "A full-array-grid-compatible wideband Tx/Rx multipack using multifunctional chips on GaN and SiGe," in *2018 15th European Radar Conference (EuRAD)*, 2018, pp. 433–436.

[4] M. Kappes, "All-digital antennas for mmwave systems," *Microwave Journal*, vol. 62, no. 6, pp. 84 –94, 2019.

[5] D. Sikri and R. M. Jayasuriya, "Multi-beam phased array with full digital beamforming for satcom and 5g," in *Microwave Journal*, vol. 62, no. 4, 2019, pp. 41–44.

[6] J. S. Herd and M. D. Conway, "The evolution to modern phased array architectures," *Proceedings of the IEEE*, vol. 104, no. 3, pp. 519–529, 2016.

[7] D. N. McQuiddy, R. L. Gassner, P. Hull, J. S. Mason, and J. M. Bedinger, "Transmit/receive module technology for x-band active array radar," *Proceedings of the IEEE*, vol. 79, no. 3, pp. 308–341, 1991.

[8] H. P. Feldle, A. D. McLachlan, and Y. Mancuso, "Transmit/receive modules for X-band airborne radar," in *Radar 97 (Conf. Publ. No. 449)*, 1997, pp. 391–395.

[9] F. Liu and C. Masouros, "Hybrid beamforming with sub-arrayed mimo radar: Enabling joint sensing and communication at mmwave band," in *IEEE International Conference on Acoustics, Speech and Signal Processing (ICASSP)*, 2019, pp. 7770–7774.

[10] A. Morsali, S. Norouzi, and B. Champagne, "Single RF chain hybrid analog/digital beamforming for mmwave massive-mimo," in *2019 IEEE Global Conference on Signal and Information Processing (GlobalSIP)*, 2019, pp. 1–5.

[11] A. Morsali and B. Champagne, "Achieving fully-digital performance by hybrid analog/digital beamforming in wide-band massive-mimo systems," in *ICASSP 2020 - 2020 IEEE International Conference on Acoustics, Speech and Signal Processing (ICASSP)*, 2020, pp. 5125–5129.

[12] P. K. Tyagi, A. Trivedi, and S. Bhadauria, "Hybrid beamforming channel estimation with reduced phase shifter numbers for massive mimo systems," in *2019 International Conference on Electrical, Electronics and Computer Engineering (UPCON)*, 2019, pp. 1–5.

[13] M. Oppermann and R. Rieger, "Multifunctional MMICs – key enabler for future aesa panel arrays," in *2018 IMAPS Nordic Conference on Microelectronics Packaging (NordPac)*, 2018, pp. 77–80.

[14] G. W. Stimson, H. D. Griffiths, C. J. Baker, and D. Adamy, *Introduction to Airborne Radar, 3rd Edition*. SciTech, 2014.

[15] H. Franz and NSWCWD, *Electronic Warfare and Radar Systems Engineering Handbook*. SciTech, 2013.

[16] P. E. Pace, *Detecting and Classifying Low Probability of Intercept Radar*. Norwood, MA: Artech House, 2009.

[17] M. L. Taylor, K. L. Virga, and R. G. Yaccarino, "Comparison of antenna transmit weighting functions for active arrays," in *IEEE Antennas and Propagation Society International Symposium. 1999 Digest. Held in conjunction with: USNC/URSI National Radio Science Meeting (Cat. No.99CH37010)*, vol. 4, 1999, pp. 2302–2305 vol.4.

[18] J. Wolf, M. Livadaru, R. A. Dana, and J. B. West, "Limitations of AESA's on monopulse beamforming," in *2016 IEEE International Symposium on Antennas and Propagation (APSURSI)*, 2016, pp. 915–916.

[19] T. Enthusiast, "AESA elements counting," http://http://www.f-16.net/forum/viewtopic.php?t=24978, Accessed: 2020-05-20.

[20] W. P. Hull and R. D. Nordmeyer, "Active-element, phased-array radar: affordable performance for the 1990s," in *NTC '91 - National Telesystems Conference Proceedings*, 1991, pp. 193–197.

[21] C. Kopp, "Active electronically steered arrays a maturing technology," https://www.ausairpower.net/technology.html, Accessed: 2020-05-22 2014.

[22] C. M. Alabaster and E. J. Hughes, "Examination of the effect of array weighting function on radar target detectability," *IEEE Transactions on Aerospace and Electronic Systems*, vol. 46, no. 3, pp. 1364–1375, 2010.

[23] H. Xiao and Y. Guang, "A compact S-band digital T/R module in multi-function MIMO system," in *2018 12th International Symposium on Antennas, Propagation and EM Theory (ISAPE)*, 2018, pp. 1–4.

[24] M. Hurtado, J. Xiao, and A. Nehorai, "Target estimation, detection, and tracking," *IEEE Signal Processing Magazine*, vol. 26, no. 1, pp. 42–52, 2009.

[25] E. Kowalski, D. Conway, A. Morris, and C. Parry, "Multifunction phased array radar advanced technology demonstrator (MPAR ATD) nearfield testing and fielding," in *2019 IEEE Radar Conference (RadarConf)*, 2019, pp. 1–4.

[26] M. Cooley, S. Essman, S. Quade, S. Geibel, T. Spence, T. Fontana, and K. Browne, "Planar-fed folded notch (PFFN) arrays: A novel wideband technology for multi-function active electronically scanning arrays (aesas)," in

2016 IEEE International Symposium on Phased Array Systems and Technology (PAST), 2016, pp. 1–6.

[27] Y. Mancuso and C. Renard, "New developments and trends for active antennas and TR modules," in *2014 International Radar Conference*, 2014, pp. 1–3.

[28] W. Gautier, W. Gruener, R. Rieger, and S. Chartier, "Broadband multifunction AESA front-ends: New requirements and emerging technologies," in *2016 46th European Microwave Conference (EuMC)*, 2016, pp. 1481–1484.

[29] Y. Mancuso and C. Renard, "New developments and trends for active antennas and tr modules," in *2014 International Radar Conference*, 2014, pp. 1–3.

[30] E. Brookner, "Metamaterial advances for radar and communications," in *2017 IEEE Radar Conference (RadarConf)*, 2017, pp. 1614–1621.

[31] W. Weng and C. Hung, "An H-fractal antenna for multiband applications," *IEEE Antennas and Wireless Propagation Letters*, vol. 13, pp. 1705–1708, 2014.

[32] A. Bakytbekov and A. Shamim, "Additively manufactured triple-band fractal antenna-on-package for ambient RF energy harvesting," in *2019 13th European Conference on Antennas and Propagation (EuCAP)*, 2019, pp. 1–3.

[33] Yashmi and B. S. Dhaliwal, "Design of a dual band crown fractal antenna," in *2017 IEEE Applied Electromagnetics Conference (AEMC)*, 2017, pp. 1–3.

[34] R. Rieger, A. Klaaßen, P. Schuh, and M. Oppermann, "GaN based wideband T/R module for multi-function applications," in *2015 European Microwave Conference (EuMC)*, 2015, pp. 514–517.

[35] S. Rathod, A. Raut, A. Goel, K. Sreenivasulu, K. S. Beenamole, and K. P. Ray, "Novel fpga based t/r module controller for active phased array radar," in *2019 IEEE International Symposium on Phased Array System Technology (PAST)*, 2019, pp. 1–5.

[36] D. Conway, M. Fosberry, G. Brigham, E. Loew, and C. Liu, "On the development of a c-band active array front-end for an airborne polarimetric radar," in *2013 IEEE International Symposium on Phased Array Systems and Technology*, 2013, pp. 198–201.

[37] D. Xiaohui and K. Minggang, "Design of T/R-module for the ultra-wide-band multi-function aesa," in *2009 2nd Asian-Pacific Conference on Synthetic Aperture Radar*, 2009, pp. 261–262.

[38] R. Cacciola, E. Holzman, L. Carpenter, and S. Gagnon, "Impact of transmit interference on receive sensitivity in a bi-static active array system," in *2016 IEEE International Symposium on Phased Array Systems and Technology (PAST)*, 2016, pp. 1–5.

REFERENCES

[39] null, "AESA applications by new technologies evolution: Invited paper," in *2019 IEEE International Symposium on Phased Array System Technology (PAST)*, 2019, pp. 1–4.

[40] P. Hindle, "Antenna technologies for the future," in *Microwave Journal*, no. 1, 2018.

[41] M. W. Elsallal, J. Hood, and I. McMichael, "3d printed material characterization for complex phased arrays and metamaterials," in *Microwave Journal*, no. 1, 2016, pp. 41–44.

[42] E. Brookner, "Metamaterial advances for radar and communications," in *2017 IEEE Radar Conference (RadarConf)*, 2017, pp. 1614–1621.

[43] Y. Wada, K. Fujita, Y. Kuji, M. Iwasaki, T. Mizuno, and M. Tanabe, "Transmit and receive module with a fully-digital interface," in *2020 14th European Conference on Antennas and Propagation (EuCAP)*, 2020, pp. 1–4.

Chapter 4

Choosing the Correct Wideband Receiver

In the previous chapter we discussed the importance of the AESAs and the digital beamforming that is critical for the DRFM developer. These types of antennas can provide significant gains and serve several simultaneous DRFM multifunctions. Digital beamforming represents a major requirement as a part of network-enabled massive input/massive output (MIMO) system. They enable advanced processing techniques, such as adaptive nulling of antenna directivities toward jammers, to effectively suppress clutter and interferences while maintaining target detection sensitivies. In this chapter, information concerning how to choose the best wideband spectrum-sensing receiver architecture is presented. Starting from the AESA antenna, methods to connect the antenna up to the receiver are discussed. Then the process of choosing the best wideband spectrum sensing architecture for a particular DRFM application is presented. Digital architectures for wideband spectrum-sensing and ultra wideband channel estimation include compressive sensing receivers such as the *modulated wideband converter*.

Employing the technique of compressed sensing stands as a promising choice for future DRFM receivers as technique of it has the potential to reduce the typically required sampling rate and the volume of data that needs to be collected and processed. The chapter also discusses bandpass sampling receivers, undersampling channelizers, and analysis/synthesis channelized receivers including a look at the ones using embedded technologies such as FPGAs and GPUs. Digital receiver/exciters or transceivers are also addressed with an example shown.

4.1 AESA ANTENNA-TO-RECEIVER INTERFACE

In Chapter 2 we saw there are basically two choices for an AESA antenna to interface to the ADC path for receive and the DAC path and transmit: the switch as shown in Figure 4.1(a) and the circulator, diplexer, or duplexer as shown in Figure 4.1(b).

The switch is shown in Figure 4.1(a). A switch can provide almost infinite isolation. Note also that the switch can be used at both the antenna interface and at the back as shown in Figure 4.1(a). The *circulator* can provide isolation up to 60 dB between paths. In addition to the isolation, other circulator considerations are the temperature variation, voltage standing wave ratio (VSWR), the insertion loss, and whether a phase match is required. A *duplexer* for single antenna operation separates a transmit and receive path based on signal direction and can be used for the same frequency signals. It consists of two bandpass filters that allow simultaneous transmission and reception within the same band and using the same antenna. There are two types of duplexers, one by using PIN diode switches and the other using circulators as shown. Both transmit and receive paths usually will have frequency bands very nearer, hence narrowband filters are used to separate these frequencies. A duplexer is often referred as 9-port RF circulator. Typical isolation transmit to receive isolation is 90 to 95 dB.

A *diplexer* is also used for single antenna operation, and separates two different frequency bands in the receive path and combines them in the transmit path. These bands are typically too far apart in the frequency domain for the diplexer to work satisfactorily. Often referred as a RF power combiner/divider it can have the added functionality of broadband filtering to pass the appropriate bands at the transmit and the receive path. Typical transmit to receive isolation is 50 to 60 dB. Their operation is not interchangeable, and a diplexer could not replace a duplexer in common circuits.

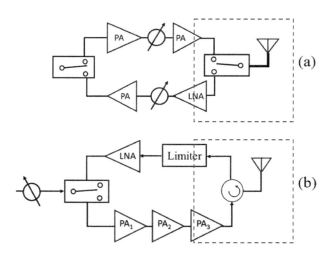

Figure 4.1 Transmit/receive module showing the use of a (a) circulator at the antenna interface and (b) a switch at the antenna interface.

4.1.1 RF CMOS Technology

RF high-speed CMOS is well established for many microwave applications–especially where very densely packed mixed-signal RF integrated circuits (RFICs) are concerned. The basic CMOS logic inverter shown in Figure 4.2 represents an important example of how two contrasting types of MOS transistors (pMOS, nMOS) are interconnected to form an inverter, a fundamental type of circuit configuration. The functional behavior of the CMOS inverter circuit can be characterized by just two cases. A low voltage on the input turns on the p-channel transistor Q2, and turns off the n-channel transistor Q1. Since the output is connected to V_{dd} 1.3V, through Q2, the output is high. When the input voltage is high, we have the opposite behavior and the output is connected to ground (0V) through Q1 and is low. This is exactly an inverter function – the output logic value is the opposite of the input logic value.

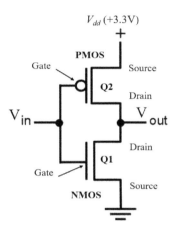

Figure 4.2 CMOS inverter using both an nMOS and a pMOS transistor.

RF CMOS and its derivatives now represent a mainstream RF technology that can be adopted for the relatively low-power portions of MMICs/RFICs. CMOS data converters are essential components for coherent transceiver design since they provide the conversion between the digital and analog domain and vice versa. In addition to semiconductor digitization technology, photonics has also made significant inroads [1]. These converters today are fabricated in technology to scale the bandwidth with power and cost efficiently. However, there is an upcoming limit in the future scalability of CMOS as the transistor nodes, today at 7 nm gate size, will have trouble meeting future optical interface data rates, even as nodes scale to 5 nm and 3 nm. To ease this scaling issue, parallel data converter architectures, better known as interleaved data converters, are enabling performance beyond the limits of CMOS. State-of-the-art

CMOS technologies are used in addition to circuits in material alternatives to silicon to scale the aggregate converter and channel bandwidth.

To further increase both the analog bandwidth and the sampling rate, bipolar CMOS (BiCMOS) technologies offer an attractive alternative if they can be integrated or interconnected with CMOS DSP through innovative techniques. Among these new technologies are silicon germanium or III/V semiconductors (e.g., indium phosphide) for which ft and fmax values >1 THz have been demonstrated, as described in [3]. High-speed ADCs are usually based on pipelined, successive approximation registers or flash architectures. To connect the AESA antenna up to the TX (transmitter) path, for the HPA chain and DAC and the RX (receiver) path for the LNA ADC (digitization), a GaN switch such as the one shown in Figure 4.3 can be used.

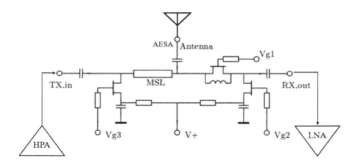

Figure 4.3 Schematic of an asymmetrical CMOS switch.

For efficiency on transmit, the path to the HPA must be low loss and the isolation of the path to the LNA should be as high as possible for its protection. For an efficient receive mode however, a minimal loss must be achieved for the LNA path to enable a low noise figure.

An asymmetrical CMOS switch such as those in Figure 4.3 can provide the critical link between the HPA on transmit and the LNA on receive [2]. The TX path uses a transmission line with low loss (0.8 dB) over the 10-GHz bandwidth and provides good isolation with low insertion loss. In Rx, a series-parallel-FET design with a large series FET is used to keep insertion loss low. Conventional GaN switch designs ask for a high negative control voltage to switch the FET devices. The design also avoids negative control voltages through introduction of a positive offset voltage [2]. This voltage is already available in the HPA drain supply (on the Tx side). The source lead for the parallel FETs is equipped with series capacitors that provide a DC blocking and RF short effect. Another advantage of a switch over a duplexer is it can be controlled using a T/R module controller such as the one discussed in [3] where the switch is used at the back of the T/R module as shown in the bottom of Figure 4.1.

An RF-MEMS switch fabricated in 0.13μm SiGe BiCMOS process technology for 240-GHz applications [4]. The fabricated RF-MEMS switch provides a high capacitance Con/Coff ratio of 8.78, 0.44 dB of insertion loss, and 24.6 dB of isolation at 240-GHz. A compact high-isolation Ku-band SPDT switch using triple-well transistors based on 0.35-μm SiGe BiCMOS process is shown in Figure 4.4 in [5].

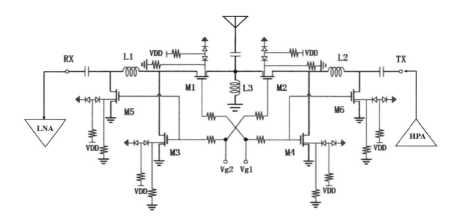

Figure 4.4 Schematic of the SPDT switch using body-floating transistor technology. (After [5].)

An improved series-shunt-shunt topology is used in this design to enhance the isolation and to reduce the insertion loss simultaneously. In order to improve the power handling capability, *body-floating transistor technology* is employed and analyzed. The full-wave simulated results show that the insertion loss of the ON state path is better than 1.76 dB, and the isolation of the OFF state path is higher than 40.8 dB in the entire designed frequency band of 14-18 GHz.

4.2 ANALOG-TO-INFORMATION SAMPLING

SDRs digitize the spectrum as far toward the antenna as possible, allowing the spectral signal processing to take place digitally, in turn allowing it to be reconfigured in many different ways. Strictly allocated resources, such as power and area however, coupled with multifunction integration requirements, have resulted in an increased role for the DRFM's front end within the unmanned and miniature air-launched vehicle. As software-defined, reconfigurable receivers continue to play an important role in communication, EMMW, and transceiver designs, there becomes a need for new configurations that move away from the traditional trade-offs.

In addition, bandwidth and reconfigurability currently play a key requirement creating systems that provide high performance under stringent data and sampling demands leading to a recent concept in the signal processing literature. The concept is sampling the environment based on the *information rate* rather than the bandwidth. When the signal environment is relatively sparse, the reduction in sample rate may be very large. Vetterli, Marziliano, and Blu show that Dirac sequences may be sampled at a finite rate of innovation (FRI) – even with extra sampling to give robustness against noise, this is still far less than the Nyquist/Shannon rate for ultrawideband impulse-like signals [6]. Although these initial results apply to a limited class of signals, recent results in compressed sensing (CS) have shown more generally that the information from a signal can be captured with far fewer measurements than the traditional Nyquist/Shannon criteria, as long as the signal has a sparse (or nearly sparse) representation in some basis or frame [6–8]. One of the more surprising aspects of CS (also called compressive sampling) is that the prescribed measurements do not require a priori signal knowledge beyond the basic sparsity or compressibility assumption. Thus it is possible, at least in principle, to design a universal CS measurement system that can be used to encode a wide range of signal types. The idea of sampling based on the signal information rate, rather than the Shannon bandwidth criteria, suggests a new approach denoted *analog-to-information* (A-to-I) as an alternative to conventional ADCs or digital receivers.

The applicability of CS theory to practical RF receivers has been somewhat limited. Original CS theory described the recovery of sparse (and perhaps very high dimensional) vectors from a set of observations in the form of projections of the signal onto random basis vectors. This model is ill-suited to RF applications for several reasons. First, it assumes that the signal could be subjected to repeated observations of the same signal vector, which is not the case with time-varying signals. Second, the discrete nature of this model implicitly assumes that the signal has already been sampled, which leads back to the original problem of the ADC limitations.

Nonetheless, some promising approaches for practical receivers have been proposed, including discrete-time random filters [9], random sampling [10, 11], and random demodulation [12, 13]. Note that A-to-I architectures based on either discrete time random filtering or random demodulation still require Nyquist-rate components relative to the maximum analog frequency at the receiver front end, thus limiting the applicability of these approaches for direct processing of high RF signals. On the other hand, A-to-I approaches based on nonuniform sampling are feasible for direct RF implementation without the need for Nyquist-rate components. In this chapter we describe a novel approach that uses structured nonuniform sampling, rather than random sampling, to implement a direct RF A-to-I receiver that is effective at recovering signals that have a sparse frequency-domain representation [14, 15]. Among the benefits of the proposed receiver is its greatly simplified signal recovery compared to random sampling.

4.3 COMPRESSIVE SENSING RECEIVERS

Consider the problem of recovering unknown, real, length N vectors that are sparse in some specified basis or dictionary. Let x be such a vector, and assume it is K–sparse, meaning that it can be expressed as a linear combination of K elements of the basis set $\{\psi_i\}$; for example,

$$x = \sum_{i=1}^{K} \alpha_i \psi_i$$

Recovery of x is, of course, possible using traditional methods – for example, by sampling each entry of x. The problem with this approach is that sampling at such a rate could be physically impossible, and even if it were possible we would expend quite a bit of energy encoding bits that will eventually be discarded in the final representation.

In [16], Donoho posed the question of whether there is a way to "just measure directly the part that won't end up being thrown away." The answer is yes, and compressive sensing describes a collection of methods by which x can be recovered using a minimal number of measurements. If the correct basis elements were known prior to sampling, one could simply observe the K relevant entries of the signal in the representation $\{\psi_i\}$. But the lack of prior signal knowledge makes this an impossible task. Instead, the CS approach prescribes collecting samples that are projections of the unknown signal onto elements from a second basis set $\{\phi_i\}$ that is *incoherent* with $\{\psi_i\}$. By incoherent, we mean that a sparse representation of any element of the basis set $\{\phi_i\}$ does not exist using the basis vectors $\{\psi_i\}$ and vice-versa [16]. A collection of such observations can be described by the linear model $y = \Phi x$, where y is a length m vector and $K < M << N$.

One of the significant results of CS theory is that the signal does not need to be known for incoherent projections to be obtained. In particular, random vectors will be incoherent with any fixed basis vectors with high probability. Given the basis vectors, the problem then reduces to solving an l_1 minimization problem, finding the estimate for x in the underdetermined set of linear equations:

$$\hat{\alpha} = \arg\min \|\alpha\|_1 \quad s.t. \quad y = \Phi x = \Phi \Psi \alpha$$
$$\hat{x} = \Psi \hat{\alpha}$$

On average, CS succeeds in recovering K sparse vectors when the number of observations greater than a small constant times $K \log_{10}(N)$.

Since CS involves discrete-time observations, one of the goals of practical A-to-I receivers is to extend and apply discrete time CS concepts to an analog continuous time signal environment. For example, a model of the discrete time random filters has been created with a Toeplitz sensing matrix allowing for efficient CS solutions.

Information recovery for locally Fourier sparse signals is performed via the sparsogram – a fast iterative greedy pursuit algorithm that includes computing a nonuniformly sampled fast Fourier transform algorithm on the sampled residual. The novel approach, described below, relies on nonuniform chirped sampling for simplified information recovery compared to other A-to-I approaches, and allows sampling high analog input frequencies without the need for high-speed components operating at the Nyquist rate for the maximum analog input frequency.

4.3.1 Nyquist Folding Analog-to-Information Receiver

The primary challenge in reconstructing a signal from its samples is that many different signals could possibly give rise to the same set of samples. For example, uniformly sampling a time-varying signal can result in aliasing and the inability to recover the original frequencies of the aliased components. The novel sampling scheme proposed here overcomes this issue by imposing a frequency-dependent signature on each component of the original signal, from which the original signal component frequencies can be obtained [17]. We now examine the proposed A-to-I receiver architecture, which folds multiple Nyquist zones into a narrower bandwidth before the ADC, in more detail. The Nyquist folding receiver (NYFR), shown in Figure 4.5, uses a wideband pre-select filter $H(\omega)$ rather than a standard anti-aliasing filter prior to sampling; thus allowing multiple Nyquist zones to be sampled and subsequently folded into a continuous time analog interpolation filter.

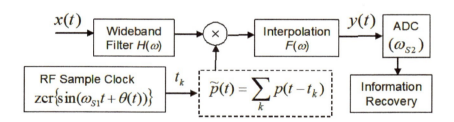

Figure 4.5 Nyquist folding A-to-I receiver architecture.

In order to provide direct RF sampling at high analog input frequencies and avoid high-speed Nyquist rate components, the RF sampling is performed using pulse-based sampling rather than using a sample and hold circuit. The RF sample clock is modulated about a carrier to provide a nonuniform sample rate $\varphi'(t)$. The phase of the RF sampling clock may be viewed as an occurrence function for a monotonically increasing function $\varphi(t)$ – the samples are taken as $\varphi(t)$ crosses multiples of 2π. That is, sample times correspond to zero crossing rising voltage times of $\sin(\varphi(t))$. Other than the fact that the sample clock is modulated and the filter $H(\omega)$ allows aliasing, the front-end portion before the ADC follows the standard impulse sampling algorithm.

Figure 4.6 shows the NYFR folding the RF input spectrum through direct RF pulse-based sampling without quantizing followed low-pass filtering.

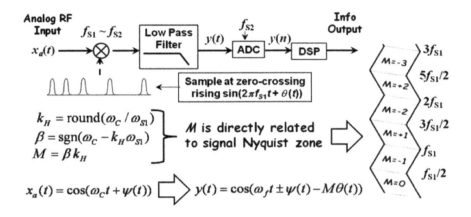

Figure 4.6 Nyquist folding A-to-I receiver architecture.

Below we examine the case where $F(\omega)$ is an ideal low-pass filter with cutoff $\omega_{s1}/2$ and where $\varphi(t)$ represents a narrowband phase of frequency modulation centered at ω_{s1}:

$$\varphi(t) = \omega_{s1}(t) + \phi(t)$$

For this narrowband frequency modulated RF sample clock, if the input is a narrowband signal with center frequency ω_c and information modulation $\psi(t)$:

$$x(t) = \cos(\omega_c t + \psi(t))$$

the normalized interpolation filter output will be:

$$y(t) \approx \cos(|\omega_c - \omega_{s1} k_H| t + \beta \psi(t) - M\theta(t)), \text{where}$$
$$M = \beta k_H \quad \beta = \text{sgn}(\omega_c - \omega_{s1} k_H)$$

$$k_H = round(\omega_c/\omega_{s1}) \tag{4.1}$$

In 4.1, $|\omega_c - \omega_{s1} k_H|$ is the intermediate frequency after bandpass down conversion sampling, β is negative for spectrally reversed bandpass sampled signals from odd Nyquist zones as shown in Figure 4.7. Here M is the modulation scale factor, and k_H is the harmonic in the Fourier series of the pulse train that corresponds to the interpolation filter output. This compressive sensing result appears at the output of the interpolation filter. A derivation of the interpolation filter output is given in Appendix 4A at the end of this chapter.

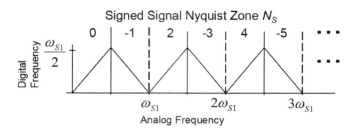

Figure 4.7 Bandpass sampled signal Nyquist zones.

From the interpolation filter output in (4.1), the received signal has an induced modulation $M\theta(t)$ of the same form as the RF sample clock modulation $\theta(t)$ with a modulation scale factor M and orientation β depending on the signal Nyquist zone N_s.

Even though multiple signals from different Nyquist zones may alias into the same band, the information from the different signals, including the original RF, can still be recovered. We take advantage of the fact that the added modulation is different for each Nyquist zone so that the folded signals are separable under the condition that the signal environment is relatively sparse. Note that the continuous time interpolation filter allows the RF sample rate to be decoupled from the ADC sample rate so that the ADC may sample at a uniform rate. This feature helps to simplify the clocking of the DSP, including local data movement between ADC and DSP.

With narrowband FM sampling in the NYFR architecture of Figure 4.6, processing can be performed directly on the folded data without performing a computationally complex l_1 minimization. For example, if the sample clock is a FM continuous wave (FMCW) periodic chirped signal, received narrowband signals will all have the same chirp pattern in the time-frequency plane, with center frequencies depending on bandpass sampling translation and frequency modulation scale factors depending on originating Nyquist zone. Signal recovery may be performed by de-modulation as opposed to using a chirp CS matrix with a minimization process.

4.3.1.1 An Illustrative Example

In Figure 4.8, an illustration of a MATLAB example of a NYFR with an ideal low-pass interpolation filter and a sinusoidal FMCW RF sample clock is shown.

4.3 Compressive Sensing Receivers 147

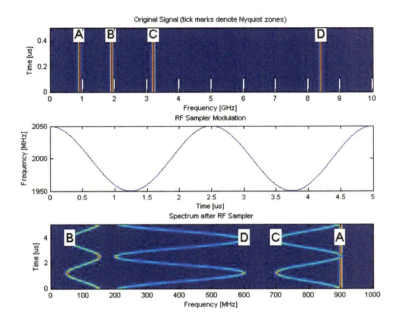

Figure 4.8 Folded narrowband signals at the output of the interpolation filter, each with a unique Nyquist zone dependent modulation.

The top panel frequency-time plot shows four narrowband signals: signal A is at 900 MHz; signal B is at 1.9 GHz; signal C is at 3.2 GHz; signal D is at 8.4 GHz. The middle panel shows the RF sample clock varying from 1950 to 2050 Msps over a 2.5-μs window, with an average sample rate of 2.0 Gsps. The bottom panel shows the folded narrowband signals at the output of the interpolation filter with unique Nyquist zone dependent modulation for each signal. Figure 4.9 shows the signal Nyquist zone, sampling harmonic (k_H), folded intermediate frequency (IF), and modulation bandwidth for each of the four signals using (4.1), and signed Nyquist zones N_s from Figure 4.8.

Signal	N_S	k_H	M	IF (MHz)	$M\Delta F$ (MHz)
A	0	0	0	900	0
B	-1	1	-1	100	-100
C	-3	2	-2	800	-200
D	8	4	4	400	400

Figure 4.9 Table of the four signals showing their Nyquist zones and frequency parameters (From [17].)

4.3.1.2 Laboratory Test

Experimental results were obtained using a DCSM 7620 sampling device from Picosecond Pulse Labs with a nominal sample aperture of about 20 picoseconds.

In this example, we consider three narrow band tones: signal A is at 7.65 GHz; signal B is at 17.22 GHz; and signal C is at 32.6 GHz. The interpolation filter bandwidth is 850 MHz. The output of the interpolation filter is uniformly sampled by an Atmel 10-bit ADC at 2 Gsps. For this experiment, we consider two cases: Uniform RF sampling at 2 Gsps and modulated RF sampling using an FMCW clock waveform (similar to the prior illustrative example) with an average sample rate $F_{s1} = 2,000$ Msps, $\Delta F = 5$ MHz, and the modulation period $t_m = 2$ μs. Figure 4.10 shows a multisignal lab test example. On the left, the results are after uniform sampling at 2 Gsps.

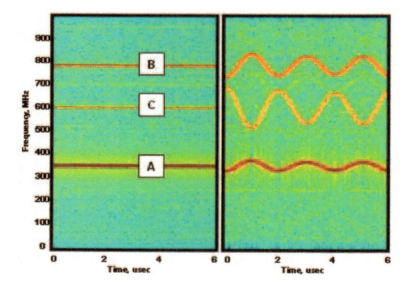

Figure 4.10 Multisignal lab test example [17].

Signals A, B, and C fold to 350 MHz, 780 MHz, and 600 MHz, respectively, as predicted by conventional bandpass sampling. The right panel of the figure shows the resulting time-frequency plot with the FMCW modulated RF sample clock. The corresponding calculated values for this lab test are presented in Figure 4.11. From this example, we can see that the signals fold to the correct locations and that the induced modulation bandwidth and modulation orientation corresponds to the expected values listed in Figure 4.11. Note that in addition to determining the original RF, it is also possible to remove the induced modulation to recover the original signal.

Signal	N_S	k_H	M	IF (MHz)	$M\Delta F$ (MHz)
A	-7	4	-4	350	-20
B	-17	9	-9	780	-45
C	32	16	16	600	80

Figure 4.11 Table showing the three signals and their Nyquist parameters [17].

In summary, the Nyquist folding receiver is an analog-to-information receiver motivated by compressive sensing that performs frequency modulated pulsed sampling directly at the RF to allow for unambiguous recovery after compressing multiple Nyquist zones into an analog interpolation filter. The RF sample clock modulation induces a Nyquist-zone dependent frequency modulation on the received signals that can be measured and removed when the RF signal environment is relatively sparse so that the folded signals do not overlap too much. With this architecture, it is possible to recover signals without performing a computationally complex l_1 minimization. In some applications, the folded signal parameters may be detected and measured directly in the compressed space, further reducing the computational complexity. Although we have limited the discussion to narrow band (Fourier-sparse) signals in this chapter, the NYFR is applicable to a broader class of signals.

4.3.2 The Modulated Wideband Converter

The modulated wideband converter (MWC) is a compressive sensing architecture able to break the compromise between bandwidth, noise figure, and energy consumption of ADCs and was originally proposed by Mishali and Eldar [18]. The technique exploits spread spectrum techniques and uses an analog mixing front end to alias the spectrum such that that a spectrum portion from each band appears in baseband. The primary design goals were efficient hardware implementation and low computational load on the supporting digital processing. The system consists of several channels, with each sampling channel containing a mixer, a low-pass filter, and a sampler. A block diagram of the MWC is shown in Figure 4.12.

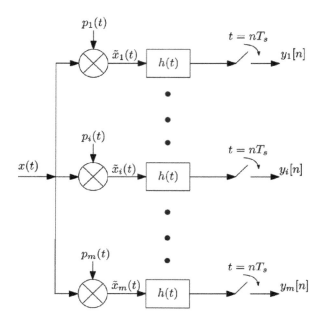

Figure 4.12 The modulated wideband converter (MWC),(©IEEE, reprinted with permission from [18].)

An input signal from a wideband antenna, for example, is processed in parallel by m channels. The mixer in each channel makes the spectrum of the multiband signal $p_i(t)$ extend periodically, piecewise smooth, and the spectrum portion from each subband can appear in the baseband. This operation is similar to spread-spectrum technology. The input signal is multiplied by a mixing function $p_i(t)$ that is T_p-periodic. The multiplier output $\tilde{x}_i(t)$ is processed by a low-pass filter with cut-off frequency $1/(2T_s)$ and the filter output, y_i is obtained by sampling with frequency $f_s = 1/T_s$ which is orders of magnitude smaller than Nyquist. In summary, the design parameters are:

- m : Number of channels;
- T_p: The period;
- $1/T_s$ The sampling rate;
- $p_i(t)$ The mixing functions for $1 \leq i \leq m$.

The mixing function $p_i(t)$ and the frequency response for the lowpass filter $h(t)$ are shown in Figure 4.13.

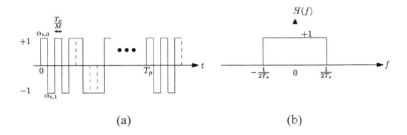

Figure 4.13 Modulated wideband converter's (a) mixing function $p_i(t)$ and (b) frequency response of lowpass filter $h(t)$. (©IEEE, reprinted with permission from [18].)

The concept is that a sufficiently large number of mixtures allows one to recover a relatively sparse multiband signal. The sequence $p_i(t) = \alpha_{ik}$ with

$$k\frac{T_p}{M} \leq t \leq (k+1)\frac{T_p}{M}, \quad 0 \leq k \leq M-1$$

with $\alpha_{ik} \in \{+1, -1\}$, and $p_i(t + nT_p) = p_i(t)$ for every $n \in \mathbb{Z}$.

Perfect recovery from the proposed samples is achieved under certain necessary and sufficient conditions. For the ith channel, the mixing function $p_i(t)$ has a Fourier expansion

$$p_i(t) = \sum_{l=-\infty}^{\infty} c_{il} \exp(2\pi l t / T_p) \quad (4.2)$$

where

$$c_{il} = \frac{1}{T_p} \int_0^{T_p} p_i(t) \exp(-j2\pi l t / T_p) dt \quad (4.3)$$

The mixing function above is applied to the multiband signal $x(t)$, obtaining the modulated signal as $\tilde{x}_i(t) = x(t) p_i(t)$. Its Fourier transform is evaluated as

$$\tilde{X}_i(f) = \int_{-\infty}^{\infty} \tilde{x}_i(t) \exp(-j2\pi l t / T_p) dt \quad (4.4)$$

$$= \int_{-\infty}^{\infty} x_i(t) \left(\sum_{l=-\infty}^{\infty} c_{il} \exp(2\pi l t / T_p) \right) \exp(-j2\pi f t) dt$$

$$\tilde{X}_i(f) = \sum_{l=-\infty}^{\infty} c_{il} X(f - lf_p) \quad (4.5)$$

where $X(f)$ is the Fourier representation of $x(t)$ and $f_p = 1/T_p$.

After the mixer operation, the spectrum of the modulated signal achieves periodic expansion, causing intentional spectrum aliasing. Subsequently, a linear combination of f_p−shift copies of $X(f)$ are then input to the low-pass filter with with cutoff frequency $1/(2T_s)$. After low-pass filtering, only the spectrum in the baseband is retained, and contains the spectrum information from each subband. Finally, the filtered signal is sampled uniformly at low rate $f_s = 1/T_s$ and is matched to the sampling rate of the low-pass filter. As a practical consideration, f_p is chosen slightly larger than B to avoid edge effects and in order to get the complete information within each subband. The period T_p determines the aliasing of $X(f)$ by setting the shift intervals to $f_p = 1/T_p$ and we choose $f_p \geq B$ so that for each subband itself, there is no aliasing.

As a practical matter, f_p is chosen to be slightly more than the B to avoid edge effects. In order to get all the information within each subband, the relationship between f_p and f_s must satisfy $f_s \geq f_s$. The simplest choice is $f_p \approx B$ as this allows the lowest sampling rate for each sampling channel. The overall sampling rate of the MWC is equal to the product of the channel number m and the single channel sampling rate f_s. That is, $f_{sys} = m f_s$. Note that this sampling rate is much lower than the Nyquist rate, so it is called a *sub-Nyquist rate*.

A digital architecture is also developed that allows either reconstruction of the analog input or processing of any band of interest at a low rate; that is, without interpolating to the high Nyquist rate. Also shown are numerical simulation results that demonstrate many interesting engineering aspects of the MWC device, including: robustness to noise and mismodeling, potential hardware simplifications, real-time performance for signals with time-varying support, and stability to quantization effects. Which also compare the MWC with two previous approaches: periodic nonuniform sampling, which is bandwidth limited by existing hardware devices, and the random demodulator, which is restricted to discrete multitone signals and has a high computational load. In the broader context of Nyquist sampling, they point out that the MWC scheme has the potential to break through the bandwidth barrier of state-of-the-art analog conversion technologies such as interleaved converters, which are discussed in Chapter 6. Since their MWC invention in 2010, many reconstruction techniques for A-to-I have grown out of this development. For example, a blind multiband signal reconstruction method, referred to as a statistics multiple measurement vectors (MMV) iterative algorithm approach, is described in [19]. By exploiting the jointly sparse property of MMV model, the supports can be obtained by statistical analysis for the reconstruction results. Simulation results show that, without the sparse prior, the statistics MMV iterative algorithm can accurately determine the support of the multiband signal in a wide range of signal-to-noise ratios by using various numbers of sampling channels.

For instance, the investigation into a new pseudorandom code sequence, the Zadoff-Chu, is examined in [20]. This code-based, real-valued sensing matrix satisfies cyclic properties with good spectral properties and also increases the robustness against noise. Second, a quasi-systematic study of the influence of code families and of row selection is carried out on different criteria. Specifically, the influence on the coherence, vital to limit the number of branches, is investigated. Additionally, an original approach that focuses on evaluating the isometric properties is established. These measures are helpful since isometry is essential to the noise robustness.

Unfortunately, the MWC also has a high hardware complexity owing to the high degree of freedom of the random waveforms constructing the measurement matrix. To reduce the complexity, the authors in [21] present a compressive circulant matrix (CCM) based generating random waveform by cyclic shift of a special sequence with unit amplitude and random phase in frequency domain. Theoretical analysis shows this scheme is optimal for signals sparse in frequency. CCM-AIC outperforms MWC and is more robust.

4.4 BLIND COMPRESSIVE SENSING WITHOUT A PRIORI BASIS MATRIX KNOWLEDGE

Since CS-based EW receivers are able to detect and locate targets using only a small number of compressively obtained samples, their computational cost is much lower than traditional ES intercept receivers. A sparse Bayesian learning (SBL) framework, is blind in the sense that the knowledge of the sparsity basis is not available. As a practical consideration, the lack of a priori knowledge on the characteristics of received signals is one of the main challenges in the design and development of modern DRFMs.

In [22], the authors develop a model of the observed signal by receiver as a linear signal contaminated by the additive noise and a term to account for the clutter and other structured noises. Let (d_e, α) denote the location in polar coordinates of the receiver, where d_e denotes the distance between the receiver and the origin, while α stands for the azimuth angle of the EW receiver. Also, consider that the adversary radar lies on the same plane. The adversary radar is at azimuth angle θ and moves with constant radial speed v. Note that the velocity remains constant only during each processing interval but the velocity can change independently from one processing interval to another. So the range is

$$d_r(t) = d_r(0) - vt$$

where $d_r(0)$ is the distance between the adversary radar and the origin at time instant zero. Under the far field assumption $d_r(t) >> d_e$ the distance between the adversary radar and the DRFM receiver, $d(t)$, can be approximated as [22]

$$d(t) \approx d_r(t) - \eta(\theta) = d_r(0) - vt - \eta(\theta)$$

where $\eta = (\theta) = d_e \cos(\theta - \alpha)$.

The work in [22] focuses on pulse compression emitters employing continuous-time phase modulated prototype waveforms $z(t)$ of duration τ. Such a pulse can be modeled as a collection of L contiguous subpulses $z_l(t)$ of duration $\tau_z = \tau/L$, each with the same frequency but a (possibly) different phase [22], or

$$z(t) = \sum_{l=0}^{L-1} z_l(t - l\tau_z) \quad \text{for } 0 \leq t \leq \tau \tag{4.6}$$

and $z(t) = 0$ zero elsewhere. For the individual subpulses $z_l(t) = \exp^{j\phi_l}$ for $0 \leq t \leq \tau_z$ and they are often referred to as a subcodes in LPI radar [23].

The goal of the proposed blind-SBL framework is to estimate the unknown pulse modulating code sequence Z, as well as finding the azimuth angle θ and movement speed v of the target. A blind SBL algorithm is formulated and a solution is derived in [22]. Numerical simulations are run to demonstrate the performance of the proposed blind-SBL framework and analyze its properties through a set of Monte Carlo simulations.

4.5 BANDPASS SAMPLING CHANNELIZER

An analog-to-information DRFM architecture based on a bandpass sampling technique is presented in this section. It presents an SoC implementation of a wideband channelized transceiver with subbands that use discrete time signal processing to process intercepted signals with a continuous frequency range from 1 GHz to 20 GHz and an instantaneous bandwidth up to 300 MHz. In addition, nonparametric power spectral density estimation is used to detect incoming signals in the noisy environment. Performance results show an input dynamic range of 40 dB ranging from −60 dBm to −20 dBm with receiver noise figure as low as 4 dB. Also, the worst case acquisition time of 20 pulse repetition intervals is achieved. The system-on-a-chip (SOC) implementation as the wideband channelized transceiver is as shown in Figure 4.14.

The receiver uses bandpass sampling as a downconversion mixer to convert the RF input to baseband using discrete time processing. For bandpass sampling, the signal spectral characteristics such as carrier frequency and bandwidth are needed to be known prior to sampling. The different subband analog filters shown in Figure 4.14 are used for detection of incoming signals and are activated by digital controls using analog multiplexers and demultiplexer circuits. These bandpass filters ensure precise portions of the entire DRFM channel spectrum are passed to the sampling mixer, thus enabling the bandpass sampling mixer to downconvert the filtered signal. Each channel also has its own antenna.

The filtered signal can be stored for processing and waveform detection in the DSP without knowing the exact signal bandwidth and carrier frequency. The bandpass filter's characteristics are designed to prevent flipping of original spectrum after sampling and downconversion and to avoid aliasing due to bandpass sampling.

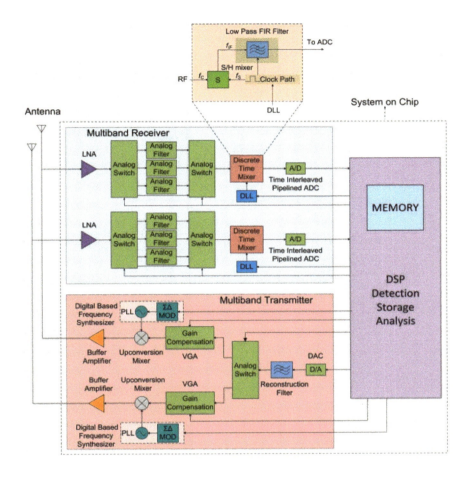

Figure 4.14 Channelized DRFM with frequency of operation 1 - 20 GHz with instantaneous bandwidth of 300 MHz. (©IEEE, reprinted with permission from [24].)

The multichannel receiver front end architecture with parallel overlapping DRFM channels as shown in Table 4.1 are for continuous coverage over the entire bandwidth. The overall DRFM system channel specifications are given in Table 4.2.

The front end architecture is divided into parallel DRFM channels but only two channels are shown in Figure 4.14 with each channel using a separate wideband antenna (bandwidth in Figure 4.1). The antenna and RF design considerations are discussed in [24] – one for each channel. All the DRFM channels work simultaneously, allowing the maximum coverage over the entire RF spectrum under consideration and for the ability of the DRFM to process more than one signal simultaneously. The DRFM channels overlap alias in the frequency domain in order to not miss any signals

Table 4.1
DRFM Channels for Overlapping Continuous Coverage (from [24])

DRFM Channels
1 – 5.5 GHz
5 – 9.5 GHz
9 – 13.5 GHz
13 – 16.5 GHz
16 – 20 GHz

Table 4.2
Overall DRFM System Specifications (from [24])

Parameter	Value
Frequency of operation	$L - K_u$ Band (1 – 20 GHz)
Instantaneous bandwidth	Up to 300 MHz
Noise figure	2 dB to 8 dB (based on DRFM channel)
Sensitivity	up to -60 dBm (at 20 dB SNR)
Input dynamic range	-60 dBm to -20 dBm (at 20 dB SNR)
ADC resolution	8 bits

at the band edges. Table 4.3 shows the subband bandpass filters upper and lower cutoff frequencies and the corresponding sampling frequencies for the first DRFM channel. The other main idea of this sampling in the receiver is to carry out discrete time signal processing as close to the antenna as possible to provide flexibility in overall operation. That is, by exploiting the bandpass nature of the signals, sub-Nyquist sampling techniques are used to in order to reduce the sampling frequency to much less than the traditional Nyquist-rate sampling of traditional DRFM architectures.

The sampling frequency f_s for detection of the bandpass signals is calculated as

$$f_s = \frac{4f_c}{2m-1} \quad (4.7)$$

where f_c is the center frequency, f_L and f_H are the band edge frequencies in the RF, $m = 1, 2, \ldots, m_{max} =$ the maximum integer value $< f_H/B$ and B is the bandwidth. The algorithm used in the detection of the incoming radar signal selects each subband filter in a DRFM channel for a predefined time interval sufficient enough to ensure occurrence of multiple signals. These filters are selected sequentially (one after the other). According to the filter properties, a corresponding, predefined sampling frequency is selected by the DSP controller and a sampling clock is generated using

Table 4.3
Subband Bandpass Filters Upper and Lower Cut-off Frequencies and the Corresponding Sampling Frequencies for the First DRFM Channel (from [24])

Subband filter	f_L(GHz)	f_H(GHz)	BW(MHz)	f_S(GHz)
1	1	1.3	300	0.92
2	1.3	1.8	500	1.24
3	1.8	2.5	700	1.72
4	2.5	3.3	800	2.32
5	3.3	4.15	850	2.98
6	4.15	4.95	800	2.02
7	4.95	5.5	550	2.32

a delay locked loop (DLL) that is used to clock the discrete time mixer. The sampled signal is digitized using a time interleaved pipelined ADC and stored for power spectral density (PSD) estimation in the DSP.

The information on the signal properties and the noise levels are extracted from the power spectral density estimation using FFTs and windowing with the presumption that the signals are corrupted with additive white Gaussian noise. The Welch method of spectral analysis was used which, consisted of windowing an FFT to highlight the signal while diminishing the noise as a function of an asynchronous spectral averaging mechanism. With this estimation, the band edge frequencies and instantaneous bandwidth of the signal can be detected. The PSD threshold required for detection is decided based on the signal characteristics. A maximum instantaneous bandwidth of 500 MHz is used by adjacent channels for reliable detection of radar pulses. The Wigner-Ville time-frequency distribution was also used for analysis of the signal intercepts.

To test their DRFM architecture, an input pulsed radar signal with liner frequency modulation pulse compression signal is applied. The chirp signal used for pulse compression has linear frequency variation from 1.4 to 1.6 GHz with 20 dB SNR. The pulse width is 100 ns and pulse repetition interval of 500 ns. The input signal is as shown in time and frequency domain in Figure 4.15 and Figure 4.16, respectively.

158   CHOOSING THE CORRECT WIDEBAND RECEIVER

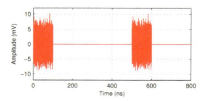

Figure 4.15   Linear frequency modulation test signal in the time domain (200-MHz modulation bandwidth). (From [24].)

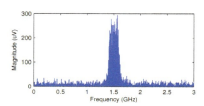

Figure 4.16   Linear frequency modulation test signal in the frequency domain (200-MHz modulation bandwidth). (©IEEE, reprinted with permission from [24].)

The received signal passes through the sub-band filter number 2 in Table 4.3 after sampling downconversion as shown in Figure 4.17.

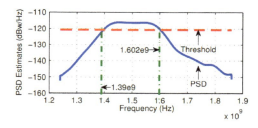

Figure 4.17   Detected linear frequency modulation signal with threshold. (From [24].)

This demonstration shows the estimated power density of the downconverted baseband signals along with threshold for detection. Once the DRFM samples, detects and stores the signal, it is able to estimate the pulse width and extract the pulse compression modulation (chirp sweep rate). Figure 4.18 shows the chirp modulation in the time domain (left plot), illustrating the increase in frequency of the waveform over time, and on the right, shows the linear increase in frequency in time as a function of frequency over the modulation bandwidth.

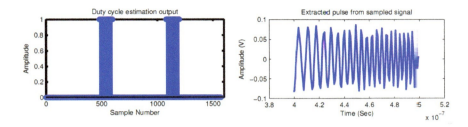

Figure 4.18 Detected linear increase in frequency modulation over time (left) and increase in frequency modulation bandwidth over time (right). (©IEEE, reprinted with permission from [24].)

Their implementation provided a path to a single-chip, silicon CMOS DRFM-solution for acquisition, storing, and time-frequency analysis of intercepted signals over a 20-GHz bandwidth.

4.6 POLYPHASE ANALYSIS SYNTHESIS CHANNELIZERS

ADCs and DACs with blazing speed are now used to enable the synthesis of high-fidelity state of the spectrum (SoS) realizations. Enabled by high capacity ADCs, the instantaneous bandwidth of the processed signals has increased rapidly to detect and extract the received signal parameters. In Chapter 1, we discussed polyphase channelizers and how they can divide the entire band into subbands with each subband processed independently (channels). Spectral channelization is performed once the analog RF signal is translated to an IF signal allowing multiple RF signals to be captured and processed by the receiver at any one time. It additionally provides a significant improvement in sensitivity relative to the original collection bandwidth and corresponding reduction in digital sample rate for follow-on processing.

For example, by taking a different approach than the channelizer in the section above, the polyphase channelizer decomposes the high sample rate data into a parallel data at a lower sample rate as it is digitized as shown in Figure 4.19.

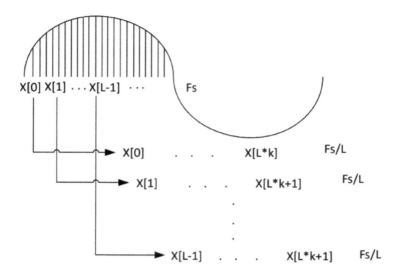

Figure 4.19 Channelized signal sampling concept converting the high-speed data samples, (sampled at Fs) into L parallel streams of lower sampled data channels (sampled at Fs/L) (©IEEE, reprinted with permission from [25].)

As the number of channels increases, the amount of resources to implement the design increases rapidly.

Signal processing algorithms, such as matched filtering, equalization, and synchronization will then require significant parallelism to accomplish their processing tasks. At high sample rates even the simplest digital filtering task, i.e., FIR filtering, may saturate the hardware's processing limit. This is because the hardware operation speed is limited by its clock rate and the number of operations required per clock interval is directly related to the signal's sampling rate or bandwidth.

For DRFM processing, a broadband frequency coverage can be covered using a channelized approach. In addition, increased efficiency in real-time amplitude, time, and frequency modulation EA processing can be accomplished through channelization. One way is the efficient implementation of resampling filters as shown in Figure 4.20(a) Rule 1: filtering with M-unit delays followed by a M:1 down sampling is equivalent to M:1 downsampling followed by filtering with 1 unit delays. Also shown in Figure 4.20(b) Rule 2: Processing with 1:M upsampling followed by filtering with M-unit delays is equivalent to filtering with 1 unit delays followed by 1:M upsampling. Consequently, it is always more efficient to apply the filter at the lower sample rate.

4.6 Polyphase Analysis Synthesis Channelizers

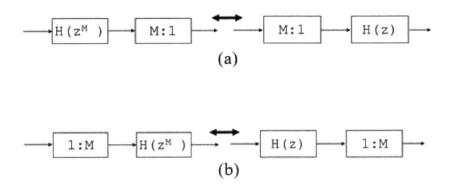

Figure 4.20 Efficient implementation of resampling filters (a) Rule 1: filtering with M-unit delays followed by a M:1 down sampling is equivalent to M:1 downsampling followed by filtering with 1 unit delays, (b) Rule 2: 1:M upsampling followed by filtering with M-unit delays is equivalent to filtering with 1 unit delays followed by 1:M upsampling.

Historically, the efficient filtering problem is mainly solved via fast convolution, whereby the linear convolution is converted into circular convolution via either overlap-and-save or overlap-and-add algorithms. The circular convolution is then efficiently computed via the fast Fourier transform (FFT) algorithm. In practice, the fast convolution is rarely used in real-time systems or communication systems (except OFDM) [26].

The channelizers use FFT-based processing as an efficient way of frequency translation and parameter extraction. In terms of FPGA resources they are also easy to implement as there are fast and efficient FFT IP (intellectual property) cores that are available from, for example, Intel, Xilinx, and Actel. Typical features include streaming single-precision floating-point or fixed-point representations, radix-4, mixed radix-4/2 implementations (for floating-point FFTs) touting reduced memory requirements, and support for 8-bit to 32-bit data and twiddle width (for fixed-point FFTs). For non-power-of-two FFTs (e.g., modular) other algorithms must be developed, such as Bluesteins [27]. However, processing speeds of the FGPAs are still lower than the sampling rates of the ADCs. With cognitive, pulsed, and CW radars with a wideband frequency coverage and a high instantaneous bandwidth requirement the design and implementation of such wideband radar intercept and EA, jamming systems is difficult.

The analysis filter bank (AFB) and synthesis filter bank (SFB) are a coupled analysis and synthesis *filter bank pair*. For the DRFM attack system, the analysis and synthesis reside at the intercept (ES) and jammer (EA) side and for the communication system the coupled filter bank pair resides at both the modulator side and at the demodulator side of the communication link. An example is shown in Figure 4.21.

162 CHOOSING THE CORRECT WIDEBAND RECEIVER

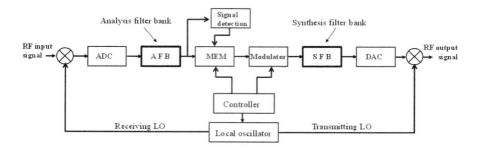

Figure 4.21 DRFM using an analysis filter bank at the front end for wideband receive (ES) and a synthesis filter bank at the output for jammer activity (EA).

4.6.1 Nonmaximally Decimated Analysis/Synthesis Filtering

The DRFM signal processing architecture described briefly in this section is based on oversampled, uniform-channel, discrete Fourier transform filter banks (DFT-FB) or DFT-nonmaximally decimated filter banks (NMDFBs), that include a pair of AFBs and SFBs, as well as the intermediate processing elements (IPEs).

The coupled filter bank pair at each end of the link enables the coupled filter bank pair to replace a high-speed DSP engine by applying a matched filter to a quadrature 1-GHz bandwidth signal with a 2-GHz sample rate. Following the example in [26], the analysis filter partitions the 1-GHz bandwidth input signal into forty 25-MHz bandwidth segments with 50-MHz sample rates. These low data rate intermediate signals are processed and modified by forty parallel low-speed processing engines to affect the spectral envelope change derived from a matched filter. The gain modified intermediate signals are recombined by the coupled synthesis filter bank to form the high bandwidth, high sample rate processed output signal.

Figure 4.22 shows a block diagram of a generalized M-path decimation by D polyphase synthesis channelizer. In the case of $D = M/2$ the complex rotators $\exp(-j2\pi mnD/M)$ followed by each polyphase component vanish if m is even or they reduce to $(-1)^n$ when m is odd. Generalized M-path decimation by D polyphase synthesis channelizer. The phase rotation can be offset by using a two-state circular shift buffer at the input to the FFT [26]. For first stage: (all channels weighted by 1) M-point IFFT is performed; second stage: (all odd channels are weighted by -1, even channels weighted by 1) phase offset absorbed by switching upper half outputs with bottom half outputs.

4.6 Polyphase Analysis Synthesis Channelizers

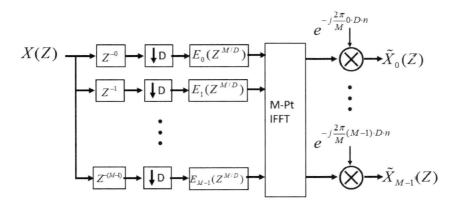

Figure 4.22 Generalized M-path decimation by D polyphase synthesis channelizer. (©IEEE, reprinted with permission from [26].)

The polyphase synthesis channelizer shown in Figure 4.22 is the dual of the analyzer shown in Chapter 3 and significantly reduces the computational complexity and allows building realizable systems that require only one FFT and one polyphase partitioned low-pass filter. For example, the prototype FIR filter in [26] has 769 taps in a masking configuration that costs $769 \times 2 = 1538$ real multiples per complex input. This filter bank approach requires approximately 144 operations with $M = 64, D = 32$, resulting in a 90% workload reduction [26]. Reconfigurable IIR masking filters [28] and FIR masking filters [29, 30] can obtain very sharp cut-off features with low power requirements and a narrow transition bandwidth [31].

In Figure 4.23 a block diagram of a generalized M- path, upsample by D polyphase synthesis channelizer is shown.

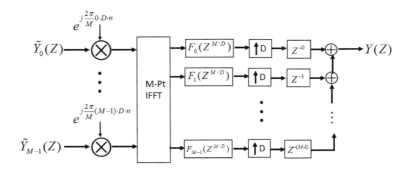

Figure 4.23 Generalized M-path upsample by D polyphse synthesis channelizer. (©IEEE, reprinted with permission from [26].)

The M-path polyphase partition of the synthesis filter for the mth channel can be written as

$$G_m(Z) = G(ZW_M^m) = \sum_{k=0}^{M-1} Z^{-k} W_M^{-km} F_K(Z^M) \qquad (4.8)$$

where $F_k(Z^M)$ is the kth polyphase component. Sliding the upsampling operation through the partitioned filter $G_m(Z)$ arrives at the polyphase representation $G_m^p(Z)$ as

$$G_m^p(Z) = \sum_{k=0}^{M-1} Z^{-k} W_M^{-km} F_K(Z^{M/D}) \qquad (4.9)$$

Similarly, the IFFT block is placed in front of the synthesis polyphase filter servicing all M channels. In summary, the significance of this work is that a cascade of perfect reconstruction (PR), nonmaximally decimated analysis, and synthesis filter banks (NMDFBs) have been applied to filtering wideband digital signals sampled at multiple GHz sample rates requiring only one FFT and one polyphase partitioned masking filter with a sharp cut-off frequency for either the analysis or synthesis section. An M-path analysis channelizer partitions an input signal with a sample rate f_s into M baseband time signals with bandwidths f_s/M operating at the reduced sample rate of $2f_s/M$. A binary mask or filter delivers a subset of the channelized time series to the synthesis channelizer that upsamples and upconverts the selected multiple streams to synthesize a reduced bandwidth output time series [32]. The binary mask or filter connecting the analysis and synthesis filter banks can synthesize filters with bandwidths kf_s/M. In [32] the authors also show that when a filter has a large sample rate-to-bandwidth ratio, the cascade of a downsampling and upsampling filters offer significant advantage over the direct implementation. That is, when a filter has a large ratio of sample rate to bandwidth, the cascade of downsampling and upsampling filters offers substantial computational advantages over the direct implementation [33].

A number of techniques use perfect reconstruction (PR) NMDFBs to implement variable bandwidth filters by taking this into account. Consider a cascade of 3 filters, an M-path input filter to reduce the sample rate $M-to-1$, an inner filter to perform the specified filtering task at the reduced sample rate, and an M-path output filter to increase the sample rate $1-to-M$. In this architecture the input and output filters do not have to be formed from the original filter but can be designed to facilitate an efficient resampling operation.

This is illustrated in the filter architecture shown in Figure 4.24(a) that shows a 2-MHz signal applied to the 120 tap 20-to-1 input, 40 tap inner, 120 tap, 1-to-20 output FIR filter and Fig. 4.24(b) shows the spectrum for architecture shown in Figure 4.24(a). The passband is 0 to 10 kHz, stopband 20 to 1000 kHz with stopband attenuation of -90 dB.

4.6 Polyphase Analysis Synthesis Channelizers

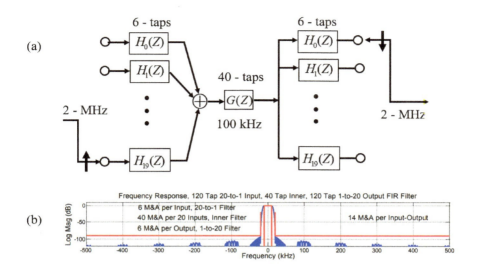

Figure 4.24 Cascade filter showing 120 tap, 20-path 20-to-1 downsample filter, 40-tap inner filter and 120 tap, 20-path 1-to-20 upsample filter and (b) Frequency response. (Adapted from [26].)

The three-section cascade is formed by a 120-tap filter partitioned into a 20-path filter with 6 coefficients per path that processes 20 input samples to form 1 output sample. The inner filter operates at 100 kHz, (1/20th of the input sample rate) and performs the spectral shaping through matching the passband and stopband specifications given above. Since the inner filter is designed to operate at $f_s/20$, its 40 tap length is 1/20th of a direct implementation designed to operate at f_s. This filter is not only shorter, it is operating at 1/20th the sample rate. This gives a workload reduction of 1/400th of the workload of the direct implementation. The third filter in the cascade is the dual of the 120-tap input 20-path filter with 6-coefficients per path that processes 1 input sample with 20 different path filters that output 20 output samples. The workload for the 3-filter cascade is a significant savings over the direct implementation (809-tap filter).

4.6.1.1 Improving the Frequency Resolution

The cascade polyphase filter banks can synthesize wide bandwidth filters with any even or odd integer multiple of the channel bandwidth. Consider a filter with $c = 21$ channels, each 20 MHz wide but now we need a filter with a bandwidth spanned by a noninteger multiple number of channel bandwidths such as $c = 20.5$, 20-MHz channels (instead of $c = 21$, 20-MHz channels). To vary the bandwidth of these filters, three tightly coupled tasks can be used as described in [33].

Option I – A number of methods have been introduced to accommodate the requirements for finer or arbitrary increments of synthesized bandwidths. One method is to alter the bandwidth of the two edge filters. This can be done in a number of ways [33].

Option II – Rather than alter the bandwidth of only the edge filters, we alter the bandwidth, but not center frequency, of all the filters in the *analysis filter bank*. We increase (or decrease) the bandwidth of the *even* indexed channel filters by an amount while decreasing (or increasing) the bandwidth of the *odd* indexed channel filters by the same amount. The channelizer formed in the same manner will have two sets of interleaved complementary bandwidth channels [33]. The complementary channel bandwidths are formed in two analysis filter bands and then interleaved and binary masked in a single filter bank.

Option III – In the previous techniques, we modified the bandwidth of the edge channel filters by modifying the bandwidth of all the channel's filters. This was accomplished by forming two filter sets with wider and with narrower bandwidths and interleaving them with an ordering that placed the narrowband channel at the band edge filter position. To change the band edge filter bandwidth, the impulse response of both prototype filters had to be altered and redesigned. This final version is a hybrid option where one channelizer, the primary, with equal bandwidth channels that form all the channel filters except the band edge pair and design a second channelizer, the secondary, forms the pair of variable band edge channel filters. The architecture of this third option is shown in Figure 4.25.

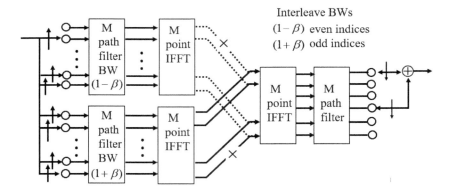

Figure 4.25 Synthesizing interleaved narrow and wide channel channelizer from dual analysis channelizers. (©IEEE, reprinted with permission from [26].)

Although intermediate processing (between analysis and synthesis) can typically take the form of spectral shaping, communication demodulation and so forth, the DRFM can take advantage of the opportunity of creating false targets, jamming signals, false clutter profiles, and so forth and can do it at real-time rates to keep up with the emitters being targeted.

4.6.2 Another Form Using the Noble Identity

4.6.2.1 Analysis Filter Bank

The channelizer has three main blocks: the analysis filter bank, the DRFM kernel, and the synthesis filter bank [34]. The channelizer divides the wideband channel into smaller parts. The channelizer using $x(n)$ as the input is shown in Figure 4.26.

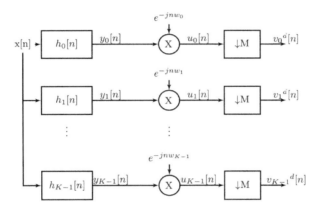

Figure 4.26 Block diagram of a channelizer.

The input is filtered using a bank of bandpass filters $h_0[n]...h_{K-1}[n]$. The outputs of the filters are then

$$y_k[n] = \sum_{m=0}^{N-1} h_0[m]x[n-m]e^{j2\pi km/K} \tag{4.10}$$

where N is the filter length and K is the number of channels. The outputs are then downsampled by a factor of M. By using the noble identity, the analysis filter bank can be transformed into a more convenient and efficient architecture as shown in Figure 4.27.

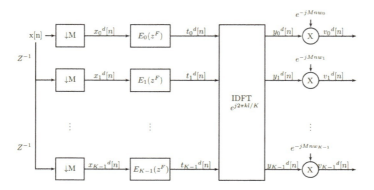

Figure 4.27 Block diagram of a channelizer (©IEEE, reprinted with permission from [34].)

With $K > 2M$, ambiguities between channels can be prevented. At the output of the channelizer, all the channels have a processing rate of f_s/M. With a proper threshold, the probability of false alarm can be maintained. The formation of 32 channels is shown in Figure 4.28.

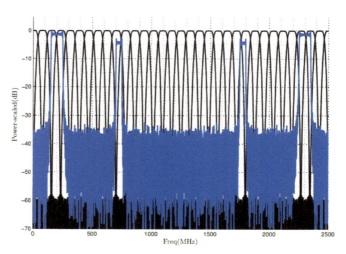

Figure 4.28 Formation of 32 channels. (©IEEE, reprinted with permission from [34].)

All the channels are symmetric about $f_s/2$ and consequently, the middle channel and the first channel can be used as guard channels. In addition, the first half of the channels can be recorded in the DRFM dual-ported memory and the second half of the channels complex conjugate can be used to provide the output of the DRFM.

4.6.3 Synthesis Filter Bank

The synthesis filter bank in direct form is shown in Figure 4.29.

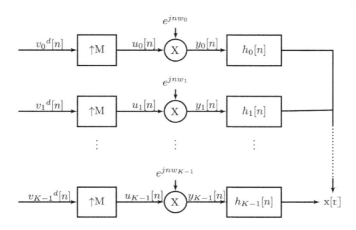

Figure 4.29 Synthesis filter bank in direct form. (©IEEE, reprinted with permission from [34].

It receives the K channels at the rate of f_s/M and generates the reconstructed signal at the rate of f_s. The output of the channelizer shown in Figure 4.27 can be written as

$$V_k^d[z] = \frac{1}{M} \sum_{p=0}^{M-1} U_k[z^{1/M} e^{-j2\pi p/M}] \qquad (4.11)$$

When the output of the channelizer shown in Figure 4.27 is connected to the synthesis bank shown in Fig. 4.29, the output of the interpolator can be written as

$$\hat{U}_k[z] = \frac{1}{M} \sum_{p=0}^{M-1} Y_k[z e^{jw_k} e^{-j2\pi p/M}] \qquad (4.12)$$

The output of the upconverter is

$$\hat{Y}_k[z] = \frac{1}{M} \sum_{p=0}^{M-1} Y_k[z e^{-j2\pi p/M}] \qquad (4.13)$$

170 CHOOSING THE CORRECT WIDEBAND RECEIVER

Summing the outputs of all the filters is then

$$\hat{X}[z] = \sum_{k=0}^{M-1} \hat{Y}_k[z] H_k[z] \tag{4.14}$$

or after a bit of algebra, the sum of the filter outputs is

$$\hat{X}[z] = \frac{1}{M} X[z] \sum_{k=0}^{K-1} H_k^2[z]$$

$$+ \sum_{p=0}^{M-1} X[ze^{-j2\pi p/M}] \sum_{k=0}^{K-1} H_k[ze^{-j2\pi p/M}] H_k[z] \tag{4.15}$$

The coefficients of the analysis filter bank and synthesis filter bank can be calculated as shown in [35] to have perfect reconstruction (PR). These are similar to the techniques above that use PR NMDFBs to implement variable bandwidth filters – the first task is, using an M-path analysis filter bank to partition the full bandwidth input time series into a set of M reduced bandwidth, reduced sample rate, intermediate time series. The second task is selecting the subset of the M channel time series whose contiguous spectra spans the design bandwidth of the digital filter [35]. The third task uses the M-path PR synthesis bank to assemble the desired output time series from the subset of selected channel time series.

The first benefit of this process is that the computational workload of the cascade analysis-synthesis filter bank filter implementation is typically an order of magnitude below that of the tapped delay line, direct implementation, of the same filter. A second benefit is that a high data rate input time series is partitioned into a set of multiple, reduced sample rate, intermediate time series processed by reduced speed parallel arithmetic processors. This process enables simple, reduced cost processing of input signals with GHz sample rates. Below another channelizer technique is discussed with similar advantages.

4.6.4 Split N–Point FFT

To address the ultrawideband spectrum sensing problem, an FPGA design of wideband FFT module that uses a split N-point FFT form of a channelizer that includes $L-$FFT channels of size K, and $L-$point parallel FFTs, is derived in [25]. They derive a new *split-N point FFT* as

$$X(Lk_2 + k_1) = \sum_{n_1=0}^{L-1} \left\{ \sum_{n_2=0}^{K-1} x(Kn_1 + n_2) W_K^{(n_2 k_2)} \right\} W_N^{(n_2 k_1)} W_L^{(n_1 k_1)} \tag{4.16}$$

4.6 Polyphase Analysis Synthesis Channelizers

where $n_2 = 0, 1 \ldots, K-1$ and $k_1 = 0, 1, \ldots, K-1$. Also, this form includes L number of parallel FFT channels of size K. As well, this form includes the same structures as a split N−point form that includes L number of FFT channels with size K, L point parallel FFTs, and twiddle factor given in (4.17) but differs however, in the ordering of input samples and the ordering of parallel L−point FFTs and pipelined K−point FFTs. This is shown in Figure 4.30.

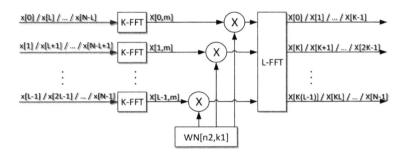

Figure 4.30 Split N-point FFT L−FFT channels of size K, L−point parallel FFTs[25].

The twiddle factor is given as

$$W_N^{kn} = e^{-j2\pi kn/N} \tag{4.17}$$

The combined N−point FFT block module is shown in Figure 4.31.

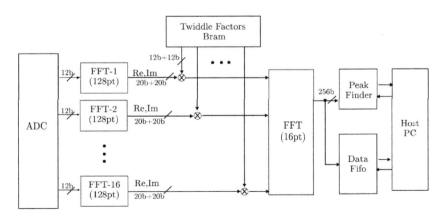

Figure 4.31 Combined N-point FFT using a smaller K-point FFT and parallel L-point FFT [25].

The FPGA module has 16 channels with each channel having a 12-bit ADC. The architecture is pipelined the 16 channels are processed using 128-point, parallel FFT blocks. After obtaining the output data from the parallel FFTs, they are complex multiplied with the real and imaginary twiddle factors that are stored in block RAM (read only memory). Next, the samples are scaled and processed by the 16-point parallel FFT blocks (that process the input samples from the 16 parallel channels at the same time [25]. The 256-bit real and imaginary result of the 16-point parallel FFTs are used then used for ELINT analysis. The technique discussed in this section, a chirp signal, a CW, pulsed signals were all digitized successfully at an ultrawideband FFT and is accomplished by using smaller size FFT structures and combining them by reordering the phase-shifted samples of the parallel received data. With this design, the high-speed data samples from the ADC are digitally processed by a low-speed FPGA. The design of combined FFT structures are implemented on an FPGA board that has a 2.5-Gsps ADC with 16 parallel data channels. The whole picture of the spectrum is acquired by combining FFT structures. Finally, in [36] four parallel ADCs sampling at 250 MHz to achieve an overall 1 GS/s sampling rate utilizing FPGA Virtex-7 technology was successfully reported in the development of a DRFM.

4.6.5 Scalable Graphical Processing Unit Architecture

The implementation of a uniformly partitioned polyphase filter bank spectral channelizer with a 500-MHz instantaneous bandwidth that operates on a high performance GPU is described below [37]. A graphics processing unit is a computer chip (invented by NVIDIA in 1999) that performs rapid mathematical calculations, mostly for the purpose of rendering images. A GPU is able to render images more quickly than a central processing unit (CPU) because of its parallel processing architecture, which allows it to perform multiple calculations at the same time. CPU and GPU architectures are also differentiated by the number of cores. Most CPUs have between four and eight cores, though some have up to 32 cores. NVIDIA GPUs, however, can have several thousand cores. Cores are responsible for various tasks related to the speed and power of the GPU. The cores are often referred to as CUDA (Compute Unified Device Architecture) cores and were developed by the NVIDIA Corporation [38]. CUDA is a parallel processing platform that allows software developers access to the NVIDIA GPU to perform parallel computing.

The parallel processing architecture of the GPU is taken advantage of to maintain the high data throughput rates that are required of a 500-MHz instantaneous bandwidth RF digital receiver. Such a device can form the front end to a reconfigurable SDR DRFM for RF spectral sensing. Channelizer implementations that rely on a GPU hosting offer a number of advantages over competing parallel computing technologies, such as multicore CPUs and FPGAs, in certain applications [37]. These advantages include the GPU's raw computational processing speed against the multicore CPUs and the ease and speed of development when compared to FPGAs.

4.6 Polyphase Analysis Synthesis Channelizers

Figure 4.32 shows the architecture that performs the channelization as a polyphase M-path filter bank.

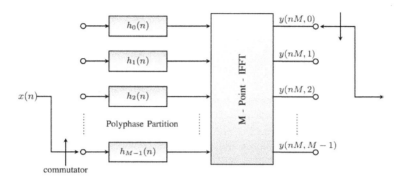

Figure 4.32 Efficient GPU M-channel polyphase channelizer architecture to perform channelization (©IEEE, reprinted with permission from [37].)

One of the main contributions of using a GPU architecture design for a DRFM is that it is scalable and can take in streaming data. That is, the implementation allows continuous channelization of raw RF intercepted data across multiple GPU cards and multiple processing streams contained within a GPU. There is also maximum flexibility in the number of channels that can be supported to account for varying signal types of signals. This ultimately means that the performance and throughput of the channelizer is consistent for the number of channels that are configured. The DRFM design can also easily incorporate configurable oversampling on the spectral outputs, which permits the synthesis of inphase and quadrature data snippets for post-processing [37]. Configurable filter lengths can also be supported in the polyphase channelizer design.

As shown in Figure 4.32, the ordering of operations should be optimized for efficiency such that filtering and shifting is performed at the decimated rate. In addition, channels should be aliased back to baseband where the same prototype low-pass filter is then applied in the polyphase M-channel partition [37]. An IFFT operation can then be efficiently used to separate the aliases. The above architecture produces M channels, each with a sample rate of f_s/M. This is often referred to as critically sampled or maximally decimated (i.e., where the output sample rate of the channelizer matches the channel spacing).

The DRFM signals of interest (communication, radar, etc.) will vary significantly in function and form. That is, their bandwidth and duration will vary widely. In the channelizer described here the oversampling process can be digitally controlled to adjust for this. The first step is to control the commutator that delivers the digital samples to the polyphase stages at the output sample rate performing a *serpentine shift*. For example, the ×2 oversampling case delivers $M/2$ samples to the polyphase stages and performs an $M/2$ serpentine shift to produce one sample for M channels at a sample rate of $2f_s/M$. In the case of *critical sampling*, the architecture delivers M samples into the filter to produce one output sample for M channels at a sample rate of f_s/M.

The second step that allows for a different output sample rate involves a *cyclic shift* of the data prior to the FFT operation [37]. This shift absorbs the phase shift that corresponds to the linear time shift of data through the polyphase filter. An example of this two state process for the 2× oversampling case is shown in Figure 4.33.

These concepts can easily be extended to support arbitrary output sample rates. The number of output channels is defined as M and the number of taps per channel in the low-pass filter is N. The output of each channel is oversampled by a integer value I such that M/I is an integer (coherent sampling) [39], resulting in a output sample rate of $I \times f_s/M$ per channel. Also, the number of new samples that are needed to provide to the channelizer commutator to produce one output time sample across all channels $D = M/I$. The output time samples across all of the channels are considered a *frame*.

To take advantage of the streaming functionalities that are provided by the GPUs, the channelizer design is split across multiple GPU streams and multiple GPU cards, thereby improving concurrency with memory transfers and kernel executions. To enable this, a CUDA program was used [38] that handles the synchronization of all data between GPU streams and across multiple GPU cards. Data packets are simply pushed to the framework and they are automatically distributed across streams and the results are returned in order.

4.6 Polyphase Analysis Synthesis Channelizers

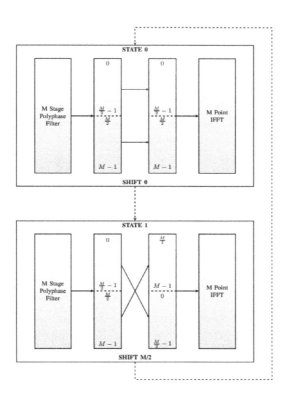

Figure 4.33 Circular shift ×2 oversampling. (©IEEE, reprinted with permission from [37].)

For the channelizer implementation example illustrated in [37] the overlap raw data buffers are passed to the framework so that they have filter history (full registers across the arms of the polyphase filter) for each stream execution. The amount of overlap required is equal to the filter length $M \times N$ minus the number of new samples per frame D. This operation is shown in Figure 4.34.

The overlapped data is limited to be a small percentage of the total data that each kernel operates on ($< 1\%$). The proposed channelizer kernels consist of a filter, a circular shift stage, and an IFFT stage. Each processing block operates on $M \times N$ data overlapped by $M \times N - D$ samples as shown in Figure 4.35. Inside each block, each thread performs the multiply and accumulate (MAC) for each output channel. This is the same as performing the filter operations in each arm of the polyphase filter bank.

176 CHOOSING THE CORRECT WIDEBAND RECEIVER

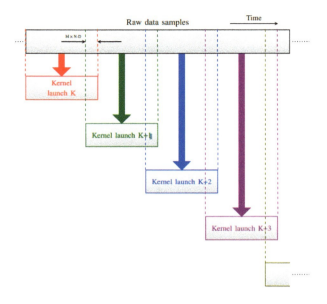

Figure 4.34 Overlapping data across multiple kernel launches. (©IEEE, reprinted with permission from [37].)

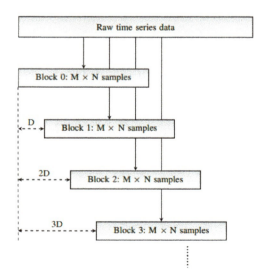

Figure 4.35 Filter kernel operation over blocks. (©IEEE, reprinted with permission from [37].)

The input data and coefficients are stored in the global memory (GM). GM access is continuous for a warp of threads, which reduces the number of read operations, which in turn minimizes transfer times. For increased performance, the MAC operations will occur in shared memory (SM). Since SM is being used for the MAC operations, this imposes a limitation on the channelizer size, as SM is a scarce resource on the GPU. However, for EW applications with a digital intercept receiver that operates with at 1.333 GSamples/s input sample rate, the SM on the GPU is sufficient to support up to 4096 output channels, or a channel bandwidth of 325.5 kHZ.

The pseudocode for the filtering operation across threads is given in [37]. The final step in this kernel is to perform the circular shift to account for the phase offset associated with the oversampling rate. Each block knows what frame it is producing and hence what state is required for the circular shift. Each thread shifts one sample from SM back to GM, and this memory access is once again coalesced for maximum performance. The cuFFT library provides a wide range of input interfaces and highly optimized FFT algorithms for computing FFTs on an NVIDIA GPU [1]

In Figure 4.36 a plot of the respective real-time performance of the GPU and the CPU channelizer implementations for the reference sample rate of 1.333 GSa/s and with equivalent processing samples for the two different channelizer sizes is shown [37].

The dashed line in red indicates the real-time requirements (786 μs) for $M = 2,048$ with 1,024 output samples and $M = 256$ with 8,192 output samples. Note that the execution time of the channelizer needs to be below the red line to satisfy the real-time constraints. In this case it can be observed that the GPU implementation meets the real-time requirement, while the CPU designs do not reach the real-time requirement. Also, the two CPU implementations are approximately 20× slower than the expected real-time target.

1 The actual operation in this step that is needed is an IFFT operation. It is often convenient to implement a FFT corresponding to one transform direction and then to obtain the other transform direction from the first.

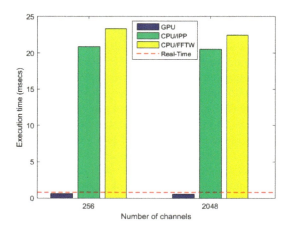

Figure 4.36 Real-time performance comparison between CPU and GPU channelizer implementations (©IEEE, reprinted with permission from [37].)

4.7 DIGITAL RECEIVER / EXCITERS

The growing trend toward digital beamforming in phased arrays for DRFMs requires highly integrated RF circuits combined with digital ADCs and DACs. The embedded technology focus for enabling the next generation DRFM modules and AESA phased array antennas is the ultimate goal of having a DREX at each array element. Dual-polarized, wideband performance is desirable for many applications and current devices are now including a RF SoC such as the XiLinx Zynq SoC.

The main features for next generation DREX technology include dual-ADC/DACs (to support both horizontal and vertical on transmit and receive) and software-defined functionality [40]. As the sensor's SOSA open-systems standard has now gained significant momentum for embedded computing, it is driving down the required cost and complexity of components, widening the bandwidth, and speeding up the ability to tune quickly across multiple bands using software with very low latency [40]. Another key feature is the required on-board storage and bulk memory for the coherent processing interval (CPI) per channel needed for digital beamforming and adaptive array signal processing. Minimization of the backplane architecture for the RF, digital and DC low-noise interconnects is also key including thermal heat dissipation.

Advances in FPGA devices, GPU devices, and ADCs and DACs have had a significant effect on the DREX technology and how they execute the new cognitive algorithms for identifying the state of the spectrum. The common approach is to bring together SiGe system-on-a-chip approaches, for example from DARPA's DAHI (Diverse Accessible Heterogeneous Integration) program, but these have been mostly for certain applications are not generic, and are not easily modified. An S-band DREX with a 550-1050 MHz baseband I/Q input was developed in [41] and consisted of an FPGA, ADC, and DAC as core components with a 5-Gbps PCIe interface, DDR3 memory, and a USB 2.0 and TCP/IP GigE interface I/O.

Another one of the key challenges in next generation DREX architectures is integrating and miniaturizing filters in order to meet the new requirements. Moreover, as DREX moves on-chip and on wafer, filters must integrate at this level to achieve the desired level of miniaturization and channelization required by today's wide bandwidth complex signal environments. Notable strides have been made across many enabling technologies such as Si/CMOS/SiGe for affordability, GaN for power and linearity, InP HEMT for speed and low noise, MEMS for low loss switching, and photonics for high-speed distribution [42].

Recently, a scalable multichannel development platform based on an integrated transceiver used as the DREX along with measured results applicable to larger array level performance estimates was described in [40]. The eight channel DREX is shown in Figure 4.37 and is based on the ADRV9009 transceiver IC that houses two waveform generators and two receivers in a single monolithic IC [43]. The transceiver has two receive channels and two transmit channels with digital signal processing and achieves 200-MHz Rx bandwidth and a tunable Tx bandwidth of 450-MHz. An application program interface (API) is provided to program and control the transceiver from the customer's platform. Gain and attenuation can be controlled using the on-chip front-end networks, and built-in initialization and tracking calibration routines provide the performance required for many communications and military applications.

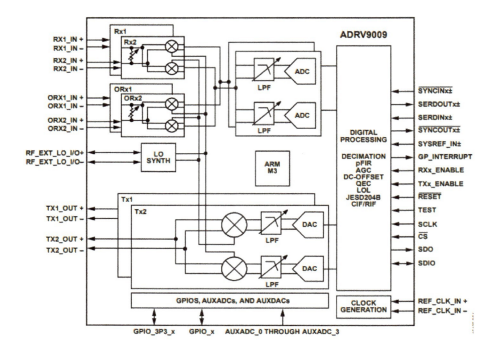

Figure 4.37 Analog Devices ADRV9009 example of an integrated transceiver that combines RF and digital functions on a single IC with 200 MHz instantaneous bandwidth [43].

The latest release offers 200 MHz instantaneous bandwidth and additional features including fast frequency hopping and provisions for synchronization across multiple ICs. As discussed in the next chapter on design concepts, spurious signals are generated that can be problematic and steps are being investigated on how to reduce these effects as they are unwanted beacons to the outside world.

Another exciting development is combining the functionality of the system-on-module SDR that combines the integrated RF agile transceiver Analog Devices AD9361 with the Xilinx Z7035 Zynq-7000 all-programmable SoC. A block diagram of the Analog Devices ADRV9361-Z7035 shown in Figure 4.38.

4.7 Digital Receiver / Exciters   181

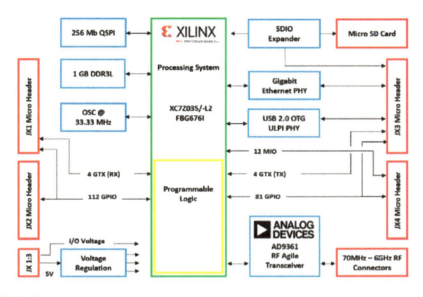

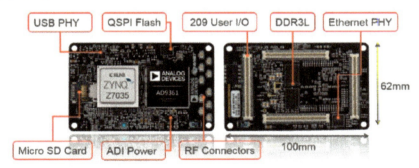

Figure 4.38   Analog Devices ADRV9361-Z7035 SDR 2x2 System-on-Module [42].

Although only tunable to 6 GHz, they provide a tunable channel bandwidth of 200 kHz to 56 MHz with a maximum output power of 8 dBm. The embedded Zynq Z-7035 provides a dual ARM Cortex -A9 MP Cores running at 800 MHz and 275K Kintex–7 logix cells with 900 DSP48 slices they provide multiple operating system support.

4.8 SUMMARY

In summary, we have covered several of the various approaches to wideband spectrum sensing. We began with bringing the antenna signal down to the receiver and looked at two the RF CMOS devices that have low insertion loss and good transmit and receive pathways with low noise. Of course the limiting factor in any of the wideband spectrum sensing techniques will be the antenna. We then discussed the software-defined receivers that digitize the spectrum and examined the promising approach of compressive sensing receivers, the first one being the Nyquist folding receiver. Also examined was another recent technique called the modulated wideband converter. Blind compressive sensing was then presented that relies on a Bayesian learning algorithm. This was followed by a discussion of several channelizer algorithms. These included the bandpass sampling channelizer algorithm, the nonmaximally decimated analysis/synthesis filtering approach, and a split N-point FFT algorithm using FPGAs and GPUs. Finally we presented a look at some examples of the current transceiver technology available. These ICs integrate multiple digital and analog functions within a single, small transceiver and can simplify phased array design and accelerate development.

APPENDIX 4A: INTERPOLATION FILTER OUTPUT (SECTION 4.3.1)

A derivation for the NYFR interpolation filter output $y(t)$ is given. First, consider that the pulse train $\tilde{p}(t)$ can be expressed as a convolution of a Dirac sequence with a pulse template. Using the Dirac scaling property to keep the impulse area normalized to 2π, we have

$$\tilde{p}(t) = p(t) * \sum_k 2\pi \delta(t - t_k) = p(t) * \varphi'(t) \sum_k 2\pi \delta(\varphi(t) - 2\pi k)$$

Noting that the Fourier series of a uniform impulse train is $2\pi \sum_k \delta(z - 2\pi k) = \sum_k e^{jkz}$ and following [44] with $z = \omega_{s1} t + \theta(t)$, we have

$$\tilde{p}(t) = p(t) * (\omega_{s1} + \theta'(t)) \sum_k e^{jk(\omega_{s1} t + \theta(t))}$$
$$\approx p(t) * \omega_{s1} \sum_k e^{jk(\omega_{s1} t + \theta(t))}$$

where the approximation follows from the assumption that the RF sample clock modulation is narrowband so that $\omega_{s1} >> \max|\theta'(t)|$.

The impact of this approximation can be appreciated by noting that the output magnitude is proportional to the sample rate. As the sample rate changes, the output amplitude varies accordingly. For example, if the sample rate varies from 1950 Msps to 2050 Msps as described in the illustrative example discussed earlier, this will result in an amplitude variation of $20\log_{10}(2050/1950)$, or slightly greater than 0.4 dB.

4.8 Summary

We now take the Fourier transform of the pulse train:

$$\tilde{P}(\omega) \approx P(\omega)\omega_{S1} \sum_k (\delta(\omega - k\omega_{S1}) * \Im\left\{e^{jk\theta(t)}\right\})$$
$$\approx \omega_{S1} \sum_k P(k\omega_{S1})(\delta(\omega - k\omega_{S1}) * \Im\left\{e^{jk\theta(t)}\right\})$$
$$= \sum_k T_k(\omega - k\omega_{S1})$$

where the narrowband modulation term $T_k(\omega)$ is given by:

$$T_k(\omega) = \omega_{S1} P(k\omega_{S1}) \Im\left\{e^{jk\theta(t)}\right\}$$

The additional approximation assumes that the sample aperture is short so that the Fourier transform of the pulse $P(\omega)$ is approximately constant over the frequency range ω where

$$k\left[\omega_{S1} - \min\left(\theta'(t)\right)\right] < \omega < k\left[\omega_{S1} + \max\left(\theta'(t)\right)\right]$$

Then using the notation shown in Figure 4.5, for input signals $x(t)$ that fall within the passband of the preselect filter $H(\omega)$, the output $y(t)$ of the interpolation filter in the frequency domain is given by:

$$Y(\omega) = \left((X(\omega)H(\omega)) * \tilde{P}(\omega)\right) F(\omega)/2\pi$$
$$\approx \left(X(\omega) * \sum_k T_k(\omega - k\omega_{S1})\right) F(\omega)/2\pi$$
$$= \left(\sum_k X(\omega - k\omega_{S1}) * T_k(\omega)\right) F(\omega)/2\pi$$
$$= (X_L(\omega - k_H\omega_{S1}) * T_{k_H}(\omega) +$$
$$X_R(\omega + k_H\omega_{S1}) * T_{-k_H}(\omega))/2\pi$$

where X_L and X_R represent the left-hand and right-hand sides of the Fourier transform, respectively (i.e., $X_L(\omega) = X(\omega)$ for $\omega < 0$ and 0 otherwise and similar for $X_R(\omega)$. In (4.10), we assume that the signal and modulation $T_k(\omega)$ are narrow enough in bandwidth so that the shifted left-hand and right-hand terms fall completely within the interpolation filter $F(\omega)$ for the bandpass sampling harmonic k_H. In particular, for the case where $x(t)$ is a narrowband signal, we have

$$X_L(\omega) = \delta(\omega + \omega_C) * \Im\left\{e^{-j\psi(t)}\right\}/2$$
$$X_R(\omega) = \delta(\omega - \omega_C) * \Im\left\{e^{j\psi(t)}\right\}/2$$

Substituting this into the equation above results in

$$Y(\omega) = (\delta(\omega + \omega_C - k_H \omega_{S1}) * \Im\left\{e^{-j\psi(t)}\right\} * T_{k_H}(\omega)$$
$$+ \delta(\omega - \omega_C + k_H \omega_{S1}) * \Im\left\{e^{j\psi(t)}\right\} * T_{-k_H}(\omega))/4\pi$$

Taking the inverse Fourier transform yields the desired time-domain result.

REFERENCES

[1] A. Bokov, V. Vazhenin, and E. Zeynalov, "Improving the accuracy of digital simulation of the radio signal propagation delay," in *2019 Radiation and Scattering of Electromagnetic Waves (RSEMW)*, 2019, pp. 348–351.

[2] P. Schuh, H. Sledzik, and R. Reber, "High performance GaN single-chip frontend for compact x-band aesa systems," in *2017 12th European Microwave Integrated Circuits Conference (EuMIC)*, 2017, pp. 41–44.

[3] S. Rathod, A. Raut, A. Goel, K. Sreenivasulu, K. S. Beenamole, and K. P. Ray, "Novel FPGA based T/R module controller for active phased array radar," in *2019 IEEE International Symposium on Phased Array System Technology (PAST)*, 2019, pp. 1–5.

[4] S. T. Wipf, A. Göritz, C. Wipf, M. Wietstruck, A. Burak, E. Türkmen, Y. Gürbüz, and M. Kaynak, "240 GHz RF-MEMS switch in a 0.13 μm sige bicmos technology," in *2017 IEEE Bipolar/BiCMOS Circuits and Technology Meeting (BCTM)*, 2017, pp. 54–57.

[5] Z. Fan, K. Ma, S. Mou, and F. Meng, "A high-isolation Ku-band SPDT switch in 0.35μm SiGe BiCMOS technology," in *2017 IEEE Electrical Design of Advanced Packaging and Systems Symposium (EDAPS)*, 2017, pp. 1–3.

[6] M. Vetterli, P. Marziliano, and T. Blu, "Sampling signals with finite rate of innovation," *IEEE Transactions on Signal Processing*, vol. 50, no. 6, pp. 1417–1428, 2002.

[7] E. J. Candes, J. Romberg, and T. Tao, "Robust uncertainty principles: exact signal reconstruction from highly incomplete frequency information," *IEEE Transactions on Information Theory*, vol. 52, no. 2, pp. 489–509, 2006.

[8] I. Maravic and M. Vetterli, "Sampling and reconstruction of signals with finite rate of innovation in the presence of noise," *IEEE Transactions on Signal Processing*, vol. 53, no. 8, pp. 2788–2805, 2005.

[9] Liang Zhongyin, Huang Jianjun, and Huang Jingxiong, "Sub-sampled IFFT based compressive sampling," in *TENCON 2015 - 2015 IEEE Region 10 Conference*, 2015, pp. 1–4.

[10] J. A. Tropp, M. B. Wakin, M. F. Duarte, D. Baron, and R. G. Baraniuk, "Random filters for compressive sampling and reconstruction," in *2006 IEEE International Conference on Acoustics Speech and Signal Processing Proceedings*, vol. 3, 2006, pp. III–III.

[11] Y. Chen, A. J. Goldsmith, and Y. C. Eldar, "Channel capacity under sub-nyquist nonuniform sampling," *IEEE Transactions on Information Theory*, vol. 60, no. 8, pp. 4739–4756, 2014.

[12] J. A. Tropp, "Random filters for compressive sampling," in *2006 40th Annual Conference on Information Sciences and Systems*, 2006, pp. 216–217.

[13] C. Luo and J. H. McClellan, "Discrete random sampling theory," in *2013 IEEE International Conference on Acoustics, Speech and Signal Processing*, 2013, pp. 5430–5434.

[14] G. L. Fudge, M. A. Chivers, S. Ravindran, R. E. Bland, and P. E. Pace, "A reconfigurable direct rf receiver architecture," in *2008 IEEE International Symposium on Circuits and Systems*, 2008, pp. 2621–2624.

[15] G. L. Fudge, H. M. Azzo, and F. A. Boyle, "A reconfigurable direct rf receiver with jitter analysis and applications," *IEEE Transactions on Circuits and Systems I: Regular Papers*, vol. 60, no. 7, pp. 1702–1711, 2013.

[16] D. L. Donoho, "Compressed sensing," *IEEE Transactions on Information Theory*, vol. 52, no. 4, pp. 1289–1306, 2006.

[17] R. Maleh, G. L. Fudge, F. A. Boyle, and P. E. Pace, "Analog-to-information and the nyquist folding receiver," *IEEE Journal on Emerging and Selected Topics in Circuits and Systems*, vol. 2, no. 3, pp. 564–578, 2012.

[18] M. Mishali and Y. C. Eldar, "From theory to practice: Sub-nyquist sampling of sparse wideband analog signals," *IEEE Journal of Selected Topics in Signal Processing*, vol. 4, no. 2, pp. 375–391, 2010.

[19] M. Mashhour and A. I. Hussein, "Sub-nyquist wideband spectrum sensing based on random demodulation in cognitive radio," in *2017 12th International Conference on Computer Engineering and Systems (ICCES)*, 2017, pp. 712–716.

[20] M. Marnat, M. Pelissier, O. Michel, and L. Ros, "Code properties analysis for the implementation of a modulated wideband converter," in *2017 25th European Signal Processing Conference (EUSIPCO)*, 2017, pp. 2121–2125.

[21] J. Zhang, N. Fu, and X. Peng, "Compressive circulant matrix based analog to information conversion," *IEEE Signal Processing Letters*, vol. 21, no. 4, pp. 428–431, 2014.

[22] S. Salari, I. Kim, F. Chan, and S. Rajan, "Blind compressive-sensing-based electronic warfare receiver," *IEEE Transactions on Aerospace and Electronic Systems*, vol. 53, no. 4, pp. 2014–2030, 2017.

[23] P. E. Pace, *Detecting and Classifying Low Probability of Intercept Radar*. Norwood, MA: Artech House, 2009.

[24] A. Kale, P. V. S. Rao, and J. Chattopadhyay, "Design and simulation of a wideband channelized transceiver for DRFM applications," in *2014 IEEE Asia Pacific Conference on Circuits and Systems (APCCAS)*, 2014, pp. 635–638.

[25] A. Orduyilmaz, M. Ispir, M. Serin, and M. Efe, "Ultra wideband spectrum sensing for cognitive electronic warfare applications," in *2019 IEEE Radar Conference (RadarConf)*, 2019, pp. 1–6.

[26] X. Chen, F. J. Harris, E. Venosa, and B. D. Rao, "Non-maximally decimated analysis/synthesis filter banks: Applications in wideband digital filtering," *IEEE Transactions on Signal Processing*, vol. 62, no. 4, pp. 852–867, 2014.

[27] P. A. Milder, F. Franchetti, J. C. Hoe, and M. Püschel, "Hardware implementation of the discrete Fourier transform with non-power-of-two problem size," in *2010 IEEE International Conference on Acoustics, Speech and Signal Processing*, 2010, pp. 1546–1549.

[28] Q. Liu, Y. C. Lim, and Z. Lin, "Design of pipelined IIR filters using two-stage frequency-response masking technique," *IEEE Transactions on Circuits and Systems II: Express Briefs*, vol. 66, no. 5, pp. 873–877, 2019.

[29] S. Roy and A. Chandra, "A new design strategy of sharp cut-off FIR filter with powers-of-two coefficients," in *2018 International Conference on Wireless Communications, Signal Processing and Networking (WiSPNET)*, 2018, pp. 1–6.

[30] A. A. Hanimol and B. Prameela, "Reconfigurable linear phase less complex sharp transition band filter using FRM method," in *2018 International Conference on Emerging Trends and Innovations In Engineering And Technological Research (ICETIETR)*, 2018, pp. 1–6.

[31] Y. Hong and Y. Lian, "Continuous-time FIR filters based on frequency-response masking technique," in *2015 IEEE International Conference on Digital Signal Processing (DSP)*, 2015, pp. 191–195.

[32] F. Harris, E. Venosa, X. Chen, and C. Dick, "Interleaving different bandwidth narrowband channels in perfect reconstruction cascade polyphase filter banks for efficient flexible variable bandwidth filters in wideband digital transceivers," in *2015 IEEE International Conference on Digital Signal Processing (DSP)*, 2015, pp. 1111–1116.

[33] ——, "Cascade non-maximally decimated filter banks form efficient variable bandwidth filters for wideband digital transceivers," in *2017 22nd International Conference on Digital Signal Processing (DSP)*, 2017, pp. 1–5.

[34] O. Ozdil, M. Ispir, I. E. Ortatatli, and A. Yildirim, "Channelized DRFM for wideband signals," in *IET International Radar Conference 2015*, 2015, pp. 1–5.

[35] W. Abu-Al-Saud and G. Stuber, "Efficient wideband channelizer for software radio systems using modulated PR filterbanks," *IEEE Transactions on Signal Processing*, vol. 52, no. 10, pp. 2807–2820, 2004.

[36] M. A. S. Aseeri, A. A. Alasows, and M. R. Ahmad, "Design of DRFM system based on fpga with high resources," in *2016 11th International Design Test Symposium (IDT)*, 2016, pp. 177–180.

[37] S. Faulkner, S. D. Elton, T. A. Lamahewa, and D. Roberts, "A reconfigurable wideband streaming channeliser for RF sensing applications: A multiple GPU-based implementation," in *2017 11th International Conference on Signal Processing and Communication Systems (ICSPCS)*, 2017, pp. 1–8.

[38] G. G. Technology, "NVIDIA developer," https://developer.nvidia.com/cuda-gpus, accessed: 2020-09-11.

[39] P. E. Pace, *Advanced Techniques for Digital Receivers*.   Norwood, MA: Artech House, 2000.

[40] P. Delos and M. Jones, "Digital arrays using commercial transceivers: Noise, spurious, and linearity measurements," in *2019 IEEE International Symposium on Phased Array System Technology (PAST)*, 2019, pp. 1–5.

[41] W. H. Weedon and R. D. Nunes, "Low-cost wideband digital receiver/exciter (DREX) technology enabling next-generation all-digital phased arrays," in *2016 IEEE International Symposium on Phased Array Systems and Technology (PAST)*, 2016, pp. 1–5.

[42] M. C. Smith and R. Dixit, "Future trends in filter technology for military multifunction systems," in *2012 IEEE International Conference on Wireless Information Technology and Systems (ICWITS)*, 2012, pp. 1–4.

[43] M. Jones and P. Delos, "Integrated transceivers simplify design, improve phased array radar performance," *Microwave Journal*, vol. 63, no. 1, pp. 88–98, 2020.

[44] F. Marvasti, *Random Topics in Nonuniform Sampling, in Nonuniform Sampling: Theory and Practice.* Norwell, MA: Kluwer, 2001.

Chapter 5

Transceiver Design and Practical Considerations

In this chapter, the main components of the DRFM-transceiver are examined. Mathematical models of the receive process are presented so a better understanding can be gained of the embedded design process as it proceeds from concept to fabrication. Figures of merit are also included for the DRFM, ADC, and DAC. Although ADCs and DACs are complex integrated circuits (ICs), usually housed in multipin packages and taken for granted, finding the best for a particular design requires some familiarity with how they function and how their performance is described in terms of key parameters and figures of merit (FOM). In addition, important memory considerations are examined. This includes the recent dual ported memory technology that allows the user to simultaneously store and readout signals from memory at the same time. Mathematical models of how to include these components into a design are also presented Direct digital sampling technology is also addressed.

5.1 MATHEMATICAL MODELS OF THE TRANSCEIVER PROCESS

A novel class of emerging digital transceiver ICs are now being targeted for wideband RF spectrum sensing. These integrated transceivers are able to combine multiple functions onto a single IC, simplifying system design and improve phased array performance [1]. However, they offer little flexibility in their design offerings. Consequently, users are designing and building their own DRFMs and incorporating their own transceiver architectures with the needed requirements, while at the same time streamlining the throughput and including boutique capabilities not offered by the big design houses. Converter technology continues to evolve every year. Today's ADCs and DACs from major semiconductor companies such as Analog Devices, Texas Instruments, National Semiconductor, and Maxim Integrated, to name just a few, are sampling at rates orders of magnitude faster than that of their predecessors five years ago at even higher resolutions.

The increased resolution of these high-speed ADCs provides DRFMs and transceivers with higher dynamic range and a wider instantaneous bandwidth. Dynamic range is a critical factor for seeing down further into the noise and dealing with high-power microwave sources so they don't cause damage. More instantaneous bandwidth provides several advantages, including increased EMS understanding and the ability to implement advanced jammer techniques. As the transistor size has continued to shrink it has enabled DAC speeds to increase and their position within the transmitter architecture has gotten closer to the antenna, consequently eliminating the need for conventional analog RF mixing. That is, these RF-DACs can synthesize RF signals directly from digital inputs and include multi-Nyquist DACs, power-mixer-DACs, and other high-frequency designs such as the $\Sigma\Delta$–DACs.

Figure 5.1 shows the major components contained in a DRFM-transceiver. They include the multiple pairs of ADC and DAC networks, a local oscillator and mixer combination for frequency conversion, dual-ported memory for reading and writing at the same time and a clock with its distribution network. The issues behind the integration of these components is discussed below. The LO and mixer combination provides the conversion of the RF band to the IF band and is discussed first.

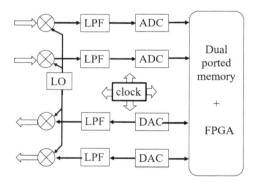

Figure 5.1 Components of an integrated transceiver.

5.1.1 Quadrature Conversion Process

Historically, the quadrature conversion process consisting of the magnitude to in-phase and quadrature channels was done in the analog domain to provide a 90° offset between the channels and also down convert the signal in frequency in order to be compatible with the ADC technology available at the time of build. Nyquist sampling for 9-10 GHz frequency systems was previously not available until recently with the time-interleaved architectures and flash ADCs available. Currently the process is to digitize the magnitude signal at the antenna without downconversion and have the quadrature conversion process take place in the digital domain. In the following development, an analog conversion is illustrated. For an easy digital representation, just replace t by nT where n is the sample number and T is the sampling period. For convenience, one may also drop the T and just replace t by n.

The quadrature conversion process is shown in Figure 5.2. The conversion process can also be done in the digital domain where the input signal shown $s(t)\cos\omega_c(t)$ is a digital signal from the ADC $s(n)\cos(\omega n)$.

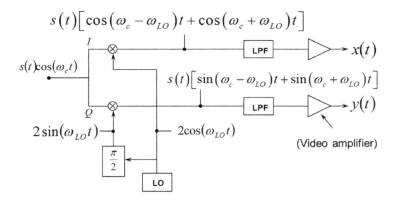

Figure 5.2 Single sideband digital RF memory architecture.

The input signal, $s(t)$, is on a carrier ω_c. The signal $s(t)\cos(\omega_c t)$ is split 3 dB to provide both an in-phase and quadrature channel. The output of the in-phase mixer is $s(t)[\cos(\omega_c - \omega_{LO})t + \cos(\omega_c + \omega_{LO})t]$. The output of the quadrature mixer is $s(t)[\sin(\omega_c - \omega_{LO})t + \sin(\omega_c + \omega_{LO})t]$. After low-pass filtering the in-phase signal is

$$x(t) = s(t)\cos(\omega_c - \omega_{LO})t \tag{5.1}$$

and the quadrature signal is

$$y(t) = s(t)\sin(\omega_c - \omega_{LO})t. \tag{5.2}$$

These signals are at an intermediate frequency $\omega_{IF} = |\omega_c - \omega_{LO}|$ and can now be digitized by the ADC. In fact, this is the only reason the signals are down converted. As discussed below, the down conversion process gives rise to many problems that could be eliminated if the signal was converted directly at the antenna (digital antenna).

Recently the development of a high-speed quadrature modulator and demodulator with an embedded digital correction block (DCB) circuit using 180-nm BiCMOS technology has been developed in [2]. The control block uses three control inputs and the includes topologies for both quadrature modulator and demodulator with a frequency range of 1800 MHz to 6 GHz with a modulation bandwidth of 700 MHz. The central nodes of quadrature modulators and demodulators are heterodyne tracking components, mixers, and output buffers as shown in Figure 5.3.

The central nodes of the quadrature modulators and demodulators are heterodyne tracking components, mixers, and output buffers. To split the signal of the heterodyne LO into in-phase LO_I and quadrature LO_Q components, a polyphase filter is used. It is connected to amplifiers-limiters, and rectangular oscillations are formed at their output. The limiting amplifiers compensate for the attenuation of the polyphase filters and reduce the amplitude imbalance between the I and Q channels because they operate in the limiting mode. The block diagram of the heterodyne tracking is shown in Figure 5.3.

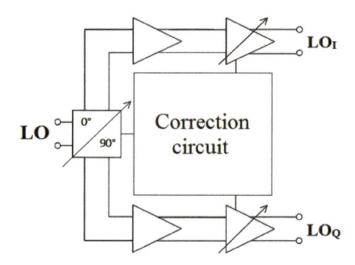

Figure 5.3 Heterodyne tracking correction circuit. (From [2].)

The standard polyphase filter is a circuit with ordered phase shifts [2], which is a regular structure consisting of resistors with equal nominal values and capacitors whose nominal costs decrease exponentially. Special varactors are used to adjust the phase difference between the *I* and *Q* signals at the outputs of the polyphase filter instead of capacitors [3]. The capacity of each varactor depends on the voltage at the outputs. Structurally, this element is a multiplier n-channel MOS transistor connected to drain and source, which lie in the N-Well. PFP is a two-stage. A 4-bit DAC (16 levels) with a resistor circuit has been developed to control the voltage on the varactor and frequency response with a minimum step size of 0.133V.

5.1 Mathematical Models of the Transceiver Process

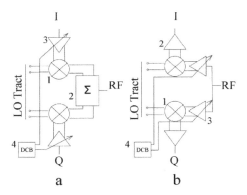

Figure 5.4 Heterodyne tracking correction circuit showing (a) block diagram of the modulator and (b) demodulator with 1-mixer; 2-output buffer; 3-limiter amplifier; 4 digital correction block (DC3). (From [2].)

Three-bit current source DACs are used to control the output signal amplitude and depending on the input control only one of the eight current sources is active at a time. Further details can be found in [2].

5.1.2 Single Sideband Output Modulation

The in-phase single sideband modulator is shown in Figure 5.5. The input signal is the baseband signal

$$x'(t) = s(t)\cos(\omega_{IF}t + \phi). \tag{5.3}$$

The output of the cosine mixer is

$$s(t)\cos(\omega_{IF}t + \phi)\cos(\omega_{LO}t) = \frac{s(t)}{2}[\cos((\omega_{IF} + \omega_{LO})t + \phi) \\ + \cos((\omega_{IF} - \omega_{LO})t + \phi)]. \tag{5.4}$$

The output of the Hilbert transform filter is

$$s(t)\cos\left(\omega_{IF}t + \phi - \frac{\pi}{2}\right) = s(t)\sin(\omega_{IF} - \phi) \tag{5.5}$$

The output of the sine mixer is

$$s(t)\sin(\omega_{IF}t + \phi)\sin(\omega_{LO}t) = \frac{s(t)}{2}[\cos((\omega_{IF} - \omega_{LO})t + \phi) \\ - \cos((\omega_{IF} + \omega_{LO})t + \phi)]. \tag{5.6}$$

194   TRANSCEIVER DESIGN AND PRACTICAL CONSIDERATIONS

The output is the difference between the two channels

$$x(t) = s(t)\cos\left((\omega_{IF} + \omega_{LO})t + \phi\right). \tag{5.7}$$

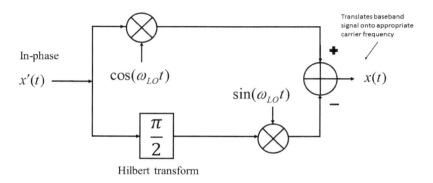

Figure 5.5 Single sideband digital RF memory architecture.

The quadrature single sideband modulator is the same as the in-phase modulator except that the position of the Hilbert transform filter and the sine mixer are interchanged as shown in Figure 5.6.

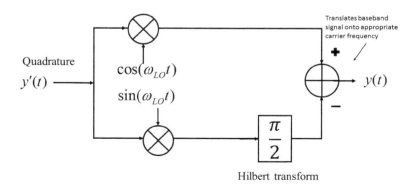

Figure 5.6 Quadrature signal sideband modulator.

The input signal is
$$y'(t) = s(t)\sin(\omega_{IF}t + \phi). \tag{5.8}$$

The output of the cosine mixer
$$s(t)\sin(\omega_{IF}t + \phi)\cos(\omega_{LO}t) =$$
$$\frac{s(t)}{2}\left[\sin\left((\omega_{IF} + \omega_{LO})t + \phi\right) + \sin\left((\omega_{IF} - \omega_{LO})t + \phi\right)\right]. \tag{5.9}$$

The output of the sine mixer is
$$s(t)\sin(\omega_{IF}t + \phi)\sin(\omega_{LO}t) =$$
$$\frac{s(t)}{2}\left[\cos\left((\omega_{IF} - \omega_{LO})t + \phi\right) - \cos\left((\omega_{IF} + \omega_{LO})t + \phi\right)\right]. \tag{5.10}$$

The output of the Hilbert transform is
$$\frac{s(t)}{2}\left[\cos\left((\omega_{IF} - \omega_{LO})t + \phi - \frac{\pi}{2}\right) - \cos\left((\omega_{IF} - \omega_{LO})t + \phi - \frac{\pi}{2}\right)\right]$$
$$= \frac{s(t)}{2}\left[\sin\left((\omega_{IF} - \omega_{LO})t + \phi\right) - \sin\left((\omega_{IF} + \omega_{LO})t + \phi\right)\right]. \tag{5.11}$$

Finally, the output is the difference
$$y(t) = s(t)\sin\left((\omega_{IF} + \omega_{LO})t + \phi\right). \tag{5.12}$$

The DRFM signal is now at the carrier frequency and ready for amplification, filtering, and retransmission.

5.1.3 Hilbert Transform

As shown by, for example (5.5), the Hilbert transform (after David Hilbert) shifts all frequencies in the signal by $\lambda/4$ in the time domain and all spectral components by 90° in the frequency domain. The Hilbert transform does not change domain however. A time domain function remains in the time domain and a frequency domain function remains in the frequency domain. The Hilbert transform of a signal $x(t)$ is

$$H[x(t)] = \frac{1}{\pi}\int_{\tau=-\infty}^{\infty} x(\tau)\frac{1}{t-\tau}d\tau = \frac{1}{\pi}x(t) * \frac{1}{t} \tag{5.13}$$

where the $*$ indicates convolution. By the convolution theorem of Fourier transforms, the Hilbert transform of a signal may be evaluated as the product of $F\{x(\cdot)\}$ (the

196   TRANSCEIVER DESIGN AND PRACTICAL CONSIDERATIONS

Fourier transform of $x(t)$) with $-j * sgn(t)$, where

$$sgn(t) = \begin{cases} -1 & t < 0 \\ 0 & t = 0 \\ 1 & t > 0 \end{cases}$$

In summary, the Hilbert transform enables the single-sideband operation for both in-phase and quadrature modulators. In the next section we discuss the signal conversion process that takes place after any downconversion baseband processing that is necessary.

5.2 RECEIVE PROCESS

To examine some of the important nonidealities in the digitization process, a real-life ADC model is shown in Figure 5.7 and consists of five major steps.

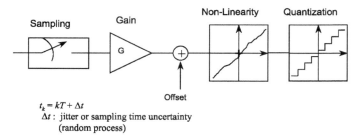

Figure 5.7   Real-life ADC model.

First the analog signal is sampled. The sampling operation obtains a representation of the analog signal at $t_k = kT + \Delta t$ where T is the sampling period and $k \in \{0, 1, 2, \cdots\}$. The additional Δt represents the jitter or sample time uncertainty, which is a random process. After sampling, a gain is applied followed by the addition of a possible offset. Nonlinearities can also be present and affect the amplitude of the sampled signal. The quantization step induces the (nonrecoverable) quantization error.

5.2.1   Sampling Operations

The three types of sampling operations are (a) the ideal sampler, (b) the zero-order hold, and (c) the track and hold. In ideal sampling, the analog signal is represented by a periodic train of impulses as

$$y_a(t) = x(t) \sum_{k=-\infty}^{\infty} \delta(t - kT) . \tag{5.14}$$

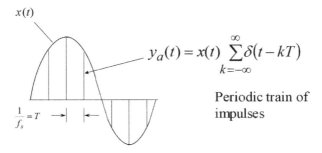

Figure 5.8 Ideal sampling operation.

This is shown in Figure 5.8 where the period between samples is $T = 1/f_s$. This causes the signal spectrum to be convolved with a train of impulses in the frequency domain, thus replicating and shifting the signal spectrum by integer multiples of T^{-1}. Taking the Fourier transform

$$\begin{aligned} Y_a(f) &= X(f) * \sum_{n=-\infty}^{\infty} \frac{1}{T} \delta\left(f - \frac{n}{T}\right) \\ &= \frac{1}{T} \sum_{n=-\infty}^{\infty} X\left(f - \frac{n}{T}\right). \end{aligned} \quad (5.15)$$

This sampling process places a copy of the spectrum at integer multiples of the sampling frequency. Although easy to analyze, this representation is not very practical. The operation of the zero-order hold amplifier is shown in Figure 5.9 [4]. The response of this amplifier is

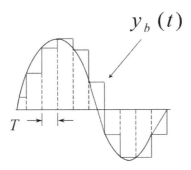

Figure 5.9 Zero-order hold amplifier operation.

$$y_b(t) = \left[x(t) \sum_{k=-\infty}^{\infty} \delta(t-kT) \right] * \prod\left(\frac{t}{T} - \frac{1}{2}\right) \quad (5.16)$$

where $\prod(t/T - 1/2)$ is a unit height pulse of duration T. The Fourier transform of the output is then

$$Y_b(f) = \left(e^{-j\pi fT}\right) \frac{\sin(\pi fT)}{\pi fT} \sum_{n=-\infty}^{\infty} X\left(f - \frac{n}{T}\right). \quad (5.17)$$

Practically, the input cannot be captured in zero-time and the amplifier requires a sufficiently narrow sampling window to provide an adequate approximation of an "ideal" zero-order hold.

The function of the track and hold amplifier is shown in Figure 5.10. As the name suggests, half of the sampling period is used to track (or follow) the input while the latter half holds the last value tracked.

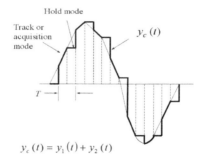

Figure 5.10 Track and hold amplifier.

The output of the track and hold amplifier can be decomposed into two modified square waves:

$$y_c(t) = y_1(t) + y_2(t) \quad (5.18)$$

where

$$y_1(t) = x(t) \left[\prod\left(\frac{2t}{T} - \frac{1}{2}\right) * \sum_{k=-\infty}^{\infty} \delta(t-kT) \right]$$

$$y_2(t) = \left[x(t) \sum_{k=-\infty}^{\infty} \delta\left(t - kT - \frac{T}{2}\right) \right] * \prod\left(\frac{2t}{T} - \frac{1}{2}\right). \quad (5.19)$$

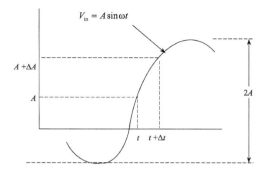

Figure 5.11 Sample-time uncertainty (Δt).

In the first square wave $y_1(t)$, the amplitude is multiplied by the input signal. In the second square wave $y_2(t)$, the amplitude is equal to the last held voltage value. The output tracks the input during sampling (acquisition or tracking mode). The voltage remains at the last value of input when it enters the hold mode. Note each mode is $T/2$ seconds in duration. Taking the Fourier transforms,

$$Y_c(f) = \sum_{n=-\infty}^{\infty} e^{-jn\pi/2} \frac{\sin\left(n\frac{\pi}{2}\right)}{n\pi} X\left(f - \frac{n}{T}\right)$$

$$+ e^{-(3j\pi fT)/2} \frac{\sin\left(\pi f \frac{T}{2}\right)}{\pi fT} \sum_{n=-\infty}^{\infty} X\left(f - \frac{n}{T}\right). \quad (5.20)$$

To summarize, we note that the zero-order hold amplifier and the track and hold amplifier have output spectra containing sinc envelopes. This will not be a problem for the ADC; however, sinc compensation techniques must be used for DACs.

5.2.2 Sample Time Uncertainty

To quantify the acceptable sample-time uncertainty or jitter (Δt), we consider a sinusoidal input voltage $V_{in} = A\sin\omega t$ as shown in Figure 5.11. The section of waveform between time t and $t + \Delta t$ corresponds to A and $A + \Delta A$, respectively. The slope of this section is

$$\text{Slope} = \frac{\Delta A}{\Delta t} = \frac{dV_{in}(t)}{dt} \quad (5.21)$$

200 TRANSCEIVER DESIGN AND PRACTICAL CONSIDERATIONS

and is called the slew rate. Solving for Δt

$$\Delta t = \frac{\Delta A}{\frac{dV_{in}}{dt}} = \frac{\Delta A}{A\omega \cos \omega t}. \qquad (5.22)$$

Also,

$$\Delta A = \frac{2A}{2^n - 1} \cong \frac{2A}{2^n} = 1 LSB. \qquad (5.23)$$

Substitution of (5.23) into (5.22) gives

$$\Delta t = \frac{2A}{2^n A \omega \cos \omega t}. \qquad (5.24)$$

The maximum slew rate occurs at $t = 0$ so the sample time uncertainty is

$$\Delta t_{max} = \frac{2}{2^n \omega} = \frac{2}{2^n 2\pi f_{in}}, \qquad (5.25)$$

or

$$\Delta t_{max} = \frac{2^{-n}}{\pi f_{in}}. \qquad (5.26)$$

For example, for a 16-bit digital audio receiver with $f_{in} = 20$ kHz, $\Delta t < 0.25$ ns.

5.2.3 Digital Sampling Clock Accuracy

Accurate clock timing is essential to high quality signal sampling (ADC) and reconstruction (DAC). The sampling clock can also have a certain amount of variation in the time difference between clock pulses. To investigate the digital sampling clock time uncertainty, we consider the squaring amplifier as shown in Figure 5.12. The squaring amplifier has a bandwidth f_b and a 1σ rms thermal noise power of $e_n^2 = 4kTR_n f_{noise}$ where R_n is the equivalent noise resistance of the squaring circuit and f_{noise} is the equivalent noise bandwidth of the system and $kT = 4(10^{-21})$ W/Hz. The first-order squaring amplifier is driven using a sampling clock oscillator that can be considered a sinusoidal source $V_{CL} = A \sin(2\pi f_{CL} t)$. Proceeding as before, the time uncertainty can be determined by calculating the derivative of V_{CL} as

$$\frac{dV_{CL}}{dt} = \frac{e_n}{\Delta t_{CL}} = 2\pi f_{CL} A \cos(2\pi f_{CL} t). \qquad (5.27)$$

The time uncertainty is then

$$\Delta t_{CL} = \frac{e_n}{2\pi f_{CL} A \cos(2\pi f_{CL} t)}. \qquad (5.28)$$

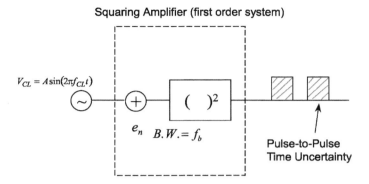

Figure 5.12 Model of digital sampling clock.

The maximum sensitivity of the squaring circuit is when $\cos(2\pi f_{CL}t) = 1$ (zero crossing of sinewave). Then

$$\Delta t_{CL} = \frac{2e_n}{2\pi f_{CL}A} = \frac{e_n}{\pi f_{CL}A} \qquad (5.29)$$

where the factor of 2 accounts for the total 1σ time uncertainty.

The above expression is a function of the equivalent noise bandwidth of the system. For a first-order system, the equivalent noise bandwidth is related to the system bandwidth (a more convenient parameter). To find this relationship, we start with the expression for the equivalent noise bandwidth for a first-order system

$$f_{\text{noise}} = \frac{\int_0^\infty |H(\omega)|^2 d\omega}{|H(0)|^2}. \qquad (5.30)$$

For example, for a low-pass filter (LPF)

$$H(\omega) = \frac{1}{1 + j(\omega/\omega_b)} \qquad (5.31)$$

and

$$|H(\omega)|^2 = \frac{1}{1 + (\omega/\omega_b)^2}. \qquad (5.32)$$

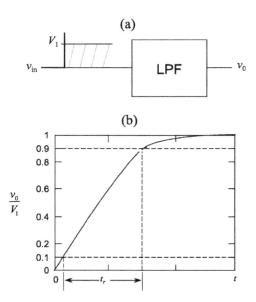

Figure 5.13 (a) Step input applied to LPF and (b) step response.

Note $|H(\omega = 0)| = 1$. From (5.30)

$$f_{noise} = \frac{1}{2\pi}\left(\int_0^\infty \frac{\omega_b^2}{\omega_b^2 + \omega^2}d\omega\right)$$

$$= \frac{1}{2\pi}\left(\omega_b \tan^{-1}\left(\frac{\omega}{\omega_b}\right)\right)\Big|_0^\infty = \frac{1}{2\pi}\left(\omega_b\left(\frac{\pi}{2}\right)\right) \quad (5.33)$$

and finally,

$$f_{noise} = f_b\left(\frac{\pi}{2}\right). \quad (5.34)$$

Substituting (5.34) into (5.29) gives the total sampling clock time uncertainty as

$$\Delta t_{CL} = \frac{(2\pi k T R_n f_b)^{1/2}}{\pi f_{CL} A}. \quad (5.35)$$

The clock time uncertainty can also be expressed as a function of the sampling pulse rise time. Again we use a first-order LPF with a step input as shown in Figure 5.13. Figure 5.13(a) shows an input pulse of height V_1 being applied to the low-pass filter. The output voltage is $v_0 = V_1(1 - e^{-t/\tau})$ where $\tau = 1/\omega_b$. Figure 5.13(b) shows the normalized step response and the rise time t_r. For v_0 to reach $0.1V_1$, $t = 0.1\tau$. For v_0

to reach $0.9V_1$, $t = 2.3\tau$. The rise time is defined as

$$t_r = (2.3 - 0.1)\tau = 2.2\tau \qquad (5.36)$$

or

$$t_r = 2.2\tau = \frac{2.2}{\omega_b} = \frac{2.2}{2\pi f_b} \cong \frac{0.35}{f_b}. \qquad (5.37)$$

That is,

$$t_r f_b \cong 0.35. \qquad (5.38)$$

Substitution of (5.38) into (5.35) gives the sampling clock uncertainty time as a function of the measured rise time as

$$\Delta t_{CL} = \frac{(0.7kTR_n)^{1/2}}{f_{CL}A\sqrt{\pi t_r}}. \qquad (5.39)$$

For example, when $R_n = 50\Omega$, $t_r = 250$ ps, pulse repetition frequency = 10 MHz and $A = 0.5$ V, then the RMS clock time uncertainty $\Delta t_{CL} = 2.7$ ps.

The key design point here is that now that the rise time of a clock pulse can be measured at different points in the circuit, an estimate of the sample time uncertainty can be obtained from Δt_{CL}. The next step is to do a comparison to the fundamental limit derived above and given by Δt_{max} and a determination of how close to ideal of a clock distribution the design is.

5.2.4 Sampling Circuit Designs

The sample and hold technique is demonstrated in the circuit shown in Figure 5.14. As the input voltage is applied (switch closed), the output voltage tracks the input voltage and the capacitor charges up. In the hold mode, the switch transitions to the open state and V_{out} remains constant.

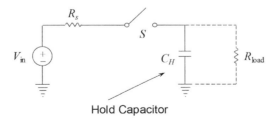

Figure 5.14 Sample and hold circuit.

The function of a sample and hold is to capture the input signal and hold it constant during the subsequent ADC conversion cycle. In this circuit C_H is usually large to minimize the switching errors. The charge-up time depends on the RC time constant set by C_H, the source resistance R_s, and switch resistance. The load resistor R_{load} discharges the capacitor when the switch is opened and also causes errors in the stored voltage. The switching operation and transient currents generated make this an unsuitable design for high-performance applications.

The open-loop sample and hold architecture shown in Figure 5.15 is more suitable for high-speed operations. It uses an input buffer B_1, a sampling circuit (switch) S, a hold capacitor C_H, and an output buffer B_2. The input buffer B_1 has a very high input impedance and provides large currents to quickly charge the capacitor C_H. B_2 is connected as a voltage follower (output follows input) to buffer the hold capacitor from the load resistance to prevent discharge. B_2 also has a low output impedance.

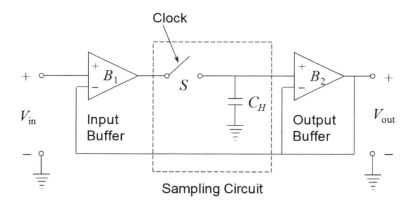

Figure 5.15 Open-loop sample and hold architecture.

When the switch is closed the circuit is in the acquisition mode and $V_{out} = V_{in}$ since the voltage across the B_1 input must be zero. Also, $V_{CH} = V_{in}$ since B_2 is connected as a voltage follower. With the switch open, V_{in} remains stored on C_H and the circuit enters the hold mode and the output is held at V_{in} by the voltage follower (output remains constant). Advantages of this configuration is that it is unconditionally stable (no feedback) and is suitable for high-speed operations. The most serious drawback is the input dependent charge that is injected by the sampling switch onto the hold capacitor (pedestal errors). This is a source of nonlinearity.

Figure 5.16 depicts a simple open-loop differential CMOS S/H circuit consisting of S/H NMOS switches, S/H (holding) capacitors, and buffers. As this architecture includes no global feedback and has unconditional stability, it can also be designed for high-speed operations.

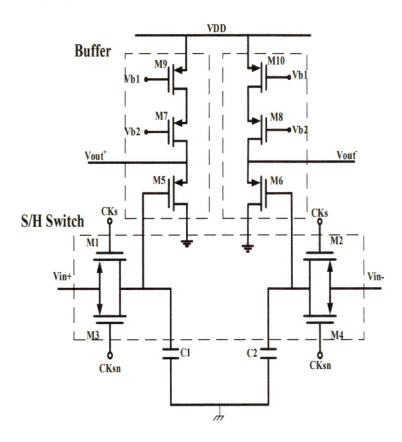

Figure 5.16 Open-loop sample and hold architecture. (From [5].)

Even considering ideal buffers, there will be problems with the voltage-dependent charge injection of the input switch [5]. To reduce this charge injection effect, circuits were developed that incorporate both NMOS and PMOS devices such that the opposite charge packets injected by the two cancel each other [5]. However, even with complementary switches to reduce charge injection, the flicker noise and thermal noise from all the CMOS switches and buffers can cause nonlinearities in the circuit. The authors in [5] address this using a unique chopper design to reduce the noise and charge injection and improve the SNR and noise floor.

A closed-loop sample and hold architecture is shown in Figure 5.17. The closed-loop sample and hold suppresses the input-dependent pedestal errors by including the sampling switch in the feedback loop so that the voltage swings are much smaller than the input and output swings. In the acquisition mode S is ON and the circuit is a two-stage op-amp compensated by C_H. This configuration resembles a unity gain buffer where the output closely follows input. Note, if A_0 is large, X is a virtual ground node allowing the voltage across C_H to track the input.

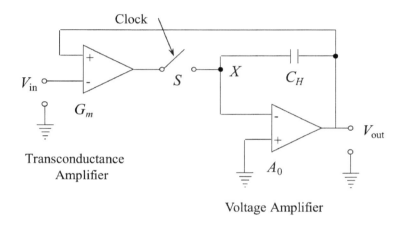

Figure 5.17 Closed-loop sample and hold architecture.

When S turns off, the instantaneous output voltage is stored on C_H and the feedback circuit consisting of A_0 and C_H retains the sampled voltage at the output.

One of the advantages of the closed-loop sample and hold is that in the sampling mode, the output voltage of G_m is close to the ground potential (virtual ground property of node X). Consequently, switch S always turns off with a constant voltage at its input and output terminals. This results in the injection of a constant charge onto C_H that exists (primarily as an offset error independent of input). This configuration is especially useful in high-precision systems. Stability and slow time response, however, are still a concern. There is also a significant hold-mode feedthrough due to the V_{in} to V_{out} signal path through the input capacitance of G_m.

Recently, a 30-GS/s sample-hold amplifier (SHA) implemented in a combined InP HBT and Si CMOS heterogeneous integration technology was reported [6]. The high-speed signal path is entirely in InP, but droop in the sampled voltage arising from HBT bias currents is suppressed by an integrated CMOS feedback circuit. Under this closed-loop control, in hold mode, the droop rate of the single-ended outputs is reduced to 20 mV/ns. The closed-loop sample and hold consists of a master track and hold amplifier (THA) and a slave THA as shown in Figure 5.18.

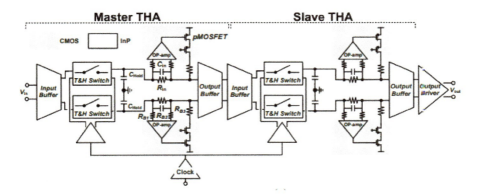

Figure 5.18 InP HBT and Si CMOS closed-loop sample and hold architecture with master THA and slave THA. (From [6].)

InP-CMOS interconnect parasitics are isolated from the high-speed signal path by isolation resistors and active bootstrapping. The SHA is a pair of track-hold amplifiers, each with input buffer, track-hold switch, and output buffer. The design includes a linearized input buffer, feedthrough cancellation, and fast base-collector diodes as switches. The output driver provides a fast 50 Ω interface. Given an 8 GHz input sampled at 32 GHz, the circuit shows an input-referred 1-dB compression point, (P1dB) and an input-referred third-order intercept point (IIP3) of 0.5 dBm and 5.8 dBm, respectively, as shown in Figure 5.19. The total power consumption reported was 2.7 W and the chip area was $815 \times 855 \ \mu m^2$.

208 TRANSCEIVER DESIGN AND PRACTICAL CONSIDERATIONS

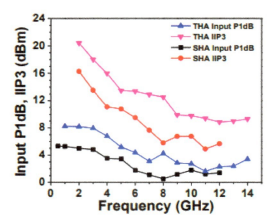

Figure 5.19 InP HBT and Si CMOS closed-loop sample and hold architecture with master THA and slave THA. (From [6].)

5.3 BASIC DEFINITIONS

As the transceiver is designed and laid out through fabrication, the digitizer (ADC) and DAC are quantified by the foundry and the specifications are presented to the developer. Below are the specifications reported that identify the quality of the components being used. The *absolute* accuracy is the worst-case difference between the actual converter output and the ideal converter output. This performance metric is often used by the National Bureau of Standards. The absolute accuracy is related to the low noise reference source used in the converter. The *relative* accuracy is the worst-case difference between the actual converter output and the ideal converter output after the gain and offset errors are removed. The least significant bit size (LSB) is defined as a fraction of the full-scale voltage, V_{FS} as [7]

$$LSB = \frac{V_{FS}}{2^n} \quad \text{(unipolar)} \tag{5.40}$$

and

$$LSB = \frac{V_{FS}}{2^{n-1}} \quad \text{(bipolar)} \tag{5.41}$$

and is sometimes known as the converter's resolution. The general *linearity error* Δ measures the deviation of the actual converter from a fitted line passing through the two end points of the measured converter output (terminal point fit). The linearity error is specified as a "fraction of an LSB" or as a "percentage of V_{FS}." A good converter typically has $\Delta \leq 1/2$ LSB. For example, if the magnitude of the largest error for $n = 3$ bits is -1.5 LSBs, the linearity error is $|-1.5 \text{ LSB}| = 1.5 \text{ LSB} = 18.75\% \, V_{FS}$.

The *differential nonlinearity* (DNL) is the maximum deviation in the output step size from the ideal value of one LSB.

$$DNL_k = V_k - V_{k-1} - LSB \tag{5.42}$$

where V_k, V_{k-1} are consecutive transition points. The step size is $V_k - V_{k-1}$. For any transition point [7]

$$V_j = \sum_{k=1}^{j} (V_k - V_{k-1}) \tag{5.43}$$

or

$$V_j = jLSB + \sum_{k=1}^{j} DNL_k \tag{5.44}$$

where $jLSB$ is the ideal level corresponding to V_j (assuming zero gain error and offset) and $V_0 = 0$ [7].

The *integral nonlinearity* (INL) is the maximum deviation of the input/output characteristic from a straight line passed, for example, through its end points

$$INL_j = \sum_{k=1}^{j} DNL_k = V_j - jLSB \tag{5.45}$$

The *offset* is defined as the vertical intercept of the straight line passed, for example through the end points. The *gain error* is the deviation of the slope of the line passed through the end points from its ideal value (usually unity). The *settling time* is the time required for the output to experience full-scale transition and settle within a specified error-band around its final value.

5.4 ANALOG-TO-DIGITAL CONVERSION

Conversion of an analog bandwidth of signals into their digital representations is important in many DRFM and front-end transceiver modules. The ADC translates a continuously variable (analog) signal into a digital signal. Digital signals propagate more efficiently than analog signals since they are mostly well defined and orderly and are easier for electronic circuits to distinguish from noise. A block diagram of an ADC is shown in Figure 5.20.

210 TRANSCEIVER DESIGN AND PRACTICAL CONSIDERATIONS

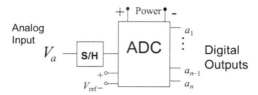

Figure 5.20 Block diagram of an ADC.

The input to an ADC consists of a voltage that varies among a theoretically infinite number of values. Examples are sinewaves, human speech, and the signals from a conventional television camera. The signals $a_1, a_2, \cdots, a_n$ now represent the digital output values corresponding to the analog input V_a. The output of the ADC has well defined levels or states. A sample and hold (S/H) circuit is often included in the input process to sample the signal and hold it until conversion into a digital number takes place. The transfer function for an n-bit (unipolar) ADC is [8]

$$V_a = V_{FS}\left(\sum_{i=1}^{n} a_i 2^{-i}\right) + q_e \tag{5.46}$$

where q_e represents the quantization error. A voltage V_{ref} is typically adjusted for a_i such that absolute value of the error is less than 1/2 of a LSB.

$$\left|V_a - V_{FS}\sum_{i=1}^{n} a_i 2^{-i}\right| < \frac{1}{2}\text{ LSB}.$$

5.4 Analog-to-Digital Conversion

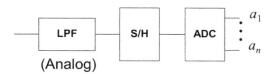

Figure 5.21 Analog LPF used as an ADC antialiasing filter.

The analog LPF needs to limit the bandwidth $< f_s/2$ (as a Nyquist converter).

$$H_{LPF}(f) = \begin{cases} 1: & |f| < \frac{f_s}{2} \\ 0: & |f| > \frac{f_s}{2} \end{cases}$$

Note that a bandpass filter can also be used here limiting the band of signals entering into the digitizer. The analog bandwidth of components on (or about) the ADC such as the SHA or comparators, can be considerably larger than the Nyquist band of the converter or the sampling frequency. For example, consider that the comparator's bandwidth, f_{comp}, has the largest bandwidth of all the components. The noise and frequency components attenuated by the input antialiasing filter between

$$f_s/2 < f << f_{comp}$$

are folded back into the baseband of the ADC. The number of Nyquist folds N_{fold}, as a function of the sampling frequency and the largest bandwidth component is

$$N_{fold} = \frac{f_{comp} - f_s/2}{f_s/2} = \frac{2f_{comp}}{f_s} - 1 \approx \frac{2f_{comp}}{f_s} \tag{5.47}$$

This adds to the quantization noise and reduces the SNR. For total foldback noise equal to quantization noise, the stopband attenuation must be increased by $\sqrt{N_{fold}}$ [9]. Consequently, the anti-alias filter's stopband attenuation must be increased to

$$A_{\text{stopband}} = 6.02n + 1.76 + 10\log(N_{\text{fold}}) \text{ dB} \tag{5.48}$$

where n represents the number of bits in the ADC.

212 TRANSCEIVER DESIGN AND PRACTICAL CONSIDERATIONS

5.4.1 Transfer Functions to Determine Linearity

To demonstrate the concepts discussed, the transfer functions (input/ output relationships) for several ADC configurations are examined.

5.4.1.1 ADC Transfer Functions

The four common characteristic transfer functions for an ADC are shown in Figure 5.22.

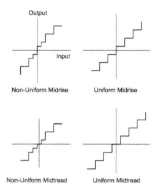

Figure 5.22 Characteristic ADC transfer functions.

These include the uniform and nonuniform midrise and the uniform and nonuniform midtread. Although the uniform quantizer is the most common means of amplitude analyzing a signal, a smaller quantization error can result by choosing smaller intervals where the signal is more likely to occur. As such a nonuniform midrise or midtread transfer function is the proper choice although post-processing is a bit more difficult.

The transfer function for an ideal 3-bit ADC is shown in Figure 5.23. The ADC output code is shown as a function of the input voltage. The staircase shows the quantized output code with the line through its center crossing the ideal code transition points [10]. Shown below the transfer function is the corresponding quantization error. As the line through the staircase crosses the ideal code transition points, note the quantization error goes to zero.

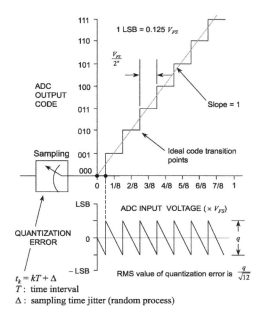

Figure 5.23 Transfer function and quantization error for ideal 3-bit ADC.

The error signal is $v_e(t) = -qt/T$ for $-T/2 < t < T/2$. The value of T here is the period of the sawtooth folding waveform. To find the RMS value of the error, we first calculate the mean squared error as

$$\begin{aligned} \overline{v_e^2} &= \frac{1}{T}\int_{-T/2}^{T/2} \left(\frac{qt}{T}\right)^2 dt \\ &= \frac{q^2}{T^3}\frac{t^3}{3}\Big|_{-T/2}^{T/2} = \frac{q^2}{3T^3}\left[\frac{T^3}{8}+\frac{T^3}{8}\right] \\ \overline{v_e^2} &= \frac{q^2}{12}. \end{aligned} \quad (5.49)$$

The value of the RMS error is then $v_{e,RMS} = q/\sqrt{12}$. The RMS quantization noise in a real ADC, in which the transfer function has long and short codes, can be calculated in a similar manner. An example of a nonideal ADC transfer function is shown in Figure 5.24. The converter has several problems including two linearity errors and a missing code. These are reflected in the quantization error as plotted on the bottom.

214 TRANSCEIVER DESIGN AND PRACTICAL CONSIDERATIONS

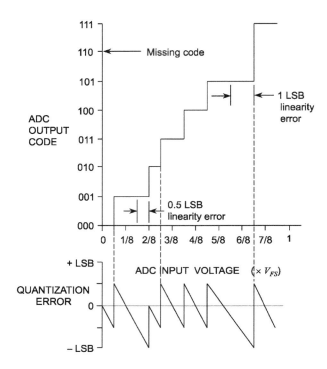

Figure 5.24 Nonideal ADC transfer function.

The converter can also have a nonmonotonic I/O relationship [7]. A good converter will be monotonic and have no missing codes over its full temperature range.

5.4.1.2 Linearity Errors

The linearity errors for the nonideal ADC in Figure 5.24 are shown in Figure 5.25. The step size calculation is the length of the transition (in volts) at each output code. That is, the step size is measured horizontally along the input voltage axis. The ideal step size is 1 LSB.

5.4 Analog-to-Digital Conversion 215

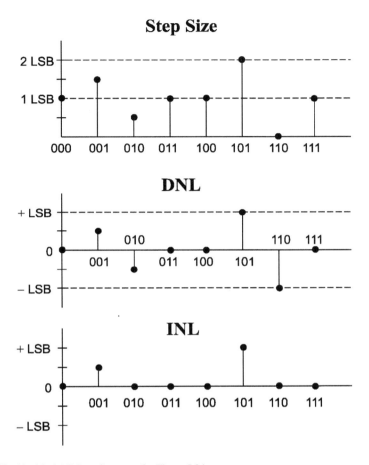

Figure 5.25 Nonideal ADC performance for Figure 5.24.

An uncalibrated converter with gain and offset errors is shown in Figure 5.26. The transition occurs at a voltage of 0.5 LSBs too high, giving a 0.5 LSB offset. All deviations are temperature dependent. Converter specifications include temperature coefficients for gain, offset, and linearity.

216   TRANSCEIVER DESIGN AND PRACTICAL CONSIDERATIONS

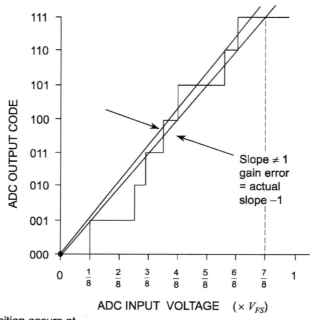

Transition occurs at
 voltage 0.5 LSBs
 too high − 0.5 LSB offset

Figure 5.26 Uncalibrated converter with gain and offset errors.

5.4.2 Operational Amplifiers

Operational amplifier circuits play a critical role in both analog-to-digital conversion and digital-to-analog signal conversion. Analog circuits that contain operational amplifiers such as the one shown in Figure 5.27(a) are easily analyzed since the input currents are both zero ($I_+ = 0 = I_-$) and $V_{in} = V_0/A \to 0$ as $A \to \infty$.

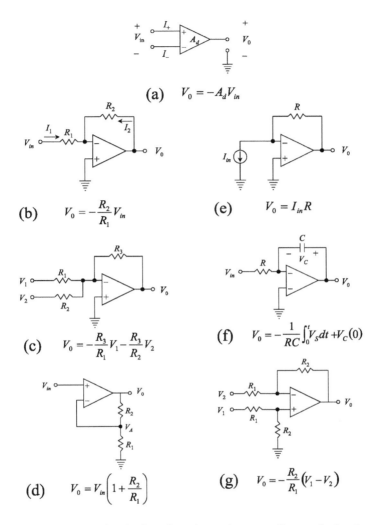

Figure 5.27 Operational amplifier circuit configurations and corresponding transfer functions: (a) single operational amplifier, (b) inverting amplifier, (c) summing amplifier, (d) noninverting amplifier, (e) current to voltage converter, (f) integrator, and (g) differential amplifier.

Several important operational amplifier circuit configurations are also shown in Figure 5.27. Figure 5.27(b) is an inverting amplifier, Figure 5.27(c) is a summing amplifier, and Figure 5.27(d) is a noninverting amplifier. Other functional configurations include a current-to-voltage converter, an integrator, and a differential amplifier in Figures 5.27(e) through 5.27(g). The transfer function of each configuration is shown as a function of the input voltage or currents and the output voltage V_0 [10].

5.4.3 Comparator Circuits

To amplitude analyze the analog waveforms and the detected optical pulses, high-speed comparator circuits are used. Comparators compare the voltage at its two inputs, producing an output voltage that is high or low when the input V_+ is greater or less than V_-, respectively. The comparator is a hybrid device in that its output is a binary signal but its inputs are continuous analog signals. The comparator symbol is identical to the symbol for the operational amplifier and any operational amplifier in analog electronics may be used as a comparator. Special purpose comparators, however, have been developed for fast switching speeds (devices such as the LM306, LM311, and LM303 have a very fast response).

In the past decade a remarkable advancement in comparator architectures has occurred in terms of both speed and power consumption. Some of those candidates for high-speed applications include the single-stage dynamic comparator preamplifier based latched comparator [11] and a two-stage dynamic regenerator [12].

The input is applied to one terminal of the device and the matching threshold voltage is applied to the other as shown in Figure 5.28. The matching voltage in this case is supplied by a resistor divider network (R_1 and R_2).

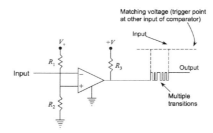

Figure 5.28 Comparator circuit showing the existence of a metastable state at the output due to the input at/or near the threshold.

Also used is a pull-up resistor R_3. The pull-up resistor ensures that an ungrounded output will float reliably high even in the presence of noise. The pull-up resistor does not interfere with the change of state as long as its value is large compared to the resistance of the closed switch. No feedback is used in this comparator circuit so for slowly varying input, the output swing can be rather slow. As shown in the figure, if the input is noisy the output may make several transitions as the input passes through the trigger point. With no negative feedback the inputs are not at the same voltage. In this configuration the differential input impedance is not bootstrapped to high values typical of operational amplifiers. Consequently, the input signal sees a changing load and changing input current as the output switches. Most comparator circuits also have a limited differential input swing (e.g., ± 5 volts) and the specifications must be checked before use.

The Schmitt trigger comparator circuit solves a number of the problems associated with the previous comparator circuit. The Schmitt trigger comparator circuit is shown in Figure 5.29. The big difference is the use of positive feedback (R_3 and C). The resistor R_3 lets the circuit have two thresholds depending on the output state. Positive feedback also ensures a rapid output transition. The capacitance C (10 – 100 pf) is used as a speed-up capacitor to help the output response.

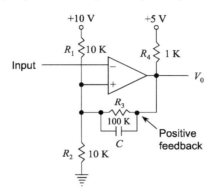

Figure 5.29 Schmitt trigger comparator circuit showing the positive feedback connection to create two thresholds thereby eliminating the metastability condition.

Several metrics are used to detail the performance of comparators. The *comparator resolution* is the minimum input difference that yields a correct digital output. The *comparison rate* is the maximum clock frequency at which the comparator can recover from a full-scale overdrive and correctly respond to a subsequent LSB input. The *dynamic range* is the ratio of a maximum input swing to a minimum resolvable input. The *kickback noise* is the power of the transient noise observed at the comparator input due to the switching of the amplifier and the latch.

5.4.3.1 Comparator Considerations

The comparator determines the overall performance of flash ADC. The characteristics of the comparator, such as the speed, gain, and delay, should be considered while designing an ADC. Different comparator designs for flash ADC architecture are compared in [13]. For N-bit flash ADC, $2^N - 1$ comparators are needed. The number of comparators required gets doubled with the increase in the resolution of flash ADC, even by a single bit. Because of the parallelism in its architecture [10], the conversion is completed within a single clock cycle.

There are basically four comparator design techniques that are variations of the switched comparator: (a) the improved double tail comparator, (b) the double tail dual rail dynamic latched comparator, (c) the double tail comparator with clock gating, and (d) the charge sharing dynamic latch comparator. In the improved double tail, the

Comparator	Technology (nm)	Power (uW)	Delay (ps)	PDP (fJ)
Improved double tail comparator [8]	180	329	210	69.09
Double tail dual rail dynamic latched	180	10.2	1012	10.3
Double tail comparator with clock gating [10]	180	35.5	5.1	0.181
Charge sharing dynamic latch comparator [11]	180	0.00133	1699	0.002

Figure 5.30 Comparison of Comparator Architectures (©IEEE, reprinted with permission from [13].)

delay time is reduced by adding a few more transistors that strengthen the positive feedback in the latch regeneration cycle. A major drawback of this orientation is the static power consumption. In the double tail dual rail dynamic latched comparator design, an inverter is present between the front end and the back end. This design exhibits less power dissipation. In the double tail comparator with clock gating, clock gating techniques are incorporated and the power as well as the delay are improved in comparison with the previous topologies. In the charge, sharing dynamic latch comparator, the good features of both resistive dividing comparator and differential comparator with current sensing are included. The sensitivity is increased and delay is reduced by using a latched comparator design. With low supply voltage, a stable output is generated here as compared to the previous cases [13]. Table 5.30 shows a comparison of the four types of comparators in 180-nm technology including power usage, delay, and power delay product (PDP).

5.5 ADC CIRCUIT CONCEPTS

Below we examine one of the most useful architectures for digitization, *the flash*. The includes the output encoders and calibration methods. The flash is often used in a number of different configurations to achieve higher resolution in devices due to the limitations as discussed below.

5.5.1 Flash ADCs

The flash ADC is the fastest type available. A flash ADC uses comparators, one per voltage step, and a string of resistors. A 4-bit ADC will have 16 comparators, and an 8-bit ADC will have 256 comparators. All of the comparator outputs connect to

5.5 ADC Circuit Concepts

a block of logic that determines the output based on which comparators are low and which are high. The conversion speed of the flash ADC is the sum of the comparator delays and the logic delay (the logic delay is usually negligible). Flash ADCs are very fast, but consume enormous amounts of IC real estate. Flash ADCs face challenges due to the large input capacitance and often have a limited bandwidth, or otherwise they require a broadband input buffer. They also suffer from the large input parasitics but also limited resolutions. Flash ADCs are typically preferred for high-speed, low-resolution applications.

An example of a $n = 3$-bit all parallel flash converter is shown in Figure 5.31. This configuration uses $2^n - 1$ comparators in a parallel array. Note also that no hold circuit is used. The matching threshold voltages are supplied to the comparator by a resistor divider network. Note that each comparator can be represented by a preamplifier, a latch, and a clock [4]. With the clock signal high, the input difference is amplified by the preamplifier while the latch is disabled. With the clock signal low the latch amplifies the difference while the preamplifier is disabled. The latched comparator outputs are recombined in a thermometer-to-binary (3 bit) output.

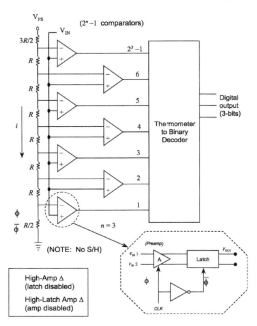

Figure 5.31 $n = 3$-bit all parallel flash converter.

In this example let $V_{FS} = 8$ mV (unipolar) with $R = 1\Omega$. The current in the resistor ladder is then

$$i = \frac{V_{FS}}{R_{\text{total}}} = \frac{8 \text{ mV}}{8\Omega} = 1 \text{ ma}. \qquad (5.50)$$

Table 3.1 shows the matching threshold voltage V_R for each comparator and the resulting input voltage range for turning the comparator on.

Table 5.1

Matching Threshold Voltages for the $n = 3$-bit Flash Converter

Comp. No.	V_R (mV)	Input Voltage Range (mV)
1	0.5	$V_{in} > 0.5$
2	1.5	$V_{in} > 1.5$
3	2.5	$V_{in} > 2.5$
4	3.5	$V_{in} > 3.5$
5	4.5	$V_{in} > 4.5$
6	5.5	$V_{in} > 5.5$
7	6.5	$V_{in} > 6.5$

For example, with $V_{in} < 0.5$, no comparators are on. The comparator outputs represent the input signal in a thermometer code (total number of comparators on). Both the thermometer code and the resulting binary representations are shown in Table 3.2. A transit time of 1 ps is required for a signal to propagate 200 μm of IC interconnect. This is important in a parallel configuration since both the clock signals and analog signals must arrive at the comparators at the same time (synchronously).

Table 5.2

Thermometer Corresponding Binary Representation

Binary Code c b a	Thermometer Code g f e d c b a
0 0 0	0 0 0 0 0 0 0
0 0 1	0 0 0 0 0 0 1
0 1 0	0 0 0 0 0 1 1
0 1 1	0 0 0 0 1 1 1
1 0 0	0 0 0 1 1 1 1
1 0 1	0 0 1 1 1 1 1
1 1 0	0 1 1 1 1 1 1
1 1 1	1 1 1 1 1 1 1

5.5.1.1 Encoder Considerations

The encoder is a device that converts one form of code to another. Encoding is the reverse process of decoding. An encoder has m number of input lines and n number of output lines. It produces at the output an n-bit binary code corresponding to the digital input number. There are four types of encoders: (a) read only memory (ROM), (b) fat tree encoder, (c) Wallace tree encode, and (d) the multiplexer-based encoder [14]. The ROM input input stage consists of logic gates to detect the thermometer code transitions. The ROM encoder schematic is shown in Figure 5.32. Based upon the transition a particular row in the ROM encoder table is activated for a binary output. Logic gates at the input stage form an $M - to - 1$ block formed using 2-input AND gates. One input is the output of comparator n and other input is the inverted output of comparator $n + 1$. The AND gate output activates a row in the ROM encoder table. Typical ROM encoder delay is 0.135 ns with a power dissipation of 432 μW. The ROM encoder drawbacks are that it consumes a lot of power and is prone to bubble errors requiring additional circuitry to prevent these from occurring.

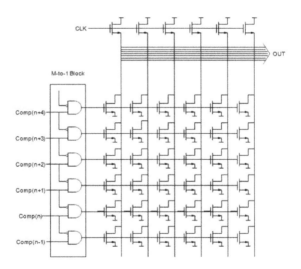

Figure 5.32 Block diagram of a read-only-memory encoder structure. (©IEEE, reprinted with permission from [14].)

The fat tree encoder is the fastest encoder converting the analog signal into n-bits of data in binary form. The encoding is carried out in two stages. The first stage converts the thermometer code to a one-out-of-N code. Then it is converted to binary code in the next stage. The fat tree encoder is less sensitive to noise and uses a relatively small area compared to a ROM encoder. It is more power-efficient than ROM encoder since it doesn't require clock or pull-up circuitry [14]. The fat tree encoder designed using OR gates is as shown in Figure 5.33.

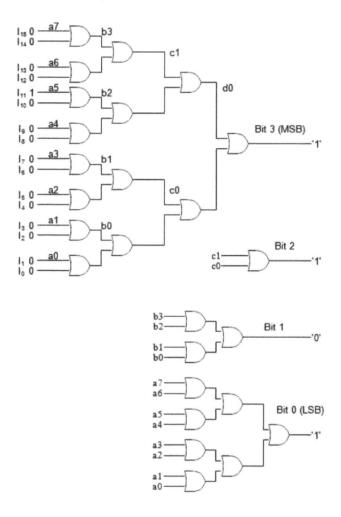

Figure 5.33 Block diagram of a fat tree encoder structure. (©IEEE, reprinted with permission from [14].)

Here the equation for bit 0 b_0 is

$$b_0 = a_0 + a_1 + a_3 + a_4 + a_5 + a_6 + a_7 \qquad (5.51)$$

and for b_1
$$b_1 = b_0 + b_1 + b_2 + b_3 \tag{5.52}$$
and for b_2
$$b_2 = c_0 + c_1 \tag{5.53}$$
and for b_3
$$b_3 = d_0 \tag{5.54}$$

The fat tree encoder structure has a power dissipation of 0.110 μW and a delay of 0.093 ns. The 4-bit Wallace tree encoder is shown in Figure 5.34 and shows the basic building block used is a full adder. The first stage counts the number of logic 1s in the thermometer code and outputs the binary value.

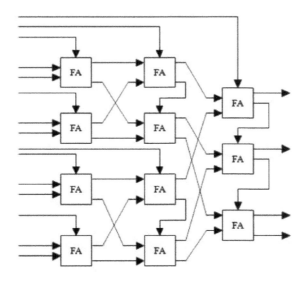

Figure 5.34 Block diagram of a Wallace tree encoder structure. (©IEEE, reprinted with permission from [14].)

The quantity of elementary adder cells required to implement the Wallace tree encoder for N-bits is given as

$$X_N = \sum_{i=1}^{N} (i-1) 2^{N-i} \tag{5.55}$$

The advantage of a Wallace encoder is that it can reduce the bubble errors. However for large values of N, the Wallace tree encoder is not practical. The architecture shown in Figure 5.34 has a delay of 4.05 ns and a power dissipation of 8.27 μW. The multiplexer encoder is shown in Figure 5.35 for a 7-to-3 multiplexer thermometer-to-binary encoder.

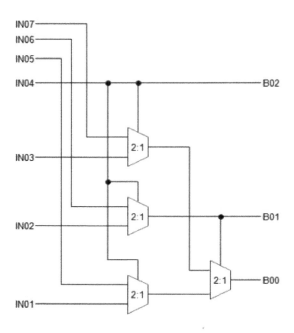

Figure 5.35 Block diagram of a multiplexer tree encoder structure. (©IEEE, reprinted with permission from [14].)

The multiplexer configuration uses a minimum die area due to the semiconductor real estate needed. For an N-bit flash ADC, the most significant bit (MSB) of the thermometer to binary encoder output is high if more than half of the outputs in the thermometer code are high. For the second most significant bit, the original thermometer code is divided into two equal parts separated by MSB. The center bit of each part is decoded by the multiplexer and its output gives the second most significant bit. Table 5.3 compares the delay, power, and power delay product of each design.

5.5 ADC Circuit Concepts 227

Table 5.3
Comparison of Delay, Power, and Power Delay Product

Type of Encoder	Power Dissip. (μW)	Delay (ns)	PDP(fJ)
ROM	431.8	0.1346	58.1
Fat tree	0.1104	0.09343	0.0103
Wallace tree	8.27	4.05	83.4
Multiplexer	0.0442	0.1325	0.0058

Digitally assisted integrated circuit design has gained popularity in recent years due to challenges associated with analog components in mixed-signal SOC and miniaturization of transistors. These challenges include process variations, mismatches, and increased demand for circuit complexity for designs in submicron fabrication process technologies. In flash ADCs, the offsets are typically random variables with normal distributions and have a direct impact on an ADC's performance, worsening its integral and differential nonlinearities (INL and DNL, respectively). On-chip calibration techniques aid to address these issues by reducing the sensitivity of analog circuit designs to variations. One particular calibration scheme is shown in Figure 5.36.

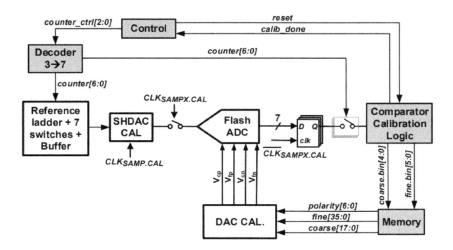

Figure 5.36 Digital offset calibration of a flash ADC. (©IEEE, reprinted with permission from [15].)

5.6 ADC NOISE FLOOR WITH WINDOWING

The dynamic range of a digitizer can be determined using a full-scale sinusoid and a subsequent examination of the noise floor. With a spectral averaging test, the digitized signal is repeatedly acquired and a DFT calculated. The magnitude response is then averaged point-by-point. The noise floor level depends on the amount of jitter and thermal noise present. Quantization noise and any harmonics that are generated are also present.

In an ideal system, the noise floor is only a function of the number of bits n, the number of points integrated into the DFT, N, and the equivalent noise bandwidth of any window function being used, E_B. In Figure 5.37, the process for examining the noise floor is shown.

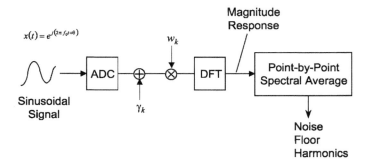

Figure 5.37 Process for examining the noise floor of an ADC.

In this section analytical expressions are presented to describe the noise floor. This is done by first examining the magnitude spectrum and evaluating the signal-to-noise. To begin this process, let x_k, $k = 0, 1, \cdots, N-1$ be the samples of a sinusoid. Also let γ_k, $k = 0, 1, \cdots, N-1$ be a white noise process, and w_k, $k = 0, 1, \cdots, N-1$ be the window samples. The DFT of the windowed sinusoid can be expressed as [16]

$$A(\ell) = \sum_{k=0}^{N-1} (x_k + \gamma_k) w_k e^{-jk\ell 2\pi/N}. \tag{5.56}$$

Since γ_k is a random process, $A(\ell)$ and $|A(\ell)|^2$ are random variables.

5.6 ADC Noise Floor with Windowing

To obtain the SNR relationship, the expected value of the magnitude squared is needed, $E\left\{|A(\ell)|^2\right\}$. The DFT of the windowed sinusoid can be broken up as

$$A(\ell) = \underbrace{\sum_{k=0}^{N-1} \gamma_k w_k e^{-jk\ell 2\pi/N}}_{\substack{\text{DFT of windowed noise} \\ \text{(no longer white sequence)}}} + \underbrace{\sum_{k=0}^{N-1} x_k w_k e^{-jk\ell 2\pi/N}}_{\substack{\text{DFT of windowed} \\ \text{sinusoid}}} \quad (5.57)$$

or

$$A(\ell) = \Gamma(\ell) + X(\ell). \quad (5.58)$$

Here $\Gamma(\ell)$ is the DFT of the windowed noise sequence and is no longer a white sequence [16]. We know, however, that $E\{\gamma_k\} = 0$, $E\{\gamma_k^2\} = \sigma^2$, $E\{\gamma_k \gamma_\ell\} = 0$ for $k \neq \ell$ because the noise is zero mean, white, and stationary. Therefore, $E\{\Gamma(\ell)\} = 0$, $\ell = 0, 1, \cdots, N-1$ and

$$E\{\Gamma(m)\Gamma^*(n)\} = \sigma^2 \sum_{k=0}^{N-1} w_k^2 e^{-j(m-n)2\pi/N}$$

and

$$E\{\Gamma(\ell)\Gamma^*(\ell)\} = \sigma^2 \sum_{k=0}^{N-1} w_k^2$$

which is the variance. Note that $\sigma^2 = e_{qns}^2$.

$X(\ell)$ is just the DFT of a windowed sinusoid. Consider $x(t)$ to be a sinusoid and $w(t)$ to be a window function with $0 < t < N/f_s$ where f_s is the sampling frequency and N is the record length. The Fourier transform is then $FT\{x(t)w(t)\} = |W[2\pi(f-f_0)]|$, where $W(f)$ is the (continuous time/analog) Fourier transform of the window function $w(t)$. That is, the sinusoidal signal translates the window spectrum in frequency. To find the spectrum of the discrete time signal, recall the time domain sampling process of an analog signal $x_a(t)$ where $t = nT = n/f_s$ and $f_s > 2$ (bandwidth of interest). Note that F is the frequency of the analog signal. The spectrum of the discrete time signal is then

$$X(f) = f_s \sum_{k=-\infty}^{\infty} X_a[(f-k)f_s] \quad \text{(periodic)} \quad (5.59)$$

where X_a is the spectrum of the analog signal and $f = F/f_s$ [5]. This can also be written as

$$X\left(\frac{F}{f_s}\right) = f_s \sum_{k=-\infty}^{\infty} X_a(F - kf_s). \quad (5.60)$$

The relationship between the spectrum of an analog signal and the spectrum of a discrete time signal is shown in Figure 5.38.

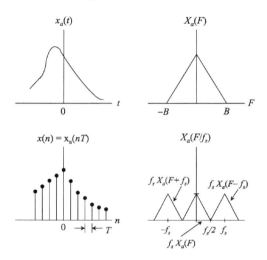

Figure 5.38 Relationship between the spectrum of an analog signal and the spectrum of a discrete time signal.

Now we can sample the spectrum $X(f)$ of the discrete time signal $x(n)$ at N equally spaced frequencies. For $f = \ell/N$,

$$X(\ell) = X(f)|_{f=\ell/N} = f_s \sum_{m=-\infty}^{\infty} X_a\left[\left(\frac{\ell}{N} - m\right) f_s\right]$$
$$\ell = 0, 1, \cdots, N-1 \tag{5.61}$$

or

$$X(\ell) = f_s \sum_{m=-\infty}^{\infty} X_a\left(\frac{\ell f_s}{N} - m f_s\right)$$
$$\ell = 0, 1, \cdots, N-1. \tag{5.62}$$

Therefore,

$$X(\ell) = f_s \left| W\left[2\pi f_s \left(\frac{\ell}{N} - \alpha\right)\right] \right| \tag{5.63}$$

5.6 ADC Noise Floor with Windowing

where we let $\alpha = f_0/f_s$. Now for $E\left\{|A(\ell)|^2\right\}$,

$$E\{A(\ell)A^*(\ell)\} = f_s\left|W\left[2\pi f_s\left(\frac{\ell}{N} - \alpha\right)\right]\right|^2 - \sigma^2\sum_{k=0}^{N-1} w_k^2 \qquad (5.64)$$

and we note that the first term on the right is the signal power and the second is the additive noise power [16]. If we average point by point, a large number of $|A(\ell)|^2$ (squares of magnitude of DFT records), the SNR converges to a value of

$$\frac{S}{N} = \frac{f_s^2\,|W\left[2\pi f_s(\ell/N - \alpha)\right]|^2}{\sigma^2\sum_{k=0}^{N-1} w_k^2}$$

$$= \frac{N}{\sigma^2}\frac{f_s^2 W(0)^2}{N\sum_{k=0}^{N-1} w_k^2}\frac{|W\left[2\pi f_s(\ell/N - \alpha)\right]|^2}{W(0)^2}$$

$$= \frac{N}{\sigma^2 E_B}\frac{|W\left[2\pi f_s(\ell/N - \alpha)\right]|^2}{W(0)^2} \qquad (5.65)$$

where E_B is the equivalent noise bandwidth of the window function $w(t)$[6]. With the magnitude spectrum of the window function fairly flat and N reasonably large, then the ℓth frequency bin, where ℓ is the integer nearest to $N(f_0/f_s)$, will contain the signal component and

$$\left|W\left[2\pi f_s\left(\frac{\ell}{N} - \frac{f_0}{f_s}\right)\right]\right| \approx W(0). \qquad (5.66)$$

Therefore,

$$\frac{S}{N} \approx \frac{N}{\sigma^2 E_B}. \qquad (5.67)$$

For a perfect n-bit ADC, and a sinusoid with a peak-to-peak value of 2, $\sigma^2 = LSB^2/12$ where

$$LSB = \frac{V_{FS}}{2^n - 1} = \frac{2}{2^n - 1}. \qquad (5.68)$$

Now in dB,

$$10\log\left(\frac{S}{N}\right) \cong 10\log\left(\frac{12N}{LSB^2 E_B}\right)$$

$$\cong 10\log\left(\frac{3N}{E_B}(2^n-1)^2\right)$$

$$\cong 10\log\left(\frac{3N}{E_B}\right) + 10\log(2^n-1)^2$$

$$\cong 10\log\left(\frac{3N}{E_B}\right) + 6.02n \text{ dB}. \tag{5.69}$$

In practice $x(t) = \sin(2\pi f_0 t + \theta)$ (a sinewave instead of $e^{j\omega t}$). For the *magnitude squared* spectrum [16]

$$10\log\left(\frac{S}{N}\right) \cong 10\log\left(\frac{3N}{4E_B}\right) + 6.02n \text{ dB}. \tag{5.70}$$

For the magnitude spectrum (i.e., $E\{|A(\ell)|\}$) assume $x(t) \gg \gamma(t)$. That is, consider the signal to be much greater than the additive noise. $A(\ell)$ is just the signal part

$$|A(\ell)| \cong f_s \left| W\left[2\pi f_s\left(\frac{\ell}{N} - \alpha\right)\right]\right|. \tag{5.71}$$

For frequencies where the signal is not present ($f \neq f_0$) the noise is given by $\Gamma(\ell)$. By the law of large numbers both Γ_{real} and Γ_{img} are Gaussian random variables, $N(\mu = 0, \sigma)$. They are also uncorrelated and independent. Therefore, $\sqrt{\Gamma(\ell)\Gamma^*(\ell)}$ is Rayleigh distributed with

$$\mu = \sqrt{\sigma^2\left(\sum_{k=0}^{N-1} w_k^2\right)\frac{\pi}{4}}. \tag{5.72}$$

Therefore, the S/N should be increased by

$$10\log\left(\frac{4}{\pi}\right) \cong 1.05 \text{ dB}. \tag{5.73}$$

The expression for the noise floor for the magnitude spectrum is then

$$10\log\left(\frac{S}{N}\right) \cong 6.02n + 10\log\left(\frac{3N}{\pi E_B}\right) \text{ dB}. \tag{5.74}$$

5.6 ADC Noise Floor with Windowing

If the noise floor is substantially higher, it can be concluded that the system's other additive noise sources (e.g., thermal noise) are dominant [16].

The averaged magnitude square spectrum of an experimental 7-bit A/D converter is shown in Figure 5.39.

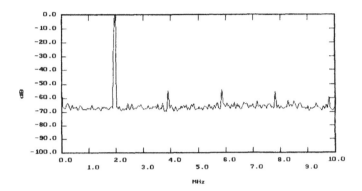

Figure 5.39 Averaged magnitude square spectrum of an experimental 7-bit A/D converter. (©IEEE, reprinted with permission from [16].)

The signal was sampled at 20-M samples per second then digitized by an ADC ($w(t)$; 4-term Blackman-Harris, 200 spectral averages, $N = 1024$). From (4.22) the expected noise floor is 68 dB and agrees nicely with Figure 5.39 [16].

To verify the expressions for the noise floor a computer simulation of a unit-amplitude sinewave sampled at 20 Ms/sec is also run using $N = 1024$ and $E_B = 2$. The ADC has $n = 7$-bits, a 4-term Blackman-Harris window is used ($E_B = 2$), and 200 spectral averages are performed. To simulate the asynchronous acquisition of the sinusoid a random phase was added. Also, white Gaussian noise was added with $\sigma = LSB/4$. The adjusted noise floor calculation is then

$$10 \log \left(\frac{S}{N}\right) \cong 10 \log \left(\frac{6.86N}{16E_B}\right) + 6.02n. \tag{5.75}$$

The expected noise floor for the simulation using (4.27) is 65.6 dB and also agrees with the results shown in Figure 5.40 [16].

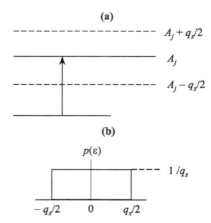

Figure 5.41 (a) Quantization of a sampled analog signal and (b) uniform distribution of ε.

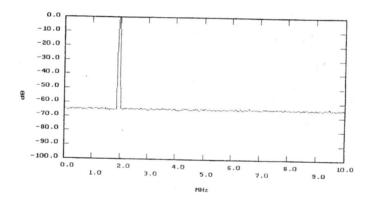

Figure 5.40 Average magnitude square spectrum of an ideal 7-bit A/D converter. (©IEEE, reprinted with permission from [16].)

5.6.1 Quantization Error

The quantization errors are irreversible errors in that the original signal cannot be recovered. If the sampled analog signal amplitude is $A_j + \varepsilon$ then the level out of the quantizer is A_j if $-q_s/2 < \varepsilon < q_s/2$ where q_s is the quantization step size as shown in Figure 5.41(a). The difference between the true value and the quantizer output level A_j is known as the quantization error. The quantization error is a noise source that is added to the signal. The mean square quantization noise power (considering $R = 1\Omega$) can be determined by assuming a uniform distribution for ε as shown in Figure 5.41(b). The mean square quantization noise power is then

$$e_{qns}^2 = E\{\varepsilon^2\} = \int_{-q_s/2}^{q_s/2} \frac{\varepsilon^2}{q_s} d\varepsilon$$

$$= \frac{1}{q_s}\left(\frac{\varepsilon^3}{3}\bigg|_{-q_s/2}^{q_s/2}\right) = \frac{1}{q_s}\left(\frac{q_s^3}{24} + \frac{q_s^3}{24}\right) \quad (5.76)$$

or

$$e_{qns}^2 = \frac{q_s^2}{12} \quad \text{watts}. \quad (5.77)$$

The RMS quantization noise voltage is then

$$e_{qns} = \sqrt{E\{\varepsilon^2\}} = \frac{q_s}{\sqrt{12}} \quad \text{volts}. \quad (5.78)$$

Recall that for n bits, $2^n - 1$ quantization levels are required. The maximum peak-to-peak signal amplitude is then $A_{pp} = 2^n q_s$ before the next quantization level is reached. However, the maximum peak-to-peak level cannot always be obtained. For a sinusoid $(2^n - 1)q_s \leq A_{pp} \leq 2^n q_s$. For $n \geq 5$, $A_{pp} = 2^n q_s$.

5.6.2 Signal-to-Noise Relationships

The SNR provides an important measure of the digitization process. To derive the SNR we start with the RMS signal value

$$A_{rms} = \frac{A_{pp}}{2\sqrt{2}} = \frac{2^n q_s}{2\sqrt{2}}. \quad (5.79)$$

The theoretically ideal SNR ($n \geq 5$) is obtained by dividing by $\sqrt{e_{qns}^2}$ or indexADC!SNR

$$SNR = \frac{A_{rms}}{\sqrt{e_{qns}^2}} = \frac{2^n q_s \sqrt{12}}{q_s 2\sqrt{2}} = 2^n \sqrt{1.5}. \quad (5.80)$$

The term theoretically ideal refers to the fact that the only noise source being considered is quantization noise. In dB,

$$SNR = n20\log(2) + \frac{1}{2}20\log(1.5) \quad (5.81)$$

or

$$SNR = 6.02n + 1.76 \text{ dB}. \quad (5.82)$$

That is, each additional bit increases the SNR by 6 dB. The ideal SNR for $n < 5$ is

$$SNR = \left(2^n - 2 + \frac{4}{\pi}\right)\sqrt{1.5}. \quad (5.83)$$

When fewer than five bits are used, the quantization error has a strong correlation with the input signal.

The dynamic range of an ADC equals the SNR of that system measured over a bandwidth equal to 1/2 the sampling frequency, f_s. The quantization noise power density is

$$e_{qns}^2(f) = \frac{e_{qns}^2}{f_s/2} = \frac{q_s^2}{6f_s}. \quad (5.84)$$

From (5.80), the SNR density is then

$$\begin{aligned} SNR(f) &= 2^{n-1}\sqrt{6}\sqrt{\frac{f_s}{2}} \\ &= 2^{n-1}\sqrt{3}\sqrt{f_s}, \end{aligned} \quad (5.85)$$

and for a system with a bandwidth $BW = f_{sys}$, the result is

$$SNR_{sys}(f) = 2^{n-1}\sqrt{3}\sqrt{\frac{f_s}{f_{sys}}}. \quad (5.86)$$

A summary of the important SNR relationships is shown in Figure 5.42. Note the step size of the quantizer is usually chosen such that the quantization noise is below the thermal noise, which may typically be $6q_s$ to $10q_s$.

5.7 FFT SPECTRUM ANALYSIS

Examination of the signal's spectrum reveals a number of important characteristics including the S/N, the S/N plus distortion (SINAD), the effective number of bits (ENOB), and a measure of the distortion that may be present in the digitization system. Below, the fundamentals of several important spectral definitions are given. Probably the most important is the SNR. The SNR is defined as

$$SNR = 20\log\left(\frac{\text{rms signal}}{\text{rms noise}}\right) \quad (5.87)$$

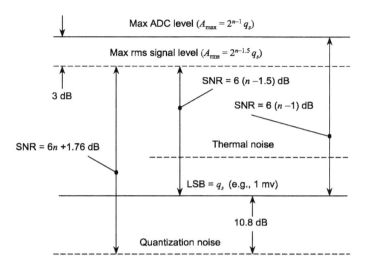

Figure 5.42 Dynamic range summary.

where the rms noise is the sum of all spectral components except the fundamental and first five harmonics. The expression for the theoretical SNR is repeated below

$$SNR(+\mathrm{dB}) = 20\log\left[2^{n-1}\sqrt{6}\right] = 6.02n + 1.76 \tag{5.88}$$

and assumes only quantization noise.

The SINAD ratio is the ratio of the rms signal (fundamental) to the rms noise at a specified input and sampling frequency. The SINAD takes into account all the noise (including harmonics) to give indication of the useful dynamic range and excludes the DC component. The expression

$$SINAD = 20\log\left(\frac{\text{rms signal}}{\text{rms noise}}\right) \tag{5.89}$$

is the same where the "rms signal" is the measured rms signal in the fundamental. The "rms noise" is the rms sum of all other spectral components *below* the Nyquist frequency (excluding DC). The expression for the SINAD is

$$SINAD(+\mathrm{dB}) = -20\log\sqrt{10^{-(SNRW/0DIST)/10} + 10^{+(THD/10)}}. \tag{5.90}$$

where the *THD* is the total harmonic distortion (defined below).

Although the digitizer amplitude analyzes the signal with n-bits, due to noise, the usable dynamic range is somewhat less. The ENOB is [9]

$$ENOB = \frac{SINAD - 1.76 + 20\log\left(\frac{\text{full scale amplitude}}{\text{actual input amplitude}}\right)}{6.02}. \quad (5.91)$$

The actual input amplitude of the analog signal is sometimes less than the full-scale level to avoid distorting the digitized waveform.

The three indicators of a converter's nonlinearity are the (1) total harmonic distortion (single tone input), (2) spurious free dynamic range (single tone input), and (3) intermodulation distortion (multiple tones). The total harmonic distortion (THD) measures the harmonics of the input signal that show up at integral multiples of the fundamental. Recall that the harmonic components beyond Nyquist bandwidth are folded back or aliased. The THD

$$THD(-\text{dB}) = 20\log\left(\frac{\text{rms noise}}{\text{rms signal}}\right) \quad (5.92)$$

where the rms noise is now the sum of the input signal's first five harmonics. That is,

$$THD(-\text{dB}) = 20\log\sqrt{\left(10^{2\text{nd}HAR/20}\right)^2 + \left(10^{3\text{rd}HAR/20}\right)^2 \cdots} \quad (5.93)$$

where the harmonics are in $-$dB. The spurious free dynamic range (SFDR) is the ratio (in dB) of the fundamental component magnitude to the largest harmonic or intermodulation product (usually a harmonic of the fundamental). The intermodulation distortion (IMD) is the change in one sinusoid that the presence of another sinusoidal input at a different frequency causes. The distortion is usually the product of two tones (f_1, f_2). The intermodulation products are $[mf_1 + nf_2]$ where $n, m \in \{0, \pm 1, \pm 2, \cdots\}$. The order of the IMD is $|n| + |m|$. As an example, consider the input tone signals

$$v(t) = A_1 \sin \omega_1 t + A_2 \sin \omega_2 t. \quad (5.94)$$

The second-order term is

$$v(t)^2 = A_1^2 \sin^2(\omega_1 t) + 2A_1 A_2 \sin(\omega_1 t)\sin(\omega_2 t) + A_2^2 \sin^2(\omega_2 t). \quad (5.95)$$

The first and last terms produce a harmonic distortion at $2\omega_1$ and $2\omega_2$. The cross product term produces the IMD. The second-order IMD is

$$2A_1 A_2 \sin(\omega_1 t)\sin(\omega_2 t) = A_1 A_2 \{\cos\left[(\omega_1 - \omega_2)t\right] - \cos\left[(\omega_1 + \omega_2)t\right]\} \quad (5.96)$$

and are the sum and difference frequencies. The third-order IMDs are $2f_1 + f_2$ and $2f_2 + f_1$ and are usually outside the filter passband. The most bothersome third-order IMD is $2f_1 - f_2$ and $2f_2 - f_1$ since it lies within the filter passband. The IMD is

$$IMD = 20\log\left(\frac{\text{rms noise}}{\text{rms signal}}\right) \quad (5.97)$$

where the rms noise includes the desired IMD products.

5.8 DIGITAL-TO-ANALOG CONVERSION

Current high-speed DACs are based on segmented current-steering architectures [17]. Their performance is mainly limited by a code-dependent output impedance, timing mismatches between individual current cells, and clock feedthrough.

Figure 5.43 shows a block diagram of a DAC with its required inputs. The input signals are typically digital and binary and represent a weighting factor for the output analog voltage. The expression for the DAC output voltage (unipolar case) is [7]

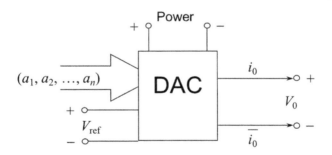

Figure 5.43 Mathematical model of a DAC.

$$V_0 = kV_{FS}\left(\sum_{i=1}^{n} a_i 2^{-i}\right) + V_{os} \quad (5.98)$$

where k is the gain of the device, V_{FS} is the full-scale output voltage, and V_{os} is the offset voltage. The binary (digital) input signal $a_1, a_2, \cdots, a_n$ is to be translated into the analog signal V_0 representing a binary weighting of the bit values. This is reflected in the 2^{-i} term in (5.98). Equation (5.98) is often referred to as the DAC transfer function. The offset voltage V_{os} induces a shift in the transfer function of the DAC

240   TRANSCEIVER DESIGN AND PRACTICAL CONSIDERATIONS

as will be discussed shortly. An analog lowpass reconstruction filter is often used at the output of a S&H/DAC combination as shown in Figure 5.44. The LPF is needed to reject repeated spectra around sampling frequency and multiples of the sampling frequency. The high-frequency components out of the DAC, can overload for example, an output amplifier if they are not eliminated. The LPF also aides in interpolation between quantization steps and to output a smoothed signal.

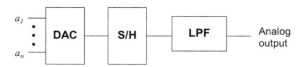

Figure 5.44   Analog low-pass filter to remove extraneous DAC output signals.

5.9   TRANSFER FUNCTIONS TO DETERMINE LINEARITY

To demonstrate the concepts discussed, the transfer functions (input/output relationships) for several DAC configurations are examined.

5.9.0.1   DAC Transfer Functions

The ideal 3-bit unipolar DAC transfer function is shown in Figure 5.45, which shows the output voltage as a function of the binary input values.

5.9 Transfer Functions to Determine Linearity

Table 5.4
Output Voltage Corresponding to Each 3-Bit Input code.

3-Bit Input Code	Output Voltage $\times V_{FS}$	
000	0.000	(0/8)
001	0.125	(1/8)
010	0.250	(1/4)
011	0.375	(3/8)
100	0.500	(1/2)
101	0.625	(5/8)
110	0.750	(3/4)
111	0.875	(7/8)

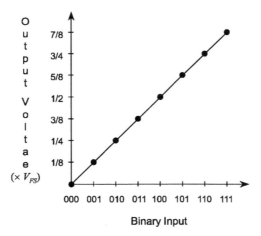

Figure 5.45 Three-bit unipolar DAC transfer function.

This is an ideal device in that the output voltage increases monotonically by one LSB for each subsequent binary input value. Also note from (4.1) the $LSB = V_{FS}/2^3 = V_{FS}/8$. The output voltage corresponding to each 3-bit input code is summarized in Table 4.1.

A 3-bit DAC with both gain and offset errors is shown in Figure 5.46. Also shown for comparison is the 3-bit ideal transfer function. Note the gain or slope of the line is not unity. The expression for the gain for the unipolar case is

$$k = (V_{2^n-1} - V_0)2^n/2^n - 1$$

or $k = 0.57$ for $n = 3$-bits. The offset voltage V_{os} is the point at which the terminal point line crosses the output axis (for the $n = 3$-bit DAC $V_{os} \approx 1/8 V_{FS}$).

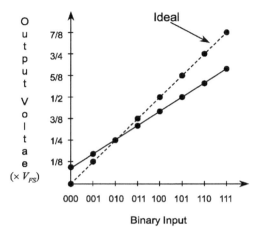

Figure 5.46 Three-bit DAC with gain and offset errors.

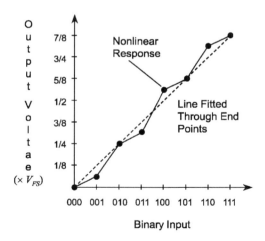

Figure 5.47 Three-bit DAC containing circuit mismatches.

Figure 5.47 shows the output voltage as a function of the input code for a 3-bit DAC containing circuit mismatches. The circuit mismatches result in an output response that is not perfectly linear [4]. A terminal point line fit (dashed line) gives a reference to quantify the performance of this device. Figure 5.48 shows the corresponding step size, the differential, and integral linearity errors for the DAC whose outputs are shown in Figure 5.47. These results are found by using (4.3) and (4.6) described above.

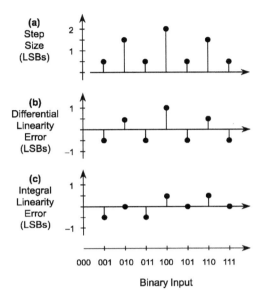

Figure 5.48 Linearity errors for 3-bit DAC shown in Figure 5.47.

Figure 5.49 shows an example of a nonmonotonic DAC transfer function with its corresponding terminal point fit. The corresponding linearity errors are shown in Figure 5.50. Note that in both Figures 5.48 and 5.50, the INL sums to zero at the largest binary input.

244   TRANSCEIVER DESIGN AND PRACTICAL CONSIDERATIONS

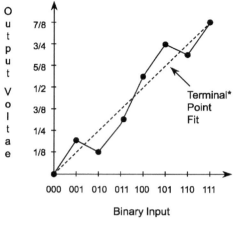

Figure 5.49   Example of a nonmonotonic DAC transfer function.

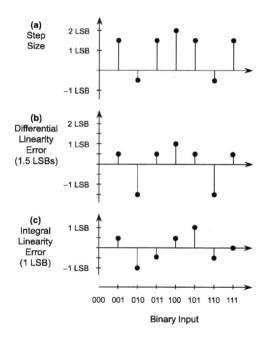

Figure 5.50   Linearity errors for the nonmonotonic 3-bit DAC shown in Figure 5.49.

5.9.1 Digital-to-Analog Conversion Models

A *weighted resistor DAC* is shown in Figure 5.51 and uses a summing amplifier in which the total input resistance is determined by the switch positions $a_1, \cdots, a_n$. For example, if $n = 3$, $a_1 = a_2 = 1$ (switch closed), and $a_3 = 0$ (switch open), $-V_r$ is across the resistance $2R$ in parallel with $4R$, or $4/3R$ and $V_0 = 3/4V_r$. This configuration can also be biased into a bipolar configuration using V_{ref} connected to the virtual ground through a resistor R_0. The current through R_0 changes the voltage at this node. Note that the bipolar configuration results in a different transfer function.

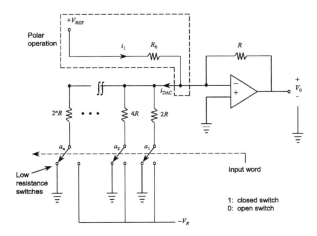

Figure 5.51 Weighted resistor digital-to-analog converter.

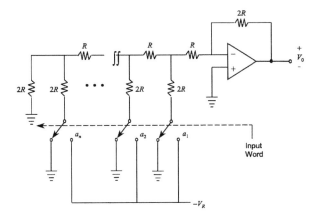

Figure 5.52 *R*-2*R* ladder DAC.

Since the output voltage for the weighted resistor DAC depends on the resistor ratios, it is important that the ratios are accurate. For example, a 10-bit DAC requires an accuracy of 1 in 1,024. Also, the current drawn from the reference supply varies depending on the binary input pattern, causing various superposition errors. An important advantage, however, of this DAC over other types is its ability to drive resistive loads with no need for a buffer. This property proves crucial if the DAC must drive a transmission line, as in wireline systems, or if the load contains a resistive component, as in displays and optical modulators, which are discussed in Chapter X.

For IC designs, the switch resistance should be low when the switch is ON. The source resistance should also be low. This low resistance is usually not a problem with good MOSFETS and JFETS. A key drawback of this architecture however, is the wide range of resistors required for moderate-to-large resolutions. This requirement precludes monolithic integration. The *R-2R ladder DAC* shown in Figure 5.52 also uses a summing amplifier and resistor values of R and $2R$. This configuration has the same transfer function but avoids requiring a wide range of resistor values [10].

In the above two DACs we note that the currents change as the binary input data changes. This change in current induces additional heat and power dissipation. To avoid the errors associated with this power dissipation, the *inverted R-2R DAC* is configured such that the currents flowing in the ladder and reference are independent of the digital input word. The inverted *R-2R* DAC is shown in Figure 5.53. More sophisticated designs can also be found but the above examples illustrate the important trade-offs that occur in DAC design.

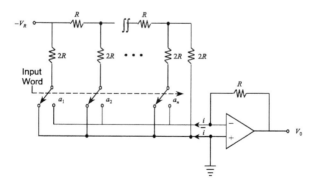

Figure 5.53 Inverted *R-2R* DAC.

5.10 TECHNICAL ISSUES AND LIMITATIONS

In this section a number of technical concepts are discussed as they relate to the DRFM. The first is the instantaneous bandwidth (IBW). The IBW is the bandwidth of the baseband processor and is subject to ADC/DAC technology as discussed before. For the SSB DRFM

$$IBW_{SSB} = a\frac{f_s}{2} \tag{5.99}$$

and for the DSB DRFM

$$IBW_{DSB} = af_s \tag{5.100}$$

where a is the available fraction of the ideal bandwidth ($0.5 < a < 1$) due to BPFs and the LPFs. The operating bandwidth (OBW) is the range of RF frequencies over which a signal may be received and processed. This is typically several times the IBW. The OBW is controlled by the LO. The fidelity is the number of bits instrumented by the DRFM using a larger number of bits minimizes spurious signals. The SNR for both the SSB and the DSB is $SNR = 6.02n + 1.76$ dB.

The bandwidth in a DRFM is extremely important. To achieve a large bandwidth, multiple (narrowband) DRFMs that are independently tunable can be used. Sometimes, however, a single DRFM with a wide operating bandwidth (OBW > 100 MHz) is required [7]. Applications include interception of frequency-hopped LPI waveforms, wideband chirped signals, and phase-coded P.C. signals. Wideband DRFMs can be a significant cost savings to EW systems by reducing the costly analog front ends.

When using a wideband DRFM, however, there is a high probability of multiple simultaneous signals being present. In jamming applications one of the problems is that some of the signals could be friendly. Also, the EA intended for one emitter could beacon to a second emitter. Another potential problem is that the signal suppression of a low-power high-priority threat can occur at the preconverter AGC amplifier or in the transmission TWT. Due to these types of problems, head-to-tail reconstruction of the received signal is (nearly) impossible. This head-to-tail reconstruction is made even more difficult by any phase noise present (raises noise floor). Other major drawbacks include the fact that any two-tone intermodulation is potentially detectable by the victim and also results in power wasted.

One potential solution to the multiple signal problem is to use a digital bandpass filter with 10–50 MHz bandwidth to isolate the desired signal. Figure 5.54 shows an example of a wideband DRFM spectrum.

248   TRANSCEIVER DESIGN AND PRACTICAL CONSIDERATIONS

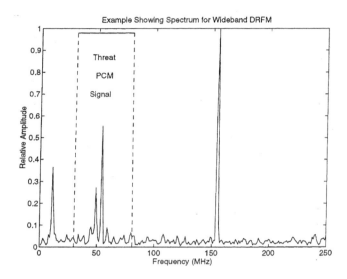

Figure 5.54   Spectrum for wideband DRFM. Note the pulse code modulated (PCM) threat signal to be eliminated by a wideband pass-reject filter [8].

In this situation, the threat signal of interest has an amplitude that is less than the signal at 150 MHz. A digital bandpass filter at 50 MHz can easily isolate this threat and exclude the higher amplitude signal. The processing rate requirements for DRFM applications using a standard FIR filter implementation are severe. Efficient filtering approaches such as multirate polyphase filters, interpolated FIR, and FFT based filters may allow practical implementations.

5.10.1   DRFM Problems

There are several problems associated with DRFM architectures. One of the most troublesome problem is the image frequencies. Image frequencies are caused by the ambiguity created in a superhet between the f_{IF} and its mirror image at $f_c - 2f_{IF}$ such that both frequencies are translated to the same location. For example, consider a carrier frequency $f_c = 850$ KHz, a local oscillator $f_{LO} = 1305$ KHz, and a third frequency $f = 1760$ KHz. The desired frequency is $f_{IF} = |850 - 1305| = 455$ KHz. However, due to the superhet process, the third frequency also resolves as $f_{IF} = |1760 - 1305| = 455$ KHz.

Unbalance between the quadrature channels can also cause image frequencies. The components that form the quadrature channels can contribute to amplitude or phase unbalance or both. Image suppression to 30 dBc requires a cumulative amplitude unbalance of less than 0.55 dB, a cumulative phase unbalance of less than 3.6 degrees or a combined phase and amplitude unbalance as shown in Figure 5.55.

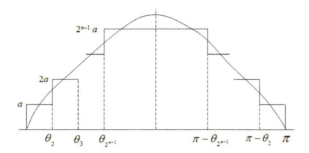

Figure 5.56 2^{n-1} amplitude levels in positive half of a sinewave.

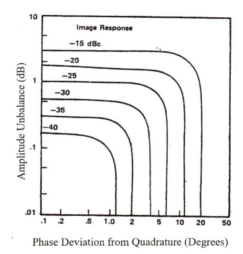

Figure 5.55 Image response as a function of the amplitude imbalance and phase deviation from quadrature.

Spurious frequencies in the DRFM output can also be due to nonlinear mixer characteristics. The spurious frequencies in this case are $f_{IF} = |Mf_{RF} - Nf_{LO}|$ where $M, N \in \{0, \pm 1, \pm 2, \cdots\}$. For the fundamental output frequency, $M = N = 1$. Another problem, especially in electronic attack applications, is LO leakage. LO leakage is characterized by a CW emission in the center of the IBW. This emission provides a beacon signal and results from inadequate isolation in the up-conversion process (the mixers only provide approximately 60 dB isolation). Other problems can be characterized as tuning errors resulting from an instability of the LO that causes spectral spreading. Phase noise and clock jitter also raise the noise floor. Another source of spurious frequencies is the digitizing process in an n-bit DRFM. The spurs (harmonics) can be detected and used to identify the DRFM presence that is preventing

250   TRANSCEIVER DESIGN AND PRACTICAL CONSIDERATIONS

the DRFM from disturbing the sensor track loop. The digitization spurs are created by the quantization of the RF signal into discrete levels. Recall that a DRFM with n-bits has 2^{n-1} amplitude levels in the positive half cycle [18]. With positive and negative half cycle symmetry, only odd harmonics are present. The 2^{n-1} amplitude levels in the positive half of a sine wave are shown in Figure 5.56. The spurious amplitudes can be calculated as [19]

$$b_m = \frac{2}{\pi} \left[\int_0^{\theta_2} a \sin mx\, dx + \int_{\theta_2}^{\theta_3} 2a \sin mx\, dx \right.$$

$$+ \cdots + \int_{\theta_{2^{n-1}}}^{\pi - \theta_{2^{n-1}}} (2^{n-1}) a \sin mx\, dx$$

$$\left. + \int_{\pi - \theta_2}^{\pi} a \sin mx\, dx \right]. \quad (5.101)$$

For a $n = 1$-bit DRFM

$$b_m = \frac{2}{\pi} \int_0^{\pi} a \sin(mx)\, dx, \quad (5.102)$$

or

$$b_m = \frac{4a}{m\pi}. \quad (5.103)$$

For $n = 2$-bits

$$b_m = \frac{4a}{m\pi} [1 + \cos m\theta_2], \quad (5.104)$$

and for $n = 3$-bits,

$$b_m = \frac{4a}{m\pi} [1 + \cos m\theta_2 + \cos m\theta_3 + \cos m\theta_4]. \quad (5.105)$$

In reducing the spurs, we are concerned with the level a and the transition angles θ. We would like to find the threshold values such that the staircase is a best fit to the reference sinewave. To do this we perform a sum squared error (SSE) minimization of the difference between the staircase and the sinewave [19]. The error between the staircase level and the sinewave is shown below:

$$0 < \theta < \theta_2 \qquad e = a - \sin\theta$$

$$\theta_2 < \theta < \theta_3 \qquad e = 2a - \sin\theta$$

$$\theta_3 < \theta < \theta_4 \qquad e = 3a - \sin\theta$$

$$\vdots$$

$$\theta_{2^{n-1}} < \theta < \frac{\pi}{2} \qquad e = (2^{n-1})a - \sin\theta$$

The SSE can be shown to be

$$SSE = \frac{1}{4}\sum e^2 = a^2\pi 2^{2n-3} + \frac{\pi}{4} - 2a$$
$$-a^2[3\theta_2 + 5\theta_3 + 7\theta_4 + \cdots + (2^n-1)\theta_{2^{n-1}}]$$
$$-2a(\cos\theta_2 + \cos\theta_3 + \cos\theta_4 + \cdots + \cos\theta_{2^{n-1}}). \quad (5.106)$$

For a minimum SSE, the partial derivatives w.r.t. $\theta_2, \theta_3, \theta_4, \cdots$ must all be equal to zero providing the following relationships:

$$\sin\theta_2 = \frac{3a}{2}$$
$$\sin\theta_3 = \frac{5a}{2}$$
$$\vdots$$
$$\sin\theta_{2^{n-1}} = \frac{(2^n-1)a}{2}. \quad (5.107)$$

The partial derivatives w.r.t. a must also be equal to zero and gives

$$a\pi 2^{2n-3} - 1 - a(3\theta_2 + 5\theta_3 + 7\theta_4 + \cdots + (2^n-1)\theta_{2^{n-1}})$$
$$-(\cos\theta_2 + \cos\theta_3 + \cos\theta_4 + \cdots + \cos\theta_{2^{n-1}}) = 0. \quad (5.108)$$

This results in 2^{n-1} transcendental equations and 2^{n-1} unknowns. An iterative method for solving the set of transcendental equations is shown below:

Step (1) Choose an initial estimate of a equal to 2^{1-n}.

Step (2) From (5.107), calculate $\theta_2, \theta_3, \theta_4, \cdots, \theta_{2^{n-1}}$.

Step (3) Solve for $(\cos\theta_2 + \cos\theta_2 + \cdots + \cos\theta_{2^{n-1}})$.

Step (4) Substitute variables into (5.108), which should produce a positive result (not zero).

Step (5) Decrease estimate of a by 20% and solve (5.108). This should produce a negative result indicating the desired root is between a and $0.8a$.

Step (6) By observing the sign of (5.108), and varying a between $0.8a$ and a it is possible to obtain an a that satisfies (5.108) to any required accuracy. Once a has been determined to the required accuracy, $\theta_2, \theta_3, \theta_4, \cdots$ can be determined using (5.107).

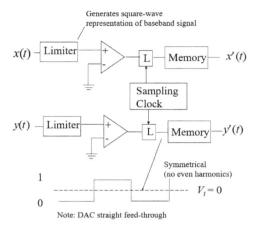

Figure 5.57 Block diagram of a 1-bit DRFM (magnitude only A/D).

5.10.2 Spurious Signals and Their Dependence on Bit Resolution

5.10.2.1 1-Bit Magnitude Example

A block diagram of a 1-bit magnitude DRFM is shown in Figure 5.57. In this configuration, the limiter at the input generates a square wave representation of the baseband signals $x(t)$ and $y(t)$. The comparator threshold $V_t = 0$ generates a symmetrical output about zero (no even harmonics). If the sampling time, however, is at unequal time intervals (sampling jitter) even harmonics will be generated. Sampling at the latch also forms intermodulation products between each square wave harmonic and multiples of the sampling clock. Every harmonic of a square wave is folded into the baseband spectrum. To examine the spurious frequencies, the Fourier series expansion of the square wave is given as

$$f(x) = \frac{4a}{\pi}\left[\sin x + \frac{1}{3}\sin 3x + \frac{1}{5}\sin 5x + \cdots\right] \quad (5.109)$$

where x is the fundamental, $3x$ is the third harmonic, $5x$ is the fifth harmonic, and so forth. From this expression it is clear that the output spectrum is monotonically decreasing with the third harmonic being the worst (-9.5 dB). The total spur power is

$$\left(\frac{1}{3}\right)^2 + \left(\frac{1}{5}\right)^2 + \left(\frac{1}{7}\right)^2 + \cdots = \frac{\pi^2}{8} - 1 = 0.234 = 23.4\% \quad (5.110)$$

5.10 Technical Issues and Limitations 253

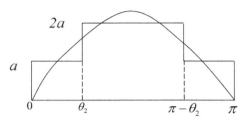

Figure 5.58 2-bit DRFM levels during a half cycle of a sine wave.

5.10.3 2-Bit DRFM Example

To examine the spurious frequencies in a 2-bit DRFM, we recall the 2-bit DRFM levels during a half cycle of a sine wave as shown in Figure 5.58. The additional bit resolution offers a better approximation to the sine wave than a square wave but by how much? To answer this we examine the relationship for the mth harmonic as

$$b_m = \frac{4a}{m\pi}[1+\cos m\theta_2]\Big|_{\substack{a=0.4188161 \\ \theta = 38.92°}}$$

or

$$b_m = \frac{0.53325}{m}[1+\cos m 38.92°]. \qquad (5.111)$$

The ratio of the mth harmonic to the fundamental is

$$\frac{b_m}{b_1} = \frac{1}{1.778m}[1+\cos m 38.92°]$$

or in dB

$$20\log\left(\frac{b_m}{b_1}\right) = 20\log\left\{\frac{1}{1.778m}[1+\cos m 38.92°]\right\} \text{ dB}. \qquad (5.112)$$

The spurious amplitudes for a 2-bit DRFM are shown in Figure 5.59. The total spur power equals 5.44% of the fundamental power (first 1000 harmonics). Note that for a 2-bit DRFM the spurious amplitudes are not monotonically decreasing. The worst spur occurs at the ninth harmonic (-18.12 dB). The levels for a 3-bit DRFM during a half cycle of a sine wave is shown in Figure 5.60. The spurious amplitudes are shown in Fig. 5.61. The total spur power contained in the first 1000 harmonics for a 3-bit DRFM is 1.17% of the fundamental. The spur amplitudes for both a 4-bit and a 5-bit DRFM are shown in Figure 5.62 and 5.63, respectively. The total spur power for the 4-bit DRFM equals 0.27% of the fundamental (first 1000 harmonics). The total spur

254   TRANSCEIVER DESIGN AND PRACTICAL CONSIDERATIONS

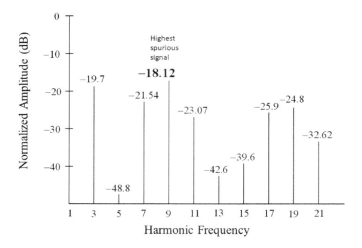

Figure 5.59   Spur amplitudes for a 2-bit DRFM.

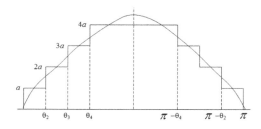

Figure 5.60   3-bit DRFM levels during a half cycle of sine wave.

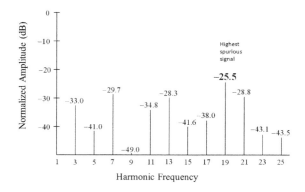

Figure 5.61   Spur amplitudes for a 3-bit DRFM.

5.10 Technical Issues and Limitations 255

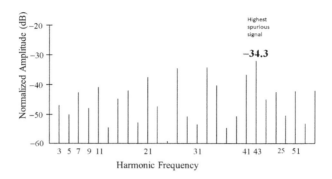

Figure 5.62 Spur amplitudes for a 4-bit DRFM.

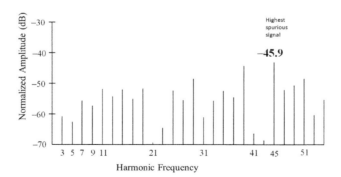

Figure 5.63 Spur amplitudes for a 5-bit DRFM.

Table 5.5
Summary of Spurious Performance

No. of Bits	Worst Spur	Worst Spur Level (dB)	Total Spur Power as % of Fundamental Power
1	3rd	−9.5	23.40
2	9th	−18.1	5.44
3	19th	−25.5	1.17
4	43rd	−34.3	0.27
5	45th	−45.9	0.063

power for the 5-bit DRFM equals 0.063% of the fundamental (first 1000 harmonics). The results for $m = 1$ to 5 for the DRFMs are summarized in Table 5.5. A general rule of thumb is that the highest spur level is approximately nine times the number of bits (in dB) down from the fundamental.

5.10.4 Reducing Harmonic Spurs

In this section we address the reduction of spurs. There are several methods to reduce the spurious response of a DRFM. One particular method involves dithering the LO/input signal with random noise. This disburses the spurs over the IBW but is a cosmetic solution. That is, the actual spurs are not reduced, just harder to observe on a spectrum analyzer. Also, the total spur power is the same. Another method is to use multiple bit DRFMs as discussed previously. The use of multiple bits, however, restricts the bandwidth of the device. The use of multiple bit DRFMs also expands the storage requirements that are needed.

Another important method of reducing the harmonic spurs is to purposely introduce a certain amount of intentional time jitter which is applied during the sampling process. That is, instead of sampling at exact discrete multiples of the sampling period T, we take the samples at $t_k = kT + \Delta$. Time-jitter sampling analysis was discussed in Chapter 4. Figure 5.64(a) shows the spectral density of a 3-bit DRFM with a full dynamic range input signal with no ADC jitter added [19]. The signal frequency is $f_T = 400$ MHz and the sampling frequency is $f_s = 1280$ MHz. Note the presence of the spurious signals, the largest being −22.9 dB. To reduce the spurious signals, a 1% jitter is added (1% of the sampling period) as shown in Figure 5.64(b). The jitter is uniformly distributed in the range ±7.8125 ps. The largest spur is not the same as that shown in Figure 5.64(a) and is −27.9 dB down. That is, the spurious response is improved by 5 dB. Note, however, that the noise floor is also higher. To use

5.10 Technical Issues and Limitations 257

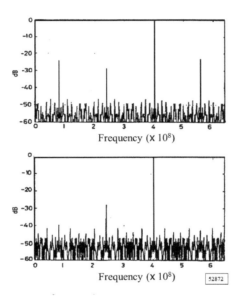

Figure 5.64 Spectrum of 3-bit DRFM with full dynamic range input sinusoid ($f_s = 1280$ MHz, $f_T = 400$ MHz) (a) without jitter and (b) with 1% jitter. (©IEEE, reprinted with permission from [19].)

this method, the designers need to choose the minimum jitter that reduces the worst harmonic power to an acceptable value.

Another technique that can be used in the case of a 1-bit DRFM is based on the relationship between the level of the odd harmonics of the rectangular wave and its duty cycle. In this method the duty cycle of the rectangular wave is adjusted to $1/M$ to eliminate any harmonic spur M at the expense of generating even harmonics. The phase of the even harmonics reverses when the duty cycle goes above 0.5 to below 0.5 (phase of odd harmonics unchanged). When the duty cycle is modulated about 0.5, the energy is transferred from the odd to the even harmonics. The even harmonics are phased out by the reverse duty cycle (over a finite period of time in the victim receiver). The total spur power is then spread over the IBW.

5.10.5 Mixers and the RF Up / Downconversion Process

RF mixers are an important signal processing component and used for many reasons. They are 3-port devices and used for microwave and millimeter wave processing of RF signals and come as *balanced, double balanced*, and *triple balanced*. The mixer is shown in Figure 5.65.

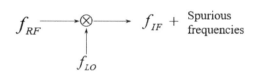

Figure 5.65 Mixer schematic diagram showing the production of spurious frequencies.

The mixer's function is to change or mix the frequency of an electromagnetic waveform while preserving every other characteristic such as phase and amplitude of the input signal. Most often the mixer is used to covert the input RF signal frequency to an intermediate frequency and also to upconvert a baseband frequency to an RF frequency for transmission. The IF frequency f_{IF} is a result of the mixer's nonlinear behavior as $f_{IF} = |Mf_{RF} - Nf_{LO}|$ where $M, N \in \{0, \pm 1, \pm 2, \cdots\}$ where $M = N = 1$ is the fundamental output frequency. The upconversion and downcoversion processes are shown in Figure 5.66.

5.10 Technical Issues and Limitations

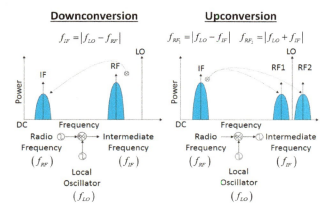

Figure 5.66 Downconversion process and upconversion process using mixers with input frequency f_{RF} with local oscillator frequency f_{LO} and intermediate frequency f_{IF}.

The downconversion process from RF to IF and the upconversion process from IF to RF1 and RF2 using mixers is realized if *direct sampling* at the antenna/LNA is not used.[1] Downconversion is typically accomplished in two stages to avoid overloading the RF and IF amplifiers used while three steps is overkill.

Figure 5.67 shows the schematic of a single diode mixer design.

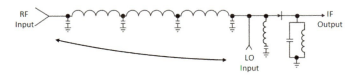

Figure 5.67 Schematic of a single diode (unbalanced) mixer design.

Note in this design, as illustrated, there is no isolation between the RF port and the LO port. Use of this mixer therefore requires external filters to isolate any applied output, LO, or RF signals or else they become beacons at the output. Use of this design is also recommended only for narrowband designs and is difficult to employ for wideband intercepts.

A balanced mixer is shown in Figure 5.68. A diode bridge in the center of the device provides a good deal of common-mode noise cancellation. In addition, a good number of the IMD products are isolated and further suppressed. The drawback to using this device is that a higher LO drive level is required.

---

1 The direct sampling at IF is the preferred method; however, if the RF bandwidth of the signal is too high and the preferred ADC, DAC technology is not available, then analog downconversion and upconversion is warranted.

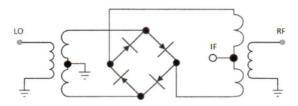

Figure 5.68 Schematic of a balanced bridge mixer design.

The first major concern is the mixer's *conversion loss*. The conversion loss, typically measured in dB, is a measure of how efficiently a mixer converts energy from the input frequency to the output frequency. It is measured as the difference in signal level between amplitude of input signal and desired output signal level and is defined as the ratio of the IF output power to the RF input power. Typical values of conversion loss range between 4 to 10 dB and are a result of transmission line losses, balun mismatch, diode series resistance, and mixer imbalance. In addition, the wider the frequency range the larger the conversion loss typically is. The *noise figure* is the next major concern and is defined in the usual manner as the amount of noise added to the output above and beyond the thermal noise of the device. For a passive mixer, the noise figure is approximately equal to the mixer's loss. Another serious concern is the signal *isolation*. The isolation is defined as the amount of power leakage from one port to another. When isolation is high, the leakage between the ports or the power that leaks from one port to another port will be small. The isolation is a measure of the circuit balance within the mixer. When the isolation is high, the amount of leakage or feed-thru between the mixer ports also will be very small. As an example, consider Figure 5.69.

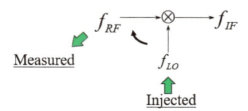

Figure 5.69 Example of LO to RF isolation calculation.

With an injection of f_{LO} of +15 dBm. If a power measurement at f_{RF} port measures -20 dBm, the LO to RF isolation is $+15 - (-20) = 35$ dB. Typical mixers today have the best isolation on the order of 60 dB. Also of concern are *image frequencies*. Image frequencies are a concern in any RF communication or

5.10 Technical Issues and Limitations

radar system. These are ambiguities created in a mixer super heterodyne architecture between f_{IF} and its mirror image at $f_c - 2f_{IF}$ such that both frequencies are translated to the same location. Recall the IF frequency after quadrature conversion of an RF frequency f_{RF} using a local oscillator frequency f_{LO} is $f_{IF} = |f_{LO} - f_{RF}|$. As an image frequency example, consider $f_c = 850$ kHz with $f_{LO} = 1,305$ kHz and consider an arbitrary signal $f = 1,760$ kHz. The desired signal would be $f_{IF} = |850 - 1,305| = 455$ kHz. The image frequency however, would also be $f_{IF} = |1,760 - 1,305| = 455$ kHz.

Another major issue that we've discussed is the 1-*dB compression point*, P1dB, as illustrated in Figure 5.70. Note the region where the output power is linearly related to the input power. However, there is a point where intermodulation products and other nonlinearities cause a divergence from linear operation. The 1-dB compression point is where the output diverges by 1 dB and is approximately 4 to 7 dB below the mixers minimum recommended LO drive level.

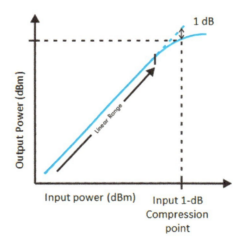

Figure 5.70 Illustration of the 1-dB compression point and the linear range of the mixer.

One of the most serious problems is the intermodulation distortion caused by two-tones called third-order intermodulation distortion. This is most serious since the products that are generated (due to the mixer's nonlinear behavior) fall within the IF bandwidth. For example, consider f_{RF_1} and f_{RF_2}. The third-order intermodulation products f_{if_1} and f_{if_2} causing interference at the IF port are

$$f_{if_1} = 2f_{RF_1} - f_{RF_2} - f_{LO}$$

and

$$f_{if_2} = 2f_{RF_2} - f_{RF_1} - f_{LO}$$

These are the most serious since they are close to the desired IF output frequency. In addition, filtering cannot remove these unwanted distortion products. However, a mixer can possess the ability to suppress these products. Suppression of these third-order intermodulation products is evidenced by the (input referred) third-order intercept point IIP3 as illustrated in Figure 5.71.

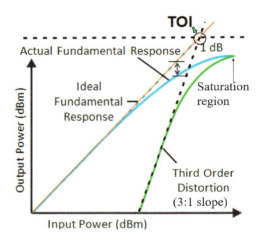

Figure 5.71 Illustration of the third-order distortion, the actual response and the third-order intercept (TOI) point.

Prediction of the mixer's nonlinear behavior can be accomplished by changing the amplitude. That is, any 1 dB change the input amplitude will cause a 3-dB jump in the third-order products.

5.11 PHASE ANGLE SAMPLING SYSTEM: QUANTIZATION NOISE

The quantization noise performance of a phase sampled system is very dependent on the nature of the input signal. For sinusoidal input signals the quantization noise appears as spurious signals at harmonic frequencies aliased into the system bandwidth. For an ideal n-bit system, spurious signals occur at frequencies $\omega = (1 \pm m2^n)\omega_c = H\omega_c$ where m is any integer $\neq 0$. The amplitude of the spectral components has a $\sin x/x$ relationship is given by

$$\frac{F}{H\omega_c} = \frac{\sin\left(\frac{H\pi}{2^2}\right)}{\left(\frac{H\pi}{2^n}\right)}. \tag{5.113}$$

5.11 Phase Angle Sampling System: Quantization Noise

Table 5.6
Peak Spurious Signal for a Range of Phase Quantization Schemes

n-bits	H_{max} (No. of the largest harmonic present)	$F(\omega_c)$ (amplitude of the fundamental)	$F(H_{max}\omega_c)$ (amplitude of largest harmonic)	Ratio dB
1	−1	0.6366	0.6366	0
2	−3	0.9003	0.3001	−9.5
3	−7	0.9745	0.1392	−16.9
4	−15	0.9936	0.0662	−23.5
5	−31	0.9984	0.0322	−29.8
6	−63	0.9996	0.0159	−36.0
7	−127	0.9999	0.0079	−42.1
8	−255	1.0000	0.0039	−48.1

Table 5.6 indicates the level of the largest spurious signal when a sinusoidal signal of frequency ω_c is represented by an ideal phase sampled system of n-bits. Here H_{max} is the number of the largest harmonic present, $F(\omega_c)$ is the amplitude of the fundamental component, and $F(H_{max}\omega_c)$ is the amplitude of the largest harmonic. The primary source of spurious signals is likely to be the simultaneous reception of multiple signals.

5.11.1 Comparison Between Amplitude and Phase Sampling

A comparison between the *phase sampling transceiver* architectures and the *amplitude analyzing* transceiver architectures is shown in Table 5.7. The percent power in the spurious signals is shown as a function of the number of bits. Also compared is the SNR

Table 5.7
Spurious Signal Level Summary

n-bits	No. levels	dB amplitude quantization	% power spurs **amplitude** quantization	SNR (dB) phase quantization	% power spurs **phase** quantization
1	2	6.3	18.9		
2	4	13.1	4.47	6.3	18.9
3	8	19.5	1.11	12.8	5.04
4	16	25.6	0.275	18.9	1.28
5	32	31.7	0.0684	24.9	0.321
6	64	37.7	0.0170	31.0	0.0803
7	128	43.8	0.00421	37.0	0.201
8	256	49.8	0.00104	43.0	0.00502

5.12 DIGITIZATION FIGURES OF MERIT

Architectures continue to morph and refine to address application pull and process technology push. The two main performance specifications are conversion speed and dynamic range. Secondary to these are the power requirements and power dissipation. The flash, pipeline, SAR, and $\Delta\Sigma$ (and their hybrids) are the architectures with the most in-roads to applications with interleaving at the forefront of making them go as fast as possible (discussed in the next chapter). The FOM lumps several performance metrics into a single number, thereby creating a proxy for the overall efficacy of a circuit or device. The main parameters of data converters for DRFM transceiver applications are sampling rate, analog bandwidth, spurious free dynamic range, ENOB, and power consumption.

For CMOS data converters, the analog bandwidth and the ENOB are bottlenecks for increasing the data rate while interleaving can provide an in-road to giving the speed necessary for the applications at hand. To compare the various approaches, two figures of merit have been proposed and are consistently reported typically on the Walden chart after R. H. Walden's publication [20]. The FOM_W typically reported as the Walden FOM can be expressed as

$$FOM_W = \frac{Power}{f_s 2^{ENOB}} \quad \text{Joules/conversion step} \quad (5.114)$$

where f_s is the sampling frequency and *Power* is the power used by the device. This is sometimes written as

$$FOM_W = \frac{Power}{2 \times ERBW} \quad \text{Joules/conversion step} \quad (5.115)$$

where *ERBW* represents the effective resolution bandwidth [21]. The FOM_W typically lies between $100 < FOM_W < 10$ femto-joules/conversion step.

A second FOM is called the Schreier FOM, FOM_S and has two different forms. The first is for the DR (dynamic range) and the second is for the SNDR. For the DR

$$FOM_S(DR) = \text{DR (dB)} + 10\log\left(\frac{BW}{Power}\right) \quad \text{dB} \quad (5.116)$$

and typically lies between $160 < FOM_S < 140$ dB. The second FOM for SNDR is [22]

$$FOM_S(SNDR) = \text{SNDR (dB)} + 10\log\left(\frac{f_s/2}{Power}\right) \quad \text{dB} \quad (5.117)$$

It is important to note that stand-alone commercial ADCs have far worse FOMs than IP blocks that are reported in the academic literature and in the updated Walden tables (now the Murmann tables from Stanford University) shown in Figure 5.72. The FOM plot clearly shows that there is an inflection point just below 1 GHz where the FOM rises nearly one order of magnitude for each order of magnitude of increasing input frequency. It is not fair to compare the FOM of an ADC core with that of a commercially available part [23]. The example given is the AD9213 with a sample rate of 10 GS/s and an SNDR of 52 dB and ENOB of 8.3 bits for a 4-GHz incoming frequency, but consumes 4 W of power. That power as noted is for the entire packaged device with all of the internal and external support circuits that are loading it down.

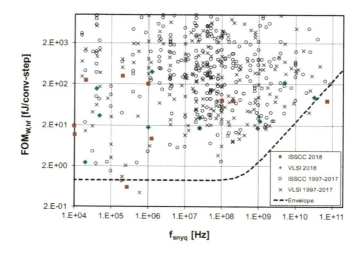

Figure 5.72 The Walden/Murmann FOM plot from 2017. (©IEEE, reprinted with permission from [24].)

5.12.1 DRFM ADC Requirements

For intercepting the pulsed/beamforming radar system the receiver requires ADCs with a $\geq$8-bit resolution, a $\geq$2-GS/s sampling rate, and a large input bandwidth ae required. The backplane/Ethernet receiver requires even higher sampling rates (e.g., 28 GS/s for PAM4 56-Gbps systems) with BER. Although time interleaving is inevitable to achieve such sampling rates, higher speed per sub-ADC is still preferred to avoid massive interleaving and to simplify the clock distribution and necessary calibration of the interleaving artifacts. The successive approximation register (SAR) and flash ADCs are two of the most common candidates in this regime. For the SAR ADC, its excellent power and area efficiencies are the main reasons why they are suitable for such applications. Meanwhile, the input loading of the SAR ADC is usually smaller than that of the flash ADC, making it easy to drive. However, its low conversion speed due to the looped operation and poor BER due to metastability result in severe challenges for SAR ADC arrays.

In contrast, the flash ADCs show advantages in speed and BER performance in that its higher conversion speed obviates massive interleaving while open-loop latch pipelining effectively reduces the BER of the flash ADC without significant speed degradation. What makes the flash ADC less attractive is its large power consumption, area, and input loading due to the exponential dependence of the comparator number on the resolution. For conventional gigahertz flash ADCs, the resolution is usually limited to 6–8 bits, even when signal folding is applied. Digitally assisted integrated circuit design has gained popularity in recent years due to challenges associated with analog components in mixed-signal SOC and miniaturization of transistors.

These challenges include process variations, mismatches, and increased demand for circuit complexity for designs in submicron fabrication process technologies. In flash ADCs, the offsets are typically random variables with normal distributions and have a direct impact on an ADC's performance, worsening its integral and differential nonlinearities (INL and DNL, respectively). On-chip calibration techniques aid to address these issues by reducing the sensitivity of analog circuit designs to variations.

5.12.2 Direct Sampling Signal-to-Noise Considerations

Figure 5.73 shows the direct RF sampling architecture used with one particular configuration of a DRFM AESA antenna interface. The first digital (direct sampled, no analog downconversion) UHF receiver deployed was for the E-2C Hawkeye airborne radar [8]. New efforts now are also investigating HBTs in InGaP/GaAs and SiGe. In an active array, each antenna element requires its own ADC and DAC directly sampling at the instantaneous bandwidth. The antenna output is first processed by a band-select filter to place the sampling at the proper frequency band within the EMS. The output of the band-select filter is then amplified by a LNA. Prior to the ADC is a anti-aliasing filter with a sample-and-hold (shown together).

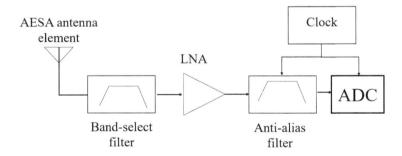

Figure 5.73 Direct RF sampling architecture.

If direct sampling cannot be done, each T/R module then requires its own upconversion and downconversion stage leading to increased design costs, size, and variation in performance. That is, you can reduce costs, size, and complexity by using a direct RF sampling architecture by not using the mixer and LO. With such a large array of transmitters and receivers, direct RF sampling architectures can significantly increase channel density and reduce the cost per channel. Below we discuss a basic model to estimate the dynamic range of the direct receiver as a function of the ADC sample rate and basic parameters of the critical components.

The main factors limiting the SNR include the jitter on the sampling clock (typically derived from a crystal oscillator) and the phase noise on the local oscillator, noise in the ADC converter and jitter in the ADC sample and hold circuit. The best achievable ADC resolution depends on its power consumption and sample rate [8]. The useful FOM that relates all three variables (repeated here) is

$$FOM_W = \frac{P}{2^{ENOB} f_s}$$

where P is the power consumption of the ADC, $ENOB$ is the effective number of bits in the ADC, and f_s is the ADC sampling rate. The above FOM has units of joules per conversion and a smaller number indicates a higher-performance ADC. Current state-of-the-art ADC converters achieve a FOM as shown in Figure 5.72. Values between 4 pJ/conversion to less 1 pJ/conversion for sample rates below 100 MHz have been achieved with a power consumption under 1W. The SNR in a signal bandwidth of B Hz can be expressed as

$$SNR_{ADC} = 20\log_{10}\left(\frac{1}{FOM_W}\right) + 1.76 - 20\log_{10}(f_s) + 10\log_{10}\left(\frac{f_s}{2B}\right)$$

The phase noise in the receiver is due to the LO/frequency synthesizer combination. The phase noise on the sampling clock is modulated onto the incoming signal after being amplified by the subsampling factor. Consequently, the dynamic range of the receiver will be limited by the SNR of the sampling clock signal according to

$$SNR_S = \left(\frac{f_s}{f_i}\right)^2 SNR_{clk} \qquad (5.118)$$

where f_i is the ADC input frequency and SNR_{clk} is the SNR of the sampling clock. The noise power on the clock signal is integrated over the entire bandwidth of the ADC, which can be several GHz (e.g, $B = 2$ GHz).

The noise floor from a high-quality crystal oscillator (e.g., 100 MHz) is typically around -180 dBc/Hz. For higher sampling frequencies the crystal oscillator has to be multiplied ($\times N$) increasing the phase noise by $20\log(N)$. Consequently, the overall SNR of the clock is given by [25]

$$SNR_{clk} = 180 - 10\log_{10}(2 \times 10^9) - 20\log_{10}(f_s/100 \times 10^6) \qquad (5.119)$$

assuming a 2-GHz bandwidth. The SNR limitation imposed by the phase noise on the sampled signal is then

$$SNR_s = 20\log_{10}\left(\frac{f_s}{f_i}\right) + 247 - 20\log_{10}(f_s) + 10\log_{10}(f_s/2B) \qquad (5.120)$$

or
$$SNR_s = 247 - 20\log_{10}(f_i) + 10\log_{10}(f_s/2B). \tag{5.121}$$

The sample-hold circuit will degrade the SNR by introducing timing jitter that modulates the noise onto the signal. The SNR of the sampled signal is related to the standard deviation of the sample-hold jitter σ_t as $-20\log_{10}(2\pi f_i \sigma_t)$ resulting in the dynamic range of the sampled signal as

$$SNR_s = -20\log_{10}(2\pi f_i \sigma_t) + 10\log_{10}(f_s/2B). \tag{5.122}$$

Consequently, the SNR of the clock depends on the sampling frequency.

5.12.2.1 Key Considerations

Flash ADCs are key components in DRFMs and many applications require very high sampling rates and low resolutions such as data-storage read channels and wired communication systems [26]. As the demand for high data rates increases, receiver front ends that depend on high-speed (multi-GS/s) low-resolution ADCs are being considered in order to fully exploit the benefits of sophisticated digital signal processing techniques. However, the large area and high power consumption of fast ADCs are major concerns.

At mm-wave clock frequencies, skew due to mismatch in the clock and data distribution paths is a significant challenge for both flash and time-interleaved converter architectures. A full-rate front-end THA may be used to reduce the effect of skew [27]. However, it is found that the THA output must then be distributed to the comparators with a bandwidth greater than the sampling frequency in order to preserve the flat regions of the track and hold waveform. Instead, if the data and clock distribution have very low skew, the THA can be omitted thus obviating the associated nonlinearities resulting in improved performance SNDR when the input is a sinusoid. The SNDR degradation at high input frequencies is dominated by the following:

1. Signal-frequency-dependent spurious tones at the THA output;

2. Aperture jitter limitation including external clock jitter (approximately 500 fs) and internal clock distribution and thermal jitter;

3. Timing misalignment at the encoder logic;

4. Increasing second harmonic distortion, which may come from the single-ended output measurement and the phase error from a 180° hybrid coupler used for differential input at the test setup.

5.13 DRFM TRANSCEIVER DESIGN CHALLENGES

5.13.1 Electronic Support

Today with a steady increase in sophistication of radar systems the DRFM transceiver designer faces a lot of challenges. Pulse-to-pulse PRI agility, frequency agility, and the numerous complex modulations make it difficult to detect, identify, and classify specific emitters. Although wideband spectrum sensing capabilities are possible, many reports can be generated for a single emitter appearing in the bandwidth. Now with many communication CW signals within the radar spectrum, the typical intercept receiver can be blinded. That is the emitters that the DRFM was designated to intercept, they are now no longer able to handle.

In addition, interference from modern radars transmitting from friendly platforms is a significant issue for wideband EMS sensing transceivers. In addition, when the transceiver processes signals outside its bounds or near its internal thresholds, this often can cause a reset, bringing the transceiver system off-line for several minutes.

Another significant problem is the failure to intercept the threat signal that is present. The DRFM transceiver as an ES asset must be able to intercept both pulsed radar signals and CW signals within a very wide band (e.g., .5 to 300 GHz). The presence of ultrawideband sources can present a problem by raising the noise floor of the transceiver. Spread spectrum communication signals, impulse jammers, and impulse radar systems can pose a significant problem. One of the most important considerations is that the transceiver must also have the capability to detect high power microwaves (HPM) and include in the design high power detection and protection circuits at the front end tuner. The transceiver front end must protect itself against microwave weapons, directed energy weapons, and electromagnetic impulse weapons such as an HPM E-bomb warhead. Depending on the type of HPM device used, the weapon could be integrated with an existing cruise missile, but could also be used with an existing glide bomb kit such as the JDAM as shown in Figure 5.74.

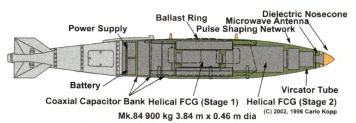

Figure 5.74 HPM E-bomb warhead in a GBU MK-84 form factor. (From [28].)

If the power density is extremely high, burnout will occur. Burnout is the physical damage to the electronics within the system. This level of energy will not just cause a reset but will render the DRFM inoperable. If a large power density just below burnout occurs, this will cause an *upset*. An upset is the temporary disruption of memory or logics in the electronic system. This level of power will critically fail the system where it will be of little or no use until reloaded. The next level below upset is EA. This effectively blinds the receiver to any microwave or RF radar communication signals in the area. The level below that is considered *deception*. Deception is the spoofing of the system into mission failure. These are discussed later in the text. The DRFM must have the ability to reject these unwanted signals and this is just as important of a capability as it is to process the signals of interest. The use of adaptive notch band reject filters at the front end along with limiters can be used to protect the weak components (such as the LNA) from stray signals as shown in Figure 5.75(a). Yttrium iron garnet (YIG) filters such as those used in microwave, acoustic, optical, and magneto-optical applications can be used.

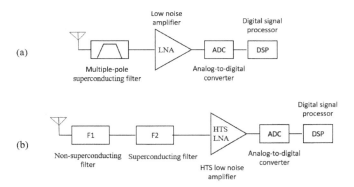

Figure 5.75 Transceiver showing (a) front end using a HTS superconducting filter for protection against HPM weapons and (b) use of a HTS and non-HTS filter combination.

The effects of HPM attacking the back door of an LNA have been examined in [29]. They conclude that if no shielding measures are taken, the performance of the receiver will be degraded under HPM. The degradation is caused by the deterioration of the noise factor (NF) and gain of the LNA. The gate terminal of the LNA is more sensitive to interference signals than the drain terminal so it's necessary to adopt shielding measures and pay particular attention to the LNA in the protection of the gate terminal. In addition, other active microwave RF circuits and digital circuits, such as active mixers, IF amplifiers, power amplifiers, and operational amplifiers should also be protected. As well, significant protection should be put on the signal input ports.

Adaptive thresholds can also increase the receiver's sensitivity. The use of high temperature superconductivity (HTS), low temperature superconductivity (LTS), and cryogenic components at the front end is a viable alternative as shown in Figure 5.75(a). In addition, combining HTS filters and non-HTS filters can also be useful as shown in Figure 5.75(b). In this manner the input to the DRFM will only pass the desired signals, reject interference from nearby sources at frequencies close to the desired frequency resonators and tunable filters and resonators coupled together can present a unique design. The filter network combining nonsuperconducting and superconducting filters is a good design challenge.

5.14 SUMMARY

Commercial technology is currently bringing into play the latest advantages such as modularity and the utilization of integrated FPGA devices to provide low-latency, real-time digital signal processing. The signal can then be converted back to analog and transmitted. This enables real-time processing to mitigate electronic threat emitters and jammers. In addition, the integration of open-systems architectures (e.g., OpenVPX) allows the platform user the ability to leverage their own intellectual property and algorithms. The semiconductor technology that is leveraged onto the transceiver design components allows a growth capability in addition to enabling mission-specific customization and ruggedization. The open systems architecture allows minimal integration time and insertion of rapid upgrades to mezzanine cards when the next-generation technology becomes available.

Future DRFM transceiver missions of signal collection and processing will require AESA architectures that can directly sample at the front end with a transceiver at each antenna element. Clearly, this is an enormous amount of heat and DC power that must be supplied to the antenna. The key question here is can critical technology such as those that can facilitate direct sampling at multigigahertz rates directly at the antenna such as photonics and superconductivity provide the answer? Or can undersampling, interleave processing and channelizers relax the ADC and DAC requirements on these transceiver designs for wideband spectrum sensing.

The ADC is one of the critical building blocks in any DRFM transceiver. This also includes the companion DAC and the dual-ported memory block in between them. The demanding requirements of a large SFDR and wide bandwidth capability have led to the development of HBT integrated circuit technology. The HBT fabrication based on GaAs and InP have lead to state-of-the-art devices for direct sampling at the antenna/LNA combination. These are discussed in the next chapter.

REFERENCES

[1] M. Jones and P. Delos, "Integrated transceivers simplify design, improve phased array radar performance," *Microwave Journal*, vol. 63, no. 1, pp. 88–98, 2020.

[2] R. Shabardin, N. Shabardina, I. Mukhin, D. Morozov, and L. Nedashkovskiy, "The development of quadrature modulators and demodulators 1800 mhz – 6 ghz with digital correction of parameters," in *2019 IEEE Conference of Russian Young Researchers in Electrical and Electronic Engineering (EIConRus)*, 2019, pp. 1612–1615.

[3] E. H. ñigo Guitierrez, Juan Melendez, *Design and Characterization of Integrated Varactors for RF Applications.* New York, NY: John Wiley and Sons, 2006.

[4] B. Razavi, *Principles of Data System Design.* Princeton, N.J.: IEEE Press, 1995.

[5] H. Liu, N. Ghaderi, D. Yu, Y. Jin, and K. Han, "A novel low-noise open-loop sample and hold using the chopper technique," in *2019 IEEE 4th Advanced Information Technology, Electronic and Automation Control Conference (IAEAC)*, vol. 1, 2019, pp. 760–763.

[6] Seong-Kyun Kim, S. Daneshgar, A. D. Carter, Myung-Jun Choe, M. Urteaga, and M. J. W. Rodwell, "A 30 GSample/s InP/CMOS sample-hold amplifier with active droop correction," in *2016 IEEE MTT-S International Microwave Symposium (IMS)*, 2016, pp. 1–4.

[7] Aadil Rafeeque K.P. and A. Sahu, "A cost effective static linearity testing scheme for ADCs," in *2015 IEEE International Conference on Electrical, Computer and Communication Technologies (ICECCT)*, 2015, pp. 1–4.

[8] P. E. Pace, *Advanced Techniques for Digital Receivers.* Norwood, MA: Artech House, 2000.

[9] R. van de Plassche, *Integrated Analog-to-Digital Converters and Digital-to-Analog Converters.* Boston, MA: Kluwer Academic Publishers, 1994.

[10] S. Max, "Optimum DAC and ADC testing," in *2008 IEEE Instrumentation and Measurement Technology Conference*, 2008, pp. 573–578.

[11] A. Khorami, M. B. Dastjerdi, and A. F. Ahmadi, "A low-power high-speed comparator for analog to digital converters," in *2016 IEEE International Symposium on Circuits and Systems (ISCAS)*, 2016, pp. 2010–2013.

[12] Q. lia and Z. Lib, "A 8-bit 2 gs/s flash ADC in 0.18 μm cmos," in *IWIEE*, vol. 29, 2012, pp. 693–698.

REFERENCES

[13] L. Chacko and G. Tom Varghese, "Comparator design for low power high speed flash ADC-a review," in *2019 3rd International Conference on Computing Methodologies and Communication (ICCMC)*, 2019, pp. 869–872.

[14] M. P. Ajanya and G. T. Varghese, "Thermometer code to binary code converter for flash ADC - a review," in *2018 International Conference on Control, Power, Communication and Computing Technologies (ICCPCCT)*, 2018, pp. 502–505.

[15] M. Zlochisti, S. A. Zahrai, N. Le Dortz, and M. Onabajo, "Comparator design and calibration for flash ADCs within two-step ADC architectures," in *2019 IEEE International Symposium on Circuits and Systems (ISCAS)*, 2019, pp. 1–5.

[16] Y. . Jenq, "Measuring harmonic distortion and noise floor of an A/D converter using spectral averaging," *IEEE Transactions on Instrumentation and Measurement*, vol. 37, no. 4, pp. 525–528, 1988.

[17] T. Drenski and J. C. Rasmussen, "ADC & DAC — Technology trends and steps to overcome current limitations," in *2018 Optical Fiber Communications Conference and Exposition (OFC)*, 2018, pp. 1–3.

[18] D. C. Schleher, *Electronic Warfare in the Information Age*. Norwood, MA: Artech House, 1999.

[19] S. J. Roome, "Digital radio frequency memory," *Electronics Communication Engineering Journal*, vol. 2, no. 4, pp. 147–153, 1990.

[20] R. H. Walden, "Analog-to-digital converter survey and analysis," *IEEE Journal on Selected Areas in Communications*, vol. 17, no. 4, pp. 539–550, 1999.

[21] A. Varzaghani, A. Kasapi, D. N. Loizos, S. Paik, S. Verma, S. Zogopoulos, and S. Sidiropoulos, "A 10.3-gs/s, 6-bit flash ADC for 10g ethernet applications," *IEEE Journal of Solid-State Circuits*, vol. 48, no. 12, pp. 3038–3048, 2013.

[22] B. Murmann, "The race for the extra decibel: A brief review of current ADC performance trajectories," *IEEE Solid-State Circuits Magazine*, vol. 7, no. 3, pp. 58–66, 2015.

[23] S. Norsworthy, "Rf data conversion for software defined radios," in *2019 IEEE 20th Wireless and Microwave Technology Conference (WAMICON)*, 2019, pp. 1–4.

[24] "A/d converter figures of merit and performance trends," https://web.stanford.edu/~murmann/adcsurvey.html, stanford Univ. report by Boris Murmann, Accessed: 2020-7-24.

[25] M. Trinkle, "Snr considerations for rf sampling receivers for phased array radars," in *2006 International Radar Symposium*, 2006, pp. 1–4.

[26] M. M. Ayesh, S. Ibrahim, and M. M. Aboudina, "A 15.5-mW 20-GSps 4-bit charge-steering flash ADC," in *2015 IEEE 58th International Midwest Symposium on Circuits and Systems (MWSCAS)*, 2015, pp. 1–4.

[27] S. Shahramian, S. P. Voinigescu, and A. C. Carusone, "A 35-GS/s, 4-bit flash ADC with active data and clock distribution trees," *IEEE Journal of Solid-State Circuits*, vol. 44, no. 6, pp. 1709–1720, 2009.

[28] "E-bombs vs. pervasive infrastructure vulnerability," https://http://www.ausairpower.net/PDF-A/AOC-PACOM-Kopp-Oct-2012-A.pdf, air Power Australia by Carlo Kopp, Accessed: 2020-7-24.

[29] C. Guo, Z. Lv, X. Shi, L. Xu, and M. Cai, "Effects of low noise amplifier under high power microwave back-door coupling," in *2012 International Conference on Microwave and Millimeter Wave Technology (ICMMT)*, vol. 4, 2012, pp. 1–4.

Chapter 6

High-Performance Transceiver Technologies

In this chapter, high-performance transceiver component technologies and architectures for DRFMs are presented. ADCs, DACs, and dual-ported memory and materials are key blocks discussed and examples of recent designs are examined.

In Chapter 4, we analyzed the flash architecture and derived expressions for the INL and the DNL as a means to understanding the linearity of the device as usually reported and discussed its importance and the requirement of the linearity being $< 1/2$ LSB. Also presented were the FOMs for comparison of these technologies.

In this chapter, we also identify the current component technologies and describe the methods of constructing these ADC digitizers that lead to their extreme performance. We begin by taking a look at the state of the art in comparator designs, which represent a mixed-signal technology (analog and digital) and are the interface (or transducer) between the analog world and the digital technology domain. Also addressed are direct digital synthesis modules (DDSs) and the local oscillator phase noise and a discussion of its limitations in the minimum signals that can be detected in the overall goal of wideband spectrum sensing.

6.1 COMPARATOR DESIGNS

Comparators play a vital role in analog-to-digital conversion and having a high-speed comparator is critical. Flash converters often suffer from kickback noise, the existence of sparkles and bubbles within their conversion. Many comparators and their parallel arrangements within the flash architectures serve to eliminate these problems [1]. Since comparators are a mixed-signal component most of the noise and most of the power is generated and consumed, respectively, in this part of flash. There are several different comparator designs that provider trade-offs for increase in speed for power consumption, gain, delay, and other parameters.

6.1.1 Two-Stage Open-Loop Design

The two-stage, open-loop comparator is a nonclocked design (clock applied only at the latch) that resembles an operational amplifier and operates in the saturation region on the applied differential inputs (input and output phase). The two-stage open loop comparator design is shown in Figure 6.1. The input is applied to M1 and the reference voltage is applied at M2. The output (e.g., to the latch) is taken from

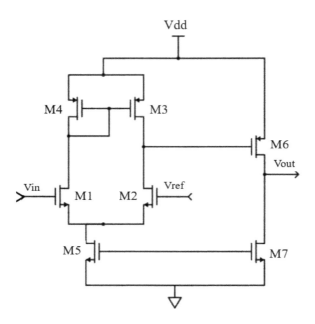

Figure 6.1 Schematic diagram of a two-stage open-loop comparator design.

M6. The switched comparator version turns off after the evaluation phase is complete and turns off after the maximum speed of operation is not needed. These states are added both to save power and additional circuitry is required.

6.1.2 Conventional Double-Tail Dynamic Design

The conventional double-tail dynamic comparator, proposed in [2], is a modified version of the traditional latch type, voltage sense amplifier comparator but in this case the comparator uses two tail current sources, one for the input stage (or preamplifier stage), and the other for the latching stage. The double-tail dynamic comparator schematic is shown in Figure 6.2. When CLK is a logic 1 and CLKn is a logic 0, the circuit operates in *a reset phase*.

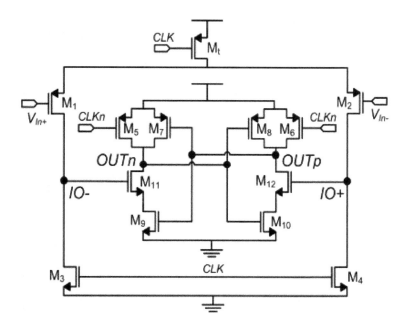

Figure 6.2 Schematic diagram of the double-tail dynamic comparator (©IEEE, reprinted with permission from [3, 4].)

In this phase, Mt (at VDD) is OFF and M3,4 are ON [4]. Consequently, the intermediate output nodes IO+/IO- discharge to ground. As M5,6 transistors are ON, output nodes (OUTn,p) of the latch stage charge to VDD. When the CLK Logic is 0 and CLKn is a logic 1, it enters *a decision phase*. In this phase, Mt is ON and M3,4 are OFF. IO+/IO- nodes start charging. An intermediate differential voltage ΔVIO+(−) is developed due to the differential input ΔVin [4].

6.1.3 New Double Tail with Clock Gating

Figure 6.3 shows the schematic for a new double-tail comparator with clock gating.

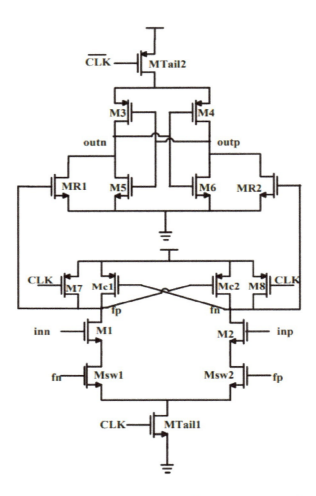

Figure 6.3 Schematic diagram of the proposed double-tail dynamic comparator (©IEEE, reprinted with permission from [3].)

The parameters that affect the total delay time are the latch effective trans-conductance and ΔV_0. The proposed comparator design exploits this dependence of latch regeneration time on ΔV_0. By adding two key transistors to the latch circuit the effective transconductance of intermediate stage transistors is enhanced. This in turn enhances ΔV_0, which leads to increased latch regeneration speed or reduced delay time. Detailed operation of the comparator is given in [3].

In [5], the delay of a single-tail comparator, a double-tail comparator, and double-tail comparator with low power are analyzed in CMOS using Cadence simulation tools. In addition, a new double-tail comparator design is proposed that reduces the leakage power and reduces the delay. An improved version of the conventional dynamic comparator design for high-speed and low power consumption has also been proposed in [6, 7]. Designed in 180-nm CMOS technology with a constant supply voltage of 0.8V used, a conventional double-tail comparator is designed by adding transistors but without hindering the functionality but by providing a faster, more efficient modification of the comparator design. In this new design, a dynamic double-tail comparator is modified to use a new, regenerative clock gating technique that further reduces the power consumption and provides higher speed by reducing the delay time of the circuit.

In [8], a dynamic comparator that consists of a low gain amplifier connected to a latch circuit is examined. The inputs are amplified during the evaluation period and the outputs are latched during the regeneration time. At a clock frequency of 1.25 GHz and 100 mV ΔV_{in}, the delay is 176 ps with an average power consumption of 119 μW for a 1.8V supply. What is unique about the proposed work in [8] is that the design is compared against five previous designs. The dynamic comparators are compared by using 180-nm SCL technology. NMOS and PMOS lengths are taken as 180nm and widths are taken as 420nm. Clock frequency is 1.25 GHz and input frequency is 625 MHz. The voltage supply is 1.8V with common mode input voltage of 0.9V and 100-mV ΔVin.

6.1.4 Low Power Ultrahigh-Speed Design

The latest design examined here is a low power, ultrahigh-speed, dynamic latched comparator that uses a NAND latch circuit that has been used in the latching stage and results in both high-speed operation and low power dissipation [9]. The schematic diagram of the comparator design is shown in Figure 6.4.

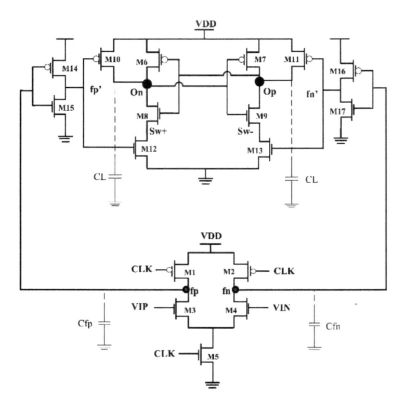

Figure 6.4 Schematic diagram of the low power ultrahigh-speed dynamic latched comparator.(©IEEE, reprinted with permission from [9].)

Due to its maximum three stacked transistors in both preamplifier and latch stage, it works at low supply voltage operation for full swing output [9]. In operation, only one clock pulse is enough to complete both the reset and comparison phase, which reduces the comparison time and saves power consumption. By decreasing the regeneration process time, it also improves the comparison performance, which is the main concern of the proposed architecture [9].

Their proposed comparator design has two phases of full operation, (i) precharge or reset phase and (ii) comparison or decision-making phase. During the reset phase (CLK = 0), the tail transistor (M5) of the precharged stage gets turned off and stops current flow from VDD to ground (which avoids static power dissipation) while both pull-up transistors (M1 and M2) being turned on pulls both Vfn and Vfp nodes to VDD. This in turn brings the output voltages of both intermediate stage inverters, Vfn' and Vfp', down to zero. Sequentially, pull-up transistors (M10, M11) of the latching stage will turn on and output nodes On and Op are precharged up to VDD while both pull-down transistors (M12 and M13) remain off [9]. During the comparison phase (CLK = VDD), transistors M1 and M2 are off condition while the tail transistor of precharging stage M5 is on. Thus, each fn and fp nodes precharged capacitive voltage begin to discharge from VDD to ground with a different discharging rate with respect to their associated applied input voltages VIP and VIN. Further details on the operation are given in [9]. Figure 6.5 shows the performance of the comparator design.

Item	Value
Technology	90-nm CMOS
Supply voltage range	0.6 V – 1.5 V
Maximum clock frequency	4 GHz
Best case delay (VDD = 1.2 V, Vcm = 1 V, Vin = 100 mV)	89 ps
Worst case delay (VDD = 0.6 V, Vcm = 0.4 V, Vin = 1 mV)	424 ps
Delay time (VDD = 0.8 V, Vcm = 0.6 V, Vin = 5 mV)	224 ps
Average power consumption per conversion at frequency = 1 GHz	8.42 µW

Figure 6.5 Performance of the proposed comparator. (©IEEE, reprinted with permission from [9].)

6.2 FLASH ADC EXAMPLES

The examples below illustrate the advantages, architecture trade-offs, and performance levels for the flash ADC. The first example is a 24 GS/s, single-core flash with a $n = 3$-bit resolution in 28 nm low-power digital CMOS. A block diagram of the device is shown in Figure 6.6. The ADC is capable of delivering its full sampling rate without using any time interleaving [10].

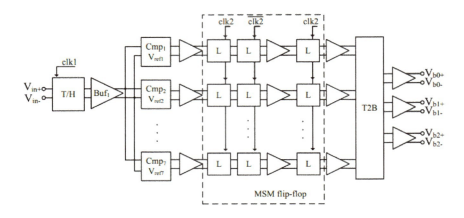

Figure 6.6 Block diagram of a 24 GS/s single-core flash ADC with 3-bit resolution in 28 nm low-power digital CMOS. (©IEEE, reprinted with permission from [10].)

The architecture consists of a clocked track-and-hold (T/H) stage and subsequent buffer (Buf_1) at the input, followed by the individual comparators (Cmp_i) with follow-on amplifiers and latches (L) in each of the parallel data processing paths. A thermometer-to-binary conversion logic (T2B) creates the full-speed output signals. By utilizing a modern CMOS process with a transit frequency substantially above 100 GHz, it is possible to achieve sampling rates of tens of GHz with circuit structure sizes in the range of hundreds of μm. A standard thermometer-to-binary conversion logic is used for generating the output [10].

A 25 GS/s 4-bit single-core flash ADC in 28 nm fully depleted silicon on insulator (FDSOI) CMOS was recently developed and reported in [11]. Transistors manufactured using FDSOI use a planar process technology that delivers the benefits of reduced silicon geometries while simplifying the manufacturing process. Figure 6.7 illustrates the flash ADC architecture.

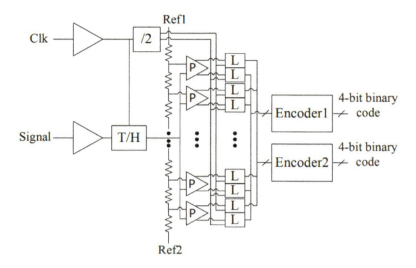

Figure 6.7 Schematic diagram of 4-bit 25 GS/S single-core flash ADC designed in 28-nm FDSOI CMOS process (©IEEE, reprinted with permission from [11].)

An input buffer with a differential 100 Ω termination resistor drives the input analog signal to the T/H stage [11]. The T/H samples the analog signal, which is then quantized by a bank of comparators. After the quantization, the thermometer code is converted to binary data using a fat-tree encoder. Current mode logic (CML) latches are used for the comparator design for high-speed and a divide-by-2 circuit is used to generate a differential 12.5 GHz on-chip clock for the 1:2 demux from an external 25-GHz clock signal since the comparator needs to work at 25 GS/s [11].

Each comparator consists of a preamplifier and two StrongARM latches as shown in Figure 6.8. To achieve sufficient gain and also to maintain a proper output common mode voltage, two current steering transistors P1 and P2 are utilized.

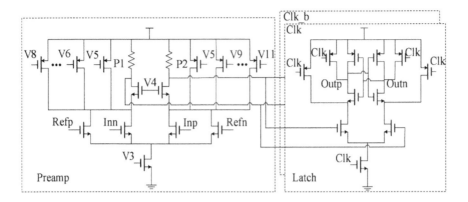

Figure 6.8 Comparator design consisting of a preamplifier and two StrongARM latches are used. To achieve sufficient gain and also to maintain a proper output common mode voltage, two current steering transistors P1 and P2 are utilized. A cascode structure in the preamplifier is used to suppress kickback noise. (©IEEE, reprinted with permission from [11].)

The cascode structure is utilized in the preamplifier to suppress the kickback noise. A 3-bit current DAC is implemented to calibrate the compactor offset voltage down to 0.2 LSB. Body biasing in FDSOI is also utilized to lower the VTH of the input transistors of the preamp and the StrongARM latches.

The thermometer-to-binary encoding is carried out in two stages in the fat-tree encoder. The first stage converts the thermometer code to one-out-of-N code. This code conversion is done in N bit parallel using N 3-input AND gates. The second stage converts the one-out-of-N code to binary code using the multiple trees of OR gates. The encoder only consists of AND and OR gates and they are implemented using static CMOS helping to achieve low power consumption. Simulation results are summarized in Figure 6.9 and show the ADC achieves a SNDR of 25.3 dB and a SFDR of 35.9 dB with the 1.07-GHz input and the SDNR and SFDR with the 12.4-GHz input are 19.5 dB and 22.9 dB, respectively.

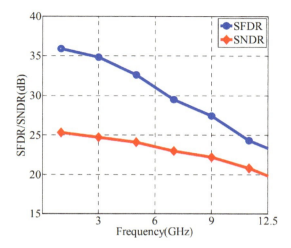

Figure 6.9 Spurious free dynamic range and signal to noise plus distortion ratio for comparator design shown in 6.8. (©IEEE, reprinted with permission from [11]).

6.2.1 SiGe BiCMOS

In order to increase the operating speed (hence also frequency) it is also possible to add one (or occasionally two) bipolar transistors to a CMOS circuit stage. Where such a BJT is NPN type, the base is then often doped with germanium (Ge)–which serves to shrink the bandgap substantially. This silicon-germanium alloy is always termed "SiGe" and the resulting overall technology is called SiGe BiCMOS. Compared to silicon alone, the addition of Ge has several advantages, including:

(a) A much higher value of the current gain ($\beta = I_c/I_B$), where I_c is the collector current and I_B is the base current;
(b) Also leads to a much larger transition frequency, (f_T).

The fastest flash ADCs are typically fabricated in SiGe. For example, the design of a 10-GS/s, 3-bit analog-to-digital converter in a 0.13-μm SiGe BiCMOS technology for optical communication applications is discussed in [12]. To achieve the maximum sampling frequency for the design, the SiGe HBT were chosen for improving the high-speed performance of the circuit. A wide-bandwidth THA that included a capacitor bandwidth enhancement was applied to achieve the wide input bandwidth. As described in [12] the comparators shown in Figure 6.10 adopt a fully differential architecture and use differential input signals and differential reference voltage signals.

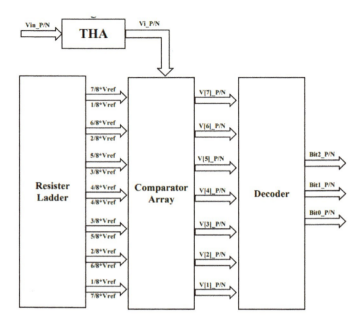

Figure 6.10 Simplified block diagram of 3-bit flash ADC (©IEEE, reprinted with permission from [12].)

The thermometer to 3-bit decoder includes a bubble-correct module, an XOR-gate-array module, and a fully differential ROM-decoder module for differential signaling. Figure 6.11 shows the simulated SFDR of THAs with a conventional output buffer and the modified output buffer for a differential 0.6Vpp sine-wave input. The results are shown as a function of the input frequency F_{in}.

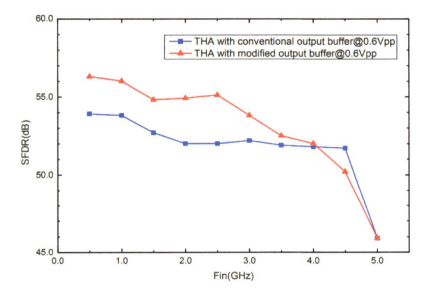

Figure 6.11 SFDR of the THAs with respect to different input frequency.(©IEEE, reprinted with permission from [12].)

A low power 6-bit, 20 GS/s Nyquist-rate flash ADC is presented in [13]. It operates without a track-and-hold or without time interleaving as well as without any kind of correction or calibration. The circuit is intended to demonstrate the energy efficiency limits achievable with a plain SiGe bipolar transistor (analog) design. Figure 6.12 shows the ADC block diagram with binary signal distribution tree (BSDT) and differential reference ladder.

A unique comparator placing concept along with a differential reference ladder is used to take advantage of the differential signaling with a pseudo-differential comparator architecture. A passive input signal distribution tree is employed to lower the power consumption. The 10-GHz bandwidth ADC data input signal is buffered in continuous time by an emitter follower (EF) directly at the input. No track-and-hold is employed to minimize the power consumption and demonstrate the inherent high-speed capability of the technology. This input EF drives a passive binary signal distribution tree (BSDT) of differential transmission lines to distribute the full speed signal to 16 EFs.

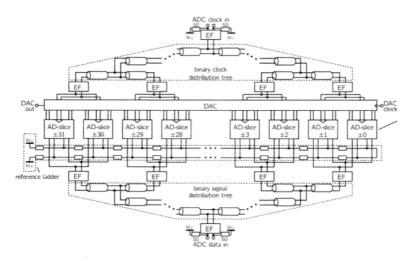

Figure 6.12 ADC block diagram with BSDT and differential reference ladder. (©IEEE, reprinted with permission (©IEEE, reprinted with permission from [13].)

Each EF drives two complementary AD-slices in parallel; each complementary AD-slice performs the comparison of its differential input against a positive and the corresponding negative reference level. At 20 GS/s sampling and 10 GHz signal frequency a high ENOB=3.7 is achieved. At a core power consumption of only 1.0 W it leads to a FOM_W =3.9 pJ/cs under Nyquist conditions [13].

A different approach in [14] develops a 5-bit 20-GS/s flash ADC realized in 0.18-μm SiGe BiCMOS technology. The ADC includes a THA that incorporates linear distortion compensation, a double-interpolation preamplifier, current bias-weighted comparators, and a high-speed logic encoder. A block diagram of the 40-Gb/s 10-GBd coherent QPSK optical receiver architecture is shown in Figure 6.13. At 4V, the THA has a bandwidth that exceeding 23 GHz and an IIP3 of 24 dBm. The ADC achieves a SINAD ratio of 28.6 dB and a spurious-free dynamic range SFDR of 36 dB with a 1-GHz input sinusoid sampled at 20 GS/s. The performance specifications are shown in Table 6.1. The figure of merit $FOM_W = 9.54$ pJ/conversion-step. The SNDR at Nyquist frequency and the SFDR versus sampling frequency is illustrated in Figure 6.14.

Table 6.1
Performance Specifications of 20-GS/s Flash ADC Realized in 0.18-μm SiGe BiCMOS Technology BiCMOS.

Specifications	Value
Input voltage range	1 V$_{p-p}$ differential
Resolution	4-bits
Resolution bandwidth	7 GHz
Signal bandwidth	20 GHz
Sampling rate	20 GS/s
Static performance	1 LSB
Supply voltage	4V and 3V
Power budget	3W
FoM_{Walden}	10 pJ/conv-step

From: [14]

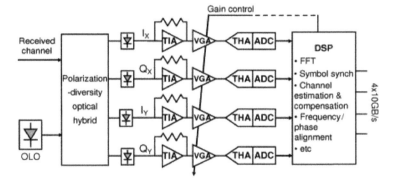

Figure 6.13 Block diagram of a 40-Gb/s 10-GBd coherent QPSK optical receiver. (©IEEE, reprinted with permission from [14].)

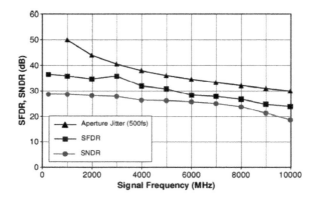

Figure 6.14 Measured SNDR and SFDR versus analog input frequency at $f_s = 20$ GS/s. (©IEEE, reprinted with permission from [14].)

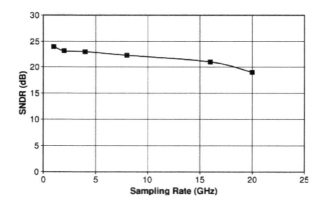

Figure 6.15 Measured SNDR versus sampling frequency at Nyquist frequency $f_s/2$. (©IEEE, reprinted with permission from [14].)

The SNDR is better than 24 dB at 1 GS/s, but it is degraded to 19 dB at 20 GS/s. The SNDR at Nyquist frequency versus sampling frequency is illustrated in Figure 6.15. The ADC has a wide resolution bandwidth of 7 GHz, and the figure of merit is $FOM_W = 9.54$ pJ/conversion-step. The ADC consumes 3.24 W from 4- and 3-V supplies when sampled at 20 GHz. The prototype ADC occupies 8.68 mm of silicon area.

As a final example, a 40 GS/s 4-bit flash ADC in 0.13 μm SiGe BiCMOS technology on a single-core was described in [15]. The architecture is shown in Figure

6.2 Flash ADC Examples

6.16(a) and (b) and was developed as a sub-ADC part of an optical communication 100 Gbit/s wireless system.

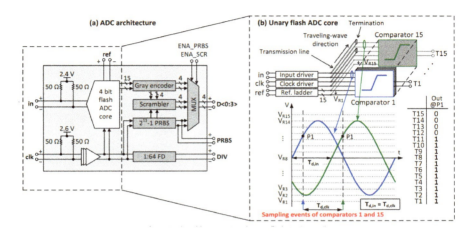

Figure 6.16 A 40 GS/s 4-bit SiGe BiCMOS flash ADC showing (a) the ADC architecture and (b) the unary flash ADC core. (©IEEE, reprinted with permission from [15].)

The main building blocks are a unary flash ADC core and a bubble-error suppressing Gray encoder. To omit the need for a front-end track-and-hold amplifier with fast settling time, the flash core exploits a traveling-wave topology, where analog input and clock signal travel synchronously from comparator to comparator.

Figure 6.16(b) illustrates this *traveling-wave signal distribution* concept to the comparators. A linear input driver feeds the analog input signal with help of a transmission line (TL) to the bank of parallel comparators. A high-gain clock driver does this same distribution for the clock signal. As the comparators are spatially separated from each other, the input signal and clock signal do not arrive at the same time at all the comparators, but rather successively with small time delays [15]. The idea exploited in traveling-wave ADCs is to keep the delays of the input and clock signals equal between adjacent comparators. This ensures that every comparator quantizes the same input signal value at each sampling event, as illustrated in Figure 6.16(b). Even though the comparators operate asynchronously due to the clock delays, the same results can be obtained as with synchronous flash signal distribution approaches, where the input and the clock signal arrive at all comparators[1]. Only two transmission lines are required– one for the input and one for the clock– and as such, the traveling wave concept significantly consumes less area and has lower complexity.

---

[1] This is the same concept as the traveling-wave velocity matched optical modulator in titanium diffused, lithium niobate (Ti:LiNbO$_3$) as discussed in the next chapter when optical sampling circuits are investigated.

The DSP thermometer outputs of the core are Gray-encoded to minimize the encoding errors as shown in Figure 6.17.

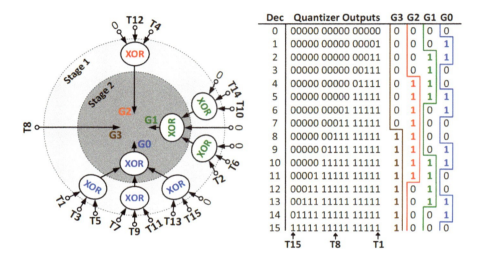

Figure 6.17 Thermometer 4-bit-to-Gray encoder with truth table. (©IEEE, reprinted with permission from [15].)

For speed enhancement and power reduction of the encoder, each XOR gate in Figure 6.17 is replaced by a pseudo-exclusive-OR (PXOR) gate. The PXOR gate is a folded-cascode differential logic and is based on the comparator outputs being unary coded. That is, the XOR operations in the encoder can alternatively also be fully emulated with the help of a folding amplifier [15]. This approach allows the ADC to realize the XOR operation with minimum circuitry with only three folded-cascode differential pairs required. The PXOR enables operation at high speed and low power dissipation. Due to the low systematic jitter, several PXORs can directly be cascaded without the need for data retiming flip-flops in between.

Also as a digital communication interface the device consists of a scrambler, a multiplexer (MUX), and a 1:64 frequency divider to enable the synchronous storage of the samples on a FPGA. The ENOB and the SFDR, as formulated in the last chapter, are shown below for the 4-bit device in Figure 6.18 as a function of the sampling frequency in GHz for several sampling frequencies. Also shown in Figure 6.19 is the INL and the DNL as a function of the output code.

6.2 Flash ADC Examples 295

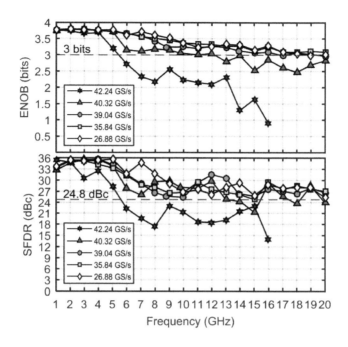

Figure 6.18 Measured ENOB and SFDR as a function of sampling frequency (GHz). (©IEEE, reprinted with permission from [15].)

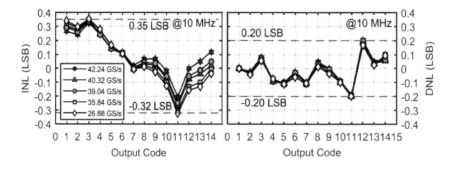

Figure 6.19 Measured INL and DNL as a function of sampling frequency (GHz). (©IEEE, reprinted with permission from [15].)

The ADC dissipates 3.5 W while operating from a 3.5 V and 3 V power supply. The unary ADC core consumes 1.4 W and the encoder with the FPGA interface 2.1 W, whereby 57 % of the latter power is dissipated in the six retiming output drivers.

6.3 TIME-INTERLEAVING FLASH TECHNIQUES

Many techniques are available to upgrade a flash ADC resolution. However, the hardware complexity of higher-resolution flash ADCs achieved by adding on comparators for increased resolution is a major drawback resulting in high power consumption, high input capacitance, and the creation of sparkles [16]. Time-interleaving (TI) is a major trend and a good choice for creating a high-speed ADC design. Figure 6.20 shows a block diagram of a time-interleaved architecture composed of M ADC modules in a parallel configuration. There is a delay τ in the sampling clock between each ADC in the parallel array. The master sampling clock samples the parallel array at f_s/M. This is the sampling clock of subsystem m for $m = 0, 1, \ldots, M-1$. The overall sampling frequency of the digitizer is f_s The output of each ADC is time multiplexed into memory for subsequent processing.

The phenomenon of time interleaving itself is far more complex than simply placing sub-ADCs in parallel. Numerous repercussions such as timing mismatch, offset error, and gain error should be considered when designing a TI system. Moreover, the design of the S&H circuit for a TI-ADC is quite different from that of a noninterleaved ADC. In modern designs, a digital calibration block is usually employed to account for all mismatch calibrations [17].

The timing errors in TI ADCs often generate undesirable spurs, and hence, degrade the SFDR. Degradation can occur for two main reasons. The first is the gain variations that can occur among the different modules. These gain variations cause amplitude variation in the sampled signal and tends to raise the noise floor of the digitizers. The second problem is the nonuniform sampling effect due to the difficulty of maintaining uniform delay between two consecutive modules. This effect also tends to raise the noise floor. Note that Jenq, by using nonuniform sampling theory, has developed analytic expressions to examine the spectrum of time-interleaved digitizers as a function of their mismatch offset nonlinearities and the number of ADC modules in the parallel arrangement [18, 19].

6.3 Time-Interleaving Flash Techniques

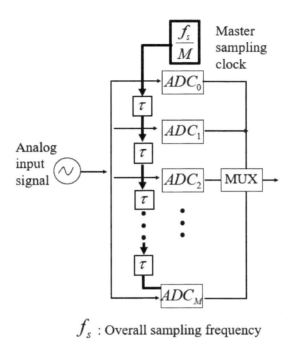

f_s : Overall sampling frequency

Figure 6.20 Block diagram of a time-interleaved ADC with M ADCs in parallel.

Calibration techniques have been actively pursued to compensate the mismatches. Multi-ADC time interleaved sampling requires providing the sampling clock of different phases for each ADC. Time interleaving requires the sub-ADCs to operate on sampling clocks that have the same sampling speeds, but different phase angles, such that only one of the M sub-ADCs is active at a given time. The generation of a precise clock phase for all channels is imperative to the system performance, as any skew in the generated phases can result in the sampling of incorrect analog values.

Even though 2× interpolation techniques utilizing dynamic latches have achieved competitive flash ADC performance, it is not so straightforward to realize a higher interpolation factor than two because of the nonlinear behavior of dynamic latches [20]. It is also possible to reach high sampling rates with basic circuit structures by applying time-interleaving. As long as the overhead of multi-phase clock generation is negligible, it is theoretically possible to increase the sampling rate with no penalty in terms of required energy per conversion [10]. For this reason, time-interleaving topologies are widely used for high-speed ADCs and have been extensively exploited to achieve low figures of merit with low energy per conversion. Problems however include the jitter in multiphase clock generation and distribution, clock transition times, input capacitance, and the requirements on the THAs.

Multiple techniques have been proposed for the generation of precise clock phases. A digital-background calibration technique is often used to minimize the mismatch effects. One particular technique is based on digital interpolation. Digital interpolation estimates the correct output values from the output samples that suffer from timing errors [21]. Since this technique requires an accurate estimation of the timing errors of the individual channels, a digital-background timing-error measurement technique is also examined.

6.3.1 Time Interleaved Flash Examples

A 6-bit, 10-GS/s device in 65-nm CMOS that only draws 63 mW is described is described below [20]. The major contribution here is that the architecture reduces the flash ADC hardware burden with the use of a voltage-to-time converter (VTC) and a time-domain interpolator array. The technique also results in a reduced power consumption, reduced hardware and footprint.

Figure 6.22 shows the schematic of the $8\times$ time-domain interpolating unit block.

6.3 Time-Interleaving Flash Techniques

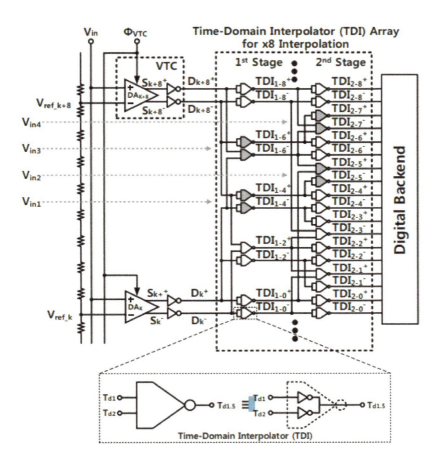

Figure 6.21 Block diagram of the 6-b 10-GS/s 4x TI flash ADC using the sub-ADC. (©IEEE, reprinted with permission from [20].)

It consists of a resistor ladder string to serve as a voltage reference, the VTC and an inverter based time-domain interpolator (TDI) array. The VTC enables a linear zero-crossing interpolation in the time domain with an interpolation factor of eight [20].

The VTC is composed of a dynamic amplifier (DA) whose output slope in the time-domain is proportional to the input voltage difference [20]. It also has two inverters that convert the input voltage difference into the logic level delay difference. This then results in a voltage-to-time conversion where each VTC generates delays D_k^+, D_k^-, D_{k+8}^+ and D_{k+8}^- which is based on the difference between the reference voltages ($V_{ref_k}, V_{ref_{k+8}}$) and the input voltage (V_{in}).

The TDI array is made up from two stages of inverters. Through the first stage of TDI, three additional ZX points are generated resulting in an increase of resolution by 2-bits more (4× interpolation). The second stage adds an additional 1-bit resolution by the (2× interpolation). Each TDI has two inverters. The output $T_{d1.5}$ is an interpolated time delay that is in the middle of the two time delay inputs (T_{d1}, T_{d2}) as shown. The TDI and the interpolation operation is further explained (with examples) in [20]. In addition, a single-channel 6-bit, 2.5-GS/s flash ADC with the proposed interpolation scheme is described.

A block diagram and timing diagram of a prototype 6-bit 10-GS/s, 4× TI flash ADC is shown in Figure 6.22. The architecture uses a sub-ADC that consists of a single-channel, 6-bit 2.5-GS/s flash ADC with the proposed interpolation scheme.

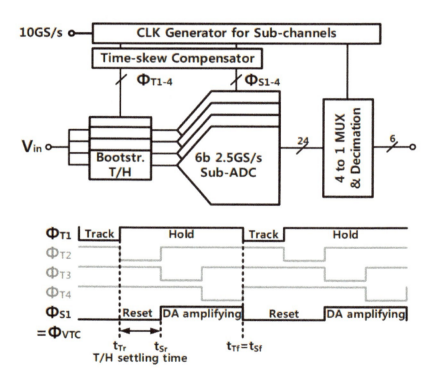

Figure 6.22 Block diagram of the 6-b 10-GS/s 4x TI flash ADC using the sub-ADC. (©IEEE, reprinted with permission from [20].)

The sub-ADC consists of a sub-clock generator, bootstrapped THA, voltage reference string, VTC array, TDI array, a NAND-based SR-latch, thermometer-gray binary encoder, and a VTC offset calibration block. The calibration architecture uses a sequential slope-matching offset technique that reduces the offset of the VTC itself, and also reduces the offset of the interpolated zero-crossing (ZX) between adjacent VTCs for improved linearity. The sub-clock generator replicates the signal path from the VTC to the SR-latch (SRL) array via TDIs, and defines the time-delay from Φ_{VTC} to Φ_{EN}. Only ten VTCs are used including two dummies for interpolation scheme. This reduces the power consumption, and input capacitance and the core area of the ADC making it a compact flash ADC suitable for multi-channel TI-ADCs.

Experimental results of the 4 channel, 6-bit 10 GS/s ADC using 65-nm CMOS technology were taken. The measured DNL and INL profiles after VTC offset calibration are shown in Figure 6.23.

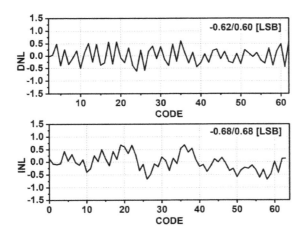

Figure 6.23 Measured DNL and INL after offset calibration (©IEEE, reprinted with permission from [20].)

The peak DNL and INL are $+0.60/-0.62$ LSB and $+0.68/-0.68$ LSB, respectively. The magnitude spectrums before and after calibration at 10-GS/s conversion rate with a Nyquist input signal is shown in Figure 6.24.

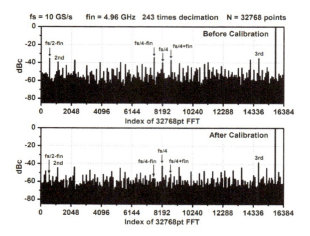

Figure 6.24 Measured magnitude spectrums (©IEEE, reprinted with permission from [20].)

A plot of the dynamic performances at 10-GS/s for various input frequencies is shown in Figure 6.25. Shown are both the SNDR and SFDR both with and without calibration.

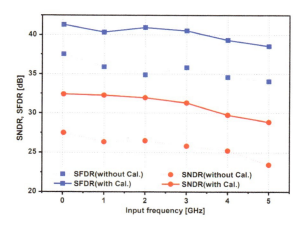

Figure 6.25 Measured SNDR and SFDR before and after calibration at 10-GS/s. (©IEEE, reprinted with permission from [20].)

The performance roll-off at higher frequencies is due to the limited bandwidth of the THA. The SFDR and SNDR are improved from 34.1 dB and 23.4 dB to 38.5 dB and 28.9 dB, respectively, by the offset calibration. The ENOB is 4.51-bits and the figure of merit excluding the output multiplexer power is $FOM = 277$ fJ/c-s.

6.3 Time-Interleaving Flash Techniques

The dependency of the SINAD on the combination of several different channel mismatch effects is investigated in [22]. By using either explicitly given mismatch parameters or given parameter distributions, a derivation in closed-form to calculate the explicit or the expected SINAD for an arbitrary number of channels is presented. They also extend the explicit SINAD by the impact of timing jitter.

To demonstrate the trade-off in bit resolution for sampling frequency, another example of time-interleaving using the flash ADC is shown in Figure 6.26. Here a low-power 4-bit 20-GS/s ADC using a 4× time-interleaved flash ADC in the same 65-nm CMOS is reported [23]. The architecture block diagram with a shared reference resistor ladder is shown in Figure 6.26.

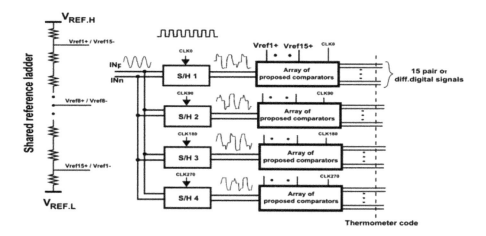

Figure 6.26 Block diagram of the low power ADC using charge steering showing the 15 pair of differential digital signals at the output of each stage. Together, the channel outputs represent the input signal in a time-interleaved manner. (©IEEE, reprinted with permission from [23].)

A new ultra-low-power comparator is used that merges a charge steering preamplifier with an embedded regenerative latch into one block and relies on the charge redistribution between different phases of operation to achieve preamplification and regeneration.

The schematic diagram of the new comparator shown with an embedded regenerative latch is shown in Figure 6.27(a), which is followed by an SR latch to complete a flip flop. The comparator works in three phases as shown in Figure 6.27(b). Phase 1 is a reset phase when the ClK3 is high and CLK1 is low. This is followed by the preamplification and regenerative latch stage.

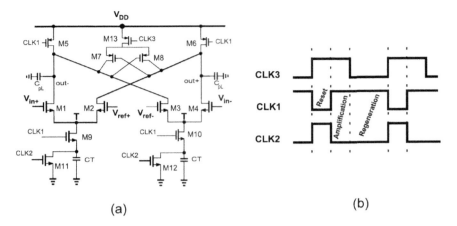

Figure 6.27 New ultra-low power comparator (a) schematic diagram merging a charge steering preamplifier with an embedded regenerative latch into one block and relies on the charge redistribution between different phases of operation to achieve (b) preamplification and regeneration. (©IEEE, reprinted with permission from [23].)

Sharing the reference ladder between the interleaved branches while keeping the kickback noise at an acceptable level is also a feature resulting in a significant power reduction as well. This proposed comparator relies on the concept of charge redistribution and reusing the stored charges to obtain the required output voltage at the end of each phase [23]. The power consumption is 15.5 mW with a peak SNDR of 23 dB for a 9.84-GHz input frequency with a SFDR of 34 dB. As a final TI flash example, an extremely wide bandwidth and high sampling-rate (128 GS/s) TI-flash ADC is discussed. The TI flash is a 5-bit architecture in 55-nm SiGe BiCMOS that takes advantage of the SiGe HBT (larger bandwidth and lower thermal noise than the most advanced FinFETs). It also takes advantage of the novel minimum-power combined SiGe HBT and MOS-HBT topologies that achieves in silicon a better energy-per-bit than previously reported in the SiGe BiCMOS ADCs [24]. A block diagram of a 2 × TI flash architecture integrating two track and hold amplifiers, each driving a 5-bit flash sub-ADC sampled at 64 GHz in antiphase, is shown in Figure 6.28 [24].

6.3 Time-Interleaving Flash Techniques   305

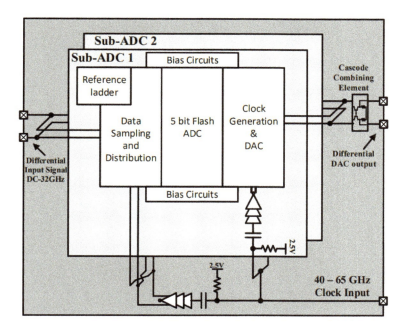

Figure 6.28 Block diagram of a 2 × time-interleaved architecture integrating two track and hold amplifiers, each driving a 5-bit flash sub-ADC sampled at 64 GHz in antiphase. (©IEEE, reprinted with permission from [24].)

The power consumption and layout footprint of the ADC was critical for operation at 128-GS/s, and were minimized by employing novel 1-mA Cherry-Hooper comparators [25] and quasi-CML MOS-HBT latches with active peaking.

The time interleaving approach is employed to increase the sampling rate while keeping the required speed of the flash devices moderate. The time interleaved sub-ADCs include data sampling and distribution to 5 bits for the flash ADC. The clock generation is implemented by an active distribution network. The delay matching is added to assure the right sampling time for the output bits.

Each flash ADC generates a thermometer code using 31 lanes. The lanes include the comparator and the required buffer and latches for recording the bits. This work combines the output bits of the ADC using a time interleaved current steering DAC. This action eliminates the need for an on-chip memory and storing the bits and also eases the dynamic testing of the ADC. Each sub-ADC lane includes a differential Cherry-Hooper comparator, a SiGe-HBT inverter, and a latch. The on-die DAC and ADC building blocks are shown in Figure 6.29. The block diagram of the linear, broadband data sampling, and distribution network are also shown.

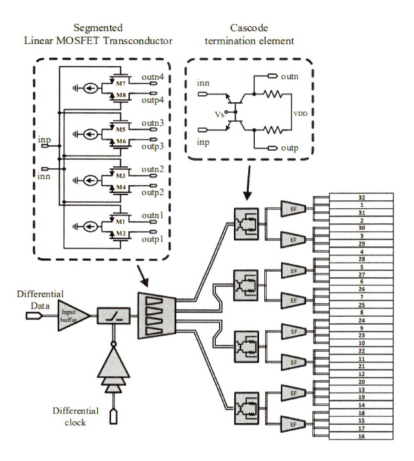

Figure 6.29 Block diagram of the data sampling and distribution network in each sub-ADC. (©IEEE, reprinted with permission from [24].)

Each block was designed for a full-scale input signal of 300 mVpp per side with a fanout-of-2 load. The overall gain of the entire network is approximately 0 dB with a bandwidth of 50 GHz from the input pad of the ADC to the input of the comparator in each sub-ADC lane [24].

SFDR and ENOB measurements are shown in Figures 6.30, and 6.31, respectively. The measurements were conducted for different clock frequencies of 45, 55, and 64 GHz, corresponding to 90, 110, and 128-GS/s sampling rate with no calibration or postprocessing of the measured data beyond integrating the spectrum overt the intended 32-GHz input signal bandwidth [24]. Measurements show an ENOB better

6.3 Time-Interleaving Flash Techniques   307

than 4.1 up to 32-GHz input frequencies for sampling rate of 128 GS/s. The ENOB degradation is less than 1 bit over the entire bandwidth of interest [24].

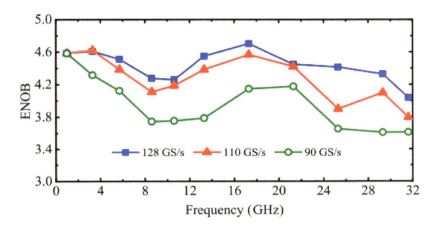

Figure 6.30 Measured ENOB as a function of frequency for sampling rates of 128 GS/s, 110 GS/s, and 90 GS/s. (©IEEE, reprinted with permission from [24]).

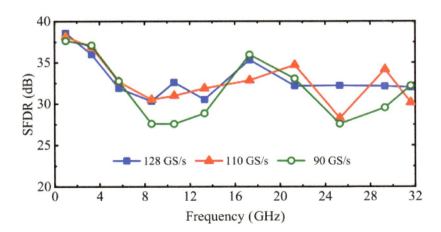

Figure 6.31 Measured SFDR (dB) as a function of frequency for sampling rates of 128 GS/s, 110 GS/s, and 90 GS/s. (©IEEE, reprinted with permission from [24]).

6.4 TIME-INTERLEAVING PIPELINE TECHNIQUES

Time interleaving a pipelined architecture also has its advantages. A pipelined ADC consists of several stages with each stage resolving only a few bits and the output is combined in an output register. Figure 6.32 shows a block diagram of a pipelined ADC that has k−stages.

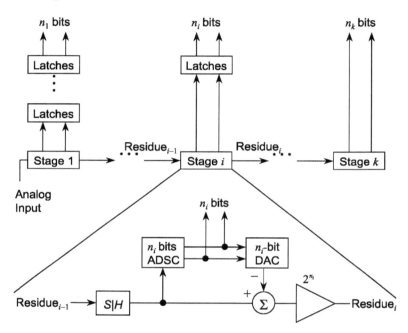

Figure 6.32 Block diagram of a k-stage pipeline ADC.

The total number of bits resolved is

$$n = \sum_{i=1}^{k} n_i \quad (6.1)$$

where subscript i is the stage index. As shown, each pipeline stage consists of a S/H (the pipeline latch), a low-resolution n_i-bit ADC subconverter (ADSC) and DAC, a subtracter, and an interstage amplifier. Typically, n_i is in the range of 2-4. In operation, each stage initially samples and holds the output from the previous stage [35]. Next the held input is converted into a digital code by the first stage ADSC and back into an analog signal by the first stage DAC. The difference between the DAC output and the held input is the *residue* that is amplified and sent to the next stage where this process is repeated. At any time, the first stage operates on the most recent sample while all other stages operate on the amplified residues from previous samples. To synchronize the outputs from the k stages, digital latches are used.

The main advantages of pipelined ADCs are that they can provide high throughput rates and occupy small die areas. If the ADSCs are done with flash converters, pipelined architectures require only two main clock phases per conversion. Also, the number of stages used to obtain a given resolution is not constrained by the required throughput rate. Due to the small size and low power consumption, the pipelined architecture is more suitable for high-resolution applications than the full-flash but is susceptible to circuit imperfections such as offset, gain error, and nonlinearities. Therefore, under the constraint of a total desired resolution, the number of stages may be chosen to minimize the required die area [16]. As expected, any errors generated in one stage will be passed on down the chain.

6.4.1 Time-Interleaved Pipeline Examples

A demonstration of a 12-bit, 3.072-GS/s TI pipelined ADC is developed in [26] specifically for applications in a digital wideband receiver architectures. The design of a fully digital wideband receiver that produces targeted signal descriptor words requires (i) high-input instantaneous bandwidth and (ii) high-speed, real-time signal processing. A block diagram of the wideband receiver is shown in Figure 6.33.

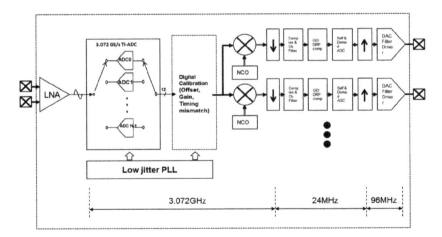

Figure 6.33 Wideband fully digital receiver. (©2020 IET, reprinted with permission from [26].)

The design must incorporate a high-speed ADC that can provide high accuracy, along with a large SFDR at the same time lower-power consumption. Instead of using a single ultra-high-speed ADC, a parallel sampling approach is employed, in which multiple low data-rate ADCs are time interleaved using phase-shifted sampling clocks to achieve a high overall sampling rate (GS/s). In this device, 32 pipelined ADCs each with a sampling rate of 96 MS/s are interleaved in the time domain to achieve the overall sampling rate. Figure 6.34 shows the working concept of the TI pipelined wideband receiver concept.

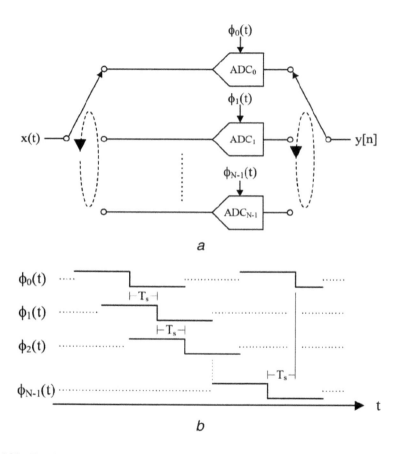

Figure 6.34 Time interleaved wideband fully digital receiver concept showing (a) time interleave concept and (b) the sampling time waveform T_s. (©2020 IET, reprinted with permission from [26].)

The basic structure is shown with $x(t)$ as the input signal and $y[n]$ as the digital output. Considering the number N of sub-ADCs, with T_s being the sampling time for a single ADC, the sampling time of N sub-ADCs can be expressed as

$$\hat{T} = NT_s \tag{6.2}$$

and the kth sub-ADC will trigger at

$$t_k(n) = n\hat{T}_s + kT_s = (nN+k)T_s \qquad (6.3)$$

Considering the input signal to be sampled uniformly (ideally), the output of the kth sub-ADC can be expressed as [26]

$$\hat{y}_k(n) = x(t_k(n)) = x((nN+k)T_s) \qquad (6.4)$$

where $(n-k)/N$ is an integer so

$$y_k(n) = \hat{y}\left[\frac{n-k}{N}\right] \qquad (6.5)$$

The output from all of the ADCs is then multiplexed to achieve $y(n)$ or

$$y(n) = \sum_{k=0}^{N-1} y_k(n) = x(nT_s) \qquad (6.6)$$

and is the output of the TI-ADC when all of the sampling periods are considered ideal. Problems such as mismatch in gain, offset, and timing skew occur, which affect system performance. Taking these effects into account the output can be expressed as

$$y_k(n) = (G_k x(nT_s - \tau_k) + O_k) \qquad (6.7)$$

where the gain error of the ADC is G_k, the offset is O_k, and the timing skew is τ_k. In the proposed design, 32 channels of the 12-bit 96-MS/s pipelined ADC are TI to achieve an overall bandwidth of 3.072 GS/s [26].

Time interleaving requires the sub-ADCs to operate on sampling clocks that have the same sampling speeds, but different phase angles, such that only one of the M sub-ADCs is active at a given time [26]. The generation of a precise clock phase for all channels is imperative to the system performance, as any skew in the generated phases can result in the sampling of incorrect values.

The conversion performance of a TI-ADC cannot surpass the performance of the sub-ADC used. Thus, the precise selection of a sub-ADC based on application is an important task. As the number of interleave channels increases, problems such as mismatch for example, in timing skew, bandwidth, offset, and gain are increased [26]. In addition, a larger number of channels results in a higher power requirement so it is necessary to balance the trade-off and select an ADC that offers high speed with an optimal compromise in terms of power consumption, accuracy, and have a high-input dynamic bandwidth [26]. For these reasons the pipeline ADC was chosen as the sub-ADC.

Each stage contributes a certain number of bits to the final converted output, and the process continues until the signal reaches the final stage where the m-bit ADC (usually low-resolution flash ADC) resolves the final bits of the pipelined ADC. That is, the final output is

$$x = Q_1 + \frac{1}{A_1}Q_2 + \frac{1}{A_1 A_2}Q_3 + \frac{1}{A_1 A_2 A_3}Q_4 +, \cdots, \tag{6.8}$$

where Q_i is the quantization output and A_i is the residue amplification gain. All of the stages in the pipeline pass their bits to the time alignment and error correction block, which time aligns the output bits gathered at the different time instances and performs necessary calibrations for error correction [26].

Generally, the number of bits per stage varies from 1 to 4 (with a redundancy of 0.5 bit for error correction). The most popular among these is 1.5 bit per stage, the use of which simplifies the stage design requirements. That is, the requirement for a residue gain amplifier is $A_i = 2^n$ (where n is the number of bits per stage and i is the stage number) for the residue to acquire a full stage range for the approaching next stage. However instead of $A_i = 2^n$ for the full-scale range, to account for errors $A_i = 2^{(ni-1)}$ is used for amplification [26]. The total resolution of the pipelined ADC can be calculated as

$$Resolution_{ADC} = \#_{Stages}(n-1) + m$$

where n is the number of bits per stage and m is the number of bits in the final stage. When all of the bits are resolved, the digital error correction and time alignment block combines the bits from all of the stages, provides them with appropriate weighting, aligns them in time, and subtracts the redundant bits to attain target resolution. The same reference voltages are used across the entire chain for the quantizers and DAC circuits to simplify the layout implementation and eliminate the path mismatch. This decreases the parasitic capacitance and for this reason, the traditional 1.5-bit architecture is used in their proposed design. A total of ten, 1.5-bit stages are cascaded with a 2-bit flash quantizer to achieve 12-bit resolution [26].

A 1.5-bit architecture has two main parts: the sub-ADC (in the pipeline stage) and a multiply DAC (MDAC). The comparator candidates considered were the (a) single-stage dynamic comparator [27],(b) the preamplifier, latch-based comparator [28], (c) the two-stage dynamic regenerator comparator, and (d) the dual-tail, dual-rail dynamic latched comparator [26]. Since the preamplifier, based latched comparators have high speeds, with low power consumption and low offset in their outputs they were found to be the most appropriate.

The preamplifier-based latched comparator completes its conversion in three stages: (1) *preamplification*, (2) *regenerative latch*, and (3) *output latch*. The complete structure is described in [26]. A block diagram is shown in Figure 6.35.

6.4 Time-Interleaving Pipeline Techniques   313

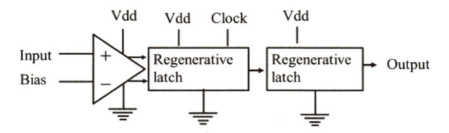

Figure 6.35 Block diagram of the preamplifier-based latched comparator showing the preamplifier, the clocked regenerative latch, and the output latch.

The design of the three stages including the clocked stage is described below with the design equations referencing the components shown on the schematics. Figure 6.36; schematic diagram of the latched comparator showing the preamplifier section.

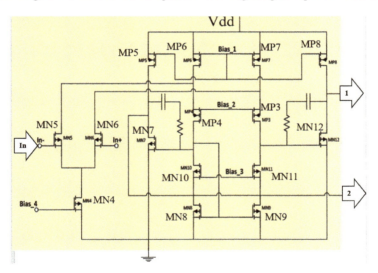

Figure 6.36 Schematic diagram of the preamplifier section of the comparator showing the input and the bias. (©2020 IET reprinted with permission from [26].)

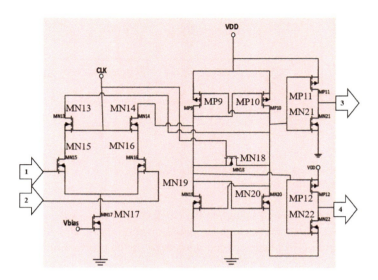

Figure 6.37 Schematic diagram of the comparator showing the regenerative latch section, and the clock, and the two inputs and two outputs. (©2020 IET reprinted with permission from [26].)

6.4 Time-Interleaving Pipeline Techniques

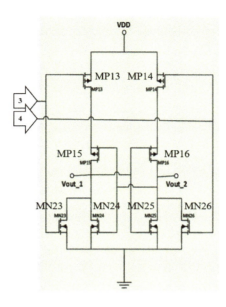

Figure 6.38 Schematic diagram of the comparator showing the output latch. (©2020 IET reprinted with permission from [26].)

A comparator that uses a preamplification in the first stage has the advantage of increased conversion speed as it only amplifies the difference between the two input signals reducing the comparison time for the regenerative latch. It also reduces the offset error of the comparator. For high-speed designs, it is also necessary to have a high gain-bandwidth (GBW) in each stage. Consequently, a number of amplification stages can be used. However the decision on the number of amplification stages is critical as it increases the transmission delay. The transmission delay of a preamplifier can be expressed as [26]

$$t = n\tau = \frac{nA^{1/n}}{G(GBW)} \qquad (6.9)$$

where t is the transmission delay, n is the number of amplifiers, τ is the delay of each amplifier, A is the total gain of the preamplifier, G is the gain of the individual stage amplifiers and GBW is the gain-bandwidth of each stage. In the reported design, the two-stage folded cascode amplifier is used for preamplification of the input signal. The circuit provides an effective gain of 58.98 dB with a phase margin of 66.39 deg and the effective GBW is 1.602 GHz with the amplifier output then passed on to the regenerative latch stage [26].

The regenerative latch is the *decision making stage* of the comparator and is always the preferred choice due to its high speed. The speed depends on the recovery time constant $\tau_{recovery}$ and the regeneration time constant τ_{regen} as

$$\tau_{recovery} = 2R_{on,MN18}C_{tot} \qquad (6.10)$$

$$\tau_{regen} = \frac{C_{tot}}{g_{MP9,10}g_{MN19,20}} \qquad (6.11)$$

where $R_{on,MN18}$ is the on-time impedance of $MN18$, C_{tot} is the total parasitic capacitance and $g_{MP9,10}$ and $g_{MN19,20}$ are the transconductances of MP9-MP10 and MN19-MN20, respectively. From (6.10), the recovery time of the regenerative latch can be decreased by decreasing $R_{on,MN18}$ (i.e., increasing the width of MN18). The regeneration time usually remains constant, because broadening the widths of MP9–MP10 and MN19–MN20 will certainly increase $g_{MP9,10}$ and $g_{MN19,20}$ but at the same time increase the value of C_{tot} by the same amount. The final regenerative latch stage is an SR latch that buffers the data at the output and keeps it stable [26]. While the regenerative latch is in reset mode the output latch will maintain the value of the last clock period and only generate an output when the regenerative latch is in the compare mode.

The next most important component in a pipeline ADC is the multiplying DAC (MDAC). The MDAC samples the residue from the previous stage and reconstructs the analog equivalent of the digital output from the sub-ADC, thereby generating the residue for the next stage by subtracting the reconstructed signal from the stage output and maintaining residue amplification. Details on the MDAC are given in [26].

6.5 TI-SUCCESSIVE APPROXIMATION REGISTER TECHNIQUES

A SAR is a converter that takes continuous analog waveform and converts it into a discrete digital representation using a binary search through all possible quantization levels before finally converging upon a digital output for each conversion. A block diagram showing how a SAR converter operates is given in Figure 6.39.

6.5 TI-Successive Approximation Register Techniques

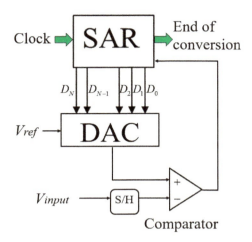

Figure 6.39 SAR showing its basic operation.

A S/H circuit acquires the input voltage V_{input} and a comparator amplitude analyzes the input comparing it to a reference voltage from an internal DAC output–the result of a comparison on a previous clock signal. The SAR subcircuit is designed to supply an approximate digital code of V_{input} to the internal DAC. The internal reference DAC uses V_{ref} to generate the analog voltage the comparator needs for amplitude analyzing the input analog voltage using the digital code output from the clocked SAR component in Figure 6.39.

A more detailed SAR schematic shows, along with the comparator operation, counting logic to perform the conversion as shown in Figure 6.40. The first step in the conversion is to see if the input is greater than half the reference voltage. If it is, the MSB of the output is set. This value is then subtracted from the input, and the result is checked for one quarter of the reference voltage. This process continues until all the output bits have been set or reset. A successive approximation ADC takes as many clock cycles as there are output bits to perform a conversion. That is, the process repeats with progressively smaller steps (according to the binary weights), and all bits are resolved as the voltage is nulled to within a small residual quantization error over time [29].

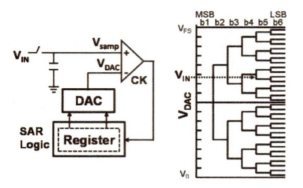

Figure 6.40 SAR ADC and its time trellis operation [30].

An example of the charge redistribution binary weighted capacitor array DAC is shown in Figure 6.41. It consists of switches, capacitors, a voltage comparator, and digital logic. This composition aligns the converter particularly well with the strengths of modern nano-CMOS, which is first and foremost optimized for digital logic (and switch) performance.

In the first step, the SAR ADC samples the input voltage V_{in} on all of the capacitors. The register in the SAR logic holds the estimated digital value of the input voltage and the DAC reproduces the corresponding analog value (V_{DAC}). If the sampled input voltage V_{samp} is lower than the V_{DAC} then estimated digital value is increased, otherwise it is decreased [30]. Normally a binary weighted capacitor array is used to generate V_{DAC} with the trellis shown on the right-hand side in Figure 6.40.

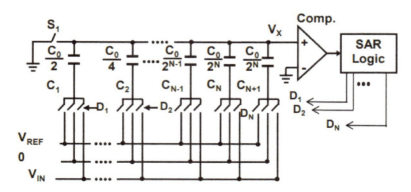

Figure 6.41 Charge redistribution SAR-ADC[30].

The four advantages of a SAR digitizer include:

- By using a dynamic comparator (regenerative dynamic latch), no static current is required and therefore, only low power is consumed as there is no need of an operational amplifier [29, 30].
- Suitable to nano-CMOS process[29, 30].
- Good lattice matching to recent metal-oxide-metal layout density with about 10-bit matching accuracy [30].
- Sampling kT/C noise advantage: total binary weighted, metal fringe capacitors $\sum C$ are used to sample input voltage. Therefore, kT/C is small compared with other ADCs [29, 30].

The four major disadvantages include:

- Long conversion time that is difficult to make smaller since the bits are determined from the MSB to the LSB each clock cycle. For n-bit resolution, $n+2$ operating clock cycles required to complete conversion [30].
- Limited precision by comparator decision error due to binary weighted capacitor array from MSB to LSB [30].
- Limited precision by comparator thermal noise, 1/f noise. The problem is that SNR is limited, when low supply voltage used in fine process [30].
- Limited precision by capacitor matching to about 10-bit or 0.1% relative precision level. If greater than 10 bits are required, a method is needed to improve precision [30].

The speed of a SAR ADC operation can be improved by the use of asynchronous clocking. In this scheme, by equipping the ADC with a comparison completion detector, after completing one comparison, the next bit comparison automatically starts, which can shorten the conversion time by half. Also by the use of deciding multiple bits, each clock can greatly reduce the time in going from the MSB to the LSB. Pipelined SAR ADCs can also help [30].

6.5.1 Time-Interleaved SAR Examples

The biggest improvement for SAR devices is through time interleaving. Time interleaving has emerged as the most common method of achieving ultra-fast quantization at reasonably high resolution. Time interleaving is a widely used option (as we have shown above) to push ADC performance boundaries to higher throughputs by exploiting arrays of reduced speed digitizers operating on the signal.

The TI SAR technique however, is not without its challenges [31]. The first example we show is an 11-bit 1 GS/s time-interleaved (×2) SAR-assisted pipeline ADC specifically for wideband direct-sampling radio-frequency receivers. The ADC architecture combines the speed advantage of the pipeline algorithm and the structural simplicity of the SAR structure. Figure 6.42 shows the system architecture with the input voltage V_{in} divided into between the two channels to ADC1 and ADC2. The front-end S/H is employed to mitigate the timing mismatch between two channel ADCs [32].

Subsequent to the front-end S/H, a high-precision front-end comparator is employed to first predetermine the MSB. In this manner, the ensuing channel ADCs only need to resolve the subsequent 10 bits instead of the full 11 bits, hence reducing the hardware and power overheads. Consequently, both the structure and the operation of the pipeline stages are simplified, thereby enhancing the conversion rate and accuracy [32].

In particular, the advantages include that the proposed ADC eliminates the multiplying DAC in the conventional pipeline ADC, and hence makes it compatible with process portability. By eliminating the multiplying DAC, the accuracy limitation of the conventional pipeline ADC, especially when the process technology scales to smaller nodes, is also removed and the operation of the pipeline stages is simplified to a single-iteration successive approximation (SA), thereby enhancing the conversion rate [32]. In addition, the digital transmission between the pipeline stages features a high noise tolerance and improves the overall ADC accuracy.

In addition, 2-bit/step conversion scheme is employed to reduce the number of S/Hs [32]. In this manner and the speed requirement of these S/Hs is significantly relaxed by the time-interleaved technique and the hardware is reduced by extracting the quantized MSB at the front of the ADC as shown in Figure 6.42.

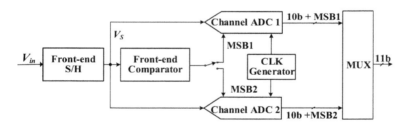

Figure 6.42 Charge redistribution SAR-ADC[32].

The prototype ADC fabricated in 65-nm CMOS process and achieves SNDR$\geq$ 56-dB across 500 MHz Nyquist bandwidth at 1-GS/s conversion rate with a 230 mV power dissipation leading to a $FOM_W = 449.2$ fJ/conv.-step.

Figure 6.43(a) depicts the measured SNDR and SFDR versus input frequency at 1 GS/s. The SAR-assisted pipelined ADC achieves an SNDR $\geq$ 55.9 dB (equivalent to 9 ENOB) up to 475 MHz. The peak SNDR is 59.1 dB, equivalent to 9.52 ENOB. The SFDR remains above 60.5dB up to 475 MHz, and exhibits a peak value of 66 dB at low frequency. Figure 6.43(b) depicts the measured SNDR/ENOB versus conversion rate. The measurement is performed with normalized input frequency

6.5 TI-Successive Approximation Register Techniques

(Input frequency/Conversion rate= 0.475). It can be observed that the SNDR remains largely constant (< 1 dB degradation) with the conversion rate up to 1-GS/s.

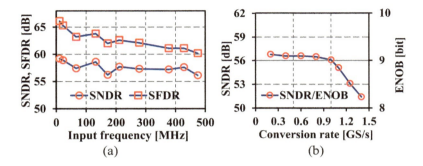

Figure 6.43 Time-interleaved SAR-assisted pipelined 11-bit ADC showing (a) measured SNDR and SFDR versus input frequency, and (b) measured SNDR/ENOB versus conversion rate[32].

The last time-interleaved SAR example shows an ultra-high-speed architecture that directly connects the input to four sampling and interleaving slices that feed buffered samples to 16 sub-ADCs resulting in 64 sub-ADCs that convert the analog samples. The aggregated digital output is captured by a large high-speed memory block storing the 8192 digitized samples. Figure 6.44 shows the top-level overview of the ADC showing the four samplers, 1:16 interleavers and the clocks showing their respective offset phase shifts to interleave the samples. Within each bank, the ADC slices process the incoming samples in a round-robin fashion [33]. The aggregate throughput of this example is thus 64 times the sampling rate of each ADC slice (e.g., 64 GS/s when 1 GS/s SAR slices are employed).

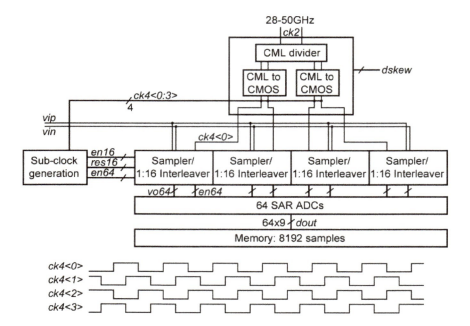

Figure 6.44 Highly interleaved SAR ADC showing (a) the top-down architecture with CML to CMOS divider to the four samplers and their 64 SAR ADCs each. Also shown is the memory housing the 8192 [33].

Consequently, this architecture requires only four timing-critical clock phases, namely those connected to the first input sampling switches [29]. These critical phases are derived from a half-rate differential clock ck2 of up to 50 GHz, which is divided by 2 in CML and converted to CMOS levels leverages 64 SAR ADCs running in parallel using a hierarchical 4-4-4 interleaving approach. Figure 6.45 shows the SNDR vs. sampling frequency for different supply voltages.

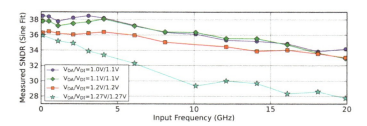

Figure 6.45 Measured SNDR vs. sampling frequency for different supply voltages [33].

The ADC supply on the interleaver, including the sampler, CML clock divider, and CML to CMOS stages (VDI), is 1.2V for 90GS/s, and the SAR ADC supply (VDA) is set to the save level for 90GS/s. To save power, VDA can be lower than VDI for lower conversion rates. The best FoM is achieved at 70 GS/s with 121 fJ/conversion-step. The FoM at 90 GS/s is higher with 203 fJ/conversion-step, (due to increased voltage on the SAR ADCs). The SFDR of 41.4dB at 19.9 GHz input frequency and full-scale input amplitude is limited by the third-order harmonic distortion.

In summary, the above sections have examined several of the key digitizer techniques available for DRFM transceiver design and development. Examples were given showing the architecture concepts and results.

Below we show some of the FOM graphs and plots to get a better understanding of the current capability of these digitizer designs and performance levels.

6.6 FIGURE OF MERIT PLOTS FOR ADCS

Figure 6.46 shows the FOMs versus conversion rate (f_s). Here the FOM plotted is $FOM_S(SNDR)$ (6.46). The red markers indicate data reported after 2010. Points with solid fill mark time-interleaved designs.

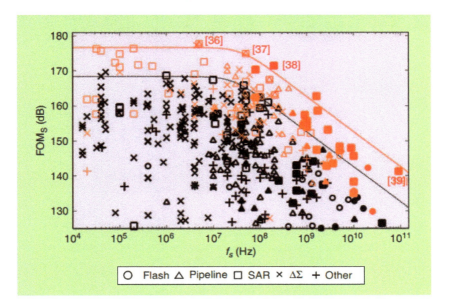

Figure 6.46 FOMs versus conversion rate (f_s). The red markers indicate data reported after 2010. Points with solid fill mark time-interleaved designs. (©IEEE, reprinted with permission from [34].)

First, note that the achieved FOMS is highest for low conversion rates. This is expected since it is generally harder to make a fast ADC energy efficient. Second, between 10 and 100 MHz, the efficiency begins to deteriorate and rolls off with a slope of approximately −10 dB per decade, which indicates that the power dissipation scales with the square of the conversion speed in this area [34]. Third, the recent pipelined SAR designs set the peak performance near the corner while the time-interleaved SAR ADC reported marks the fastest design and lies almost exactly on the −10 dB/decade roll-off portion of the envelope. The overall progress rate comes out almost exactly to 1 dB per year (or doubling of energy efficiency every three years) [34].

Architecture trends are reflected in 6.47. More details can be found in [34]. The SAR ADCs range from ultralow-power to ultrahigh-speed designs (using time interleaving). Much of the progress in SAR converters is enabled by technology scaling, and there have been a number of important circuit and architecture innovations as well, including the combination of SAR with pipelining and the use of dynamic residue amplification in hybrid topologies.

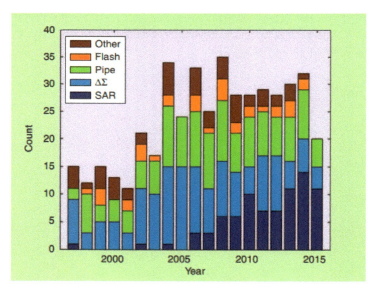

Figure 6.47 Architectures of ADCs described in the literature (ISSCC 1997–2015 and VLSI Circuits Symposium 1997–2014). (©IEEE, reprinted with permission from [34].)

With pipelined ADCs notable are the energy efficiency and scaling robustness of SAR converters. A low stage count leads to significant reductions in power and complexity. With proper calibration, the pipelined architecture can be pushed to 1 GS/s at 14 bits.

Flash ADCs have regained interest due to the time interleaving of the design. A time-interleaved flash design is reported that operates at 10.3 GS/s and thereby enables a wideband transceiver. The speed can also be extended to 20 GS/s while maintaining power efficiency. The key to high efficiency in flash ADCs is to identify a proper offset calibration/mitigation scheme and to minimize the circuit complexity. For example, using dynamic interpolation at the comparators' regenerative nodes. Another approach is to adaptively control the decision levels to minimize the system's bit error rate. Analog-to-information or compressed sensing receivers are also an alternative.

Oversampling $\Delta\Sigma$ converters are for high resolution and have reached 6 GHz in recent work [34]. Signal bandwidth have scaled proportionally, exceeding 100 MHz. Digitally assisted design concepts have been applied to all of the architectures mentioned above, including a time-interleaved pipeline design that leverages two million logic gates to reach a performance level of 14 bits at 2.5 GS/s. Another area is voltage-controlled oscillator–based Nyquist converters. Here, digital calibration is required to make these approaches practical.

Figure 6.48 show recently reported SNDR and power efficiency (P/f_s) for several ADC architectures. The green triangle points show the SAR ADCs and power efficiency is very good for the SAR ADCs but high-SNR/SNDR cannot be achieved by the SAR architecture. High SNDR can only be achieved by sigma-delta ADCs, which are shown as yellow circles.

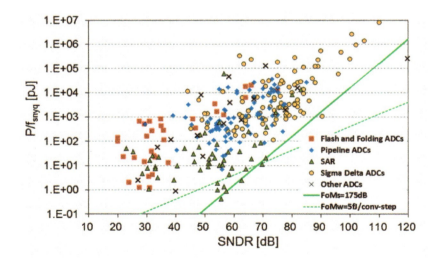

Figure 6.48 Power efficiency (P/f_s) vs. SNDR with ADC architectures. (Modified from B. Murmann "ADC Performance Survey 1997-2015," [online] http://web.stanford.edu/murmann/adcsurvey.html.)

Wideband architectures pushing high-resolution digitizers out of the audio band and into the RF band are now coming of age. The majority of new results published, for example, in the *IEEE Journal of Solid State Circuits*, will have these FOMs presented. Figure 6.49 shows a recent snapshot of the technologies outlined below quantifying their FOM (fJ/c.s.) versus Nyquist sampling frequency (GHz) [35]. Shown are the relative positions for the flash and time-interleaved architectures including TI=flash, TI-successive approximation, TI-pipeline, and TI-binary search.

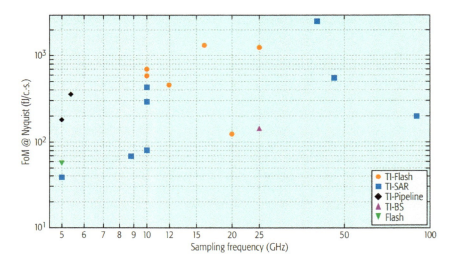

Figure 6.49 FOM (fJ/c.s.) versus Nyquist sampling frequency (GHz). (©IEEE, reprinted with permission from [35].)

Another look at Nyquist rate ADCs comparing the flash, the two step, folding, pipeline, and SAR techniques are shown in what is referred to as an *aperture plot* in Figure 6.50. This plot illustrates that the conversion region covered by $\Sigma\Delta$ converters covers much the same region as the SAR converters by the different ADC architectures. Note that a wide conversion region is covered by ADCs, comparable with that achieved by SAR converters.

6.6 Figure of Merit Plots for ADCs

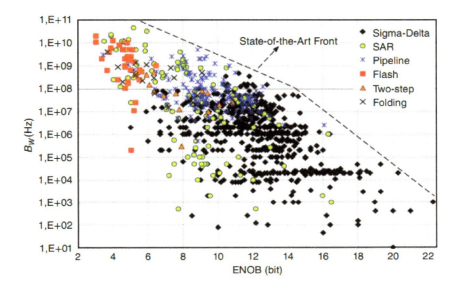

Figure 6.50 Bandwidth versus effective number of bits, ENOB. (©IEEE, reprinted with permission from [36].)

Figure 6.51 shows a graph of the ADC power efficiency (quantified via the Schreier FOMs vs. sampling frequency (f_s). Time-interleaved designs are marked with a bold outline (for SAR) or gray shading (for all other architectures) [29]. From this figure we see that SAR-based designs (in orange) deliver leading-edge performance for sampling rates from tens of kilohertz to tens of gigahertz. Finally, in the ultra-high-speed regime, the groundbreaking 90 GS/s design of demonstrated that a sea of 64 SAR ADCs running in parallel can be an attractive solution for emerging optical and electrical data links.

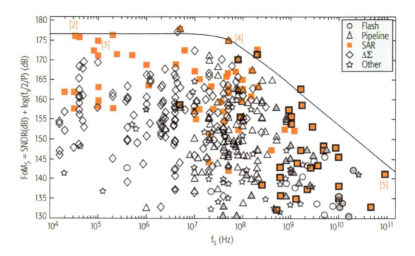

Figure 6.51 ADC power efficiency (quantified via the Schreier FOMS vs. sampling frequency (f_s). Time-interleaved designs are marked with a bold outline (for SAR) or gray shading (for all other architectures). (©IEEE, reprinted with permission from [29].)

In the next section a key technology that is currently on the rise and being investigated as a new type of transistor is discussed.

6.7 FINFET TRANSISTOR TECHNOLOGIES

With scaling for advanced devices beginning to approach physical limits, new architectures using *vertical transistor structures* are being investigated. FinFETs use tall fin-like structures to achieve higher chip speed and lower power. A comparison of the planar transistor structure shown in Figure 6.52(a) and the FinFET architecture shown in Figure 6.52(b) shows the key differences in the channel region.

6.7 FinFET Transistor Technologies

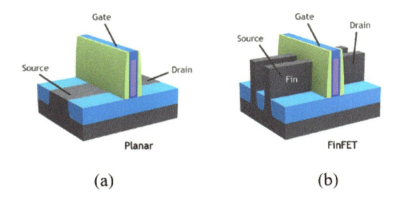

Figure 6.52 Comparison of the (a) planar transistor and the (b) FinFET architecture.

Advantages of using FinFET transistors include:

- Better performance and reduced power consumption compared to planar transistors;
- A 16-nm/14-nm FinFET process can offers 40%-50% performance increase or 50% power reduction compared to a 28-nm process;
- Can achieve higher frequencies for a given power budget;
- Can exhibit more drive current per unit area than planar devices due to the height of the fin that can be used to create a channel with larger effective volume but still take advantage of wraparound gate;
- Power reduction due to reduced supply voltage requirement.

The main disadvantage is the cost. This is due to building the FinFET as it uses a number of additional steps in the manufacturing flow process. On a bulk-silicon process, the control over the fin depth is more difficult to maintain.

For the current high-speed 16-nm FinFET CMOS DACs and ADCs achieve sampling rates of up to 128 GS/s with analog bandwidths of around 35-GHz and 8-bit nominal resolution at an average ENOB of 5.5 bits [37]. Both DACs and ADCs are integrated with the DSP to avoid the data interface bottleneck; for example, an interface rate of 3.2 Tb/s is required between four 100 GS/s 8-bit data converters and the DSP high-speed digital and mixed-signal circuits were fabricated using indium phosphide double heterojunction bipolar transistor (InP-DHBT) technology, which is capable of producing circuits with high ft and fmax while also providing high breakdown voltage leading to large output swing circuits. While 7-nm and 14-nm FinFET technology and highly parallelized SAR architectures have been favored in ADCs because of the low power consumption and small layout footprint facilitated by the continued shrinkage of the minimum feature size, FinFETs suffer from poor f_{MAX} and large parasitic metal capacitance and resistance, which limit the ADC bandwidth.

Below we examine some of the recent dual-port memory advancements being developed today that can both read and write to the memory at the same time.

6.8 EMBEDDED DUAL-PORT MEMORY

For the DRFM to process the digital signals of interest, the ADC/digitizers strobe the bits into a dual-port memory using a memory controller. These embedded memories can often dominate the total area and power budget of entire SoCs. Various blocks that are integrated on these SoCs, such as ADCs, DACs, communication modems, signal processing units, and computational cores, require multiport RAM embedded memories to improve the DRFM bandwidth by enabling multiple operations using the same memory bank in parallel. For example, the dual-port memory shown in Figure 6.53 is well utilized for EA processing because several read and write access operations are possible at the same time.

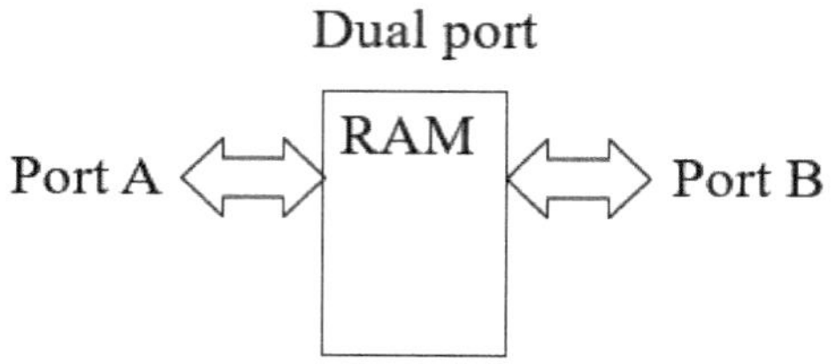

Figure 6.53 Block diagram of a dual-port random access memory.

That is, dual-port memory provides a common memory accessible to both processors that can be used to share and transmit data and system status between the two processors.

The controller enables the data to be read simultaneously as bits are streamed out at the same time. It does this by providing the memory address and control signals for the signal storage and recall operations. A number of user controls are also available. The storage enable function is used to designate the pulse to be stored. The store address designates where the leading edge of the pulse is to be stored. The recall initiates a recall cycle at the next word clock time. The recall output triggers must be synchronized with the DRFM clock for coherent output. The recall address is the address where recall begins and the delay time is the throughput delay impressed upon the stored signal.

6.8 Embedded Dual-Port Memory

Dual-port (DP) static random access memory (SRAM) is useful for a DRFM design because it can offer simultaneous read and write operations with two asynchronous clocks. The dual-port memory cell is a multiport, SRAM cell (DP-SRAM) realized using flip-flops, which provide the required flexibility needed by multiprocessor applications. Typically, each cell uses 6 to 10 transistors (6T to 10T). Note that the circuit design is more complex than for a single-port (SP) SRAM in that resistance and capacitance degradation caused by process scaling and the read-disturb-write (RDW) problem in DP-SRAM cells are a major challenge for DP-SRAM design.

6.8.1 FinFET Technology for SRAM

Figure 6.54 shows five different transistor configurations for implementing DP-SRAM. Each SRAM cell is accessed by dual (two) ports with devoted word and bit lines to each. Single port SRAM only allows access to a single address of a memory cell at a time during each clock pulse. However, dual port overcomes this drawback and allows concurrent read or write access at different addresses. Thus the efficiency is almost doubled by using DP-SRAM.

Figure 6.54(a) shows the schematic of a 6T (6-Transistor) SRAM commonly used for on-chip memory. The smallest size of 6T SRAM cell is called as the 'high density' (HD) cell. In an HD cell, all transistors use one fin each. In order to obtain a larger cell current, a 'high current' (HC) cell is prepared and consists of 2 fins for the PG and PD transistors.

Figure 6.54(b) shows the HD, HC 14-nm FinFET cell schematic [38]. In FinFET technology, the same design concept as the straight type cell is used but an integer number of fins are engaged for each transistor. An integer number of fins are included instead of adjusting the gate width as is the case for bulk CMOS technology.

In Figure 6.54(c) an 8T SRAM is shown that has two ports; the read-out port with stacked nMOS transistors apart from the write port.

In Figure 6.54(d) shows a 10T 'differential-signal' (DS) SRAM. The 10T SRAM cell shown in Figure 6.54(d) is basically enhanced from an 8T cell with read-out transistors from both storage nodes as dual wing.

Figure 6.54(e) shows an 8T DP SRAM and indicates a DP.

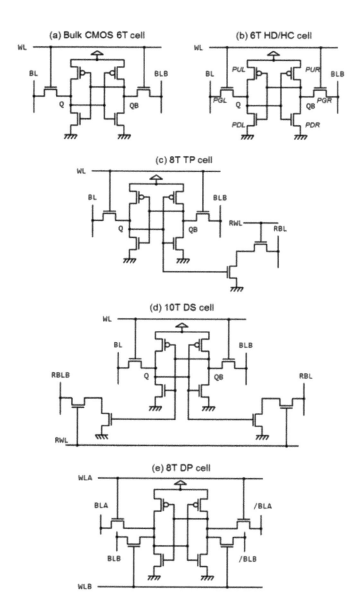

Figure 6.54 Five dual-port SRAM configurations showing (a) 6T SRAM commonly used for on-chip memory, (b) 14-nm 6T HD, HC FinFET SRAM cell, (c) an 8T TP SRAM, (d) 10T DS SRAM, and (e) 8T DP SRAM (©IEEE, reprinted with permission from [38].)

Since the readside output of an 8T SRAM cell is a single-ended signal, a large-swing sense scheme is typically employed, although a pseudodifferential signal scheme could be utilized with the penalty of additional area and complexity. The 10T SRAM cell in 14-nm FinFET technology has a larger number of transistors required for one cell. This results in larger cell area and parasitic capacitances. The 10T SRAM shows performance improvement over the 8T cell with the drawback of larger cell size. However, an increased number of read-out fins beyond 2 on the 10T do not provide much speed improvement due to the resulting larger parasitic capacitance.

6.8.2 MOSFET Technology

Figure 6.55 shows the block diagram of a dual-ported SRAM device consisting of four blocks: the bit cell array, the I/O circuitry, the row decoder, and the control circuitry. A single bit cell is designed initially for successful read and write operations.

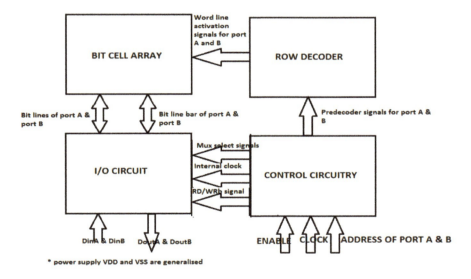

Figure 6.55 Block of SRAM. (©IEEE, reprinted with permission from [39].)

Bit line charges depending on the input given such as 0 or Vdd. It comprises word lines on Port A and B, respectively. The function of word lines is to select a particular row in an array of memory during read and write operation. During read operation, both the bit lines of Port A and Port B are precharged first. In dual-port simultaneous write operation cannot be done using both the ports of the same bit cell in the same clock pulse. Either Port A or Port B can be used to write "1" but simultaneous write "1" operation can be made to a bit cell from Port A and can write "1" to another bit cell from Port B, but for read operation, the data stored in a single bit cell is read using both ports [39].

The I/O circuit consists of subblocks such as precharge circuit, column mux, sense amplifier, and latches that are integrated into one complete block. The area of the bit cell layout is measured in the bit cell array layout. Each bit cell needs an area of 0.45×2.7 μm area. So the total 128 X 128 bit cells need 311.04 μm^2. The power consumed by the chip when the SRAM is working is measured to be 71.8 μW.

6.8.3 NEMFET Technology

The dual-ported memory technology and its design has advanced significantly in the last few years. Based on a nano-electro-mechanical field effect transistor (NEMFET), a volatile low-power dual port (one port for write and one for read) multitier 3D stacked hybrid NEMFET-CMOS memory cell has been developed [40]. The cell has the appealing properties of the NEMFET transistor; that is, ultra-low leakage currents, abrupt switching and hysteresis plus the versatility of CMOS technology.

The cell relies on an energy-effective NEMFET-based inverter, designed such that no short-circuit current can occur during switching to store the data. The design is also based on adjacent CMOS logic to allow for read and write operations, and for data preservation. The 3D stacked hybrid memory approach relies on the hysteretic NEMFET inverters to store the data. As well, included adjacent to the NEMFET is the CMOS-based logic arrays to allow for read/write operations and data preservation. A NEMFET suspended gate cross-section illustration showing the two states, i.e., pull-out (OFF) and pull-in (ON), is shown in Figure 6.56.

6.8 Embedded Dual-Port Memory

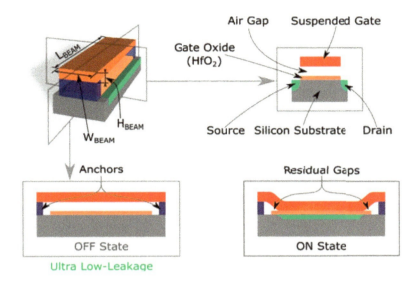

Figure 6.56 NEMFET suspended-gate illustrative cross section showing the two states, pull-out (OFF) and pull-in (ON). (©IEEE, reprinted with permission from [40].)

Here the gate-oxide is the thickness of the gate oxide, H_{BEAM} is the thickness of the suspended gate, W_{BEAM} is the width of the beam, L_{BEAM} is the length of the beam, and Air Gap (gap) is the gap between the oxide and the suspended gate.

The NEMFET inverter transfer characteristic is depicted in Figure 6.57 and shows that when the nNEMFET beam is pulled-in, the pNEMFET beam is pulled-out, and vice versa, eliminating the short-circuit current.

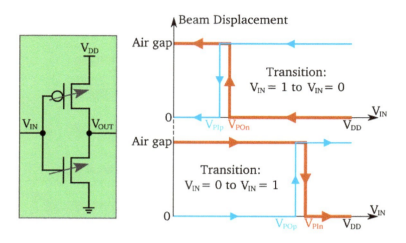

Figure 6.57 NEMFET inverter schematic and its hysteretic transient behavior. (©IEEE, reprinted with permission from [40].)

The fast switching behavior also mitigates the subthreshold leakage (SL), while NEMFET's gate leakage (GL) is mitigated in the OFF state due to the gap between the suspended gate and the oxide. The short-circuit current (SCC) contributes to dynamic energy dissipation, while the SL and GL contribute to the static energy consumption. The NEMFETs are also able to diminish these three components and they can be an energy-effective MOSFETs replacement in CMOS-like Boolean gate implementations. Other nice NEMFET properties include ultra-low leakage currents in addition to the fast switching. They also have the advantages of CMOS technology versatility.

The device relies on an energy effective NEMFET-based inverter, designed so that no short-circuit current can occur during switching to store the data, and on adjacent CMOS-based logic to allow for read and write operations and for data preservation. It makes use of the 3D integration in order to enable the potential co-integration of NEM and MOS transistors, which can be challenging on the same die.

A model of the NEMFET-based inverter transfer characteristics is shown in Figure 6.58 and shows the hysteresis behavior influence on the NEMFET inverter and current consumption. The NEMFET's inverter current consumption consists of the current required to charge and discharge the load and the internal capacitances, while SCC, GL, and part of SL have all been eliminated.

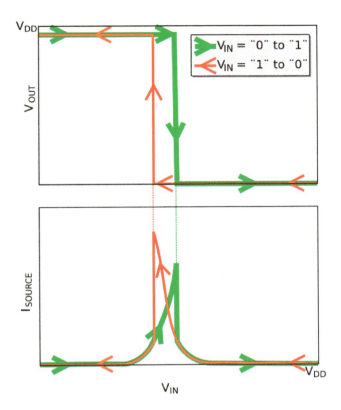

Figure 6.58 NEMFET-based inverter transfer characteristics. (©IEEE, reprinted with permission from [40].)

The stability of an inverter is given by its noise margin. The noise margin is the maximum noise voltage on the inverter input that does not disturb its output. The low and high noise margins (i.e., NML and NMH, respectively are defined as)

$$NM_L = V_{IL} - V_{OL} \qquad (6.12)$$

and

$$NM_H = V_{OH} - V_{IH} \qquad (6.13)$$

V_{IH} is the minimum HIGH input voltage, V_{IL} is the maximum LOW input voltage, V_{OH} is the minimum HIGH output voltage, and V_{OL} is the maximum LOW output voltage. The NEMFET-inverter hysteresis behavior showing these voltage levels is shown in Figure 6.59.

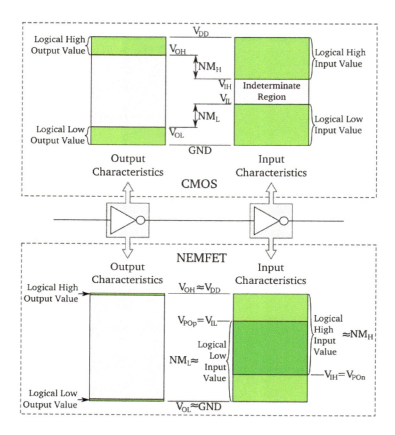

Figure 6.59 CMOS versus NEMFET inverter noise margin. (©IEEE, reprinted with permission from [40].)

One can see that NM_L and NM_H are equal with V_{PO_p} and the $VDD - V_{PO_n}$ respectively, because V_{OH} is approximately equal with logic "1." That is, VDD and V_{OL} are approximately equal with logic "0" or ground GND.

As a final note, a high-density alternative for two-ported functionality, gain-cell embedded DRAM, GC-eDRAM has also been examined. However, as a dynamic memory element, periodic refresh operations are required for data retention, limiting the data availability.

In summary, the dual-port SRAM in 14-nm FinFET technology and also MOS-FET technology was discussed. In the next section we look at the practical considerations for the DAC and some examples of the correct design approaches.

6.9 PRACTICAL CONSIDERATIONS FOR DACS

6.9.1 Time-Interleaved ΣΔ DACs

Time-interleaved ΣΔ DACs have the potential for high-speed wideband spectrum monitoring operation. Their SNR is limited by their timing skew between the output delays between the channels to the output. In a two-channel interleaved ΣΔ DAC, the channel skew arises from the duty cycle error in the half-sample rate clock. The effects of timing skew error can be mitigated by hold interleaving, digital prefiltering, or compensation in the form of analog postcorrection or digital precorrection.

A general structure of two-channel time-interleaved ΣΔ DAC with effective sample rate of F_s is shown in Figure 6.60. This type of twofold ΔΣ time interleaved DAC is of interest in flexible radio transmitters as they have the potential for wideband and high-speed operation. They are also of interest because they can relax the analog complexity by moving a part of the signal processing to the digital domain and can reduce the order of analog reconstruction filter after the DAC [41].

Figure 6.60 shows a general two-channel ΣΔ DAC that implements an overall noise transfer function $1 - H(z)$, where $H_1(z)$ and $H_2(z)$ are the two polyphase components of $H(z)$ obtained through block digital filtering [42]. The two outputs y_0 and y_1 are multiplexed to a single output y by using a single clock $F_s/2$ [41].

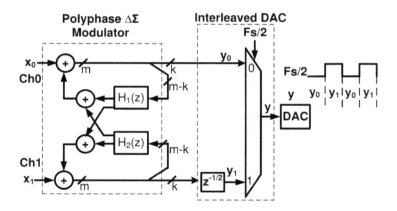

Figure 6.60 A general structure of two-channel time-interleaved ΔΣ DAC with effective sample rate of F_s. (©IEEE, reprinted with permission from [41].)

The overall analog performance of the time-interleaved ΣΔ DAC is limited by the multiplexing at the output. Hence, two-channel DACs are of special interest because they relax the logic speed while keeping the complexity of the multiplexing and the clocking scheme to a minimum. The timing skew in the output delays of the two channels in Figure 6.60 arise from an error in the $Fs/2$ clock duty cycle [41]. Figure 6.61 shows the effect of a 49% duty cycle on the spectrum of the TIDSM DAC.

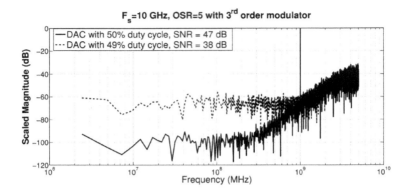

Figure 6.61 Effect of channel timing skew on the $\Delta\Sigma$ DAC SNR. (©IEEE, reprinted with permission from [41].)

This 1% duty cycle error results in a 9-dB SNR loss in a 4-bit 10 GS/s DAC with OSR=5 and third-order noise shaping. The duty cycle error modulates the shaped noise at the higher frequencies and folds back into the band of interest, reducing the SNR.

In discrete time, the timing error e_t can be expressed as a difference equation as [41]

$$e_t(n) = [y(n) - y(n-1)]\frac{\Delta t(n)}{T_s} \qquad (6.14)$$

where $\Delta t(n)/T_s = 2d_e$ and where $\Delta t(n)$ is the absolute timing error for the nth output $y(n)$, T_s is the effective sampling time period and d_e is the duty cycle error (deviation in duty cycle from 50%). This above equation identifies three ways to reduce the timing error [41]:

- Reduce $\Delta t(n)$ through duty cycle correction;
- Reduce the high-frequency content in the shaped output relative to the frequency of interest (i.e., $y(n) - y(n-1)$);
- Compensate for the error $e_t(n)$ by injecting a quantity $-e_t(n)$ into the DAC to cancel the error.

Note that the simplest solution is to reduce $y(n) - y(n-1)$ is to introduce more bits in the DAC. Every additional DAC bit yields about 6-dB improvement in SNR as the cost of increased mismatch requirement of the DAC cells [41].

The current-switching structure shown in Figure 6.62 suffers from dynamic errors. That is, when a switch turns off, the top terminal voltage of its corresponding current source collapses to zero. Thus, the next time that this branch is enabled, the (nonlinear) capacitance at this terminal must charge up, drawing a significant transient current from the output node. Moreover, since switching actions change the total current carried by the array, the ground voltage experiences large fluctuations in the presence of parasitic series inductances, such as those due to bond wires. Both of these effects can be greatly suppressed through the use of a current-steering DAC.

Among various DAC realizations, the current-steering DAC topology offers the highest speed and becomes the de facto solution at gigahertz frequencies, especially if the analog output must be delivered to a resistive load. One drawback of current-steering DACs is their limited output voltage compliance [43]. That is, the differential-pair transistors must operate in saturation, and, therefore, at least two drain-source voltages are subtracted from the supply voltage. Another difficulty in the design is the choice of the digital input voltage excursions. The most convenient design relies on rail-to-rail swings of the input voltage. This issue can be avoided by segmentation to give moderate swings. For an N-bit DAC, we still incorporate $2^N - 1$ unit cells but apply a different switching sequence. The segmented architecture avoids the jumps because, at the major carry transition, it simply turns on one more LSB cell rather than turn off one group of current sources [43].

A systematic approach to finding an optimized switching sequence for a current steering DAC to reduce the static nonlinearity is examined in [44]. As methodology for calculating the error patterns of a particular process is explained in detail. Also, a new algorithm is proposed to find an INL bounded optimized switching sequence for any gradient profile that can even compensate for the linear gradient errors due to the voltage drop in the biasing line. A new generalized symmetric routing technique for the common centroid placement that can be used for any switching sequence is also discussed.

In Figure 6.62, the signal chain of an I/Q transmitter with a mixing DAC with good signal frequency values is shown [45]. Shown is a standard Cartesian transmitter architecture based on mixing-DACs. This architecture realizes a direct step from digital to analog RF modulated signal. As a single unit, the mixing-DAC features much more architecture choices compared to just combining a separate DAC and mixer.

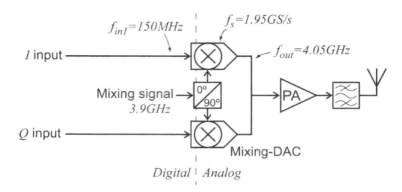

Figure 6.62 Signal chain of I/Q transmitter with mixing-DAC, with good signal frequency values. (©IEEE, reprinted with permission from [45].)

Potential advantages of these new architecture choices include higher linearity and larger clean bandwidth. For efficiently generating a large output power, typically a dedicated (non-CMOS) PA is used. The total system efficiency is therefore dominated by this PA, hence the mixing-DAC efficiency is a secondary concern, after spectral purity and bandwidth [45]. By cointegrating a current-steering DAC with a passive cascoded mixer, high spectral purity can be achieved for a wide signal band at a high RF carrier frequency [45]. The key techniques that enable such a performance are local mixing per current cell, multilevel cascoding with double bleeding currents and elevated bulk voltage, supply-isolated LO driver, sort-and-combine calibration, and digital dither. The two main linearity limitations are the mismatch between the unit cells and the coupling of the output signal to the internal nodes. The presented work features the highest output frequency f_{out} = 5.26 GHz the largest usable reduced bandwidth of 300 MHz and the lowest intermodulation distortion (IMD) < -82 dBc at 1.9 GHz and IMD of < -62 dBc at 4.1 GHz at the time of the research.

It is also worth noting that a 12-bit 20 MS/s SAR ADC incorporating a window-switching technique is examined in [46], which also includes a DAC. The proposed fast-binary-window (FBW) DAC switching scheme can effectively remove the major capacitor-DAC transition error to improve the DAC linearity and suppress DAC switching errors to improve the SNR. To maintain a good production yield, a dual-reference capacitor-DAC is applied to have a small total capacitance. The ADC was implemented in 180-nm CMOS. It consumes 1.22 mW from a 1.5-V supply. The measured peak SNDR and SFDR are 61.9 and 81 dB, respectively. The peak ENOB is 10-bits with an $FOM_W = 59.6$ fJ/cs. Figure 6.63(a) shows the proposed ADC architecture.

6.9 Practical Considerations for DACs 343

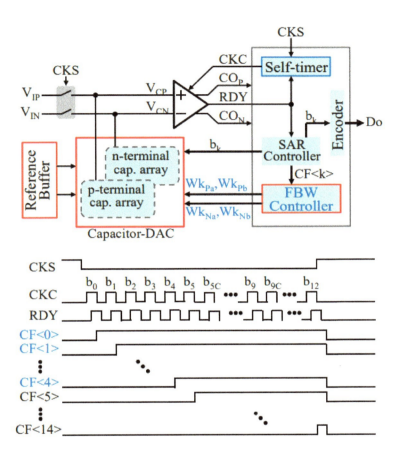

Figure 6.63 (a) Proposed ADC architecture and (b) timing diagram showing the clock signals. (©IEEE, reprinted with permission from [46].)

It consists of a sampling switch, a differential capacitor-DAC (C-DAC), a dynamic comparator, a self-timer, a FBW DAC controller, a SAR controller, and an output encoder. A self-timed clocking scheme was applied to speed up the ADC operation and avoid using a high-frequency clock signal. Figure 6.63(b) show the digital signals. After the sampling clock (CKS) goes low, the self-timed loop is activated with the recursive operation (CKC /RDY) until the last conversion cycle. In the C-DAC, two redundant capacitors were inserted to tolerate DAC settling errors and thus speed up ADC operation. The output data with two resultant bits are encoded by the digital error correction scheme. To reduce DAC switching errors, the FBW DAC switching is applied to the first four MSB capacitors. After window-switching operation, the monotonic DAC switching scheme is applied for the following LSB cycles.

In Figure 6.64, the DAC schematic using the p-terminal C-DAC as an example is shown. The capacitor-DAC schematic with two redundant capacitors is controlled by b5C and b9C. To accommodate the process limitation, a subreference voltage (VR2 = VR / 16) was applied to achieve the dual-reference capacitor-DAC. The MSB DAC uses VR for the first seven capacitors and the LSB DAC uses VR2 for the last four capacitors. Therefore, only 150 unit capacitors are used in the p-terminal capacitor array. The unit capacitance (C) is 6 fF to have a total capacitance of 1 pF including the parasitic capacitance of 100 fF. The main reference voltage (VR) is 1.5 V. With the gain-down factor of 0.7, the differential input range is 2.1 VP-P to have one LSB voltage (VLSB) of 0.5 mV. In Figure 6.65 the FBW DAC switching operation for the first four MSB cycles and the window encoder are shown.

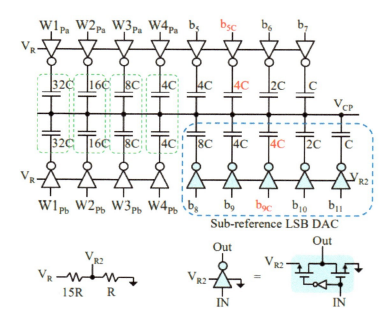

Figure 6.64 The DAC schematic (using the p-terminal C-DAC as an example). (©IEEE, reprinted with permission from [46].)

6.9 Practical Considerations for DACs 345

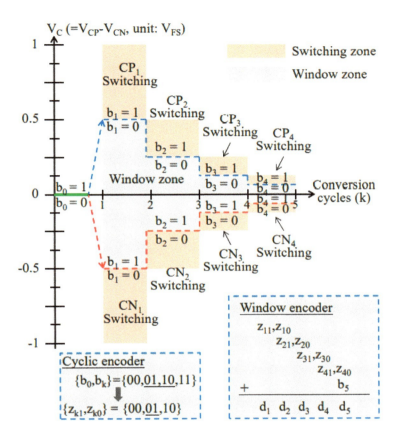

Figure 6.65 The FBW DAC switching operation for the first four MSB cycles and the window encoder. (©IEEE, reprinted with permission from [46].)

Finally, a compact DAC for digital-intensive transmitter architectures is presented in Figure 6.66.

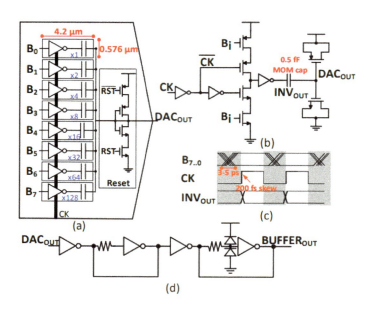

Figure 6.66 DAC architecture and building blocks: (a) switched-capacitor core, (b) SC DAC unit-cell, (c) timing diagram, and (d) inverter-based buffer. (©IEEE, reprinted with permission from [47].)

To minimize area and to leverage the strengths of FinFET CMOS, the implementation departs from the traditional current steering approach and consists mainly of inverters and subfemtofarad switched capacitors. The 14 GS/s 8-bit design occupies only 0.011 mm2 and supports up to 0.32 Vpp signal swing across its differential 100 Ω load. It achieves IM3 < −45.3 dBc across the first Nyquist zone while consuming 50 mW from a single 0.8 V supply.

The DAC's frequency response and measured output spectra as a function of frequency in GHz is shown in Figure 6.67. The DAC's SFDR and IM3 versus output frequency at 14 GS/s are shown in Figure 6.68.

6.10 Machine Learning Calibration Technology

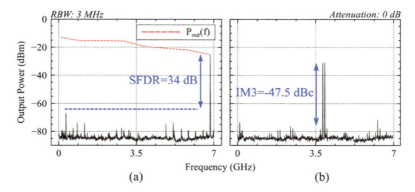

Figure 6.67 SFDR and IM3 versus output frequency at 14 GS/s. (©IEEE, reprinted with permission from [47].)

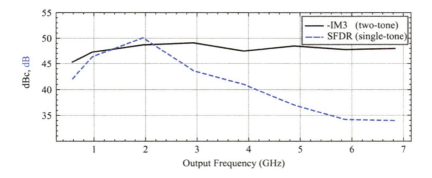

Figure 6.68 SFDR and IM3 versus output frequency at 14 GS/s. (©IEEE, reprinted with permission from [47].)

The observed roll-off at high frequencies is mainly due to package losses that were not de-embedded in these measurements. For a single-tone near Nyquist, the DAC achieves 34-dB SFDR. A midband two-tone test is shown in Figure 6.67(b), indicating IM3 at −47.5 dBc. The measured noise spectral density (NSD) is −147 dBm/Hz and mainly limited by quantization noise [47].

6.10 MACHINE LEARNING CALIBRATION TECHNOLOGY

As a final example, an image calibration algorithm for the twofold $\Delta\Sigma$ time-interleaved DAC (TIDAC) discussed above is presented. The algorithm is based on simulated annealing concept (included), which is used in the field of machine learning discussed in a later chapter to solve *derivative-free optimization* (DFO) problems. The DAC under consideration is part of a DRFM / digital transceiver core that contains a high-speed ADC, microcontroller, and digital control via a serial peripheral interface (SPI) [48].

A common theme throughout the text so far is replacing the RF front-end solutions with high-speed ADCs, DACs, and DSP, which perform frequency conversion and filtering operations in the digital domain. Placing the sampling and data converters closer to the antenna significantly reduces the system cost and power consumption. In addition, high-speed converters have their thermal and quantization noise power spread across a wide Nyquist zone, which enhances the dynamic range after any processing gain has been achieved. As we have seen, the time interleaving of several lower-speed converters is able to achieve multigigahertz digital conversion rates. The high speed of the TIDAC coupled with the area efficiency inherent in 14-nm CMOS presents an ideal use case for phased array systems.

The big problem, however, is timing errors and mismatch among the low speed converter slices which results in images (or spectral replicas), which corrupt the converter output spectrum. For example, the clock and data misalignment aggravate an image at $f_s/2$ and therefore, calibration schemes are often necessary in order to avoid considerable loss in dynamic range. As discussed earlier, the clock duty cycle error (timing uncertainty) is understood to be the limiting impairment regarding dynamic range and loss in SNR and calibration schemes are proposed.

Some solutions have been investigated such as the twofold $\Delta\Sigma$ TIDAC, sampling at an aggregate sample rate of 10 GS/s [42]. The author's solution involved digital prefiltering or equivalently increasing the DAC's resolution and tightening the matching requirements. Then with the postcorrection scheme proposed, an accurate measurement of clock duty cycle was required but this proves to be increasingly challenging at higher sampling rates as pointed out in [48].

These are used as tools for designing an algorithm that suppresses the interleave image to the noise floor. The algorithm is supported with experimental results in silicon on a 10-bit two-fold TI-DAC operating at a sample rate of 50 GS/s in 14-nm CMOS technology. The clock duty cycle error is understood to be the limiting impairment regarding dynamic range, and calibration schemes are proposed.

However, the recommended solution involves digital prefiltering, which is essentially equivalent to increasing the DAC resolution and tightening the matching requirements [48]. Although an analog postcorrection scheme is proposed, an accurate measurement of clock duty cycle is required, and this proves to be increasingly challenging at higher sample rates [48]. In this work they consider a 10-bit twofold, TIDAC with current steering architecture operating at an aggregate rate of 50 GS/s using two 25 GS/s sub-DAC slices in 14-nm CMOS technology. The DAC is part of a digital transceiver core from Jariet Technologies that contains an on-chip high-speed ADC, microcontroller, and digital control via an SPI interface. For the DAC under consideration, there is an image that appears at half of the aggregate sample rate $f_s/2$.

The closed-loop configuration shown in Figure 6.69 is used to design an algorithm which suppresses the interleave image to the noise floor. The algorithm also does not depend on the sub-DACs being gain balanced and does not rely on the circuits being bandwidth limited.

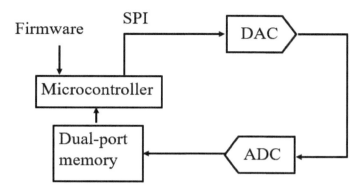

Figure 6.69 Block diagram of the TIDAC in a closed loop configuration (©IEEE, reprinted with permission from [48].)

A block diagram of the general M-bit TIDAC operating at a sampling frequency of f_s is shown in Figure 6.70. A PLL generates the clock at $f_s/2$ that is uniformly distributed to the blocks by a serializer, sub-DAC A, sub-DAC B, and an analog multiplexer (AMUX) [48]. The serializer contains a clock tree with several 2-to-1 multiplexers that serialize the N low-speed parallel lanes into two high-speed ones at *half the sampling rate* $f_s/2$. The sub-DAC slices employ current drivers for each bit to convert the M-bit code presented at the input to an analog output current. The drivers are binary weighted current sources. The AUX controls the current being dumped to a dummy load (not shown) when the switches are inactive.

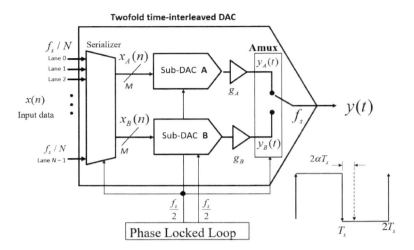

Figure 6.70 Simplified block diagram of the TIDAC, the fractional timing error, α, shown in lower right. (©IEEE, reprinted with permission from [48].)

When sub-DAC A undergoes a data transition, there is a settling window of τ_{settle} as shown in Figure 6.71. During this time, sub-DAC A is dumping current to the dummy node (AMUX controlled) while sub-DAC B is driving current to the output. The ideal scenario corresponds to the case where the clock edges are equidistant from the data transitions as illustrated in Figure 6.71. In any other scenario, one sub-DAC has a longer (or shorter) τ_{drive} than the other [48].

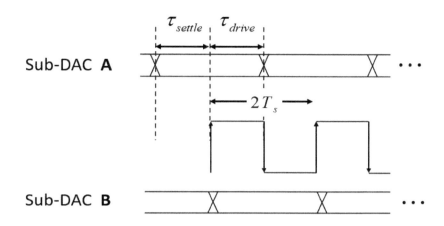

Figure 6.71 Illustration of clock and data alignment for a twofold TIDAC (After [48].)

As a consequence of the mismatch and misalignments, a ML, image calibration algorithm for the twofold TIDAC is proposed and verified in silicon on a 10-bit 50 GS/s DAC in 14nm CMOS. The important contribution is that the algorithm does not aggravate the matching requirements and does not assume the sub-DAC gains are balanced. As well, bandwidth limited calibration circuitry is also not required [48].

6.11 DIRECT DIGITAL SYNTHESIS

The DDS is a method of producing an analog waveform, for example, a sine wave or a triangular waveform, by generating a time-varying signal in digital form and then using a DAC to perform the conversion followed by a low-pass filter to smooth the transitions out. Because the DDS, also known as a *numerically controlled oscillator* (NCO), is mostly digital, it can offer fast switching between output frequencies, fine frequency resolution, and can operate over a broad spectrum drawing little power.

The basic digital mapping architecture, introduced by Tierney et. al. in 1971 [49], used a phase accumulator, a phase to amplitude converter block, and a DAC, and is shown (for historical purposes) in Figure 6.72.

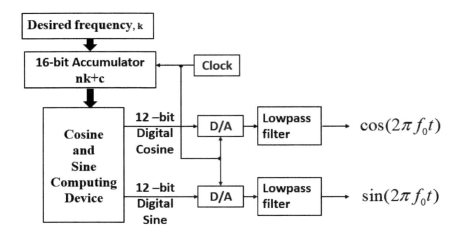

Figure 6.72 Block diagram of the original digital frequency synthesizer to produce quadrature outputs proposed by J. Tierney et. al. (After [49].)

An input frequency control word k is stored in a register and used to update an accumulator every T seconds. Each time the accumulator is changed, the value $nk+C$ is used to compute the real and imaginary parts of $\exp[j(2\pi/N)Y]$ where Y can be represented by a binary number $0 \leq Y \leq N-1$ and N is a frequency design parameter $f_0 = 1/NT$ as described in [49]. The computed values of $\sin(2\pi/N)Y$, $\cos(2\pi/N)Y$ are used to drive a pair of DACs with the proper word length to produce analog samples [49].

By generating analog waveforms digitally, it is able to take advantage of programmability and higher levels of integration and lower cost. Additionally, it is a technique for using digital data processing blocks as a means to generate a frequency- and phase-tunable output signal referenced to a fixed-frequency precision clock source, making it a primary generator for advanced digital modulation techniques such as binary frequency shift keying (BFSK), binary phase shift keying (BPSK) and spread spectrum, as well as the engine of interference mitigation techniques such as frequency hopping. The DDS offers many advantages over more complex solutions. A few of these are listed below [50]:

- Microhertz tuning resolution of output frequency and subdegree phase tuning capability, all under complete digital control;
- Extremely fast hopping speed tuning output frequency (or phase), phase-continuous frequency hops, no over/undershoot or analog-related loop settling time anomalies;

- DDS digital architecture eliminates need for manual tuning, tweaking associated with component aging, and temperature drift in analog synthesizer solutions;
- Digital control interface of the DDS architecture facilitates an environment where systems can be remotely controlled, and minutely optimized, under processor control;
- When utilized as a quadrature synthesizer, DDS affords unparalleled matching and control of I/Q synthesized outputs.

The main disadvantages of the DDS are (a) an insufficiently high maximum frequency of the synthesized signal, (b) a comparatively high amplitude noise level, and (c) a high level of spurious spectral components [51]. These factors are play a major limitation in the direct application of the DDS.

6.11.1 Early DDS Architectures

One of the fundamental early DDS architectures is shown in Figure 6.73 and consists on an input clock, an address counter, a lookup table (LUT) that stores a sine wave, a register, and a DAC. The clock drives a programmable-read-only-memory (PROM) that stores one or more integral number of cycles of a sinewave (or other arbitrary waveform, for that matter). As the address counter steps through each memory location, the corresponding digital amplitude of the signal at each location drives a DAC [52]. This in turn generates the analog output signal with its spectral purity then determined primarily by the DAC and phase noise basically determined by the reference clock.

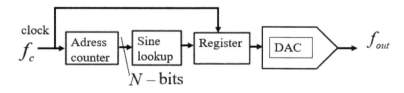

Figure 6.73 Elementary version of a direct digital synthesizer using an address counter, sine lookup table, register, and a DAC.

The output frequency of this implementation is dependent on [52]

1. Frequency of the reference clock;
2. Sinewave step size programmed into the PROM.

A fundamental problem with this simple DDS system is that the final output frequency can be changed only by changing the reference clock frequency or by reprogramming the PROM, making it rather inflexible [53].

The components of a typical DDS today (such as the AD9833) produce a sine wave, square wave, or triangular waveform at a given frequency. The block diagram of a DDS is shown in Figure 6.74.

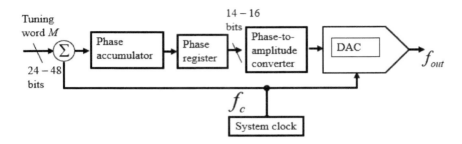

Figure 6.74 Direct digital synthesizer components showing the input tuning M and system clock f_c to produce an output waveform f_{out}.

The DDS in Figure 6.74 produces a waveform at a given frequency that depends on the reference-clock frequency f_c and the binary number programmed into the frequency register called the tuning word, M. Note that for a system reference clock f_c, the maximum fundamental output frequency is $f_c/2$ since sampling theory requires that at least two samples per cycle are required to reconstruct the output waveform. The binary number in the frequency register provides the input to the phase accumulator. With a sine lookup table, the phase accumulator computes a phase (angle) address for the LUT, which sends the digital value of amplitude (corresponding to the sine of that phase angle) to the DAC. The DAC then converts that number to a corresponding value of analog voltage or current. To generate a fixed-frequency sine wave, a constant value (the phase increment—which is determined by the binary number) is added to the phase accumulator with each clock cycle. If the phase increment is large, the phase accumulator will jump quickly through the sine lookup table and thus generate a high-frequency sine wave. If the phase increment is small, the phase accumulator will take many more steps, accordingly generating a slower waveform.

The sinusoidal signals are continuous and exhibit a repetitive nature in angular phase over the range of 0 to 360 degrees. The concept for the phase accumulator block can be as depicted by the digital phase wheel shown in Figure 6.75 [52–54].

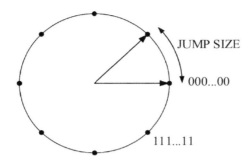

Figure 6.75 Digital phase wheel showing a jump of a single step.

For an n-bit with the frequency resolution of the DDS equal to $f_c/2^n$. In addition, all of the n-bits are *not* passed on to the LUT but are truncated, reducing the size of the LUT. However, this does not affect the resolution but adds a small amount of (acceptable) phase noise to the output [53].

The phase accumulator is a modulus M counter that increments its stored number each time it receives a clock pulse. The magnitude of the increment is determined by a digital word M contained in a delta phase register and represents the amount the phase accumulator is incremented each clock cycle. Then if f_c is the clock frequency, the frequency of the output waveform is equal to:

$$f_s = \frac{f_c M}{2^n} \quad \text{Hz}$$

This is known as the tuning equation where M is the binary tuning word, f_c is the system clock frequency, and n is the bit-length of the phase accumulator. Changes to the value of M result in immediate and phase-continuous changes in the output frequency with no loop settling time incurred [50].

6.11.2 ROM-Based DDS Architecture

A ROM version of the DDS is shown in Figure 6.76. The input is a digital frequency control word to a phase accumulator that is read by a ROM-based LUT. The size of the ROM module is a function of the bits incoming from phase accumulator module that establishes a direct relationship between the size of the ROM module and the LUT. As the magnitude of the phase accumulator grows, the size of ROM module or LUT also grows.

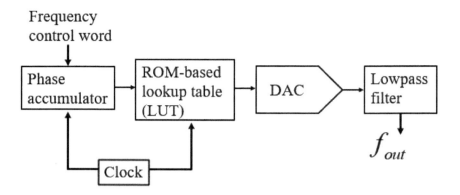

Figure 6.76 Direct digital synthesizer ROM-based version.

In order to achieve a high spectral purity in the output reconstructed waveform, the LUT entries should be increased as well as the size of ROM module. For the above trade-off, phase truncation can also be used. The size of ROM module depends on the number of the bits input to the phase accumulator. In the phase information, some of the bits are removed that are not significant, leaving only the most significant bits. This leads to phase truncation. The output frequency is dependent on the three parameters; namely, the frequency of the reference clock, the size of accumulator, and the frequency control word.

6.11.3 CORDIC DDS Architecture

In a CORDIC DDS the phase-to-sine waveform mapping is implemented by the CORDIC module. A CORDIC can also perform a complex-to-phase angle conversion. It does this through a series of rotations using only shifts and add operations. It also uses a pipelined architecture that optimizes the throughput [55]. A block diagram of the CORDIC DDS is shown in Figure 6.77. The I and Q coordinates are shown in Figure 6.78.

6.11 Direct Digital Synthesis

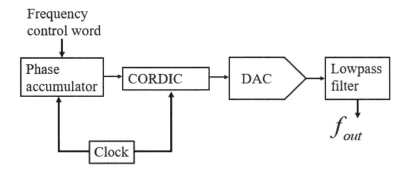

Figure 6.77 Direct digital synthesis using a CORDIC.

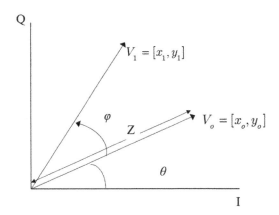

Figure 6.78 A vector being rotated counterclockwise by angle φ.

In Figure 6.78, a vector with magnitude Z has a phase angle of θ and is then rotated by an angle φ in the counterclockwise direction. At its first position, the components of the vector can be expressed as

$$x_o = Z \cos \theta$$

and

$$y_o = Z \sin \theta$$

After the rotation, the components of the resulting vector can be expressed as [55]

$$x_1 = Z\cos(\theta + \varphi) = \cos\varphi(x_o - y_o \tan\varphi) \tag{6.15}$$

and
$$y_1 = Z\sin(\varphi + \theta) = \cos\varphi(y_o - x_o \tan\varphi) \qquad (6.16)$$

By conforming $\tan\varphi$ to take on values such as $\pm 2^{-i}$ where i is the rotation index, the multiplication of $\pm 2^{-i}$ is the same as a shift operation that can be implemented. The $\cos\varphi$ term can be expressed

$$\cos\varphi = \cos\left(\tan^{-1}\left(2^{-i}\right)\right) = \frac{1}{\sqrt{1+2^{-2i}}}.$$

With the above being an even function, it can be treated as a constant independent of the direction of rotation. The expressions for the x and the y components of the resulting vector after every subsequent rotation are x_i and y_i and can be expressed as

$$x_{i+1} = \frac{1}{\sqrt{1+2^{-2i}}}(x_i - y_i d_i 2^{-i})$$

and
$$y_{i+1} = \frac{1}{\sqrt{1+2^{-2i}}}(y_i + x_i d_i 2^{-i})$$

where $d_i = \pm 1$ depending on the direction of rotation. The constant scale factor for every rotation $1/\sqrt{1+2^{-2i}}$ can be treated as the system gain as

$$A_n = \prod_n 1/\sqrt{1+2^{-2i}} \qquad (6.17)$$

and this value approaches 1.647 as the number of rotations increases. The design is synthesized with a clock frequency of 1 GHz there by yielding a sinusoid of frequency 500 MHz at the output.

6.11.4 A Comparative Study

A comparative study of the ROM-based version and a DDS that uses a CORDIC based version was reported in [54]. The CORDIC eliminates the usage of the large ROM based look up table. The SFDR for both the ROM direct digital frequency synthesizer (DDFS) and the CORDIC DDFS are shown below in Figure 6.79. Note the CORDIC sidelobes are fairly monotonically decreasing except for the few harmonics that show themselves.

6.11 Direct Digital Synthesis

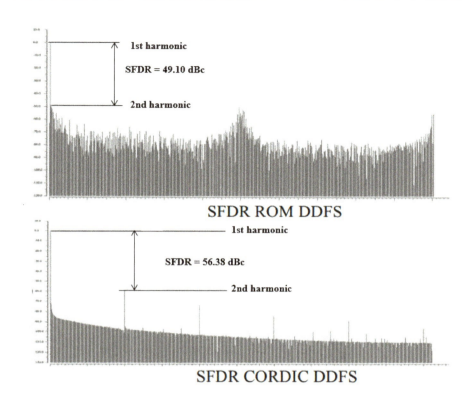

Figure 6.79 An SFDR comparison between the ROM DDFS and the CORDIC DDFS (©IEEE, reprinted with permission from [54]).

Comparison of ROM and CORDIC DDS architectures by their area, power, and SFDR is shown in Figure 6.80.

Parameter	ROM DDS	CORDIC DDS
Area (μm^2)	41,731	159,513
Power (mW)	24.18	71.78
SFDR (dBc)	49.10	56.38

Figure 6.80 Comparison of ROM and CORDIC DDS architectures by their area, power, and SFDR.

As a final note, a brief mention of the Analog Devices group AD983X where X is 1–5 with evaluation boards also available. As a result, DDS devices are rapidly replacing or augmenting traditional PLLs and other analog RF sources, while still offering high stability and signal purity making ideal sources for deception and suppression jamming systems.

In fact, the question arises whether a DRFM or a DDS is a more effective (smart noise) jammer against a coherent RF emitter. This question was initially answered in a (unclassified) masters thesis by Capt. Charles Watson, U. S. Army at U. S. Naval Postgraduate School [56]. This question of whether a DRFM or DDS is a more effective smart-noise jammer against and a tactical 3D air surveillance radar (the AN/TPS-70) was initially investigated Capt. Watson under the guidance from the late Professor D. Curtis Schleher and myself.

The results of this study are discussed in Chapter 10 on electronic attack but to summarize we learned that the ability to meet this challenge lies largely in the expertise, creativity, and stubbornness of the EA designer. A brief overview of the threat is also given in Chapter 10. The effectiveness requires that a number of issues be considered from both operational and technical perspectives with an eye on the changing technology.

6.12 DRFM OSCILLATORS AND PHASE NOISE

In this section we address the modern oscillators that are used within the DRFM transceiver and specifically the phase noise and its limitations on the DRFM performance. Oscillators represent an essential part of an RF system. They are commonly used to provide frequency and timing synchronization and supply the signal used for frequency conversion (upconversion and downconversion). Phase noise can affect channel and adjacent channel performance, especially in wideband receivers used in EW and EA. Unfortunately, the outputs of the oscillators are not perfectly periodic and they suffer from many imperfections, making it difficult for them to give a precise periodic timing reference. Assessing the effect of the noisy oscillators on the detection performance of DRFMs can help the development of algorithms to mitigate these unwanted distortions [57, 58].

6.12.1 Oscillator Phase Noise Limitations

As discussed in Chapter 1, the DRFM serves to detect an RF signal from an adversary emitter, digitally process the signal, and retransmit the signal. The system delays each subsequently received pulse to intentionally delay retransmission. This causes the tracking radar to, for example, slowly move away from the target. The DRFM has to determine the direction of arrival (DOA) and the frequency of the RF signal. An oscillator is an important component in the system's transceiver collection of components and drives the signal processing elements such as the clock and the mixers. A crystal oscillator is an oscillator that uses the resonance property of a quartz crystal to create an electric signal at a particular frequency. Selecting a crystal oscillator for a DRFM the frequency stability versus temperature is one of the key parameters that must be examined. Several design approaches can be used here.

6.12.2 Oscillator Choices

There are many things to consider with choosing an oscillator type. With a quartz crystal oscillator (XO), the frequency versus stability is provided *only by the quartz resonator itself* and is influenced primarily by the cut of the quartz crystal. Frequency stability ($\Delta f / f_0$) versus temperature for an XO can reach ± 10 to $\pm 15 \times 10^{-6}$ from -40^oC to $+85^oC$.

With a temperature compensated quartz oscillator (TCXO), additional components apply a control voltage to a varactor diode that compensates for temperature effects on the frequency. Frequency stability versus temperature for a TCXO can reach ± 1 to $\pm 3 \times 10^{-7}$ from -40^oC to $+85^oC$.

An oven controlled quartz crystal oscillator (OCXO) design places the quartz resonator and all basic circuits inside an oven at constant temperature. Frequency stability versus temperature for an OCXO can reach ± 1 to $\pm 5 \times 10^{-11}$ from -40^oC to $+85^oC$. Of these three designs, only the OCXO, which has the highest frequency stability versus temperature, is discussed here.

A PLL, is a feedback system designed to closely track and match the phase of two RF sources: a reference oscillator and voltage-controlled oscillator (VCO). A phase locked oscillator (PLO) is a stable frequency source with inherently low phase noise and spurious signals. The PLO is meant to function as a stable microwave source or LO and often leverages an analog or digital PLL to impart onto the VCO the frequency stability and phase noise of a crystal oscillator or other low noise reference [59].

A simple PLL is shown in Figure 6.81. In a PLL the output of a VCO is fed in a feedback loop through a divider, where the divider generates one output pulse for every N input pulses. The output of the divider is fed to a phase or error detector that detects the phase between the output of the divider and the output of the reference [59]. The phase detector produces a pulsed error signal representing the phase difference between the reference and the frequency divided VCO. This error is filtered in a low-pass filter, also known as the loop filter, and controls the VCO, ultimately locking it to a specific frequency, which is related to the phase.

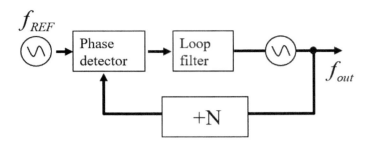

Figure 6.81 A simple phase locked loop (PLL).

The reference can also be generated through direct synthesis, indirect synthesis or through DDS. A frequency synthesizer is a mixed signal device that allows a designer to generate a variety of output frequencies as multiples of a single reference frequency. A PLO can be considered a fixed frequency synthesizer that generates a stable output frequency regardless of temperature drift, vibration, and aging. Frequency synthesizers can be implemented using one of three topologies: direct synthesis, indirect synthesis, or using a DDS.

6.12.3 Definition of Phase Noise

The concept of phase noise is broadly used and its definition is subject to much treatment recently. Phase noise, typically indicated by $\mathscr{L}$ in the literature, is defined as the ratio of the power of the signal in a 1-Hz bandwidth at offset f from the carrier, divided by the power of the carrier or

$$\mathscr{L}(f) = \frac{S'_v(f_0 + f)}{P} \qquad (6.18)$$

where $S'(v)$ is the PSD of a periodic one-sided voltage signal with frequency f_0 and power P that is affected by random excess phase modulation [60]. The phase noise $\mathscr{L}$ is defined over positive frequencies only $f \geq 0$. In 1999, the IEEE released the Standard 1139, revised in 2008, in which the phase noise is defined as the two-sided PSD S_ϕ and is one half of the one-sided PSD S'_ϕ of the excess phase of the carrier. So the definition and the preferred quantity $\mathscr{L}(f)$ (pronounced "script el") to $S_\varphi(f)$ where

$$\mathscr{L} = \frac{1}{2} S_\varphi(f) \qquad (6.19)$$

and is given in dBc/Hz. That is, the quantity used in practice is $10\log_{10}[S_\varphi(f)] - 3\text{dB}$.

6.12 DRFM Oscillators and Phase Noise

A model of the phase noise PSD $S_\varphi(f)$ that has been found to be the most useful is the composite power-law function

$$S_\phi(f) = K_\alpha |f|^{-\alpha}, \quad \alpha \in \{0, 1, 2, 3, 4\}, \quad K_\alpha \in \mathbb{R} \qquad (6.20)$$

where f denotes offset frequency from actual carrier frequency and is almost always plotted on a log-log scale. Phase noise PSD modeled as [61]:

$$S_\varphi(f) = K_0 + \frac{K_1}{|f|} + \frac{K_2}{|f|^2} + \frac{K_3}{|f|^3} + \frac{K_4}{|f|^4}$$

where weights $K_i, i = 0, 1, 2, 3, 4$ make up a combined power-law spectrum and produce an infinite value at 0 Hz. Use of the models are subject to max PSD, usually around 5- to 10-Hz offset frequency; the coefficients K_0 to K_4 describe contributions from (1) white phase noise, (2) flicker phase noise, (3) white frequency noise, (4) frequency flicker noise, (5) random walk frequency noise [60].

The spectral density $S_\Phi(f)$ is often called the *Leeson model* and is shown in Figure 6.82.

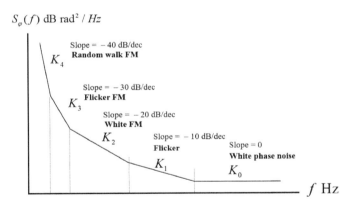

Figure 6.82 Power spectral density diagram of the Leeson oscillator phase noise model.

For convenience, the phase noise in bistatic and monostatic radar systems is often modeled by a discrete white process (see for example, [62, 63]). However, this has been shown to be somewhat inadequate and a more correct approach for the coherent oscillators and stable local oscillators is modeling the undesired phase fluctuations via a *multivariate circular distribution* and the phase noise PSD through a composite power-law model [64, 65].

In designing EA or EW systems, great emphasis is placed on the frequency plan, with the goal of reducing receiver and/or transmitter spurs to at least −45 dBc below the minimum true signal level for reliable detection and signal processing. Less consideration is given to the LOs selected for the system. However, LO phase noise affects the signal jitter, the SNR, and BER in the communication receivers and causes phase and frequency measurement errors. Phase noise can also affect adjacent channel performance, especially in wideband multichannel DRFMs used in EW and EA.

Since the EA and ES components within a DRFM accomplish different functions that a bistatic, monostatic radar system, a different approach has to been used. For example, the EA systems discussed are digital systems that detect friendly radar signals from adversary radars, digitally process the signals, and retransmit the signals with EA or some type of deception and/or suppression. The ES system components are used to detect and then geolocate adversary radars, generally using phase interferometer techniques that measure the angles of arrival (AOA). After detection and geolocation, the signals are classified and clustered (or grouped) into specific categories, as discussed in Chapter 9.

The phase noise, however, will affect these systems as well in terms of accuracy of geolocation, frequency, and emitter classification. An ES wideband DRFM for passive interception and passive coherent detection should include the phase noise analysis limitation.

Following the method presented in [57, 58, 66] we consider that the DRFM frequency conversions in the receiver are accomplished with high-frequency PLOs. The oscillators used in multiple frequency conversion receiver, multirate wideband systems such as those discussed in Chapter 4 are assumed to be externally phase-locked to a preferred 100 MHz, stress compensated (SC) cut, ultra-stable OCXOs.

To find the upper bound on the phase noise for the conversion circuits considered here, the LO power levels are assumed to be $+13$ dBm at the input to the mixer and that the noise floor of the PLO is -161 dBm/Hz. The frequency and phase fluctuations are measured in terms of instantaneous frequency. The phase variations, normalized to a nominal frequency f_0, are defined as phase or time jitter as

$$y(t) = \frac{1}{2\pi f_0} \frac{d\varphi(t)}{dt} = \frac{\dot{\varphi}}{2\pi f_0} \tag{6.21}$$

$$\mathscr{J} = \frac{\Phi}{2\pi f_0} \tag{6.22}$$

where $y(t)$ is the frequency departure from the nominal frequency f_0 or phase noise and $\mathscr{J}$ is the root-mean-square time jitter in units of seconds.

The PLO signals are represented by

$$e_{LO_i}(t) = \sqrt{2P_{C_i}} \cos[\omega_i t + \Phi_i(t)]$$

where $i = 1, 2$, or 3 for the first, second, or third LO stage in the DRFM circuit. Also, P_{C_i} is the ith LO power in units of watts, $\omega_i = 2\pi f_i$ where f_i is the ith LO frequency, and

$\Phi_i(t)$ is a zero-mean low-pass stationary Gaussian random process with a mean square phase deviation Φ_i^2 that phase modulates the ith LO signal. The amplitude-generated noise is usually much smaller than the phase noise and is cancelled in balanced mixers. The phase noise spectrum was described above by the composite power law.

6.12.4 Estimation of the Composite Power Law Parameters

Following the procedure outlined in [64, 65], the parameters can be estimated for the composite power-law model.

1. Let $f_i, i = 1, 2, \ldots, N_1$ be discrete set of offset frequencies where PSD has been measured;
2. Also let $S_m(f_i), i = 1, 2, \ldots, N_1$ be the phase noise PSD measurement in dBc/Hz obtained corresponding to f_i;
3. Estimation of parameters K_0, K_1, K_2, K_3, K_4 obtained via least-square fitting of model:
$$S_{dB}(f) = 10\log_{10}(S_\phi(f))$$
with PSD measurements $S_m(f_i), i = 1, 2, \ldots, N_1$
4. Defined as optimal solution to:
$$\min_{K_0, K_1, K_2, K_3, K_4} \sum_{i=1}^{N_1} |S_{dB}(f_i) - S_m(f_i)|^2$$
5. Nonconvex (nonlinear) optimization problem;
6. Solution by using (a) search of discrete grid of point or (b) search grid algorithm;
7. First, need good initial point to compute optimized solution;
8. Fix the set of four frequencies $\bar{f}_1, \bar{f}_2, \bar{f}_3, \bar{f}_4$ and evaluate estimate according to:
$$K_0 = 10^{N_{floor}/10}, \quad \text{with} \quad N_{floor} = \frac{1}{M} \sum_{i=N_1-M+1}^{N_1} S(f_i)$$
9. This is average of phase noise PSD measurements over last M offset frequencies;
10. The values of K_1, K_2, K_3, K_4 are then the solution to the following system of linear equations:
$$K_1 \frac{1}{\bar{f}_1} + K_2 \frac{1}{\bar{f}_1^2} + K_3 \frac{1}{\bar{f}_1^3} + K_4 \frac{1}{\bar{f}_1^4} = 10^{S_m(\bar{f}_1)/10} - K_0$$

$$K_1 \frac{1}{\bar{f}_2^1} + K_2 \frac{1}{\bar{f}_2^2} + K_3 \frac{1}{\bar{f}_2^3} + K_4 \frac{1}{\bar{f}_2^4} = 10^{S_m(\bar{f}_2)/10} - K_0$$

$$K_1 \frac{1}{\bar{f}_3^1} + K_2 \frac{1}{\bar{f}_3^2} + K_3 \frac{1}{\bar{f}_3^3} + K_4 \frac{1}{\bar{f}_3^4} = 10^{S_m(\bar{f}_3)/10} - K_0$$

$$K_1 \frac{1}{\bar{f}_4^1} + K_2 \frac{1}{\bar{f}_4^2} + K_3 \frac{1}{\bar{f}_4^3} + K_4 \frac{1}{\bar{f}_4^4} = 10^{S_m(\bar{f}_4)/10} - K_0$$

As an example the phase noise power spectral density (dBc/Hz) comparison of the model with phase noise values as a function of frequency offset $f_0 = 150$ MHz ($f osc = 100$ MHz) is given in Figure 6.83. Note the scale and close match.

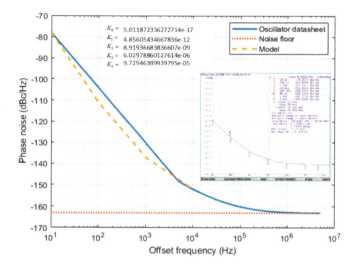

Figure 6.83 Example of phase noise power spectral density (dBc/Hz) comparison of model with phase noise values as a function of frequency offset $f_0 = 150$ MHz ($f_{osc} = 100$ MHz).

6.13 SUMMARY

An examination of the high-performance all-electronic transceiver technologies were presented. Starting with the analog comparator designs, a look at the high-speed ADCs including the SiGe BiCMOS technologies which are typically the fastest. Time-interleaved flash designs and time-interleaved pipeline methods are also emphasized. This is followed by the time-interleaved SAR techniques. Several figures of merit plots were examined to demonstrate the state-of-the-art. FinFET technology was then presented including its advantages. Embedded dual port memory architectures were then investigated and several FinFET SRAM configurations examined. Practical DAC considerations were then presented including time-interleaved DACs. Machine learning transceiver calibration techniques were then investigated. Direct digital synthesis technologies were then presented. Finally, DRFM oscillators and their phase noise characteristics and limitations were emphasized including a method to estimate the composite power law profile parameters.

REFERENCES

[1] H. Molaei and K. Hajsadeghi, "A low-power comparator-reduced flash ADC using dynamic comparators," in *2017 24th IEEE International Conference on Electronics, Circuits and Systems (ICECS)*, 2017, pp. 5–8.

[2] D. Schinkel, E. Mensink, E. Klumperink, E. van Tuijl, and B. Nauta, "A Double-Tail Latch-Type Voltage Sense Amplifier with 18ps Setup+Hold Time," in *2007 IEEE International Solid-State Circuits Conference. Digest of Technical Papers*, 2007, pp. 314–605.

[3] R. Jain, A. K. Dubey, V. Varshney, and R. K. Nagaria, "Design of low-power high-speed double-tail dynamic CMOS comparator using novel latch structure," in *2017 4th IEEE Uttar Pradesh Section International Conference on Electrical, Computer and Electronics (UPCON)*, 2017, pp. 217–222.

[4] S. R. Vemu, P. S. S. N. Mowlika, and S. Adinarayana, "An energy efficient and high speed double tail comparator using cadence EDA tools," in *2017 International Conference on Algorithms, Methodology, Models and Applications in Emerging Technologies (ICAMMAET)*, 2017, pp. 1–5.

[5] ——, "An energy efficient and high speed double tail comparator using cadence EDA tools," in *2017 International Conference on Algorithms, Methodology, Models and Applications in Emerging Technologies (ICAMMAET)*, 2017, pp. 1–5.

[6] S. Aakash, A. Anisha, G. J. Das, T. Abhiram, and J. P. Anita, "Design of a low power, high speed double tail comparator," in *2017 International Conference on Circuit, Power and Computing Technologies (ICCPCT)*, 2017, pp. 1–5.

[7] L. Chacko and G. Tom Varghese, "Comparator design for low power high speed flash adc-a review," in *2019 3rd International Conference on Computing Methodologies and Communication (ICCMC)*, 2019, pp. 869–872.

REFERENCES

[8] S. Hussain, R. Kumar, and G. Trivedi, "Comparison and Design of Dynamic Comparator in 180nm SCL Technology for Low Power and High Speed Flash ADC," in *2017 IEEE International Symposium on Nanoelectronic and Information Systems (iNIS)*, 2017, pp. 139–144.

[9] F. Rabbi, S. Das, Q. D. Hossain, and N. Sad Pathan, "Design of a Low-Power Ultra High Speed Dynamic Latched Comparator in 90-nm CMOS Technology," in *2018 International Conference on Computer, Communication, Chemical, Material and Electronic Engineering (IC4ME2)*, 2018, pp. 1–4.

[10] G. Tretter, M. Khafaji, D. Fritsche, C. Carta, and F. Ellinger, "A 24 GS/s single-core flash ADC with 3-bit resolution in 28 nm low-power digital CMOS," in *2015 IEEE Radio Frequency Integrated Circuits Symposium (RFIC)*, 2015, pp. 347–350.

[11] Y. Feng, Y. Tang, Q. Fan, and J. Chen, "A 25-GS/s 4-bit Single-core Flash ADC in 28 nm FDSOI CMOS," in *2018 IEEE Asia Pacific Conference on Circuits and Systems (APCCAS)*, 2018, pp. 30–33.

[12] Y. Shen, Y. Zhang, L. Yang, Y. Yang, Y. Guo, X. Li, and Y. Zhang, "A 10 GS/s, 3-bit ADC with novel decoder in SiGe BiCMOS technology," in *IEEE 3rd International Conference on Integrated Circuits and Microsystems (ICICM)*, 2018, pp. 196–200.

[13] P. Ritter, S. Le Tual, B. Allard, and M. Möller, "A SiGe bipolar 6 bit 20 GS/s nyquist-rate flash ADC without time interleaving," in *2013 IEEE Bipolar/BiCMOS Circuits and Technology Meeting (BCTM)*, 2013, pp. 65–68.

[14] J. Lee, J. Weiner, and Y. Chen, "A 20-gs/s 5-b sige adc for 40-gb/s coherent optical links," *IEEE Transactions on Circuits and Systems I: Regular Papers*, vol. 57, no. 10, pp. 2665–2674, 2010.

[15] X. Du, M. Grözing, M. Buck, and M. Berroth, "A 40 GS/s 4 bit SiGe BiCMOS flash ADC," in *2017 IEEE Bipolar/BiCMOS Circuits and Technology Meeting (BCTM)*, 2017, pp. 138–141.

[16] P. E. Pace, *Advanced Techniques for Digital Receivers*. Norwook, MA: Artech House, 2000.

[17] R. H. Walden, "Analog-to-digital converter survey and analysis," *IEEE Journal on Selected Areas in Communications*, vol. 17, no. 4, pp. 539–550, 1999.

[18] Y. C. Jenq, "Digital spectra of non-uniformly sampled signals: Fundamentals and high speed waveform digitizers," in *IEEE Transactions of Instrumentation and Measurement*, vol. 37, no. 6, 1988, pp. 245–251.

[19] ——, "Digital spectra of non-uniformly sampled signals: A robust sampling time offset estimation algorithm for ultra high-speed waveform digitizers using interleaving," in *IEEE Transactions of Instrumentation and Measurement*, vol. 39, no. 2, 1990, pp. 71–75.

[20] D. Oh, J. Kim, M. Seo, J. Kim, and S. Ryu, "A 6-bit 10-GS/s 63-mW 4x TI time-domain interpolating flash ADC in 65-nm CMOS," in *ESSCIRC Conference 2015 - 41st European Solid-State Circuits Conference (ESSCIRC)*, 2015, pp. 323–326.

[21] Huawen Jin and E. K. F. Lee, "A digital-background calibration technique for minimizing timing-error effects in time-interleaved adcs," *IEEE Transactions on Circuits and Systems II: Analog and Digital Signal Processing*, vol. 47, no. 7, pp. 603–613, 2000.

[22] C. Vogel, "The impact of combined channel mismatch effects in time-interleaved adcs," *IEEE Transactions on Instrumentation and Measurement*, vol. 54, no. 1, pp. 415–427, 2005.

[23] M. M. Ayesh, S. Ibrahim, and M. M. Aboudina, "A 15.5-mw 20-gsps 4-bit charge-steering flash adc," in *2015 IEEE 58th International Midwest Symposium on Circuits and Systems (MWSCAS)*, 2015, pp. 1–4.

[24] A. Zandieh, P. Schvan, and S. P. Voinigescu, "A 2x-Oversampling, 128-GS/s 5-bit Flash ADC for 64-GBaud Applications," in *2018 IEEE BiCMOS and Compound Semiconductor Integrated Circuits and Technology Symposium (BCICTS)*, 2018, pp. 52–55.

[25] ——, "Design of a 55-nm SiGe BiCMOS 5-bit Time-Interleaved Flash ADC for 64-Gbd 16-QAM Fiberoptics Applications," *IEEE Journal of Solid-State Circuits*, vol. 54, no. 9, pp. 2375–2387, 2019.

[26] W. H. Siddiqui and G. S. Choi, "12 bit 3.072 GS/s 32-way time-interleaved pipelined ADC with digital background calibration for wideband fully digital receiver applications in 65 nm complementary metal oxide semiconductor," *IET Circuits, Devices and Systems*, vol. 14, no. 2, 2020.

[27] A. K. M. Sharifkhani, "High-speed low-power comparator for analog to digital converters," in *International Journal of Electronic Communications (AEU)*, vol. 70, no. 7, 2016, pp. 886–894.

[28] Z. L. Q. Lia, "A 8 bit 2 Gs/s flash ADC in 0.18 μm CMOS," in *International Workshop of Information and Electronic Engineering (IWIEE)*, vol. 29, no. 7, 2012, pp. 693 – 698.

[29] B. Murmann, "The successive approximation register ADC: a versatile building block for ultra-low- power to ultra-high-speed applications," *IEEE Communications Magazine*, vol. 54, no. 4, pp. 78–83, 2016.

[30] T. Matsuura, "Integrated CMOS ADC — Tutorial review on recent hybrid SAR-ADCs," in *2017 MIXDES - 24th International Conference "Mixed Design of Integrated Circuits and Systems*, 2017, pp. 29–34.

[31] A. Buchwald, "High-speed time interleaved adcs," *IEEE Communications Magazine*, vol. 54, no. 4, pp. 71–77, 2016.

[32] Q. Liu, W. Shu, and J. S. Chang, "A 1-GS/s 11-Bit SAR-assisted pipeline ADC with 59-dB SNDR in 65-nm CMOS," *IEEE Transactions on Circuits and Systems II: Express Briefs*, vol. 65, no. 9, pp. 1164–1168, 2018.

[33] L. Kull, T. Toifl, M. Schmatz, P. A. Francese, C. Menolfi, M. Braendli, M. Kossel, T. Morf, T. M. Andersen, and Y. Leblebici, "22.1 A 90GS/s 8b 667mW 64× interleaved SAR ADC in 32nm digital SOI CMOS," in *2014 IEEE International Solid-State Circuits Conference Digest of Technical Papers (ISSCC)*, 2014, pp. 378–379.

[34] B. Murmann, "The race for the extra decibel: A brief review of current ADC performance trajectories," *IEEE Solid-State Circuits Magazine*, vol. 7, no. 3, pp. 58–66, 2015.

[35] S. Palermo, S. Hoyos, A. Shafik, E. Z. Tabasy, S. Cai, S. Kiran, and K. Lee, "Cmos adc-based receivers for high-speed electrical and optical links," *IEEE Communications Magazine*, vol. 54, no. 10, pp. 168–175, 2016.

[36] J. M. de la Rosa, *Frontiers, Trends and Challenges: Towards Next-generation $\Delta\Sigma$ Modulators*, 2018, pp. 389–461.

[37] T. Drenski and J. C. Rasmussen, "ADC & DAC — Technology trends and steps to overcome current limitations," in *2018 Optical Fiber Communications Conference and Exposition (OFC)*, 2018, pp. 1–3.

[38] M. Ichihashi, Y. Woo, M. A. Ul Karim, V. Joshi, and D. Burnett, "10T Differential-Signal SRAM Design in a 14-nm FinFET Technology for High-Speed Application," in *2018 31st IEEE International System-on-Chip Conference (SOCC)*, 2018, pp. 322–325.

[39] S. Srinivas and A. B. Gudi, "Design and implementation of dual port SRAM memory architecture using MOSFET's," in *2017 International Conference on Smart grids, Power and Advanced Control Engineering (ICSPACE)*, 2017, pp. 357–362.

[40] M. Enachescu, M. Lefter, G. R. Voicu, and S. D. Cotofana, "Low-Leakage 3D Stacked Hybrid NEMFET-CMOS Dual Port Memory," *IEEE Transactions on Emerging Topics in Computing*, vol. 6, no. 2, pp. 184–199, 2018.

[41] K. D. Choo, J. Bell, and M. P. Flynn, "27.3 Area-efficient 1GS/s 6b SAR ADC with charge-injection-cell-based DAC," in *2016 IEEE International Solid-State Circuits Conference (ISSCC)*, 2016, pp. 460–461.

[42] A. Bhide and A. Alvandpour, "Timing challenges in high-speed interleaved $\delta\sigma$ DACs," in *2014 International Symposium on Integrated Circuits (ISIC)*, 2014, pp. 46–49.

[43] B. Razavi, "The current-steering dac [a circuit for all seasons]," *IEEE Solid-State Circuits Magazine*, vol. 10, no. 1, pp. 11–15, 2018.

[44] R. K. Srivastava, A. Vellathu, S. V. R. Kaipu, H. S. Jattana, and A. Rampal, "A systematic method to find an optimized quad-quadrant random walk sequence for reducing the mismatch effect in current steering DAC," in *2017 International conference on Microelectronic Devices, Circuits and Systems (ICMDCS)*, 2017, pp. 1–6.

[45] E. Bechthum, G. I. Radulov, J. Briaire, G. J. G. M. Geelen, and A. H. M. van Roermund, "A Wideband RF Mixing-DAC Achieving IMD < −82 dBc Up to 1.9 GHz," *IEEE Journal of Solid-State Circuits*, vol. 51, no. 6, pp. 1374–1384, 2016.

[46] Y. Chung, Y. Lin, and Q. Zeng, "A 12-bit 20-MS/s SAR ADC With Fast-Binary-Window DAC Switching in 180nm CMOS," in *2018 IEEE Asia Pacific Conference on Circuits and Systems (APCCAS)*, 2018, pp. 34–37.

[47] P. Caragiulo, O. E. Mattia, A. Arbabian, and B. Murmann, "A Compact 14 GS/s 8-Bit Switched-Capacitor DAC in 16 nm FinFET CMOS," in *2020 IEEE Symposium on VLSI Circuits*, 2020, pp. 1–2.

[48] D. Beauchamp and K. M. Chugg, "Machine learning based image calibration for a twofold time-interleaved high speed DAC," in *2019 IEEE 62nd International Midwest Symposium on Circuits and Systems (MWSCAS)*, 2019, pp. 908–912.

[49] J. Tierney, C. Rader, and B. Gold, "A digital frequency synthesizer," *IEEE Transactions on Audio and Electroacoustics*, vol. 19, no. 1, pp. 48–57, 1971.

[50] "Ask the application engineer—33: All about direct digital synthesis, by E. Murphy and C. Slattery," https://www.analog.com/en/analog-dialogue/articles/all-about-direct-digital-synthesis.html, accessed: 2020-10-22.

[51] E. S. Klyuzhev, S. V. Tolmachev, I. V. Ryabov, and N. V. Degtyarev, "Direct digital synthesis of signals in radiolocation, communication and telecommunication systems," in *Systems of Signal Synchronization, Generating and Processing in Telecommunications (SYNCHROINFO)*, 2018, pp. 1–4.

REFERENCES

[52] Analog Devices Inc., "A technical tutorial on digital signal synthesis," *Analog Devices, Inc.*, 1999.

[53] ——, "Fundamentals of digital signal synthesis (DDS)," *Analog Devices, Inc.*, vol. MT-085 Tutorial, no. 10, pp. 1–9, 2008.

[54] R. Suryavanshi, S. Sridevi, and B. Amrutur, "A comparative study of direct digital frequency synthesizer architectures in 180nm CMOS," in *2017 International conference on Microelectronic Devices, Circuits and Systems (ICMDCS)*, 2017, pp. 1–5.

[55] P. S. Ang, "DRFM CORDIC processor and sea clutter modeling for enhanching structured false target synthesis," in *U. S. Naval Postgraduate School Masters Thesis (U)*, 2017, p. 137.

[56] C. J. Watson, "A Comparison of DDS and DRFM Techniques in the Generation of Smart Noise Jamming Waveforms(U)," in *U. S. Naval Postgraduate School Masters Thesis (U)*, 1996, p. 150.

[57] S. J. Caprio, "Specifying upper bounds on phase noise in phase-locked oscillators in electronic warfare systems–part i [application notes]," *IEEE Microwave Magazine*, vol. 12, no. 7, pp. 96–112, 2011.

[58] ——, "Specifying upper bounds on the phase noise in phase-locked oscillators in electronic warfare systems—part ii [application notes]," *IEEE Microwave Magazine*, vol. 13, no. 1, pp. 152–160, 2012.

[59] T. Galla, "Selecting Phase-Locked Oscillators for Frequency Synthesis," *Microwave Journal*, vol. 62, no. 9, 2019.

[60] N. D. Dalt and A. Sheikholeslami, *Uunderstanding Jitter and Phase Noise*. Cambridge, UK: Cambridge University Press, 2018.

[61] E. Rubiola, *Phase Noise and Frequency Stability in Oscillators*. Cambridge, UK: Cambridge University Press, 2009.

[62] J. Yu, J. Xu, and Y. Peng, "Upper bound of coherent integration loss for symmetrically distributed phase noise," *IEEE Signal Processing Letters*, vol. 15, pp. 661–664, 2008.

[63] M. A. Richards, "Coherent integration loss due to white gaussian phase noise," *IEEE Signal Processing Letters*, vol. 10, no. 7, pp. 208–210, 2003.

[64] A. Aubry, V. Carotenuto, A. Farina, and A. De Maio, "Radar phase noise modeling and effects-Part II: pulse doppler processors and sidelobe blankers," *IEEE Transactions on Aerospace and Electronic Systems*, vol. 52, no. 2, pp. 712–725, 2016.

[65] A. Aubry, A. D. Maio, V. Carotenuto, and A. Farina, "Radar Phase Noise Modeling and Effects-Part I : MTI Filters," *IEEE Transactions on Aerospace and Electronic Systems*, vol. 52, no. 2, pp. 698–711, 2016.

[66] V. Carotenuto, A. Aubry, A. De Maio, and A. Farina, "Correction," *IEEE Microwave Magazine*, vol. 13, no. 1, pp. 160–160, 2012.

Chapter 7

Microwave-Photonic Transceiver Technologies

With the increasing demands on performance of DRFM transceivers, a prominent trend is to (a) increase the input bandwidth, and (b) move the digital signal processing closer to the antenna. Consequently, digital microwave tranceivers and DRFMs are expected to digitize the signals directly at the antenna and eliminate the requirement for downconversion to intermediate frequencies.

Microwave-photonics technology and the development of advanced optical-electronic (OE) transceivers is playing an important role in the development of these DRFMs and is the subject of this chapter. Microwave-photonics merges the worlds of RF and photonics to enable new techniques in the processing of high-bandwidth signals. Besides its inherent advantages of extremely large time-bandwidth product and natural immunity to EM interference, a fundamental benefit of photonic techniques for processing wideband EM systems is that the entire RF/mm-wave spectrum constitutes only a small fraction of the optical carrier frequency. This often enables dynamic reconfiguration of the processor characteristics over multi-GHz bandwidths, and offers the ability to realize functions in microwave domain that are either very complex or perhaps not even possible in the RF domain.

These OE techniques offer not only new capabilities but improved performance. As an example, in tunable filtering, optical-microwave mixers with very high port-to-port isolation are now possible [1, 2]. Consequently, filtering and down-conversion tasks over a wide frequency band are possible due to the tiny fractional bandwidth required to cover the RF range at optical frequencies. Other DRFM applications have also been changing the competition and include ultrawide bandwidth wideband spectrum sensing and signal processing.

In this chapter, a tutorial of the important OE photonic component technologies is given and their role is studied in the development of microwave-photonic DRFM receivers. Important components include the laser (and more specifically, the mode-locked fiber laser), Mach-Zehnder modulators (MZMs), detectors, and photonic RF memories. Microwave-photonics is gaining significant ground for use in optical sampling and digitization, wideband spectrum sensing, frequency measurement receiving, and the photonic NYFR (pNYFR) is described in this chapter.

7.1 PHOTONIC RECEIVER DIGITAL ANTENNA COMPONENTS

The design of a DRFM microwave-photonic transceiver begins at the antenna and opens up the possibility to efficiently couple a much wider bandwidth of RF signals directly into the digital domain. Figure 7.1 illustrates the concept of a digital antenna and brings together five of the most important components used today. The majority of microwave-photonic transceiver solutions for DRFMs are typically not bulk optic components (e.g., lenses) but integrated optical processors.

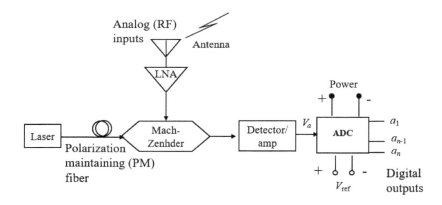

Figure 7.1 Schematic diagram of a microwave-photonics digital antenna showing the laser sampling the antenna signal at the output of a low noise amplifier at the MZM. A detector then serves to envelope detect the laser signal (at the output of the MZM) prior to being amplitude analyzed by an ADC.

Integrated optical components use wavelengths at 850 nm, 1300 nm, and 1550 nm for single-mode propagation. The reason for this choice is because the attenuation and scattering of the light (bouncing off atoms or molecules in the crystal) is much less at those wavelengths. There is also absorption occurring in several specific water bands outside these particular wavelengths, due to minute amounts of water vapor in the crystal. At longer wavelengths ($\lambda \geq 1550$ nm), the ambient temperature becomes background noise and disturbs the signals. Below 880 nm, one approaches the visible part of the spectrum. Consequently, the three major wavelengths for integrated optical processing are 850, 1300, and 1550 nm (but mostly at 1550 nm). Fortunately, we can also build lasers and photodetectors at these wavelengths. Current crystal technologies that are playing a key role in the microwave-photonic transceiver investigations include lithium niobate (LiNbO$_3$) [3], hybrid silicon (Si) and LiNbO$_3$ modulators [4, 5], polymer materials [6], and indium phosphide (InP) [7]. However, polymers and InP materials have higher loss and less long-term stability among other factors. The centerpiece processor in the diagram is the MZM with its low half-wave drive voltage V_π (for example, see [8]). The MZM has been the mainstay of integrated optical signal processing and is the most often used component in integrated optical designs. The MZM is an interferometer, typically fabricated within a LiNbO$_3$, lithium niobate crystal (index of refraction $n_e = 2.21$ at 1550 nm). The optical waveguides are placed by traditionally diffusing titanium (Ti) into the LiNbO$_3$, creating a higher index of refraction (i.e., 2.25) to guide the light at a wavelength at first, $\lambda = 1300$ nm [9] and then around 1995 to 1999 a gradual change took place toward a more efficient $\lambda = 1500$ nm [10]. This was also about the time when other active and nonlinear waveguide materials were starting to be diffused into the LiNbO$_3$ crystal [11].

Disregarding the particular technology used, it goes without saying that the MZM has been the major workhorse in the microwave-photonics industry and has been increasing in performance now for decades. Lithium niobate-based modulators have demonstrated high performance, but the material cannot be easily integrated with other photonic components and electronics due to the lattice mismatch between the LiNbO$_3$ and the other materials. The integration of active devices and detectors (also shown in Figure 7.1) along with the MZMs on a hybrid substrate is one of the exciting investigations going forward. For example, a silicon thin-film lithium niobate MZM has been fabricated to have a 3-dB bandwidth of 30.6 GHz and a half-wave voltage of 6.7 V×cm [4].

At the output of the antenna is a LNA, as the antenna signals are typically in the microvolt range. Depending on the antenna architecture, the LNA serves to boost the signal power, but as an active component also adds a certain amount of noise. It is the LNA noise factor (NF) that also affects the sensitivity of the receiver. The LNA can also be used to modulate the half-wave voltage of the MZM (discussed below) for tuning and calibration. The laser pulses or CW at 1550 nm propagate through single-mode fiber into the MZM. The antenna voltage $V(t)$ appearing on the MZM's electrodes modulates the laser's input intensity I_i that propagates through both arms of the MZM in a push-pull phase change. That is, the voltage is applied in equal and opposite polarities on each arm. The laser intensity at the MZM output, I_o, is coupled into the photodetector that transforms the impinging photons into an analog voltage and a small preamplifier amplifies the output voltage V_a. Having a low noise equivalent power is important as it also contributes to the detection threshold and overall sensitivity of the digital antenna. In fact, it is the largest value between the NF of the LNA and the NEP of the detector determines the sensitivity. Below we give an introductory discussion of these photonic components and then explore the newest concepts in microwave-photonic transceivers.

7.2 LASERS

The term laser is an acronym for light amplification by stimulated emission of radiation. Laser radiation, or more appropriately, oscillation, requires three components for operation: an active medium in which a gain can be maintained, a *pumping process* or mechanism where the energy is input to the maintain the gain in the medium (in the form of a population inversion), and a *resonator* to form the cavity to facilitate the feedback necessary for the oscillation. For most lasers, the active medium is a quantum system and the resonator is a mirror structure. The pump may operate by one of several phenomena depending on the type of laser (optical absorption, electron impact, chemical reaction, etc.) The quantum system can exist in any number of stationary discrete states that can be represented as different energy levels.

Transitions occur when the system changes from one stationary state to another. The three types of transitions that occur are (1) spontaneous emission from an excited state (finite lifetimes), (2) absorption, where the incident photon disappears and the system goes into an excited state, and (3) simulated emission which is the inverse of absorption, a downward transition caused by an incident photon (2 photons leave). As in the case of absorption, certain selection rules limit the possible emissive transitions. Conservation of energy and momentum are required [12]. Thus for a direct transition the energy difference between the initial and final states must be absorbed or emitted as a single photon. That is,

$$E_{upper} - E_{lower} = h\nu = \frac{hc}{\lambda}$$

where $E_{upper} - E_{lower}$ is the energy difference between the upper (initial state) and lower (final state) energy levels causing the transition, h is Planck's constant ($h = 6.62607015 \times 10^{-34}$ J s), ν is the frequency of the photon in s^{-1}, $c = 3 \times 10^8$ m/s is the speed of light, and λ is the wavelength of the photon. There are also several types of pump actions for lasers depending on the type of laser being used. This also depends upon the many types of laser materials. A resonator is a typically a pair of mirrors – any light ray along the axis is trapped within the system. When all of the oscillating frequencies are in phase, they will gain enough power at that frequency ν to exceed a threshold to begin a lasing action which is a process in which stimulated emission builds ups into a strong standing wave in the resonator, producing coherent collimated output that dominates the spontaneous output noise. The spectral line width may be up to 10^9 Hz compared to the operating frequency 10^{15} Hz of the laser (that is relatively monochromatic). Figure 7.2 shows a schematic diagram of a laser producing a CW light at the output of a semitransparent mirror. The cavity consists of two mirrors (one partially transmitting) separated by an active medium of length L with an index of refraction n_0.

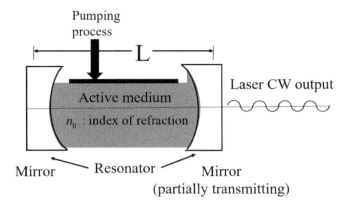

Figure 7.2 Schematic diagram of a laser resonating cavity producing CW light at the output.

The active medium serves to provide amplification to certain modes of light within the medium. When the mode power is great enough it is passed through the partially transmitting mirror as a CW output at a single frequency. Microwave-photonic receivers sample the antenna signal appearing on the MZM and require a stable, coherent laser pulse train. There are three ways to generate a laser pulse train using a CW laser: (1) directly modulate the laser, (2) externally modulate the CW laser to generate the laser pulse, and (3) use a mode-locked laser. Of these choices only the last two choices are worth considering due to the stringent pulse-to-pulse time uncertainty (time jitter) and pulse-to-pulse amplitude uncertainty (amplitude jitter).

Creating the very short, stable laser pulse train, for example, can be done using method (1) by *gain switching* or turning the laser pump on and off. Other techniques include Q-switching (laser output is turned off by increasing the resonator loss with a modulated absorber inside the cavity) and cavity dumping (a technique based on storing photons in the resonator during the off-times and releasing them during the on-times). Switching or direct modulating the laser can cause a number of unwanted effects such as chirping and overshoot along with a large amount of unwanted jitter. It also requires a setup of complex equipment for stable operation. Creating the very short, stable laser pulse train in a DRFM using methods (2) and (3) are the preferred methods and are considered below.

Many airborne DRFMs require lightweight transceivers and the stable laser must also be lightweight. As such there has been a growing interest over the last ten years in *fiber* mode-locked-lasers (MLL). This active mode-locking mechanism is discussed below.

7.2.1 Active Mode-Locking Mechanism

The laser consists of an optical resonator formed by two coaxial mirrors and some laser gain medium within this resonator. The frequency band where the laser oscillation occurs is determined by the frequency region where the gain of the laser medium exceeds the resonator losses. There are many modes of an optical resonator that fall within this oscillation band, and the laser output consists of radiation at a number of closely spaced frequencies. The total output of such a laser (as a function of time) will depend on the amplitudes, frequencies, and relative phases of all of these oscillating modes. The oscillating modes must be forced to maintain equal frequency spacings with a fixed phase relationship to each other, so the output as a function of time will vary in a well-defined manner. When this happens the laser is then said to be mode-locked or phase-locked [12]. This more useful technique, mode-locking, couples together the modes of the laser and locks their phases together inside the cavity. When the phases of these modes are locked together, they behave like the Fourier series components of a periodic pulse train $p(t)$ with pulsewidth $\tau = 2t_a$ and pulse repetition interval T or

$$p(t) = \frac{2t_a}{T} + \frac{4t_a}{T} \sum_{n=1}^{\infty} \frac{\sin(2\pi n t_a)/T}{(2\pi n t_a)/T} \cos(2\pi n n t/T)$$

The Fourier series expansion snapshot-in-time of a periodic pulse train concept is shown in Figure 7.3. Here the formation of a narrow pulse with full width half maximum (FWHM) Δt is formed from the combination of many sinusoids mode-locked together.

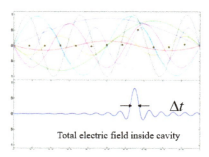

Figure 7.3 Concept of mode locking showing the formation of a narrow pulse with FWHM Δt formed from the combination of many sinusoids mode-locked together.

If a finite number of modes are locked in phase, they form a narrow pulse of photons that reflect back and forth between the mirrors of this resonating laser cavity. Each time the pulse reaches the output mirror, it transmits a short pulse. It is easy to see from the above Fourier series that the envelope of the laser output pulse can be expressed as

$$E(t) = A \frac{\sin(M\pi t)/T_F}{(M\pi t)/T_F} \quad (7.1)$$

where $T_F = 2Ln/c$ is the period of the pulse train for a single round-trip within the resonator, M is the number of modes locked together, and L is the length of the laser cavity. The optical intensity is expressed as

$$I(t) = |A|^2 \frac{\sin^2[M\pi(t-z/c)/T_F]}{\sin^2[\pi(t-z/c)/T_F]} \quad (7.2)$$

and is a periodic function of time, and z is the direction of propagation. The resulting pulsewidth $PW = T_f/M$ s and the period between pulses is $2L$.

A modulator inside the resonator cavity is used to actively lock the modes together. The modulating mechanism is introduced into the laser cavity to actively couple or force the oscillating frequencies to lock together in phase. The active modulation frequency f_{mod} must be an integer multiple q of the mode spacing frequency interval $(1/T_F)$ or

$$f_{mod} = \frac{qc}{2Ln} \quad (7.3)$$

For an actively mode-locked laser the sampling rate is then the modulation frequency or integer multiple of the modulation frequency. The pulsewidth of the actively mode-locked laser is

$$PW_a = (2M^{1/2}f_{mod})^{-1} \quad (7.4)$$

an expression known as the Kuizenga-Segman equation.

7.2.2 Mode-Locked Fiber Lasers

Actively mode-locked lasers can produce short-duration pulses in the picosecond range at high pulse repetition frequencies (PRFs) in the gigahertz regime. Mode-locked lasers can be constructed using solid-state lasers, semiconductor diode lasers, and fiber lasers. Fiber lasers are lightweight and use glass fibers common to the telecommunications industry and other fiber optic components to form a laser cavity. These fiber mode-locked lasers have demonstrated PRFs of 20 GHz at pulsewidths of one picosecond [4] and are ideal sources for optically sampling wideband signals directly at the antenna. Reducing the power requirements for these devices be-comes an important issue when considering mobile platform applications such as space-based signal sampling, and stand-in collection using UAVs.

The demand for stable, actively mode-locked fiber laser sources with extremely small timing jitter is driven by the development of high-speed signal sampling. In optical communications, the presence of jitter degrades the BER. For high resolution optical sampling, the jitter degrades the SNR. An example is given below of a mode-locked laser that uses a 980 nm pump light at its input. Figure 7.4 shows the schematic diagram of the erbium-doped fiber laser capable of generating fs wide laser pulses.

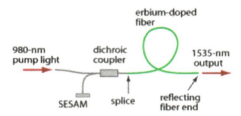

Figure 7.4 Fiber laser with a 1530-1535 nm output using a 980-nm pump laser input.

It has a broad bandwidth of 1530-1535 nm and is terminated by a semiconductor - saturable absorber mirror (SESAM). Bare fiber is used which provides a 4% reflection and the low power, 980-nm pump laser light is used at the input to pump the laser cavity.

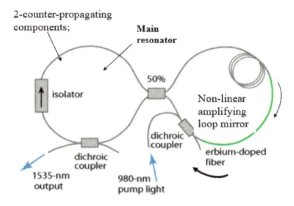

Figure 7.5 Fiber laser with a 1530-1535 nm output using a 980-nm pump laser input.

The pulsewidth and PRF are both determined by the fiber length. Only one pulse exists within the cavity at a time since it is not actively mode-locked.

By using an erbium-doped fiber amplifier, the pulsed fiber source shown in Figure 7.5 is able to generate ultrashort ps pulses or fs pulses by including two counter-propagating resonator components. The main resonator contains two counterpropagating components; the isolator and a dichroic coupler from which the output is taken. A 50% splitter in the middle separates the two resonators. The resonator on the right contains a nonlinear amplifying loop mirror that reflect the propagating light and an erbium-doped fiber amplifier and a dichroic coupler where a 980-nm pumped light source is applied.

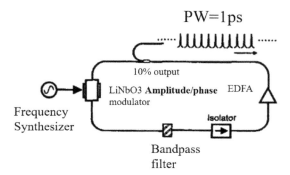

Figure 7.6 Schematic diagram of a fiber ring laser.

The schematic diagram of a *fiber ring laser* is shown in Figure 7.6. The ring laser is locked using a frequency synthesizer that drives a LiNbO3 amplitude, phase modulator (discussed below). The cavity propagation is counter-clockwise and to strengthen the mode characteristics, a bandpass filter in combination with an isolator. The isolator is used to prevent the light from reflecting off the next component backward into the cavity. The next component is a EDFA (erbium doped fiber amplifier) that amplifies the signal within the cavity. A small coupler (10%) takes a small amount of light out that might be needed. The pulsed light can be on the order of 1 ps.

Any amplitude modulation or timing fluctuations are damped out by the amplitude modulator as the mistimed pulses are first frequency shifted by the phase modulator before being damped out. The frequency shifts are converted to retiming shifts via group-velocity dispersion (GVD). The timing restoration depends on the product of frequency shifts exerted by the modulator and GVD.

Another actively mode-locked laser often used is the miniature fiber Fabry-Perot laser (FFPL) shown in Figure 7.7. The FFPL shown uses a 10.17-GHz source to actively mode lock the laser. Also within the cavity is a BPF, an optional EDFA and dispersion shifted fiber for narrowing up the mode, and a coupler to allow a 980-nm pump laser diode to help amplify the cavity.

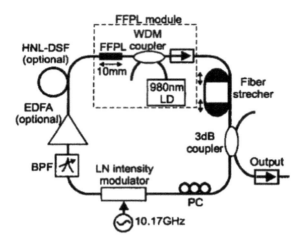

Figure 7.7 Schematic diagram of an actively mode-locked miniature FFPL.

Also included to tune the cavity to the correct resonance is a *fiber stretcher*. This is often built using a ceramic core wrapped with many turns of fiber and hooked up electrically to a feedback control circuit that uses a small amount of the output light from the laser. If the cavity is not tuned correctly (e.g., too small) then the control circuit applies a voltage to the ceramic core to expand it slightly until it comes into the correct mode.

7.2.3 Sigma Mode-Locked Fiber Laser

Probably one of the best performing fiber lasers for microwave-photonics as well as optical communications and optical comb generation is the *fiber sigma MLL*. The high repetition rate mode-locked fiber lasers have operations up to hundreds of GHz and have made practical sources for high-speed signal sampling as well as communications [13].

Considering the high level of harmonic mode locking ($N \approx 10^4$) at its GHz repetition rates, their low observed pulse dropout ratio of $< 10^{-14}$ was not previously explained. A theoretical model indicates that pulse dropouts may be avoided by the use of strong dispersion management within the cavity [14]. A schematic diagram of the sigma laser is shown in Figure 7.11 and consists of several types of fiber. These include polarization maintaining (PM); Yb-Er co-doped gain fiber (YEFA); dispersion-shifted fiber (DSF); and dispersion compensating fiber (DCF).

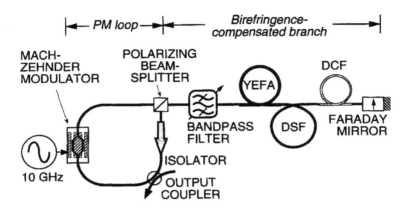

Figure 7.8 Schematic diagram of of the dispersion-managed sigma laser. Fiber types are polarization maintaining (PM); Yb-Er co-doped gain fiber (YEFA); dispersion-shifted fiber (DSF); and dispersion compensating fiber (DCF). (After [14].)

The sigma has PM fiber within the sigma loop and birefringence compensating fiber within the segment from the polarizing beam splitter to the Faraday mirror where total internal reflection takes placed. There is 13.5m of dispersion-compensating fiber (DCF) - average dispersion $D_{avg} \cong 0.8$ ps/(nm·km) giving a dispersion-managed strength parameter of $\gamma \approx 6.6$ for $\tau = 1.3$-ps pulsewidths [14].

The light's polarization, when injected into the non-PM branch changes, about in a random manner but is transformed into an orthogonal state by the Faraday mirror at the end [14]. Linearly polarized light is injected into the branch by the polarizing beamsplitter and returns to the beamsplitter also linearly polarized but rotated by 90°.

Theory from the model shown in Figure 7.11 was used to verify that at the lowest average powers, where *soliton formation* cannot be sustained, noisy $\tau \geq 5$ ps pulses are formed. With higher power, combinations of $\tau = 4.4$-ps solitons and missing pulses are formed. At higher power still, the soliton pulse train becomes completely filled (no dropouts) – an important property of the fiber laser in signal sampling. At the highest available pump power, 1.1-ps Gaussian pulses are produced [14]. For the sigma-type MLL cavity structure, the optical pulses propagate bi-directionally in the non-PM linear section and unidirectionally in the PM ring section. Under the assumption of small round-trip changes, such a hybrid mode-locked fiber laser can be described by the *master equation model* given by [15, 16]:

$$\frac{\partial u(T,t)}{\partial T} = \left(\frac{g_0}{1+(1/E_s)\int |u|^2 dt} - I_0\right)u + (d_r + jd_i)\frac{\partial^2 u}{\partial t^2} + jk_i|u|^2 u + jM\cos(w_m t)u \tag{7.5}$$

where $u(T,t)$ is the complex field envelope of the pulse, g_0 is the linear gain, E_s is the gain saturation energy, I_0 is the linear loss, d_r represents the effect of filtering, d_i is the group velocity dispersion coefficient, k_i is the self-phase modulation coefficient, M is the phase modulation depth, w_m is the angular modulation frequency, T is the large time scale denoted by the number of cavity roundtrips, and t is the short time scale. Only the Kerr nonlinear term is included and there is no nonlinear saturable absorption term because there is no equivalent saturable absorption action in the sigma-type cavity configuration [15].

To gain physical insight about the observed stability/instability of the two mode-locking states, a nonlinear eigenstate analysis was developed to study the problem in [16]. The steady-state solution of (7.5) can be reformulated as a nonlinear eigenstate problem as

$$jM\cos(w_m t)u + (dr + jdi)\frac{\partial^2 u}{\partial t^2} + +jk_i|u|^2 u = \lambda u \qquad (7.6)$$

with the constraint

$$\frac{g_0}{1 + (1/E_s)\int |u|^2 dt} - I_0 + \Re[\lambda] = 0 \qquad (7.7)$$

where the constraint in (7.7) comes from the balance of the net gain and loss at the steady state. Equations (7.6) and (7.7) can be solved together numerically by using the root-finding algorithm like Newton's method. The physical meaning of $\Re[\lambda]$ is the effective gain/loss coefficient produced by the combined effects of phase modulation, optical filtering, dispersion, and Kerr nonlinearity [16]. Or equivalently, is the required steady state lasing gain (relative to the constant loss) for the considered eigenstate.

The simultaneous generation of short pulses at multiple wavelengths is of particular interest to wavelength-division-multiplexed (WDM) communications systems, photonic microwave systems, and pump-probe applications. Multiple-wavelength sources can have the advantage of eliminating the need for synchronization between the different wavelengths. A configuration for a two-wavelength erbium-doped fiber laser based on the sigma configuration was developed at the Naval Research Laboratory [17]. Figure 7.9 shows the setup of the multiple-wavelength sigma laser. The design is similar to the single-wavelength design they describe in [18], allowing a single polarization operation with a more easily obtainable, non-polarization maintaining components shown to the right of the polarization beam splitter. In this design, a 3-dB coupler is added after the erbium-doped fiber and each of the 3-dB coupler output ports is used to complete the laser cavity for one wavelength. A 1-nm bandpass filter for selecting the wavelength, a fiber delay line for matching the cavity round-trip times, and an attenuator for balancing the gain and loss for the particular wavelength is included after each output port of the 3-dB coupler. A terminating Faraday mirror completes the cavity, providing *birefringence compensation* returning the mode to the polarization beam splitter in the orthogonal polarization. The saturation power of the erbium-doped fiber is approximately 15 mW.

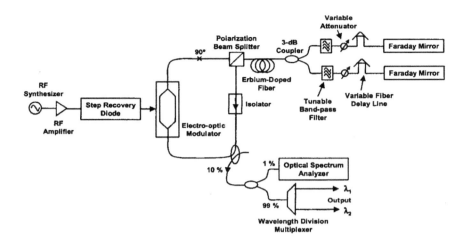

Figure 7.9 Schematic diagram of an actively mode-locked fiber sigma laser. (After [17].)

The laser is harmonically mode-locked at 500 MHz using a RF synthesizer that generates the 500-MHz signal that is amplified and sent to a step recovery diode that drives an electro-optic modulator within the cavity [17] (electro-optic modulators are described below). The BPF are set to the wavelengths that can be selected by the wavelength-division-multiplexer at the output of the laser. The attenuators are adjusted to equalize the gain and loss of both wavelengths and the fiber delay lines are tuned so that the drive frequency is a harmonic of the fundamental laser PRF.

To measure the pulsewidths, autocorrelators are used (described later). The pulses at 1538 nm and 1560 nm have pulsewidths, FWHM 10 ps and 14 ps, respectively, with no dropouts and are shown in Figure 7.10.

388 MICROWAVE-PHOTONIC TRANSCEIVER TECHNOLOGIES

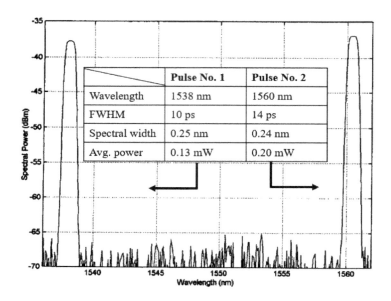

Figure 7.10 Schematic diagram of an actively mode-locked fiber sigma laser. (After [17].)

The NRL fiber laser was built with the goal of high-speed communications in mind. With the goal of signal sampling, a different set of criteria and performance measurements had to be undertaken. For example, although the sample time uncertainty (or time jitter) was of concern to both architectures, the concept of *amplitude jitter* or amplitude uncertainty was not a concern to the communications community when OOK was being transmitted. The mode-locked fiber sigma laser shown in Figure 7.11 shows the detailed construction of the laser including the feedback to the controller for the piezo-electric cavity fiber stretcher for tuning the cavity.

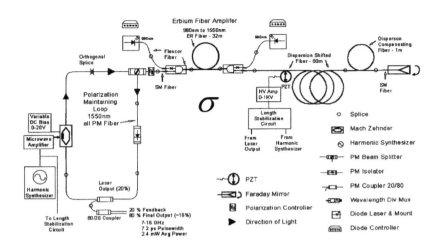

Figure 7.11 Schematic diagram the NPS mode-locked fiber sigma laser. (After [19].)

The Naval Postgraduate School (NPS) sigma laser is a fiber laser that also uses active mode-locking techniques to promote pulse shortening. The Sigma laser was originally invented at the Naval Research Laboratory (NRL) for a 100-GHz time-domain multiplexed fiber optic communication system. The NRL laser uses an erbium/ytterbium (Er/Yb) co-doped fiber amplifier pumped by a 1 W neodymium (Nd) solid-state laser. The NRL fiber amplifier has a saturated output power of 200 mW.

The NPS Sigma laser (shown in Figure 7.11) is similar in design to the original NRL Sigma laser except the fiber amplifier has a lower saturated output power. For the NPS laser, an erbium-doped fiber amplifier pumped by 200 mW of optical power from two 980 nm pump laser diodes is used to replace the Er/Yb co-doped fiber amplifier in the original NRL design. The use of only laser diodes instead of a solid-state laser results in a reduced saturated output power from the amplifier of 51 mW (one-fourth the power of the NRL amplifier). Although this reduction in fiber amplifier power results in a wider linewidth, a lower peak power output and not as much pulse compression, the decreased power requirement makes the device attractive for small, mobile platforms where power is at a premium. The ability of this lower power design to directly sample antenna signals was the subject of this investigation with the result given in [20].

7.3 DETECTORS

There are several types of detectors used in photonic transceivers. These can include photodiodes, photodiode arrays, light-dependent resistors, and avalanche photodiodes.

Requirements for detectors in DRFM transceivers include high sensitivity, short response time, minimum noise, linear and stable response over a wide range of intensities, and a low bias voltage requirement. They must also be easily coupled to the photonic system and low cost.

7.3.1 Detector Physics and Detection Mechanism

The photodetector converts the light intensity (power) into electrical signals (photocurrent) using a photodetection mechanism. Photodiodes are devices in which p-type, n-type, and intrinsic semiconductors are arranged as shown in Figure 7.12.

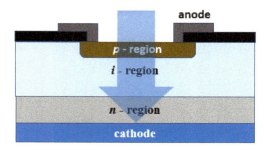

Figure 7.12 Schematic diagram of a photodiode device showing the arrangement of the p-type, n-type, and intrinsic semiconductors.

This structure is known as a p-i-n type diode, where "i" represents the intrinsic undoped layer between the n and p layers as shown in Figure 7.12. The intrinsic layer absorbs photons and transfers their energy to electrons in the atoms. This transfer of photon energy frees electrons from the atoms and simultaneously generates holes. The electrons migrate to the p-type region where they are collected by the anode (positive terminal) of the device while the holes move toward the n-region and the cathode (negative terminal). In this way, the incoming photons are converted to an electrical current if the proper wavelength is utilized. A current known as *dark current* is the relatively small electric current that flows through the photodiode (or charge-coupled device) even when no photons are entering the device. It consists of the charges generated in the detector when no outside radiation is entering the detector.[1] Dark current is a main source for noise and approximately doubles for every 10°C increase in temperature, and shunt resistance tends to double for every 6°C rise. The important selection of the semiconductor material will make a difference in the wavelength that

---

1 Dark currents are present in all semiconductor diodes. Physically, this dark current is due to the random generation of electrons and holes within the depletion region of the device and is referred to as the reverse bias leakage current.

Table 7.1

Photonic Detector Materials Used and Their Wavelength and Dark Current Relative Magnitude

Semiconductor Material	Wavelength	Dark Current
Silicon (Si)	200 - 1100 nm	Medium
Germanium (Ge)	800 - 1600 nm	High
Gallium arsenide (GaAs)	400 - 900 nm	Low
Indium gallium arsenide (InGaAs)	500 - 1800 nm	Lowest

can cause the photodetection mechanism in the photonic transceiver to work properly as well as the magnitude of the dark current. The semiconductor materials and their associated wavelength are shown in Table 7.1.

The schematic of a photodiode is shown in Figure 7.13 and includes the presence of the photodetection current I_{PD}, when light at the proper wavelength is absorbed in the depleted region of the junction semiconductor photodiode (also shown) [21].

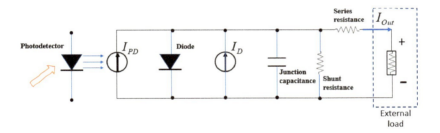

Figure 7.13 Schematic diagram of a photodiode device showing the presence of the junction capacitance, shunt resistance, series resistance and the induced photo detection current I_{PD}, and the presence of the dark current I_D. (After [21].)

In semiconductor detectors, the principle of photodetection starts with the schematic of a conduction band and a valence band separated by a bandgap as shown in Figure 7.14. The electrons in the valence band are bound while the electrons in the conduction band are free. When an incident photon having energy $E = h\nu \geq E_g$ (the bandgap energy), the photon (light) is absorbed in the semiconductor and electron-hole pairs are generated. Here ν is the frequency of the light or $\nu = c/\lambda$. As the electron is raised to a conduction band, the hole falls to a valance band. The free electron then travels down the barrier and free hole will travel up the barrier constituting the current flow that can be accounted for in an external measuring circuit.

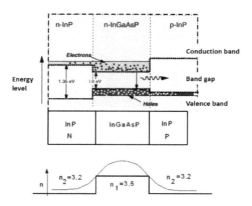

Figure 7.14 Schematic diagram of a double heterojunction PIN photodiode.

The most popular detector for integrated optical systems is the p-into-n or PIN photodetector. At 1550 nm, the most common detector architecture is the *double heterojunction* InGaAsP (indium gallium arsenide phosphide) photodetector. This advanced junction design is to reduce the diffraction loss in the optical cavity and is accomplished by modification in the material to control the index of refraction of the cavity and the width of the junction [22].

When a layer of material with a lower bandgap energy is sandwiched between layers of material with a higher energy bandgap, a double heterojunction architecture is formed (two heterojunctions present, one on each side of the active material). The schematic diagram of a InGaAs photodetector heterojunction architecture is shown in Figure 7.14.[2]

7.3.2 Detector Figures of Merit

Detector noise can be classified into three distinct groups: the photon noise, the detector noise, and the postdetector noise. The signal source as well as the background sources that are present during the sampling of the signal spectral noise equivalent power (NEP) is the amount of rms monochromatic signal power incident upon a detector that produce a SNR of unity in a detector. The expression for the NEP is

$$NEP = \frac{\sqrt{A_D \Delta f}}{D^*} \quad W \tag{7.8}$$

---

[2] In a thermal detector the energy of an incident, absorbed photon is used to heat the entire detector and thus, indirectly causes externally measurable effect such as a change in the resistance voltage (or volumetric length in a liquid thermometer). Since thermal detectors are sensitive to the amount of incident energy absorbed their spectral detectivity is relatively flat. In contrast, the detectivity of quantum detectors, being a function of the incident number of photons absorbed, increases as the energy per photon decreases.

where D^* represents the specific detectivity, A_D is the area of the detector and Δf is the noise equivalent bandwidth (Hz). A plot of the common detector materials D* (cm Hz$^{1/2}$/W) is shown in Figure 7.15.

The spectral *responsivity* $\mathfrak{R}$ of a detector is defined as the ratio of rms signal voltage, measured at the detector, to the incident rms infrared signal power. In the spectral irradiance H_λ (W/cm^2), the responsivity is expressed as

$$\mathfrak{R} = \frac{V_{rms}}{H_\lambda A_D} \quad \text{V/W} \tag{7.9}$$

The responsivity also varies with wavelength and diode material.

As an example, the responsivity curve for the Newport 818-BB-35F fiber-optic PIN photodiode detector, InGaAs that has a bandwidth of $\Delta f = 15$ GHz for $1000 < \lambda < 1650$ nm, and is battery biased, is shown in Figure 7.16. The detector has a diameter of 0.032 mm. At 1550 nm, the responsivity was measured to be 0.88 A/W with and $NEP < 0.04$ pW/Hz$^{1/2}$. PIN photodiode detectors use a photovoltaic effect to convert optical signals into electrical signals. These detectors consist of p-type and n-type doped silicon separated by an intrinsic region of high resistivity material. Absorption of photons in the silicon material causes electrons to drift and creates current [22].

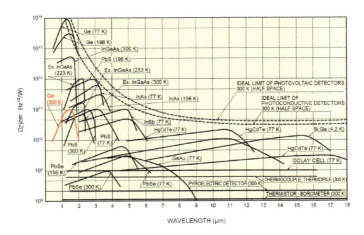

Figure 7.15 Detector D* curves (cm Hz$^{1/2}$/W) for various detector materials used including their operating temperature in Kelvin (K).

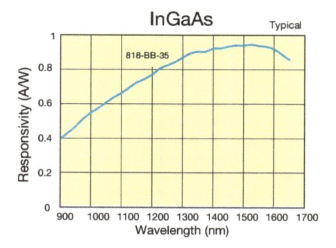

Figure 7.16 Typical responsivity curve for Newport detector 818-BB-35 PIN photodiode [22].

The pNYFR photodiodes that were used for example, are matched to the 1550-nm wavelength of the CW laser. The two most important factors for PIN diodes are gain and frequency response. The NYFR photodiodes have an internal LNA with a 10-dB gain and a frequency response bandwidth of 20 GHz. This response is primarily limited by the transit time of charge carriers separating from the substrate and moving to opposite ends of the depletion region. An internal shunting effect is also present and is calculated as [22]

$$\omega_p = \frac{1}{R_i C_i} \qquad (7.10)$$

where ω_p is the photodiode cutoff frequency in rad/s and R_i and C_i are the internal resistance and capacitance of the diode, respectively.

7.4 OPTICAL LINK AND COMPONENTS

A block diagram of an optical link is shown in Figure 7.17. It consists of a data input to a drive circuit. The drive circuit is typically a laser that can be either internally modulated or externally modulated, which is the preferred method, using a MZM (the second block). After propagation over the optical fiber (or free space), the data is detected with a photoreceiver or detector. The output of the detector is then amplified to an appropriate level.

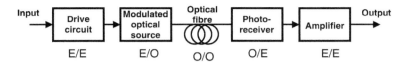

Figure 7.17 Block diagram of a photonic link.

If the propagation is through free space, a certain amount of delay is incurred and also loss. The propagation loss for 1 μs (in dB) for several materials is shown in Figure 7.18. Note the log-log scale. This includes microstrip.

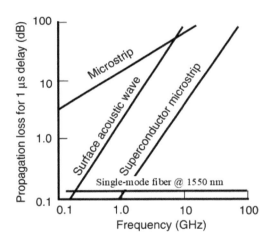

Figure 7.18 Propagation loss for a 1-μs delay for various materials (in dB).

The broad bandwidth of the optical media, such as $LiNbO_3$, is the main enabler for microwave-photonics. The main components that facilitate microwave-photonics are the electrodes that can be laid about each side of an optical waveguide media in order that the electrical signals that are applied can be *efficiently coupled* into the optical domain. The electrodes that are laid about the optical guide form contacts that create a transmission line for the electrical signals to travel on.

Another component is the optical fiber, more specifically, the *single-mode fiber*. The major advantage of single-mode fiber is the low loss of < 0.2 dB/km at 1550 nm and the large time-bandwidth $> 10^6$. However, the signal degradation due to chromatic dispersion, polarization-mode dispersion, and the nonlinearities are also present.

7.5 PHOTONIC LOCAL OSCILLATOR

A limiting component in microwave systems is the local oscillator (LO) being limited by the inherent phase noise that is present. For commercially available 10-GHz electronic LOs, the phase noise can reach $S_\varphi(f) < -105$ dBc/Hz @ 1-kHz frequency offset or $S_\varphi(f) < -115$ dBc/Hz at 10-kHz offset [23].

A measure of frequency stability can be quantified using the Allan variance and Allan deviation. The Allan variance, named after David W. Allan, also known as the two-sample variance, is a measure of frequency stability in oscillators, clocks, and amplifiers. It can be expressed as

$$\sigma_y^2(\tau) = \frac{1}{2\tau^2} \langle (x_{n+2} - 2x_{n+1} + x_n) \rangle$$

where τ is the observation period and the samples are taken with no dead-time between them. The Allan deviation is the square root of the Allan variance and is also known as *sigma-tau*. The long-term frequency stability of an LO can be evaluated using the Allan frequency deviation. The Allan deviation for a typical 10-GHz commercial RF source is $< 1 \times 10^{-11}$ at 1s.

Optical electronic oscillators are considered one of the most promising photonic LO generation approaches as shown in Figure 7.19.

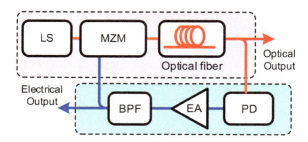

Figure 7.19 Functional block diagram of an OEO. LS: laser source; MZM: Mach-Zehnder modulator; PD: photodetector; EA: electrical amplifier; BPF: band-pass filter. (©IEEE, reprinted with permission from [23].)

A CW light from a laser source passes through a Mach-Zehnder modulator and a long optical fiber, and is then converted into an electrical signal at a PD. The generated electrical signal is amplified, filtered, and finally fed back to the RF port of the MZM, forming an oscillation loop. The phase noise of a single-loop OEO is given by [23]

$$S_\varphi(f) = noise floor - 10 \log \left[\left(1 - \frac{f_{osc}}{f_{osc} + j2Qf} e^{-j2\pi f n_r L/c} \right)^2 \right] \quad (7.11)$$

where f_{osc} is the oscillation frequency, f is the frequency offset from f_{osc} and Q, L, and n_r are the quality factor of the electric filter, fiber length and fiber refractive index, respectively. To achieve low phase noise oscillation with relatively short length fiber a coherent noise cancellation method to reduce the noise floor of the OEO is shown in Figure 7.20 where a pair of cascaded phase modulators (PMs) is applied to expand the output optical spectrum and keep the optical power constant. This reduces the intensity noise induced by the nonlinear effects in the optical fiber.

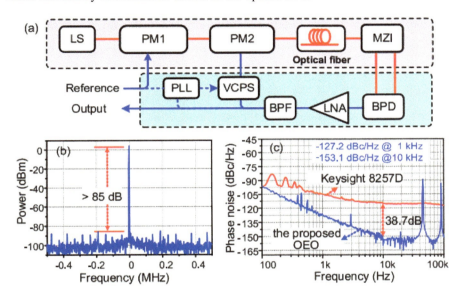

Figure 7.20 Experimental results of the OEO based on coherent noise cancellation. (a) Schematic diagram, (b) electrical spectrum, and (c) phase noise of the generated 10-GHz signal. LS: laser source; PM: phase modulator; MZI: Mach-Zehnder interferometer; LNA: low-noise amplifier; BPF: bandpass filter; PLL: phase-locked loop; VCPS: voltage-controlled phase shifter. (©IEEE, reprinted with permission from [23].)

Because of the complementary intensity modulation and the balanced detection, the common–mode intensity noise of the link will be largely suppressed. Using this approach, a 10-GHz signal with a phase noise of $S_\varphi(f) < 153$ dBc/Hz @ 10 kHz is achieved using a 4.4-km optical fiber which is 38.7-dB lower than that of a commercially available Keysight 8257D signal generator. The sidelobe suppression ratio reaches 85 dB (injection locking) with a frequency stability of 10^{-12}.

7.6 ELECTRO-OPTICAL MODULATORS

The use of photonics and integrated optical (IO) devices is a main thrust in wideband signal sampling technology. The first major issue in optical sampling technology concerns the optical modulators that are used to externally modulate the optical carrier from the laser. The most widely used I/O component is the Mach-Zehnder interferometer or MZM fabricated by diffusing titanium into LiNbO$_3$. The input RF signal is applied to electrodes that are placed about the optical waveguide and velocity matched such that the optical pulse travels down through the guide as the voltage does across down the electrode. This is similar to a transmission line. These devices can efficiently couple very wideband signals (up to 40 GHz) into the optical domain and function on the principle of the linear electro-optic Pockels effect.

Figure 7.21 shows the schematic diagram of a MZM with electrodes of length L across about one arm. The RF voltage applied to the electrode is sampled using an optical pulse (e.g., from a mode-locked laser). The input laser pulse I_{in} is split by a 3-dB splitter (or a directional coupler) and recombines at the output I_{out}. The RF voltage incident on the integrated optical interferometer's electrode changes the index of refraction of the guide generating a change in the propagation of the laser pulse. This causes a phase difference between the two arms. If the path length difference is zero (e.g., $V = 0$), the pulse recombines in phase for a maximum output. Otherwise, the pulses recombine out-of-phase and the destructive interference causes a decrease in the output pulse amplitude. That is, the applied RF voltage amplitude modulates the laser pulse.

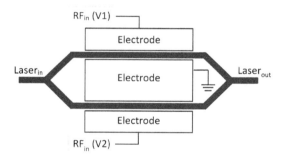

Figure 7.21 Schematic diagram of a Mach-Zehnder modulator.

The LiNbO3 MZM is the most critical component in the microwave-photonic phase detector design. Unlike conventional super-heterodyne receivers, the electro to optical conversion process with the MZM allows for optical sampling as high as 1 THz by distributed feedback lasers. Therefore, it has the potential to provide very wideband front-end phase detection without the need for signal downconversion. The schematics of the MZM is shown in Figure 7.21.

A typical MZM transfer function is shown in Figure 7.22. To understand how the MZM can couple the voltage into the optical domain, Figure 7.22 shows an antenna voltage $V(t)$ being applied to an MZM (one) electrode and a linear ramp being applied in time. The mode-locked laser is sampling the electrode voltage as the ramp is being applied in time. The antenna voltage starts just below zero volts and rises up, crossing zero as the MZM transfer function peaks at zero volts representative of the $\cos^2(\Delta\phi/2)$. Note the samples of the MLL and the V_π characteristic of the modulator build, determining the voltage it takes to go from a minimum to a maximum extinction or a maximum to minimum extinction. This value is not 100% and is also a function of frequency and degrades further with increasing frequency.

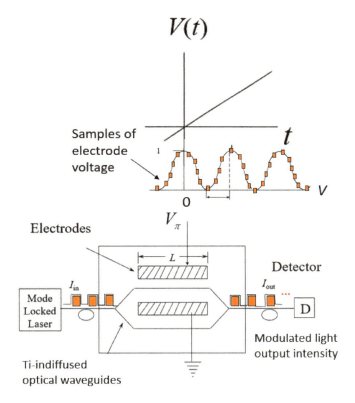

Figure 7.22 Mach-Zehnder modulator sampled transmission function (transfer function) showing the position of input voltage V_π (single electrode).

The input voltage V_π is the range of the transfer function and is the offset voltage that will be coupled onto the RF inputs to ensure that the system operates in the linear region of the MZM's transfer function.

$$H_{MZI} = \frac{I_{out}}{I_{in}} = \left[\frac{1}{2} + \frac{1}{2}\cos\Delta\phi(V) + \theta\right] \quad (7.12)$$

where θ is the constant phase change due to a DC bias voltage that may be applied to the guide. The DC bias can be used to move the transfer function laterally back and forth.

$$H_{MZI} = \cos^2\left(\frac{\Delta\phi(V)+\theta}{2}\right). \qquad (7.13)$$

Note that the transfer function is also called the *optical transmission* T_{MZ} and expressed as a function of the MZM's contrast ratio C (where $C=1$ corresponds to infinite on/off extinction), and V_π, the half-wave voltage as

$$T_{MZ} = \frac{1}{2}\left(1+C\sin\left(\frac{\pi V(t)}{V_\pi}+\theta\right)\right) \qquad (7.14)$$

The phase difference between the arms is a function of the applied voltage $V(t)$ as

$$\Delta\phi(V(t)) = \frac{2\pi n_e^3 r \Gamma L V(t)}{G\lambda} \qquad (7.15)$$

where n_e is the index of refraction, r is the pertinent electro-optic coefficient, G is the interelectrode gap, Γ is the electrical/optical overlap parameter, and λ is the free space wavelength. Figure 7.22 shows the normalized transfer function of the device for a sampled input voltage $V(t)$. Note the $\cos^2$ pattern. Another important performance specification for an integrated optical interferometer is the amount of electrode voltage needed to transition the normalized output from a minimum ($\Delta\phi = 0$) to a maximum ($\Delta\phi = \pi$). This half-wave voltage V_π is

$$V_\pi = \frac{G\lambda}{2L_i n_e^3 r \Gamma}. \qquad (7.16)$$

Another device constraint is the maximum voltage that may be applied to the electrodes. Applied voltage beyond the rated maximum (minimum) will spark across the electrode structure and damage the device. The specification of V_π and V_{max} reveal the maximum number of folds available from the device. As shown in Figure 7.22(b), a complete fold is $2V_\pi$. The maximum number of folds F available from a device is therefore

$$F = \frac{2V_{max}}{2V_\pi}. \qquad (7.17)$$

For typical LiNbO3 device parameters, $V_\pi = 0.6$ v and $F = 60$.

The dominate optical material for these devices in the future will be LiNbO$_3$. It currently has a strong foothold in high bandwidth applications (40 GHz) and will be difficult to replace. The insertion loss in these devices typically is 3-7 dB. III-V semiconductors are also somewhat popular due to their demonstrated higher bandwidth (60 GHz) and hybrid integration capabilities. This technology, however,

7.6 Electro-Optical Modulators

has a somewhat higher insertion loss (10 dB) due to mismatches. The third most popular material is polymers due to their low cost and ease of packaging. Currently, the major focus is to reduce the cost of these devices. As well, the bandwidths are being increased. Their performance is also extremely sensitive to polarization and temperature. The LiNbO3 Mach-Zehnder interferometer is and will be the most important photonic device due to its range of signal processing capabilities, its extremely large bandwidth, and its increasingly wide temperature stability. It is currently being investigated for many commercial applications as well (e.g., cable TV).

The second issue in optically sampling wideband signals concerns the lasers that produce the pulsed waveforms. Wavelengths of 1.3 μm and 1.55 μm currently dominate the technology and will do so into the near future. The use of diode pumped solid-state lasers are now being investigated to replace the Nd:YAG laser. These lasers are currently being ruggedized, reduced in size, and capable of higher power (greater than 500 mW) at a lower cost. Also, mode-locked fiber lasers (e.g., ring laser, Sigma laser) currently can produce pulse repetition rates greater than 20 gigasamples per second with power levels approaching 1 watt (considering the pump diodes supplying the optical amplifier are of sufficient power).

The third issue concerns the optical link itself (from laser to modulator and modulator to detector). Since the length of the fiber being used in a sampling system is not long, power limitations due to, for example, Brillion scattering are not a major problem. However, with polarization maintaining (PM) fiber, the cost is very expensive since the core is constructed to be asymmetrical. PM fiber is also more difficult to use since rotational alignment is always necessary. Also, there are a number of practical difficulties in the physical arrangement for PM fiber (e.g., coils). More efficient fiber splicing techniques are also being researched both for PM and non-PM fiber. Fiber optic connectors have not yet been standardized and the multipin connectors are not as reliable as they should be. The most likely connectors to be standardized are the ST for non-PM and the FC for PM fiber.

The fourth issue is the dynamic range and power-handling capabilities of the detectors. High bandwidth (3-4 GHz) PIN detectors are currently pushing 60-dB dynamic range with maximum input powers of 1W. Depending on the implementation, the optical sampling might pose a problem. Most recently, the development in MZMs are using thin film LiNbO$_3$ (TFLN) modulators with bandwidths over 100 GHz. For these high-speed traveling-wave thin-film designs, there are three major factors that need to be fulfilled: (1) the velocity match between the microwave and optical wave, (2) the impedance match of the traveling-wave electrode, and (3) low microwave attenuation [24]. However, for modulators with 2-cm-long phase shifters, the modulation bandwidth is limited by velocity mismatch but only in reported platforms with bandwidths lower than 100 GHz. Combined with other limiting factors such as lossy conductor and impedance mismatch, the experimental results showed limited bandwidth of 26–40 GHz. The development of the TFLN MZMs and the cause of the velocity mismatch is discussed with references in [24].

7.6.1 Electro-Optical Sampling Error

Optical sampling techniques using pulsed lasers, electro-optic modulators, and detectors have also been combined with comparators to amplitude analyze microwave signals into discrete levels (quantization). Henry Taylor recognized that the periodicity of the output of an electro-optic modulator with applied voltage was similar to the periodic variation of a binary representation of an analog quantity and could therefore be the basis for a folding ADC architecture. This architecture is discussed in more detail in Section 7.8. To discuss some of the limitations involved with the sampling, however, consider the maximum error due to the finite duration of the electro-optic interaction (RF voltage and laser pulse). Let the signal waveform applied to the electrodes be (see Figure 7.22)

$$V(t) = A\sin(2\pi f_m t) \tag{7.18}$$

where f_m is the maximum frequency. The error in voltage for a sample at time t_i is

$$\delta V = \frac{1}{\Delta T} \int_{t_i-(\Delta T/2)}^{t_i+(\Delta T/2)} A\sin(2\pi f_m t)dt - A\sin(2\pi f_m t_i) \tag{7.19}$$

where ΔT is the duration of the interaction (pulsewidth + optical transit time). Using the Taylor series expansion about the point $t = t_i$

$$\begin{aligned}\sin(2\pi f_m t) &= \sin(2\pi f_m t_i) + 2\pi f_m \cos(2\pi f_m t_i)(t-t_i) \\ &\quad -(2\pi f_m)^2 \sin(2\pi f_m t_i)(t-t_i)^2/2 + \cdots\end{aligned} \tag{7.20}$$

and evaluating the integral gives

$$\delta V = (\pi f_m \Delta T)^2 A \sin(2\pi f_m t_i)/6. \tag{7.21}$$

The maximum error is then

$$|\delta V|_{\max} = (\pi f_m \Delta T)^2 A/6. \tag{7.22}$$

It is required that error in an amplitude analyzing ADC be less than half the level spacing; that is

$$|\delta V|_{\max} < \frac{2A}{2^n}\left(\frac{1}{2}\right) = \frac{A}{2^n}. \tag{7.23}$$

The duration of the electro-optic interaction must satisfy

$$\Delta T < \left(\frac{3}{2^{n-1}}\right)^{1/2} / \pi f_m. \tag{7.24}$$

Here ΔT is understood to represent the sum of the pulsewidth and the optical transit time in the crystal.

For example, for $n = 6$ and $f_m = 300$ MHz, then $\Delta T < 325$ ps. The pulse transit time in LiNbO$_3$ is approximately 73 ps/cm. For a modulator length = 3.4 cm, the total transit time = 247 ps. Therefore, the maximum allowable laser pulsewidth is 78 ps. The wideband signal applied to the interferometer is sampled by the laser pulse. The laser pulse output is then detected by high-speed detectors.

7.6.2 Maximum Amplitude Jitter

Amplitude jitter in the sampling laser pulse is a problem if its variation is significant enough to exceed an ADC step size or least significant bit. A binary-weighted converter with n bits of resolution has $2^n - 1$ quantization levels that can be approximated by $2n$ (for most useful values of n). To avoid exceeding a step size, the amplitude variation ΔA must be less than or equal to

$$\Delta A \leq \frac{2A}{2^n}$$

where $2A$ is the peak-to-peak voltage of the sampled signal [18]. Since amplitude jitter, σ_A, is measured as a function of percentage of peak value, a maximum amplitude jitter of

$$\sigma_{Amax} = \frac{100 \Delta A}{A} \leq \frac{100}{2^{n-1}}$$

as a percentage. These equations can be used to plot the maximum tolerable amplitude jitter and compare against the measured amplitude jitter from the laser in order to understand the limits of the laser in an optical sampling ADC in terms of the bits of resolution.

7.6.3 Modulator Bias Controller

The widespread use of microwave-photonic transceivers and electro-optic modulators has created a great need for cost effective solutions that are reliable. However great the number of these devices in use, the fact that remains, all types of photonic modulators suffer from thermal drift problems and require efficient monitor and control circuits that can provide reliable operation [25]. There have been limited publications and commercial solutions for available for controllers to keep the V_π at a stable position. The solution in [25] is shown in Figure 7.23 and uses a feedback control to monitor in two steps only the optical average power. The monitor and control circuit is implemented in standard CMOS IC technology and has been successfully demonstrated with commercial 1550-nm MZM along with external Ge PD.

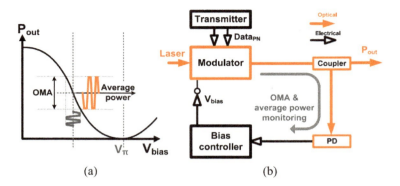

Figure 7.23 (a) MZM transfer curve with OMA and average power and (b) block diagram of our MZM bias controller.

In the first step, it searches the modulator bias voltage that generates the largest OMA by scanning the bias voltages within the predetermined range. In the second step, the circuit provides the optimal bias voltage by monitoring only the digitized average modulated output power and controlling the bias voltage so that the average power remains at the condition determined as optimal in the first step [25]. This two-step approach has been shown to greatly reduce the power consumption.

Maximum Temporal Jitter To prevent an amplitude error in a photonic ADC, the maximum temporal jitter is limited to $\sigma_{tmax} = \Delta t_{max}$. The total rms temporal jitter measured in the NPS laser, for example, was 386 fs (best measurement). Because of the low amplitude jitter, the temporal jitter measurements at 14.1, 15, and 16 GHz are believed to be lower but the second harmonic spectrum was not available due to the limits of the RF spectrum analyzer. Figure 7.24 shows the maximum temporal jitter.

7.6 Electro-Optical Modulators 405

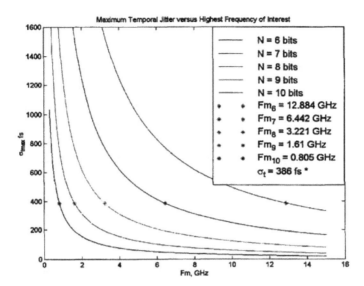

Figure 7.24 Laser temporal jitter allowed σ_{tmax} for N=6 through 10 bits (* indicates NPS laser measured at $\sigma_t = 386$ fs).

The curves are plotted for $n = 6$ to 10 bits. Using $\sigma_{tmax} = 386$ fs for comparison purposes, the maximum antenna frequencies that can be sampled are listed as Fm_6 through Fm_{10} on the graph. The maximum PRF of this laser is measured at 16 GHz. Since Nyquist sampling requires a PRF of at least twice the highest maximum frequency, this laser can be used to sample a signal as high as 8 GHz as indicated by the dashed line. At 6 bits, the laser temporal jitter is well below the maximum value allowed. For $n = 10$ bits, the maximum antenna frequency of interest is limited to 805 MHz.

Maximum Pulsewidth The maximum electro-optic interaction time (pulsewidth) allowable in an optical sampling system is also a function of the highest frequency of interest and the number of bits n [26]. The maximum electro-optic interaction time is given by ΔT. Figure 7.25 is a plot of ΔT versus the maximum antenna frequency to be sampled for resolutions between 6 and 10 bits. The smallest pulsewidth obtained from the NPS Sigma laser was 7.2 ps and was used to estimate the maximum RF frequencies that can be sampled. These maximums are listed as Fm_6 through Fm_{10} on the graph. Since the NPS Sigma laser is PRF limited to Nyquist sampling an 8-GHz signal, sampling is limited to 6 or 7 bits. At 10 bits, the pulsewidth limits the maximum frequency of interest to 3.38 GHz.

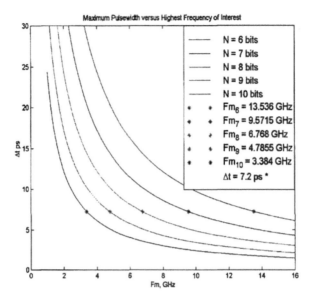

Figure 7.25 Maximum pulsewidth allowed vs. highest antenna frequency.

7.7 SIGNAL PROCESSING

Discussed in Section 7.4, the single-mode fiber offers a time-bandwidth product greater than 10^6 making it ideal for the processing of wideband signals. An optical transversal filter is shown in Figure 7.26 [27]

7.8 Photonic RF Memory

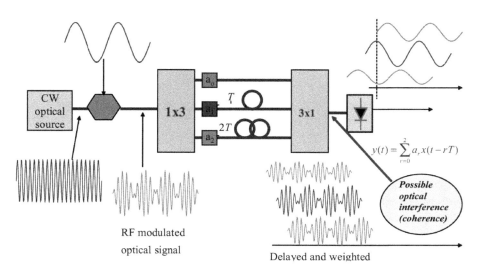

Figure 7.26 Block diagram of microwave-photonic 3-tap transversal filter. (From [27].)

The noise figure NF (dB) of a microwave photonic filter is defined as the ratio between the total noise power spectral density at the device output N_{out} and the noise power due to only the thermal noise spectral density applied to the input typically through a LNA at the reference temperature including the gain.

$$NF(\text{dB}) = 10\log_{10}\left[\frac{N_{out}}{4kT_0 T_{RF}/R}\right] = 10\log_{10}\left[\frac{N_{RIN} + N_{shot} + N_{sig-ASE} + N_{PIIN} + N_{th}}{4kT_0 T_{RF}/R}\right] \quad (7.25)$$

where k is Boltzmann's constant, T_0 is 298 K, and R is the load resistance at the RF source applied to the filter (or the LNA). The total noise spectral density at the output of the filter is composed of different sources of noise generated within the filter. These include N_{RIN}, the laser's relative intensity noise (RIN), N_{shot} is the shot noise within the detector and $N_{sig-ASE}$ is the noise power spectral density due to signal to ASE beating within the detector that depends on the quantum efficiency and the population inversion parameter of the detector's amplifier and its gain [27]. N_{PIIN} is usually the dominant noise source in single laser, non-coherent microwave-photonic signal processors. The term N_{PIIN} arises since the non-coherence due to the laser having a wide linewidth (typically to obtain a robust transfer characteristic).

7.8 PHOTONIC RF MEMORY

Optical waveguides such as fiber are one of the best delay line media for microwave and millimeter-wave modulated signals. Their unique capabilities can overcome the inherent bottlenecks caused by limited sampling speeds in conventional electrical signal processors since their structures can be designed to have a wide dynamic range. As such, time-coincident multiple signals, with each at a different RF frequency, can be stored and retransmitted.

One such approach for a photonic RF memory (PRFM) structure is shown in Figure 7.27 that can realize a long storage time, wide instantaneous bandwidth (IBW), and high dynamic range [28]. The memory structure uses frequency shifting recirculating delay line structures.

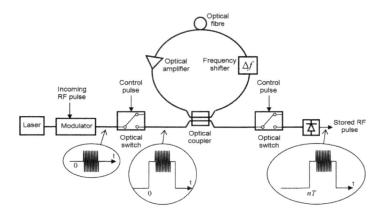

Figure 7.27 New photonic RF memory that can realize a long storage time, wide instantaneous bandwidth, and high dynamic range, and is based on the use of an optical frequency shifter inside a recirculating delay line loop to overcome a lasing problem and to enable multiple recirculations and long storage time to be realized. (©IEEE, reprinted with permission from [28].)

The modulated light from a laser is injected into a frequency shifting recirculating delay line (FS-RDL). The loop contains an optical frequency shifter and an optical amplifier. Each recirculation imposes a frequency shift Δf on the light. Consequently, it prevents the significant problem of oscillation or lasing in the cavity of conventional single-wavelength loops that operate with a closed-loop gain close to unity, because here all the spectral components are continuously shifted and they cannot resonate. This permits a large number of recirculations to be obtained without pulse power degradation. Optical switches at the input switch the pulse into the loop and at the output switch out the selected RF pulse after the desired storage time for subsequent detection by a photodetector. Important performance parameters include the maximum storage time, the SNR degradation as a function of the number of circulations and the phase distortion caused by the fiber dispersion [29].

The authors have shown high-fidelity RF signal storage after 150 μs equivalent to 100 circulations [28]. Their measurements have also demonstrated that the structure can store an RF microwave pulse with only small changes in the pulse amplitude for

280 μs. Note this is well enough time for EA techniques to be employed considering a round-trip delay of 12.4 μs / nmi (nautical mile) or 2 ns per foot.

One approach for moving to apply and compensate for the Doppler frequency shift (DFS) in the received RF signals and for meeting the moving target countermeasures demand and increasing further the storage time, a coherent frequency shift and time-gated amplifying recirculating delay line is included. This approach can simultaneously control the RF storage time and DFS, realizing long storage times greater than 500 μs and random Doppler frequency shift within ±3 MHz of the high-frequency RF signal, and its spurious suppression ratio greater than 36dB. This architecture is shown in Figure 7.28 [30]. A CW light comes from the laser with narrow linewidth (less than 50 kHz) and wavelength of 1550 nm. It is divided into two parts by a 50:50 optical coupler, with one part of the light modulated by the input RF signal through the dual parallel MZM (DPMZM) with single-sideband carrier suppression (SSB-CS) modulation undergoing a fixed 80MHz frequency shift by acousto-optic modulator AOM1. The other part of input light shifts a frequency that is decided by the demand of the DFS compensation and its maximum variable range is 80±3 MHz decided by AOM2.

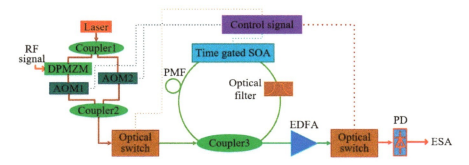

Figure 7.28 Experiment setup for simultaneous control of the storage time and random DFS of RF pulses. (From [30].)

The two optical signals combine in the 50:50 optical coupler2. Tuning the center frequency of AOM2 could randomly achieve a frequency shift within the central frequency of the RF signal. In order to maintain the stability of the suppression effect at different modulation frequencies, especially linear frequency modulation (FM) signals, the DPMZM was controlled by automatic bias feedback [30].

The coupler2 was connected to an optical switch that has an ultrafast response (rising edge was less than 2 ns) and high extinction ratio greater than 50 dB. Using an extinction ratio greater than 50 dB effectively suppressed the coherent lasing of the optical signal and the cofrequency light leaking into the recirculating loop. The optical switch was controlled by a trigger signal synchronized with the input RF pulse in a manner that it only passed the RF pulse-modulated optical signal while blocking the unmodulated light for the rest of the time.

The time-gated optical signals were then injected into a fiber recirculating loop to generate delay. The fiber recirculating loop was made up of a 50:50 coupler, a time-gated SOA, an optical bandpass filter with narrowband, and a long fiber (time delay was 9.5 us). In order to ensure that the optical gain remains unchanged after multiple cycles, an EDFA or SOA was required to compensate the power attenuation in the recirculating loop.

Figure 7.29 shows the measured signal spectrum of the 10-GHz carrier with random tuning of Doppler frequency shift within ±3-MHz. Figure 7.30 shows a measured signal spectrum of the RF pulse after 3 MHz Doppler frequency shifting and being stored for 209 μs.

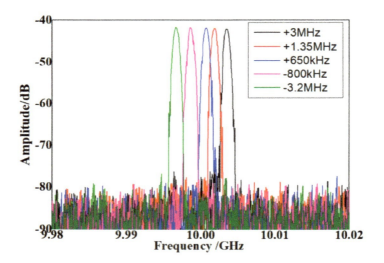

Figure 7.29 Measured signal spectrum of the 10-GHz carrier with random tuning of Doppler frequency shift within ±3MHz. (©IEEE, reprinted with permission from [30].)

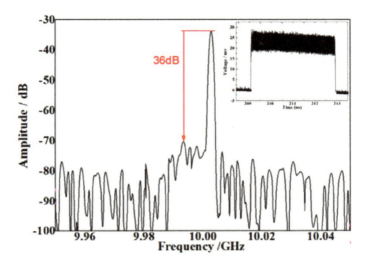

Figure 7.30 Measured signal spectrum of the RF pulse after 3-MHz Doppler frequency shifting and being stored for 209 μs. (©IEEE, reprinted with permission from [30].)

7.9 DESIGNING MICROWAVE-PHOTONIC ANTENNAS

DRFMs and transceivers taking advantage of microwave-photonic transmitters start at the antenna with a design approach that allows them to model the integration of the technologies together. One of these methods is described below for the generation of microwaves (millimeter-waves) over fiber using photonic frequency upconversion technology. Microwave-photonic transmitter technology can provide high-frequency, multiband local oscillator source and high-definition, large bandwidth waveform generation. The antenna, based on microwave-photonic technology, can effectively overcome several technical bottlenecks of traditional all-electronic devices, improving the bandwidth and quality of the signal processing techniques that follow.

Photonic monolithic antennas have the following advantages [31]:

1. Lightweight and small size since a photonic antenna does not require RF cables and connectors;

2. The possibility of remote antenna control due to low loss in optical fiber (below 0.2 dB/km);

3. Wide bandwidth, limited only by the antenna radiator itself;

4. Immunity to EM interference (important for large antenna systems);

5. Possibility to use optical signal processing and optical generation of microwaves in the antenna system.

7.9.1 Microwave-Photonic Antenna Modules

A transmitting photonic antenna can be constructed using a InGaAs/InP p-i-n photodiode as shown in Figure 7.31. The photodiode has an 8-GHz 3-dB bandwidth and a sensitivity (responsivity) of 1 A/W at 1310 nm [31].

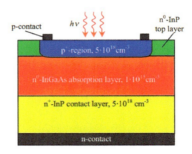

Figure 7.31 Cross section of a InGaAs/InP p-i-n photodiode.

The main disadvantages of the transmitting photonic antenna are the relatively low output microwave power that is limited by the maximum photocurrent that can be generated by the photodiode. This is typically not more 10-20 mA. The output power is also limited by the optical to electronic conversion loss, which can exceed 10 dB.

The microwave signal to be transmitted (5.8 GHz) is routed to an E-shaped microstrip patch antenna via a single mode fiber. A photodiode equivalent circuit is shown in Figure 7.32.

7.9 Designing Microwave-Photonic Antennas

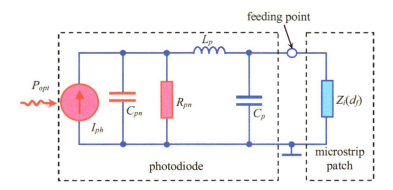

Figure 7.32 Equivalent electrical circuit of an active integrated photodiode-microstrip patch photonic antenna as a high frequency-transmitter.

The microstrip patch is represented as complex impedance Z_i. The photodiode is represented by the current source I_{ph}, the p-n junction capacitance C_{pn} (0.07 pF), and resistance R_{pn}, as well as the package capacitance C_p (0.5 pF) and inductance L_p (2 nH). The effective radiated power of a photonic antenna depends on the impedance matching between the photodiode and the microstrip patch. Since input impedance Z_i of the microstrip patch radiator depends on the feeding point position d_f, there is an *optimal feed point* providing the highest antenna efficiency within the widest frequency band.

To find the optimal feed point, a detailed knowledge of input impedance of both the photodiode and the microstrip patch radiator at the desired frequency band is required [31]. The model shows the input resistance Z_i increases for higher frequency (unfavorable condition for the photodiode) and the antenna efficiency is frequency-dependent. With the feed point at $d_f = 2.3$ mm (resistance increases with frequency monotonically) and power supply point $d_{DC} = 7.1$ mm [31]. Experimental scattering results $|S_{21}|$ dB, are shown for a conventional antenna, a hybrid photodiode and an integrated photodiode for the using a probe are shown in Figure 7.33.

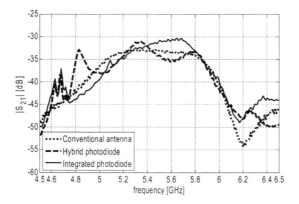

Figure 7.33 Scattering measurements S_{12} versus frequency for a conventional antenna, a hybrid photodiode, and a integrated photodiode (©IEEE, reprinted with permission from [31].)

A model that enables the electromagnetic properties of the antenna (input impedance, and antenna efficiency) to be readily incorporated into a system simulation tool can enable full link optimization and be useful in the design of integrated photonic antennas. A two-port, S-parameter model that allows the accurate representation of an antenna to be incorporated into a system simulation to enable parameters such as NF, system gain, and dynamic range properties is shown in Figure 7.34.

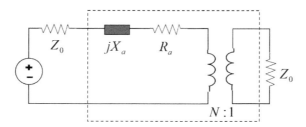

Figure 7.34 Schematic diagram of a two-port antenna representation simulation of the input impedance of the antenna and its overall efficiency (which includes dielectric, conductor, and surface wave losses).

The model can assist the design of an efficient integrated microwave-photonic into the antenna and also include simulation of the input impedance of the antenna and its overall efficiency (which includes dielectric, conductor, and surface wave losses) [32]. The Friis equation is then used to determine the overall performance of the integrated module, which allows for the accumulation of errors which can explain some of the discrepancies observed between measured and predicted results. The equivalent circuit

7.9 Designing Microwave-Photonic Antennas

of the antenna using the model shown in Figure 7.34 is defined as

$$Z_a = R_a + jX_a = R_r + R_1 + jX_a \tag{7.26}$$

$$R_r = \eta R_a \tag{7.27}$$

$$R_1 = (1-\eta)R_a \tag{7.28}$$

$$N = \sqrt{R_r/Z_0} \tag{7.29}$$

where R_r is the radiation resistance, R_1 is the dissipative loss resistance, X_a is the antenna reactance, η is the antenna efficiency, N is the turns ratio of the transformation of the power delivered to/from the antenna and free-space, and Z_0 is the system impedance.

Three examples are shown in [32]. The first example considers a 3-GHz directly modulated RF over a fiber link. Cascaded preamplifiers before the link are used to improve the NF and gain. The gain, noise figure, and SFDR at 1.5 GHz are 10.0 dB, 10.7 dB, and 107 dB/Hz, respectively. Next a highly efficient *stacked patch antenna* that operates over 1–2 GHz is incorporated with the resulting link gain, NF, and SFDR or 9.8 dB, 11.1 dB, and 107 dB/Hz, respectively [33]. A more commonly used technique to achieve a wideband operation of an antenna is a *printed flared monopole structure* with a matching circuit integrated with the antenna assembly. In this case the integrated antenna/link parameters are 6.9 dB (gain), 13.8 dB (NF), and 107 dBHz$^{2/3}$ (SFDR) [33].

The second example considers the hybrid integration of nonterminated photodiode with the antenna in particular, working to improve the impedance response of the photodiode over a wide band. As shown in Figure 7.35, by conjugate matching over a selected band, good efficiency was achieved.

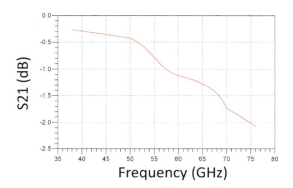

Figure 7.35 Predicted S-parameters of integrated photodiode/antenna.

The third example investigated concerns the simultaneous transmit and receive (STAR) photonic-based modules that have been successfully reported in [34] and are of high interest for use in CW emitters. The main issue here is their sensitivity to reflections and transmissions from interconnecting devices such as antennas connected to the modulator [32]. The model described above was used to design an endfire STAR antenna connected to a modulator with a low reflection coefficient (better than -30 dB over the frequency band of 9.5–10.5 GHz). The antenna (quasi-Yagi) was designed for an input impedance that matched the optimum efficiency of the electro-optic modulator with the results shown in Figure 7.36.

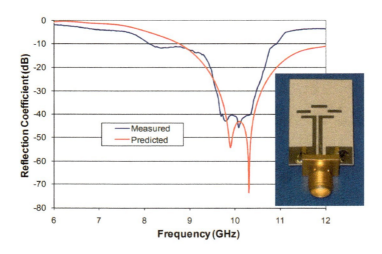

Figure 7.36 Reflection coefficient of quasi-Yagi printed antenna (©IEEE, reprinted with permission from [32].)

7.9.2 Microwave-Photonic Antenna Arrays

The design of a microstrip 4-cell antenna array for optical processing is described below. On the premise of ensuring good performance of the microstrip antenna unit, the array spacing, distribution form, excitation amplitude, and phase are adjusted to control the radiation characteristics [35]. Selecting the proper dielectric substrate material and its thickness directly affect the antenna radiation performance. That is, an *increase* in the antenna's material dielectric constant can reduce the size of the antenna and reduce the radiation efficiency. Increasing the thickness of the plate can also increase the antenna gain.

7.9 Designing Microwave-Photonic Antennas

The antenna substrate material used is AD255 with a relative permittivity $\varepsilon_r = 2.55$ with a thickness of 0.8 mm and width of [35]

$$w_{mp} = \frac{c}{f}\left(\frac{\varepsilon_r+1}{2}\right)^{-1/2} \tag{7.30}$$

or $w_{mp} = 5.6$ mm. After determining the width, the *equivalent* dielectric constant is calculated as

$$\varepsilon_e = \frac{\varepsilon_r+1}{2} + \frac{\varepsilon_r-1}{2}\left(1 + \frac{12h}{w_{mp}}\right)^{-1/2} \tag{7.31}$$

The length of the microstrip is theoretically one-half wavelength. Taking into account the half-wavelength, the actual length of the antenna should be subtracted from the *extension length* Δl [35]. The extension length can be calculated as

$$\Delta l = 0.412h\left(\frac{\varepsilon_e+0.3}{\varepsilon_e-0.258}\right)\left(\frac{w_{mp}/h+0.264}{w_{mp}/h-0.8}\right) \tag{7.32}$$

giving the overall length of the antenna calculated as

$$L = \frac{c}{2f\sqrt{\varepsilon_e}} - 2\Delta l \tag{7.33}$$

Using simulation and modeling to minimize mutual coupling, the array spacing was determined to be $d = 7.5$ mm. The antenna schematic is shown in Figure 7.37(a) and a micrograph of the printed circuit antenna is shown in Figure 7.37(b). The dimensions of the antenna array are given in Table 7.2.

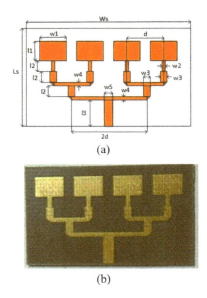

Figure 7.37 Microwave-photonic 4-cell Antenna Array Prototype (©IEEE, reprinted with permission from [35].)

Table 7.2
microwave-photonic 4-Cell Antenna Array Parameters.

Parameter	Dimension (mm)	Parameter	Dimension (mm)
Ws	32.00	w5	2.20
Ls	19.60	l1	3.90
w1	5.50	l2	2.15
w2	0.83	l3	2.05
w3	1.45	h	0.76
w4	0.80	d	7.50

The feed network design must be impedance matched (no or minimum reflections) and must have a wide bandwidth.

To generate the microwave (millimeter-wave) over fiber, photonic frequency upconversion can be used as shown in Figure 7.38.

7.9 Designing Microwave-Photonic Antennas

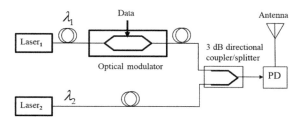

Figure 7.38 Block diagram of a photonic frequency upconversion using two wavelengths λ_1 and λ_2 combined with a directional coupler prior to being sent to a photo detector (PD).

The development of photonic antennas uses laser diodes, optical modulators, directional couplers and photo detectors as shown in Figure 7.38. For this transmitter, the microwave data signal is first applied to an electro-optical MZM. The MZM is a optical guided-wave, interferometric device fabricated in materials that have a strong electro-optic (*Pockels'*) effect such as lithium niobate ($LiNbO_3$), gallium arsenide (GaAs), or indium phosphide (InP). The waveguides are fabricated by a diffusion process using, for example, titanium (Ti), which creates a guide (Ti:$LiNbO_3$) having a higher index of refraction which then confines the optical mode within the guide. A pair of electrodes are placed about the guide on each side during fabrication. When the data signal is applied to the electrodes, an electric field is created through the guide and directly changes the index of refraction, coupling the data signal efficiently into the optical domain. The modulator output from the first laser, λ_1 (from laser 1), is combined with the light from the second laser, λ_2, within the 3-dB directional coupler and sent to the SiGe photodiode (PD).

For example, with $\lambda_1 = 1553.44$ nm (1 mW) and $\lambda_2 = 1553.60$ nm, the output frequency from the PD, Si-based antenna is 20 GHz. A 2.5-Gbps binary on-off keying (OOK) data signal is applied to the modulator. Simulation of the Si-based lateral contact structure avoids the harmful ohmic contact in the Ge region and drastically increases the RC bandwidth. The 3-dB bandwidth can approach 50 GHz. The measured 20-GHz results are shown in Figure 7.39.

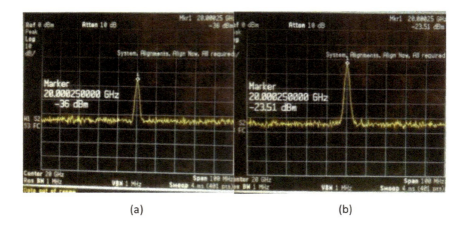

Figure 7.39 Measured 20-GHz signal applied to optical modulator, generated by beat frequency of both lasers, transmitted by optical fiber, detected by PD, and emitted by the photonic antenna showing (a) lower peak power and (b) higher peak power. (©IEEE, reprinted with permission from [35].)

The antenna was placed in their EM chamber and the gain of the microwave-photonic antenna was measured as $G_{mp} = 12.5$ dB, [35] which was different than the predicted results (by 3 dB). This was due to factors such as fabrication accuracy, bias of the interface, and inaccuracies of the test.

7.9.3 Microwave-Photonic Beamforming Phased Arrays

The use of integrated microwave-photonic technology in phased-array antennas has attracted strong interest because of the many advantages it offers. These include a wide bandwidth operation, the ability to provide antenna remoting, immunity to electromagnetic interference, excellent isolation, and the potential benefits of small size and low weight, which is ideal for DRFMs and transceivers on UAVs. A microwave-photonic broadband true time delay (TTD) continuously tunable beamformer module for phased array antenna applications is discussed in this section. For phased array antennas the application of TTD beamformers is advantageous as this results in the absence of beam squint and offers large bandwidths [36]. Traditionally, TTD beamformers are usually realized purely in the microwave domain; however, they suffer from high RF crosstalk and high propagation losses and are very bulky.

To get the best performance in the design, several photonic components should be integrated into the optical beamformer with the RF signal interface. For the processing, an ultra-low loss Si_3N_4-based optical waveguide platform called TriPleX has been used [36]. The active components used for the high-speed modulation and photo detection are realized in indium phosphide (InP). As mentioned above, InP is

7.9 Designing Microwave-Photonic Antennas

a III-V semiconductor material that has excellent electro-optic properties making it suitable for the fabrication of modulators and detectors. The pump laser is created using a hybrid cavity approach between InP and TriPleX, resulting in high optical power lasers with narrow linewidths. A schematic overview of the microwave-photonic beamformer developed in [36] is shown in Figure 7.40.

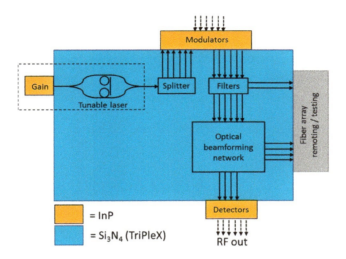

Figure 7.40 Schematic overview of a microwave-photonic beamformer (©2019 EuMC, reprinted with permission from [36].)

To design the beamformer, an analog photonic link is used as shown in Figure 7.41. The analog link uses a laser with optical output power P_0 into a single-mode fiber pigtailed phase modulator with half-wave voltage, V_π, which processes the input RF signal. The responsivity of the photo detector is R_{pd}, H is the response of the photodiode and the circuit between the output of the diode and the load. For lossy impedance matching $H = 1/2$.

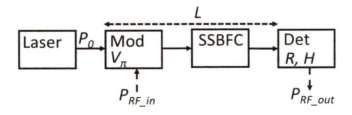

Figure 7.41 Analog photonic link with phase modulation and single-sideband filtering with full carrier (©2019 EuMC, reprinted with permission from [36].)

The optical link loss is L. The RF link gain with phase modulation and *single-sideband full carrier* (SSBFC) filtering is given by [36]

$$G_{PM-SSBFC} = \frac{P_{ml_{RF}}}{P_{in}} = \frac{I_{pd_{DC,q}}^2}{V_\pi^2} \pi^2 R_L^2 |H|^2 \qquad (7.34)$$

or

$$G_{PM-SSBFC} = \left(\frac{R_{pd} P_0 R \pi |H|}{L V_\pi} \right)^2 \qquad (7.35)$$

where the DC photocurrent for the link is given by

$$I_{pd_{DC,q}} = R_{pd} \frac{P_0}{L} \qquad (7.36)$$

The four key components are described below and include the laser, the modulator, and the beamforming network.

7.9.3.1 Laser

The laser parameters that directly affect the link performance are its output power $P_0 \geq 40$ mW and its relative intensity noise $RIN = -165$ dBc/Hz [36]. The wavelength of the tunable laser was 1530-1560 nm (optical C band).

7.9.3.2 Phase Modulator

The important parameter of the phase modulator is its half-wave voltage that is a function of wavelength $3.5V \leq V_\pi \leq 5.5V$. The half-wave voltage is shown in Figure 7.42 from 0 to 20-GHz.

7.9 Designing Microwave-Photonic Antennas

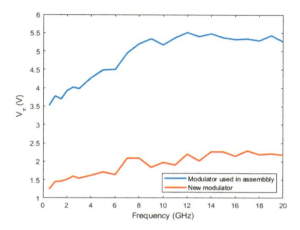

Figure 7.42 Phase modulator measured half-wave voltage, V_π, as a function of modulation frequency. (©2019 EuMC, reprinted with permission from [36].)

7.9.3.3 Photodetector

The most important parameters of the photodiode detector are the responsivity $R_{pd} = 0.8$ A/W and the bandwidth $B \geq 40$ GHz [36].

7.9.3.4 Optical Beamforming Network

The fourth component is the optical beamforming network design for TTD beamforming which uses optical ring resonators in addition to an optical sideband filter (OSBF) in order to create the single sideband full carrier (SSBFC) signal. When a signal is fed through a series of optical ring resonators, the signal is delayed and in this way a TTD can be created and the individual delays can be tuned by controlling the coupling coefficients via thermo-optic tuners. An example of TTD beamforming is shown in Figure 7.43.

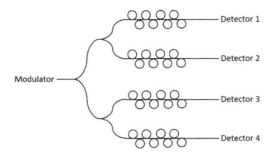

Figure 7.43 Schematic diagram of a binary tree optical beam forming network (OBFN). (©2019 EuMC, reprinted with permission from [36].)

7.9.3.5 Measured Results

The measured link gain S_{21} of the beamforming network over the frequency range from 10 MHz to 40 GHz is shown in Figure 7.44. The results are measured using a single sideband full carrier modulation with an optical sideband filter of 20 GHz passband and a free spectral range of 50 GHz. The carrier is located 3 GHz from the edge of the passband and the result is measured from the RF input of the modulator to the RF output of the detector. (The dip around 20 GHz in Figure 7.44 is due to the OSBF.) The achieved link gain of −23 dB at 4 GHz is the highest measured in an integrated optical beamforming network reported to date [36].

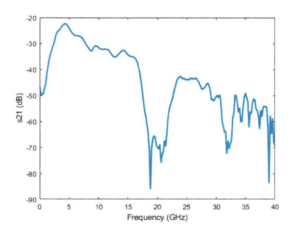

Figure 7.44 Measured link gain results S_{21} for one branch of the optical beamformer. (©2019 EuMC, reprinted with permission from [36].)

A flexible microwave-photonic image-reject mixer based on a widely tunable microwave-photonic filter (MPF), that is used to select the desired signal and suppress the image. The MPF is realized by using stimulated Brillouin scattering, which offers significant advantages of wideband tunability, high selectivity, and large image rejection. In this work, a theoretical model is established to describe the operation principle of the proposed mixer at first. Then, in the experiment, a 7.8-GHz radio frequency (RF) signal is successfully downconverted to an intermediate frequency signal between 0.2 and 1.5 GHz, with an image rejection of larger than 40 dB. Additionally, when the RF frequency is tuned from 1.8 to 19.8 GHz, the image rejection remains above 40 dB.

7.10 PHOTONIC ANALOG-TO-DIGITAL CONVERTERS

Recently, there has been concern that ADC performance might have reached a saturation point within the deep, submicron CMOS technology. However, the recent innovation into photonic ADCs have refuted such fears [37]. By leveraging the unique benefits of photonic technology, we are able to build ADC systems capable of far greater speed, bandwidth, and accuracy than the current all-electronic methods. Consequently, these photonic methods allow us to directly sample the RF signals *at the antenna*. By pushing the ADC close to the antenna, the many problems and complications of downconversion are eliminated allowing us to accurately capture and process the incoming signals without requiring high-speed electronics [20]. Direct digital systems, featuring upsamplinng and downsampling of information content directly to- and from- the intended carriers now offer the best combination of agility and performance for photonic transceivers [38].

7.10.1 Jitter - the Limiting Factor

Thermal noise appears to dominate the performance of high-resolution low-bandwidth ADCs while aperture jitter and comparator ambiguity become important at high sampling rates. Figure 7.45 illustrates the allowable timing jitter for a Nyquist ADC (i.e., maximum input frequency 1/2 sampling rate) as a function of sampling rate and effective bits, assuming that jitter is the only error source.

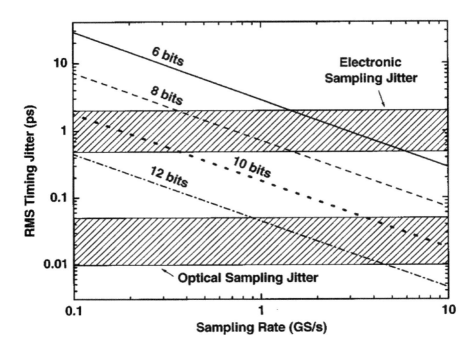

Figure 7.45 Timing-jitter requirement for ADCs as a function of sampling rate and number of effective bits. The calculation assumes that timing jitter is the only noise source. The present-day electronic and optical sampling-aperture jitter ranges (hatched areas) are from [39, 40]. (©IEEE, reprinted with permission from [41].)

Estimates of aperture jitter in state-of-the-art electronic ADCs range from 0.5 to 2 ps are shown in the hatched region in Figure 7.45. This reveals that that electronic sampling jitter is presently a major limit to the performance of high-speed, high-resolution ADCs. For example, the 50-fs timing jitter required to implement a 1-GS/s 12-bit converter is far below that of present electronic jitter estimates. In contrast, sampling-aperture jitter less than 50 fs has been attained in optical sampling systems [41].

7.10.2 The Quantum Limit

Recently, the jitter performance of different photonic sampling techniques have been investigated in detail [42]. Here the MLL with negligible jitter σ_{MLL} is considered since it is the basic motivation for photonic assisted ADCs where other mentioned noise sources in the system are dominant. The MLL with this negligible jitter, emits ultrashort pulses and a following erbium doped fiber amplifier (EDFA) amplifying these pulses then adds noise due to amplified spontaneous emission (ASE). This causes the pulse center to shift in time also inducing a slight change in the pulse center frequency leading to a group velocity change with a corresponding change in the arrival time at the fiber end [42]. The impact of a varying center frequency and fiber dispersion is known as Gordon-Haus jitter after its discovery in [43]. Either the center of mass or the rising edge of the detected pulses, emitted by an MLL, can be used as the sampling instant. The quantum noise model takes into account the shot noise, the thermal noise in the receiver, the ASE noise of an (optional) EDFA and the pulse-to-pulse energy fluctuations resulting in quantum jitter limits of 1 to 2 fs.

7.10.3 The Taylor Scheme: How It All Began

In the mid-1980s, Henry F. Taylor recognized that the output of a MZM was periodic and had the form of a *raised* $\cos(\theta)$ or $\cos^2$ waveform and that this periodicity could be controlled with the electrodes (placed about the optical waveguides) where the wideband RF voltage was applied as well as a DC electrode (where a tuning bias could be applied) [44]. This periodic MZM transfer function could be used to perform mathematical calculations such as an analog multiplier as shown in Figure 7.46.

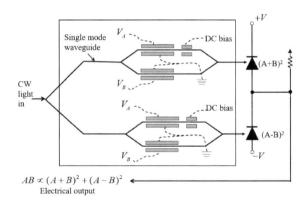

Figure 7.46 Schematic diagram of a wideband photonic multiplier. (©IEEE, reprinted with permission from [44].)

As the CW light enters into the modulators the analog voltages V_A and V_B to be multiplied are applied to the electrodes as shown. The output is detected with the properly biased diode detectors, resulting in $(A+B)^2$ and $(A-B)^2$. Subtracting the photocurrents in a common load resistor will yield an output voltage that is proportional to the product AB. Based on the performance of today's MZMs and high-speed detectors, multiplication of signals with bandwidths in excess of 40-50 GHz should be readily achievable.

With the periodic transfer function of the MZMs, a CW laser could be applied and the output detected, amplified and thresholded at the mid-point, he also recognized that with a properly tuned bias, an n-bit ADC with an output Gray code would result. His schematic architecture is shown in Figure 7.47.

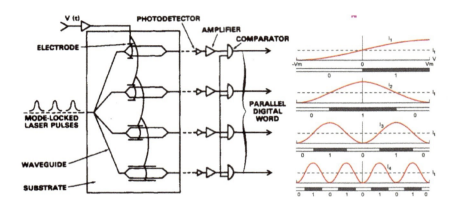

Figure 7.47 Schematic diagram of the Henry Taylor scheme for a 4-bit, guided wave ADC. (After [44].)

Note that each MZM electrode length doubles for each additional bit where the shortest length electrode provides the largest folding period of the device or V_π. The four waveforms and the Gray code that results are also shown in Figure 7.47.

Following this, the demonstration of a $n = 2$-bit 1-GS/s guided-wave ADC was published [45]. The pulses were generated with a diode laser driven with a comb generator. The pulses were sent over a single-mode fiber, sampling the RF on the MZM device shown in Figure 7.48 at 1 GHz. Separate DC biases were used to offset the transfer functions.

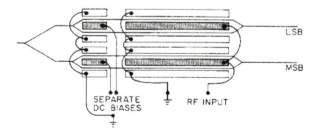

Figure 7.48 Schematic diagram of a 2-bit LiNbO3 guided-wave ADC. (©IEEE, reprinted with permission from [45].)

7.10 Photonic Analog-to-Digital Converters 429

The output was detected with an avalanche photodiode and amplified with an amplifier before driving a comparator chip. A 1-to-8 serial to parallel converter resulted in the 125-MHz output shown in Figure 7.49.

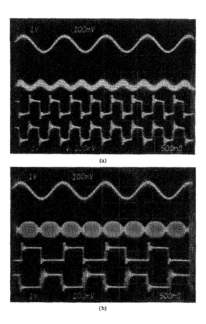

Figure 7.49 Beat-frequency test of electrooptic A/D converter. The input RF test signal frequency is 499.2 MHz and the sampling rate is 1 GHz. Shown are a derived beat signal (top trace), detector output (second trace), and comparator outputs (bottom two traces) for the (a) least significant bit, and (b) most significant bit. The relative phase of the beat signal has not been synchronized to that of the detector and comparator. (©IEEE, reprinted with permission from [45].)

Pulsed diode lasers can also be used for signal sampling and high-speed frequency downconverters, comparators and demultiplexers used to produce the digital output.

These architectures make use of the periodic dependence of the interferometer MZM optical output power on the applied voltage to the electrodes V (input voltage to be digitized) and the electrode lengths L. The secondary DC electrodes add the additional DC bias and hence a static phase shift ψ to the output intensity to provide a Gray code. The output intensity I of each interferometer can be expressed in a form similar to (7.13) as

$$I = I_o \cos^2\left(\frac{\phi(VL_n)}{2} + \frac{\psi}{2}\right) \quad (7.37)$$

where I_o is the intensity of the input laser pulse. For an increase in the number of significant bits in the photonic ADC, for example, the electrode lengths L_n decrease by

factors of 2 or
$$L_n = 2^{n-N} L_N$$
where L_N corresponds to the electrode length of the LSB.

7.10.4 Limitations of the Taylor Scheme

One of the major limitations associated with this 1-bit per MZM type of electro-optical ADC is the exponential electrode length relationship for an increasing number of bits. For example, an 8-bit ADC (8 MZMs) using the above scheme, would require electrode lengths $L, 2L, 4L, \cdots, 128L$ with $128L$ corresponding to the LSB. This requirement ultimately constrains the achievable resolution of these converters to moderate bit resolutions and limited frequency ranges due to the transit time of the optical pulse as it traverses the various sized interferometers in the ADC array. In addition, driving a parallel array of MZMs with long electrodes presents a large capacitance to the input RF, restricting the input bandwidth considerably.

By folding the waveform in a little bit smarter fashion and amplitude analyzing the output with > 1 comparator, greater than 1-bit per MZM can be achieved [46]. Valley provides a good overview of the progress since Taylor, including time-stretched, time-interleaved, $\Sigma - \Delta$, multiple quantum well, and self electro-optic device techniques, among others [47].

7.10.5 Optical (Phase-Encoded) Sampling

In Figure 7.50 an optical sampling architecture that uses a digital linearization technique referred to as *phase-encoded optical sampling* that combines the energy from the complementary outputs of a dual-output modulator to invert the MZM transfer function is illustrated. If a pulse of energy is transmitted through the dual-output modulator at time t_0, the pulse energies at the output ports will be

$$E_A = \frac{E_0}{2}\left(1 + C\sin\left(\frac{\pi V_{in} t_0}{V_\pi}\right)\right) \quad (7.38)$$

$$E_B = \frac{E_0}{2}\left(1 - C\sin\left(\frac{\pi V_{in} t_0}{V_\pi}\right)\right) \quad (7.39)$$

where $V_{in}(t_0)$ is the voltage applied to the modulator at the instantaneous sampling time t_0. That is, the pulsewidth is negligible as is the time-of-propagation through the modulator considering the electrodes are *velocity matched*, meaning the voltage travels down the electrode at approximately the same speed as the velocity of optical sampling pulse.

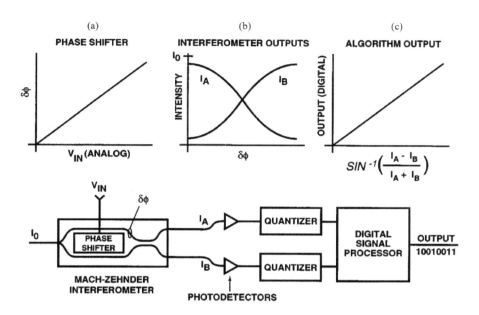

Figure 7.50 Phase-encoded optical sampling technique. (a) Electro-optic effect provides a linear phase shift with respect to the applied voltage. (b) Complementary interferometer output intensities are sinusoidal with respect to the induced phase shift. (c) The phase shift and, therefore, the applied voltage is recovered by inverting the interferometer transfer function. (©IEEE, reprinted with permission from [41].)

After measuring E_A and E_B, the applied voltage can be determined as [41]

$$V_{in}(t_0) = \frac{V_\pi}{\pi} \left[\sin^{-1}\left(\frac{1}{C}\frac{E_A - E_B}{E_A + E_B}\right) - \theta \right] \tag{7.40}$$

Note that the linearity is limited by [41]:

1. Accuracy to that E_A and E_B can be measured and to the extent that the MZM transmission characteristics can be measured by 7.14;
2. Linearity of E_A and E_B must be maintained through photodetection, signal conditioning, and quantization;
3. Estimate of $V_{in}(t_0)$ independent of input pulse energy E_0;
4. Small deviations of modulator bias point from quadrature $\theta \neq 0$ cause only dc offset error and do not degrade linearity.

As mentioned, this technique requires twice the number of calibrated electronic detection and quantization channels as a single-output intensity sampling technique.

7.11 HIGH-RESOLUTION ENCODING PROCESS FOR PHOTONIC ADCS

ADC architectures utilizing the MZM are of special interest for digital receivers due to their capability of directly digitizing high frequency, wideband signals. This is due to the high pulse repetition frequency (PRF) of mode locked fiber lasers ($PRF > 300$ Gb/s) that can be used for sampling and the large bandwidth available for the MZM ($B > 50$ GHz) [48]. These sampling sources also have much lower sampling jitter than all-electronic sampling circuits.

In the traditional photonic ADC, N MZMs arranged in a parallel array symmetrically fold the RF voltage with the folding period between MZMs being a successive factor of 2. Folding the input signal and applying a single threshold to implement a Gray code was described above and shown in Figure 7.47. Here, the folded output voltage from each detector is quantized with a single comparator. Together the comparator outputs represent the RF voltage in a binary Gray code format where any two successive values differ in only one bit. However, one of the major limitations with this ADC approach is the achievable resolution. For the folding periods to be a successive factor of 2, the length of each electrode must also be doubled, which adversely affects the device capacitance and ultimately constrains the feasible resolution. For these 1-bit per MZM approaches, resolution is limited to less than 4 bits.

A modular preprocessing technique based on the optimum symmetrical number system (OSNS), which extends the resolution of the photonic MZM approach *beyond* 1-bit per MZM, was presented in [49] and experimental results using a pulsed laser were shown in [50]. This architecture for moduli $m \in \{3, 4, 5\}$ is shown in Figure 7.51 (a).

7.11 High-Resolution Encoding Process for Photonic ADCs

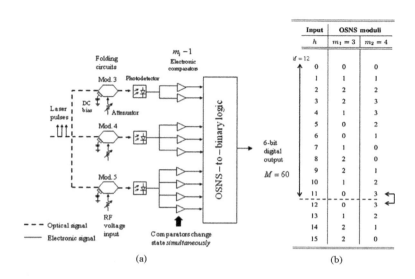

Figure 7.51 (a) $N = 3$ Mach-Zehnder modulator ADC using the OSNS with $m_i - 1$ comparators in each channel (dynamic range $M = \prod m_i = 60$) and (b) OSNS integers for $m_1 = 3$ and $m_2 = 4$ with dynamic range $M = 12$ ($0 \leq h \leq 11$). (©IEEE, reprinted with permission from [48].)

However, as demonstrated in Figure 7.51(b), consider for $N = 2$, $m_1 = 3$, and $m_2 = 4$, ($M = \prod m_i = 12$) for each code transition multiple comparator thresholds are crossed. For example, consider the transition between $h = 1$ to $h = 2$ in Figure 7.51(b). In the $m_1 = 3$ channel, the number of comparators whose threshold has been exceeded changes from 1 to 2. Similarly, the same change is observed in the $m_2 = 4$ channel. If these transitions do not occur simultaneously, a large encoding error will occur at the output of the OSNS-to-binary logic network. The OSNS presents a problem at each of these specific voltage levels since the comparators within each channel must switch states *simultaneously*. For example with $m_1 = 3, x_1 = 2, m_2 = 4, x_2 = 1, h = 9$ instead of $h = 2$ and consequently, a large error is incurred. Additional interpolation circuitry can be added to trap and correct these errors; however, this adds a good deal of complexity and overhead to the implementation.

7.11.1 Robust Symmetrical Number System

To overcome these problems, a new preprocessing technique based on the robust symmetrical number system (RSNS) is described. The RSNS is also a modular scheme consisting of $N \geq 2$ integer sequences with each sequence associated with a coprime modulus m_i [51, 52]. Due to the presence of ambiguities, the set of integers within each RSNS sequence do not form a complete residue system.[3] The ambiguities within each modulus sequence are resolved by the use of additional moduli and considering the vector of paired integers from all N sequences.

The RSNS is based on the following sequence

$$\{x'\} = [0, 1, 2, \ldots, m-1, m, m-1, \ldots, 2, 1]$$

To form the N-sequence RSNS, each term in the expression above is repeated N times in succession. The integers within one folding period of a sequence are then

$$\{x_h\} = [0, 0 \cdots, 0, 1, 1, \cdots, 1, \cdots, m-1, \cdots, m-1,$$
$$m, m, \cdots m, m-1, \cdots, m-1, \cdots, 1, \cdots, 1]$$

Each sequence corresponding to m_i, is right (or left) shifted by $s_i = i - 1$ places. The chosen shift values $s_1, s_2, \ldots, s_N$ must also form a complete residue system modulo N. The integer values within each modulus (comparator states), when considered together, change one at a time at the next code position resulting in an integer Gray code property. That is, only one comparator changes state between any two code transitions. Although the dynamic range of the RSNS $\hat{M}$ is less compared to the OSNS, its inherent Gray code property makes it particularly attractive for error control. Compare for moduli $m_i \in \{3, 4, 5\}, \hat{M} = 43$ for the RSNS, and $M = 60$ for the OSNS. With the RSNS preprocessing, the encoding errors due to comparator thresholds not being crossed simultaneously are eliminated and the interpolation circuitry can be removed.

Further details on the RSNS and the dynamic range $\hat{M}$ computation are given in the references. The RSNS has recently gained attention in a wide range of other applications, such as, code division multiple access transceiver architectures with inherent error detection and correction [53], all-electronic ADCs [54], RF direction finding interferometer antennas [55], and microwave-photonic direction finding systems for LPI waveforms [56].

7.11.2 Lessons from a Photonic RSNS ADC Prototype

To demonstrate the concept, commercially available photonic and electronic components were integrated in a $N = 3$ prototype configuration with $m_1 = 3, m_2 = 4, m_3 = 5$ as shown in Figure 7.52. The dynamic range for this configuration is $\hat{M} = 43$ (or $n > 5$-bits). There are also m_i comparators in each MZM channel.

---

[3] A set of integers $b_0, b_1, \ldots, b_{m-1} \pmod m$ form a complete residue system if $b_k \equiv k \pmod m$ for $k = 0, 1, \ldots, m-1$ [51].

7.11 High-Resolution Encoding Process for Photonic ADCs 435

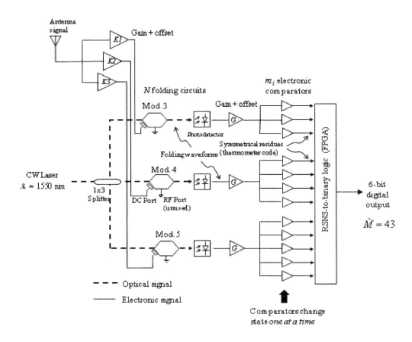

Figure 7.52 $N = 3$ Mach–Zehnder modulator ADC using the RSNS with m_i comparators in each channel. (©IEEE, reprinted with permission from [48].)

To determine the linearity of the ADC, the characteristic transfer function is obtained using a 2-kHz triangular waveform from 0 to V_{FS}. Figure 7.53 shows the input voltage and the decimal output voltage. Also shown is the quantization error. Since there is not a one-to-one correspondence in the input voltage to output voltage, a gain is present in the device, which can also be seen in the quantization error.

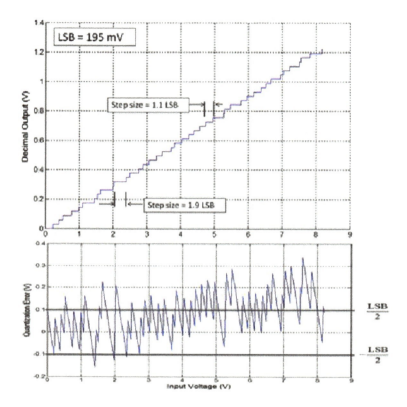

Figure 7.53 MZM amplitude analyzing the detector output into a mod5 format. (©IEEE, reprinted with permission from [48]).

7.12 CONFIGURING PHOTONIC COMPRESSIVE SAMPLING SYSTEMS

As we have pointed out, compressive sensing (CS) is a novel sampling algorithm that goes against the common framework put forth by Nyquist. Ever since Donoho's initial works [57], compressive sampling has been able to exploit the sparsity property of signals and then faithfully reconstruct them from far fewer measurements than traditional methods use, as first pointed out by Donoho. This has attracted considerable attention in wideband spectrum sensing. CS makes it possible to acquire multiband signals whose Nyquist rates may exceed the specifications of state-of-the-art ADCs by orders of magnitude with low-speed digitizers. Thus, no prior knowledge of RF signals is required and the bandwidth requirements of the digitizers can be significantly relaxed. Photonic technology provides an even larger advantage over electronics for implementing wideband compressive sensing systems. The low jitter and large bandwidths required of compressive sampling waveforms are readily available through photonic systems. Compressive sensing relies on signal sparsity in a domain, typically time, frequency or space depending on the application. Sampling of the signal by an uncorrelated sampling pattern in the appropriate domain is also required.

7.12.1 Multicoset Wideband Compressive Sensing Concepts

CS relies on two fundamental premises: sparsity and incoherence. When expressed in a convenient basis, many natural signals have concise representations whose most coefficients are small, and the relatively few large coefficients capture most of the information. A signal is considered sparse when over a certain observation duration, the occupied spectrum is a small fraction of the total observed spectrum. A measure of signal sparsity is given by [58]

$$\Omega = \frac{f_{occupied}}{f_{max} - f_{min}} \qquad (7.41)$$

where f_{min} is the lowest frequency of the space and f_{max} is the highest frequency of the frequency space and $f_{occupied}$ is the sum of all of the occupied frequency content of the signal. A signal is sparse if $\Omega \ll 1$, and compressive, *periodic nonuniform sampling* is suitable. Multicoset sampling can be described as first sampling the signal at the base rate of $1/T_b$ and then keeping only p samples from every block of L samples. In particular, a multicoset sampling is equivalent to sampling with p uniform samplers each with a period of $T = LT_b$ and with the input signal shifted in time by τ_i before being sampled by the ith uniform sampling stream. Each sampled stream is written as $x_i[n] = x(nT + \tau_i)$ where $x(t)$ is the original input signal [58].

Decomposing the multicoset sampling stream into uniformly sampled sequences allows the compressive sampling and reconstruction problem to be understood in the frequency domain. The goal is to construct the model

$$Y(f) = \Psi S(f) \qquad (7.42)$$

where $Y(f)$ is the multicoset measurement matrix, Ψ is the matrix related to the multicoset time shifts, and $S(f)$ is the matrix of the signal that we would like to reconstruct [58].

Recall the discrete time fourier transform of the signal $x_i[n]$ above with sample rate $1/T$ is

$$X_i(e^{j2\pi fT}) = \sum_{n=-\infty}^{\infty} x_i[n] e^{-j2\pi fnT} \qquad (7.43)$$

or rewriting for a bandlimited signal

$$X_i(e^{j2\pi fT}) = e^{j2\pi\tau_i}\frac{1}{T}\sum_{m=-L/2}^{L/2} x_i[n]e^{-j2\pi\tau_i m/T} X(f-k/T) \tag{7.44}$$

and now (7.44) can be expressed in matrix form as

$$Y(f) = \Psi S(f) \tag{7.45}$$

and $Y(f)$ is a vector of length p whose ith element is $X_i(e^{j2\pi fT})e^{-j2\pi f\tau_i}$ and Ψ is a $p \times (L+1)$ matrix whose ith, mth element is given by $(1/T)e^{-j2\pi\tau_i m/T}$ and $S(f)$ is the $(L+1) \times 1$ sparse matrix vector that contains the L unknown sections of frequency content that we are trying to find governed by the total observed spectrum and observation period [58].

To reduce the signal subspace on the multicoset measurement matrix $Y[n]$, a *correlation matrix*, R, is first computed to measure the correlation between multicoset channels. Next, an *eigenvector decomposition* is performed on R to provide an eigenspace to represent the matrix. Eigenvectors with eigenvalues below a defined threshold are discarded, and the remaining eigenvectors are used in the multiple measurement vectors orthogonal matching pursuit (MOMP) algorithm. The MOMP is a greedy approach that estimates the frequency bins with energy content and records the indices k of these bins. The rows of S that don't contain any frequency content are removed reducing $S(f)$ to a vector $Z(f)$ allowing the solution to the original problem we derived in (7.45). Then Ψ is modified to only contain columns indexed by k. With $\Psi^\dagger$ being the pseudoinverse of Ψ, $Z(f)$ is then found as

$$Z(f) = \Psi^\dagger Y(f) \tag{7.46}$$

Then once the unknown $Z(f)$ is found, the rows contain frequency segments of width $1/T$. The frequency segments have corresponding indices k that give the elements' locations on the full frequency interval of $[-1/2T_b, 1/2T_b]$. Frequency segments outside of the occupied space are considered the noise subspace and are set to zero. Spectral reconstruction is complete after the segments of $Z(f)$ are correctly shifted and summed [58].

7.12.2 Creating a Wideband Compressive Nonuniform Sampler

The wideband photonic compressive sampling architecture is shown in Figure 7.54. The system uses optical pulses to sample the incoming RF signal, but first creates a nonuniform pulse stream: A 10-GHz pulse source operating in the optical C-band (1530-1565 nm) and uses a nonlinear pulse compression to form pulses with pulsewidths of 2.5 ps. In the pulse source, a 5-GHz electronic oscillator with a drive

7.12 Configuring Photonic Compressive Sampling Systems

voltage of 2 Vpp drives a MZM biased at null to generate a 10-GHz pulse stream. This pulse stream then undergoes nonlinear compression to form narrow pulses.

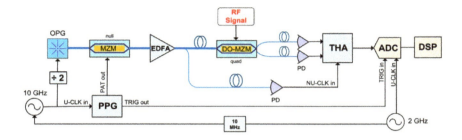

Figure 7.54 Photonic compressive sampling system diagram. Blue lines represent fiber, and thick blue lines are polarization maintaining. Black lines represent electrical cables. Abbreviations: OPG: optical pulse generator, MZM: LiNbO3 Mach-Zehnder modulator, DO-MZM: dual-output MZM, EDFA: erbium-doped fiber amplifier, PPG: pulse pattern generator, 2: frequency divide-by-2 circuit, U-CLK: uniform clock, NU-CLK: nonuniform clock, PAT: pulse selecting pattern, TRIG: trigger, PD: photodiode, THA: track-and-hold amplifier, ADC: electronic analog-to-digital converter, DSP: digital signal processing. (©IEEE, reprinted with permission from [58].)

Next, the high-rate uniform pulse stream passes through a MZM that gates the pulses to create a nonuniform pulse stream at a lower average rate. A pulse pattern generator (PPG) outputs a pattern that drives the MZM to pass specific sample pulses and suppress all other pulses according to the multicoset pattern. The PPG is phase-locked to the optical pulse source.

To relate the physical system to the mathematical description, the nonuniform pattern consists of 16 multicoset channels ($p = 16$) at a pattern repetition rate of 50 MHz ($T = 20$ ns). This results in a mean sampling rate of $f_s = 800$ MHz, which can be compared to the required Nyquist sampling rate of $f_{Nyquist} = 10$ GHz ($T_b = 100$ ps) for a 5-GHz bandwidth reconstruction. The modulator has a $V_\pi = 3.6$V at 1 GHz, a $V_\pi = 6.4$V at 40-GHz, and an insertion loss of 3.8 dB. Each photodiode has a 3-dB bandwidth of 12 GHz and a responsivity of 0.65 A/W [58]. The THA has a 3-dB bandwidth of 11.5-GHz, which roughly matches the bandwidth of the photodiodes used and they are synchronously clocked with the nonuniform pulse stream, which allows it to sample the peak of each pulse and hold its sample value until the next pulse arrives. The ADC has a 3-GHz front-end bandwidth and 6.9 ENOB, and operates at 2 GS/s using a clock that is phase locked to the sampling optical pulse source. Since the multicoset pattern mean sampling rate is not equal to the ADC sampling rate, extraneous samples of THA voltages are removed before processing [58].

The frequency response of the system is characterized with both uniform and multicoset sampling. For uniform sampling, the pulse gating modulator passes every tenth pulse of the 10-GHz pulse stream to create a 1-GS/s photonic sampling system.

The held samples of the THA are aligned in time with the optical pulse stream, and the ADC is synchronously clocked at 1-GS/s to sample the held values.

An input sinusoid is swept in frequency from 1 to 50 GHz, and the output amplitude of the tone is measured by taking the discrete Fourier transform of the captured samples [58]. The system was then configured to nonuniformly sample the RF signals. The frequency response was again measured from 1 to 50 GHz, and the tone was reconstructed. The response matched the uniform sampling results to within 0.6 dB, which demonstrated that nonuniform sampling did not adversely impact the response of the system.

Amplitude modulated signals and multiband, QPSK signals with raised-cosine filtering communication signals spanning a 5-GHz instantaneous bandwidth were collected and are shown in Table 7.3 and the results are shown in Figure 7.55.

Table 7.3
Characteristics of Communications Signals Used to Test the System in a Multi-Signal Environment.

Carrier (GHz)	Symbol Rate (MHz)	Roll-off factor	Power (dBm)
25.3	10	1.0	4.1
27.6	24	0.2	0.4
29.5	22.5	0.5	−0.4

By using nonuniform sampling, the system unambiguously identified signals from DC to 50 GHz in a 5-GHz instantaneous band using an average nonuniform sample rate at 8% of the Nyquist rate of multiple RF signals with bandwidth, showing correct frequency identification and spectral shape reconstruction [58].

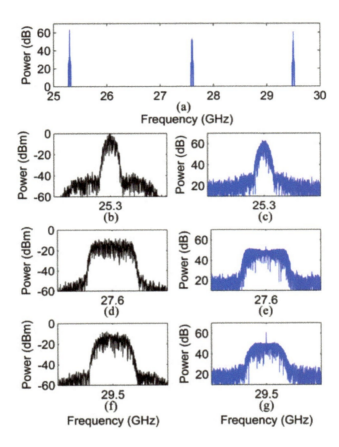

Figure 7.55 Spectrum analyzer trace (black) and system reconstruction (blue) of multiple communications signals. (a) Shows the reconstructed signals in the 5-GHz instantaneous bandwidth. The characteristics of individual signals are listed in Table 7.3. Subfigures (b)–(g) have spans of 50 MHz and cover 60 dB. (©IEEE, reprinted with permission from [58].)

7.13 DESIGNING A PHOTONIC NYQUIST FOLDING RECEIVER

The Nyquist folding receiver (NYFR) described earlier for wideband spectrum sensing is able to provide pulse rate, carrier frequency, phase variation, and frequency variation information while utilizing a compressed sensing methodology. The primary motivation for this section is to apply compressed sensing techniques to the design of a new receiver concept implemented with wideband integrated photonic components.

Progress with photonic devices is rapidly surpassing conventional electronic systems. Digital EW receivers are taking advantage of these advances in order to digitize and process wide spectrum bandwidths. Specifically, optical receivers have an advantage in bandwidth, sampling speed, and immunity to electromagnetic interference. Photonic LiNbO3 modulators now have bandwidths on the order of 40 GHz and can efficiently couple RF-antenna signals directly into the optical domain. Mode-locked lasers used for sampling now have tunable femtosecond-wide pulsewidths with pulse-repetition frequencies (PRF) on the order of 300 GS/s to allow oversampling of complex modulations.

The NYFR functional block diagram is shown in Figure 7.56(a) along with an illustration of the signed Nyquist zones in Figure 7.56(b). The photonic Nyquist folding receiver (pNYFR) has been designed and also experimentally demonstrated using a semiconductor laser. The specific design is a buried heterostructure laser consisting of a GaAs bandgap junction embedded in an AlGaAs substrate. A wavelength of 1550 nm was used to limit attenuation over an connecting single mode fiber-optic cable. The laser operates by applying current above a threshold value I_t, increasing the probability of carrier recombination in the GaAs bandgap and resulting in stimulated emission. The emitted optical power is written as

$$P_e = \frac{(I-I_t)\eta_i}{e}h\nu \qquad (7.47)$$

where I is the applied electric current, h is Planck's constant, ν is the frequency, e is elementary charge, and η_i is the probability of carrier recombination in the gap region [12]. The output of the laser is then turned on and off by an MZM to create the pulse train.

The NYFR uses a wideband front-end filter to select RF frequencies of interest from the environment. The passband must be selected so that the bandwidth is less than or equal to the NYFR maximum intercept frequency. Signals from the wideband filter are sampled by a FM swept clock that is coupled to a narrowband interpolation LPF. Lastly, a uniform sampling analog-to-digital converter digitizes the signal. Unlike traditional receivers that use a uniform Nyquist sampling rate, the NYFR undersampling rate may be many times lower than the bandwidth of interest [57] and only related to the information bandwidth.

7.13 Designing a Photonic Nyquist Folding Receiver

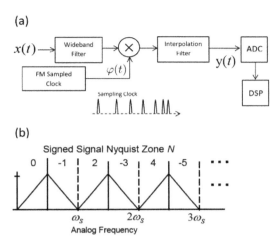

Figure 7.56 Analog-to-information receiver showing (a) NYFR operation and (b) signed Nyquist zones (see Chapter 4).

The FM sampling clock generates a sampling signal centered around an undersampling frequency ω_s with a RF sampling clock modulation $\theta(t)$ as

$$\varphi(t) = \omega_s t + \theta(t) \qquad (7.48)$$

and is used to sample the RF spectrum. An example target signal is modeled as

$$x(t) = \cos(\omega_c t + \psi(t)) \qquad (7.49)$$

where ω_c is the center frequency and $\psi(t)$ represents the RF signal information content. Convolving the two signals through the LPF yields the interpolation filter output

$$y(t) = \cos(\omega_f t + \beta \psi(t) - M\theta(t)) \qquad (7.50)$$

where ω_f is the output folded frequency in the Nyquist zone $N = 0, \beta = sgn(\omega_c - \omega_s k)$ and is the folded spectral orientation and $M = \beta k$ and is the modulation scale factor. Also, $\lfloor \omega_c/\omega_s \rfloor$ and is called the sampling harmonic factor. The induced modulation scale factor M is measurable and θ is known, so the original signal Nyquist zone can be estimated. Expanding ω_f from (7.50) results in the folded output frequency

$$\omega_f = |\omega_c - \omega_s k| \qquad (7.51)$$

that shows the relationship between the sampling modulation and the RF center frequency.

In order to extract original frequency information from the aliased signals in Nyquist zone $N = 0$, they must be unfolded and the ambiguities resolved. From (7.50) and (7.51) we see that the modulated RF signal frequency is a function of the original center frequency and the applied chirp modulation. The minimum difference between the signal modulation and clock modulation scaled by the Nyquist zone indicates the signal's original frequency region. The target frequency is extracted from the aliased signal in the $N = 0$ Nyquist zone by first generating an array $Z[i]$ of the modulated signal slope S_m minus the FM clock slope S_c scaled by the signed Nyquist zones $M(i)$. The array is modeled as

$$Z[i] = (S_m - M(i)S_c)^2 \qquad (7.52)$$

where S_c is the slope of the clock modulation, S_m is the slope of the pulse train in time-frequency analysis, and $M(i)$ is the signed Nyquist zones $N = [0, -1, 2, -3...]$, respectively. The next step is to find the index i of the minimum value of the array so that $i = min(Z[i])$. Lastly, the unfolded Nyquist zone is defined by

$$N = sgn(M(i))(i-1) \qquad (7.53)$$

where N is the original signed Nyquist zone of the RF signal. Adding the center frequency from Nyquist zone $N = 0$ yields ω_c [59]. This is shown below for a 2.75-GHz test signal applied to the photonic NYFR simulation.

7.13.1 Understanding the Double-Modulator Design Concept

The double-modulator NYFR (DM-NYFR) architecture is an implementation of a photonic compressed sensing receiver. It is designed around two MZI modulators. The first functions as a sampling impulse train generator while the second incepts the RF bandwidth. In the photonic NYFR architecture, the CW laser functions as a carrier for the signal and is analogous to electric current on a wire.

The overall block diagram of the DM-NYFR is shown in Figure 7.57. The front end of the architecture is a fiber-coupled CW laser at 1550 nm. This is connected to the first MZI that converts the CW laser into a sampling pulse train with a linearly swept PRF. An electrical signal $P(t)$ is connected to the single-arm electrodes of MZI1 and drives the modulation scheme by linearly sweeping a short electrical impulse train. The output optical pulse intensity, represented as I_o, is a function of the frequency and amplitude of the electric signal as well as the MZI transmission. The front-end optical pulse generation is discussed in detail below.

7.13 Designing a Photonic Nyquist Folding Receiver

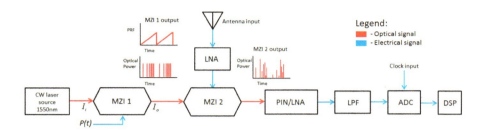

Figure 7.57 DM-NYFR block diagram showing the overall architecture of the receiver.

The front end optical pulse generation using MZI1 shown in Figure 7.57 is the most important part of the DM-NYFR architecture. Generating FM optical pulses with small pulsewidths enables the NYFR to differentially alias signals to lower Nyquist zones for detection. In the DM-NYFR, optical pulses are generated by applying a pulse signal $P(t)$ to MZI1 so that the coupled CW laser is selectively switched at a linear chirp rate. The electrical pulse signal at the input is modeled as an FM Dirac comb impulse train comprised of n number of $\delta(t)$ delta functions. This is an approximation, because the Dirac comb impulses have zero pulsewidth, which is impossible to achieve in actual operation. If the pulsewidth is much smaller then the RF signal's period, a zero-width impulse train is a good approximation. As a result, $P(t)$ is represented as a sum of impulses

$$P(t) = V_\pi \sum \delta(t - n(t)) \tag{7.54}$$

where V_π is the voltage amplitude of the MZI switching voltage and n is the impulse trigger. In order to model the variable PRF of $P(t)$, the delta functions must be triggered from a linear FM sinusoid. In the DM-NYFR architecture, this is modeled as a chirp signal $S_c(t)$, with $n(t)$ set on every positive-to-negative zero crossing such that

$$n(t) = ZCR^- (S_c(t)) \tag{7.55}$$

where ZCR^- is the notation for a single-ended zero crossing. For the DM-NYFR, two types of linear chirp signals are used, a sawtooth waveform $S_{c1}(t)$ and a triangular waveform $S_{c2}(t)$. The sawtooth chirp is given by

$$S_{c1}(t) = \sin 2\pi \left(f_c t + \frac{\Delta F}{2t_m} t^2 \right) \tag{7.56}$$

where f_c is the center frequency, ΔF is the chirp bandwidth, and t_m is the chirp period. The triangular waveform is defined as

$$S_{c2}(t) = \sin 2\pi [S_1(t) + S_2(t)] \tag{7.57}$$

with S_1 as the rising chirp sweep and S_2 as the falling chirp sweep. The chirp sweep functions are given by

$$S_1(t) = \left(f_c + \frac{\Delta F}{2}\right)t - \frac{\Delta F}{2t_m}t^2 \qquad (7.58)$$

for the positive slope and

$$S_2(t) = \left(f_c - \frac{\Delta F}{2}\right)t + \frac{\Delta F}{2t_m}t^2 \qquad (7.59)$$

for the negative slope, where f_c is the center frequency, ΔF is the chirp bandwidth, and t_m is the chirp period [26]. The linear chirp sinusoid plotted along with the pulse function is detailed in Figure 7.58. Setting the amplitude of the pulse signal to V_π switches the optical modulator at the PRF and coverts the input CW laser into a sampling impulse train. This illustrates how the front-end pulse generation block works in the DM-NYFR architecture.

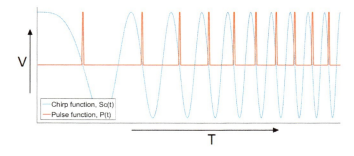

Figure 7.58 An example linear chirp used to generate a FM Dirac comb every positive-to-negative zero crossing

7.13.1.1 Living Within Design Constraints and Limitations

The photonic NYFR's primary operating boundaries and constraints come from the optical sampling with interferometers. The optical pulsewidth and ADC resolution determine the maximum frequency that can be modulated onto a laser source. Both pulsewidth and sampling limitations are examined in detail in this section. The performance of potential hardware builds is also analyzed in order to determine the maximum RF-signal intercept frequencies.

7.13.1.2 Pulsewidth Limitations

The photonic NYFR relies on signal sampling by optical pulses where the instantaneous power of the pulse train is proportional to the sampled waveform. In order to effectively sample a target RF signal, the pulsewidth must be smaller than the temporal variation of the modulating signal [?]. The 3-dB bandwidth B_p of an optical pulse is the inverse of the pulsewidth T_p or $B_p = 1/T_p$. From the Nyquist criterion, the sampling frequency f_s for an arbitrary signal with bandwidth B is $f_s = 2B$, written in terms of pulsewidth as

$$T_p < \frac{T_s}{2} \tag{7.60}$$

where T_p is the largest allowable pulsewidth that can effectively sample a signal with a period of T_s. For the photonic NYFR design, this means that optical pulsewidths limit the maximum frequency that can be sampled by MZI2, and pulses must be smaller than half the largest RF signal period.

7.13.1.3 Electro-Optical Sampling Limitations

The optically sampled signal from MZI2 is coupled into a PIN photodiode that converts optical power into an electrical signal. The output voltage must then be quantized by a sampling ADC with finite resolution. Digitizing an analog waveform imparts a sampling error that constrains the photonic NYFR's operating limits. Let the signal applied to the MZI electrodes be

$$V(t) = A \sin(2\pi f_m t) \tag{7.61}$$

where f_m is the maximum signal frequency and A is the signal amplitude. For a sampling pulsewidth T_p, the error in voltage δV for an arbitrary signal at time t_i in a traveling wave optical device is represented by

$$\delta V = \frac{1}{T_p} \int_{t_i + \frac{T_p}{2}}^{t_i - \frac{T_p}{2}} A \sin(2\pi f_m t) dt - A \sin(2\pi f_m t_i). \tag{7.62}$$

The voltage error is the difference in instantaneous voltage integrated over the width of a sampling pulse. Using a Taylor series expansion to evaluate the integral (3.15) for the maximum voltage error yields

$$|\delta V|_{max} = \frac{(\pi f_m T_p)^2 A}{6}. \tag{7.63}$$

For a n-bit sampling ADC, the voltage error must be less than half of the level spacing, which is

$$|\delta V|_{max} < \frac{2A}{2^n}\left(\frac{1}{2}\right) = \frac{A}{2^n}. \tag{7.64}$$

448 MICROWAVE-PHOTONIC TRANSCEIVER TECHNOLOGIES

This results in a T_p pulse limitation of

$$T_p < \frac{\left(\frac{3}{2^{n-1}}\right)^{1/2}}{\pi f_m} \qquad (7.65)$$

where f_m is the maximum target frequency and n is the resolution of the ADC (7.24).

7.13.1.4 Understanding Implications for the Photonic NYFR Design

The electro-optical sampling error associated with the ADC is the dominant limitation for the photonic NYFR architecture. Design implications are shown in Figure 7.59 and Figure 7.60, which are plots of the maximum intercept frequency as a function of pulsewidth for 6-bit and 8-bit ADCs, as well as for two separate hardware designs. Figure 7.59 is an illustration of a prototype build using inexpensive off-the-shelf components with pulse widths in the nanosecond range. The upper limit build illustrating picosecond and subpicosecond pulsewidths, which are achievable with either high-quality electrical pulse generators or mode-locked fiber lasers, is shown in Figure 7.60. There is a clear design trade-off between maximum intercept frequency and ADC resolution, with higher resolutions corresponding to lower target RF signals. Additionally, higher frequencies require shorter pulsewidths, which increases costs.

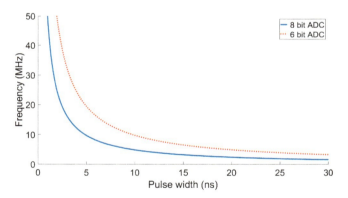

Figure 7.59 Maximum intercept frequency for nanosecond pulse-widths illustrating a low-cost prototype build.

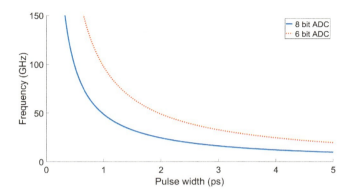

Figure 7.60 Maximum intercept frequency for picosecond and sub-picosecond pulse-widths illustrating a potential high-cost build.

Below a test of the simulation and evaluation of the results are shown for both configurations. Test and evaluation (T&E) is an important part of photonic hardware development.

7.13.2 pNYFR T&E Simulation Results

7.13.2.1 Double Modulator-NYFR Simulation Results

The simulated DM-NYFR architecture is tested with target signals in order to see if the model can successfully undersample and modulate signals for recovery by the DSP. The simulation is conducted with a sawtooth FM modulation. The modulation period $t_m = 5.0$ μs with $f_{min} = 1$ GHz and $f_{max} = 1.15$ GHz, resulting in a ΔF sweep bandwidth of 150 MHz. This gives Nyquist zones every 500 MHz, with zone $N = 0$ from 0 to 500 MHz. The test results show the first two Nyquist zones ($N = 0, -1$) as well as the modulation signal. Scope six, the sampling ADC, has a $2f_s$ sampling rate of 2.0 GHz, and the LPF cutoff frequency is $2f_s$. For all conducted tests, the SNR was 10 dB.

7.13.2.2 Test One: No Modulation

The first test checks the architecture's response with no FM sweep applied to the first MZI. The VCO was disconnected, so the sampling pulse train had a constant frequency of 1.0 GHz. The applied signal is 2.7 GHz, so the expected results when examining the DSP output are aliased signals centered around 500.0 MHz. The output shown in Figure 7.61 confirms the architecture with no modulation. Aliased signals at 0.3 GHz and 0.7 GHz are seen along with the 1.0-GHz sampling signal. Without modulation the undersampled signals cannot be resolved to their original frequencies.

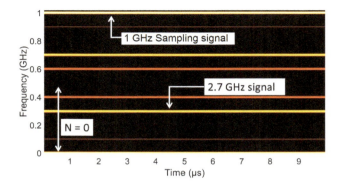

Figure 7.61 DM-NYFR Test One: no modulation. Disconnected VCO results in aliased signals that are unresolvable to their original frequencies.

7.13.2.3 Test Two: 2.75-GHz Signal

The second test is the architectures response with the FM sweep enabled. The optical pulse train has an FM chirp and successfully undersamples the signal. The target signal is 2.75 GHz, chosen because that is a common aerial search radar frequency. The received signal is shown in Figure 7.62 in the $N = 0$ Nyquist zone with a modulated slope from the MZI1 driver signal. The results are a time-frequency spectrogram computed in MATLAB. Two Nyquist zones, $N = 0, -1$, are shown completely, and are mirror images of each other. Also visible is sampling signals from 1.0 to 1.15 GHz.

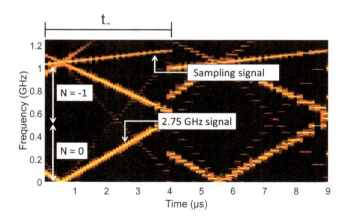

Figure 7.62 DM-NYFR Test Two: 2.75-GHz signal. Sampling FM pulses successfully modulating a signal using NYFR architecture.

7.13 Designing a Photonic Nyquist Folding Receiver

The original frequency can be extracted from the sampling signal and the folded signal. A MATLAB program is used to unfold the signal. The lowest difference is Nyquist zone $N = -5$, which corresponds to signals between 2.5 and 3.0 GHz. Estimating a center frequency of approximately 0.15 GHz and adding it to the lower frequency bound of the Nyquist zone yields the correct original frequency of 2.75 GHz.

7.13.2.4 Test Three: 5.1-GHz and 5.65-GHz Signals

The next test was a 5.1-GHz signal and a 5.65-GHz signal, which are some typical C-band radar frequencies. Each signal was applied separately and the results analyzed in MATLAB. The results are shown in Figure 7.63, which displays both signals in the $N = 0$ Nyquist zone. Comparing the signal slopes and modulation signals, we can extract the original frequency ranges. Adding the center frequencies from the results yields the original frequencies of the signals. Visible in Figure 7.63 are weaker images that appear to be signals; however, these are actually aliased signals from higher Nyquist zones folding down due to the wide bandwidth of the interpolation LPF. Reducing the LPF cutoff frequency to the upper limit of Nyquist zone $N = 0$ removes these artifacts.

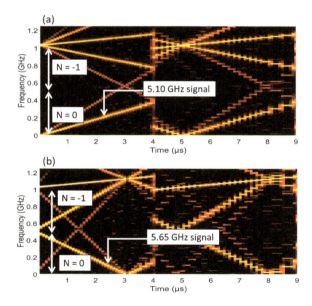

Figure 7.63 DM-NYFR Test Three: Results displaying (a) 5.10-GHz and (b) 5.65-GHz signals. Both target waveforms are successfully undersampled and recoverable from the spectrogram.

7.13.2.5 Test Four: 2.7-GHz and 9.1-GHz Combination Signal

This test was conducted to determine how multiple target signals are resolved by the DM-NYFR simulation. Two target waveforms were applied to the antenna input: 2.7 GHz at 100 mV_{pp} and 9.1 GHz at 50 mV_{pp}. The uniform noise block was disconnected for this test. A spectrogram of the instantaneous optical power at MZI2's output is shown in Figure 7.64. Visible is the FM signal at 1.0 GHz and the modulated target signals in Nyquist zone $N = 0$. The antenna signals exhibit nonuniform aliasing expected from the NYFR, and two distinct signals are clearly identifiable. Given the slopes and center frequencies from the spectrogram, the equations above can be used to successfully unfold the input RF signals to their original values.

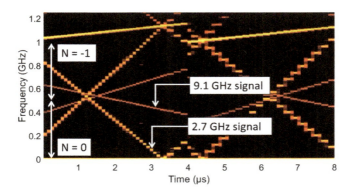

Figure 7.64 Test Four: Results displaying the spectrogram of folded 2.7-GHz and 9.1-GHz signals.

7.13.3 Receiver Hardware Prototype Results

After successfully confirming the photonic NYFR designs in rSoft's OPTSIM design suite, a prototype NYFR was built to demonstrate physical operation. The principal challenge in constructing the hardware prototype was integrating commercial of-the-shelf (COTS) components in unique configurations that were not considered when the equipment was being manufactured. The hardware build makes use of components never intended for SIGINT or ELINT applications.

Additionally, component cost and funding were also a deciding factor during the prototype build. Due to funding limitations, lower-cost equipment was used that did not meet the intended performance specifications of the designed or simulated architecture. Consequently, the prototype build is a scaled-down design that functions as a technology demonstrator and proof-of-concept rather than a fieldable receiver. Specifically, although the NYFR was designed and simulated to function in the GHz range with Nyquist zone bandwidths of 500 MHz, the hardware prototype is only able to operate in the kHz/MHz range with zone sizes of 50 kHz. Despite these constraints,

7.13 Designing a Photonic Nyquist Folding Receiver 453

the prototype is able to successfully undersample at sub-Nyquist rates and extract the original frequency information.

7.13.3.1 Signal Processing Methodology

The signal processing methodology, specifically noise removal and thresholding, that is required due to the high noise floor of the optical signals is covered in this section. This is due to the low optical power on the input of MZI2 as well as characteristics associated with the MZI functionality. Each MZI has roughly a 3-dB loss between the input optical intensity and the output optical intensity. As the modulators are added in series, power loss increases, resulting in low incident power on the detectors. Another factor for the high noise floor when using MZIs is the imperfect extinction ratio that prevents total destructive interference. In actual operation of an MZM, it is impossible to completely extinguish the output intensity to zero and as a result some laser output intensity is always present. Lastly, relative intensity noise (RIN) arises from power instabilities and vibrations within the CW laser cavity. With a low sampling signal duty cycle, the time-averaged noise power is larger than the signal power, and de-noising is a necessity. The optical noise is shown in Figure 7.65.

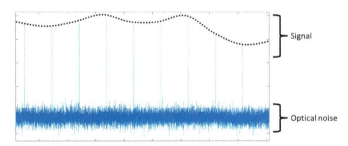

Figure 7.65 Example NYFR capture illustrating both the signal and optical noise.

The folded signal includes optical noise that needs to be filtered before analysis can take place. Additionally, thresholding and edge detection are implemented to make the undersampled signals distinguishable. The image processing block diagram is shown in Figure 7.66 and details the DSP steps to refine the NYFR output.

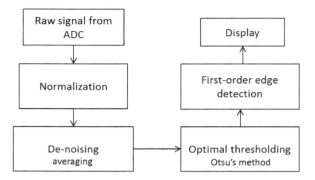

Figure 7.66 Block diagram of NYFR image processing steps.

The raw ADC signal is first normalized and then passed into a noise removal block that uses squared averaging to compute a cutoff value below that, the signal is set to zero. The noise power P_{noise} is defined as

$$P_{noise} = \frac{\sum_{i=1}^{n} \left(S[i] - \bar{S}\right)^2}{n} \tag{7.66}$$

where $S[i]$ is the n sized signal array and i is the index corresponding to a data point. The signal is de-noised by setting the signal array to

$$S[i] = \begin{cases} S[i] & \text{if } S[i] \geq P_{noise} \\ 0 & \text{if } S[i] < P_{noise} \end{cases} \tag{7.67}$$

which removes much of the optical noise. After initial processing, the signal is thresholded in order to facilitate edge detection. Otsu's method of optimal thresholding is used to isolate the signal from the background noise by identifying the 0th and 1st-order cumulative moments of an n-sized histogram. The maximum variance of the class separability between the two histogram moments is the optimal thresholding point [60]. This is shown graphically in Figure 7.67, which is a histogram of a NYFR capture with Otsu's point labeled.

Lastly, a simple edge detection method used binarization to convert the signal into a two-state array depending on the average power after thresholding. Anything less than the average power after thresholding was set to zero, which highlighted the folded signal slope. All signal processing steps were implemented with MATLAB. As a result of the signal processing steps, both the MZI optical noise and RIN was effectively removed from the ADC captures and the signals were able to be correctly unfolded.

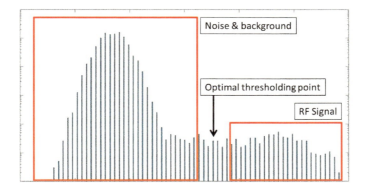

Figure 7.67 Example NYFR capture histogram illustrating otsu's method of optimal thresholding.

7.13.4 Single-Tone Test Results

Selected DM-NYFR tests using a single-frequency tone as the target RF signal are shown in this section. Target RF signals are successfully undersampled and extracted using DSP, confirming the prototype's functionality.

7.13.4.1 Selected Tests

Select tests with full DSP methodology are presented in this subsection in order to highlight the NYFR functionality. The first selected test was a 225-kHz RF signal with preset three as the modulation scheme. The results are shown in Figure 7.68, which details both the final signal and the image processing steps. All displays are time-frequency spectrograms computed in MATLAB showing a full ΔF frequency sweep and Nyquist zones $N = 0, -1$. The capture in Figure 7.68(a) is the raw received signal from the ADC block of the NYFR. The 225-kHz signal is masked by noise and is not discernible from the image. In Figure 7.68(b) we see the DM-NYFR signal after preliminary de-noising and normalization. At this stage the target signal is faintly visible from the background. The next capture, Figure 7.68(c), is after the application of optimal thresholding, which removes almost all of the clutter and leaves the target signal clearly identifiable. Figure 7.68(d) is an illustration of the final output of the DSP block and is used in conjunction with the unfolding methodology described in Section 4.1 using (2.6) and (2.7). This is shown in Figure 7.69 as a red line fit to the slope of the signal, which was used to successfully extract the correct original frequency of the target RF waveform.

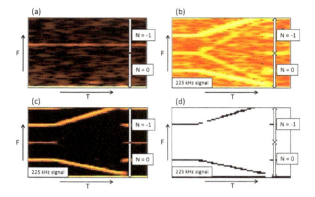

Figure 7.68 A 225-kHz test with preset three showcasing the DSP steps: (a) is the raw ADC signal, (b) is the initial de-nosing and normalization, (c) is the optimal thresholding, and (d) is the final signal after edge detection.

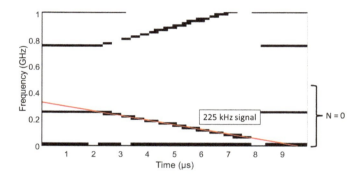

Figure 7.69 A 225-kHz signal with a fitted slope for unfolding.

The second selected test was a 525-kHz RF signal with preset four as the modulation scheme and is shown in Figure 7.70 as a series of time-frequency spectrograms. The results confirm the NYFR's operation by replicating the 225-kHz test. Clear progression is shown through each DSP block culminating in a distinguishable signal that is unfolded into its original Nyquist zone. Interestingly, in Figure 7.69(a) we see the same weak tone signal in Nyquist zone $N = -1$ that is present in Figure 7.68(a). This single-frequency anomaly is not associated with any target signal and disappears when the noise is removed. This tone likely arises from a periodic oscillation in the optical noise stemming from the MZI. The modulators have a cosine-squared transfer function that causes spurious frequencies. Another explanation is laser RIN vibrations resonating in the laser cavity. Although this does not appear to affect the DM-NYFR

7.13 Designing a Photonic Nyquist Folding Receiver 457

performance, these tones did not appear in the rSoft simulations, and this underscores the importance of hardware prototyping.

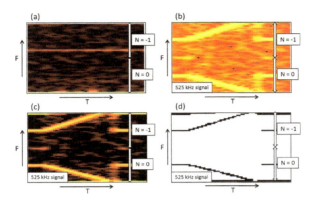

Figure 7.70 A 525-kHz test with preset four showcasing the DSP steps: (a) is the raw ADC signal, (b) is the initial de-nosing and normalization, (c) is the optimal thresholding, and (d) is the final signal after edge detection.

The modulation sweep bandwidth ΔF has a direct effect on the output modulation slope and can adversely alter the results. A common occurrence when dealing with a large modulation slope is Nyquist zone overflow. The imparted modulation S_m is modeled as

$$S_m = f_c + M \Delta F \qquad (7.68)$$

where f_c is the center frequency of modulation, ΔF is the sweep bandwidth, and M is a scale factor from the original signed Nyquist zone. As a result, S_m grows as either the target RF frequency or modulation sweep increases. This can cause an overflow effect where a signal modulation crosses the bounds of a Nyquist zone, changing the slope value and sign. Depending on the capture size, we can detect and account for this overflow. This is shown in Figure 7.71, where a 425-kHz signal is modulated with a 10-kHz ΔF. The overflow is evident at the 1.75-ms point, when the slope reflects back into the $N = 0$ Nyquist zone; however, due to the capture size of 4.0 ms, this was detected and corrected. It is recommended that the capture time t_c is at least twice the modulation time t_m such that $t_c > 2t_m$.

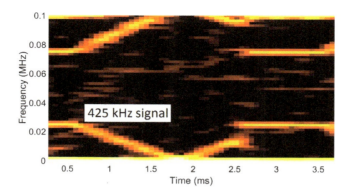

Figure 7.71 Nyquist zone overflow at 1.75 ms.

7.13.4.2 Summary

In summary, photonic NYFR designs and their operating methodologies have been discussed in this section. The constraints, limitations, and design trade-offs for the photonic NYFR architecture have been now examined, and modeling show that the optical sampling error and the resulting optical pulsewidth is the primary limiting factor. Additionally, the integrated optical components used in the NYFR were discussed.

The computer simulations and results for both photonic NYFR designs were carried out with a sampling frequency of 1 GHz the sampling pulse train had a constant frequency of 1.0 GHz. The applied signal is 2.7 GHz, so the expected results when examining the DSP output are aliased signals centered around 500.0 MHz. The output confirms the architecture with no modulation. Aliased signals at 0.3 GHz and 0.7 GHz are seen along with the 1.0-GHz sampling signal. Without modulation the undersampled signals cannot be resolved to their original frequencies.

The NYFR architecture implemented in the photonic domain takes advantage of the increased bandwidths, sampling speeds, sensitivities, and EMF noise immunity that optical devices possess.

7.14 DESIGN OF PHOTONICS COMPRESSIVE SAMPLING SYSTEMS

CS theory asserts that one can recover certain signals and images from far fewer samples or measurements than traditional methods use such as sampling by the Nyquist criteria. To make this possible, CS relies on two principles: sparsity and incoherence.

7.14 Design of Photonics Compressive Sampling Systems

The concept of sparsity pertains to the signals of interest and the fact that the information bandwidth of interest may be considerably smaller than the signal bandwidth. CS exploits that fact that most signals are sparse and can be compressed. Mathematically, they have concise representations when expressed using the proper basis function Ψ since the information bandwidth is smaller then the signal's bandwidth [61].

The concept of incoherence pertains to the sensing mode that is used [62]. It extends the duality between time and frequency and expresses the idea that objects having a sparse represenation in Ψ must be spread out in the domain in which they are acquired [61]. That is, incoherence says that unlike the signal of interest, the sampling and sensing waveforms have an extremely dense representation in ψ. The CS corresponding mathematical expression is as follows:

$$\mathbf{x} = \Psi\theta \qquad (7.69)$$

where $\mathbf{x}$ is an N-dimensional signal and $\Psi \in \{N \times N\}$ orthornormal basis matrix in which $\mathbf{x}$ can be sparsely represented. Here θ is a K-sparse vector with at most K non-zero entries. Then the sampling process is modeled by interacting $\mathbf{x}$ with a sensing matrix $A \in \{M < N\}$ that can be expressed as

$$\mathbf{y} = A\mathbf{x} + \mathbf{e} \qquad (7.70)$$

where $\mathbf{y}$ is a M-dimensional vector of measurements and $\mathbf{e}$ is a stochastic or deterministic unknown error term bounded by a known amount $||\mathbf{e}||_2 \leq \sigma$. The sensing matrix A must be incoherent with the sparse basis Ψ and random matrices are demonstrated to be good choices.

Then $\mathbf{x}$ can be reconstructed faithfully with only $M = O(K \log(N/K))$ measurements if the restricted isometry property (RIP) is satisfied. Signal reconstruction is performed by approximating the solution to the problem posed as

$$\min \frac{1}{2}||\mathbf{y} - A\Psi\theta||_2^2 + \tau||\theta||_1 \qquad (7.71)$$

Wideband spectrum sensing is a main application of the above CS theory and the wideband nature of photonics can play a key role in its implementation.

The photonics-assisted CS concept transfers all the spectral components of a RF signal to the low-frequency region by mixing them with a pseudorandom binary sequence (PRBS) of signals in the optical domain. The spectrum of the random signals or PRBS is a frequency comb as shown in Figure 7.72. The comb spacing is denoted as f_p. The input RF signals should be frequency sparse (e.g., multiband signals). This means that their spectral content (or support) occupies only a small portion of the wider input spectrum.

When mixed with the PRBS signals, all the spectral components of the RF signal will be mixed with their nearest combs and transferred to the low frequency region. Each f_p-width slice of the mixing signal's spectrum is a linear combination of f_p-shifted copies of the RF spectrum as shown in Figure 7.72. The coefficients of the combination ($c_{i,j}$) are determined by the Fourier coefficients of the PRBS signal where i is the mixed signal spectrum number, j is the frequency number, and only the low-frequency components need to be acquired so the sampling rate of the back-end digitizer can be orders of magnitude lower than the Nyquist rate of signals. In addition, a parallel architecture enables a further reduction of acquisition bandwidth.

The PRBS signals in different channels are provided by introducing *pre-designed time delays*. The delayed PRBS in each channel will be mixed with the same RF signal and each mixing signal contains all the frequency information of the input signal. As the number of measurements required for signal reconstruction remains fixed, the parallel architecture leads to a reduction of each channel bandwidth.

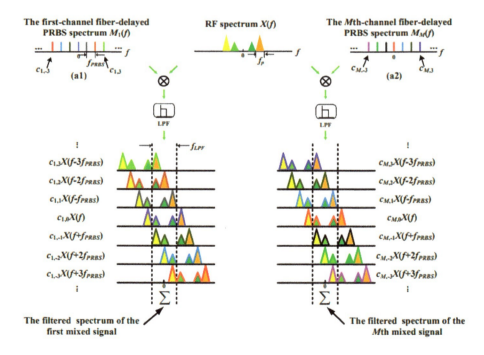

Figure 7.72 Principle of photonics-assisted compressed sensing receiver (©IEEE, reprinted with permission from [63].)

7.14.1 Single-Channel CS With Double-Parallel MZI

Single-channel and multichannel photonics-assisted CS systems are presented and demonstrated to enable accurate reconstruction of frequency-sparse signals (multiband signals and multitone signals). The use of a parallel structure of two MZM being driven by a laser diode is shown in Figure 7.73. The wideband RF is incident on one electrode and the PRBS is incident on the other electrode.

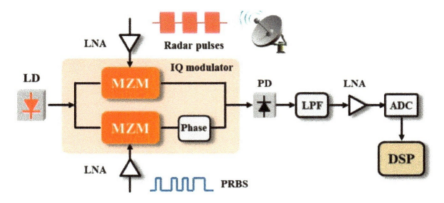

Figure 7.73 The architecture of the single-channel photonics-assisted compressive sampling system. LD: laser diode, PRBS: pseudorandom binary sequence, MZM: Mach-Zehnder modulator, PD: photodetector, LPF: low pass filter, LNA: low noise amplifier, ADC: analog-to-digital converter, DSP: digital signal processor. (©IEEE, reprinted with permission from [63].)

A CW laser (1550-nm center wavelength, 100-kHz linewidth, and 15-dBm output power) is used as the light source. A 127-bit PRBS is generated by a pulse pattern generator (PPG) driven by a 10.16-GHz clock signal. An arbitrary waveform generator (AWG) (Tektronix 7122B) is used to generate numerous radar pulses for spectrum estimation. The bit rate of the PRBS limits the system bandwidth to 5 GHz. A 12.5-GHz IQ modulator is driven by the radar pulses and the PRBS simultaneously, followed by a 1-GHz photodetector. A low-pass filter is used to extract the effective spectral information. The output signal is then acquired by an off-the-shelf ADC with a 500-MHz bandwidth. Finally, signal reconstruction is performed offline by a DSP.

A rectangular pulse (3.21-GHz carrier frequency, 2-μs pulse duration) and a linear frequency modulated (LFM) pulse (3.22-GHz center frequency, 10-MHz bandwidth and 1-μs pulse duration) are tested to evaluate the performance of the single-channel photonics-assisted CS system. Figures 7.74(a) and 7.74(b) show the normalized power spectral density of the input signals and Figures 7.74(c) and 7.74(d) give the corresponding reconstruction results. The recovery errors of the carrier frequencies are less than 100 kHz, which indicates an exact reconstruction.

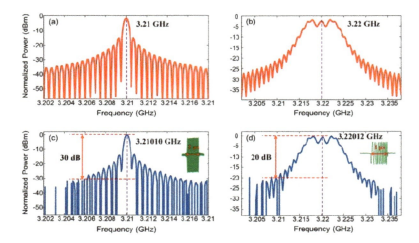

Figure 7.74 Reconstruction results of a rectangular pulse and a LFM pulse. The normalized power spectral density of (a) the input rectangular pulse and (b) LFM pulse. The power spectral density of (c) the reconstructed rectangular pulse and (d) LFM pulse. Inset in (c): the temporal waveform of the reconstruct rectangular pulse. Inset in (d): the temporal waveform of the reconstruct LFM pulse. (©IEEE, reprinted with permission from [63].)

The demonstrated system acquired approximately 100 radar pulses spanning from 500 MHz to 5 GHz in 0.1 ms. The reconstruction results are presented in Figure 7.75 The spectrogram of the input radar pulses is shown in Figure 7.75(a). The horizontal axis denotes frequency and the vertical axis denotes time. The reconstructed spectrogram is shown in Figure 7.75(b).

A close-up view of a 5-μs fraction of the spectrogram in Figure 7.75 (b) is shown in Figure 7.76. In Figure 7.76(a), the power spectral density of the input signal is shown and in Figure 7.76(c) the reconstructed signal is shown. Figure 7.76 shows the spectrogram of (c) the input signal and (d) the reconstructed signal [63].

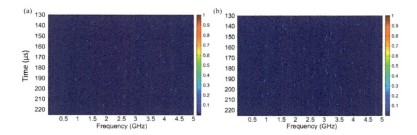

Figure 7.75 Reconstruction results of numerous radar pulses. The spectrogram of (a) the input pulse stream and (b) the reconstructed pulse stream. (©IEEE, reprinted with permission from [63].)

7.14 Design of Photonics Compressive Sampling Systems

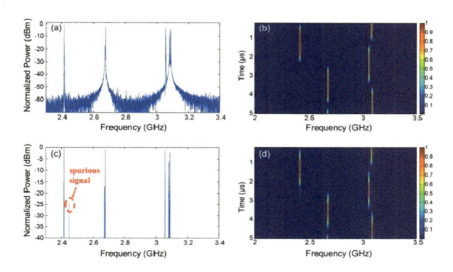

Figure 7.76 A zoom view of a fraction of the spectrogram in Figure 7.75(b). (a) The power spectral density of (a) the input signal and (c) the reconstructed signal. The spectrogram of (c) the input signal and (d) the reconstructed signal. (©IEEE, reprinted with permission from [63].)

7.14.2 Multichannel Photonics-Assisted CS System

To further reduce the acquisition bandwidth (or increase the compression ratio), a multichannel photonics-assisted CS system is proposed as shown in Figure 7.77. The architecture consists of a four-channel CS system that has a 5-GHz bandwidth. A four wavelength distributed-feedback (DFB) laser array with wavelengths of 1546.12, 1547.72, 1550.12, and 1551.72 nm is used as the light source [63]. A 10.16-Gbps PRBS signal, generated by the PPG, modulates the four optical waves simultaneously via a MZM. Then, a wavelength demultiplexer (DEMUX) is used to separate the four-wavelength PRBS signals, and different time delays are individually introduced to each wavelength by using fibers of different lengths. After that, four delayed signals are combined together by a wavelength multiplexer (MUX). The relative time delays among the PRBS signals in different channels are approximately 0, 0.34, 2.5, and 8.41 ns, respectively.

Another MZM is used to mix the input RF signal with the four-channel PRBS signals. The mixing signal is demultiplexed and then detected by a narrowband PD in each channel. After passing through a LNA and a LPF, the resultant signals are quantized by a four-channel digital phosphor oscillator (DPO) (Tektronix 72004B). The bandwidth of each ADC is set to 120 MHz. Finally, spectrum estimation is performed offline.

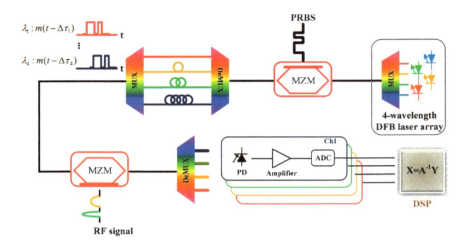

Figure 7.77 Experimental setup of the four-channel CS system. (DFB: distributed-feedback laser; MUX: multiplexer; DEMUX: demultiplexer; PD: photodetector; DSP: digital signal processor.) (©IEEE, reprinted with permission from [63].)

To demonstrate the four-channel CS system, a twenty-tone signal is tested [64]. Figure 7.78 (a) shows the input RF spectrum, measured by a spectrum analyzer (Anritsu MS2668C) with a 10-kHz resolution. The spectrum of four-channel mixing signal is shown in Figure 7.78(b). Note that the in-band SNR of the mixing signal degrades to about 26 dB due to the noise folding during the mixing process. To demonstrate the effects of different acquisition bandwidths on the reconstruction performance, the recovery results for 120- and 200-MHz bandwidths are presented in Figure 7.78(c) and (d), respectively. The twenty-tone signal is exactly reconstructed in both cases, and their SNRs are increased by about 10.0 and 11.5 dB, respectively.

7.14 Design of Photonics Compressive Sampling Systems

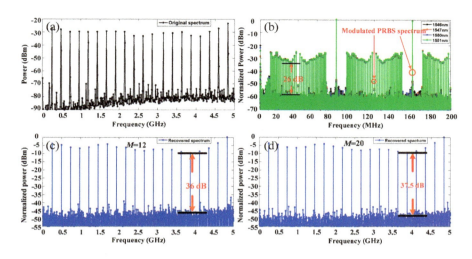

Figure 7.78 (a) The input RF spectrum, (b) the spectrum of the mixing signal, (c) and (d) reconstructed spectrum for 120- MHz and 200-MHz acquisition bandwidth. (©IEEE, reprinted with permission from [64].)

7.14.3 Extending the Bandwidth of the System

The bandwidth of the above system is limited by the bit rate of the PRBS. To obtain better system performance the system bandwidth is extended to 20 GHz by using *time-interleaved* optical sampling techniques [64].

The architecture of the ultra-wideband photonics-assisted CS system is shown in Figure 7.79 and consists of four parts: the generation of four wavelength PRBS modulated optical pulses with pre-designed time delays, signal mixing, acquisition of compressed signals and the signal reconstruction [64]. First, four optical frequency combs (OFCs) (centered at 1546.12, 1547.72, 1550.12, and 1551.72 nm) are generated by using cascaded intensity and phase modulators driven by a tone at 10.16 GHz. Then, four time-delayed PRBS pulses are recombined by an MUX and mixed simultaneously with an RF signal by another MZM biased at quadrature. The four PRBS-modulated pulses are separated and time delayed. Then the four time-delayed PRBS pulses are recombined by a MUX and mixed together with an RF signal by another MZM biased at quadrature. Then the four mixing signals are separated by a DEMUX and each one is detected by a PD, filtered by a LPF, amplified and acquired by an ADC. The DSP is finally used to reconstruct the spectrum from the compressed spectrum with a 360-MHz bandwidth [64].

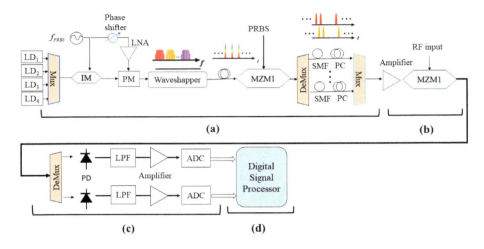

Figure 7.79 (a) Generation of 4-wavelength PRBS modulated pulses w/non-linear time delays, (b) signal mixing, (c) acquisition of compressed spectrum, (d) signal recovery, and reconstructed spectrum. (After [64].)

To evaluate the performance of the wideband spectrum sensing system, two groups of multiple tones are tested. The frequencies of the first multi-tone signal are 2.418, 3.291, 4.976, and 18.405 GHz. The frequencies of the second group are 2.418, 4.976, 12.024, and 18.405 GHz. Figure 7.80 shows the measured results with a spectrum analyzer with 100-kHz resolution bandwidth and shown in Figures 7.80 (a) and (b).

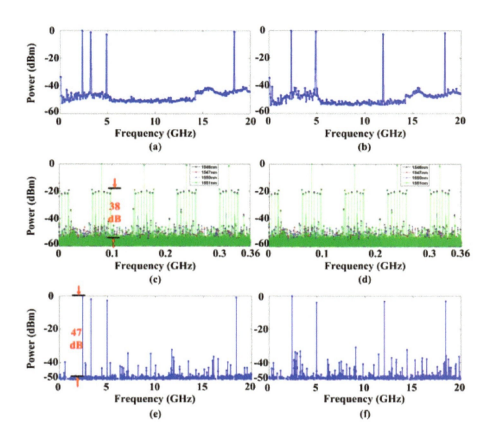

Figure 7.80 (a) The input RF spectrum, (b) the spectrum of the mixing signal, (c) and (d) reconstructed spectrum for 120- MHz and 200-MHz acquisition bandwidth. (©IEEE, reprinted with permission from [64].)

After sub-sampling, the spectrum of the signals used for mixing (four channels) with 50-kHz resolution and a 360-MHz bandwidth are acquired and used as the measurements for signal reconstruction (see Figures 7.80(c) and (d)). Finally the reconstructed RF spectrum is given in Figures 7.80(e) and (f) showing the reconstructed power spectrum matches well with the input in both cases. The frequency error is determined by the sampling depth and is 25-kHz. As shown, the architecture with an extended bandwidth of 20-GHz is achieved by time interleaved optical sampling.

7.15 WIDEBAND SPECTRUM SENSING AND ANALYSIS

Complete situational awareness in the EW domain implies the ability to detect, characterize, and continuously monitor emitters that are present within, for example, 0.5 GHz to 100 GHz and beyond. In a realistic operating environment it is required to be able to measure and classify multiple, simultaneous frequencies. Signal classification begins with electromagnetic wideband spectrum sensing. The receiver must tune to where energy is present for capture, measurement, and analysis. Signal classification by measurement or calculation of the emitter's characteristics, such as carrier frequency, bandwidth, spectral envelope, modulation type, and periodicity, and subsequent comparison to known (and unknown) emitters within a database enables the continuous monitoring for emerging threats [65]. The ideal spectrum sensing DRFM approach is a wideband-software defined receiver, replacing all the hardware functionalities with advanced DSP techniques.

The concepts being embedded into software-defined receivers are continuously evolving. Very short RF *transient events* are a subset of signals that are very important to situational awareness. The short duration and occurrence over extremely wide bandwidths make them very difficult to capture and characterize. The capture of these transient events is straightforward when center frequency information is known beforehand. However, it is a rather complex chore to perform the same measurement over an extremely wide bandwidth without advance knowledge of the center frequency of input signal. The ability to capture a transient event in its entirety implies a means of buffering the input signal for a duration greater than the time required to detect its arrival. The *detection latency* is the primary determinant of the signal storage requirements. However, without a means of input delay, the combined detection and cueing latency that exceeds the duration of the incident signal will result in complete failure to capture the transient event. That is, the signal will be gone by the time the system responds.

An instantaneous frequency measurement (IFM) receiver is a fully integrated, wideband EW system that uses phase discrimination approach to measure the intercepted signal frequency. The conventional analog IFM receiver can cover a very wide RF bandwidth (e.g., 16 GHz) and uses a correlator as the basic unit [66]. The receiver can report fine frequency measurements using a 1-bit ADC clocking at a high sampling rate. In addition, they have good sensitivity and dynamic range. The major drawback however, is that erroneous errors are generated if more than one signal is present. When there is no signal present at the input, the IFM receiver reports frequencies based on noise measurements [66].

The block diagram of a conventional IFM receiver showing the analog correlator architecture is shown in Figure 7.81. The digital IFM receiver is also based on the analog approach and typically uses a FFT-based system to digitize the received signals and compute the corresponding Fourier transforms.

7.15 Wideband Spectrum Sensing and Analysis

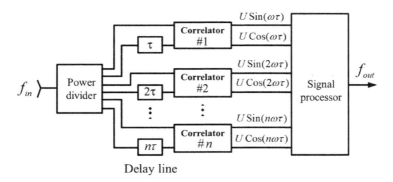

Figure 7.81 Block diagram of an IFM receiver showing the details of the analog correlator architecture.

The IFM receiver correlator uses a power divider and a number of delay lines with different lengths in order to obtain a fine frequency resolution. The interferometric principle of splitting the RF input and delaying one path before being compared by a phase comparator has inherent bandwidth limitations imposed by the microwave delay line, the power divider, and the phase comparator.

7.15.1 Channelizer Using Fiber Bragg Gratings

The most straightforward technique for implementing an IFM receiver with photonics is a channelized approach. The received RF signal is modulated on a laser using a MZM and the modulated spectrum is then split and applied to a bank of filters (e.g., a grating array), and an array of PDs then detects the power at the output of each filter within the filter bank within the filtering window [67]. Here they show the capability of resolving signals in the 2–18 GHz range into 2-GHz bands in a simple optical topology while maintaining a high probability of intercept.

Another practical photonic-assisted channelizer example is described in [68]. Here the photonic-assisted RF channelization scheme is based on dual coherent OFCs. The input broadband RF signal is upconverted and multicast by optical lines in the first OFC. The copies are then physically separated by a regular optical de-mux, each of which is mixed with the other channelized OFC. Due to the difference of the free spectrum ranges of the two combs, different spectrum slices of the RF signal are extracted and downconverted to baseband by the following I/Q demodulators.

Figure 7.82 shows the schematic diagram of an RF photonic channelizer that uses a CW laser diode and a MZM to couple the wideband antenna signal directly into the optical domain [69]. The receiver's coupler divides the light into a parallel bank of progressively incremented transmission notches formed by using a parallel array of phase-shifted chirped fiber Bragg gratings (FBG).

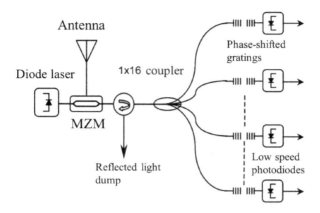

Figure 7.82 Block diagram of a photonic wideband channelizer architecture.

To understand how this architecture works, we first examine a single fiber Bragg grating as shown in 7.83(a). Here $E^+(l)$ and $E^-(l)$ are the complex amplitudes of the forward and backward propagating EM fields, respectively, and l is the grating length.

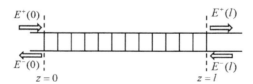

Figure 7.83 Schematic diagram of a fiber Bragg grating structure.

The relationship of the fields at the beginning $(z=0)$ and at the end $(z=l)$ is given by

$$\begin{bmatrix} E^+(0) \\ E^-(0) \end{bmatrix} = [g] \begin{bmatrix} E^+(l) \\ E^-(l) \end{bmatrix} \quad \text{or}$$

$$\begin{bmatrix} E^+(0) \\ E^-(0) \end{bmatrix} = \begin{bmatrix} m & n \\ n^* & m^* \end{bmatrix} \begin{bmatrix} E^+(l) \\ E^-(l) \end{bmatrix} \quad (7.72)$$

where

$$m = \left[\cosh(\gamma l) + \frac{i\delta\beta l \sinh(\gamma l)}{\gamma l} \right] \exp\left(\frac{i2\pi l}{\lambda_B} \right) \quad (7.73)$$

and
$$n = \frac{-\kappa l \sinh(\gamma l) \exp(-i2\pi l/\lambda_B)}{\gamma l} \quad (7.74)$$

with
$$\gamma^2 = \kappa^2 - (\delta\beta)^2 \quad (7.75)$$

where m and n are the complex conjugates of m and n respectively, κ is the coupling coefficient of the grating, and $\delta\beta = 2\pi(\lambda^{-1} - \lambda_B^{-1})$ is the detuning from the Bragg wavelength λ_B that is related to the grating period Λ as $\lambda_B = 2n_{eff}\Lambda$. Here n_{eff} is the effective mode index [70]. The reflection coefficient is then $r = E^-(0)/E^+(0) = n^*/m$ and the transmission coefficient is given by $t = E^+(l)/E^+(0) = 1/m$ and are derived by computing the fields by including the boundary conditions.

Extending the FBG the filter structure can have N linear gratings $(g_1,\ldots,g_N)$ as shown in Figure 7.84(a). The grating structures are separated by $N-1$ phase-shifts $p_1,\ldots,p_{N-1}$. The transfer matrix model is shown in Figure 7.84(b).

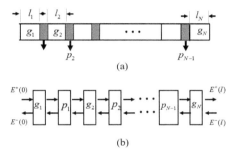

Figure 7.84 Schematic diagram of (a) compound fiber Bragg grating structure with g_N gratings and p_{N-1} phase shifts and (b) the transfer matrix model for the compound grating structure.

The transfer matrix of the compound FBG filter can be now written as

$$\begin{bmatrix} E^+(0) \\ E^-(0) \end{bmatrix} = \begin{bmatrix} A & B \\ C & D \end{bmatrix} \begin{bmatrix} E^+(L) \\ E^-(L) \end{bmatrix} \quad (7.76)$$

where
$$\begin{bmatrix} A & B \\ C & D \end{bmatrix} = [g_1][p_1]\cdots[p_i]\cdots[p_{N-1}][g_N] \quad (7.77)$$

and L is the total length of the compound FBG filter and

$$[p] = \begin{bmatrix} e^{-i\phi} & 0 \\ 0 & e^{i\phi} \end{bmatrix}$$

The amplitude transmission is then $t = 1/A$.

The incorporation of phase shifts gives rise to a narrowband transmission window inside the stopband of an FBG. By adjusting the magnitudes and locations of the phase-shift regions, one can tailor the transmission spectrum for specific needs [70]. The Bragg gratings are characterized by a transmission notch within the reflection spectrum. By adjusting the spatial position of the phase shift along the length of the grating it is possible to control the offset of the transmission notch from the center wavelength of the reflection band.

The properties of the transmission notch can be controlled precisely. The width of the transmission notch can be controlled and used to adjust the frequency resolution. The stopband attenuation can also be controlled also by adjusting the chirp, length, and size of the phase shift. The *transmission intensity* can be expressed as a function of the reflectivity of the grating stopband r and the phase shift of the region ϕ as [69]

$$I(\phi, r) = \frac{(1 - r^2)^2}{1 + r^4 - 2r^2 \cos(\phi)} \tag{7.78}$$

The demonstrated system is capable of resolving signals in the 2-18-GHz range into 2-GHz bands while maintaining a high probability of intercept. It is worth underlining that with this kind of receiver the specifics of the RF signal such as precise frequency and phase information is lost [71]. These details, however, are intended to be obtained from other signal processing modules within the DRFM.

7.15.2 Spectrum Scanning Approach

The scanning receiver shown in Figure 7.85(a) was proposed in 1999 and based on a tunable narrowband filter that scans the modulation sideband of a laser carrying the RF spectrum so that the optical power detected by a PD can be associated to the position of the filter (i.e., to the analyzed RF frequency). The receiver had a bandwidth of 40 GHz and used a piezoelectrically scanned Fabry-Perot etalon. Its drawbacks included poor sensitivity and SFDR and low probability of intercept. Details of this are given in [71].

7.15 Wideband Spectrum Sensing and Analysis

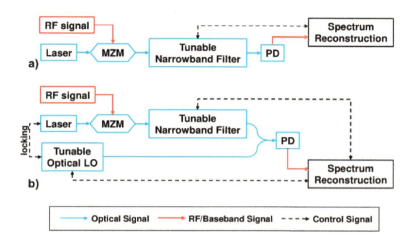

Figure 7.85 Schematic diagram of a (a) scanning receiver based on a tunable narrowband filter. MZM: Mach-Zehnder modulator, PD: photodiode. The spectrum reconstruction requires associating the power detected by the PD to the position of the tunable filter and (b) scheme of a scanning receiver based on a tunable narrowband filter and a tunable optical local oscillator (LO). The spectrum reconstruction can measure precisely the frequency of the detected RF signals, if the frequency of the LO is known. (©IEEE, reprinted with permission from [71].)

In Figure 7.85(b) the photonic sensing!spectrum scanning filtered sideband is heterodyned with a tunable laser acting as an optical local oscillator (LO). This way, the filtered spectrum portion is down-converted to an intermediate frequency (IF) that can be precisely measured by a digital acquisition and processing system after the PD [71]. Then, the original frequency of the analyzed spectrum portion can be calculated knowing the frequency of the optical LO. This approach also has the capability of analyzing both the amplitude and phase of the detected RF spectrum, and to study its time behavior.

A *photonics-assisted* IFM system is presented in [72]. A block diagram is shown in Figure 7.86.

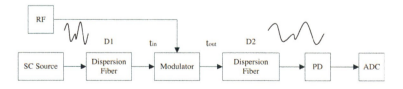

Figure 7.86 Photonic time-stretched analog-to-digital converter. (From [72].)

The concept uses a laser that is passed through a dispersion shifted fiber to generate a chirped optical signal by using the group velocity dispersion effect of the fiber.

The RF signal then modulates the chirped signal, using a MZM and passed through dispersion shifted fiber again, stretching it out before being detected by a photodiode and sampled by a slower ADC. Then the stretched signal is sent into the receiver where the measurement is made of the frequency using a Hilbert transform.

7.15.3 Direct Conversion

An example of a direct conversion receiver is shown in Figure 7.87 [73]. Here the tunable laser (TL) provides a tone centered at v_{TL} (A) that is amplified and split into two arms to drive the architecture. The following description of operation is taken from [73].

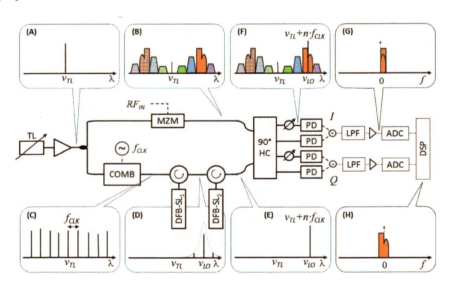

Figure 7.87 Block diagram of a direct conversion photonic receiver. Insets: optical or electrical signals at different points in the receiver. (©2016 EuRAD, reprinted with permission from [73].)

On the upper arm the TL acts as the optical carrier, and the wideband RF input signal is upconverted to the frequency v_{TL} through double-sideband suppressed-carrier modulation in a MZM (B). On the lower arm, a stable RF tone at a low frequency f_{CLK} drives a coherent optical comb generator (COMB), modulating the TL in a cascade of intensity and phase modulators generating a comb of optical lines at f_{CLK} from each other that are all phase-locked with the TL (C). Then, the optical frequency comb injects a distributed-feedback slave laser (DFB-SL$_1$) with the cavity resonance thermally tuned to v_{LO}, through an optical circulator. The injection forces the DFB-SL$_1$ to phase-lock to the comb line closest to its resonance frequency, and also suppresses

7.15 Wideband Spectrum Sensing and Analysis

the other injected comb lines. Therefore, the injection locking results in the selection and amplification of the comb line at frequency v_{LO} (D).

In order to further increase the suppression of the unwanted comb lines, the output of DFB-SL$_1$ injects another slave laser (DFB-SL$_2$). The injected DFB-SL$_2$ therefore is phase-locked to the TL, and acts as the optical LO of the direct conversion receiver (E). The frequency detuning between the TL and the optical LO is controlled by changing the optical frequency of the TL v_{TL} in discrete steps, so that

$$v_{LO} = v_{TL} + n f_{clk} \quad n = 1, 2, \ldots, (n_{max} - 1/2) \quad (7.79)$$

with n_{max} the total number of lines in the comb. The comb line at frequency v_{LO} locks the LO to the phase of the TL at any tuning step.

A 90° optical hybrid coupler (HC), two balanced photodiodes (BPD), and two ADCs then provide the optical I/Q demodulation for downconversion to baseband, antialias filtering, and digitization. The unique contribution of this receiver is that it exploits optical domain upconversion and photonic I/Q demodulation to perform a wideband scanning of the RF input spectrum thereby avoiding the use of multiple crystal oscillators and bandpass filters. In addition, their approach provides a total immunity to local oscillator self-mixing and to RF-to-baseband feedthrough. Their results show a DC offset lower than 50 μV, a second-order input intercept point (IIP2) of 57 dBm, and an operating BW of 0.5-10.5 GHz with an IBW of 1 GHz. A linear dynamic range of 141 dB/Hz and a spurious-free dynamic range of 107 dB/Hz$^{2/3}$ outperformed current pure electronic direct conversion receivers [71].

A spectrum analyzer with a resolution bandwidth of 100 kHz was used to acquire the downconverted spectrum. With a noise floor at -120 dBm a minimum detectable signal results at -87 dBm with the measured results shown in Figure 7.88. Shown are the linear dynamic range (LDR) = 110 dB, 1-dB compression point $P_{1dB} = 23$ dBm, the IIP3=28 dBm, IIP2 = 57 dBm, and the SFDR = 83 dB.

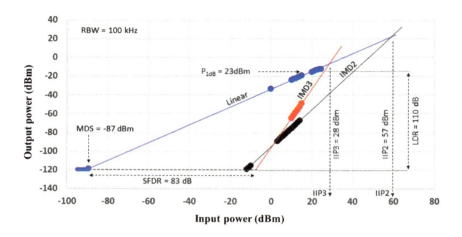

Figure 7.88 Measurement of IMD3. (©2016 EuRAD, reprinted with permission from [73].)

An improvement on this technique was reported in [71]. Here they were able to demonstrate an extended spectrum coverage up to 40.5 GHz and a fast tuning capability with a channel switching time of 10 μs, allowing a full 40 GHz-bandwidth scan to be executed in the impressive time of 20 ms with a resolution of 3 kHz. The reported results are shown in Figure 7.89.

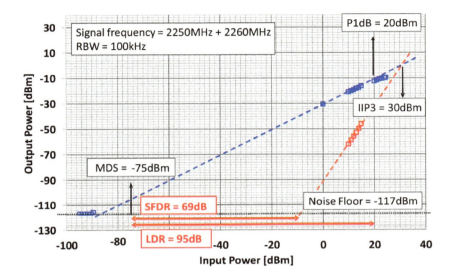

Figure 7.89 Measures of sensitivity, linearity, and distortions of the photonics-based scanning receiver. RBW: resolution bandwidth; MDS: minimum detectable signal; P1dB: input power at 1-dB compression; IIP3: input third-order intercept point; SFDR: spurious-free dynamic range; LDR: linear dynamic range. (©IEEE, reprinted with permission from [71].)

7.15.4 Cryogenic Spatial-Spectral Crystal Photonics

One photonic RF receiver technology that is able to provide a variety of EMW functions with an inherent wide instantaneous bandwidth stare is based on spatial-spectral (S2) photonic crystals. The hardware solution, with all associated components is referred to as an extreme bandwidth analysis and correlator (EBAC). The approach for RF signal processing is based on the transference of wideband analog RF signals as voltages to modulated laser light with an electro-optical modulator (EOM), then having that modulated laser light absorbed by an S2 crystal. Typical S2 crystals offer 10-200 GHz of absorption, so as to cover broadband RF signals and agile and transient energy events with 100% probability of detection.

The S2 broadband crystal has been making significant contributions for over 20 years in microwave engineering [74] and spectral analysis [75, 76]. As well, contributions in microwave holography and spatial-spectral imaging have been reported [77–82]. More recently, contributions in optical signal processing have been emphasized [83].

Currently electromagnetic spectrum (EMS) awareness can be provided in ∼ 40 GHz bandwidth segments, with the S2 approach compatible with and enabled by wideband RF front ends to feed signals, including a wideband RF antenna, low noise amplifier and power amplifiers, and the EOM. These signal input components also set the limits on instantaneous bandwidth (IBW) and spur free dynamic range (SFDR) of the overall device. So, extending beyond the current bandwidth coverage up to 110 GHz [84, 85] depends on a combination of RF front components to measure frequencies potentially up to THz, and approaching DC to daylight.

Instantaneous wideband power spectrum analysis is the straightforward implementation. This capability, along with advanced embodiments of analog correlation of instantaneous wideband phase/amplitude/frequency modulated signals, for signal processing, time difference of arrival, enables advanced applications of geolocation of RF signals in multi-receiver arrangements, and detection and identification of complex, wide bandwidth waveform signals [86]. Using these aspects enables EMS awareness, complex DRFM operation such as orientation of an antenna for sensitive detection and efficient transmission, as a cueing receiver and for detection of low probability-of-intercept signals.

7.15.4.1 S2 Photonic Crystals

The core component of the EBAC is a cryogenically cooled S2 crystal. The dopant ions in the S2 crystal in the ground state are ready to absorb resonant laser-light, thus acting as a highly efficient channelized spectral energy detectors, making millions of continuous and parallel frequency selective energy measurements. An example spectrum profile of an S2 crystal is shown in Fig. 7.90, with energy levels and identified ranges of parameters for thulium (Tm) dopant ions, that vary based on the dopant, host crystal, temperature, applied magnetic field, and optical operating parameters [82].

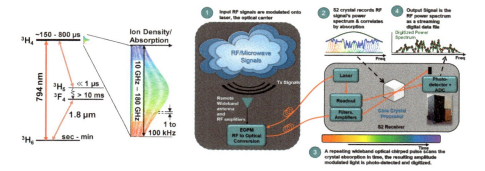

Figure 7.90 LEFT: Absorption profile and energy level diagram and properties of a thulium doped spatial-spectral (S2) crystal including linewidth, absorption bandwidth, and storage lifetimes. RIGHT: EBAC system attributes, showing a remoted antenna and key hardware components.

The typical temperature of operation is ~4K. A single crystal channel uses an irradiated volume ~1 mm3 (approximately the size of a human hair over a few cm),

7.15 Wideband Spectrum Sensing and Analysis

in interacting with modulated laser light does analog signal processing, implementing physical Fourier transforms via spectral channelized absorption and atomic energy storage. Each frequency bin has a distribution of ~1-10 billion Tm ions across the irradiated crystal volume, that will each selectively record the light intensity within the RBW of its resonance frequency channel, typically where RBW ~10-100 kHz. Thus for wide bandwidths spanning over 10-100 GHz in the optical band, an S2 crystal has ~1-10 million channelized frequency bins.

In the time domain, the crystal's coherence time is referred to as T2, where ideally T2~1/RBW. While the IBW and RBW of the crystal are key features, the persistence time T1 of the crystal absorption into an excited state or bottleneck state, makes possible two aspects of the EBAC operation, 1) to integrate energy detection results to increase energy detection sensitivity and 2) to interrogate the crystal after absorption has taken place, so that the results can be read-out before the storage power spectrum decays fully back to the atomic ground state. Applied magnetic fields can extend T1~seconds, while for most EMW applications, the typical short-term memory T1~1-10 ms of the crystal atomic states with no applied magnetic field is a feature that allows a balance between energy integration and readout persistence.

7.15.4.2 S2 EBAC Hardware Components

While the core component is the small ~1 mm3 volume of cooled S2 crystal, a full EBAC system relies on several components to enable the S2 crystal to provide persistent continuous RF spectral monitoring and signal processing. Fig. 7.90 also shows the hardware approach where EMS signals, as received by an antenna, are recorded in the crystal by first modulating the signals onto a laser carrier with an electro-optic modulator, to transfer the RF signal to the optical carrier, and the modulated laser beam is then propagated through fiber to the S2 crystal.

The main components of an S2 EBAC system include the wideband RF front end, the laser and optical amplifiers, along with single mode optical fiber cable between antenna crystal, a cryostat to cool the S2 crystal, balanced photodetectors, conventional analog-to-digital converters, and a low-latency digital processing system for signal detection, and event reporting. A ~4K crystal operation point is achieved with a small commercial closed-cycle cryocooler running on ~1 kW power, that circulates He gas, and offering long duration continuous operation for several years without maintenance, as used in hospital magnetic resonance imaging (MRI) systems. S2 EBACs with these have been field tested in several sites in various configurations, including as described in [78]. The stored spectral energy information is uniquely retrieved, asynchronously with continuous writing, by illuminating the S2 crystal's modified absorption profile with a novel frequency chirped readout laser. The full spectrum update rates can vary from kHz to MHz, and the choice of a readout bandwidth, frame rate, and balanced photo-detection bandwidth determine the displayed frequency pixel width. Readout frame rates used typically are much faster than conventional spectroscopy [77].

For the RF to optical conversion at the antenna, one benefit of the S2 absorption approach is that electro-optical phase modulators (PMs) can be used rather than of Mach Zendher amplitude modulators (MZMs), although MZMs can be used as well. Typically, PMs are lower loss, more straightforward to manufacture, have wider bandwidth and flatter responses than MZMs, and are easier to operate with no power and no bias controller needed.

Using long fiber cables gives the EBAC the same advantage of remoted antennas with analog photonic links, such as shown in Figure 7.90 where the S2 crystal replaces the photodetector in the receiver. Such fiber optical feeds are an advantage over other digital systems that situate spectrum analysis hardware such as digitizers and computers at the antennas, particularly for higher frequency operation. The EBAC technology also scales to multiple processing feeds from multiple antennas, enhancing spectral coverage and awareness over large areas with distributed antennas.

7.15.4.3 S2 EBAC Wideband Spectral Monitoring

As a spectrum monitoring device, the S2 EBAC is capable of continuous spectral recording and storage over wide bandwidths. For an example, Fig. 7.91 shows spectrogram of dynamic microwave spectral activity recorded over a 20 GHz bandwidth with a 2 kHz full spectrum update rate. The top of Fig. 7.91 shows the most recent 20 GHz spectrum captured, and below is the spectral data history displayed in a waterfall plot. Every tone burst of a random frequency hopper, operating first in the 1-6 GHz range and then in the 5-10 GHz range was captured, demonstrating the S2 EBAC's 100% probability of detection of incoming signals, which is in stark contrast to spectrum analyzers based on scanning filters or digitizers that typically record 1% of the incoming signals (or missing over 99% of the incoming signals). This is along with a variety of other spectral activity between 10-20 GHz.

7.15 Wideband Spectrum Sensing and Analysis 481

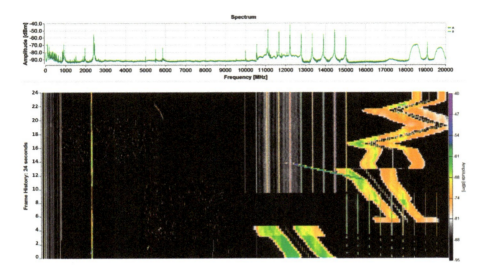

Figure 7.91 Demonstration of continuous spectral monitoring of 20 GHz instantaneous bandwidth of a dynamic microwave environment with an S2 EBAC. The top shows a current 20 GHz spectrum frame, recording is continuous at a 5 kHz frame rate, and the bottom is a waterfall plot of the last 24 seconds of frame data. Spectral resolution here was 0.24 MHz, or with ∼83k frequency pixels per frame.

Three zoomed spectrograms are shown in Fig. 7.92. The left shows a zoomed in spectrogram tracking the evolution in, a modulated frequency hopped source over 2.25 s in a 5.3 to 9.2 GHz band. The zoomed spectrogram shows a microwave signal's initial 6.88 GHz carrier with 0.7 GHz wide sidebands that were square wave modulated at 75 MHz. The S2 EBAC captured the initial shift in the sideband center frequency, followed by the stepping of the carrier frequency over several GHz. At the same time the modulated signal was being recorded, the EBAC captured two other signals. The middle zoomed spectrogram in Fig. 7.92 shows Wi-Fi (the broader features) and Bluetooth (the narrower pulses features) sharing the same spectrum in the ISM band around 2.5 GHz. The right zoomed spectrograms show a frequency hopping synthesizer that is having trouble initially locking onto its assigned frequency around 3.21 GHz. The start of the synthesizers tone slides about 7 MHz for 2 ms before locking onto one of its assigned frequency, a feature that can be used to identify the frequency hopping source, useful information in choosing the parameters of the responding DFRM.

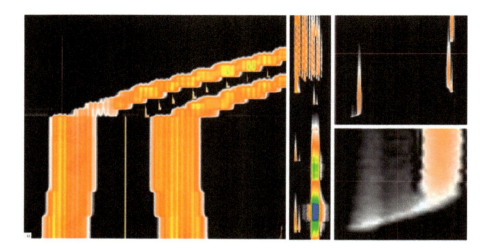

Figure 7.92 Zoomed spectrograms from the spectral data in Fig. 7.91. All were captured simultaneously by the S2 EBAC without a priori knowledge of the signals.

A critical specification of a spectral monitoring system is its spur free dynamic range (SFDR), or the inverse ratio of the smallest signal that can be detected in the presence of large signals without any spurs created by the system itself. The SFDR is typically measured by input of two tones, and monitoring the fundamental and spurious signals produced from this input. A feature of the S2 crystal is where spectral saturation in the crystal does not create intermodulation terms, so that the EBAC system is SFDR limited by the intermodulation terms created in the RF input devices and the electro-optic conversion process. The S2 EBAC directly capturing signals also eliminates frequency spurs often introduced by systems requiring RF down conversion. An example SFDR for the S2 EBAC in Fig. 7.93 shows 62 dB of SFDR with an RF front end providing 58 dB of RF gain, and down to <-120 dBm sensitivity for a 1 ms pulse [79]. Other variations and tradeoffs between SFDR and sensitivity can be achieved.

7.15 Wideband Spectrum Sensing and Analysis

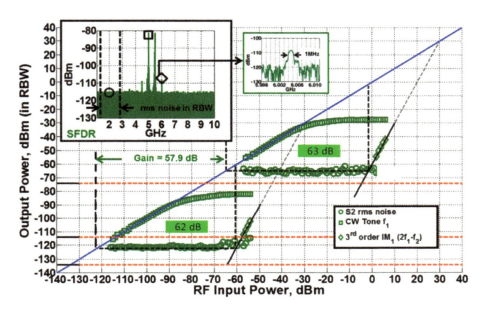

Figure 7.93 Measurements of the SFDR and Sensitivity of the EBAC, with a wideband RF front end over 10 GHz, with SFDR two tone third order limited at >62 dB, and sensitivity below -120 dBm. Higher sensitivity with less SFDR can also been achieved.

7.15.4.4 S2 Photonic Cueing Receiver Architecture

The S2 EBAC enables a novel system for assured RF signal capture of transient and agile RF signals with other components. The block diagram of an S2 photonic enabled cued digital receiver architecture is shown in 7.94 showing the combination of A) the S2 energy capture process B) fast digital signal analysis, C) a long stable optical fiber buffer delay line, and D) a fast tuning digital receiver [84]. The low-latency high-performance digital spectrum feature analysis system is enabled with fast readout of typically ∼20-50 microseconds, and the subsequent fast digital processing implemented in Field Programmable Gate Arrays (FPGAs) so that each recorded spectra frame is processed in series and identified for key spectral events, in a few microseconds. The fast logic then cues one or more RF receivers (or an optical receiver) to fast frequency tune to the energy event's center frequency.

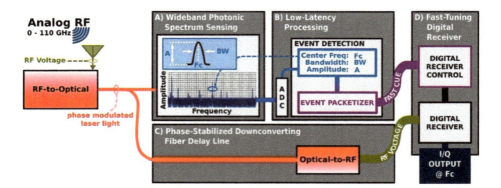

Figure 7.94 Block diagram of the S2 crystal photonics wideband receiver for assured capture of transient and agile RF events. (From [84].)

The actual phase sensitive RF energy is buffered optically in the phase-stabilized optical fiber delay, followed by coherent photo detection with a conversion to voltage and then digitization occurs with the I/Q digital receiver. Such a system would enable single event capture, and a DRFM to respond to sparse, transient, agile, single events, and random radar and RF signals.

7.15.4.5　S2 Signal Processing, Time-Difference-of-Arrival and Cross-Correlation

The correlation of analog, wideband, complex waveforms in an analog fashion is required for more advanced applications. The correlative attribute of S2 crystals is their full bandwidth phase-sensitive multi-photon absorption over the crystal's coherence time T2, and then integrated over the persistence time T1. This provides analog correlation based signal processing, where the input is the incidence of any two modulated RF signals with bandwidth IBW, delay within T2, and duration within T1, so the S2 crystal records the holographic spectral interference pattern of photons from each instance of the signal. Thus, the power spectrum is the frequency domain recording of the multiplication of physical Fourier transforms of each of the signals.

One application of S2 analog signal processing is the highly precise determination of the time delay τ between the same or similar RF signals, known as time difference of arrival (TDoA). If the same signal is received at two antenna nodes, the interference between them can lead to a power spectrum of the waveforms which is modulated a periodic component $1/\tau$. The hardware configurations are varied and can include angular separation of laser beams as well as multiple spatial volumes.

The TDoA is revealed by digital frequency analysis of the read output signal. The fiber optical feed and components from the antennas to the crystal are frequency stabilized to ensure accurate correlation and integration over $\sim$T1. In general, highly accurate measurements of time delays depend on the bandwidth of the signals, the signal-to-noise of the signal, along with the spectral resolution of the crystal, and stability of the optical fiber link. The short-term stability requirement is an advantage, not long term or absolute stability. Measurements for time delays varying over $\sim$1-5 μs with stability of $\sim$10 ps for $>>$ 10 ms have been demonstrated [86]. Other variations of this correlative approach can also be used for applications such as detection of complex RF waveforms, the use of wideband complex and varying RF waveforms for radar and RF communications, and the signal classification and identification processes.

7.16 SUMMARY

This chapter begins with a look at the design of microwave-photonic transceiver architectures starting with an overview of the technology used. This includes the mode-locked laser, photo detectors, optical links and the photonic oscillator. Electro-Optical MZMs, photonic memories suitable for DRFM transceivers, and microwave-photonic antennas were also examined. As well, photonic analog-to-digital converters were emphasized in particular, the RSNS high-resolution encoding with an inherent Gray code was presented as a smarter method of folding and quantizing the analog waveforms. The theory and applications in compressive sensing were examined including the photonic Nyquist folding receiver. Wideband spectrum sensing was presented and included a channelizer that uses a fiber Bragg grating. Finally we took a look at the wideband S2 crystal technology with its extreme bandwidth capability and its uses in spectrum sensing and DF signal processing.

REFERENCES

[1] A. C. Lindsay, G. A. Knight, and S. T. Winnall, "Photonic mixers for wide bandwidth RF receiver applications," *IEEE Transactions on Microwave Theory and Techniques*, vol. 43, no. 9, pp. 2311–2317, 1995.

[2] Y. Gao, A. Wen, W. Zhang, W. Jiang, J. Ge, and Y. Fan, "Ultra-wideband photonic microwave i/q mixer for zero-if receiver," *IEEE Transactions on Microwave Theory and Techniques*, vol. 65, no. 11, pp. 4513–4525, 2017.

[3] S. Sun, M. He, S. Yu, and X. Cai, "Hybrid silicon and lithium niobate mach-zehnder modulators with high bandwidth operating at c-band and o-band," in *2020 Conference on Lasers and Electro-Optics (CLEO)*, 2020, pp. 1–2.

[4] N. Boynton, H. Cai, M. Gehl, S. Arterburn, C. Dallo, A. Pomerene, A. Starbuck, D. Hood, D. C. Trotter, T. Friedmann, C. T. DeRose, and A. Lentine, "A heterogeneously integrated silicon photonic/lithium niobate travelling wave electro-optic modulator," *Opt. Express*, vol. 28, no. 2, pp. 1868–1884, Jan 2020.

[5] S. Sun, M. He, M. Xu, X. Zhang, Z. Ruan, L. Liu, and X. Cai, "120 Gb Hybrid Silicon and Lithium Niobate Modulators with On-Chip Termination Resistor," in *2020 Optical Fiber Communications Conference and Exhibition (OFC)*, 2020, pp. 1–3.

[6] X. Li, X. Liu, Y. Qin, D. Yang, and Y. Ji, "Ultra-low index-contrast polymeric photonic crystal nanobeam electro-optic modulator," *IEEE Photonics Journal*, vol. 12, no. 3, pp. 1–8, 2020.

[7] V. Artel, I. Bakish, T. Kraus, M. Shubely, Y. Ben-Ezra, E. Shekel, S. Zach, A. Zadok, and C. N. Sukenik, "Low temperature wafer bonding of silicon to InP and silicon to $LiNbO_3$ using self-assembled monolayers," in *OFC/NFOEC*, 2012, pp. 1–3.

[8] Y. Pan, S. Sun, M. Xu, M. He, S. Yu, and X. Cai, "Low fiber-to-fiber loss, large bandwidth and low drive voltage lithium niobate on insulator modulators," in *2020 Conference on Lasers and Electro-Optics (CLEO)*, 2020, pp. 1–2.

[9] R. C. Alferness, V. R. Ramaswamy, S. K. Korotky, M. D. Divino, and L. L. Buhl, "Efficient single-mode fiber to titanium diffused lithium niobate waveguide coupling for lambda = 1.32 μm," *IEEE Transactions on Microwave Theory and Techniques*, vol. 30, no. 10, pp. 1795–1801, 1982.

[10] A. H. Tehranchi and N. Granpaeh, "Optical beam propagation in ti:linbo/sub 3/ mach-zehnder modulator," in *Proceedings of LFNM 2002. 4th International Workshop on Laser and Fiber-Optical Networks Modeling (IEEE Cat. No.02EX549)*, 2002, pp. 128–130.

[11] W. Sohler, "Waveguide lasers and nonlinear devices in lithium niobate," in *Technical Digest. CLEO/Pacific Rim '99. Pacific Rim Conference on Lasers and Electro-Optics (Cat. No.99TH8464)*, vol. 4, 1999, pp. 1265–1266 vol.4.

[12] R. G. Hunsperger, *Integrated Optics and Technology, 5th Ed.* New York, New York: Springer, 2002.

[13] C. Luo, P. Yen, and Y. Lai, "Environmentally stable 100 ghz hybrid mode-locked burst-mode fiber laser with enhanced autocorrelation contrast," in *2019 24th OptoElectronics and Communications Conference (OECC) and 2019 International Conference on Photonics in Switching and Computing (PSC)*, 2019, pp. 1–3.

[14] T. F. Carruthers, M. L. Dennis, I. N. Duling, M. Horowitz, and C. R. Menyuk, "Enhanced stability of a dispersion-managed, harmonically mode-locked fiber laser," in *Technical Digest. Summaries of papers presented at the Conference on Lasers and Electro-Optics. Postconference Edition. CLEO '99. Conference on Lasers and Electro-Optics (IEEE Cat. No.99CH37013)*, 1999, pp. 101–102.

[15] M. Horowitz, C. R. Menyuk, T. F. Carruthers, and I. N. Duling, "Theoretical and experimental study of harmonically modelocked fiber lasers for optical communication systems," *Journal of Lightwave Technology*, vol. 18, no. 11, pp. 1565–1574, 2000.

[16] C. Luo, S. Wang, and Y. Lai, "Bound soliton fiber laser mode-locking without saturable absorption effect," *IEEE Photonics Journal*, vol. 8, no. 4, pp. 1–9, 2016.

[17] J. W. Lou, T. F. Carruthers, T. R. Clark, and M. Currie, "Multiple-wavelength mode-locked erbium-doped fiber sigma laser," in *Technical Digest. Summaries of papers presented at the Conference on Lasers and Electro-Optics. Postconference Technical Digest (IEEE Cat. No.01CH37170)*, 2001, pp. 413–414.

[18] T. F. Carruthers and I. N. Duling, III, "10-ghz 1.3-ps erbium fiber laser employing soliton pulse shortening," in *Optics Letters*, vol. 21, 1996, pp. 1927–1929.

[19] J. M. Butler, P. E. Pace, and J. P. Powers, "Experimental results of a low-power sigma mode-locked laser for applications in mobile sampling of wideband antenna signals," in *Proceedings of the 9th Annual DARPA Symposium on Photonic Systems for Antenna Applications (PSAA)*, vol. Naval Postgraduate School, no. Feb, 1999.

[20] P. E. Pace, *Advanced Techniques for Digital Receivers*.   Norwood, MA: Artech House, 2000.

[21] T. LABS, "Photodiodes and photoconductors tutorials," https://www.thorlabs.com/newgrouppage9.cfm?objectgroup_id=9020, 2020, accessed: 2020-12-11.

[22] (Jan. 11, 2021) Optical detection systems. [Online]. Available: https://www.newport.com/n/optical-detection-systems

[23] S. Pan and Y. Zhang, "Microwave photonic radars," *Journal of Lightwave Technology*, vol. 38, no. 19, pp. 5450–5484, 2020.

[24] X. Huang, Y. Liu, Z. Li, H. Guan, Q. Wei, Z. Yu, and Z. Li, "Advanced electrode design for low-voltage high-speed thin-film lithium niobate modulators," *IEEE Photonics Journal*, vol. 13, no. 2, pp. 1–9, 2021.

[25] M. Kim, B. Yu, and W. Choi, "A Mach-Zehnder modulator bias controller based on oma and average power year=2017, volume=29, number=23, pages=2043-2046,," *IEEE Photonics Technology Letters*.

[26] P. E. Pace, *Detecting and Classifying Low Probability of Intercept Radar*.   Norwood, MA: Artech House, 2009.

[27] E. Iszekiel, Stravros, *Microwave photonics : devices and applications*, ser. Wiley - ieee.   Chichester, U.K. ;: Wiley, 2009.

[28] R. A. Minasian, "Ultra-wideband and adaptive photonic signal processing of microwave signals," *IEEE Journal of Quantum Electronics*, vol. 52, no. 1, pp. 1–13, 2016.

[29] T. A. Nguyen, E. H. W. Chan, and R. A. Minasian, "Photonic radio frequency memory using frequency shifting recirculating delay line structure," *Journal of Lightwave Technology*, vol. 32, no. 1, pp. 99–106, 2014.

[30] Z. Ding, F. Yang, H. Cai, Y. Weng, M. Wang, and D. Wang, "Photonic radio frequency memory with controlled doppler frequency shift," in *2019 International Topical Meeting on Microwave Photonics (MWP)*, 2019, pp. 1–3.

[31] Y. Yashchyshyn, A. Chizh, S. Malyshev, and J. Modelski, "Technologies and applications of microwave photonic antennas," in *2010 International Conference on Modern Problems of Radio Engineering, Telecommunications and Computer Science (TCSET)*, 2010, pp. 11–14.

[32] R. Waterhouse and D. Novak, "Efficient antennas and their impact on microwave photonics signal processing," in *2018 IEEE Photonics Conference (IPC)*, 2018, pp. 1–2.

[33] K. S. et. al., "High power integrated photonic W-band emitter," in *IEEE Trans. on Microwave Theory and Techniques*, vol. 66, no. 3, 2018, pp. 1668–1678.

[34] C. Cox and E. Ackerman, "Tiprx: A transmit-isolating photonic receiver," in *IEEE Journal of Lightwave Technology*, vol. 32, no. 10, 2014, pp. 3630–3636.

[35] L. Wang, J. Yu, and Y. Li, "Microwave photonic antenna for fiber radio application," in *2018 IEEE 3rd Optoelectronics Global Conference (OGC)*, 2018, pp. 122–125.

[36] R. Grootjans, C. Roeloffzen, C. Taddei, M. Hoekman, L. Wevers, I. Visscher, P. Kapteijn, D. Geskus, A. Alippi, R. Dekker, R. Oldenbeuving, J. Epping, R. B. Timens, R. Heuvink, E. Klein, A. Leinse, P. van Dijk, and R. Heideman, "Broadband continuously tuneable delay microwave photonic beamformer for phased array antennas," in *2019 49th European Microwave Conference (EuMC)*, 2019, pp. 812–815.

[37] B. Murmann, "The race for the extra decibel: A brief review of current ADC performance trajectories," *IEEE Solid-State Circuits Magazine*, vol. 7, no. 3, pp. 58–66, 2015.

[38] T. R. Clark, J. H. Kalkavage, N. Bos, R. Schmid, J. W. Zobel, A. J. Goers, M. L. Dennis, and E. J. Adles, "Progress in photonic digital radio transceivers," in *2020 International Topical Meeting on Microwave Photonics (MWP)*, 2020, pp. 95–97.

[39] P. W. Juodawlkis, J. C. Twichell, J. L. Wasserman, G. E. Betts, and R. C. Williamson, "Measurement of mode-locked laser timing jitter by use of phase-encoded optical sampling," vol. 26, 2001.

[40] R. H. Walden, "Analog-to-digital converter survey and analysis," *IEEE Journal on Selected Areas in Communications*, vol. 17, no. 4, pp. 539–550, 1999.

[41] P. W. Juodawlkis, J. C. Twichell, G. E. Betts, J. J. Hargreaves, R. D. Younger, J. L. Wasserman, F. J. O'Donnell, K. G. Ray, and R. C. Williamson, "Optically sampled analog-to-digital converters," *IEEE Transactions on Microwave Theory and Techniques*, vol. 49, no. 10, pp. 1840–1853, 2001.

[42] E. Krune, B. Krueger, L. Zimmermann, K. Voigt, and K. Petermann, "Comparison of the jitter performance of different photonic sampling techniques," *Journal of Lightwave Technology*, vol. 34, no. 4, pp. 1360–1367, 2016.

[43] J. Gordon and H. Haus, "Random walk of coherently amplified solitons in optical fiber transmission," *Optics Letters*, vol. 11, no. 10, p. 665–667, 1986.

[44] H. Taylor, "Application of guided-wave optics in signal processing and sensing," *Proceedings of the IEEE*, vol. 75, no. 11, pp. 1524–1535, 1987.

[45] R. Becker and F. Leonberger, "2-bit 1 Gsample/s electrooptic guided-wave analog-to-digital converter," *IEEE Journal of Quantum Electronics*, vol. 18, no. 10, pp. 1411–1413, 1982.

[46] P. Pace and D. Styer, "High-resolution encoding process for an integrated optical analog-to-digital converter," *Optical Engineering*, vol. 33, no. 8, p. 2638—2645, 1994.

[47] G. Valley, "Photonic analog-to-digital converters," *Optics Express*, vol. 15, no. 5, pp. 1755–1982, 2007.

[48] M. R. Arvizo, J. Calusdian, K. Hollinger, and P. E. Pace, "Robust Symmetrical Number System Preprocessing for Minimizing Encoding Errors in Photonic ADCs," in *Optical Engineering*, vol. 50, no. 8, 2011.

[49] P. E. Pace and D. Styer, "High-resolution encoding process for an integrated optical analog-to-digital converter," in *Optical Engineering*, vol. 33, no. 8, 1994, pp. 2638—2645.

[50] P. E. Pace, R. D. Walley, R. J. Pieper, and J. P. Powers, "5 bit guided-wave SNS transfer characteristics," in *IET Electronics Letters*, vol. 31, no. 21, 1995, pp. 1799–1800.

[51] B. L. Luke and P. E. Pace, "N-sequence RSNS redundancy analysis," in *2006 IEEE International Symposium on Information Theory*, 2006, pp. 2744–2748.

[52] P. E. Pace, P. Stănică, B. L. Luke, and T. W. Tedesso, "Extended closed-form expressions for the robust symmetrical number system dynamic range and an efficient algorithm for its computation," *IEEE Transactions on Information Theory*, vol. 60, no. 3, pp. 1742–1752, 2014.

[53] Y. Jakop, A. S. Madhukumar, and A. B. Premkumar, "A robust symmetrical number system based parallel communication system with inherent error detection and correction," *IEEE Transactions on Wireless Communications*, vol. 8, no. 6, pp. 2742–2747, 2009.

[54] P. Pace, D. Styer, and I. Akin, "A folding adc preprocessing architecture employing a robust symmetrical number system with gray-code properties," *IEEE Transactions on Circuits and Systems II: Analog and Digital Signal Processing*, vol. 47, no. 5, pp. 462–467, 2000.

[55] P. Pace, D. Wickersham, D. Jenn, and N. York, "High-resolution phase sampled interferometry using symmetrical number systems," *IEEE Transactions on Antennas and Propagation*, vol. 49, no. 10, pp. 1411–1423, 2001.

[56] P. E. Pace, C. K. Tan, and C. K. Ong, "Microwave-photonics direction finding system for interception of low probability of intercept radio frequency signals," *Optical Engineering*, vol. 57, no. 2, pp. 1411–1423, 2018.

[57] D. L. Donoho, "Compressed sensing," *IEEE Transactions on Information Theory*, vol. 52, no. 4, pp. 1289–1306, 2006.

[58] T. P. McKenna, J. H. Kalkavage, M. D. Sharp, and T. R. Clark, "Wideband photonic compressive sampling system," *Journal of Lightwave Technology*, vol. 34, no. 11, pp. 2848–2855, 2016.

[59] P. E. Pace, A. Kusmanoff, and G. L. Fudge, "Nyquist folding analog-to-information receiver: Autonomous information recovery using quadrature mirror filtering," in *2009 Conference Record of the Forty-Third Asilomar Conference on Signals, Systems and Computers*, 2009, pp. 1581–1585.

[60] M. A. B. Siddique, R. B. Arif, and M. M. R. Khan, "Digital image segmentation in matlab: A brief study on otsu's image thresholding," in *2018 International Conference on Innovation in Engineering and Technology (ICIET)*, 2018, pp. 1–5.

[61] E. J. Candes and M. B. Wakin, "An introduction to compressive sampling," *IEEE Signal Processing Magazine*, vol. 25, no. 2, pp. 21–30, 2008.

[62] E. Candes, N. Braun, and M. Wakin, "Sparse signal and image recovery from compressive samples," in *2007 4th IEEE International Symposium on Biomedical Imaging: From Nano to Macro*, 2007, pp. 976–979.

[63] Q. Guo, H. Chen, M. Chen, S. Yang, and S. Xie, "Photonics-assisted compressive sampling systems," *SPIE Proceedings, Real-Time Photonic Measurements, Data Management and Processing II;*, vol. 10026E, no. 11, pp. 1–14, 2016.

[64] Q. Guo, M. Chen, Y. Liang, H. Chen, S. Yang, and S. Xie, "Photonics-assisted compressive sampling system for wideband spectrum sensing," vol. 15, no. 1, 2017, pp. 1–14.

[65] J. Hogan, W. R. Babbitt, C. Benko, S. Bekker, C. Stiffler, K. Winn, R. Price, and K. Merkel, "Assured capture of transient RF events across extremely wide bandwidths," in *2019 IEEE Research and Applications of Photonics in Defense Conference (RAPID)*, 2019, pp. 1–4.

[66] J. B. Tsui, *Special Design Topics in Digital Wideband Receivers*. Norwood, MA: Artech House, 2010.

[67] D. B. Hunter, L. G. Edvell, and M. A. Englund, "Wideband microwave photonic channelised receiver," in *Proc. International. Topical Meeting Microwaves and Photonics*, 2005, pp. 249–252.

[68] X. Xie, Y. Dai, K. Xu, J. Niu, R. Wang, L. Yan, and J. Lin, "Broadband photonic rf channelization based on coherent optical frequency combs and i/q demodulators," *IEEE Photonics Journal*, vol. 4, no. 4, pp. 1196–1202, 2012.

[69] D. B. Hunter, L. G. Edvell, and M. A. Englund, "Wideband microwave photonic channelised receiver," in *2005 International Topical Meeting on Microwave Photonics*, 2005, pp. 249–252.

[70] Li Wei and J. W. Y. Lit, "Phase-shifted bragg grating filters with symmetrical structures," *Journal of Lightwave Technology*, vol. 15, no. 8, pp. 1405–1410, 1997.

[71] P. Ghelfi, F. Scotti, D. Onori, and A. Bogoni, "Photonics for ultrawideband RF spectral analysis in electronic warfare applications," *IEEE Journal of Selected Topics in Quantum Electronics*, vol. 25, no. 4, pp. 1–9, 2019.

[72] C. Zhang, L. Yang, and S. Li, "A photonics-assisted instantaneous frequency measurement receiver," in *2017 Progress in Electromagnetics Research Symposium - Fall (PIERS - FALL)*, 2017, pp. 823–827.

[73] D. Onori, F. Laghezza, F. Scotti, A. Bogoni, P. Ghelfi, M. Bartocci, A. Zaccaron, A. Tafuto, and A. Albertoni, "A DC offset-free ultra-wideband direct conversion receiver based on photonics," in *2016 European Radar Conference (EuRAD)*, 2016, pp. 374–377.

[74] K. Merkel, R. Krishna Mohan, Z. Cole, R. Reibel, T. Harris, T. Chang, and W. Randall Babbitt, "Analog optical signal processing of baseband codes in Tm:YAG up to 10 Gb/s," in *2004 IEEE International Topical Meeting on Microwave Photonics (IEEE Cat. No.04EX859)*, 2004, pp. 138–141.

[75] W. R. Babbitt, K. D. Merkel, S. H. Bekker, C. R. Stiffler, P. B. Sellin, Z. W. Barber, R. K. Mohan, and C. H. Harrington, "Extreme bandwidth analyzer and correlator for spectrum analysis and direction finding," in *2012 IEEE Aerospace Conference*, 2012, pp. 1–8.

[76] K. D. Merkel, S. H. Bekker, C. R. Stiffler, A. Traxinger, A. Woidtke, P. B. Sellin, M. Chase, W. R. Babbitt, Z. W. Barber, and C. Harrington, "Continuous wideband spectrum analysis over 10 GHz bandwidth with 61 dBc spur-free dynamic range," in *2013 IEEE International Topical Meeting on Microwave Photonics (MWP)*, 2013, pp. 297–300.

[77] W. R. Babbitt, Z. W. Barber, S. H. Bekker, M. D. Chase, C. Harrington, K. D. Merkel, R. K. Mohan, T. Sharpe, C. R. Stiffler, A. S. Traxinger *et al.*, "From spectral holeburning memory to spatial-spectral microwave signal processing," *Laser Physics*, vol. 24, no. 9, p. 094002, 2014.

[78] S. H. Bekker, A. S. Traxinger, C. R. Stiffler, A. J. Woidtke, M. D. Chase, W. R. Babbitt, K. D. Merkel, and Z. W. Barber, "Testing of 10 GHz Instantaneous Bandwidth RF Spectrum Monitoring at Idaho National Labs," in *2014 National Wireless Research Collaboration Symposium*, 2014, pp. 21–26.

[79] K. D. Merkel, S. H. Bekker, A. S. Traxinger, C. R. Stiffler, A. J. Woidtke, M. D. Chase, W. R. Babbitt, C. H. Harrington, and Z. W. Barber, "Extreme bandwidth spectrum analysis," in *2016 IEEE Photonics Conference (IPC)*, 2016, pp. 487–488.

[80] Z. Barber, C. Harrington, R. K. Mohan, C. Stiffler, T. Jackson, P. Sellin, and K. Merkel, "100 Gbps pattern matching using spatial-spectral holographic time domain correlation," in *2016 Conference on Lasers and Electro-Optics (CLEO)*, 2016, pp. 1–2.

[81] ——, "Efficient spectral and correlative processing with spatial-spectral holography," in *2016 IEEE Photonics Society Summer Topical Meeting Series (SUM)*, 2016, pp. 112–113.

[82] Z. W. Barber, C. Harrington, K. Rupavatharam, C. Thiel, T. Jackson, P. Sellin, C. Benko, and K. Merkel, "Spatial-spectral materials for high performance optical processing," in *2017 IEEE International Conference on Rebooting Computing (ICRC)*, 2017, pp. 1–4.

[83] Z. W. Barber, C. Harrington, R. K. Mohan, T. Jackson, C. Stiffler, P. Sellin, and K. D. Merkel, "Spatial-spectral holographic real-time correlative optical processor with¿ 100 Gb/s throughput," *Applied optics*, vol. 56, no. 19, pp. 5398–5406, 2017.

[84] J. Hogan, W. R. Babbitt, C. Benko, S. Bekker, C. Stiffler, K. Winn, R. Price, and K. Merkel, "Assured capture of transient rf events across extremely wide bandwidths," in *2019 IEEE Research and Applications of Photonics in Defense Conference (RAPID)*, 2019, pp. 1–4.

[85] A. J. Mercante, S. Shi, P. Yao, L. Xie, R. M. Weikle, and D. W. Prather, "Thin film lithium niobate electro-optic modulator with terahertz operating bandwidth," *Optics express*, vol. 26, no. 11, pp. 14 810–14 816, 2018.

[86] K. D. Merkel, J. T. Jackson, R. M. Price, W. R. Babbitt, C. Benko, S. H. Bekker, K. N. Winn, C. R. Stiffler, A. J. Woidtke, J. Oset, A. S. Traxinger, J. Salveson, M. D. Chase, P. B. Sellin, R. K. Mohan, and Z. W. Barber, "A Photonics-based Broadband RF Spectrum Analysis and Geolocation System," in *2018 IEEE Research and Applications of Photonics In Defense Conference (RAPID)*, 2018, pp. 1–4.

Part II

Modern ES and EA Techniques Using Deep Learning

Chapter 8

Modern Spectral Sensing and Detection

In this chapter, we provide an introduction to the DRFM signal processing techniques for cognitive and cooperative spectral sensing. These allow the exploitation of under-utilized segments of the spectrum allowing the DRFM transceiver to maneuver about the highly defended and difficult spectral terrain for exploitation and exploration. Intercepting all the emitters within the operational arena (both threat and non-threat) can then lead to good coordination and execution of an EA response management leading to EMS dominance.

Modern *net-centric* spectrum sensing strategies including centralized and decentralized distributed concepts are discussed. Sequential detection, quickest detection, single and multiple spectrum sensing and access formulations are emphasized. EMS detection techniques using Cohen's *time-frequency methods*, with the Choi-Williams distribution, the Wigner-Ville distribution, and quadrature mirror wavelet decomposition filters are presented. Also presented is the use of *bifrequency* detection methods by computing the cyclostationary spectral correlation density function on the bi-frequency plane. Also discussed is the atomic decomposition method of detection. The above techniques provide significant subnoise visibility, increasing the processing gain of the EMS detection activities while leading to good spectrum maneuver agility.

8.1 PERSISTENT SPECTRUM SENSING

Mounted on unmanned platforms, each DRFM transceiver uses AI and machine learning and deep learning to mine the consistently large amounts of raw sensor data. Persistent surveillance activities are conducted with limited a priori information about the EMS terrain and the networked sensors being sourced in order to establish a situation awareness. Classification of all the signal modulations that are present (both known and unknown) is necessary for a fast, reactive EA response management. It is this distributed spectrum sensing that is crucial for creating awareness of the state of the spectrum (SoS) since the EMS is nonstationary and varies with time, location, and frequency. The DRFM must perform the tasks of spectrum sensing, detection, classification, parameter extraction, and in some cases manage one or more EA responses in a complex environment of high noise interference with multiple signals. Detecting security vulnerabilities within AI embedded applications are also of major concern. As such, AI also means a stronger cyber defense, countering numerous automated attacks [1].

The DRFM device today is a network-enabled wideband intercept transceiver as shown in Figure 1.36 in Chapter 1, consisting of a system-of-systems able to autonomously collect, store, detect, and classify (known and unknown) threat emitters in order to efficiently exploit an idle spectrum band as well as manage any interference. Harvesting the state-of-the-art today from the commercial market, it is the GPUs and FPGA that are driving the low-power sensor technology.

Eliminating the limitations inherent in a platform-centric configuration comes from a distributed system of systems as shown in Figure 8.1. The ability for DRFMs to acquire, track, and locate conventional threat emitters and targets, and share this information among stand-off platforms such as SOJs (e.g., for weapons targeting), is an example of a *network-centric* architecture, and represents a fundamental shift from a platform-centric approach. In Figure 8.1, a network of collaborating DRFMs perform a stand-in ES detection of an LPI emitter and another one (e.g., within a swarm) comes by and disables it.

8.1 Persistent Spectrum Sensing

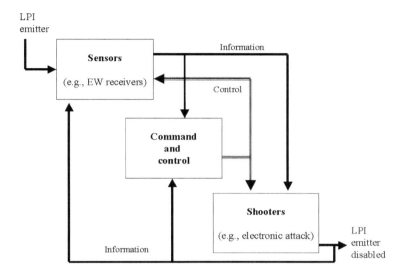

Figure 8.1 Disabling an LPI emitter within a network-centric architecture.

The wideband nature of the threat emitters can force the DRFM receiver to have a significant processing gain having to implement sophisticated wideband spectrum sensing architectures, as discussed in Chapter 3, and modern spectrum exploration and exploitation algorithms [2]. Use of time-frequency, bifrequency signal processing algorithms are also used to detect emitters, determine the waveform parameters, and often perform a specific emitter identification and perform a response.

Distributed spectrum sensing and interference management starts with understanding and scheduling of resources for detection. To quantify the detection performance within the EMS, we can define a *spatial diversity order* (D) as function of the probability of detection and the number of DRFMs cooperating together as

$$D(N) = \max \frac{\partial}{\partial \rho_{dB}} P_D(\rho_{dB}, N) \tag{8.1}$$

where $\rho_{dB} = 10\log_{10}\rho$ and $\partial/\partial\rho_{dB}$ denotes the partial derivative of ρ with respect to ρ and N is the number of DRFMs that are networked. Figure 8.2 shows that the spatial diversity order may be defined as the *maximum slope* of the probability of detection curve as a function of the logarithmic SNR ρ_{dB} [2].

500 MODERN SPECTRAL SENSING AND DETECTION

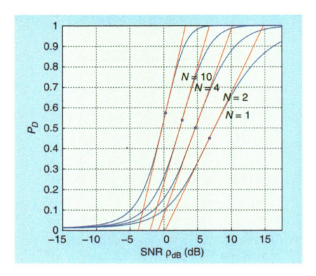

Figure 8.2 The spatial diversity order may be defined as the maximum slope of the probability of detection curve as a function of the logarithmic SNR ρ_{dB}. In this case, there are N i.i.d. Rayleigh channels (e.g., DRFMs), and the employed detection scheme is energy detection with $P_{FA} = 0.01$. (©IEEE, reprinted with permission from [2].)

In this case, there are N i.i.d. Rayleigh channels (e.g., to N distributed DRFMs), and the employed detection scheme is energy detection (radiometry using equal-gain) with $P_{FA} = 0.01$. It can be seen that the slope of the probability of detection curve grows as the number of cooperating nodes, or DRFMs, sensing the same band, increases. Also note that adding more DRFM sensors to observe a subband yields diminishing returns. The achieved again also depends on the fusion rule used in combining the local sensing information as well as the local propagation conditions surrounding the DRFMs that are distributed about the area of regard.

8.2 CENTRALIZED OR DECENTRALIZED EMS SENSING

The distributed DRFMs sensing the EMS can either be *centralized or decentralized* depending on how they transmit their SoS to the main fusion center. In a centralized DRFM network, all of the DRFMs transmit all of the locally observed (raw) data to a central node that performs the data fusion. In a decentralized DRFM network, the DRFM is able to process the (raw) data locally, computing an SoS, before transmitting the SoS to the fusion center. That is, the fusion center only receives a partial post-processed set of the data. Typically this is preferred since the bandwidth required is much less, whereas in the centralized case the DRFM networks must have significant bandwidth to transmit all of the raw data.

As the sensors in a decentralized system are processing the data locally before transmitting the results to a fusion center, a local decision or other compact statistic characterizing the SoS is carried out first to process the raw data. In this decentralized distributed spectrum sensing system, the DRFM nodes will then typically collaborate in identifying and exploiting the underutilized spectrum. The fusion center only receives a sufficient statistic from the DRFMs characterizing the SoS. The most common topologies used for the DRFMs to fusion center are serial, parallel, and tree. The serial topology is not a robust topology as a single DRFM hardware failure for example, will then take down the entire fusion center connection. In the parallel scheme, the DRFMs do not communicate with each other and no feedback is provided from the fusion center to the DRFMs.

The DRFMs use mapping rules $u_i = v_i(y_i), i = 1, \ldots, N$ where $y_i = [y_{i,1}, \ldots, y_{i,N_s}]$ is an N_s-dimensional vector of observations made by DRFM i and N is the number of DRFMs and then pass the local mappings, u_i to the fusion center. The mapping compresses the data by sending a local binary decision to the fusion center. DRFMs that are close to each other also form a cluster that sends summary information about the SoS to the fusion center. In an ad hoc configuration, there is no dedicated fusion center and the information is distributed to *all* the nodes (using multiple hops) so they all can react to the idle spectrum if needed. Note the multiple hops may cause undesired delays making the SoS information stale. The distributed DRFM sensing problem is typically modeled using binary hypothesis testing. The selected decision-making strategies such as Neyman-Pearson, minimax, or Bayes can be used, which are also related to controlling interference.

For example, the Neyman–Pearson detection scheme for distributed detection can be stated as follows: For a predefined global probability of false alarm level P_{FA}, find the (optimum) local and global decision rules $\Gamma = \{v_0(\mathbf{u}), v_1(\mathbf{y}_1), \ldots, v_N(\mathbf{y}_N)\}$, where $v_0(\cdot)$ is the global decision rule and $v_i(\cdot), i = 1, \ldots, N$ are the local rules that minimize the global probability of missed detection P_{MD}. Considering that the observed sensor data is independent, the null hypothesis H_0 at the fusion center is spectrum band idle (noise only), and H_1 at the fusion center is signal present. The decision is made by comparing the likelihood ratio $\Lambda(\mathbf{u})$ to a threshold ψ_0. In factored form $\Lambda(\mathbf{u})$ may be given as

$$\Lambda(\mathbf{u}) = \prod_{i=1}^{N} \frac{p(u_i|H_1)}{p(u_i|H_0)} \qquad (8.2)$$

where $p(\cdot)$ is the probability density function of its argument(s). Conditional independence assumption must be valid or the optimal tests are no longer simple. However, conditional independence is typically the case. If the local decision variables are *binary*, the global decision rule is a Boolean rule and the log-likelihood ratio at the fusion center can be written as a weighted sum of local sensor decisions [2]

$$\log \Lambda(\mathbf{u}) = \sum_{i=1}^{N} \left[u_i \log \frac{1 - P_{MD,i}}{P_{FA,i}} + (1 - u_i) \log \frac{P_{MD,i}}{1 - P_{FA,i}} \right] \quad (8.3)$$

where u_i is the binary local decision of the ith DRFM (i.e., 0 or 1), $P_{FA,i}$ is the probability of a false alarm at the ith DRFM, and $P_{MD,i}$ is the probability of a missed detection at the ith DRFM. The weights depend on the performance of the individual DRFMs that may not be known. Computationally simpler decision rules can be used like the well-known Boolean K-out-of-N rules such as OR, AND, and MAJORITY. The detector is commonly designed so that the levels of P_{FA} and P_{MD} at the fusion center are controlled [2].

Censoring reduces power consumption by sending only sufficiently informative decision statistics to the fusion center. A DRFM sends its test statistic to the fusion center only when its test statistic, given here as $\log \Lambda_i$, is *above* a censoring threshold defined by the network's communication rate or

$$P(\log \Lambda_i > \psi_i | H_0) \leq \kappa_i \quad i = 1, \ldots, N \quad (8.4)$$

where $\kappa_i \leq 1$ is the network communication rate being used by the DRFM of user i and ψ_i is the upper limit (no-send) region for user i. The upper limit ψ_i is selected such that the probability of DRFM i transmitting the test statistic to the fusion center under hypothesis H_0 is κ_i. For cooperative sensing, the censoring test statistic L-out-of-N users transmit is often used [3].

$$D_N = \sum_{i=1}^{L} \log \Lambda_i + \sum_{i=L+1}^{N} d_i \quad (8.5)$$

where it is assumed that the first L users with informative decision statistics transmit and $d_i = E[\log \Lambda_i | \log \Lambda_i \leq \psi_i, H_0]$ is the conditional mean of the local log-likelihood ratio of the ith DRFM in the no-send region under the null hypothesis. Note that the no-send region is not ignored. It is captured by a single quantity; that is i.e., *the conditional mean of the log-likelihood* in the no-send region, which is optimal in the minimum mean-square error sense. Practically speaking, the DRFM does not need to transmit d_i as it can be calculated at the fusion center. This is because the limits of ψ_i are determined by the DRFM's network communication rate constraints κ_i.

By using binary (hard) local decisions in a distributed decentralized detection system the communication cost is reduced at the expense of loss of information. Using soft decisions, such as log-likelihood ratios or quantized versions of them will, typically lead to an improvement in performance but, will significantly increase the amount of data to be transmitted and, maybe the power consumption. Studies suggest that the loss is negligible by using ≥ 4 bits to quantize the log likelihood ratios. A phenomenon known as the *bit-error probability* (BEP) *wall* was reported in [4] for both hard- and soft-decision-based cooperative sensing. If the BEP of the reporting channel exceeds the BEP wall value, constraints on the detector performance cannot be met at the fusion center. Soft-decision-based systems are more robust in the face of reporting errors.

8.2.1 Sequential Spectrum Sensing Methods

Sequential sensing methods are important for detecting changes in the state of the spectrum (i.e., rapidly identifying new spectral opportunities as well as vacating a specific frequency band quickly when a friendly emitter requires the band). Sensing time is an important parameter in finding idle spectrum bands. Minimizing the sensing time, fewer resources are spent sensing, and more time is allowed for transmission. Sequential detection aims to minimize sensing time and varies the sample size depending on the size of the data. Classical sequential detection aims at distinguishing between two hypotheses from a sequence of i.i.d. random observations with the objective of being able to make a decision as quickly as possible given specified error levels. An alternative formulation is the *quickest detection problem* in which the objective is to detect a change in the distribution of the data (i.e., find the change point, with minimal detection delay) [2].

For example, let $y_1, y_2, \ldots$ be a sequence of i.i.d. random observations with a common distribution F_0 or F_1. The binary test of hypothesis H_0 against H_1 can be expressed as

$$H_0 : y_n \sim F_0, \quad n = 1, 2, \ldots,$$
$$H_1 : y_n \sim F_1, \quad n = 1, 2, \ldots, \quad (8.6)$$

Let p_0 and p_1 be the pdf associated with F_0 and F_1, respectively. Spectral detection attempts to choose between these two hypotheses by attempting to minimize the number of observations given the constraints on the type I or II errors. The *sequential probability ratio test* (SPRT) of Wald requires the minimal average number of observations under both hypotheses among all tests with equal (or smaller) error probabilities [5].

The sequential sensing *stop time* when using the Wald SPRT is based on the acceptable false alarm probabilities and the probability of missed detections. The stopping time of the SPRT is given by

$$N_S = \inf\{n \geq 1 | \Lambda_n \leq B \text{ or } \Lambda_n \geq A\} \quad (8.7)$$

where $\Lambda_n = \prod_{k=1}^{n} p_1(y_k)/p_0(y_k)$ is the likelihood ratio and A and B are upper and lower stopping boundaries, respectively. After each sample, the SPRT accepts H_1 if $\Lambda_n \geq A$ or accepts H_0 if $\Lambda_n \leq B$. If $B < \Lambda_n < A$, an additional observation is required. The stopping boundaries A and B are chosen based on the target levels of the probability of false alarm and the probability of missed detection as illustrated in Figure 8.3.

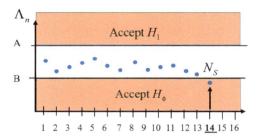

Figure 8.3 Illustration of the spectrum sensing sequential probability test for Λ_n using two stopping boundary thresholds A and B to determine N_S ($N_S = 14$).

We note that the hypotheses have to be simple and the distributions completely specified under both hypotheses for the SPRT to be optimal [2]. In cognitive spectrum sensing situations, there are often unknown parameters related to transmit powers, propagation conditions, adaptive modulation schemes, and frequency coding schemes used. There have been a number of efforts to design sequential detection tests for the case of *composite hypotheses*, such as sequential generalized likelihood ratio tests and minimax tests, making sure that a decision is reached within a certain time. In addition, the spectrum sensing within a band may be terminated within a certain time if criteria are not met. Then the band can be revisited at a later time.

8.2.2 Quickest Spectrum Detection Methods

To describe the quickest spectrum detection methods, let $y_1, y_2, \ldots$, be a sequence of independent random spectrum observations with an unknown change point m such that $y_1, \ldots, y_{m-1}$ have a common distribution F_0 and $y_m, y_{m+1}, \ldots$ have another common distribution F_1, respectively. Let p_0 and p_1 denote probability density functions of F0 and F1, respectively. Both Bayesian and non-Bayesian quickest detection approaches have received considerable attention in the research community. Page's cumulative sum (CUSUM) test is the most commonly used non-Bayesian quickest detection method. The stopping time of the CUSUM test for detecting a change is given by [6]

$$N_S = \inf\{n \geq 1 | U_n \geq \psi\} \tag{8.8}$$

where $U = \max_{1 \leq k \leq n} \prod_{i=k}^{n} p_1(y_i)/p_0(y_i)$ and $\psi \geq 0$ is the test threshold. The CUSUM test statistic U_n may be updated recursively using

$$U_n = \max\{1, U_{n-1}\} p_1(y_n)/p_0(y_n), n \geq 1, U_0 = 1$$

and the threshold value may be chosen suitably depending on the detection strategy. The optimum quickest detection algorithms are sensitive to uncertainty in distribution parameters and many of the approaches dealing with unknown distribution parameters use nonparametric approaches or generalized likelihood ratio based approaches.

8.2.3 Single DRFM: Spectrum Sensing and Access

This section considers that each DRFM can sense and access only part of the spectrum at a time. In such single-DRFM, multiband spectrum sensing and access problem, the decision maker (e.g., a UAV or fusion center) has to decide which frequency bands to sense or access at each time. Hence these are *stochastic sequential decision* problems that can be modeled as Markov decision processes (MDPs).

A Markov decision process consists of a sequence of discrete time steps, $n = 0, 1, 2, \ldots$, a decision maker i, and finite set of possible states of the environment $s \in S$. In each state there is a finite set of possible actions $a \in A$ and a state transition function $\varphi : S \times A \times S \to [0, 1]$ that defines the transition probability $P(S_{n+1}|S_n, a_n)$.

Figure 8.4 shows the Markov decision process where in each state s_n, the decision maker chooses an action a_n that results in a reward r_{n+1} and the new state s_{n+1} with the goal of maximizing the rewards.

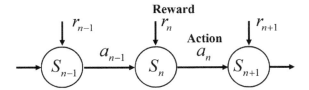

Figure 8.4 Illustration of a Markov decision process. In each state s_n, the decision maker chooses an action a_n that results in a reward r_{n+1} and new state s_{n+1}. The goal of the decision maker is to maximize a given function of the discounted rewards over a finite or infinite horizon.

A suitable objective function for spectrum sensing and access problems is the *expected sum of discounted rewards* over an infinite horizon given an initial state $s_n = s$

$$J_n = E\left[\sum_{k=0}^{\infty} \gamma^k r_{n+k+1}|s_n=s\right] \tag{8.9}$$

where r_n is the reward at time n and γ, $0 \leq \gamma < 1$ is the discount rate. This objective function gives decreasing weight to future rewards [4]. For a single-DRFM, multiband spectrum sensing and access problems, the decision maker (e.g., a fusion center) has to decide which frequency bands to sense or access at each time. These are stochastic sequential decision problems that can be modeled as MDPs.

Consider that the decision maker is the fusion center. The fusion center chooses which frequency band each DRFM senses at each time. Thus, actions correspond to sensing a particular set of frequency bands by the DRFMs. The states of the environment would be formed by the occupancy of the frequency bands. Thus, if each frequency band can be either vacant or occupied, there are in total 2^{N_B} different states, where N_B is the number of different frequency bands. Finally, the fusion center would get a reward equal to one for each frequency band sensed vacant and zero for the other frequency bands. There are many ways to formulate this problem; by defining the variables differently, by considering for example, that there is feedback from the spectrum access, the rewards could depend on the obtained throughput. The goal of the decision maker is to find an optimal policy that determines in each state the optimal action so that a particular objective function is maximized [2].

8.2.4 Multi-DRFM: Spectrum Sensing and Access

There are several ways that have been formulated for the multiuser, multi-DRFM spectrum sensing, and access formulation.

Game theory has been formulated to model the interaction of multiple users in spectrum sensing and

access problems providing models for both noncooperative and cooperative game theory. The main difference between noncooperative and cooperative games is that in noncooperative games, the players act independently and cooperation cannot be enforced, while, in cooperative games, the players act as groups and cooperation within the group can be enforced.

Dynamic programming refers to an optimization approach in which an original multistage sequential decision problem is broken down into smaller, simpler single-stage decision problems that are then solved in a recursive manner. Here, we consider solving finite MDPs using dynamic programming. Therefore, dynamic programming can be used to solve spectrum sensing and access problems formulated as finite MDPs [2]. Consider for example, the maximization of 8.9, the expected sum of discounted rewards given an initial state $s_n = s$.

The value of a policy π starting from state s is defined by

$$v^\pi(s) = E_\pi[\sum_{k=0}^{\infty} \gamma^k r_{n+k+1} | s_n = s] \tag{8.10}$$

where $E_\pi[\cdot]$ denotes the expectation when the policy is followed. Simplifying,

$$v^\pi(s) = E_\pi[r_{n+1} + \gamma v^\pi(s_{n+1}) | s_n = s] \tag{8.11}$$

we get what is known as the *Bellman equation* for v^π where the optimal value function v^* is equal to the expected return of the best action in the state ($s_n = s$) or

$$v^*(s) = \max_a E[r_{n+1} + \gamma v^*(s_{n+1}) | s_n = s, a_n = a] \tag{8.12}$$

which is called the *Bellman optimality equation* from reinforcement learning [7]. Reinforcement learning is a trial-and-error machine-learning approach. Machine learning is discussed below. Reinforcement learning is where the decision maker, called the agent, observes the state of the operational and spectral environment and chooses the actions that lead to rewards and new states. Actions leading to desired outcomes are given higher rewards, which reinforce these actions. This makes the actions more likely to be chosen again in similar situations in the future. Consequently, in reinforcement learning, the agents are faced with *exploitation versus exploration*. The agents must decide whether to trade off, that is, exploit the current best action or to explore other actions in hope of finding a better option.

8.2.5 Bandit Problems

Bandit problems are also sequential decision problems in which the name originates from the similarity to the traditional slot machines used in casinos, called one-armed bandits. Bandit problems can also be modeled as MDPs and are appropriate models for single-user multiband spectrum sensing and access problems. In the multiarmed bandit formulation, the arms of the multiarmed bandit correspond to *different frequency bands* and choosing an arm to play corresponds to sensing or accessing a particular frequency band. The multiarmed bandit problem is a classic version of the problem of optimal allocation of activity under certainty [8]. One can phrase it by saying that one has n projects, the state of project i being denoted by x_i (or by $x_i(t)$ if one wishes to emphasize its dependence on time, (t)). One can operate only one project at once: if one operates project i then one receives reward $g_i(x_i(t))$ in the time-interval $(t, t+1)$ and the transition $x_i(t)...x_i(t+1))$ follows a Markov rule specific to project i. The unused projects neither yield reward nor change state; current states of all projects are known at any time. The problem is to so choose the project at each moment that the expected discounted reward over an infinite future is maximal[8].

In a Markovian multiarmed bandit problem, the conditional state transition probability depends only on the current state, and, thus, they can be modeled as MDPs. Consider there are K arms and a decision maker controlling the selection of the arms to play. In each state s_n, the decision maker selects one arm, a_i, to play and receives a reward, r_{n+1}, for it. In the classical multiarmed bandit problems, the state of the nonplayed arms does not change and the stochastic processes for the different arms are independent of each other [2]. The optimal policy maximizing the expected sum of discounted rewards with known rewards and state transition probabilities is given by an index structure in which a priority index, called *the Gittins index*, is calculated for each state of each arm of the multiarmed bandit [9, 10]. The Gittins index for the kth arm is given by

$$v_k(s_0^k) = \max_{\tau > 0} \frac{E\left[\sum_{n=0}^{\tau-1} \gamma^n r_{n+1}^k(s_n^k) | s_0^k\right]}{E\left[\sum_{n=0}^{\tau-1} \gamma^n | s_0^k\right]} \tag{8.13}$$

where the maximization is over the set of all stopping times $\tau > 0$. Here, $r_{n+1}^k(s_n^k)$ denotes that the reward of the kth arm r_{n+1}^k depends only on the state of the kth arm s_n^k. Also, if an arm is not played, its reward is zero. The Gittins index of each arm depends only on that arm's underlying stochastic process so the original K-dimensional optimization problem is reduced to K, one-dimensional optimization problems. Once the Gittins indices have been calculated, the optimal policy reduces to selecting the arm with the largest Gittins index at each time.

Consequently, the Gittins index finds for each arm the optimal stopping time in terms of maximizing the *expected discounted reward* normalized by the expected discounted number of plays. Note that index calculations require full knowledge of the rewards and state transition probabilities for the arms. The *restless multiarmed bandit problem* is a generalization of the classical multiarmed bandit problem in which the decision maker may simultaneously play multiple $L \leq K$ arms and the nonplayed arms may change state and give rewards [11]. Here, Whittle proposed an index policy based on a Lagrange multiplier approach that is optimal for the average expected reward over the infinite horizon criterion under a relaxed constraint that the average number of played arms is equal to L but was difficult to calculate.

The frequency band occupancy can also be modeled as a two-state Markov process is shown in Figure 8.5.

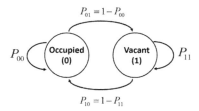

Figure 8.5 A two-state Markov chain model for a frequency band occupation.

There are two possible states: vacant (1) and occupied (0) and their known state transition probabilities. (P_{01}^i, P_{11}^i) where $i = 1, \ldots, N_B$. The user observes the frequency band states only after sensing and, hence, needs to infer the state from its past decisions and observations to make decisions. The conditional probability that a frequency band is in state 1 given all past decisions and observations is a sufficient statistic.

A *myopic policy* chooses at each time the action maximizing the expected immediate reward while fully ignoring the impact on any future rewards [12]. That is, there is no exploration as a myopic policy is always exploiting the action that gives the highest expected reward. Consider the myopic action $\hat{a}_n = [\hat{a}_n^1, \ldots, \hat{a}_n^{N_B}]$, $\hat{a}_n^i \in \{0, 1\}$

8.3 Detection Methods: Time-Frequency

for sensing M frequency bands at a time n. This is given by

$$\hat{a}_n = \arg\max_{a_n} \sum_{i=1, a_n^i \neq 0}^{N_B} \omega_n^i r_i \quad \text{s.t.} \sum_{i=1}^{N_B} a_n^i = M \tag{8.14}$$

where ω_n^i is the belief state of frequency band i, $a_n^i = 1$ denotes that frequency band i is sensed and $a_n^i = 0$ does the opposite. If the arms are i.i.d., the myopic policy is then a queue structure that depends only on the ordering of P_{11} and P_{01} time, the M frequency bands at the head of the queue are sensed [2].

Consequently, the belief states do not need to be updated, and the exact values of state transition probabilities are not needed. Only the ordering of P_{11} and P_{01} has to be known. This makes the myopic policy computationally very efficient to employ in practice for the DRFM. Figure 8.6 shows the queue structure of the myopic policy.

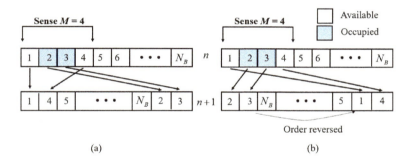

Figure 8.6 The structure of the myopic sensing policy for i.i.d. frequency bands when (a) $P_{11} \geq P_{01}$ and (b) $P_{11} \ll P_{01}$. (Adapted from [2].)

For positively correlated ($P_{11} \geq P_{01}$) i.i.d. frequency bands, the myopic sensing policy is optimal for any M [13]. This optimality holds for discounted expected reward over finite and infinite horizons and for average expected reward over the infinite horizon. For negatively correlated ($P_{11} < P_{01}$) i.i.d. frequency bands, the myopic sensing policy has been shown in [2] to be optimal for $N_B = 2$ and $N_B = 3$ and $M = 1$ but, in general, not optimal for $N_B > 4$.

8.3 DETECTION METHODS: TIME-FREQUENCY

Over the last several years, there has been an explosion of interest in alternatives to traditional signal representations. Instead of just representing signals as superpositions of sinusoids (the traditional Fourier representation) we now have available alternate methods of representation such as time-frequency and bifrequency 2D planes. These planes enhance the signal characteristics and give a good representation of the signal's properties not available with the standard Fourier transform. Using sophisticated machine learning, AI, multiagent systems, and distributed intelligence routines to capture the signal space and calculate these 2D planes enables the DRFM as the single, most important device technology in the EM operational environment.

DRFMs using the extremely limited sets of data from the many distributed and disparate DRFMs are able to detect, classify, and even fingerprint the intercepted emitters using the highly sensitive detection, fusion, and correlation technologies now available. The modulation and emitter tracking is even more difficult when the networked systems have different spatial and temporal sample rates, resolutions, and sensitivities.

In addition, these transceivers must now include precision direction finding, EA and counter-EA, control and cueing of weapon systems, enhanced radar warning, and aid in the fusion of off-board sensors and databases, with full integration of the electronic combat system. Other capabilities include emitter classification and identification, emitter-to-platform correlation, detailed analysis, and signal recording. A block diagram of a software-defined signal processor for detection and classification of signals within the environment is shown in Figure 8.7.

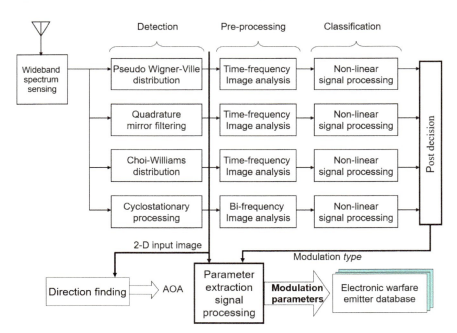

Figure 8.7 Block diagram of a software-defined autonomous classification and parameter extraction processing for a DRFM interception.

8.3 Detection Methods: Time-Frequency

The important considerations here are that the detection and classification techniques are autonomous. Also, each detection, preprocessing (or feature extraction), and classification routine are run in parallel. Further, the results of each parallel path are integrated in a final decision block that takes each classification path output and forms a final decision of the modulation type.

In this section we have discussed how the modern concept of networking sensors together can bring about a new dimension to our view of targets as well as offer a more efficient jammer configuration. Below we consider the signal processing tools for sensing and, first discuss the *time-frequency distributions*.

8.3.1 Cohen's Time-Frequency Distributions

The reason for time-frequency analysis is to give a mathematical core to the intuitive concept of the time-varying Fourier spectrum for nonstationary signals. The general class of time-frequency distributions introduced by Cohen is given by [14]

$$C_f(t,\omega,\phi) = \frac{1}{2\pi} \int \int \int e^{j(\xi\mu - \tau\omega - \xi t)} \phi(\xi,\tau) A(\mu,\tau) d\mu d\tau d\xi \tag{8.15}$$

where $\phi(\xi,\tau)$ is a kernel function and

$$A(\mu,\tau) = x\left(\mu + \frac{\tau}{2}\right) x^*\left(\mu - \frac{\tau}{2}\right) \tag{8.16}$$

and $x(\mu)$ is the time signal, and $x^*(\mu)$ is its complex conjugate. This represents a generalized class of a bilinear transformation that satisfies the marginals and has good resolution in both time and frequency spaces.

Considerable theoretical work has been carried out in this direction and has yielded many different classes of parametric solutions and other types as well, such as *adaptive diffusion* [15]. Inspired by the work on image processing by Perona and Malik [16], diffusion-based models were first investigated to improve the readability of the Cohen class time-frequency representations to rely on signal-dependent partial differential equations that yield adaptive smoothed representations with sharpened time-frequency components. Below we start with the Choi-Williams distribution followed by the Wigner-Ville distribution.

8.3.2 Choi-Williams Distribution

Choi and Williams [17] realized that by choosing the kernel in (8.15) carefully, the calculation can minimize the cross terms and still retain the desirable properties of the self-terms. The Choi-Williams distribution (CWD) uses an *exponential weighting kernel* in order to reduce the cross-term components of the distribution. The kernel function that gives the Choi-Williams distribution is

$$\phi(\xi, \tau) = e^{-\xi^2 \tau^2 / \sigma} \qquad (8.17)$$

where σ ($\sigma > 0$) is a scaling factor. By substituting this kernel into (8.15) the continuous CWD of the input signal $x(t)$ is given as [14]

$$\text{CWD}_x(t, \omega) = \int_{\tau=-\infty}^{\infty} e^{-j\omega\tau} \left[\int_{\mu=-\infty}^{\infty} \sqrt{\frac{\sigma}{4\pi\tau^2}} G(\mu, \tau) A(\mu, \tau) d\mu \right] d\tau \qquad (8.18)$$

where

$$G(\mu, \tau) = e^{\sigma(\mu-t)^2/(4\tau^2)} \qquad (8.19)$$

and t is the time variable, ω is the angular frequency variable ($2\pi f$), and σ is a positive-valued scaling factor. The bracketed term in (8.18) is the estimation of the time-indexed autocorrelation. Just as for the Wigner-Ville Distribution (WVD), the CWD can be defined from the Fourier transform $X(\omega)$ of $x(t)$ by

$$\begin{aligned}\text{CWD}_X(t, \omega) &= \frac{1}{2\pi} \int_{\xi=-\infty}^{\infty} e^{-j\xi t} \int_{\mu=-\infty}^{\infty} \sqrt{\frac{\sigma}{4\pi\xi^2}} e^{\frac{(\mu-\omega)^2}{4\xi^2/\sigma}} \\ & \quad X\left(\mu + \frac{\xi}{2}\right) X^*\left(\mu - \frac{\xi}{2}\right) d\mu d\xi\end{aligned} \qquad (8.20)$$

and in discrete form, the Choi-Williams distribution is

$$\begin{aligned}\text{CWD}_x(\ell, \omega) &= 2 \sum_{\tau=-\infty}^{\infty} e^{-j2\omega\tau} \sum_{\mu=-\infty}^{\infty} \frac{1}{\sqrt{4\pi n^2/\sigma}} \\ & \quad e^{-\sigma(\mu-\ell)^2/(4\tau^2)} x(\mu+\tau) x^*(\mu-\tau)\end{aligned} \qquad (8.21)$$

For computational purposes it is necessary to apply the weighting windows $W_N(\tau)$ and $W_M(\mu)$ for the summations in (8.21) before evaluating the distribution at each time index ℓ. The windowed Choi-Williams distribution can then be expressed as

$$\begin{aligned}\text{CWD}_x(\ell, \omega) &= 2 \sum_{\tau=-\infty}^{\infty} W_N(\tau) e^{-j2\omega\tau} \sum_{\mu=-\infty}^{\infty} W_M(\mu) \sqrt{\frac{\sigma}{4\pi\tau^2}} \\ & \quad e^{-\frac{\sigma\mu^2}{4\tau^2}} x(\ell + \mu + \tau) x^*(\ell + \mu - \tau)\end{aligned} \qquad (8.22)$$

where $W_N(\tau)$ is a symmetrical window that has nonzero values for the range of $-N/2 \leq \tau \leq N/2$, and $W_M(\mu)$ is a rectangular window that has a value of 1 for the range of $-M/2 \leq \mu \leq M/2$. The parameter N is the length of the window $W_N(\tau)$. The

8.3 Detection Methods: Time-Frequency

length N along with the shape of the window determines the frequency resolution of the distribution. The parameter M, which is the length of the window $W_M(\mu)$, determines the range from which the time-indexed autocorrelation is estimated.

The CWD_x in (8.22) can also be expressed as

$$\text{CWD}_x(\ell, \omega) = 2 \sum_{n=-L}^{L} S(\ell, n) e^{-j2\omega n} \qquad (8.23)$$

where the kernel is

$$S(\ell, n) = W(n) \sum_{\mu=-M/2}^{M/2} \frac{1}{\sqrt{4\pi n^2/\sigma}} e^{\frac{-\sigma(\mu^2-\ell)^2}{4n^2/\sigma}} x(\mu+n) x^*(\mu-n) \qquad (8.24)$$

where $W(n)$ is a symmetrical window (such as Hamming), which has nonzero values on the interval $-L$ to L, and $W(\mu)$ is a uniform rectangular window that as a value of 1 for the range of $-M/2$ and $M/2$. The choices of N and M on these windows, respectively, determine the frequency resolution of the CWD and the range at which the function will be defined. Choi and Williams state that decreasing the size of $W(n)$ reduces the "oscillatory fluctuations of the cross terms," which at the same time decreases the frequency resolution of the distribution. In other words, there is a trade-off between the reduction of the cross terms and the frequency resolution obtained from the distribution.

When the above kernel function is expressed as

$$f_\ell(n) = x(\ell+n) x^*(\ell-n) w(n) w(-n) \qquad (8.25)$$

the reader will notice that the CWD contains an exponential term and introduces a new summation. The reader will also notice that the CWD kernel function is a series of Gaussian distributions. Barry points out that these distributions are aligned diagonally and that the mean and variance of each distribution is 1 and $2n^2/\sigma$, respectively.

As we show below, the discrete CWD can be modified to fit the standard DFT by setting $\omega = \pi k/2N$. Substituting this result into (8.23) and (8.25) above, and adding the window limits, we obtain [18]

$$\text{CWD}_x\left(\ell, \frac{\pi k}{2n}\right) = 2 \sum_{n=0}^{2N-1} S'(\ell, n) e^{-j2\pi kn/N} \qquad (8.26)$$

where the kernel function $S'(\ell,n)$ is defined as

$$S'(\ell,n) = \begin{cases} S(\ell,n), & 0 \leq n \leq N-1 \\ 0, & n = N \\ S(\ell,n-2N), & N+1 \leq n \leq 2N-1 \end{cases} \quad (8.27)$$

As an example, we consider an LPI radar is transmitting a low-power CW signal modulated by a Frank code. A Frank code is derived by a step approximation to a linear frequency modulation waveform using M frequency steps and M samples per frequency [19]. If i is the number of the sample in a given frequency and J is the number of the frequency, the phase of the ith sample of the j frequency is

$$\phi_{i,j} = \frac{2\pi}{M}(i-1)(j-1) \quad (8.28)$$

where $i = 1, 2, \ldots, M$ and $j = 1, 2, \ldots, M$. Consider a DRFM with an ADC sampling frequency of $f_s = 7$ kHz. It intercepts a Frank polyphase modulation with an SNR = 0 dB, $B = 1,000$ Hz with a code period of $T = 16$ ms and a carrier frequency of $f_c = 1,000$ kHz.

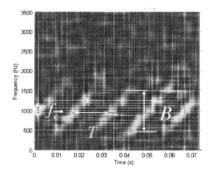

Figure 8.8 CWD for Frank code with $B = 1,000$ Hz, $T = 16$ ms, time-frequency plot for SNR = 0 dB.

Our DRFM receiver is open and intercepts the waveform shown in Figure 8.8 using the Choi-Williams distribution. Note that the bandwidth B can be measured as well as the code period T and carrier frequency f_c. Also note that the number of code periods (5) is also clearly evident.

8.3.3 The Wigner-Ville Distribution

The WVD, introduced by Wigner in 1932 as a phase representation in quantum statistical mechanics [20] and separately by Ville in 1948 addressing the question of a joint distribution function [21], simultaneously gives the representation of a signal in both time and frequency variables. The WVD has been noted as one of the more useful bilinear time-frequency analysis techniques for signal processing. The WVD has been shown to be able to detect LPI signals in [19]. The main objective is that by studying the results and correlating the signal parameters that are revealed, the user is able to learn to determine the presence of a particular LPI signal and to recognize the LPI modulation characteristics under various signal-to-noise ratios. It was also shown that the waveforms have similar time and frequency characteristics. Using the Wigner analysis tools, a DRFM can come close to having a processing gain near the LPI radar's matched filter processing gain.

8.3.4 Wigner-Ville Distribution Analysis

The WVD has been used in many fields of engineering. The WVD exhibits the highest signal energy concentration in the time-frequency plane for linearly modulated signals, but has drawbacks in the case of nonlinear frequency modulated signals. To improve the concentration where nonlinear modulations are present, various higher-order time frequency representations have been investigated. The WVD also contains interfering *cross terms* (or ghost terms) between every pair of signal components. The presence of the cross terms sometimes makes it difficult to determine the LPI modulation parameters visually. However, they can be useful for the autonomous classification of signals because they add additional information about the signals that actually improve the P_{cc} (probability of correct classification) [19].

A good review of bilinear transforms and their use in signal analysis is given in [22]. The influence that the cross-term interference has on the WVD is analyzed in [23]. The extension of the WVD to discrete time signals has been discussed in and a formulation to remove the cross terms has been reported in [24]. Below, we begin with the definition of the WVD, and then present a windowed version of the WVD, the pseudo WVD (PWVD) which is useful in the signal processing of the digital signals within the DRFM receiver.

8.3.5 Continuous WVD

The WVD of a continuous one-dimensional function (or input signal) $x(t)$ is given by

$$W_x(t, \omega) = \int_{-\infty}^{\infty} x\left(t + \frac{\tau}{2}\right) x^*\left(t - \frac{\tau}{2}\right) e^{-j\omega\tau} d\tau \tag{8.29}$$

where t is the time variable, ω is the angular frequency variable $(2\pi f)$, and the $*$ indicates a complex conjugate.

That is, the Wigner-Ville time-frequency distribution is based on (8.15) where the kernel function $\phi(\xi, \tau) = 1$. For multicomponent signals, the cross terms that are present in the Wigner-Ville distribution were demonstrated to be quite large. The cross terms cause interference that can obscure physically relevant components of the LPI signal's modulation.

The WVD is a three-dimensional function describing the amplitude of the signal as a function of time and frequency. Since the emitter modulations vary the compression of the waveform as a function of time, these types of time-frequency distributions give a higher probability of detecting the modulation parameters. The WVD can also be defined from the Fourier transform $X(\omega)$ of $x(t)$ by

$$W_X(\omega,t) = \frac{1}{2\pi} \int_{-\infty}^{\infty} X\left(\omega + \frac{\omega_0}{2}\right) X^*\left(\omega - \frac{\omega_0}{2}\right) e^{-j\omega_0 t} d\omega_0 \qquad (8.30)$$

From (8.29) and (8.30), the following relation is obtained:

$$W_x(t,\omega) = W_X(\omega,t) \qquad (8.31)$$

That is, the WVD of the spectra of a signal can be determined simply from that of the time functions by an interchange of the frequency and time variables. This shows the symmetry between space and frequency domain definitions.

Equation (8.29) implies that evaluation of the WVD is a noncausal operation. As such, this expression does not lend itself to real-time evaluation. This limitation is overcome by first applying the WVD analysis to a sampled time series $x(\ell)$, where ℓ is a discrete time index from $-\infty$ to ∞. The discrete WVD is defined as

$$W(\ell,\omega) = 2 \sum_{n=-\infty}^{\infty} x(\ell+n)x^*(\ell-n)e^{-j2\omega n} \qquad (8.32)$$

Windowing the data results in the pseudo-WVD and is defined by

$$W(\ell,\omega) = 2 \sum_{n=-N+1}^{N-1} x(\ell+n)x^*(\ell-n)w(n)w(-n)e^{-j2\omega n} \qquad (8.33)$$

where $w(n)$ is a length $2N-1$ real window function with $w(0) = 1$. Using $f_\ell(n)$ to represent the kernel function

$$f_\ell(n) = x(\ell+n)x^*(\ell-n)w(n)w(-n) \qquad (8.34)$$

the PWVD becomes

$$W(\ell,\omega) = 2 \sum_{n=-N+1}^{N-1} f_\ell(n)e^{-j2\omega n} \qquad (8.35)$$

The choice of N (usually a power of 2) greatly affects the computational cost as well as the time-frequency resolution, of the PWVD output. A large N gives a higher time-frequency resolution since it influences the frequency resolution in (8.35). When the continuous variable ω in (8.35) is sampled to produce a suitable form of the DFT, a larger N also gives more output samples, yielding a smoother result. The maximum value of N is limited by

$$N \leq \frac{M+1}{2} \qquad (8.36)$$

where M is the data length.

Once N is chosen, the kernel function can be generated. Since

$$f_\ell(n) = f_\ell^*(-n) \qquad (8.37)$$

only $f_\ell(n)$ needs to be computed for $n \geq 0$. A block diagram of the PWVD kernel generation for $N = 8$ is shown in Figure 8.9,

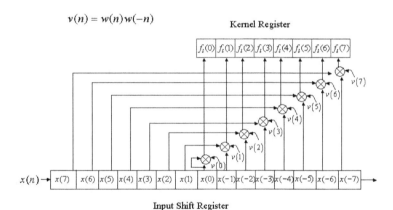

Figure 8.9 The computational structure for an $N = 8$ PWVD kernel generation[25]. (©1989 IEEE.)

where $v(n) = w(n)w(-n)$. Here the input signal enters the buffer register from the left and shifts to the right after each kernel generation. The right-most element is disposed after the next shift.

The PWVD can detect the presence of LPI signals, as well as extract the signal's modulation characteristics. For an intercept receiver, it is important that the computation be done in real time or near real time. From the PWVD expression in (8.35), we notice that it is computationally expensive to directly compute the PWVD.

Boashash et al. have presented an efficient algorithm to compute the discrete PWVD. Their algorithm is presented below [25].

To begin, the continuous frequency variable ω is sampled as

$$\omega = \frac{\pi k}{2N} \qquad (8.38)$$

where $k = 0, 1, 2, \cdots, 2N-1$ ($2N$ samples). The kernel indexes are modified to fit the standard DFT:

$$W\left(\ell, \frac{\pi k}{2N}\right) = 2 \sum_{n=-N+1}^{N-1} f_\ell(n) \exp\left(-\frac{j2\pi nk}{2N}\right) \qquad (8.39)$$

or

$$W\left(\ell, \frac{\pi k}{2N}\right) = 2 \sum_{n=0}^{2N-1} f'_\ell(n) \exp\left(-\frac{j2\pi nk}{2N}\right) \qquad (8.40)$$

where

$$f'_\ell(n) = \begin{cases} f_\ell(n), & 0 \leq n \leq N-1 \\ 0, & n = N \\ f_\ell(n-2N), & N+1 \leq n \leq 2N-1 \end{cases} \qquad (8.41)$$

Since the kernel is a symmetric function, the DFT of the kernel is always real. The resulting PWVD using $2N$ samples is

$$W(\ell, k) = 2 \sum_{n=0}^{2N-1} f'_\ell(n) \exp\left(-\frac{j\pi kn}{N}\right) \qquad (8.42)$$

Equation (8.42) is the algorithm implemented, and several examples are shown in the next section to illustrate the properties of the computation.

8.3.6 Example Calculation: Real Input Signal

Consider an example using a real input signal

$$x(\ell) = \{2, 4, 3, 6, 1, 7\} \qquad (8.43)$$

where $N = 3$ and the length of the input signal $x(\ell)$ is $2N = 6$. Here $\ell = -3, -2, -1, 0, 1, 2$ and is the discrete time index in the range $-N$ to $N-1$. Note that $x = 0$ for $\ell \leq -4$ or $\ell \geq 3$. From (8.41), with $N = 3$,

$$f'_\ell(n) = \begin{cases} f_\ell(n), & 0 \leq n \leq 2 \\ 0, & n = 3 \\ f_\ell(n-6), & 4 \leq n \leq 5 \end{cases} \qquad (8.44)$$

8.3 Detection Methods: Time-Frequency 519

From (8.34) $f_{-3}(n)$ ($\ell = -3$), for input signal $x(\ell)$ is computed as follows

$$\begin{aligned}
f_{-3}(n=0) &= x(-3)\cdot x^*(-3) = 2\cdot 2 = 4 \\
f_{-3}(n=1) &= x(-2)\cdot x^*(-4) = 4\cdot 0 = 0 \\
f_{-3}(n=2) &= x(-1)\cdot x^*(-5) = 3\cdot 0 = 0 \\
f_{-3}(n=3) &= 0 \\
f_{-3}(n=4) &= x(1)\cdot x^*(-7) = 1\cdot 0 = 0 \\
f_{-3}(n=5) &= x(2)\cdot x^*(-8) = 7\cdot 0 = 0
\end{aligned}$$

So, from (8.41), $f'_{-3} = \{4,0,0,0,0,0\}$. Similarly for f_0, ($\ell = 0$)

$$\begin{aligned}
f'_0(n=0) &= x(0)\cdot x^*(0) = 6\cdot 6 = 36 \\
f'_0(n=1) &= x(1)\cdot x^*(-1) = 1\cdot 3 = 3 \\
f'_0(n=2) &= x(2)\cdot x^*(-2) = 7\cdot 4 = 28 \\
f'_0(n=3) &= 0 \\
f'_0(n=4) &= x(-2)\cdot x^*(2) = 4\cdot 7 = 28 \\
f'_0(n=5) &= x(-1)\cdot x^*(1) = 3\cdot 1 = 3
\end{aligned}$$

and so, $f'_0 = \{36,3,28,0,28,3\}$. Repeating the above procedure, the kernel matrix for all values $\ell = -4$ to 3, and $n = 0$ to 5 is as shown in Figure 8.10.

	$n=0$	$n=1$	$n=2$	$n=3$	$n=4$	$n=5$
$l=3$	0	0	0	0	0	0
$l=2$	49	0	0	0	0	0
$l=1$	1	42	0	0	0	42
$l=0$	36	3	28	0	28	3
$l=-1$	9	24	2	0	2	24
$l=-2$	16	6	0	0	0	6
$l=-3$	4	0	0	0	0	0
$l=-4$	0	0	0	0	0	0

Figure 8.10 The kernel $f'_l(n)$ matrix for the real six input example.

The second step after the kernel transformation is to use (8.42) to calculate the Wigner distribution. As an example of the calculation, one can pick any ℓ and k to examine the values inside the PWVD matrix. For example, choose $\ell = 1$, $k = 2$, with

$N = 3$. The PWVD is

$$W(\ell = 1, k = 2) = 2 \sum_{n=0}^{2N-1} f'_\ell(n) \exp\left(-j\frac{\pi k n}{N}\right)$$

$$= 2 \sum_{n=0}^{2\cdot 3-1} f'_\ell(n) \exp\left(-j\frac{\pi 2 n}{3}\right)$$

$$= 2 \sum_{n=0}^{5} f'_\ell(n) \exp\left(-j\frac{2\pi n}{3}\right) \qquad (8.45)$$

From the kernel matrix in Figure 8.10, the kernel function for $\ell = 1$ is $f_1(n) = \{1, 42, 0, 0, 0, 42\}$. So from (8.45), the PWVD for $\ell = 1, k = 2$ (6 terms) is

$$W(1,2) = 2f'_1(0) \cdot \exp\left(-j\frac{2 \cdot \pi \cdot 0}{3}\right) + 2f'_1(1) \cdot \exp\left(-j\frac{2 \cdot \pi \cdot 1}{3}\right)$$

$$+ 2f'_1(2) \cdot \exp\left(-j\frac{2 \cdot \pi \cdot 2}{3}\right) + 2f'_1(3) \cdot \exp\left(-j\frac{2 \cdot \pi \cdot 3}{3}\right)$$

$$+ 2f'_1(4) \cdot \exp\left(-j\frac{2 \cdot \pi \cdot 4}{3}\right) + 2f'_1(5) \cdot \exp\left(-j\frac{2 \cdot \pi \cdot 5}{3}\right)$$

$$= 2 \cdot 1 \cdot (0) + 2 \cdot 42 \cdot (-0.5000 - 0.8660i)$$

$$+ 2 \cdot 0 + 2 \cdot 0 + 2 \cdot 0 + 2 \cdot 42 \cdot (0.5000 + 0.8660i)$$

$$W(1,2) = -82$$

Repeating the above procedure gives the PWVD matrix at each discrete time index $\ell = -4$ to 3 for each discrete frequency index $k = 0$ to 5. The result is a symmetric matrix about $k = 3$, as shown in Figure 8.11. An important feature of the PWVD is that all the components in the matrix are real.

l=3	0	0	0	0	0	0
l=2	98	98	98	98	98	98
l=1	170	86	-82	-166	-82	86
l=0	196	22	10	172	10	22
l=-1	122	62	-34	-70	-34	62
l=-2	56	44	20	8	20	44
l=-3	8	8	8	8	8	8
l=-4	0	0	0	0	0	0
	k=0	k=1	k=2	k=3	k=4	k=5

Symmetric

Figure 8.11 The PWVD matrix $W(\ell,k)$ for the real six input example.

8.3.7 Example Calculation: Complex Input Signal

To demonstrate the PWVD computation for a complex input, consider the signal

$$x = I + jQ \tag{8.46}$$

where

$$I = \cos(2\pi f_c t) \tag{8.47}$$

$$Q = \sin(2\pi f_c t) \tag{8.48}$$

If the carrier frequency $f_c = 1$ kHz, sampling frequency, $f_s = 7$ kHz, and $t \in \{0, 1/f_s, 2/f_s, \ldots, 7/f_s\}$, then the first eight input points for the discrete time index $\ell = -4$ to 3 is

$$\begin{aligned} x(\ell) &= \{1+0i,\ 0.62+0.78i,\ -0.22+0.97i,\ -0.90+0.43i, \\ &\quad -0.90-0.43i,\ -0.22-0.97i,\ 0.62-0.78i,\ 1+0i\} \end{aligned} \tag{8.49}$$

Consider the value when $\ell = 0$, $n = 3$. Using (8.41) with an input length $2N = 8$ or $N = 4$. The kernel is

$$f'_\ell = \begin{cases} f_\ell(n), & 0 \le n \le 3 \\ 0, & n = 4 \\ f_\ell(n-8), & 5 \le n \le 7 \end{cases} \tag{8.50}$$

or

$$f'_\ell(n) = \{f_\ell(1), f_\ell(2), f_\ell(3), 0, f_\ell(-3), f_\ell(-2), f_\ell(-1)\} \tag{8.51}$$

Since $f_\ell(n) = x(\ell+n) \cdot x^*(\ell-n)$, the kernel at $\ell = 0, n = 3$ is $f'_0(3) = x(3) \cdot x^*(-3) = 1 \cdot (0.6235 + 0.7818i)^* = 0.6235 - 0.7818i$. Repeating the same procedures as discussed in the real input case, the kernel matrix for the complex eight input example is shown in Figure 8.12.

n \ ℓ	0	1	2	3	4	5	6	7	
	1	0	0	0	0	0	0	0	3
	1	-0.2225 + 0.9749i	0	0	0	0	0	-0.2225 - 0.9749i	2
	0.9999	-0.2226 + 0.9749i	-0.9010 - 0.4339i	0	0	0	-0.9010 + 0.4339i	-0.2226 - 0.9749i	1
	1.0001	-0.2225 + 0.9749i	-0.9009 - 0.4339i	0.6235 - 0.7818i	0	0.6235 + 0.7818i	-0.9009 + 0.4339i	-0.2225 - 0.9749i	0
	1.0001	-0.2225 + 0.9749i	-0.9009 - 0.4339i	0.6235 - 0.7818i	0	0.6235 + 0.7818i	-0.9009 + 0.4339i	-0.2225 - 0.9749i	-1
	0.9999	-0.2226 + 0.9749i	-0.9010 - 0.4339i	0	0	0	-0.9010 + 0.4339i	-0.2226 - 0.9749i	-2
	1	-0.2225 + 0.9749i	0	0	0	0	0	-0.2225 - 0.9749i	-3
	1	0	0	0	0	0	0	0	-4

Figure 8.12 The kernel matrix for the complex eight input example.

Referring to Figure 8.12, we can calculate the PWVD when $\ell = -1$. The kernel is

$$f'_{-1}(n) = \{1.00, -0.22+0.97i, -0.90-0.43i, 0.62-0.78i,$$
$$0, 0.62+0.78, -0.90+0.43i, -0.22-0.97i\} \quad (8.52)$$

Consider the case when $\ell = -1, k = 4$. From (8.42), the PWVD for $N = 4$ is

$$W(\ell = -1, k = 4) = 2\sum_{n=0}^{2N-1} f'_\ell(n) \exp\left(-j\frac{\pi kn}{N}\right)$$

$$= 2\sum_{n=0}^{2 \cdot 4-1} f'_{-1}(n) \left(\frac{\pi 4n}{4}\right)$$

$$= 2\sum_{n=0}^{7} f'_{-1}(n) \exp(-jn\pi) \quad (8.53)$$

From (8.52) and (8.53)

$$W(\ell = -1, k = 4) = 2\sum_{n=0}^{7} f'_{-1}(n) \cdot \exp(-jn\pi) = -3.2073$$

8.3 Detection Methods: Time-Frequency

Again, the PWVD matrix of the complex eight input samples is real. The complete PWVD matrix is a symmetric $2N \times 2N$ matrix. Figure 8.13 shows the PWVD matrix of the complex eight input samples. Note this important feature: the PWVD is always real whether the input signal is real or complex.

								l
2.0000	2.0000	2.0000	2.0000	2.0000	2.0000	2.0000	2.0000	3
1.1099	4.1280	5.8995	5.3867	2.8899	-0.1282	-1.8997	-1.3868	2
-2.4943	2.3923	9.5036	7.1225	-0.7139	-1.8638	1.7041	0.3485	1
0.0004	-1.5882	12.6307	6.6749	-3.2073	2.1112	-1.4231	0.7965	0
0.0004	-1.5882	12.6307	6.6749	-3.2073	2.1112	-1.4231	0.7965	-1
-2.4943	2.3923	9.5036	7.1225	-0.7139	-1.8638	1.7041	0.3485	-2
1.1099	4.1280	5.8995	5.3867	2.8899	-0.1282	-1.8997	-1.3868	-3
2.0000	2.0000	2.0000	2.0000	2.0000	2.0000	2.0000	2.0000	-4
0	1	2	3	4	5	6	7	
			k					

Figure 8.13 The PWVD matrix for the complex eight input example.

Figure 8.14(a) shows a 3D mesh plot of the PWVD for the complex signal example with eight inputs. This plot shows the magnitude in both the time domain and the frequency domain. Note that it directly correlates with Figure 8.13. The peak corresponds to the 1-kHz carrier frequency. Figure 8.14(b) shows the corresponding PWVD contour plot. The contour plot is a 2D time-frequency plot that is useful for characterizing the time-frequency behavior of the signal. The magnitude is represented by a different gray scale, as shown in the legend bar.

To see the marginal details of the PWVD, Figure 8.15(a) shows a plot of the PWVD obtained by rotating the mesh plot in Figure 8.14(a) to show just the time axis with the eight samples. The time resolution is $1/f_s$. Figure 8.15(b) shows the marginal details in the frequency domain, and is obtained in the same manner as Figure 8.15(a). The carrier frequency is represented by the peak in this plot, and shows up at 900 Hz, very close to the real value 1 kHz. The frequency resolution $f_s/2/\#$ samples is also indicated.

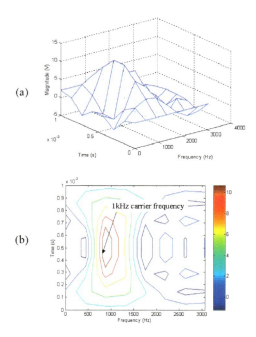

Figure 8.14 PWVD for the eight input complex example: (a) 3D mesh plot, and (b) time-frequency domain.

8.3 Detection Methods: Time-Frequency 525

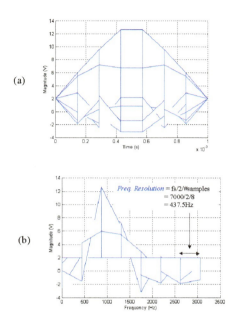

Figure 8.15 PWVD for the eight input complex example: (a) 2D mesh in time domain, and (b) 2D mesh in frequency domain.

In summary, both the real signal example and the complex signal example illustrate the mechanics of the PWVD calculation. The PWVD time-frequency results, when presented in the four different plots, give a variety of aspects so that the LPI signal and its modulation characteristics can be determined.

8.3.8 Two-Tone Input Signal Results

Now we consider the PWVD for a two-tone input (two carrier frequencies) with $f_{c1} = 1$ kHz and $f_{c2} = 2$ kHz. Now $I = \cos(2\pi f_{c1}t) + \cos(2\pi f_{c2}t)$ and $Q = \sin(2\pi f_{c1}t) + \sin(2\pi f_{c2}t)$. Figure 8.16(a) shows the PWVD results for the two-tone signal in a 3D time-frequency mesh plot. In this plot the cross terms are stronger than the signal terms and show up with many peaks. Figure 8.16(b) is the 2D PWVD time-frequency contour plot and shows the time dependence of the real signal and the cross terms.

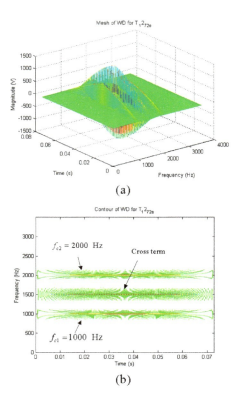

Figure 8.16 PWVD for the two-tone example, showing the (a) 3D time-frequency domain mesh plot, and (b) 2D time-frequency contour.

Figure 8.17(a) shows the marginal time domain plot obtained by rotating the 3D mesh plot in Figure 8.16(a), to show only the time axis. This reveals the cross terms as a series of positive and negative magnitude components in the time domain. Figure 8.17(b) shows the frequency domain plot obtained in the same manner, and reveals the two-carrier frequencies and the cross term. Note that the shape and magnitude of the cross term is not like the two-carrier frequency components, and can be easily identified.

8.3 Detection Methods: Time-Frequency   527

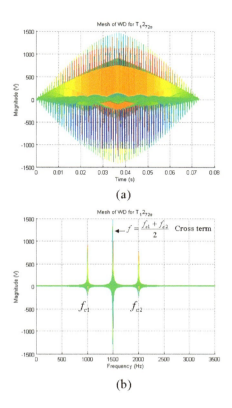

Figure 8.17 PWVD for the two-tone example showing the (a) marginal time domain plot, and (b) marginal frequency domain plot.

In summary, the PWVD, is able to compute the time frequency characteristics of an intercepted EMS and provides significant insight into the waveforms present within the signal. The drawbacks are that it contains cross terms among all frequency components contained within the signal. This however, may or may not be a disadvantage. For example, if autonomous classification is being considered, it adds a significant amount of energy at these frequencies and so can act as an advantage to actually increase the P_{cc}, doing better than for example, the Choi-Williams (without cross terms) [19].

Below we consider the quadrature mirror filter bank wavelet decomposition method of computing the time-frequency layers for signal detection.

8.4 WAVELET DECOMPOSITION USING QUADRATURE MIRRORS

Various methods of decomposing a waveform on the time-frequency plane have recently been investigated. The Wigner transform discussed in the previous section is an example of a bilinear transform.[1]

8.4.1 Wavelets and the Wavelet Transform

Wavelets are localized basis functions for time-frequency analysis of a signal. That is, the wavelet basis function is effectively nonzero for only a finite time interval and is designed to satisfy the orthonormality condition. Orthonormal and orthogonal are important considerations in wavelet transform theory and are briefly outlined next.

The basis functions are said to be orthogonal if

$$\int \Phi(t)\Phi(t-k) = E\delta(k) = \begin{cases} E & \text{if } k=0 \\ 0 & \text{otherwise} \end{cases} \quad (8.54)$$

where E stands for the energy of $\Phi(t)$. If $\Phi(t)$ is normalized by dividing by the square root of the energy $\sqrt{E}$, then the basis functions are said to be orthonormal defined by [19]

$$\int \Phi(t)\Phi(t-k)dt = \delta(k) = \begin{cases} 1 & \text{if } k=0 \\ 0 & \text{otherwise} \end{cases} \quad (8.55)$$

If the basis functions are orthonormal, there is no redundancy in the representation of the signal $f(t)$. If the signal is sampled at or above the Nyquist rate, all of the signal's information is retained. In this case, the time variable t in (8.54) and (8.55) can be considered to be discrete $t = nT$, where T is the sampling period and the integral should be replaced with summations.

From a signal processing point of view, a wavelet is a band-pass filter. In the time-frequency analysis, the wavelet filter occurs most often in pairs (a low-pass filter and a high-pass filter), and includes a resampling function that is coupled to the filter bandwidth as shown in the two-band analysis bank in Figure 8.19.

---

[1] Wigner transforms are called bilinear because the input waveform appears twice in the development of the transform. Better resolution occurs in the time-frequency plane than with linear techniques; however, the computational burden is greatly increased and the cross terms can be bothersome for some applications.

8.4 Wavelet Decomposition Using Quadrature Mirrors 529

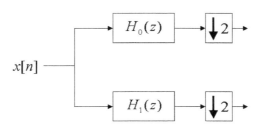

Figure 8.18 Two-band analysis bank.

Here, $H_0(z)$ is the high-pass filter and $H_1(z)$ is the low-pass filter. Like the design of conventional digital filters, the design of a wavelet filter can be accomplished by using a number of methods including weighted least squares, orthogonal matrix methods, nonlinear optimization, optimization of a single parameter (e.g., the passband edge), and a method that minimizes an objective function that bounds the out-of-tile energy.

A quadrature mirror filter (QMF) is an iteration of filter pairs with resampling to generate the wavelets. By varying the window used, resolution in time can be traded for resolution in frequency. To isolate discontinuities in signals, it is possible to use some basis functions which are very short, while longer ones are required to obtain a fine frequency analysis. One method to achieve this is to have short high-frequency basis functions, and long low-frequency basis functions. The Wavelet Transform (WT) makes this possible by obtaining the basis functions from a single prototype wavelet $h(t)$ using translation, dilation, and contraction as

$$h_{a,b}(t) = \frac{1}{\sqrt{a}} h\left(\frac{t-b}{a}\right) \tag{8.56}$$

where a is a positive real number and b is a real number. For large a, the basis function becomes a stretched version of the prototype wavelet (low-frequency function). For small a, the basis function becomes a contracted wavelet (short high-frequency function). The WT is defined as

$$X_W(a,b) = \frac{1}{\sqrt{a}} \int_{-\infty}^{\infty} h^*\left(\frac{t-b}{a}\right) x(t) dt \tag{8.57}$$

The WT divides the time-frequency plane into tiles. This is shown in Figure 8.19. Here, the area of each tile represents (approximately) the energy within the function (rectangular regions of the frequency plane).

530   MODERN SPECTRAL SENSING AND DETECTION

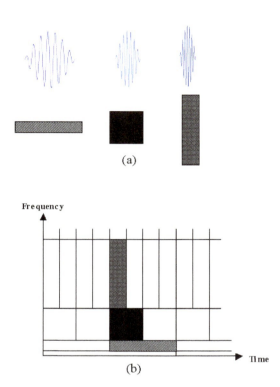

Figure 8.19 Wavelet shown in (a) the time domain and (b) wavelets shown in the time-frequency plane.

Note that not all of the signal's energy can be located in a single tile because it is impossible to concentrate the function's energy simultaneously in frequency and in time. The first wavelet in Figure 8.19(a) is longer in time than the others. Consequently, for the frequency characteristics it has the narrowest band (smallest frequency resolution) as shown in Figure 8.19(b). The WT can be interpreted as constant-Q filtering with a pair of subband filters (a low-pass filter and a high-pass filter), followed by a sampling at the respective Nyquist frequencies corresponding to the bandwidth of the particular subband of interest.

8.4.2 Wavelet Filters

FIR filters are the popular choice for the wavelet filter. To meet the requirements for a wavelet filter, the coefficients must ensure an orthogonal decomposition of the input signal, such that the energy at the input will equal the energy at the output from each filter pair. The filter pairs are designed to divide the input signal energy into two orthogonal components based on the frequency. The filter should also pass as much energy within its tile with a flat passband, and reject as much energy outside the tile as possible.

8.4.3 Sinc and Modified Sinc Filter

A function that tiles the energy perfectly in frequency would have a flat magnitude response across the passband, an infinitely narrow passband-to-stopband transition, and a zero across the stopband. From the time-domain description (inverse Fourier transform), the function is called a *sinc filter*. While it has an infinite number of coefficients, this condition can be modified by windowing. The sinc filter can be expressed as

$$\mathrm{sinc}(k) = \begin{cases} \frac{\sin(\pi k)}{\pi k} & k \neq 0 \\ 1 & k = 0 \end{cases} \qquad (8.58)$$

Since the passband ranges from $-\pi/2 < \omega < \pi/2$ or $-0.25 < f < 0.25$, the nulls of the sinc function will be at $2T$ for a sampling period of T [26]. To obtain the filter coefficients, the sinc function is sampled at the normalized sampling period of $T = 1$ for a situation similar to that shown in Figure 8.20.

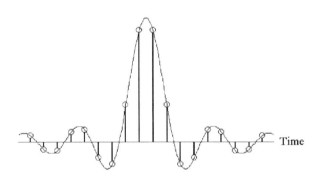

Figure 8.20 Sampled sinc filter impulse response.

One way to sample the function would be to let the main tap sample occur at the center of the main lobe. However, two main taps are needed, and their sum needs to be as large as possible. This occurs for the sinc function if both main tap samples are equally spaced about the center of the main lobe [26]. The sum of the square of the coefficients must also be unity, which is achieved by scaling the sinc by $1/\sqrt{2}$, giving

$$h(n) = \frac{1}{\sqrt{2}} \text{sinc}\left(\frac{n+0.5}{2}\right) \qquad (8.59)$$

where n is an integer.

This filter meets the criteria of wavelet filters. The only problem is that there is an infinite number of coefficients. A small amount of nonorthogonality will occur when this filter is truncated. Some cross correlation will also take place between both high-pass and low-pass filters. If the ends of the filter are simply truncated (a rectangular window in the time domain), some ripples in the passband of the frequency response will appear (Gibb's phenomena).

One solution is to use a nonrectangular window, and one whose Fourier transform has a narrower main lobe and smaller side lobes than the sinc function. The Hamming window is one that is commonly used. Multiplying the coefficients from (8.59) by this window, and using the results in an FIR filter, the frequency response needed is generated. Energy will be lost at the filter transitions, which is primarily the result of the loss of orthogonality from truncating the filter.

For detection, instead of losing the energy at those frequencies, a better trade-off would be a small amount of cross correlation between the filters so that some energy appears in more than one tile. To achieve this type of prototype filter, the impulse response can be modified to have a passband that is slightly greater than $\pi/2$. Thus, the lowpass and highpass filters are squeezed together slightly. This can be achieved by compressing the sinc envelope of (8.59) slightly. At the same time, it is desirable to rescale the coefficients slightly so the sum of the squares equals one. With these modifications, a *modified sinc filter* results as [26]

$$h(n) = \sqrt{\frac{S}{2}} \text{sinc}\left(\frac{n+0.5}{C}\right) w(n) \qquad (8.60)$$

where $-N/2 \leq n \leq (N-2)/2$, C is the compression variable, S is the scaling variable, N is the number of coefficients, and $w(n)$ is the Hamming window to suppress the Gibb's phenomena. For these filters, the greatest cross correlation occurs between tiles in the same frequency band and adjacent in time, when $N = 512$ (the number of coefficients), with values $C = 1.99375872328059$, $S = 1.00618488680080$, and a Hamming window with a cross correlation of less than 0.001 results.[2] Note the number of coefficients N can be changed using the MATLAB file tsinc_su.m.

[2] Personal communication between P. Jarpa and T. Farrell, March 20, 2002.

8.5 QMFB TREE RECEIVER

Orthogonal wavelet decomposition of the unknown signal can be implemented using QMFs, by designing filter pairs to divide the input signal energy into two orthogonal components, based on frequency. The tiles are used to refer to the rectangular regions of the time-frequency plane containing the basis function's energy. By arranging the QMF pairs in a fully developed tree structure, it is possible to decompose the waveform in such a way that the tiles have the same dimensions within each layer. Thus, every filter output is connected to a QMF pair in the next layer, as shown in Figure 8.21.

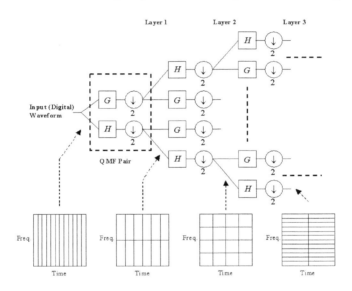

Figure 8.21 Quadrature mirror filter bank tree.

Each QMF pair divides the digital input waveform into its high-frequency and low-frequency components, with a transition centered at π. A normalized input of one sample per second is assumed, with a signal bandwidth of $[0, \pi]$. Since each filter output signal has half the bandwidth, only half the samples are required to meet the Nyquist criteria; therefore, these sequences are downsampled by two. The same number of output samples is returned. For example, if 100 samples appear at the input of the first QMF pair, 100 samples appear at the output. Each of the two resulting sequences is then fed into QMF pairs, forming the next layer, where the process is repeated, and so on down the tree.

Within the time-frequency plane, the WT is sharper in time at high frequencies. At low frequencies, the WT is sharper in frequency. That is, the tiles become shorter in time and occupy a larger frequency band as the frequency is increased. Since the WT is linear, there is a fundamental limit on the minimum area of these tiles. However, the nature of the QMFB configuration is such that each layer outputs a matrix of coefficients for tiles that are twice as long (in time) and half as tall (in frequency) as the tile in the previous layer. The outputs from each layer of the tree in Figure 8.21 form a matrix whose elements, when squared, approximately represent the energy contained in the tiles of the corresponding time-frequency diagrams shown in the figure.

8.5.1 Example Calculations

In this section two example calculations are shown for a complex input. A complex single-tone example is shown first, followed by a two-tone signal. These examples serve to demonstrate the different QMFB output layers and show the trade-off in time-frequency resolution as a function of the layer number being examined. The lower the layer number, the smaller (better) the resolution in time, and consequently the larger (poorer) the resolution in frequency. As the layer number gets larger, the resolution in time gets larger, and the resolution in frequency gets smaller.

8.5.1.1 Complex Single-Tone Signal

To demonstrate the results available from the QMFB signal processing, we again consider a complex, single-tone signal. The signal has a carrier frequency $f_c = 1$ kHz and is sampled by the ADC at a rate of 7 kHz. The results, shown in Figure 8.22, show layers 2, 3, and 4, respectively, in the time-frequency domain using gray scale plots. Figure 8.23(a, b), show layers 5 and 6.

8.5 QMFB Tree Receiver 535

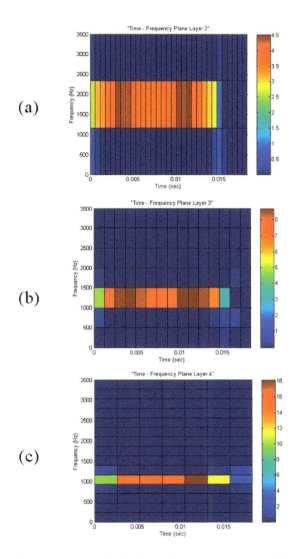

Figure 8.22 Time-frequency layers for the 1-kHz single-tone signal, showing (a) layer 2, (b) layer 3, and (c) layer 4.

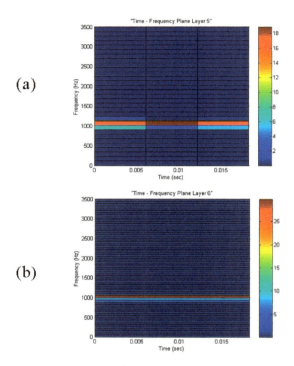

Figure 8.23 Time-frequency layers for the 1-kHz single-tone signal showing (a) layer 5, and (b) layer 6.

One of the important objectives of showing the five layers of the QMFB is to demonstrate how each layer results in a matrix of *energy values,* and the fact that the tiles are twice as long (in time) and half as tall (in frequency) as the tile in the previous layer. That is, as the layer number increases, the frequency resolution gets smaller and the time resolution gets larger. This adds quite a bit of flexibility to the analysis of nonstationary signals. Several different layers can be examined and compared, and the parameters of the signals can be extracted with high fidelity. Also, since the first and last layer in the QMFB are a single row of data, it is not useful to display them in a time-frequency format.

The input signal is zero padded with z zeros, such that the resulting number of data points is a power of 2. This resulting power of 2 is the number of layers L within the QMFB that result. That is, $N_p = 2^L$. The QMFB output resolution depends on the layer number. The frequency resolution of a layer l is

$$\Delta f = \frac{f_s}{2(2^l - 1)} = \frac{f_s}{2(N_F)} \qquad (8.61)$$

where N_F is the number of tiles displayed in frequency. For example, for layer 2 in Figure 8.22, $\Delta f = 7{,}000/2(3) = 1{,}166.67$ Hz. The resolution in time is determined by how many samples are integrated within the QMFB. For layer $l < L$

$$\Delta t = \frac{N_p}{f_s(2^{L-l}-1)} = \frac{N_p}{f_s N_T} \qquad (8.62)$$

where L is the total number of layers, and N_T is the number of tiles in time. Also, $N_p = 2^L$. For example, for layer 2 in Figure 8.22, $\Delta t = 128\ (1/7{,}000)/31 = 590\ \mu$s. Also note that the lower layers (e.g., layers 2 and 3) can be used to identify how many samples of the signal were collected (excluding zero padding). Since the sampling period for this example is $T = 0.143$ ms, from layer 2 we see that 105 samples were collected, and that 23 zeros were used to pad the signal.

Referring to layer $l = 6$ in Figure 8.23(b), the tiles have a frequency resolution of $\Delta f = 55.5556$ Hz and a resolution in time of $\Delta t = 18.286$ ms. Layer 6 shows the signal between 944.445 Hz and 1055.56 Hz, and from 0 to 18.286 ms. That is, we can say that $f_c \approx 1{,}000$ Hz with the accuracy limited by the tile resolution. Note that if more detailed time information is required, a lower layer could be examined.

8.5.1.2 Complex Two-Tone Signal

The second example consists of a signal with two frequencies, $f_{c1} = 1$ kHz and $f_{c2} = 2$ kHz, with a sampling frequency $f_s = 7$ kHz. In this example, a *contour plot* is used. Although the gray scale plot illustrated above quantifies the energy within each tile, the contour plot is useful for other types of information (such as time-domain characteristics), as illustrated in the results below. The number of signal samples collected, the time resolution Δt, and frequency resolution Δf for each layer within the QMFB are the same as for the single-tone example above. Figure 8.24 shows the contour plot for layers 2 through 4. Figure 8.25 shows the contour plot for layers 5 and 6. As before, layers 1 and 7 are not displayed, since they have only a single row vector of data.

538   MODERN SPECTRAL SENSING AND DETECTION

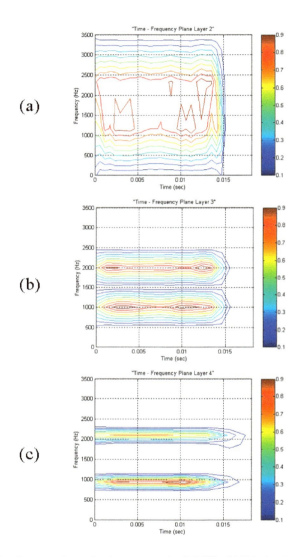

Figure 8.24 Time-frequency layers for the two-tone signal (1 kHz, 2 kHz), showing (a) layer 2, (b) layer 3, and (c) layer 4.

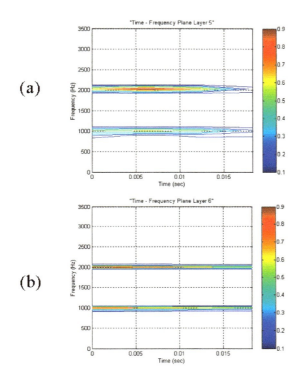

Figure 8.25 Time-frequency layers for the two-tone signal (1 kHz, 2 kHz), showing (a) layer 5, and (b) layer 6.

This example illustrates the important concepts that are evident using a contour image. First, for lower layers such as layer 2 and layer 3, the time domain characteristics of the signals can be clearly identified. In layer 2, the complex phase interaction in time, of the two signals within a single filter can also be identified. The high-frequency resolution layers such as layer 6 shown in Figure 8.25(b) reveal the frequencies contained in the input signal with a good amount of accuracy ($\Delta f = 55$ Hz).

The polyphase Frank modulation is examined next as an example. The polyphase modulation is applied both to the I and Q channels, which are 90 degrees out of phase. In this and the following sections, it is shown that the QMFB can be used to not only identify a particular type of phase modulation, but also to extract the important parameters of the signal.

The Frank phase code signal is generated with $N_c = 64$ ($M = 8$). The phase codes for $M = 8$ are shown in Chapter 5, in [19]. This is demonstrated in the QMFB $l = 2$ layer shown in Figure 8.26. The number of layers in this example is $L = 12$ ($N_p = 4,096$). For this layer $\Delta f = 1,166.67$ Hz, and $\Delta t = 571.99$ μs (small difference). In Figure 8.26(a), the additional 48 subcodes within a code period results in a longer duration signal. The five code periods have a total length of 320 ms. Figure 8.26(b) shows a close-up of the frequency characteristics within a code period.

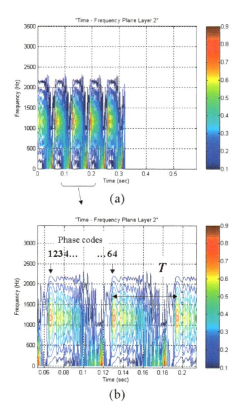

Figure 8.26 QMFB contour images for $M = 8$ Frank code with $B = 1,000$ Hz, $T = 64$ ms (signal only), showing (a) layer 2 output, and (b) close-up of layer 2, showing detailed frequency changes due to phase codes.

The linear frequency modulation characteristics are viewed in the QMFB $l = 5$ layer in Figure 8.27(a, b). Here, the bandwidth can be clearly identified, as well as the code period $T = 64$ ms. Note the wraparound characteristic within the bandwidth, similar to the 16-subcode example above. Correlation of the occurrence in time of the eight major energy concentrations within T, with the $M = 8$ Frank phase modulation waveform sections, can be easily made. The distribution of the signal energy within the nine frequency tiles within B helps in identifying the phase code, and in distinguishing between the modulation characteristics.

8.5 QMFB Tree Receiver 541

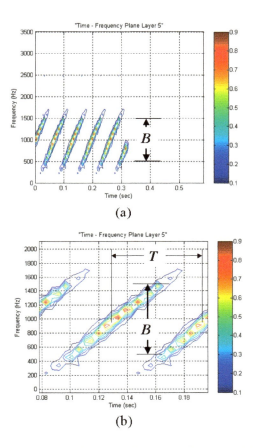

Figure 8.27 QMFB contour images for $M = 8$ Frank code with $B = 1,000$ Hz, $T = 64$ ms (signal only), showing (a) layer 5 output, and (b) close-up of layer 5, showing resulting linear frequency modulation.

The frequency characteristics for the $M = 8$ Frank code is shown in Figure 8.28. The energy is distributed about the carrier frequency in a Gaussian-type distribution, with the carrier frequency f_c centered about tile nine (the tile with the largest energy content). In fact, from Figure 8.28 the five largest energy tiles (in order from largest to smallest) are 9, 10, 7, 11, and 8. Figure 8.29 shows the $l = 6$ layer results for the $M = 8$ Frank code with the SNR $= 0$ dB. Note that the parameters can still be measured quite accurately. The QMFB results for the $N_c = 16$ Frank signal ($M = 4$) are given in Appendix K and the results for the P1, P2, P3, and P4 are given in Appendix L in [19].

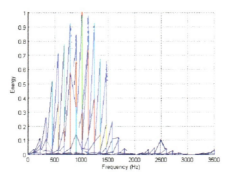

Figure 8.28 QMFB layer 5 frequency profile for $N_c = 64$ Frank code.

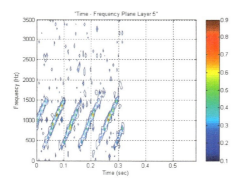

Figure 8.29 QMFB layer 6 contour image for Frank $N_c = 64$ signal with SNR = 0 dB.

The block diagram of a receiver that uses the QMFB structure is shown in Figure 8.30. A received waveform is bandpass filtered and sampled at the Nyquist rate. The digital sequence is then fed to the QMFB tree where it is decomposed. Matrices of values are output from each layer, and are then squared to produce numbers representing the energy in each tile.

8.5 QMFB Tree Receiver

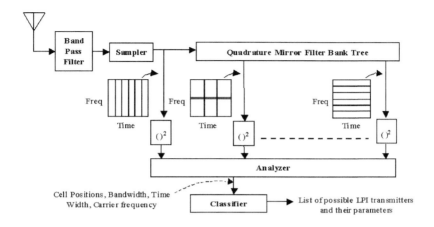

Figure 8.30 Quadrature mirror filter bank tree receiver.

Wavelet decomposition has been investigated as a tool for pattern recognition and target detection, and also as a means for identifying signal modulations. Wavelets have been used in many other types of signal identification techniques. One useful technique with dealing with very low SNRs is combining the ideas of compressive sensing (discussed in Chapter 4) in a projection-based wavelet denoising application [27]. This method consists of making orthogonal projections of wavelets (subbands) signals of the noisy signal onto an upside-down pyramid-shaped region in a multidimensional space. Each horizontal slice of the upside-down pyramid is a diamond-shaped region and it is called an l_1-ball[27]. The upside-down pyramid is called the epigraph set of the l_1−norm cost function and is shown to be a soft-thresholding denoise technique where the orthogonal projection operations automatically determine the soft-threshold values of the wavelet signals and is very efficient. Unfortunately, it is not possible to incorporate any prior knowledge about the noise PDF or any other statistical information. [27] shows results and the algorithm is outlined also.

To study and analyze the dynamic behavior of the earth's atmosphere, results showing the denoising of atmospheric mesosphere-stratosphere-troposphere (MST) radar signals using a *multiband discrete wavelet transform* are shown in [28]. The method is developed in order to clean the Doppler spectrum, and to improve detection and estimation of wind velocity parameters. The algorithm demonstrated via denoising comparisons on MST radar data with the existing method using the discrete wavelet transform (DWT). The data was collected from the National Atmospheric Research Laboratory NARL (Gadanki, Andhra Pradesh, India). When tested on the atmospheric data it was found that the algorithm produced better results than the existing method even in the presence of high degrees of noise (3-dB improvement at low SNR) [28].

8.6 CYCLOSTATIONARY SPECTRAL ANALYSIS

The Choi-Williams distribution, Wigner-Ville distribution, quadrature mirror filter bank processing, and atomic decomposition together give time-frequency results that allow the communication and radar signals to be detected and their parameters to be determined with good accuracy. This chapter presents an additional bifrequency spectral analysis technique, known as cyclostationary processing, that offers some additional capability in the detection and classification of intercepted modulations.

Cyclostationary spectral analysis is based on modeling the signal as a cyclostationary process rather than a stationary process. A signal is cyclostationary of order n if and only if one can find some nth order nonlinear transformation of the signal that will generate finite-strength additive sine wave components that result in spectral lines. For example, an $n = 2$ or quadratic transformation (like the product of the signal with a delayed version of itself, often used to detect BPSK signals) will generate spectral lines. That is, a signal $x(t)$ is cyclostationary with cycle frequency α, if and only if at least some of its delay product waveforms, $z(t) = x(t - \tau)x^*(t)$ for some delays τ, exhibit a spectral line at frequency α, and if and only if the time fluctuations in at least some pairs of spectral bands of $x(t)$, whose two center frequencies sum to α, are correlated. In contrast, for stationary signals, only a spectral line at frequency zero can be generated.

For signals with periodic features (e.g., LPI radar signals), the advantage of using a cyclostationary model is that nonzero correlation is exhibited between certain frequency components when their frequency separation is related to the periodicity of interest. Applications that use cyclostationary spectral analysis include time difference of arrival estimation, signal detection, identification, and parameter estimation.

8.6.1 Cyclic Autocorrelation

To discuss the cyclic autocorrelation, we begin with the definition of the correlation integral. The correlation integral is defined as

$$R_c(x) = \int_{-\infty}^{\infty} f(u)g(x+u)du \qquad (8.63)$$

Applying the FFT to both sides gives

$$\mathscr{F}\{R_c(x)\} = F(s)G^*(s) \qquad (8.64)$$

If $f(x)$ and $g(x)$ are the same function, the integral above is called the *autocorrelation* function and *cross correlation* if they differ. The autocorrelation function is a quadratic transformation of the signal, and may be interpreted as a measure of the predictability of the signal at time $t + \tau$ based on knowledge of the signal at time t. When considering a time series of length T, the autocorrelation function becomes the time-average autocorrelation function given by

$$R_x(\tau) \triangleq \lim_{T \to \infty} \frac{1}{T} \int_{-T/2}^{T/2} x\left(t + \frac{\tau}{2}\right) x^*\left(t - \frac{\tau}{2}\right) dt \tag{8.65}$$

The *cyclic autocorrelation* of a complex-valued time series $x(t)$ is then defined by

$$R_x^\alpha(\tau) \triangleq \lim_{T \to \infty} \frac{1}{T} \int_{-T/2}^{T/2} x\left(t + \frac{\tau}{2}\right) x^*\left(t - \frac{\tau}{2}\right) e^{-j2\pi\alpha t} dt \tag{8.66}$$

and can be interpreted as the Fourier coefficient of any additive sine wave component with frequency α that might be contained in the delay product (quadratic transformation) of $x(t)$. The nonzero correlation (second-order periodicity) characteristic of a time series $x(t)$ exists in the time domain if the cyclic autocorrelation function is not identically zero. That is, the signal $x(t)$ is said to be cyclostationary if $R_x^\alpha(\tau)$ does not equal zero at some time delay τ (any real number) and cycle frequency $\alpha \neq 0$.

8.6.2 Spectral Correlation Density

Recall that the power spectral density is defined as the Fourier transform of the autocorrelation function

$$S_x(f) = \int_{-\infty}^{\infty} R_x(\tau) e^{-j2\pi f \tau} d\tau \tag{8.67}$$

In the same manner, the spectral correlation density (SCD), or cyclic spectral density, is obtained from the Fourier transform of the cyclic autocorrelation function (8.66) as

$$S_x^\alpha(f) \triangleq \int_{-\infty}^{\infty} R_x^\alpha(\tau) e^{-j2\pi f \tau} d\tau = \lim_{T \to \infty} \frac{1}{T} X_T\left(f + \frac{\alpha}{2}\right) X_T^*\left(f - \frac{\alpha}{2}\right) \tag{8.68}$$

where α is the cycle frequency and

$$X_T(f) \triangleq \int_{-T/2}^{T/2} x(u) e^{-j2\pi f u} du \tag{8.69}$$

which is the Fourier transform of the time domain signal $x(u)$. The additional variable α (cycle frequency) leads to a two-dimensional representation $S_x^\alpha(f)$; namely, the bifrequency plane (f, α) [29]. Measurement of (8.66) and (8.68) in signal analysis constitutes what is referred to as *cyclic spectral analysis*.

Good insight is gained if we examine a second-order cyclostationary process and compare the time-domain implementation and the frequency-domain implementation. In Figure 8.31 it is shown that the time-domain implementation consists of a delay

(τ_i) and multiply operation, followed by the multiplication by the exponential cycle frequency term. The expected value then gives the cyclic autocorrelation function $\hat{R}_x^\alpha(\tau)$ and the subsequent FFT gives the spectral correlation density $\hat{S}_x^\alpha(f)$. With this perspective, it is easy to see that if the signal $x(t)$ contains a periodic component and the delay is chosen properly, a strong sinusoid will be present at the output.

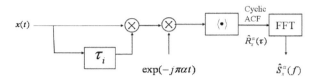

Figure 8.31 Time-domain implementation of a second-order cyclostationary process.

A frequency-domain implementation of a second-order cyclostationary process is shown in Figure 8.32. The input signal $x(t)$ with spectral representation $X(\nu)$ is split into two channels and multiplied by the two exponential factors that are a function of the cycle frequency and are complex conjugates of each other. The time-domain output signals are $u(t)$ and $s(t)$, which have spectral representations of $U(\nu)$ and $S(\nu)$, respectively. This time-domain multiplication results in a spectral shift of $u(t)$ by $-\alpha/2$ and a spectral shift of $s(t)$ by $\alpha/2$. Figure 8.33 shows the spectral representations $X(\nu)$, $U(\nu)$, and $S(\nu)$ and illustrates the narrowband spectral components of $x(t)$ being aligned at $\nu = f$. Both $u(t)$ and $s(t)$ are filtered with a band-pass filter with bandwidth B and center frequency f. Note that this captures the narrowband spectral components of $x(t)$ centered at $f + \alpha/2$ and $f - \alpha/2$. The Fourier transform is taken of both filter outputs and then the correlation of the two spectrums is computed. The expected value of the correlation output is then the spectral correlation density function $\hat{S}_x^\alpha(f)$.

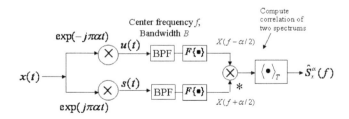

Figure 8.32 Frequency-domain implementation of a second-order cyclostationary process.

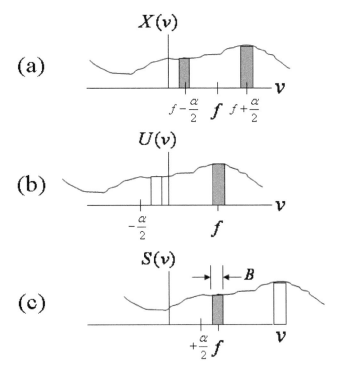

Figure 8.33 Frequency-domain representation of (a) $x(t)$ $[X(v)]$, (b) modulation of $x(t)$ by $-\alpha/2$ $[U(v)]$, and (c) modulation of $x(t)$ by $\alpha/2$ $[S(v)]$.

8.6.3 Spectral Correlation Density Estimation

Estimates of the cyclic spectral density or SCD can be obtained via time or frequency-smoothing techniques. Since the signals being analyzed are defined over a finite time interval Δt, the cyclic spectral density is only an estimate. An estimate of the SCD can be obtained by the time-smoothed cyclic periodogram given by

$$S_x^\alpha(f) \approx S_{xT_W}^\alpha(t,f)_{\Delta t} = \frac{1}{\Delta t}\int_{t-(\Delta t/2)}^{t+(\Delta t/2)} S_{xT_W}(u,f)du \qquad (8.70)$$

where

$$S_{xT_W}(u,f) = \frac{1}{T_W} X_{T_W}\left(u, f + \frac{\alpha}{2}\right) X_{T_W}^*\left(u, f - \frac{\alpha}{2}\right) \qquad (8.71)$$

with Δt being the total observation time of the signal, T_W is the short-time FFT window length, and

$$X_{T_W}(t,f) = \int_{t-(T_W/2)}^{t+(T_W/2)} x(u)e^{-j2\pi fu} du \qquad (8.72)$$

is the sliding short-time Fourier transform. Figure 8.34 shows the SCD estimation graphically for a signal $x(t)$. Here the frequency components are evaluated over a small time window T_W (sliding FFT time length) along the entire observation time interval Δt. The spectral components generated by each short-time Fourier transform have a resolution, $\Delta f = 1/T_W$. In Figure 8.34, L is the overlap (sliding) factor between each short-time FFT. In order to avoid aliasing and cycle leakage on the estimates, the value of L is defined as $L \leq T_W/4$ [29].

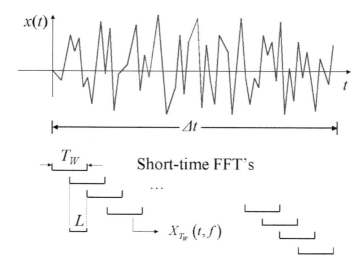

Figure 8.34 Cyclic spectral density estimation using short-time FFTs.

Figure 8.35 shows the spectral components of each short-time FFT being multiplied according to (8.71), providing the same resolution capability $\Delta f = 1/T_W$, for the cyclic spectrum estimates. Note that the dummy variable t has been replaced by the specific time instances $t_1 \ldots t_p$. Within each window (T_W), two frequency components centered about some f_0 and separated by some α_0 are multiplied together, and the resulting sequence of products is then integrated over the total time (Δt), as shown in (8.70).

8.6 Cyclostationary Spectral Analysis

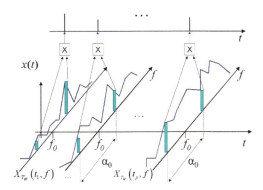

Figure 8.35 Sequence of frequency products for each short-time Fourier transform.

The estimation $S_x^\alpha(f) \approx S_{xT_W}^\alpha(t,f)_{\Delta t}$ can be made as reliable and accurate as desired for any given t and Δf, and for all f by making Δt sufficiently large, provided that (8.66) exists within the interval Δt and that a substantial amount of smoothing is carried out over Δt. This leads to the *Grenander's uncertainty condition* $\Delta t \Delta f \gg 1$ [29]. indexcyclostationary analysis!Grenander's uncertainty This uncertainty condition means that the observation time (Δt) must greatly exceed the time window (T_W) that is used to compute the spectral components. A data taper window is also used to minimize the effects of cycle and spectral leakage (estimation noise) introduced by frequency component side lobes [14]. The spectral components obtained from the short-time FFT have a resolution of

$$\Delta f = \frac{1}{T_W} \qquad (8.73)$$

The cycle frequency resolution of the estimate is related to the total observation time by

$$\Delta \alpha = \frac{1}{\Delta t} \qquad (8.74)$$

The estimation of some (f_0, α_0) represents a very small area on the bifrequency plane, as shown in Figure 8.36, and, since one needs a significant number of estimates to represent the cyclic spectrum adequately, it follows that obtaining estimates becomes very computationally demanding, and efficient algorithms are required.

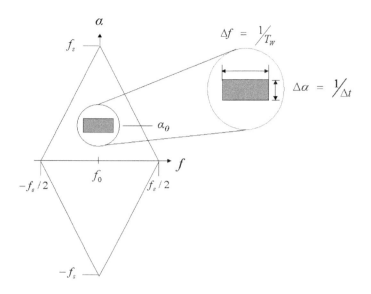

Figure 8.36 Frequency and cycle frequency resolutions on the bifrequency plane.

8.7 DISCRETE TIME CYCLOSTATIONARY ALGORITHMS

Cyclostationary signal processing can be used to extract the parameters from the sampled LPI signals in an intercept receiver when moderate to large amounts of additive noise are present. With the signal displayed on the bifrequency plane (frequency-cycle frequency), the intercept receiver or operator can examine and compare the modulation characteristics using several algorithms that estimate the SCD. Computationally efficient algorithms for implementation of time- and frequency-smoothing techniques are discussed in [16]. These are the FFT accumulation method (FAM), a time-smoothing algorithm, and the direct frequency-smoothing method (DFSM), a frequency-smoothing algorithm, as described below. The temporal and spectral smoothing equivalence is also addressed in [29].

8.7.1 The Time-Smoothing FFT Accumulation Method

The time-smoothing FFT accumulation method was developed to reduce the number of computations required to estimate the cyclic spectrum. This technique divides the bifrequency plane into smaller regions called *channel pair* regions, and computes the estimates one block at a time using the fast Fourier transform. Describing the estimated

8.7 Discrete Time Cyclostationary Algorithms

time-smoothed periodogram from (8.70) and (8.71), in discrete terms, yields

$$S^{\gamma}_{X_{N'}}(n,k) = \frac{1}{N} \sum_{n=0}^{N-1} \left[\frac{1}{N'} X_{N'}\left(n, k+\frac{\gamma}{2}\right) X^{*}_{N'}\left(n, k-\frac{\gamma}{2}\right) \right] \qquad (8.75)$$

where

$$X_{N'}(n,k) \triangleq \sum_{n=0}^{N'-1} w(n) x(n) e^{-(j2\pi kn)/N'} \qquad (8.76)$$

is the discrete Fourier transform of $x(n)$, $w(n)$ is the data taper window (e.g., Hamming window), and the discrete equivalents of f and α are k and γ, respectively. A block diagram of the FFT accumulation method is shown in Figure 8.37.

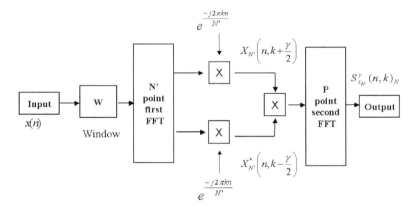

Figure 8.37 Block diagram of the FAM. (Adapted from [29].)

The algorithm consists of three basic stages: computation of the complex demodulates (divided into data tapering, sliding N' point Fourier transform, and baseband frequency translation sections), computation of the product sequences, and smoothing of the product sequences. Table 8.1 shows the relationship between the variables in (8.70), (8.71), and (8.75). The parameter N represents the total number of discrete samples within the observation time, and N' represents the number of points within the discrete short-time (sliding) FFT. In the FAM algorithm, spectral components of a sequence, $x(n)$, are computed using (8.76). Two components are multiplied (8.75) to provide a sample of a cyclic spectrum estimate representing the finite channel pair region on the bifrequency plane, as shown in Figure 8.38. There are N^2 channel pair regions in the bifrequency plane. Note the 16 small channel pair regions corresponding to a value of $M = 4$ in Figure 8.38.

Table 8.1
Comparison of Continuous and Discrete Time

Name	Continuous Time	Discrete Time
SCD	$S_{X_{T_W}}^{\alpha}(t,f)_{\Delta t}$	$S_{X_{N'}}^{\gamma}(n,k)_N$
Short FFT size	T_W	N'
Observation time	Δt	N
Time	t	n
Frequency	f	k
Cycle frequency	α	γ
Grenander's Uncertainty Condition	$M = (\Delta f/\Delta\alpha) \gg 1$	$M = (N/N') \gg 1$

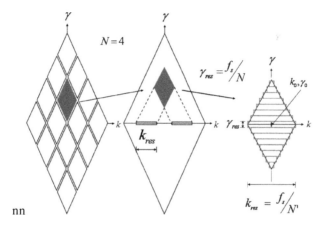

Figure 8.38 Channel pair regions within the bifrequency plane. (Adapted from [29])

A sequence of samples for any particular area may be obtained by multiplying the same two components of a series of consecutive short-time sliding FFTs along the entire length of the input sequence. After the channelization performed by an N'-point FFT sliding over the data with an overlap of L samples, the outputs of the FFTs are shifted in frequency in order to obtain the complex demodulate sequences (see Figure 8.37). Instead of computing an *average* of the product of sequences between the complex demodulates, as in (8.70), they are Fourier-transformed with a P-point (second) FFT. The computational efficiency of the algorithm is improved by a factor of L, since only N/L samples are processed for each point estimate. With f_s the sampling frequency, the cycle frequency resolution of the decimated algorithm is defined as $\gamma_{\text{res}} = f_s/N$ (compare to $\Delta\alpha = 1/\Delta t$), the frequency resolution is $k_{\text{res}} = f_s/N'$ (compare to $\Delta f = 1/T_W$), and the Grenander's uncertainty condition is $M = N/N' \gg 1$ (compare to $\Delta t \Delta f \gg 1$).

8.7 Discrete Time Cyclostationary Algorithms

Figure 8.39 reveals that the estimates toward the top and the bottom (shaded areas) of the channel pair region do not satisfy the uncertainty condition.

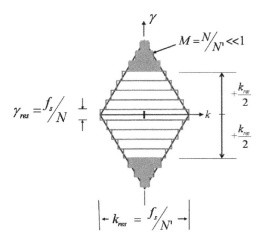

Figure 8.39 Cycle frequency and frequency resolutions of the grenander's uncertainty condition.

In order to minimize the variability of these point estimates, we can retain only those cyclic spectrum components that are within $\gamma = \pm k_{res}/2$ from the center of the channel pair region. A solution to resolve the entire area of the channel pair region without leaving gaps is to apply a data taper window on the frequency axis (such as a Hamming window) to obtain better coverage.

8.7.2 Direct Frequency-Smoothing Method

Direct frequency-smoothing algorithms first compute the spectral components of the data, and then execute spectral-correlation operations directly on the spectral components. Generally, the direct frequency-smoothing method is computationally superior to indirect algorithms that use related quantities such as the Wigner-Ville Distribution, but DFSM is normally less efficient than a time-smoothing approach [13].

The basis for the DFSM is the discrete time frequency-smoothed cyclic periodogram represented by

$$S^{\gamma}_{X_N}(n,k)_{\Delta k} = \frac{1}{N} \sum_{n=0}^{N-1} X_N\left(n, k + \frac{\gamma}{2}\right) X_N^*\left(n, k - \frac{\gamma}{2}\right) \qquad (8.77)$$

where

$$X_N(n,k) \triangleq \sum_{n=0}^{N-1} w(n) x(n) e^{-(j2\pi kn)/N} \qquad (8.78)$$

is the discrete Fourier transform of $x(n)$, $w(n)$ is the rectangular window of length N that is the total number of points of the FFT related to the total observation time, Δt, γ is the cycle frequency discrete equivalent, the frequency-smoothed ranges over the interval $|m| \leq M/2$, and $\Delta k \approx M \cdot f_s/N$ is the frequency resolution discrete equivalent. The block diagram in Figure 8.40 illustrates the implementation of the DFSM.

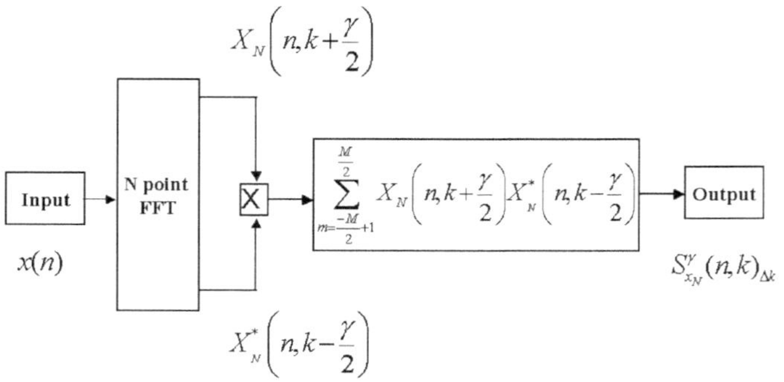

Figure 8.40 Block diagram of the direct frequency-smoothing algorithm. (Adapted from [29].)

In order to provide full coverage of the bifrequency plane with minimal computational expense, (8.77) is computed along a line of constant cycle frequency, thus spacing the point estimates by $\Delta k = M \cdot f_s/N$. This method is easier to implement, and is generally used to validate the time-smoothing approach, but may become more computationally demanding. This is especially true in the last block in which the complex demodulate product sequences are summed. Finally, we note that combinations of both time-smoothing and frequency-smoothing methods may also be advantageous for certain applications.

8.7.3 Test Signals

In order to gain an understanding of how the signals appear on the bifrequency plane, this section examines several test signals used in previous sections. The first test signal examined is a tone composed of a single carrier frequency with $f_c = 1$ kHz, and is sampled with sampling frequency $f_s = 7$ kHz. The *time-smoothing* technique to estimate the SCD is demonstrated first using the real part of the input signal. Figure 8.41 shows the time-smoothing SCD results. Figure 8.41(a) shows the bifrequency plane and reveals that the signal's frequency shows up at four separate locations. The (γ, k) frequency pairs are $(-2f_c, 0), (0, f_c), (0, -f_c)$, and $(2f_c, 0)$. Figure 8.41(b) details a close-up of the time-smoothing estimation characteristics for the signal outlined in the box in Figure 8.41(a). For these results the frequency resolution is $\Delta k = 128$ Hz. With the Grenander's uncertainty value of $M = 2$, the cycle frequency $\Delta \gamma = 64$ Hz. The overlap parameter is fixed at $L = 4$. The number of points in the first FFT N' is the next largest power of 2 value of $f_s/\Delta k$ or $N' = 64$. The number of points in the second FFT P is the next largest power of 2 value of $4f_s/\Delta N'$ or $P = 8$. The total number of signal samples integrated into the SCD are $N = PL = 128$. Note that in Figure 8.41(b), the $\gamma = 2f_c$ cycle frequency position lies at the resolved signal's centroid.

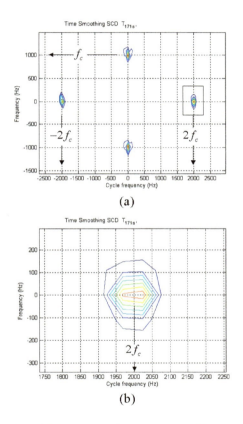

Figure 8.41 Time-smoothing SCD for a single frequency $f_c = 1$ kHz tone, showing the (a) bifrequency plane, and (b) close-up of the time-smoothing estimation characteristics.

The frequency-smoothing SCD results for the single-tone signal are shown in Figure 8.42. Figure 8.42(a) shows the bifrequency plane, and Figure 8.42(b) details a close-up of the frequency-smoothing estimation characteristics. The results serve to demonstrate the differences between the time-smoothing and frequency-smoothing techniques for estimating the SCD. For the frequency-smoothing results, $\Delta k = 128$ Hz. The number of samples integrated into the FFT is the next largest power of 2 value of $f_sM/\Delta f = 109$ or $N = 128$. Note in Figure 8.42(b), the $\gamma = 2f_c$ cycle frequency position does not lie at the resolved signal's centroid.

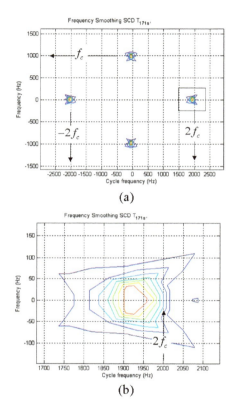

Figure 8.42 Frequency-smoothing SCD for a single frequency $f_c = 1$ kHz-tone, showing the (a) bifrequency plane, and (b) close-up of the frequency-smoothing estimation characteristics.

Next, the time-smoothing technique is used to estimate the SCD of a two-tone signal ($f_{c1} = 1$ kHz and $f_{c2} = 2$ kHz). Figure 8.43 shows the time-smoothing SCD results. Figure 8.43(a) shows the bifrequency plane and reveals that the two tones show up in the four separate quadrants along with the cross terms. Figure 8.43(b) details a close-up of the time-smoothing estimation characteristics for the signal outlined in the box in Figure 8.43(a). For these results, the frequency resolution $\Delta k = 128$ Hz, the Grenander's uncertainty value $M = 2$, and the cycle frequency $\Delta \gamma = 64$ Hz. Also, N', N, and P are the same as for the single-tone signal.

8.7 Discrete Time Cyclostationary Algorithms 557

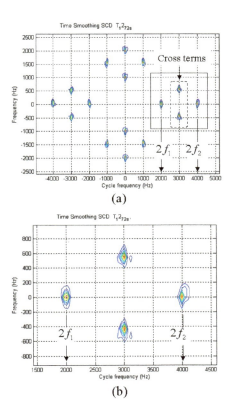

Figure 8.43 Time-smoothing SCD for a two-tone signal ($f_{c1} = 1$ kHz, $f_{c2} = 2$ kHz), showing (a) the bifrequency plane, and (b) a close-up of the time-smoothing estimation characteristics.

The frequency-smoothing SCD results for the two-tone signal are shown in Figure 8.44. Figure 8.44(a) shows the bifrequency plane and Figure 8.44(b) details a close-up of the frequency-smoothing estimation characteristics including the cross terms. As for the single-tone results, $\Delta k = 128$ Hz and $N = 128$. Note that in Figure 8.44(b), the $\gamma = 2f_{ci}$ cycle frequency positions do not lie at the signal centroids.

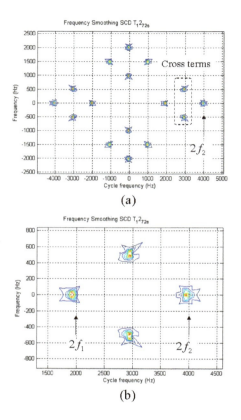

Figure 8.44 Frequency-smoothing SCD for a two-tone signal ($f_{c1} = 1$ kHz, $f_{c2} = 2$ kHz) showing (a) the bifrequency plane, and (b) a close-up of the frequency-smoothing estimation characteristics.

8.8 ATOMIC DECOMPOSITION

The time-frequency and time-scale communities have recently developed a large number of overcomplete waveform dictionaries—stationary wavelets, wavelet packets, cosine packets, chirplets, and warplets, to name a few. Decomposition into overcomplete systems is not unique, and several methods for decomposition have been proposed, including the method of frames (MOF), matching pursuit (MP), and, for special dictionaries, the best orthogonal basis (BOB) [30].

The objective of atomic decomposition is to represent the sampled signal vector from the receiver for analysis **x** as the weighted sum of a set of elementary functions called *atoms*

$$\mathbf{x} = \sum_p \hat{b}_p \mathbf{h}_{\hat{\gamma}_p} \qquad (8.79)$$

For radar detection and signal intercept by the DRFM transceiver where sensitivity is critical, atomic decomposition presents a significant result especially when appropriate dictionaries are available (e.g., for detecting chirplet families and also for recognizing signal modulations) [31]. It has also been shown to give excellent results in ISAR and SAR [32].

The atoms in (8.79) are estimated through an iterative procedure maximizing, at each iteration, the inner product of a residual signal and the atoms of the dictionary [22]

$$\hat{\gamma}_p = \arg_\gamma \max |\mathbf{h}_\gamma^H \mathbf{x}_{p-1}|^2 \tag{8.80}$$

$$\hat{b}_p = \mathbf{h}_\gamma^H \mathbf{x}_{p-1} \tag{8.81}$$

where the superscript H indicates complex conjugate transposition and $\mathbf{x}_p$ denotes the residual signal after the pth iteration [31]. This residual signal is updated for the next iteration by subtracting its *orthogonal projection* onto the estimated atom as

$$\mathbf{x}_p = \begin{cases} \mathbf{x}_{p-1} - \hat{b}_p \mathbf{h}_{\hat{\gamma}_p} & \text{for } p = 1, 2, \ldots \\ \mathbf{x} & \text{for } p = 0 \end{cases} \tag{8.82}$$

Due to the iterative and sequential character of the estimation of the atoms, the atomic decomposition is a greedy algorithm and, consequently, can estimate more energetic atoms than any of the signal components when these components are close in the time-frequency domain [33]. This greediness reduces the time-frequency resolution but it can represent an advantage for interception purposes.

The implementation of atomic decomposition (8.80) can be done in several ways. Atomic decompostion estimates each atom of the expansion by means of an optimization procedure. The objective function in this equation can present numerous local maxima and extensive plane regions [31]. These methods include exhaustive-search genetic algorithm, time-frequency, enhanced time-frequency and refinement algorithms and fast atomic decomposition. They define the sensitivity as the input SNR value required in order to obtain a detection probability $P_D = 90\%$ for a false alarm probability $P_{FA} = 10^{-6}$ and results are shown in [31] for each method.

8.9 SUMMARY

In this chapter, persistent spectrum sensing using centralized or decentralized methods were discussed. Sequential and quickest spectrum detection rules using both single and multiple DRFM architectures were emphasized. In addition, several methods were given both time-frequency and bifrequency to detect the signals of interest giving good subnoise visibility. These include the Choi-Williams, and Wigner-Ville time-frequency distributions, the quadrature mirror filter bank approach and the cyclostationary method of spectral analysis. Software for many of these methods can be found in [19].

In the next chapter, the inclusion of artificial intelligence is emphasized for the optimization of spectral dominance and automatic modulation classification using deep learning and machine learning. Deep learning concepts include auto encoders, convolutional neural networks, deep belief networks, recurrent neural networks and generative adversarial networks. Transfer learning is also discussed.

REFERENCES

[1] J. Lynch, "US Army turns to AI to prevent cyberattacks," *Defense News*, no. 1, p. 23, 2019.

[2] J. Lunden, V. Koivunen, and H. V. Poor, "Spectrum exploration and exploitation for cognitive radio: Recent advances," *IEEE Signal Processing Magazine*, vol. 32, no. 3, pp. 123–140, 2015.

[3] J. Lunden, V. Koivunen, A. Huttunen, and H. V. Poor, "Collaborative cyclostationary spectrum sensing for cognitive radio systems," *IEEE Transactions on Signal Processing*, vol. 57, no. 11, pp. 4182–4195, 2009.

[4] S. Chaudhari, J. Lunden, V. Koivunen, and H. V. Poor, "Cooperative sensing with imperfect reporting channels: Hard decisions or soft decisions?" *IEEE Transactions on Signal Processing*, vol. 60, no. 1, pp. 18–28, 2012.

[5] A. Wald and J. Wolfowitz, "Optimum character of the sequential probability ratio test," *The Annals of Mathematical Statistics*, vol. 19, no. 3, pp. 326–339, 1948. [Online]. Available: http://www.jstor.org/stable/2235638

[6] J. Oksanen, J. Lundén, and V. Koivunen, "Reinforcement learning based sensing policy optimization for energy efficient cognitive radio networks," *Neurocomputing*, vol. 80, pp. 102 – 110, 2012, special Issue on Machine Learning for Signal Processing 2010. [Online]. Available: http://www.sciencedirect.com/science/article/pii/S092523121100600X

[7] R. S. Sutton and A. G. Barto, *Reinforcement Learning, An Introduction*. Cambridge, MA: MIT Press, 1998.

[8] P. Whittle, "Restless bandits: Activity allocation in a changing world," *Journal of applied probability*, pp. 287–298, 1988.

[9] J. C. Gittins and D. M. Jones, "A dynamic allocation index for the sequential design of experiments," *Progress in Statistics*, pp. 241–266, 1974. [Online]. Available: https://ci.nii.ac.jp/naid/10027289393/en/

[10] M. Malekipirbazari and O. Cavus, "Risk-averse allocation indices for multi-armed bandit problem," *IEEE Transactions on Automatic Control*, p. 1, 2021.

[11] P. Whittle, "Restless bandits: Activity allocation in a changing world," *Journal of applied probability*, pp. 287–298, 1988.

[12] K. Wang, L. Chen, Q. Liu, and K. Al Agha, "On optimality of myopic sensing policy with imperfect sensing in multi-channel opportunistic access," *IEEE Transactions on Communications*, vol. 61, no. 9, pp. 3854–3862, 2013.

[13] Q. Zhao, B. Krishnamachari, and K. Liu, "On myopic sensing for multi-channel opportunistic access: structure, optimality, and performance," *IEEE Transactions on Wireless Communications*, vol. 7, no. 12, pp. 5431–5440, 2008.

[14] L. Cohen, "Time-frequency distributions-A review," *Proceedings of the IEEE*, vol. 77, no. 7, pp. 941–981, 1989.

[15] J. Gosme, C. Richard, and P. Goncalves, "Adaptive diffusion as a versatile tool for time-frequency and time-scale representations processing: a review," *IEEE Transactions on Signal Processing*, vol. 53, no. 11, pp. 4136–4146, 2005.

[16] P. Perona and J. Malik, "Scale-space and edge detection using anisotropic diffusion," *IEEE Transactions on Pattern Analysis and Machine Intelligence*, vol. 12, no. 7, pp. 629–639, 1990.

[17] H. Choi and W. J. Williams, "Improved time-frequency representation of multicomponent signals using exponential kernels," *IEEE Transactions on Acoustics, Speech, and Signal Processing*, vol. 37, no. 6, pp. 862–871, 1989.

[18] D. T. Barry, "Fast calculation of the Choi-Williams time-frequency distribution," *IEEE Transactions on Signal Processing*, vol. 40, no. 2, pp. 450–455, 1992.

[19] P. E. Pace, *Detecting and Classifying Low Probability of Intercept Radar*. Norwood, MA: Artech House, 2009.

[20] E. P. Wigner, "On the quantum correction for thermodynamic equilibrium," *Physics Review*, vol. 40, pp. 749–759, 1932.

[21] J. Ville, "Theorie et applications de la notion de signal analytique," *Cables et Transmission*, vol. 2A, pp. 61–74, 1948.

[22] V. C. Chen and H. Ling, *Time Frequency Transforms for Radar Imaging and Signal Analysis*. Norwood, MA: Artech House, 2002.

[23] L. Stankovic and S. Stankovic, "On the wigner distribution of discrete-time noisy signals with application to the study of quantization effects," *IEEE Transactions on Signal Processing*, vol. 42, no. 7, pp. 1863–1867, 1994.

[24] S. Kadambe and T. Adali, "Application of cross-term deleted wigner representation (cdwr) for sonar target detection/classification," in *Conference Record of Thirty-Second Asilomar Conference on Signals, Systems and Computers (Cat. No.98CH36284)*, vol. 1, 1998, pp. 822–826.

[25] B. Boashash and P. Black, "An efficient real-time implementation of the Wigner-Ville distribution," *IEEE Transactions on Acoustics, Speech, and Signal Processing*, vol. 35, no. 11, pp. 1611–1618, 1987.

[26] T. Farrell and G. Prescott, "A nine-tile algorithm for LPI signal detection using qmf filter bank trees," *Proceedings of MILCOM '96 IEEE Military Communications Conference*, vol. 3, pp. 974–978, 1996.

[27] A. E. Cetin and M. Tofighi, "Projection-based wavelet denoising [lecture notes]," *IEEE Signal Processing Magazine*, vol. 32, no. 5, pp. 120–124, 2015.

[28] G. Chandraiah and T. Sreenivasulu Reddy, "Denoising of MST radar signal using multi-band wavelet transform with improved thresholding," in *2018 Second International Conference on Inventive Communication and Computational Technologies (ICICCT)*, 2018, pp. 1026–1030.

[29] W. A. Gardner, *Statistical Spectral Analysis–A Nonprobabilistic Theory*. Cambridge, MA: Prentice Hall, 1989.

[30] S. S. Chen, D. L. Donoho, and M. A. Saunders, "Atomic decomposition by basis pursuit," *SIAM Review*, vol. 43, no. 1, pp. 129–159, 2001. [Online]. Available: https://doi.org/10.1137/S003614450037906X=

[31] O. A. Yeste-Ojeda, J. Grajal, and G. Lopez-Risueno, "Atomic decomposition for radar applications," *IEEE Transactions on Aerospace and Electronic Systems*, vol. 44, no. 1, pp. 187–200, 2008.

[32] Y. Gao, K. Wang, X. Liu, and W. Yu, "Atomic decomposition-based sar imaging technique," in *2011 IEEE International Geoscience and Remote Sensing Symposium*, 2011, pp. 629–631.

[33] G. López-Risueño, "Atomic decomposition-based radar complex signal interception," *IET Proceedings - Radar, Sonar and Navigation*, vol. 150, pp. 323–331(8), August 2003.

Chapter 9

Machine Learning in Electromagnetic Warfare

Unmanned platforms (space, air, surface, subsurface) and AI are now commanding center stage in any discussion of future military requirements and objectives. Current (TTPs) must now incorporate AI with each mission objective giving each platform its own unique capabilities [1]. Unmanned movement (e.g., flight) is one matter; however, autonomous or automatic movement (or flight) is quite another. Full autonomy requires the use of a AI and is a major technology challenge replete with its own controversies–not only technologically, but also ethically, morally, and legally [2].

In this chapter, AI is presented as an electromagnetic technology method of EMS exploration and exploitation. Understanding the difference between deep learning and machine learning is emphasized with applications to automatic sensing, detection, and classification of emitters, including unknown emitters that have not previously been encountered. We also walk through how to construct a *feature vector* from a radar and/or communication signal's (T-F) and bifrequency (B-F) detection image. The use of principal component analysis is used to reduce the size of the feature vector. Machine learning classification algorithms are used to identify, recognize, and specify the particular emitter. A detailed description of a *perceptron* and a *multilayer perceptron* classification nonlinear neural network (NN) representing the human brain is given.

Deep learning concepts are then presented including autoencoders (AEs), convolutional NN (CNN), recurrent NN, the long-, short-term memory (LSTM) networks, and gated recurrent units. Boltzmann machines including the restricted and deep Boltzmann machines are described. Deep belief networks and generative adversarial networks (GAN) are presented. Transfer learning is also emphasized for the objective of projecting a strong ES. To provide for secure distributed networks, the concept of Federated Learning is emphasized.

9.1 MODERN AI CONCEPTS FOR DISTRIBUTED SENSING

It is important to begin with the following excerpt from Richard Hamming [3]:

> After quite a few years, the field of the limits of intellectual performance by machines acquired the dubious title of *artificial intelligence* (AI) that does not have a *single meaning*. First it is a variant on the question,
>
> Can machines think?
>
> While this is a more restricted definition than is artificial intelligence, it has a sharper focus and is a good substitute in the popular mind. This question is important to you because if you believe computer cannot think then as a prospective leader you will be slow to use computers to advance the field by your efforts, but if you believe of course computers can think then you are very apt to fall into a *first class failure*! Thus you cannot afford to either believe or disbelieve - you must come to your own terms with the vexing problem, "*To what extent can machines think?*"

Although the legal, moral, and ethical issues surrounding fully autonomous systems continues to be debated (especially if the platform is weaponized), integrating DRFM transceivers with fully autonomous capability seems to be less of a problem.

With the DRFM-AI being a *multilayered concept*, ranging from machine learning, to adaptive reasoning, cognitive computing to full AI, each has a role to play. A DRFM transceiver, can perform as a perfect embedded playground for future multi-function applications no matter what the discipline. Developing defense AI it is important to keep in mind the three major principles: (1) abiding by international law, (2) maintaining sufficient human control, and (3) ensuring the permanent responsibility of the chain of command [4].

For military TTPs and concepts of operations (CONOPS), activities such as EMS sensing, the detection and classification and identification of threat (and friendly emitters), and the EA disruption of an adversary's radar and communication activities are all interleaved autonomously, and cognitively operating within the DRFM [5]. On the ground, the Army's AI focus is on three primary categories: (1) data poisoning, (2) counterclassification, and (3) inference attacks (configuring boundary layers for machine learning), all DRFM-centric EA functions [6]. More about this in the next chapter.

From space, AI embedded with remote sensing is helping the forest and trees by using complicated machine learning algorithms for processing remotely sensed forest imagery. Using 10 TB (terabytes) of satellite data, AI built a list of the trees standing including the species of each tree and its diameter as measured 1.4m off the ground, tree height, and total carbon storage. Previously, this high resolution imagery could not exist [7].

As shown in Figure 1.10, the DRFM represents the observation or ES stage within the OODA loop. One of the DRFM's significant tasks is the observation of the wideband EMS within an oscillating OODA loop. The operational tempo of the OODA loop depends on how fast the spectrum can be sensed, analyzed, and the information distributed. It also depends on the decision maker (computer or human) and the decision maker's tempo. In addition, it counts greatly on the action, reaction tempos, and the influence they have on the resulting spectral activity (EA, modify band spectral sensing, start/stop, times etc.).

The spectral information is gathered using AI to enhance the speed at which the autonomous classification, identification, recognition, of known (and unknown) signals takes place. Depending on the decision maker's *Sheridan level*, the OODA loop oscillations could be entirely autonomous (decision maker: computer only) or be entirely up to the human decision maker. Below, we describe these autonomous actions (i.e., no human-operator intervention), and the dichotomy they raise.

The concept of classification is modeling based on delineating the classes of output based on some set of input features. If regression analysis can give us an outcome of *how much*, then classification gives us an outcome of *what kind*. In general, there are two methods for autonomous classification of signal modulations: decision theoretic techniques and pattern recognition techniques. In particular, research on this topic has been typically applicable to military systems. Now however, with the advent of software-defined cognitive radar and cognitive radios, research on autonomously recognizing signal modulations has resulted in a boom in the realization of reconfigurable and adaptive wireless transceivers for automobiles and communications.

Most often, the autonomous recognition of communication modulations is an easier problem than the autonomous recognition of modern, quiet type radar modulations due to the fact that there are only a finite number of modulation techniques used for communication. On the other hand, there are an infinite number of modulation techniques that can be used for the quiet radar. In fact this is why the noncooperative intercept receiver has such a difficult time!

9.1.1 Reasons to Use Neural Networks/AI in Electromagnetic Warfare

There are four main situations in which neural networks are good candidates [8]:

1. When closed-form solutions do not exist, and trial-and-error methods are the main approaches to tackling the problem at hand;
2. When an application requires real-time performance;
3. When faster convergence rates are required in the optimization of large systems;
4. When enough measured data exists to train a nonlinear network for prediction purposes, especially when no analytical tools exist.

The design of a highly accurate emitter modulation classification network is an example where there is no closed for solution. When the signal is detected, the determination of the emitter modulation can help in fingerprinting the radar. This is especially useful for many reasons, but most importantly, it can aid in determining the correct EA response. The design of intelligent beamforming solutions is also an example where no closed-form solution exists.

For the second category, the high-speed neural networks using GPUs and the electronic hardware accelerators, for example, have made real-time processing of complicated, nonlinear EW solutions a reality. The use of photons rather than electrons is also spurring advanced hardware from companies such as Lightmatter. They are currently marketing a neural accelerator chip that calculates with light. This is a refinement of the prototype that the company showed off at the virtual Hot Chips conference [9].

For the third category, there are many examples such as the control of EA algorithms in avionics, to coordinate the launch chaff, decoys, and missile systems that benefit from the use of neural networks. The optimization of a large scale system using neural networks and fuzzy optimal control strategies is given in [10]. Here an intelligent coordination system is shown that utilizes a decomposition/coordination framework to reduce the complexity and computation time. Future battlefield strategies are being planned with AI as a vital component in analysis of battlefield situations on land, at sea and in the air leading the way for the development of *smart weapons*.

For the fourth category, training complex, nonlinear networks for the prediction of complicated responses to EA is just one example where measured data can be used to train the neural network. The amounts of data from well-established systems and technologies such as radar and sonar are too much for any person to process, encouraging the development of smart weapons on the battlefield that can share in the decision making. This chapter shows how AI within the DRFM can be applied for EMS and DRFM active mode configuration. Especially critical is the speed at which these solutions can formulated. Further examples as well, will be discussed in the chapter and in Chapter 10 and 11.

9.1.2 Automation and the Human-Computer Interface

In an embedded system, cost, size, power, and complexity are limited, so the HCI must be easy to use without sacrificing accuracy in the analysis capability. Human operators are often one of the biggest sources of error in any embedded system and many operator errors are attributed to a poorly designed HCI.

The need for human analysis of the T-F results limits these techniques to ELINT receivers where the emitter information derived is not time-critical. High-level automation of the classification decision, parameter extraction, and response management are, however, justified in highly time-critical situations in which there is insufficient time for a human operator to respond and take appropriate action. This is the case for ES receivers and RWRs. Human beings are often still needed to be the fail-safe in an otherwise automated system.

The Sheridan level shown in Table 9.1 is a system of eight levels to indicate the amount of automation that is incorporated in the response, its level of autonomy, and whether the response execution authority is assigned to the system or to the operator. The Sheridan levels or levels of authority (LoA) vary from level 1: "Computer offers no assistance, human does all" to level 8: "Computer selects method, executes task, and ignores human." In levels 1 to 4 the operator has authority over function execution; in levels 6 to 8 authority has moved to the system. In level 5 the authority is shared between the system and the operator.

Table 9.1
Sheridan Levels of Authority

Level	Computer Task	Human Task
1	No assistance	Does all
2	Suggests alternatives	Chooses
3	Selects way to do task	Schedules response
4	Selects and executes	Must approve
5	Executes unless vetoed	Has limited veto time
6	Executes immediately	Informed upon execution
7	Executes immediately	Informed if asked
8	Executes immediately	Ignored by computer

After [11].

Figure 9.1 shows an example where the intercept receiver calculates the T-F results from an intercepted LPI signal and must then administer a jamming waveform response. The figure shows the EW response management detailing the interaction between automation, autonomy, and authority for the jamming waveform.

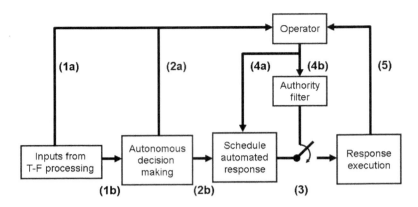

Figure 9.1 Interaction between automation, autonomy, and authority.

Depending on the Sheridan level of the response, the T-F data is presented to the operator (arrow 1a) or used by the system part "Autonomous decision making" to decide what LPI modulation is present and what the modulation parameters are (1b), given these T-F inputs [12]. Then, the system can suggest the particular modulation type to the operator (2a), who then schedules the jamming response execution (4a) or the system can select and schedule an automated response (2b). Whether the execution of the scheduled jamming response must be acknowledged by the operator depends on the LoA assigned to the response (4b). This is realized by the "Authority filter" and the switch below the filter that determines whether the scheduled response is executed (3). Depending on the setting of the "Authority filter," the operator does or does not receive feedback upon response execution (5). The interaction can be summarized as: Autonomy schedules automated responses, while authority allows or blocks response execution [12].

9.1.3 Autonomous Modulation Classification

In Chapter 8 we showed the parallel steps that can be used to autonomously classify the radar and communication signal modulations. In [11], we showed the significance of having in parallel, two or more techniques for this task, so that the measured results can be fused.

The signal is intercepted with an AESA and digital receiver that digitizes the intercepted signal. The signal is processed, in parallel, by both T-F and B-F detection techniques as illustrated in Figure 8.7. After the T-F, B-F detection processing, the resulting image planes are used by the autonomous decision making autonomous signal processing to identify the modulation *type*.[1] The autonomous decision making

---

1 Note that the signal's *parameters* are extracted from the T-F and B-F planes after the modulation *type* has been determined.

using neural networks such as the multilayer perceptron consists of a feature extraction algorithm that is used to derive the feature vector from the T-F, B-F image plane. A nonlinear classification network is then used to recognize the signal modulation type from the feature vector.

The most important step of this pattern recognition scheme is how the feature vector is formed and how it is presented to the nonlinear classification (neural) network. Note that if a high-performance reconfigurable computer is used, several T-F and B-F detection/classification algorithms can be executed quickly and in parallel. Below we first discuss the nonlinear classification multilayer perceptron (MLP) network used to identify the modulation type. Example steps in how to form a feature extraction image processing routine is then discussed and example results are shown.

9.2 NONLINEAR CLASSIFICATION NETWORKS

Nonlinear classification networks use a set of processing elements (or nodes) loosely analogous to neurons in the brain (hence the name artificial neural networks). The nodes are interconnected in a network that can then identify patterns in data as it is exposed to the data. In a sense, the network learns from experience just as people do. This distinguishes neural networks from traditional computing programs that simply follow instructions in a fixed sequential order. The architectures are specified by (1) the network topology, (2) the node characteristics, and (3) the training or learning rules used to configure the weights on each connection [20].

The classification networks can be either static or dynamic. Static networks are characterized by node equations that are memoryless. That is, their output is a function of only the current input and not of past or future inputs or outputs. Dynamic networks are systems with memory. The dynamic neural networks are characterized by differential equations or difference equations [13].

9.2.1 Single Perceptron Networks

An example of a static network is the Rosenblatt perceptron as shown in Figure 9.2.

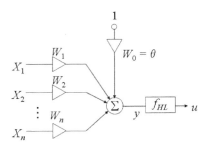

Figure 9.2 Single Rosenblatt perceptron.

Here X_n represents the n-dimensional input vector and W_n represents the n-dimensional weighting vector. The Rosenblatt perceptron forms a weighted sum of n-components of the input vector and adds a bias value, θ. The result y is passed through a nonlinear activation function to give the output value u. The activation function shown in Figure 9.2 is a *hard-limiting* nonlinearity f_{HL}. An example of a hard-limiting nonlinearity is shown in Figure 9.3 where

$$f_{HL}(y) = \begin{cases} 1 & y > 0 \\ 0 & y \leq 0 \end{cases} \tag{9.1}$$

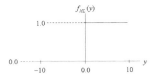

Figure 9.3 Hard limiting nonlinearity.

Another popular activation function is the *sigmoid*. The sigmoid nonlinearity is given by the expression

$$f_s(y) = \left(1 + e^{-\beta y}\right)^{-1} = \frac{1}{1 + e^{-\beta y}} \tag{9.2}$$

and is continuous. The nonlinearity varies monotonically from 0 to 1 as y varies from $-\infty$ to ∞. The β value represents the gain of the sigmoid.

One of the key attributes of the sigmoid nonlinearity $f_s(y)$ is that it is a differentiable function. This also makes it well suited to our application of pattern recognition since the output is between 0 and 1. Note that this can be interpreted as a probability distribution. The value of the output y is a weighted sum and is the inner product

9.2 Nonlinear Classification Networks

between the augmented input vector and the weight vector or [21]

$$y = \overline{W}^T \overline{X} \tag{9.3}$$

or

$$y = \underbrace{[W_0, W_1, \cdots, W_n]}_{1 \times n} \underbrace{\begin{bmatrix} 1 \\ X_1 \\ X_2 \\ \vdots \\ X_n \end{bmatrix}}_{n \times 1} \tag{9.4}$$

and then the output

$$u = f_{HL}(y) \tag{9.5}$$

A single Rosenblatt perceptron can be used to build several important logic units. One example is the AND function as shown in Figure. 9.4(a) [21]. With the weights shown the summation output $y = 2X_1 + 2X_2 - 3$. The output u for binary values of X_1 and X_2 and the value of y are as shown in the truth table. The binary logic unit OR can also be implemented with one perceptron as shown in Figure 9.4(b).

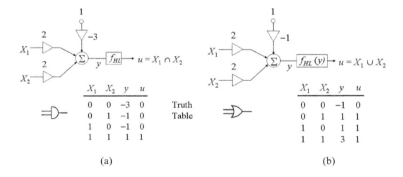

Figure 9.4 Binary logic unit: (a) AND network and (b) an OR network [21].

The summation is $y = 2X_1 + 2X_2 - 1$. The complement or NOT function can also be implemented with a single perceptron with one input as shown in Figure 9.5 [21]. The equation for the NOT summation is $y = -2X_1 + 1$.

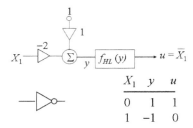

Figure 9.5 Binary logic unit: NOT [21].

Note that a single perceptron cannot implement an exclusive OR (XOR) or an exclusive NOR (XNOR).

To recognize how the perceptron can be used to recognize patterns, we examine the general two input (three weights) perceptron shown in Figure 9.6. A critical threshold occurs when the linear output $y = 0$ or

$$y = X_1 W_1 + X_2 W_2 + W_0 = 0 \tag{9.6}$$

Therefore, in slope intercept form we have

$$X_2 = -\frac{W_1}{W_2} X_1 - \frac{W_0}{W_2} \tag{9.7}$$

which is a linear separable function. That is, a linear line is formed to separate two regions of a plane as shown in Figure 9.7.

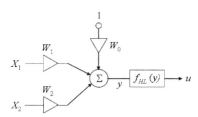

Figure 9.6 General two-input perceptron.

9.2 Nonlinear Classification Networks 573

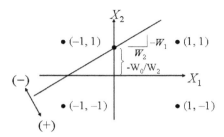

Figure 9.7 General two-input perceptron as a linear separable function.

With each additional weight, a new dimension is added to the separation boundary. That is, with four weights, the separation boundary becomes a plane, and with five weights, the separation boundary becomes a hyperplane.

9.2.2 Multilayer Perceptron Networks

In an MLP network the perceptrons (neurons or nodes) are the information processing units and they are cascaded in layers to create the complex decision regions. The inputs propagate through the network in a forward direction on a layer-by-layer basis. Most often the input set of nodes is not considered a layer. A model of a three-layer perceptron network is shown in Figure 9.8. In this model there are four neurons at the input, two hidden layers with five and four neurons, respectively, and an output layer with two neurons.

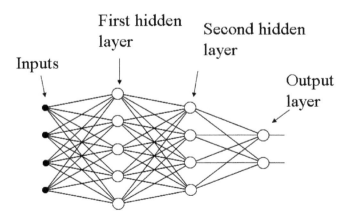

Figure 9.8 Three-layer perceptron model.

Within the MLP is a set of synapses or connecting links, each of which is characterized by a weight of its own. Each neuron has an adder for summing the input signals weighted by the respective synapses of the neuron. The activation function then limits the amplitude of the output of each neuron. The neuron may also include the externally applied bias that has the effect of increasing or lowering the net input to the activation function, depending on whether it is positive or negative, respectively. The network exhibits a high degree of connectivity determined by the synapses of the network. Most often the nodes are fully connected with every node in layer i connected to every node in layer $i+1$.

In an MLP network the inputs propagate through the network in a forward direction on a layer-by-layer basis. Training algorithms include gradient search, backpropagation, and temporal difference. The measure of how well the network performs on the actual problem, once training is complete, is called *generalization*. It is usually tested by evaluating the performance of the network on new data that is outside the training set. Parameters that can affect the generalization are (a) the number of data samples and how well they represent the problem at hand, (b) the complicity of the underlying problem, and (c) the network size. In general, a large number of weights adversely affects generalization and the time required to learn the solution. It is also worth noting that the feature vector derived from the T-F and B-F images has a significant impact on both (a) and (b).

An MLP with I input nodes, and H hidden layers can be described in general as [22]

$$y_k(\ell) = f_s \left[\sum_{h=1}^{H} w_{kh} f_s \left(\sum_{i=1}^{I} w_{hi} x_i(\ell) \right) \right] \tag{9.8}$$

where y_k is the output, x_i is the input, ℓ is the sample number, i is the input node index, h is the number of hidden layers index, and k is the output node index. Here w_{kh} and w_{hi} represent the weight value from neuron h to k and from neuron i to h, respectively, and f_s represents the sigmoid *activation function*. All weight values in the MLP are determined at the same time in a single, global (nonlinear) training strategy involving supervised learning.

The activation function f_s may vary for different layers within the network. The activation function can be any type of function that fits the action desired from the respective neuron and is a design choice that depends on the specific problem. Log-sigmoid and hyperbolic tangent (tanh) sigmoid functions are commonly used in multilayer neural networks since they are differentiable and can form arbitrary nonlinear decision surfaces [23]. The network activation function, f_s, that is popular for pattern recognition classification is the log-sigmoid discussed previously defined as

$$f_s(y) = 1/(1 + e^{-\beta y}) \tag{9.9}$$

Figure 9.9 shows a selection of some commonly used activation functions for artificial neurons, however, the hyperbolic tangent can also be expressed in the general form

9.2 Nonlinear Classification Networks

with slope parameter a as

$$g(z) = \frac{\exp(az) - 1}{\exp(az) + 1} \tag{9.10}$$

Changing the slope parameter a in the hyperbolic tangent, will then give a different look [8]. Allowing an activation function of the sigmoid type (9.10) to assume negative values, has the benefit of having a derivative.

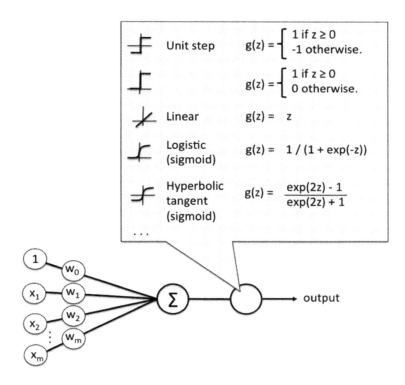

Figure 9.9 A selection of commonly used activation functions for artificial neurons.

When supervised learning is used, the input-output examples are used to train the network and derive the network weights. Since the network design is statistical in nature we can improve the network generalization during the supervised learning process by minimizing the trade-off between the reliability of the training data and the goodness of the model. This trade-off is realized during the supervised learning process through the network regularization R

$$R = gM_{SE} + (1-g)M_{SW} \tag{9.11}$$

where g is the Tikhonov's regularization parameter $0 < g < 1$ [23]. The term M_{SE} is a *performance measure* and is the mean sum of squares of the network errors. The performance measure depends on both the network design and the training data. The term M_{SW} is the mean sum of squares of the network weights and biases and is sometimes referred to as the *complexity penalty*. From (9.11), the regularization parameter g directly influences the trade-off between the complexity penalty and the performance measure. The optimum values to minimize R are found and the process is carried out for all the training examples on an epoch-by-epoch basis. Note that if $g = 1$, the network design is *unconstrained* with the solution depending only on the input-output training examples.

For most applications, a three-layer network with $H = 2$ hidden layers should be sufficient. Note that when more hidden layers are included, the convergence of the weight values becomes more difficult and significantly more time is required to complete the global training. Further, there is a much larger chance that an overgeneralization will be provided that degrades the ability of the network to correctly identify the modulation type present.

The number of output neurons reflects the number of modulation types that are expected. For example, if 12 modulation types were expected in the theater of operations, then the output layer should have 12 neurons, each of which corresponds to a modulation type. The output neurons can be hard limiting (0 or 1) or can be sigmoidal, which gives more of a modulation type probability. The input feature vector is extracted from the T-F or B-F detection processing image. The feature vector dimension $D \times 1$ is determined by feature extraction signal processing.

The supervised training of the feed-forward MLP network uses the gradient of the performance function to determine how to adjust the weights. The gradient is determined using a technique called backpropagation [14–16]. The backpropagation algorithm is a generalization of the least mean square algorithm used for linear networks, where the performance index is the mean square error. Basically, a training sequence is passed through the multilayer network, the error between the target output and the actual output is computed, and the error is then propagated back through the hidden layers from the output to the input in order to update the weights and biases in all layers.

Different modifications of training algorithms may improve the convergence speed of the network. One of these modifications is the *variable learning rate*. With the standard steepest descent algorithm, the learning rate is held constant throughout the training. The performance of the algorithm is very sensitive to the proper setting of the learning rate. When a variable learning rate is used and the learning rate is allowed to change during the training process, the performance of the steepest descent algorithm is improved. For an excellent discussion on multiple layer perceptron networks, radial basis networks and recurrent neural networks, refer to [8].

9.2.3 Modified Feature Extraction Signal Processing

Calculation of the cropping region using the marginal frequency distribution allows the low-frequency LPI modulation to be retained and the remaining T-F regions to be discarded. It is important that the size of the cropping region be adaptive and only contain the modulation energy so that the derived feature vector is consistently correlated with the modulation type. In the cropping technique described above, the presence of high-frequency noise within the T-F image, however, can vary the size of the cropping window. To minimize this effect, the use of a low-pass filter prior to the calculation of the marginal frequency distribution is investigated to help achieve the most consistent cropping of the modulation energy. To make the threshold calculation more robust, the marginal frequency distribution is smoothed using a Wiener filter before normalization.

The Wiener filter takes the form of a linear adaptive filter that adjusts its free parameters in response to the statistical variations in the marginal frequency distribution. As an alternative to directly using the feature vector as input to the classifier, this section also examines the use of *principal components analysis* (PCA) in order to develop a lower dimensional feature vector for use by the classifier. A block diagram of the modified autonomous T-F cropping and feature extraction algorithm is shown in Figure 9.10.

9.2.4 Low-Pass Filtering for Cropping Consistency

The detect and delete no-signal region is followed by an LPF applied to the T-F image. This ensures that the low-frequency LPI modulation energy is preserved and the high-frequency noise is removed. The filtering can easily be performed in the frequency domain. Frequency domain filtering using the 2D Fourier transform is fast and efficient. Let $f(k_1, k_2)$ for $k_1 = 0, 1, 2, \ldots, M-1$ and $k_2 = 0, 1, 2, \ldots, N-1$ denote the $M \times N$ T-F image. The 2D DFT of f denoted by $F(u, v)$ is [33]

$$F(u,v) = \sum_{k_1=0}^{M-1} \sum_{k_2=0}^{N-1} f(k_1, k_2) e^{-j2\pi(uk_1/M + vk_2/N)} \qquad (9.12)$$

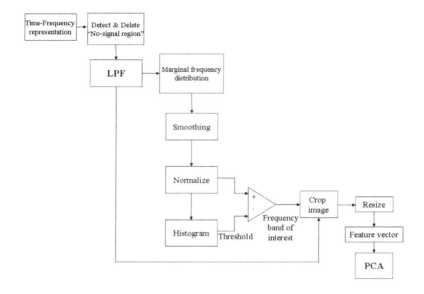

Figure 9.10 Modified T-F autonomous cropping and feature extraction algorithm.

for $u = 0, 1, 2, \ldots, M - 1$ and $v = 0, 1, 2, \ldots, N - 1$. The $M \times N$ rectangular region $F(u, v)$, defined by u and v, is often referred to as the frequency rectangle and is the same size as the input image. Note that the frequency rectangle can also be defined by digital frequencies as shown in Figure 9.11 where $\omega_1 = 2\pi u/M$ and $\omega_2 = 2\pi v/N$.

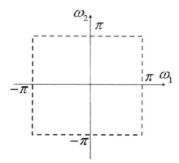

Figure 9.11 Frequency rectangle for $F(u, v)$.

Given $F(u, v)$, $f(k_1, k_2)$ can be obtained by means of the inverse DFT. Both DFT and inverse DFT are obtained in practice using a fast 2D Fourier transform algorithm

[33]. The convolution theorem, which is the foundation for linear filtering in both spatial and frequency domains, can be written as follows:

$$f(k_1,k_2) * h(k_1,k_2) \Leftrightarrow H(u,v)F(u,v) \tag{9.13}$$

and conversely,

$$f(k_1,k_2)h(k_1,k_2) \Leftrightarrow H(u,v) * F(u,v) \tag{9.14}$$

Filtering in the spatial domain consists of convolving an image $f(k_1,k_2)$ with a filter mask, $h(k_1,k_2)$. According to the convolution theorem, the same result can be obtained in the frequency domain by multiplying $F(u,v)$ by $H(u,v)$, which is referred to as the filter transfer function. A block diagram of the frequency domain filtering process is shown in Figure 9.12.

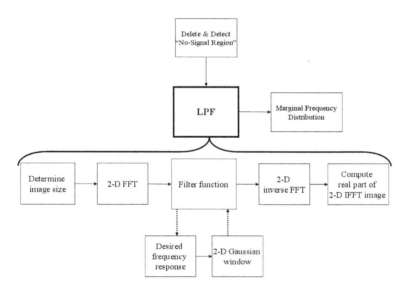

Figure 9.12 Frequency domain filtering.

The filter transfer function $H(u,v)$ can be obtained in three steps. First, the desired frequency response (ideal low-pass filter) $H_d(u,v)$ is created as a matrix. An ideal low-pass filter has the transfer function [33]

$$H_d(u,v) = \begin{cases} 1 & if\, D(u,v) \leq D_0 \\ 0 & if\, D(u,v) \geq D_0 \end{cases} \tag{9.15}$$

where D_0 (cutoff parameter) is a specified nonnegative number and $D(u,v)$ is the distance from point (u,v) to the center of the filter. D_0 can also be defined as the normalized value of digital frequencies ω_1, ω_2 by π. Second, a two-dimensional Gaussian window is created with a standard deviation $\sigma = N \times D_0/8$ where N is the number of columns in the image. The standard deviation of the window is related to D_0, and the structure becomes adaptive to the changes in the desired frequency responses. For the detection of LPI emitter modulations, both the frequency response matrix and the Gaussian window have dimensions of $M \times N$, which is equal to the image dimension $f(k_1, k_2)$ and the 2D FFT output dimension $F(u,v)$. The last step is to multiply $H_d(u,v)$ by the Gaussian window.

The transfer function of the Gaussian low-pass filter obtained by this multiplication process is then given by [34]

$$H(u,v) = e^{D^2(u,v)/2\sigma^2} \tag{9.16}$$

These steps are illustrated in Figure 9.13. Figure 9.13(a) shows the desired frequency response with $D_0 = 0.3$ (where $|D_0| \in [0,1]$) or $\omega_1 = \omega_2 = 0.3\pi$, and Figure 9.13(b) shows the Gaussian window with $\sigma = N \times D_0/8 = 33.825$. The dimension of both the frequency response matrix and Gaussian window is $M = 1{,}024, N = 902$. Figure 9.13(c) shows the resultant Gaussian low-pass filter and Figure 9.13(d) shows the Gaussian low-pass filter as an image. Several values of ω_1, ω_2 can be tested during the simulation process to find an optimum value for each distribution. For each trial the digital cutoff frequencies should be set to $\omega_1 = \omega_2$.

9.2 Nonlinear Classification Networks 581

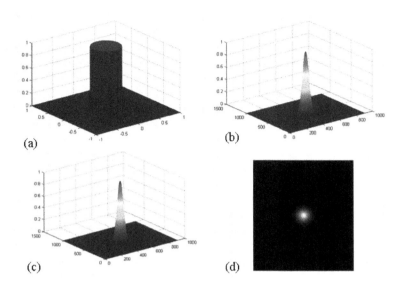

Figure 9.13 Implementation of filter function (a) desired frequency response, (b) Gaussian window, (c) Gaussian low-pass filter, and (d) Gaussian low-pass filter as an image [32].

After obtaining the low-pass filter, the frequency domain filtering can be implemented by multiplying $F(u,v)$ by $H(u,v)$. This operation is followed by shifting the frequency components back and taking the inverse FFT of the filtered image. The last step is obtaining the real part of the inverse FFT.

9.2.5 Calculating the Marginal Frequency Distribution

After the LPF is used to eliminate the high-frequency noise, the marginal frequency distribution of the T-F image is calculated. The marginal frequency distribution gives the instantaneous energy of the signal as a function of frequency. The steps for determining the modulation frequency band from the T-F plane are shown in Figure 9.14. The operations are applied to the MFD of the T-F plane. The MFD gives the instantaneous energy of the signal as a function of frequency. This is obtained by integrating the time values for each frequency in the T-F image, resulting in an $M \times 1$ vector **A**.

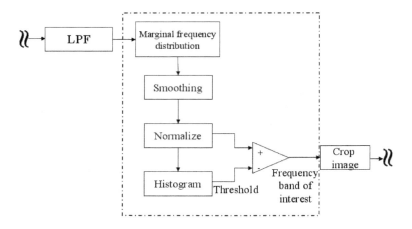

Figure 9.14 Modified method for determining the cropping region [32].

As an example, the marginal frequency distribution of a Frank coded signal with $f_s = 7$ kHz, $f_c = 1,495$ Hz, $N_c = 36$, and $cpp = 1$ ($B = 1,495$ Hz) with an SNR = 0 dB is shown in Figure 9.15(a).

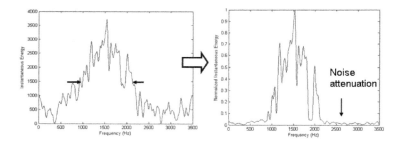

Figure 9.15 Frank code signal with $N_c = 36$ (a) MFD and (b) MFD after thresholding [32].

The higher-energy interval corresponds to the frequency band of interest and contains the modulation energy. The goal is to isolate and crop the LPI modulation as accurately as possible. This is done by computing the threshold from the histogram as before. As the noise level changes however, the cropping window set by the threshold may change as a function of noise (from one SNR to another). In order to minimize this effect, a smoothing operation is applied on **A** [32].

The smoothing of the marginal frequency distribution can be applied in a number of different ways. One of the most efficient methods is to apply a linear adaptive filter to attenuate the noise followed by a moving average filter to smooth the edges and local peaks. The smoothing operation is then followed by a normalization.

9.2.6 Wiener Filtering

An adaptive filter is a filter that changes behavior based on the statistical characteristics of the input signal within the filter. A Wiener filter is a good choice. The Wiener filter is applied to **A** using the local neighborhood of size m-by-1 to estimate the local image mean and standard deviation. The filter estimates the local mean and variance around each vector element. The local mean is estimated as [34]

$$\mu = \frac{1}{m} \sum_{n \in \eta} \mathbf{A}(n) \quad (9.17)$$

and the local variance is estimated as

$$\sigma^2 = \frac{1}{m} \sum_{n \in \eta} \mathbf{A}^2(n) - \mu^2 \quad (9.18)$$

where η is the m-by-1 local neighborhood of each element in the vector **A**. The processed image within the local neighborhood can be expressed as

$$b(n) = \mu + \frac{\max(\sigma^2 - v^2, 0)}{\sigma^2} (\mathbf{A}(n) - \mu) \quad (9.19)$$

where v is the noise variance estimated using the average of all the local estimated variances. When the variance is large, the filter performs little smoothing and when the variance is small, it performs more smoothing. For PWVD and CWD images a local neighborhood of $\eta = 10$ is used and for the QMFB images $\eta = 4$ is used. Figure 9.15(b) shows the output of the adaptive filter for the input MFD of the Frank signal with $N_c = 36$. Note the considerable noise attenuation.

Although the adaptive noise attenuation gives promising results, the threshold determination may be affected by the local noise peaks that could not be reduced by the adaptive filter. To avoid this problem a moving average filter is applied to the output of the adaptive Wiener filter. As a generalization of the average filter, an averaging over $N + M + 1$ neighboring points can be considered. The moving average filter is represented by the following difference equation [35]

$$y(n) = \frac{1}{N+M+1} \sum_{k=-N}^{M} x(n-k) \quad (9.20)$$

where $x(n)$ is the input and $y(n)$ is the output. The corresponding impulse response is a rectangular pulse.

For PWVD and CWD images a window length of $N + M + 1 = 10$ is used and for QMFB images $N + M + 1 = 4$ is used. The moving average filter output, $\tilde{\mathbf{A}}_{avg}$ is then normalized as

$$\mathbf{A}_n = \frac{\tilde{\mathbf{A}}_{avg}}{\max(\tilde{\mathbf{A}}_{avg})} \qquad (9.21)$$

where $\mathbf{A}_n$ is the normalized smoothed MFD. After normalization a histogram of 100 bins is generated for PWVD and CWD images and a histogram of 30 bins is generated for QMFB images. Using these histogram bins a threshold is determined. Threshold determination is illustrated in Figure 9.16(a) using the histogram of $\mathbf{A}_n$ shown in Figure 9.16(b) for $n=30$ bin. Note that the corresponding value to the 30th bin $T_h = 0.2954$ is selected as the threshold. For the simulation purposes the histogram bin numbers are optimized using a range of values for each detection technique and each network. The bin number that provides the best Pcc is selected.

Once the threshold is determined, the values of $\mathbf{A}_n$ below the threshold are set to zero. Then the beginning and ending frequencies of the frequency band of interest are determined as shown in Figure 9.16(c). Using the lowest and highest frequency values from the frequency band of interest, the modulation energy can now be cropped from the image.

After the determination of the modulation band of interest, the energy is autonomously cropped from the LPF output containing the noise filtered image. The cropping is illustrated in Figure 9.17. Figure 9.17(a) shows the LPF output that was obtained previously, Figure 9.17(b) shows the cropped region, and Figure 9.17(c) shows the contour plot where the signal energy can easily be seen.

Once the LPF output is cropped, the new image is resized to 50×400 pixels for the PWVD and CWD images. The QMFB images are resized to 30×120 pixels. Resizing is done in order to obtain as much similarity as possible between the same modulation types. Following the resizing operation the columns of the resized image are formed with the feature vector of size $50 \times 400 = 20{,}000$ for PWVD and CWD images, and of size $30 \times 120 = 3{,}600$ for the QMFB images.

9.2 Nonlinear Classification Networks 585

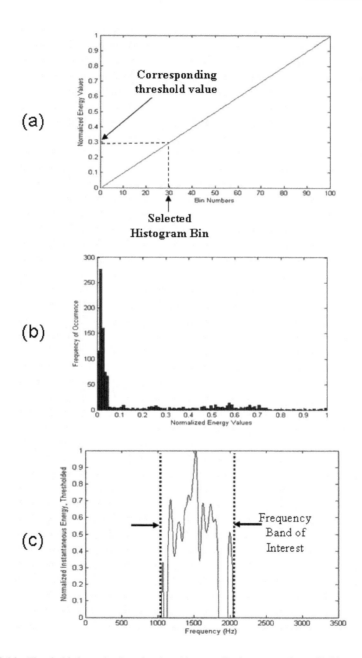

Figure 9.16 Threshold determination showing (a) normalized energy values, (b) histogram of energy values, and (c) cropped frequency band of interest using $n = 30$.

586 MACHINE LEARNING IN ELECTROMAGNETIC WARFARE

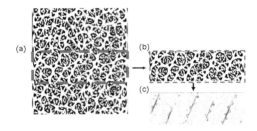

Figure 9.17 (a) LPF output, (b) cropped region, and (c) contour plot of the cropped region showing the Frank modulation.

9.3 PRINCIPAL COMPONENTS ANALYSIS

Principal components analysis (PCA) is mathematically defined as an orthogonal linear transformation that transforms the data to a new coordinate system such that the greatest variance by any projection of the data comes to lie on the first coordinate (called the first principal component), the second greatest variance on the second coordinate, and so on [36]. In other words, PCA is a rotation of the existing axes to new positions in the space defined by the original variables, where there is no correlation between the new variables defined by the rotation. PCA is theoretically the optimum transform for a given data set in least square terms. That is, the method projects the high-dimensional data vectors onto a lower-dimensional space by using a projection that best represents the data in a mean square sense. Using PCA the given data vector is represented as a linear combination of the eigenvectors obtained from the data covariance matrix. As a result, lower-dimensional data vectors may be obtained by projecting the high-dimensional data vectors onto a number of dominant eigenvectors [37].

PCA can be used for dimensionality reduction of the feature vector by retaining those characteristics of the cropped modulation that contribute most to its variance, by keeping lower-order principal components and ignoring higher-order ones. This assumes of course, that the low-order components contain the most important features of the LPI modulation within the cropped (and resized) T-F data. To facilitate the PCA, we form a training matrix $\mathbf{X}$ as shown in Figure 9.18 where N is the length of the feature vector and P is the number of training signals, which is 50 for our results. It is important to note that the mean has been subtracted from the data set.

The PCA maps the ensemble of P N-dimensional vectors $\mathbf{X} = [\underline{x}_1, \underline{x}_2, \cdots, \underline{x}_p]$ onto an ensemble of P D-dimensional vectors $\mathbf{Y} = [\underline{y}_1, \underline{y}_2, \cdots, \underline{y}_p]$ where $D < N$ using a linear projection. This linear projection can be represented by a rectangular matrix $\mathbf{A}$ so that [37]

$$\mathbf{Y} = \mathbf{A}^H \mathbf{X} \qquad (9.22)$$

9.3 Principal Components Analysis

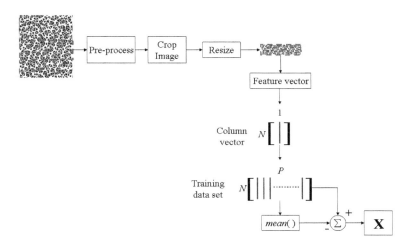

Figure 9.18 Forming the training matrix **X**.

where **A** has orthogonal column vectors, $i = 1, 2, \cdots, P$ and H is the Hermitian operation. The matrix **A** is selected as the $P \times D$ matrix containing the D eigenvectors associated with the larger eigenvalues of the data covariance matrix $\mathbf{X}^H \mathbf{X}$. With this choice of transformation matrix **A**, the transformed data vectors **Y** have uncorrelated components.

The matrix **X** is obtained first to form the training data set. The feature extraction algorithm is applied to the images in the "Training" folder for each detection technique. The cropped images are resized and a column vector is formed to represent the signal modulation. These column vectors are stacked together to form the training data set matrix. The mean of the training matrix is calculated column-wise and the mean is subtracted from the training data set matrix giving the matrix **X**. This operation is illustrated in Figure 9.18 where P is the number of training signals, which is 50 for this example, and N is the length of the feature vectors. For PWVD and CWD **X** is of dimension $20{,}000 \times 50$ (50 training signals) and for the QMFB **X** is of dimension $3{,}600 \times 50$. Figure 9.19 shows a block diagram of the PCA signal processing. In order to obtain the eigenvectors of **X**, singular value decomposition (SVD) may be performed. SVD states that any $N \times P$ matrix **X** can be decomposed as [37]

$$\mathbf{X} = \mathbf{U} \sum \mathbf{V}^H \tag{9.23}$$

where **U** is the $N \times N$ unitary matrix, **V** is the $P \times P$ unitary matrix, and $\sum$ is the $N \times P$ matrix of nonnegative real singular values. Note that

$$\mathbf{X}^H \mathbf{X} = \mathbf{V} {\sum}^H (\mathbf{U})^H \mathbf{U} \sum \mathbf{V}^H = \mathbf{V} ({\sum}^H \sum) \mathbf{V}^H \tag{9.24}$$

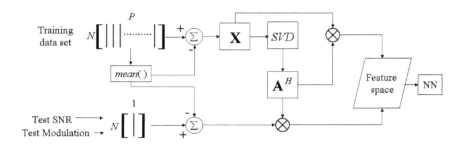

Figure 9.19 Principal components analysis.

indicates that the eigenvectors of $\mathbf{X}^H\mathbf{X}$ are contained in the $\mathbf{V}$ matrix and the eigenvalues of $\mathbf{X}^H\mathbf{X}$ are the squared singular values of $\mathbf{X}$, which are the diagonal elements of the matrix $\Sigma^H \Sigma$. It can similarly be shown that the eigenvectors of $\mathbf{X}\mathbf{X}^H$ are contained in the $\mathbf{U}$ matrix.

If $p = \min(P,N)$, both $\mathbf{X}\mathbf{X}^H$ and $\mathbf{X}^H\mathbf{X}$ will have the same p nonzero eigenvalues. The product of $\mathbf{X}$ and $\mathbf{V}$ gives

$$\mathbf{XV} = \mathbf{U}\Sigma \mathbf{V}^H \mathbf{V} = \mathbf{U}\Sigma \quad (9.25)$$

since $\mathbf{V}$ is unitary and the eigenvectors associated with nonzero eigenvalues can be extracted by

$$\mathbf{U} = \mathbf{XV}\Sigma^{-1} \quad (9.26)$$

As a result, the nonzero eigenvalues of the higher-dimensional covariance matrix $\mathbf{X}\mathbf{X}^H$ may be computed by computing the SVD of the smaller-dimensional covariance matrix $\mathbf{X}^H\mathbf{X}$.

Following the SVD of the data matrix and determination of the eigenvector matrix $\mathbf{U}$, dimensionality reduction is performed using the projection (transformation) matrix $\mathbf{A}$. The matrix $\mathbf{A}$ is composed of D eigenvectors selected from the eigenvector matrix $\mathbf{U}$ corresponding to D largest eigenvalues. In order to find the D largest eigenvalues, the biggest eigenvalue is multiplied by a threshold constant and the eigenvalues above the product are taken. Let T_{h_λ} be the *eigenvalue selection threshold* constant. In our example, three values are used as $T_{h_\lambda} = [0.001, 0.005, 0.01]$. For each case, once the eigenvalues are found, four variations of eigenvector selection are used. Let these variations be Δ_i, where $i = 0, 1, 2, 3$. The variations are defined by the i index as follows:

- Δ_0: All the eigenvectors corresponding to the eigenvalues above T_{h_λ} are used to form the matrix $\mathbf{A}$.

- Δ_1: All the eigenvectors corresponding to the eigenvalues above T_{h_λ} are selected initially; all of them except the eigenvector corresponding to the eigenvalue with the highest value are used to form the matrix **A**.

- Δ_2: All the eigenvectors corresponding to the eigenvalues above T_{h_λ} are selected initially; all of them except the two eigenvectors corresponding to the two eigenvalues with the highest values are used to form the matrix **A**.

- Δ_3: All the eigenvectors corresponding to the eigenvalues above T_{h_λ} are selected initially; all of them except the three eigenvectors corresponding to the three eigenvalues with the highest values are used to form the matrix **A**.

Once the projection matrix **A** is generated, both the training matrix **X** and the test signals are projected onto a smaller-dimensional feature space. The data set is reduced in dimension to D using the projection process. The projected data is then used for classification.

9.3.1 Classification Using Modified Feature Extraction

The classification results in this section use an extended database to determine the performance of the modified feature extraction technique as a function of the SNR. After the database is described, steps to optimize the MLP and RBF are discussed. Classification results are then shown for both the TestSNR signals (same signals used in training but with varying SNR) and the TestMod signals (different modulations and varying SNR).

9.3.2 Extended Database

To investigate the detailed performance of the modified feature extraction and classification process, a more extensive database is developed that consists of 12 LPI modulation techniques, each having 21 SNR levels (-10 dB, -9 dB, $\cdots$, 9 dB, 10 dB). The LPI modulation techniques include Costas frequency hopping, Costas frequency hopping plus a Barker phase shift keying, FMCW, PSK, and FSK. PSK signals include polyphase (Frank, P1, P2, P3, P4) and polytime (T1, T2, T3, T4) codes.

This database allows a detailed look at the Pcc as a function of the SNR. The signals are generated using the LPIT and placed in the "Input" folder within the proper subfolder (TestSNR, TestMod, Training, Signals). Note that the "Signals" folder should contain only *one* signal from each modulation *type* being used. This folder is used to correlate the modulation prefix (F for FMCW, FR for Frank, and so forth) to build the confusion matrix.

The output T-F and B-F images from the detection signal processing (Wigner-Ville, Choi-Williams, quadrature mirror filtering, cyclostationary processing) are automatically placed in the corresponding output folder (e.g., QMFB_output). Before the feature extraction and nonlinear classification signal processing algorithms are run, the detection output signals within the "TestMod" and "TestSNR" folders that have the same SNR must be collected and put into a folder that designates the SNR (e.g., TestMod-10, and TestSNR4). A sorting algorithm is included that does this collection.

The folder structure should be as shown in Figure 9.20. Note that the SNR = 10 dB signals for each modulation are used for training. This is a choice that the user can make. Training the LPI feature extraction and classification networks with only "signal only" waveforms however, is not realistic since any received signal will have a noise component related to the thermal noise present in the intercept receiver and the range of the LPI emitter.

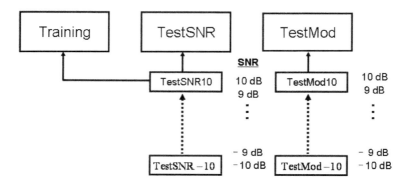

Figure 9.20 Folder structure for TestSNR, TestMod, and Training (10-dB TestSNR only) [32].

9.3.3 Signals Used for TestSNR and TestMod

The signals used to test the performance of the feature extraction and classification signal processing for various values of SNR are described below. This database is used to generate the results shown in this section. Supervised training of the autonomous classification process is done with the signal modulations below using SNR = 10 dB. The FMCW signal parameters are shown in Table 9.2. The polyphase signals (Frank, P1–P4) used for testing the performance as a function of the SNR are as shown in Table 9.3. These signals are used to evaluate the performance of the autonomous classification Pcc when the received signal has the same modulation parameters but different SNR.

9.3.4 Optimizing the Feature Extraction and Classification Network

Using the initial nonlinear network parameters two feature extraction parameters, LPF cutoff frequency and histogram bin, must be optimized. Using the optimum values derived, the PCA network parameters are then optimized. The Pcc results shown are with the final optimum values. The optimization is performed using the test signals with SNR = 10 dB. The optimum parameter selection is based on the highest average Pcc.

For each detection technique, the MLP network configuration starts with a default set of values for the epochs, the number of neurons in the first and second hidden layers S_1, S_2, the eigenvalue selection threshold constant Th_λ and eigenvector selection variations V_i. Once the initial values are set, an optimization is performed to determine optimum values for the LPF digital frequencies $\omega_1 = \omega_2$ and histogram bin number. After these two values are found, a second optimization for epochs, S_1, S_2, Th_λ, and V_i is performed.

Once all the values are found and set the classification network is tested. For the classification of PWVD images the initial values used are $epochs = 6{,}000$ $S_1 = S_2 = 50$, $Th_\lambda = 0.001$, and $V_i = V_0$. After optimization $\omega_1 = \omega_2 = 0.1\pi$ and the histogram bin number is 45. Using these values, the remaining parameters giving optimum Pcc are $S_1 = S_2 = 80$, $Th_\lambda = 0.001$, V_1 and $epochs = 5{,}000$. The optimization is repeated for the Choi-Williams, the quadrature mirror filtering, and the Wigner-Ville distribution detection techniques. Tables 9.2 and 9.3 show the modulation parameters for the FMCW and polyphase modulations, respectively.

Table 9.2
FMCW Modulation Parameters for TestSNR

Signal Modulation	Carrier Frequency f_c (Hz)	Modulation Bandwidth ΔF (Hz)	Modulation Period t_m (ms)
FMCW	1,495	250	15
	2,195	800	15

Table 9.3
Polyphase Modulation Parameters for TestSNR

Signal Modulation	Carrier Frequency f_c (Hz)	Code Length N_c	Cycles per Subcode cpp
Frank	1,495	9	5
		25	2
		36	1
....	....	....	....
	2,195	16	6
		25	3
P4	1,495	9	5
		25	2
		36	1
....	....	....	....
	2,195	16	4
		16	5

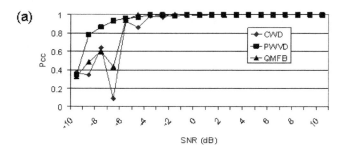

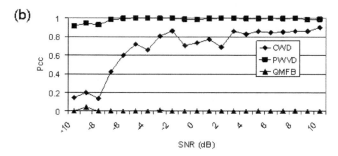

Figure 9.21 FMCW classification results using the MLP for (a) TestSNR and (b) TestMod.

9.3 Principal Components Analysis 593

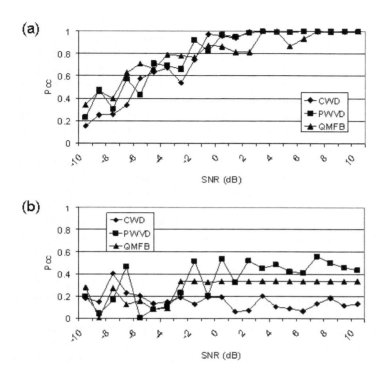

Figure 9.22 Frank classification results using the MLP for (a) TestSNR and (b) TestMod.

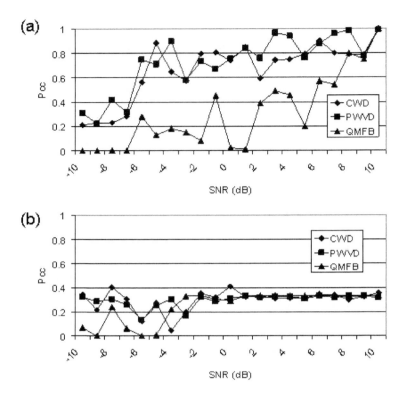

Figure 9.23 P4 classification results using the MLP for (a) TestSNR and (b) TestMod.

Concerning the results all the detection techniques show similar results on the TestSNR case. Most of the modulations are classified with more than 80% classification rate for SNR > 0 dB. There is a considerable stability in classification of signals with SNR > 0 dB. This stability indicates that the autonomous modulation energy isolation and cropping becomes more sensitive to noise variations below 0 dB. The Pcc of Frank, FMCW modulations with PWVD, and CWD techniques exhibit 100% for most of the SNR levels above 0 dB.

Concerning the TestMod case, the best results are obtained in the classification of FMCW, while the worst results are obtained in the classification of polyphase codes. Note that most of the results for Frank and P4 modulations are below Pcc = 0.4. Classification of P4 modulations with PWVD and CWD techniques exhibit similar results. Overall, the classification results with the PWVD technique outperform the other detection techniques. Overall the QMFB technique performs worse than the other techniques. Recall that the QMFB images have a very low resolution compared to the PWVD and CWD images, which becomes a disadvantage for modulation discrimination.

9.4 ARTIFICIAL INTELLIGENCE

AI is any technique that aims to enable computers to mimic human behavior, including machine learning, natural language processing, language synthesis, computer vision, robotics, sensor analysis, optimization, and simulation. As well, AI is adding new capabilities to traditional electronic design automation (EDA) bringing new emphasis to the word *automation*.

DRFM transceivers on-board UASs are now using AI to perform all segments of the OODA loop activities starting with wideband spectrum sensing. Actually, the development of AI is inherently tied to the exploitation of sensed data and is revolutionizing embedded computing for sensor processing and making a big difference in RF and image analysis.

EMW capabilities using AI, are increasingly being added to unmanned platforms, whether they are in space, the sky, on the ground, and even under the water. AI can also facilitate the OODA activities with a human more efficiently, speeding up the decision making process. Taking humans out of direct conflict is now saving lives and with the addition of AI, the UAS DRFM can retain the ability to discern nuances that in the past, only well-trained personnel could detect. In fact, AI is being efficiently coupled into all parts of the OODA loop activities for optimizing the OODA operational tempo Λ, and, as pointed out in Chapter 1, the DoD has endorsed a set of principles for the use of AI and AI-enabled technology to provide clarity.

The AI concept is an overarching umbrella for two important major types of learning: machine learning and deep learning. These enable (in different ways), the DRFM to perform the EMS computations, networking distribution, and thinking the way a human decision maker would but in a better, faster way.

9.4.1 Machine Learning

Machine learning (ML) is a subset under the AI umbrella of techniques that enables computer systems to learn from previous experience (i.e., data observations) and improve their behavior for a given task [17]. For example, if we consider solving $Ax = b$, then machine learning is based on algorithmic techniques to minimize the error in this equation through *optimization*. That is, we are focused on changing the numbers in the x parameter column vector until we reach a good set of values that gives us the closest outcomes to the actual value. That is, each weight in the weight matrix will be adjusted after the loss function calculates the error produced.

The field of ML has undergone radical transformations during the last decade. There are four key branches of machine learning:

1. *Supervised learning*: Learning to map input data to known targets given a set of examples;

2. *Unsupervised learning*: Consists of finding transformations of the input data without the help of any targets for the purposes of data visualization, data compresssion, data denoising, or to better understand the correlations present in the data;

3. *Self-supervised learning*: Supervised learning without human annotated labels. Here the labels are generated by the input data, typically using a heuristic type algorithm;

4. *Reinforcement learning*: Learning process in which an *agent* receives information about its environment and learns to choose actions that will maximize some reward criteria. Examples of this type are Google's DeepMind successfully playing Atari games, and learning to play Go at the highest level.

Some even breakouts also list *Transfer Learning* and *Generative Adversarial Networks* in this list. Since optimization forms the basis of most ML architectures, *distributed optimization* is expected to also play a key role in the ML DRFM systems onboard unmanned vehicles that are being trained, for example, using the distributed data sets being distributed to the UAVs for stand-in sensing.

Distributed optimization takes many forms. For example distributed gradient methods can be used for convex learning problems [18]. Optimization for nonconvex learning problems are discussed in [19]. For example, in distributed ML, the DRFM nodes jointly learn a model based on local data (e.g., sensed spectrum on each device). To accurately represent the local data, the nodes are often required to use nonconvex loss functions, such as those that compose multiple nonlinear activation functions through collaborative deep learning. The network topology used to distribute the data distributed stochastic optimization.

These transformations take advantage of the algorithmic and analytical tools from a number of research disciplines. Training machine learning models on a DRFM that employs deep neural networks by using petabytes of data does not scale well on a single DRFM machine. Consequently, significant research has gone into *distributing* the training of such neural networks across a cluster of machines. This is done by partitioning on both the data and the model itself giving rise to two important concepts in distributed machine learning; data-parallelism and model-parallelism.

9.4.2 Data Parallelism

In a data-parallel method of machine learning, the entire model is deployed to multiple nodes (DRFMs) a cluster and the data is shared (horizontal training). Each instantiation of the model works on a portion of the data as shown in Figure 9.24. It is the easiest type of distributed training to implement [20].

9.4 Artificial Intelligence

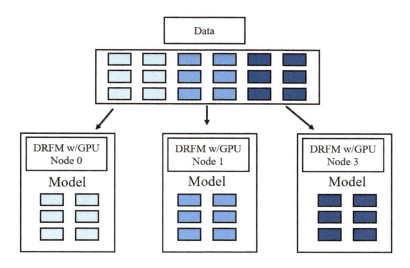

Figure 9.24 Data-parallel training entire model is deployed to multiple DRFM nodes of a cluster.

For example, in the data-parallel method the gradient is calculated for a small batch of the data at each DRFM node (for example, 50 images at once). At the end of round one of forward-backward passes by the network, the updated deep learning network weights are sent back to the initiating node. The weighted average of the weights from each node, is applied to the model parameters. The updated model parameters are sent back to the nodes for the next round of iteration.

9.4.3 Model Parallelism

The model parallel method is shown in Figure 9.25. Here, a layer (or group of layers) for the model is deployed on one node of a *cluster*, and the whole data set is copied to the node. Each DRFM node then trains on the complete dataset.

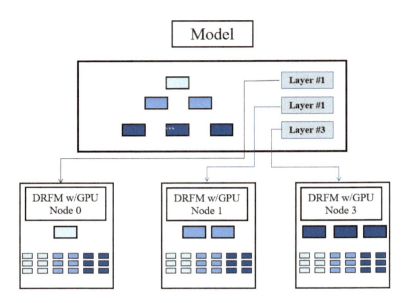

Figure 9.25 Model-parallel training.

The most common (and easiest) type of training to implement is data parallelism, shown in Figure 9.24. The main issue here is how the model parameters are stored and updated in the data-parallel approach. In both the data-parallel and model-parallel training, the concern is how the parameters are initialized, and the weights and biases updated.

For the communication of the weights and biases over the network, there are two methods; the centralized training and the decentralized training. In centralized training shown in Figure 9.26(a) a node or group of nodes are responsible for synchronizing the model parameters - these are the parameter servers. The advantage here is it is easy to synchronize the model parameters [20]. However a disadvantage is that a parameter server can itself become a bottle neck for a huge cluster and can be a single point of failure. In addition, to go along with this, the bandwidth of the network can also be an issue.

9.4 Artificial Intelligence

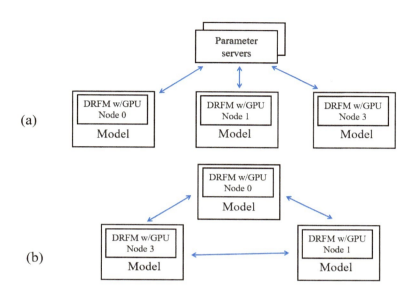

Figure 9.26 Methods of training showing (a) centralized training and (b) decentralized training.

In the decentralized approach, shown in Figure 9.26 (b), each node communicates with every other node to update the parameters. That is, the parameter-server is omitted. The advantage here is that peer-to-peer updates can be much faster. Sparse updates can be made by exchanging only the data that has changed, resulting in no single point of failure. Consequently, the bandwidth is rarely an issue.

The tools, such as decentralized computing (sensing), stochastic approximation, and distributed optimization, enable learning from data that is both *streaming and distributed*. Consequently, this has fueled our ability to collect and generate tremendous volumes of training data and in turn has enabled the leveraging of massive amounts of low-cost computing power leading to an explosion in research activity in the field [21]. For example, machine learning systems that involve distributed, decentralized learning, the interconnected network of DRFM transceivers is such that each DRFM receives its own set of training data with the goal to train a global model that is as accurate as if it had been trained on a single machine that has access to the "entire" collection of data samples.

9.4.4 Multi-Task Learning

In the distributed, decentralized machine learning task, we are typically concerned with optimizing a DRFM model, for example, for *emitter modulation classification*. In order to do this the model is trained that is actually an ensemble of models (i.e., for a number of other emitter attributes to be classified). However, in our effort to fine-tune these models we ignore information that might help us do even better on the metric we care about. Specifically, this information comes from the training signals of related tasks. That is, the multi-task learning objective is to solve multiple different tasks at the same time, by taking advantage of the *similarities* between different tasks. By sharing representations between related tasks, we can enable our model to *generalize better* on our original task. This approach is called Multi-Task Learning (MTL). There are several names for MTL: joint learning, learning to learn, and learning with auxiliary tasks are some of the other names.

The problem of simultaneously learning several related tasks has received considerable attention in several domains, especially in machine learning, with MTL [22]. In general, as the multi-domain training data is received and as you begin optimizing more than one loss function, you are effectively doing multi-task learning (in contrast to single-task learning). MTL improves generalization by leveraging the domain-specific information contained in the training signals of related tasks [23].

Hard parameter sharing is the most commonly used approach to MTL in neural networks. It is applied by sharing the hidden layers between all tasks, while keeping several task-specific output layers. Figure 9.27 shows the concept of MTL and hard parameter sharing where multi-domain training data is networked to a *composite* machine learning neural network. The data is useful for optimizing the shared layers when finally being split off to the task specific layers [24].

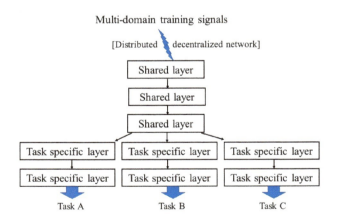

Figure 9.27 Method of MTL and hard parameter sharing.

The increasing ability for the DRFM to collect data in a distributed and streaming manner, requires the design of new strategies for jointly learning multiple tasks from streaming data (over the DRFM network). An overview of multitask strategies for learning and adaptation over networks is given in [22]. The working hypothesis for these strategies is that agents are allowed to cooperate with each other to learn distinct, though related, tasks. This cooperation steers the network-limiting point and by the network-limiting rules, allow the promotion of different task-related models.

In a massively connected world, the four key elements of problems, data, communication, and computation, enable scalable distributed processing and real-time intelligence. All of these elements are tied closely to one another. The question is, how should these elements work together in the most effective and coherent manner to realize scalable processing, real-time intelligence and contribute a vision of a smart highly connected world.

9.5 DEEP LEARNING ARCHITECTURES AND EMW APPLICATIONS

Deep learning (DL) is a subset of techniques under the umbrella of AI that consist of NNs that make computational multilayer NNs more accurate, for example, than what we showed in Section 9.3. Deep learning or Deep Neural Network (DNN) is a subset of machine learning in artificial intelligence that has networks capable of imitating the workings of the human brain in processing raw data and learning patterns for effective decision making [25]. The DL architectures discussed in this chapter include:

- Feedforward neural networks (FFNN);
- Deep belief networks (DBNs);
- Auto encoders (AEs);
- Convolutional neural networks (CNNs);
- Recurrent neural networks (RNNs);
- Generative adversarial networks (GANs).

The deconvolutional NN is a relative newcomer to the deep learning networks. Schematic block diagrams of these approaches (except the FFNN) are shown in Figure 9.28.

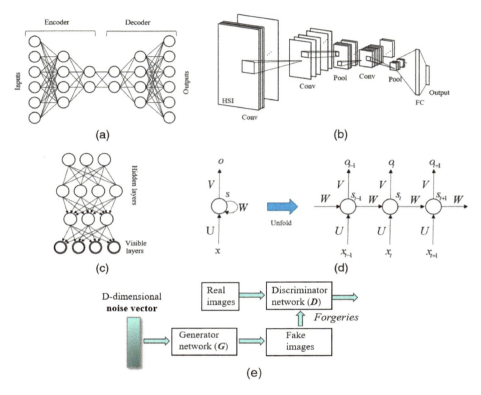

Figure 9.28 Block diagrams of deep learning architectures: (a) AE, (b) CNN, (c) DBN, and (d) RNN and (e) GAN. (Adapted from [26].)

The initial EMS gathering stage results in big data that can only be sorted with distributed, centralized computing, using autonomous and intelligent, machine learning. It is intelligent in that it adapts and gets better and better with experience with the newly collected data.

To identify the differences between deep learning and machine learning, the differences between these two was determined with the aid of clustering techniques in [27]. The machine learning programs the software to optimize a perfomance criterion (e.g., extract emitter modulation and classify with probability of correct classification $P_{cc} > 0.98$). That is, ML relates to the study, design, and development of the algorithms that give computers the capability to learn without being explicitly programmed. For example, ML is often used for *data mining*, or extracting unstructured data from the spectrum sensing contacts (discussed above) to extract knowledge or emitter modulations and patterns from it. By understanding the spectral domain and taking advantage of prior knowledge and the goals at hand, ML can identify appropriate features and build compressed representations of the data (preprocessing).

Applications for DRFMSs include emitter association, supervised learning, unsupervised learning, or clustering. Regression or defining a model assignment function is also an application. When configuring a DRFM for modulation autonomous classification, one has to decide, depending on the situation, whether it is a supervised or unsupervised algorithm or a reinforcement learning example.

Although the NN approaches above can perform extremely well in the pattern recognition task using supervised training, the deep learning approaches in contrast, (a) learn from the data itself, meaning that the feature extraction engineering is replaced either partially or completely, (b) has state-of-the-art results in many disciplines that are usually significantly better, and (c) can outperform humans and human-coded features.

9.6 AUTOENCODER

An autoencoder is a type of artificial neural network that can be used to learn efficient codings of unlabeled data (unsupervised learning). Autoencoders are similar to dimensionality reduction techniques like the PCA we studied above. They project the data from a higher dimension to a lower dimension using a linear transformation and try to preserve the important features of the data while removing the non-essential parts. The encoding is validated and refined by attempting to regenerate the input from the encoding. Consequently, the AE is typically used for dimensionality reduction by training the network to ignore insignificant data ("noise").

The output of the AE network is a reconstruction of the input data in the most efficient form. A block diagram of an AE is shown in Figure 9.29.

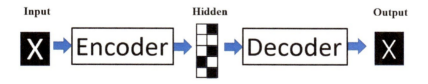

Figure 9.29 Block diagram of an AE.

The AE consists of four main parts:

- *Encoder*: Model learns how to reduce the input dimensions and compress the input data into an encoded representation.

- *Bottleneck*: Contains the compressed representation of the input data and is lowest possible dimensions of the input data.

- *Decoder*: Model learns how to reconstruct the data from encoded representation to be as close to the original input as possible.

- *Reconstruction Loss*: Method that quantifies how well the decoder is performing and how close the output is to the original input.

In essence, the AE is an unsupervised artificial neural network that learns how to efficiently compress and encode data and then learns how to reconstruct the data back (from the reduced encoded representation) to a representation that is as close to the original input as possible. For example, if you have input data that is *correlated* the AE method will work very well. That is because the encoding operation compresses the data by using the correlated features. Another advantage of using an AE is that the data do not need to be labeled.

The AE's encodings often reveal useful features from the unsupervised data. An AE can be used to detect anomalies and outliers. An AE can also be trained to remove noise from images (denoising). AEs are fundamental structures in deep networks because they are often used as part of larger networks. Like many other networks, they serve that role but can be used as a standalone network as well. The training involves any number of techniques but most often used is backpropagation.

In this section, a couple of examples are given of some of the recent AEs. One of the first AEs was described in [28], for the task of dimensionality reduction. The AE network for dimensionality reduction first receives the high-dimensional data and then converts it to low-dimensional codes by training a multilayer neural network with a small central layer to deconstruct high-dimensional input vectors. The weights can be initialized in the AE network to allow it to learn low-dimensional codes. The AE has been shown to work much better than PCA as a tool to reduce the dimensionality of the data.

The AE maps the input through an encoder function f to generate an internal (latent) representation, or code, h. The AE also has a decoder function, g, that maps h to the output $\hat{x}$. Let an input vector to an AE be $\mathbf{x} \in \mathbb{R}^d$. In a simple one hidden layer case, the function $h = f(\mathbf{x}) = g(\mathbf{W}'\mathbf{x} + \mathbf{b})$ where $\mathbf{W}$ is the learned weight matrix and $\mathbf{b}$ is a bias vector. A decoder then maps the latent representation to reconstruction or approximation of the input as $\mathbf{x}' = f'(\mathbf{W}'\mathbf{x} + \mathbf{b})$ where $\mathbf{W}'$ and $\mathbf{b}'$ are the decoding weight and bias, respectively. Usually, the encoding and decoding weight matrices are tied (shared), so that $\mathbf{W}' = \mathbf{W}^\mathbf{T}$ (transpose) [26].

A loss function $\mathscr{L}$ measures how close the AE can reconstruct the output: $\mathscr{L}$ is a function of $\mathbf{x}$ and $\mathbf{x}' = f'[f(\mathbf{x})]$. One popular loss function is the *mean squared error*, which penalizes the approximated output from being different from the input: $\mathscr{L}(x,x') = ||\mathbf{x} - \mathbf{x}'||^2$.

Consider the simple linear AE where both f and g are linear with a shared weight matrix $\mathbf{W}$.

$$\hat{\mathbf{z}} = f(\mathbf{x}) = \mathbf{W}\mathbf{x}$$

$$\mathbf{x} = g(\hat{\mathbf{z}}) = \mathbf{W}^\mathbf{T}\hat{\mathbf{z}}$$

One way to train this model is minimize the squared error $\sum_j ||\mathbf{x}_j - g(f(\mathbf{x}_j))||^2$ so that $\mathbf{x} \approx g(f(\mathbf{x}))$. Here $\mathbf{W}$ is trained so that a low-dimensional $\hat{\mathbf{z}}$ will retain as much information as possible to reconstruct the high-dimensional data $\mathbf{x}$. This linear AE turns out to be closely connected to the classical PCA discussed previously. In other words, when $\mathbf{z}$ is m-dimensional, the matrix $\mathbf{W}$ should learn to span the m principal components of the data. That is, the set of m orthogonal directions where the data has the highest variance - or equivalently, the m eigenvectors of the data covariance matrix that have the largest eigenvalues, exactly as in the PCA. It is this correspondence that suggest that there may be a way to capture more complex kinds of generative models using more complex AEs [29].

The AE is usually constrained either through its architecture or through a sparsity constraint (or both) or through a modulation mapping function. One particular approach that is worthy of mention is from [30] as it:

- Addresses the *automatic modulation classification* steps based on a deep learning AE network;
- Training introduces a nonnegativity constraint algorithm into the training of the AE to learn a sparse, part-based representation of the input data;
- Trainings can be stretched to train deep network with stacked-AEs and softmax classification layer;
- The learning algorithm constrains the negative weights;
- Is able to disentangle hidden, meaningful signal structure from partial representations of the data.

A three-layer fully connect DNN based on auto encoders for the automatic modulation classification is shown in Figure 9.30 [30].

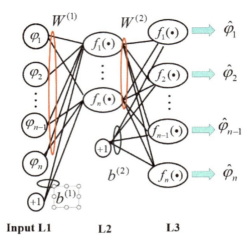

Figure 9.30 Schematic diagram of the autoencoder-based DNN. (©IEEE, reprinted with permission from [30].)

The auto encoder attempts to learn a function

$$\hat{\psi} = f_{\mathbf{W},b(\psi)} \approx \psi \qquad (9.27)$$

where $\mathbf{W} = \{W_1, W_2\}$ and $\mathbf{b} = \{b_1, b_2\}$ represents weight and biases of both layers, respectively. To optimize the parameters of the model in (9.27), for example, the average reconstruction error is used as the cost function:

$$J(\mathbf{W}, b) = \frac{1}{m} \sum_{z=1}^{m} \frac{1}{2} ||\hat{\psi}^z - \psi^z||^2 () \qquad (9.28)$$

where m is the total number of training samples.

9.6.1 Regularization Function

A regularization function $\Omega(h)$ can also be added to the loss function to force a more sparse solution [26]. The regularization function can involve penalty terms for model complexity, model prior information, penalizing based on derivatives, or penalties based on some other criteria such as supervised classification results. Regularization is most often used with a deep encoder, not a shallow one. Regularization encourages the AE to have other properties than just reconstructing the input, such as making the representation sparse, robust to noise, or constraining derivatives in the representation.

A *contractive* AE is an unsupervised deep learning technique that helps a neural network encode unlabeled training data. Autoencoders in general are used to learn a representation, or encoding, for a set of unlabeled data, usually as the first step towards dimensionality reduction or generating new data models. Denoising and contractive AEs and ones that also contain activations such as sigmoid or rectified linear units (ReLUs), (found in many AEs), satisfy conditions sufficient to encourage sparsity.

One common practice is to add a weighted regularizing term to the optimization function, such as the l_1 norm or

$$||\mathbf{h}||_1 = \sum_{m=1}^{K} |h(i)| \tag{9.29}$$

where K is the dimensionality of $\mathbf{h}$ and λ is a term which controls how much effect the regularization has on the optimization process. This optimization can be solved using alternating optimization over $\mathbf{W}$ and $\mathbf{h}$.

In conventional networks, two restrictions are imposed to the cost function in (9.28) to produce a more meaningful hidden representation of the input. The first restriction added to (9.28) is known as the *sparsity constraint*, which results in a rapid convergence of training using the backpropagation algorithm. Here, sparsity is imposed by a limitation to the activation of hidden units $\mathbf{h}$ using a Kullback–Leibler (KL) divergence function [31]. To impose sparsity, $\hat{p}_j = p$ where p is the positive sparsity parameter chosen close to 0.

In summary, the training is then stretched to train a deep network with *stacked auto encoders* and a softmax classification layer. The weights of the overall network were constrained to be nonnegative. The part-based representation of the data helped unravel the hidden structure of the data, producing a better reconstruction of the data [30].

9.6.2 Variational Auto Encoder

A variational auto encoder (VAE) based on a Bayesian construction has been introduced where the output data are conditioned on the input data in a probabilistic way. The VAE is a generative model of a set of data that simulates how the data are generated and is a directed probabilistic graphical model. The model approximates posteriors by a neural network. The generative model $p(x,z)$ can be decomposed as $p(x,z) = p_\theta(x|z)p(z)$, where x is a datum and z is a hidden variable [30]. The data are generated as independent and identically distributed (iid) data. The VAE does the inference of $p(z)$ by a *variational Bayesian* approach, and $p_\theta(x|z)$ is assumed to be parametric, where for estimating the unknown θ one can exploit the maximum likelihood method.

The VAE was also shown to have a bottleneck-shaped network able to capture the lower dimensional information efficiently. This suggests that that the VAE could be used to provide suitable distributions for the *adaptive importance sampling schemes* (AIS). It is a directed probabilistic graphical model that approximates posteriors by a neural network and is described in more detail in [32]. The introduction of an AIS method that exploits the VAE for constructing proposal distributions is described in [32]. Further, they introduce a banana-shaped target distribution as a benchmark for evaluating their importance sampling methods whose goodness-of-fit can be measured analytically.

9.7 CONVOLUTIONAL NEURAL NETWORK

The CNN consists of multiple layers of operations such as convolution, pooling, nonlinear activation functions, and normalization functions. The first part of the CNN is the *feature extractor*. The last part is a MLP that computes the probabilities of a given class being present in the input [26]. CNNs have been constructed and simulated for feature extraction, classification, and can be used to identify antenna pulse Doppler scan modulations, pulsed emitter pulse repetition intervals, and low probability of intercept (LPI) radar modulations and represent good examples for embedded DRFM applications [11, 33]. CNN deep learning has been used in 5G networks to predict mobility based on traffic flows between base stations [34]. Here the CNN architecture is explicitly tailored to the problem of anticipatory resource orchestration in mobile networks. Thorough empirical evaluations with real-world metropolitan-scale data show the substantial advantages granted by DeepCog over state-of-the-art predictors and provide a first analysis of orchestration costs at heterogeneous network slices and data centers. In addition, they can be combined with other techniques.

Convolutional neural networks (CNNs) are specialized DNN models to appropriately handle large, high-resolution images as inputs and exploit the relationship in nearby data (such as pixels in the image or location-based measurements). The CNNs use linear operations that compute a weighted average of nearby samples and pooling to summarize the data in a region. Typically, max-pooling or a similar operation is used depending on the objective. In addition, a CNN consists of multiple layers of operations called convolutions, pooling, nonlinear activation functions, and normalization, to name a few.

Most layers in the CNN have parameters to estimate, such as the convolution filter parameters, or the scales and shifts in batch normalization. For example, a CNN that analyzes grayscale imagery employs two-dimensional (2-D) convolution filters, whereas a CNN that uses red–green–blue (RGB) imagery uses three-dimensional (3-D) filters. In general, it is easy, in theory, to support any N-dimensional signal, such as in hyperspectral imagery, as it really only affects the dimensionality of the first layers of filters. Through training, these filters learn to extract hierarchical features directly from the data, versus traditional machine learning approaches that use hand-crafted features.

Consider convolution on a single grayscale image using a $m \times m$ mask. If the input data have dimensions of $K \times K$, then the convolution output will be $(K-m+1) \times (K-m+1)$ assuming a stride of one and no padding. Many times the CNNs start with the smaller mask sizes such as 3×3 or 5×5 and may have different sizes per layer.

9.7.1 Unmanned System Example

Unmanned systems track the moving objects and structures around them allowing coexistence with humans and other autonomous vehicles in the environment. Recognition and tracking of these objects in a shared location can simplify the design of autonomous navigation and obstacle avoidance algorithms and the ability to operate within multi-agent formations.

In [35], a CNN is used to solve the image processing, object recognition, and classification problems on an unmanned system. Here they discuss target recognition and target tracking. The input to the CNN is a $128 \times 128 \times 3$ image. It is transferred to three following convolutional layers with the same size. After this, a *max pooling* layer with size $64 \times 64 \times 10$ is used. Following this, three convolutional layers each with size $64 \times 64 \times 10$ are used. The second max pooling layer has $32 \times 32 \times 10$ and is followed by two convolutional layers of size $32 \times 32 \times 20$. The third max pooling layer has size $16 \times 16 \times 20$ and is followed by two convolutional layers of size $16 \times 16 \times 20$. The fourth max pooling layer is of size $8 \times 8 \times 20$ with the following two convolutional layers $8 \times 8 \times 30$. The fifth and final max pooling layer has size $4 \times 4 \times 30$. The output of the fifth max pooling layer is connected to a 1×64 fully connected layer with its output then supplied to a second fully connected layer of size 1×2. In summary, each convolutional layer is a rectified linear unit and the network has 12 convolutional layers and five max pooling layers.

Every one convolutional layer is followed by a ReLU activation. The ReLU activation is defined as

$$f(x) = \max(0, x) \tag{9.30}$$

The stride between the filters in horizontal and vertical directions is equal to one. The same padding adjustment is used for each one of the convolution layers. The usage of the same padding provides equality between the size of the input and the output matrices. Max pooling layers apply max filters to the sub regions of the input matrix. For each of the regions represented by the filter, the algorithm takes the max of that region and creates a new, output matrix where each element is the max of a region in the original input [35]. The size of the max filters is 3×3 and the stride between filters is 2. Each max pooling layer reduces the size of the matrix and the number of parameters to learn. Max pooling is used also to help overfitting. The activation function after the last fully connected layer is a softmax layer or

$$\sigma(x_j) = \frac{e^{x_j}}{\sum_{i=1}^{k} e^{x_i}} \qquad (9.31)$$

The probability distribution of the event over the n different events are calculated by the softmax function. The output from the softmax function varies from 0 to 1 with the sum of all probabilities equal to one.

For the multiclass classification problems, the CNN assigns each input to one of the k mutually exclusive classes. The loss function $E(\theta)$ for this case is the cross entropy function for a $1-of-k$ coding scheme as

$$E(\theta) = -\sum_{i=1}^{n}\sum_{j=1}^{k} t_{ij} \ln y_j(x_i, \theta) \qquad (9.32)$$

where θ is the parameter vector, t_{ij} is the indicator that the ith sample belongs to the jth class and $y_j(x_i, \theta)$ is the output for sample i and can be interpreted as the probability of association of the ith input with the class j by the network.

In this example, stochastic gradient descent with momentum (SGDM) is used as the optimization method. The updated value of the parameters is calculated according to

$$\theta_{l+1} = \theta_l - \alpha \nabla E(\theta_l) \qquad (9.33)$$

The CNN is trained with images sized 128×128 having 3 layers of RGB. The accuracy and loss during the training process is shown in Figure 9.31 using only 9 epochs.

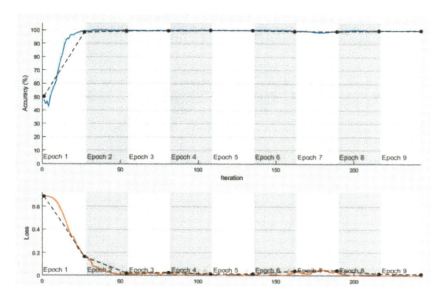

Figure 9.31 Accuracy and loss during the training process consisting of 9 epochs. (Adapted ©IEEE, reprinted with permission from [35].)

The number of convolution filters and the filter sizes, the pooling sizes and strides, and all other layer operators and associated parameters need to be learned. In general, it has been demonstrated that the lower layers of a CNN typically learn basic features, and as one traverses the depths of the network, the features become more complex and are built up hierarchically. Most layers in the CNN have parameters to estimate, for example, the convolution filter parameters, or the scales and shifts in batch normalization.

Figure 9.32 shows the detailed schematic of a convolutional neural network.

612   MACHINE LEARNING IN ELECTROMAGNETIC WARFARE

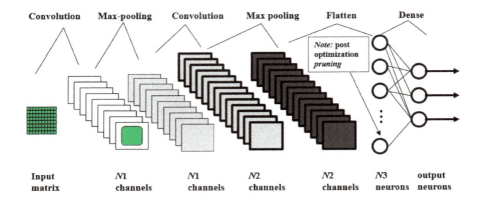

Figure 9.32 Schematic of a convolutional neural network. (©IEEE, reprinted with permission from [35].)

The CNN works by applying filters to the input data (e.g., time-frequency data) and tunes the filters as the training happens in real time. That is, the big difference between a CNN and a regular neural network like a MLP is that with the CNNs, convolutions are used to handle the filter designs behind the scenes. A convolution is used instead of matrix multiplication in at least one layer of the CNN. Convolutions take two functions and return a function. Usually the network will initially find the edges of the picture. Then this slight definition of the image will get passed to the next layer. That layer will then start detecting things like the corners, slopes, and color groups. Then that image definition will get passed to the next layer and the cycle continues until a prediction is made.

The *final layers* typically output class labels, estimates of the posterior probabilities of the class label, or some other quantities such as softmax normalized values. Convolution layers perform filtering (e.g., enhancement, denoising, detection, etc.) on the outputs of the previous layers. Most CNNs then apply a nonlinear activation function σ to the linear convolution result. A legitimate question is what hyperparameters to choose? Unfortunately, there is no easy answer, just rules of thumb and knowing from the literature what has worked for other researchers. Often experimentation and knowledge of the problem at hand are required [26].

For example, Figure 9.33 shows how an input image is filtered using a 3×3 filter mask that is convolved across the image to give a resultant to be passed along. The design of a filter mask (for example, a *Sobel* edge detection filter) and the computation of its frequency response as well has how to convolve the filter in the time-domain across the image is left as an image processing homework assignment for the reader.

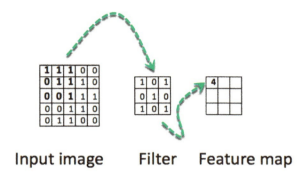

Figure 9.33 Convolution of an input image passing along the result to the feature map. (©IEEE, reprinted with permission from [36].)

9.7.2 Modulation Recognition Example

A radar signal intrapulse modulation recognition method based on convolutional denoising autoencoder (CDAE) and deep convolutional neural network (DCNN) is described in [37]. Using *denoising autoencoders* can help the network extract useful features from data, and improve the network noise immunity of the network. Cohen's time-frequency distribution (CTFD) is first used to convert radar signals into time-frequency images (TFIs), and then a bilinear interpolation and amplitude normalization is used to preprocess the TFIs.

A design for a CDAE for pretraining a classification network is used, which can directly extract the robustness features of TFIs and effectively remove noise from TFIs, and to avoid loss of information caused by filtering and binarization processing. Finally, the design of a deep convolutional neural network based on an inception architecture to identify the processed TFIs is applied. Figure 9.34 shows the data processing and recognition for identifying the signal modulations using deep CNN.

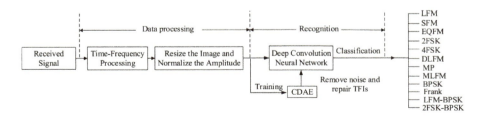

Figure 9.34 An example of the data processing and deep convolutional neural network for classification.

The autonomous identification of radar and communication intercepted modulation information provides the presence of threat emitters within the EMS. The first step after receiving the signal is to perform a Choi-Williams T-F processing. The 12 types of radar intra-pulse modulation signals are transformed into T-F images as discussed above. The 12 signals include LFM, sinusoidal frequency modulation (SFM), even quadratic frequency modulation (EQFM), 2FSK, 4FSK, dual-linear frequency modulation (DLFM), monopulse (MP), multiple linear frequency modulation (MLFM), BPSK, Frank, LFM-BPSK, and 2FSK-BPSK. After this, an image resizing and amplitude normalization is performed on all of the data with good results.

A denoising autoencoder shown in Figure 9.35 is applied to reduce the impact of noise on signal recognition at low SNR. However, there is a chance that this might result in a loss of signal information contained in the details of the image.

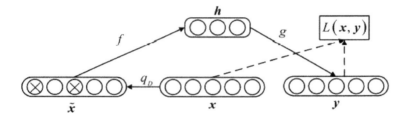

Figure 9.35 The basic structure of denoising autoencode.

The convolutional denoising autoencoder was designed to remove image noise, repair the distortion caused by noise in the image, and make the restored image closer to the original image. As shown in Figure 9.35 the purpose of the traditional autoencoder is that the output $y \approx x$, the input. The noise, q_D, is added to the input x to get $\tilde{x}$ as the input and then $y \approx x$ (input is noisy data and output is denoised data). The learning process of the denoising autoencoder can be expressed as

$$\min L(x,y) = \min L(x, g(f(\tilde{x}))) \tag{9.34}$$

9.7.3 Antenna Scan Modulation Example

In this section a CNN is used to extract the scan modulation features of a radar using the emitter's time-amplitude images. The proposed approach consists of three stages: preprocessing of signal, extracting feature, and classifying, as shown in Figure 9.36

9.7 Convolutional Neural Network

Figure 9.36 The basic structure of denoising autoencode.

First the received radar signal is converted into time-amplitude images that describe the signal strength at different time slots. Second, the convolution and pooling layers are applied for feature extraction. In the final step, the fully connected network is used to classify the scan modulation period [38].

The CNN consists of three convolutional layers, two pooling layers and two fully connected layers, to recognize the antenna scan period. A schematic of the CNN is shown in Figure 9.37

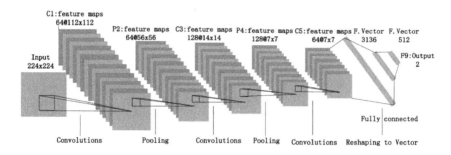

Figure 9.37 The CNN architecture used to detect the radar's scan period. (©IEEE, reprinted with permission from [38].)

The parameters for each convolutional layer are shown beginning with the input layer consisting of a 224 × 224 pixel image containing the radar signal's time-amplitude. Each convolutional kernel can be represented as a 4-D array $out_{num}, in_{num}, k_h, k_w$. Here k_h height of the kernel and k_w represents the width. Each convolutional kernel performs a weighted summation operation on a local area of the image. After a convolutional kernel operation, the result of an image is called a feature map. For instance, the convolution of the mth layer is calculated as

$$out_m = \sigma out_{m-1} * W_m + b_m \tag{9.35}$$

where σ represents the activation function, while W_m and Ab_m represent the weight and bias of the mth convolutional layer.

The deep learning for the CNN is the optimization of the objective function. For this network, a cross entropy loss function is used as optimization objective function. The gradient descent method is used to update the parameters in the network to minimize the objective function. The simulation results for a signal only; signal plus noise with amplitude 45 and noise amplitude 120 are shown in Figure 9.38.

Signal	Simulation Result			
	Number of training samples	Number of testing samples	Number of correct classification	Number of correct classification
pure signal	4200	600	600	100
signal with noise which amplitude is 45	4200	600	600	100
signal with noise which amplitude is 120	4200	600	595	99.17

Figure 9.38 Simulation results for recognizing the radar antenna scan period using a CNN.

Simulation results show that the proposed approach is effective to recognize the radar's scan period modulation.

9.7.4 Determining the Radar Pulse Repetition Interval Example

In this section, the CNN architecture is used to determine the pulse repetition interval (PRI) of an intercepted waveform [39]. Seven PRI configurations were tested [39]. To take advantage of CNN's characteristics of invariance to small shifts, the use of local filtering and max-pooling was used. In this way, practical PRI sequences with noises can be well recognized.

These seven PRI modulation modes have different characteristics described as follows:

- *Fixed PRI*: Fixed PRI sequence has a stable value that will does not change under ideal circumstances;
- *Jittered PRI*: The value of jittered PRI sequence jitters around a certain PRI value, and the range of jitter is always between 1% and 30% of the PRI value and the jittered value which is random but typically Gaussian or uniformly distributed;
- *Staggered PRI*: Several (2 to 7 typically) stable PRI values appear in cyclic order for staggered PRI sequences;
- *Sliding PRI*: Sliding PRI is the sequence whose PRI values monotonically increase (or decrease) to an extreme value. When it reaches the maximum (or minimum) value, it suddenly goes down (or up) to the minimum (or maximum) value;
- *Agile PRI*: Agile PRI has several certain PRI values like staggered PRI, but the difference is that these values are not changing in order, but in a random sequence;

- *D&S PRI*: Some constant values of PRI are used in dwell and switch PRI sequences. However, the difference between D&S PRI and staggered PRI is that the PRI remains on one certain value for a short duration, and then switches to another value in D&S PRI sequences;
- *Wobulated PRI*: Wobulated PRI sequence has the similar shape as a sinusoidal function, and it comes up periodically.

The seven types of PRI range from 200 to 2,000 to simulate a tactical situation. In this situation, the sequences can have a high ratio of lost and spurious pulses due to the influence of noise. The CNN is able to leave out complex data pre-processing, extract the features automatically, and adapt to the environment with missing and false information. These characteristics of the CNN make it ideal for recognizing the PRI modulation modes.

The CNN architecture is shown in Figure 9.39 and contains four convolution layers, 2 max-pooling layers and 2 fully connected layers. The output from the CNN are processed subsequently by a softmax function for classification.

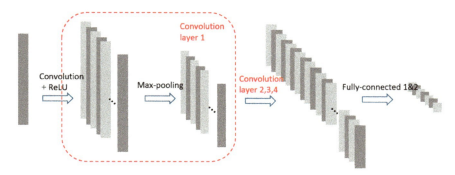

Figure 9.39 Simulation results for recognizing the radar antenna scan period using a CNN. (©IEEE, reprinted with permission from [39].)

For the input, let $x \in \mathbb{R}^{h \times \omega \times c}$ be the input PRI data where $h \times \omega \times c$ represent the height, width, and number of channels. Since the PRI data are 1-D, $h = 1$. Also, ω is a constant indicating the record length is always the same [39]. A block takes x and a set of parameters W as input and produces a vector $y \in \mathbb{R}^{h \times \omega \times c}$ as output where x is the input and W is the weight matrix. That is, $\mathbf{y} = f(\mathbf{x}, W)$.

The CNN described in [39] for PRI identification uses three kinds of layers including four convolutional layers, two pooling layers, and two fully connected layers. The first convolution layer C_1 uses 32 filters with kernel size 3, the second layer C_2 has 64 filters with kernel size 3 and the third and fourth C_3, and C_4 have 128 filters with kernel size 2. Convolution layers C3 and C4 are of the same structure with 128 filters and kernel size 2. The computation of convolutional layers can be formulated as

$$\mathbf{y} = f(\mathbf{b}_{k'} + \sum_i^h \sum_j^\omega \sum_d^c W_{ijdk'} \times \mathbf{x}_{i'+i,j'+j,d}) \quad (9.36)$$

where input **x** with the number of filters $W \in R^{h' \times \omega' \times c'}$ also called as kernels and adds a bias $\mathbf{b}\ R^{c'}$.

The max-pooling layers are used after convolution to perform down-sampling and the size of feature maps through a max-pooling layer will be reduced by 1/2. There are also two max-pooling layers that follow the first two convolutional layers [39]. The 2 fully connected layers that follow all the convolution and pooling layers all every kernel in these layer to be connected to all the feature maps in the previous layers. The first fully connected layer has 64 neurons while the other has 7 neurons in order to do the PRI classification.

For the simulation results, in training phase, the weights are initialized using random initialization ranging from 0 to 1. In all the experiments, we use 90% of data for training and 10% of data for validation. For the amount and length of the dataset, there are two parameters including the number of the training samples and the length of the signal. The process of PRI modulation recognition is shown in Figure 9.40(a) while the results are shown in Figure 9.40(b).

Three methods are compared, the traditional method, common neural network method (FC), and CNN method. The traditional method always compare the different changing regulations among the PRI sequences. Common neural networks method uses fully connected networks to recognize PRI modulation modes [39].

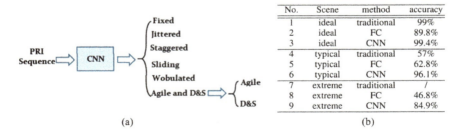

(a)   (b)

Figure 9.40 Simulation results showing (a) process for PRI modulation recognition and (b) recognition for results under the different test circumstances where FC is using a common neural network method. (Adapted from [39].)

In summary, the results show that using convolution and max-pooling significantly improves the recognition performance. Simulation results show the overall ratio of recognition is 96.1% with the upper most 50% lost pulses and 20% spurious pulses.

9.7.5 Recognition of LPI Radar Waveforms Example

In this section, a LPI waveform recognition technique using a CNN is described. Figure 9.41 illustrates the block diagram of the proposed LPI waveform recognition technique.

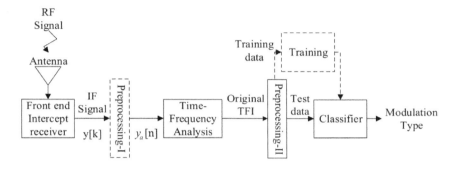

Figure 9.41 LPI waveform recognition technique [33].

The intercepted signal is down-converted to an intermediate frequency (IF) f_I and then sampled at $f_s = (1/T_s)$ to yield $y[k]$, where k is the sample index. A set of consecutive N signal samples are collected over τ_{pw} seconds (i.e. $N = \tau_{pw} f_s$) and processed by the preprocessing-I block for the proposed sample averaging technique (SAT).

This technique focuses on classifying LPI radar waveforms and assumes that the signal samples are collected for a signal pulse interval (τ_{pw}) and the coarse estimate of the carrier frequency is obtained. The preprocessing-I block is an optional function used only when $f_s/(f_I + B_s/2)$ is a multiple times larger than N_{sc} that is the required number of samples per cycle of a carrier wave of the highest frequency allowed by the receiver bandwidth B_s. Also $N_{sc} \geq 4$ does not cause an SNR loss larger than 1.0 dB [33].

The LPI modulations are shown in Figure 9.42 and are the same as those discussed in [11]. Note that this reference also quantifies the detection and classification of these exact modulations however, it only uses multi-layer perceptrons. In addition [11] also emphasizes that fact that the modulations must be identified first prior to any of the modulation parameters being extracted.

Modulation type	$f[k][Hz]$	$\phi_{i,j}[k][rad]$ for subcode
LFM	$f_0 + \frac{B}{\tau_{pw}}(kT_s)$	constant
Costas	f_j	constant
BPSK	constant	0 or π
Frank	constant	$\frac{2\pi}{M}(i-1)(j-1)$
P1	constant	$-\frac{\pi}{M}[(M-(2j-1))][(j-1)M+(i-1)]$
P2	constant	$-\frac{\pi}{2M}[2i-1-M][2j-1-M]$
P3	constant	$\frac{\pi}{\rho}(i-1)^2$
P4	constant	$\frac{\pi}{\rho}(i-1)^2 - \pi(i-1)$
T1	constant	$\mathrm{mod}\left\{\frac{2\pi}{N_{ps}}\left\lfloor (N_g(kT_s) - j\tau_{pw})\frac{jN_{ps}}{\tau_{pw}}\right\rfloor, 2\pi\right\}$
T2	constant	$\mathrm{mod}\left\{\frac{2\pi}{N_{ps}}\left\lfloor (N_g(kT_s) - j\tau_{pw})\left(\frac{2j-N_g+1}{\tau_{pw}}\right)\frac{N_{ps}}{2}\right\rfloor, 2\pi\right\}$
T3	constant	$\mathrm{mod}\left\{\frac{2\pi}{N_{ps}}\left\lfloor \frac{N_{ps}B(kT_s)^2}{2\tau_{pw}}\right\rfloor, 2\pi\right\}$
T4	constant	$\mathrm{mod}\left\{\frac{2\pi}{N_{ps}}\left\lfloor \frac{N_{ps}B(kT_s)^2}{2\tau_{pw}} - \frac{N_{ps}B(kT_s)}{2}\right\rfloor, 2\pi\right\}$

$\mathrm{mod}\{a, b\}$: remainder after division between a and b

$\lfloor a \rfloor$: biggest integer smaller than or equal to a

Figure 9.42 LPI modulations to be extracted. (From [11, 33].)

Figure 9.43 shows the CNN architecture used in the classification of the LPI waveform modulations [33].

9.7 Convolutional Neural Network

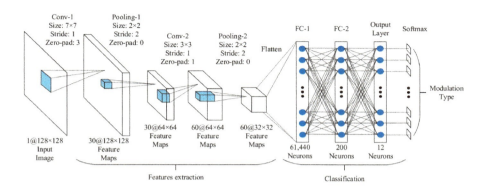

Figure 9.43 CNN architecture for the classification of the LPI modulations (©IEEE, reprinted with permission from [33].)

The CNN consists of a (image) feature extraction block and a classification block. Because the feature extraction block is integrated inside, CNN does not require any prior feature extraction function. In addition, the convolution and pooling processes in the CNN make the CNN robust to the geometrical distortions, such as scaling, shift, and rotation, and to the noise in the input image.

The basic structure of the CNN can be described in a sequence of functions as Input - Conv - ReLU - Pooling - Conv - ReLU - Pooling - FC - Dropout - FC where Conv represents the convolution layer, ReLU is the rectified linear unit, Pooling is the pooling layer, FC denotes the full connected layer, and Dropout is the dropout layer. Based on the basic structure, the hyperparameters are designed such as the input size, convolution filter size, the number of Conv feature maps, and the number of neurons in the FC, to find the optimal structure for the LPI waveform classification problem based on numerous Monte Carlo simulations for various conditions.

The confusion matrix for this example for $SNR = -6$ dB is shown in Figure 9.44. As shown, polyphase codes have low pcc, because the Choi-Williams time frequency images of the P1 and P4 codes can be confused with that of the LFM, and those of Frank and P3 codes can be confused when the SNR is low.

	BPSK	Costas	LFM	Frank	P1	P2	P3	P4	T1	T2	T3	T4
BPSK	99	0	0	0	0	0	0	0	1	0	0	0
Costas	0	99	0	0	0	0	0	0	1	0	0	0
LFM	0	0	94	0	4	0	0	2	0	0	0	0
Frank	0	0	0	89	0	0	11	0	0	0	0	0
P1	0	0	2	1	86	0	0	10	0	0	0	1
P2	0	0	0	0	0	100	0	0	0	0	0	0
P3	0	0	1	7	0	0	91	1	0	0	0	0
P4	0	0	4	0	9	0	0	85	1	0	0	1
T1	1	0	0	0	0	0	0	0	97	0	2	0
T2	1	0	0	0	1	0	0	2	0	94	0	2
T3	0	0	0	0	0	0	3	0	1	0	95	1
T4	0	0	0	0	1	0	0	2	0	3	0	94

Figure 9.44 Confusion matrix of the proposed technique for the classification of the LPI modulations for $SNR = -6$. (©IEEE, reprinted with permission from [33].)

9.7.6 Electromagnetic Warfare Example

In [40], a similar CNN was used for the classification of EMW emitters is investigated using a large data set with 58 independent emitter sources (for training and testing). The CNN was formed similar to Lenet-5 [41]. The signal preprocessing creates 3-dimensional images with a feature space composed of pulse width (PW), radio frequency (RF), and pulse repetition interval (PRI), referenced with respect to time of arrival (TOA). The images are created and randomly assigned to the training and validation set. The images are normalized between [0, 1]. Figure 9.45 shows the architecture for the classification of the EMW emitters.

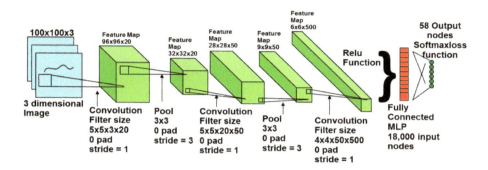

Figure 9.45 CNN architecture used for the classification of EMW emitters. (©IEEE, reprinted with permission from [40].)

The image representation has proven to be the most effective, consistently producing classification accuracies approaching 98.7%.

9.8 RECURRENT NEURAL NETWORKS

The RNN is a deep NN where the connections between nodes form a *directed graph* along a *temporal sequence* allowing it to exhibit temporal dynamic behavior. The RNN attempts to capture the temporal dynamic behavior of the observed imaging sequences, facilitating the classification on the temporal dynamics of the different inputs. They contain an internal state memory to process input sequences of variable length, making them ideal for applications such as unsegmented, connected handwriting recognition or speech recognition as well as remote sensing [42].

RNNs are based on David Rumelhart's work in 1986 with Hopfield networks–a special kind of RNN–discovered by John Hopfield in 1982. RNNs can be adjusted to the one-to-many mode used when a single input is mapped onto multiple outputs. There are also second-order RNNs and gated recurrent units.

RNNs work well with inputs that are in the form of sequences. A recurrent neural network is a type of deep learning neural network that remembers the input sequence, stores it in memory states/cell states, and predicts the future words/sentences [43]. Applications include speech recognition, machine translation, timeseries forecasting, and sales forecasting. They can be configured to allow for both sequential and parallel computation and, in principle, can compute anything a traditional computer can compute.

Figure 9.46 shows the block diagrams of two RNNs. Figure 9.46(a) shows a simplified version of a RNN in the Keras database.

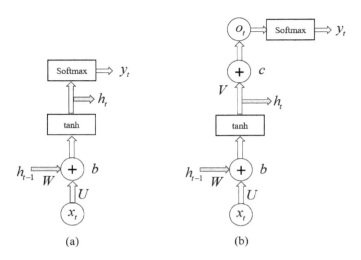

Figure 9.46 Block diagram of two RNNs showing (a) a simple RNN layer in the Keras database and (b) the true RNN.

The simple RNN expression is

$$h_t = \tanh(W h_{t-1} + U x_t + b) \tag{9.37}$$

where the subscript t represents the position within the input sequence h_t. The output expression for the true RNN shown in Figure 9.46(b) is the same as in (a) but only the softmax layer output is different since it works on a altered form of h_t or $o_t(V h_t + c)$.

Softmax is an activation layer (similar to tanh, sigmoid, etc.) and is a generalization of the logistic regression. It scales the input samples into a probability vector and can be applied to give a continuous output. The softmax activation function is

$$y(t) = \frac{e^{h_t}}{\sum_{k=1}^{K} e^{h_k}} \tag{9.38}$$

and shows the output y_t as the corresponding probability distribution for the different mutually exclusive output classes. That is, y_t is the probability that the sample belongs to certain class where the N inputs sum to 1.

There are many varieties of these networks. Most of them configured to handle temporal information in different ways to include a short term memory. Unlike traditional computers, however, RNN are similar to the human brain, which is a large feedback network of connected neurons that somehow can learn to translate long sensory input streams into a sequence of useful motor outputs. That is, the RNN feeds its intermediate or final outputs back into its own inputs, forming a dynamical system that has an internal state or memory. Consider the architecture shown in Figure 9.47. Consider x_{11}, x_{12}, x_{13} as inputs and O_1, O_2, O_3 as outputs of the hidden layers 1, 2, and 3 respectively. The inputs are sent to the network at different time intervals. That is, x_{11} is sent to the hidden layer 1 at time t_1, x_{12} at t_2, and x_{13} at t_3.

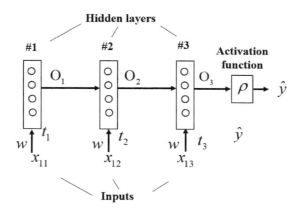

Figure 9.47 Architecture of a RNN example with three hidden layers and an output activation layer. (After [43].)

The weights are the same in the forward propagation and the output O_3 is dependent on O_2 that in turn is dependent on O_1 as

$$O_1 = \rho(x_{11} * w)$$
$$O_2 = \rho(O_{11} + x_{12} * w)$$
$$O_3 = \rho(O_{22} + x_{13} * w) \tag{9.39}$$

where w is the weight and ρ is the activation function. Finally the output is indicated by $\hat{y}$ as [43]

$$\hat{y} = \rho((x_{11} * w) + (x_{12} * w) + (x_{13} * w)) \tag{9.40}$$

The architecture shown above is a version of a simple recurrent network.

9.8.1 Loss Function

A *loss function* can be considered as the difference between the desired output y and the RNN output from the activation function $\hat{y}$. The loss function can be calculated as $L = (y - \hat{y})^2$. The goal is to reduce L to the point $L = 0$ in order to reach the global minima that establishes the appropriate weight that has to be added in the network. This is achieved in backpropagation by using *optimizers* to adjust the weights.

An example of the application of the chain rule of differentiation during backward propagation, we can calculate the loss function with respect to w_3 as:

$$\frac{\partial L}{\partial w_3} = \frac{\partial L}{\partial \hat{y}} \left(\frac{\partial \hat{y}}{\partial O_4} \right) \frac{\partial O_4}{\partial w_3} \cdots \qquad (9.41)$$

Training can also be done using gradient descent, which is a first-order iterative optimization algorithm for finding the minimum of a function.

The concept of *bidirectional RNN* is coupling two hidden layers which have the input and producing output. The invention is that the output we get for a particular hidden layer of interest will have information from the past and also the future. See the architecture in Figure 9.48 below [43].

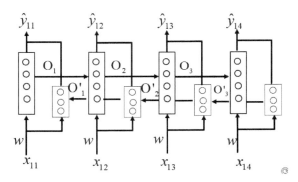

Figure 9.48 Architecture of a simple bidirectional RNN. (©IEEE, reprinted with permission from [43].)

To predict the output of $\hat{y}_{13}$, we have O_1, O_2 (from forward direction), and also O'_3 (from reverse direction).

9.8.2 Disadvantages of RNN

The major disadvantage in using the RNN for processing sequential, time-series data is what is known as the *Vanishing Gradient* problem, which occurs when the temporal contingencies present in the input/output sequences span long intervals [44]. At this point, a new weight added will become equal to old weight, thus, no change, and training the network becomes difficult. There is also the exploding gradient problem where the weight updating is so huge that the network cannot learn from training data hence global minima can never be reached.

In summary, training a RNN to learn long range input/output dependencies is a hard problem and gradient based methods appear inadequate. In [45], a modified partial RNN and its algorithm are investigated for recognition of high resolution ship targets.

9.8 Recurrent Neural Networks

In a partial RNN, the connections are mainly feedforward, but include a carefully chosen set of feedback connections as well.

There have been several RNN architectures proposed however, the two most popular types are the Elman network and the Jordan network The Elman network is shown in Figure 9.49.

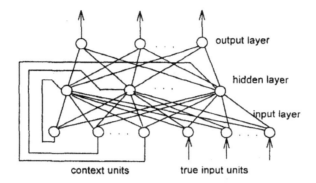

Figure 9.49 The Elman RNN architecture.

Note the similarities to the MLP studied earlier in the chapter, except for the feedback of a few of the nodes to a few of the other hidden layer nodes. The input layer is divided into two parts: the true input units and the context units. The context units simply hold a copy of the activations of the hidden units from the previous time step [45]. The modifiable connections are all feedforward, and can be trained by conventional backpropagation methods each step in time during the training of a particular sequence.

Figure 9.50 shows the Jordan architecture.

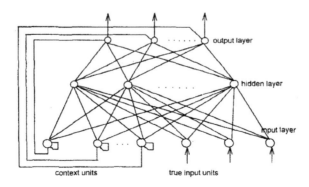

Figure 9.50 The Jordan RNN architecture.

It differs from that of the Elman by having the context units fed from the output (instead of the hidden layer). The self-connections give the context units some individual memory.

However, for a sequence recognition problem, the Elman network and Jordan network have some disadvantages. For instance, the Jordan network has feedback from the output layer, therefore its training is sensitive to the desired output. Its desired output can be set as classification result, but the convergence of training is not so easy to reach. For the Elman architecture, it is trained to output, the $(i+l)$th pattern in the sequence upon receiving at the input the ith pattern in the sequence. Usually the number of output units is not equal to that of the target class [45].

To overcome these disadvantages, a partially modified RNN was constructed as shown in Figure 9.51. Figure 9.50 shows the Jordan architecture.

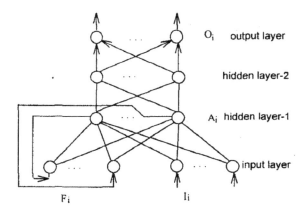

Figure 9.51 The new RNN architecture that combines the advantages of both the Elman and the Jordan network.

This new architecture combines the advantages of the above two networks. Its context units are fed from the hidden layer, that is the same as Elman network, but the number of its output units can be set equal to that of target class, that is just like a Jordan network. The two hidden layers are employed for the benefit of training speed. The feedback is taken from the first hidden layer.

Training the network can be accomplished with conventional backpropagtion. The connection weights are updated once at each time step during the training of a particular sequence. At each time step, feedback connections are ignored and the network is viewed as a multi-layer feedforward network with input units consisting of two parts Fi and Ii.

9.8.3 Naval Ship Class Example

The network was trained to recognize 8 ship-class targets within an angle ranging from 0° to 30° using 100 noisy waveforms for each target within each 1° angle. The waveforms of a particular target are relatively stable in a given direction but change considerably with angle. Figure 9.52 shows three of the training sequences.

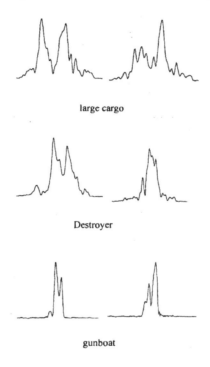

Figure 9.52 Three ship training sequences. (©IEEE, reprinted with permission from [45].)

The average waveform is taken in each 3° angle for each target to form the training set. One difficulty is if the sequence becomes too long in length. In this case, it was found that the training becomes difficult due to the gradient descent mechanism in weight update algorithm (as discussed above). Genetic algorithm-based training methods may be a useful approach to the long sequence recognition problem. The results of the ship recognition using the new RNN are shown in Table 9.4. The modified partially RNN architecture method reaches above 90% average recognition rate for 8 high-resolution radar ship-class targets. It is also tolerant of time shift up to a certain degree.

In [46] an RNN is used for pulse radar detection with both 13-bit and 35-bit barker codes used as input signal codes. The results of the simulation are compared

Table 9.4

RNN Ship Recognition Performance.

Target	Destroyer	Frigate	Gunboat	Chaser	Large cargo	Med. cargo	Small cargo	Passenger
Recog.	89%	88%	95%	93%	92%	87%	92%	86%

(From [45].)

with the results obtained from the simulation of pulse radar detection using an MLP network. The number of input layer neurons is same as the length of the signal code and three hidden neurons are taken in the present systems. Their conclusions were that the RNN based pulse detection system converges faster than the MLP based system and from the simulations, they also observed that RNN gives significant improvement in noise performance and range resolution ability. Finally under Doppler shift conditions, the RNN gives much better signal-to-sidelobe ratio of 27.68 dB as compared to the MLP which is only 16.38 dB for a 13-bit barker code.

In [47], an LSTM classifier is used to categorize land-cover using a multitemporal interferometric SAR (InSAR). Specifically, the Sentinel-1 InSAR is used as it preserves the spatial context of the land-cover by using grey-level spatial dependencies and morphological profiles. In addition, they were able to train the LSTM classifier to capture the temporal dynamics of the InSAR coherence. The Sentinel-1 was used to capture 39 interferometric coherence pairs acquired over Donana, Spain to evaluate the performance of the method [47].

9.9 LONG-, SHORT-TERM MEMORY

Fortunately, advancements in the types of neurons (or activation units) can correct the disadvantages apparent in the RNN. It is through the addition of a couple of states: a hidden state and a cell state, so they have a *long-, short-term memory* (LSTM). RNNs only have hidden states. The LSTM networks were introduced in 1997 by Hochreiter and Schmidhuber [48].

The LSTMs forget some information that is not important when the context changes, thus working very efficiently even for long sentences, which is not the case with RNN [43]. LSTM networks are useful when the data is the combination of both time and image (e.g., video). That is, they are able to capture the temporal aspect of change within the video images over time. LSTM networks are the most commonly used variation of RNNs and are specifically known for better update equations and better backpropagation [49]. Concerning the training, the computational complexity of the forward and backward pass operations scale linearly with the number of timesteps in the input sequence.

The LSTM applications include:

- Classifying time series;
- Speech recognition;
- Handwriting recognition;
- Polyphonic music modeling;

9.9.1 LSTM Architecture

The main components of the LSTM architecture are the (a) memory cell, (b) input gate, (c) forget gate, and (d) output gate. Figure 9.53 shows a schematic diagram of the LSTM architecture.

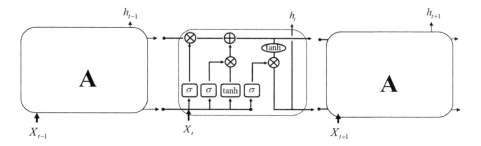

Figure 9.53 LSTM architecture schematic showing the gates of operation.

The first gate (or σ operation) within the center block is the forget gate and is shown in Figure 9.54.

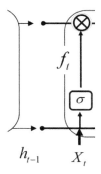

Figure 9.54 LSTM architecture schematic showing the forget gate operation.

The inputs are h_{t-1} and X_t. They are passed to the sigmoid activation function that outputs values between 0 and 1.0. Here a "0" means *completely forget* information and a "1" means *completely retain* information. Letting f_t represent the output of the sigmoid activation function σ then

$$f_t = \sigma\left(W_f[h_{t-1}, X_t] + b_f\right) \tag{9.42}$$

Here b_f is the bias and W_f is the combined weight of the two inputs.

The next gate over is the input gate and it consists of the next sigmoid and the hyperbolic tangent function as shown in Figure 9.55.

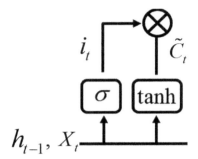

Figure 9.55 LSTM architecture schematic showing the input gate operation.

The input gate design objective is to identify new information and add to the cell state. The new information is added in two steps:

Step 1: The sigmoid layer outputs a value between 0 and 1 based on the inputs ht-1 and xt, as seen in the diagram above. At the same time, these inputs will be passed to the tanh layer, which outputs values between -1 and 1 and creates vectors for the inputs:

$$i_t = \sigma\left(W_i[h_{t-1}, X_t] + b_i\right) \tag{9.43}$$

$$\tilde{C}_t = \tanh\left(W_C[h_{t-1}, X_t] + b_C\right) \tag{9.44}$$

Step 2: The output of the sigmoid layer and tanh layer is multiplied:

$$C_t = f_t C_{t-1} + i_t \tilde{C}_t \tag{9.45}$$

which updates the state of the cell from C_{t-1} (the previous state of the cell) to C_t the current state of the cell.

The last gate is the output gate and is shown in Figure 9.56.

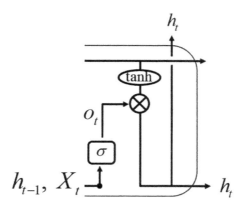

Figure 9.56 LSTM architecture schematic showing the output gate operation.

The cell state first passes through a tanh function and then simultaneously inputs h_{t-1} and X_t are passed to the sigmoid function layer. Then multiplication takes place and h_t is the output of the memory cell and it is then passed to the next cell and so on.

The input gate and the forget gate make up the various behaviors of the LSTM architecture. A truth table of the LSTM behaviors as a function of the states of the input gate and the forget gate is shown in Figure 9.57.

Input gate	Forget gate	Behavior
0	1	remember the previous value
1	1	add to the previous value
0	0	erase the value
1	0	overwrite the value

Figure 9.57 LSTM cell behaviors as a function of input gate and forget gate binary values {0,1}. (After [50].)

Alternate detailed description of the LSTM neuron development is given in the tutorial series [50].

9.9.2 Recognition of Unknown Radar Emitters Example

The first application of LSTMs and MCs for *unknown emitter recognition* is discussed below [51]. For ELINT operations, it is important to recognize if an intercepted signal belongs to an unknown radar emitter (as investigated above) often termed open-set recognition. One of the most important target sets in the ELINT area is being able to recognize radar *unknown emitters* or emitters that have not been seen before. The effort described here compares two general approaches, which are the "memoryless"

Markov chain and the LSTM recurrent neural network, which is specially designed to "remember" the past. The performance is demonstrated with two evaluation metrics in ten scenarios that contain different combinations of known and unknown emitters. An evaluation with corrupted data provides an estimate on the method's accuracy under challenging conditions.

Figure 9.58 shows a visualization of the deinterleaving classifier. The input to the classifier are pulses from various known and unknown emitters. The classifier deinterleaves the waveform pulses.

Figure 9.58 The input to a classifier that deinterleaves the PDWs that belong to different emitters.

After this step, separated PDW sequences are obtained, but no information about the emitters' identities is given.

The work in [51] presents an investigation of six approaches in several configurations to recognize unknown emitters based on the hierarchical emission model. Four of the methods employ LSTMs and two are based on Markov Chains (MCs). The LSTMs are implemented with three different loss functions, which are cross-entropy (CE), a loss function called the entropic open-set (EOS), and an approach called deep open classification (DOC). The All considered classifiers for unknown emitter recognition are trained with five training cases, which differ in the contained known unknown data.

The different types of input that a classifier for open-set recognition needs to handle are:

- *Known classes*: Classes that the classifier should recognize and identify $\mathbb{K}$;
- *Known unknown classes*: Should be classified as unknown/rejected $\mathbb{V}$;
- *Unknown unknown classes*: Data of unknown unknown classes should be classified as unknown but is unavailable at training time and only encountered at test time $\mathbb{U}$.

The evaluation is performed with ten different test cases containing several combinations of classes from the $\mathbb{K}, \mathbb{V}$ (known unknown emitters), and $\mathbb{U}$. A compromise between the true and false rejection rate, as well as the accuracy for discriminating between known classes is found by selecting the configuration of the classifier, that consists of the training case and the value of the threshold δ (if a threshold is used) [51]. In addition, estimates of the expected performance for the distinction between known and unknown input when choosing the best configuration for the identification of known classes and vice versa is presented.

Also determined is whether a single classifier is enough for a good accuracy or a hierarchical combination of different classifiers is preferred [51]. An evaluation with corrupted data further provides estimates on the robustness of the classifiers. The emitter encoding is quite unique. A hierarchical emission model that considers the emitters as systems that speak a language is the basis for the presented methods. In analogy to natural language, the different modelling levels are defined as follows:

- *Letters*: Correspond to the emitters pulses defined e.g., by PRI, RF, and PW;
- *Syllables*: Correspond to emitter bursts and are common combinations of letters;
- *Words*: Syllables can be combined to form words - resemble radar dwells;
- *Commands*: Describe word types or classes on a higher level of abstraction;
- *Functions*: Different purposes of the emissions map to functions e.g., searching or tracking targets;

Once the emitters are intercepted, they are sorted by common properties into two separated sequences, such that each sequence only contains the PDWs of a single emitter. After deinterleaving, the symbol extraction maps the PDW sequences to the symbols defined by the emission model. For example, the result might be a sequence of blue and a sequence of green syllables. These symbol sequences are the input to the open-set recognition method. Also, a low signal-to-noise ratio (SNR) can lead to a missed/wrong detection of symbols, but does not influence the processing after a correct detection.

The emitter symbols are encoded into a numerical representation using word embeddings, namely *word2vec* with each emitter e, having its individual dictionary Ω^{l_e} at model level $<$ where $l \in$ {letters, syllables, words, commands, functions}. There is also a global dictionary Ω^l containing all symbols. The embedding *word2vec* introduces a symbol called UNK that is a placeholder for a new input *without* an embedding [52]. The identification accuracies for letters, commands, and functions are not satisfactory for the LSTMs even without unknown unknown emitters, only syllables and words are considered, i.e. $l \in$ {syllables, words}.

Figure 9.59 gives an overview of the methods employed for unknown emitter recognition. Here Rv1= Rules-v1 and Rv2=Rules-v2. For each method, the case and dictionary are given and whether a threshold δ is used.

Method	Case	Dictionary	Thres.	Classes
LSTM - Cross-Entropy	0	Ω^l	yes	QoS, Rv1, Rv2
	I	Ω^l	no	QoS, Rv1, Rv2, unk
	II-V	$\Omega^l_{QoS} \cup \Omega^l_{Rv1}$	no	QoS, Rv1, unk
LSTM - Entropic Open Set	I	Ω^l	yes	QoS, Rv1, Rv2
	II-V	$\Omega^l_{QoS} \cup \Omega^l_{Rv1}$	yes	QoS, Rv1
LSTM - Deep Open Class.	I	Ω^l	yes	QoS, Rv1, Rv2
	II-V	$\Omega^l_{QoS} \cup \Omega^l_{Rv1}$	yes	QoS, Rv1
LSTM - Unknown Gate	I	Ω^l	no	knwn, unk
	II-V	$\Omega^l_{QoS} \cup \Omega^l_{Rv1}$	no	knwn, unk
MC	0	Ω^l	yes	QoS, Rv1, Rv2
	I	Ω^l	no	QoS, Rv1, Rv2, unk
	II-V	$\Omega^l_{QoS} \cup \Omega^l_{Rv1}$	no	QoS, Rv1, unk
MC - Unknown Gate	I	Ω^l	no	knwn, unk
	II-V	$\Omega^l_{QoS} \cup \Omega^l_{Rv1}$	no	knwn, unk

Figure 9.59 Overview of the methods employed for unknown emitter recognition. (From [51].)

The implementation uses Python and TensorFlow. Training and evaluation are performed on a computer with two Intel Xeon E5-2637 v2 CPUs and two GPUs (NVIDIA GeForce RTX 2080 Ti and GTX TITAN) including 130 GB of RAM. This configuration allowed the investigation to be conducted using parallel processing.

The generation of the training and testing emitter data is described in [51]. The architectures for the different LSTM approaches being tested are shown in Figure 9.60. Further details are provided in [52]. The architectures of the different MC variant approaches being tested are shown in Figure 9.61. In contrast to the LSTM-based approaches, the MCs use the index of the symbol Ω^l in the dictionary instead of the embedding vector. Therefore, the symbols are mapped to an integer.

Figure 9.62 provides the symbols and parameters for each LSTM-based method. For the LSTM and the MC-based approaches, the dictionary only contains the symbols of $\mathbb{K}$. This includes the symbols of $\mathbb{V}$ as the data is generated from the known emitters [51]. For all methods, including the MC-based approaches, the dictionary only contains the symbols of $\mathbb{K}$ and includes the symbols of $\mathbb{V}$ as the data is generated from the known emitters.

9.9 Long-, Short-Term Memory 637

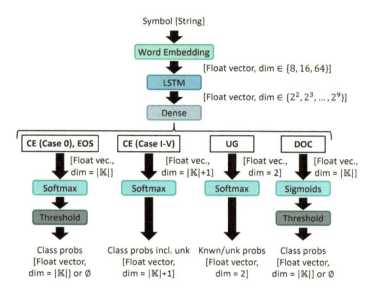

Figure 9.60 LSTM approaches being tested for the detection and recognition of unknown emitters. (©IEEE, reprinted with permission from [51].)

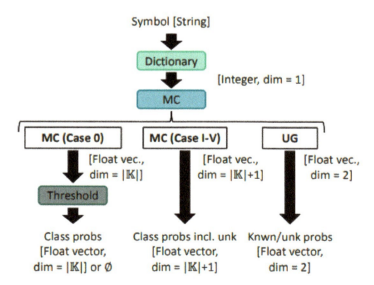

Figure 9.61 MC approaches being tested for unknown emitters. (From [51].)

			\multicolumn{6}{c}{Training Case}
	Symbol	Parameter	0	I	II	III	IV	V
LSTM-CE	Syllables	# layers # cells/layer	1 8	1 16	1 16	1 8	2 32	1 16
	Words	# layers # cells/layer	1 4	1 8	1 8	1 16	1 16	1 8
LSTM-EOS	Syllables	# layers # cells/layer	– –	1 16	1 16	1 16	1 16	2 16
	Words	# layers # cells/layer	– –	1 8	1 8	1 32	1 8	1 8
LSTM-DOC	Syllables	# layers # cells/layer	– –	2 64	1 64	2 16	1 64	1 128
	Words	# layers # cells/layer	– –	1 256	1 256	1 512	1 512	1 512
LSTM-UG	Syllables	# layers # cells/layer	– –	1 16	1 16	1 8	1 16	1 16
	Words	# layers # cells/layer	– –	1 4	1 8	1 16	1 32	1 8

Figure 9.62 LSTM architectures and their parameters. (©IEEE, reprinted with permission from [51].)

The classifiers are tested with several combinations of emitters in the $\mathbb{K}$ known, $\mathbb{V}$ (known unknown emitters), and $\mathbb{U}$ unknown unknown class. Figure 9.63 presents the mean distinction accuracies of the best configurations, averaged over all test cases "a" and "b." The MC outperforms all methods for syllables especially with only one symbol. The LSTM-UG however, provides higher accuracy than the MC-UG with longer sequences. For words, the MC is most accurate with only one symbol, but the MC-UG comes close.

9.9 Long-, Short-Term Memory

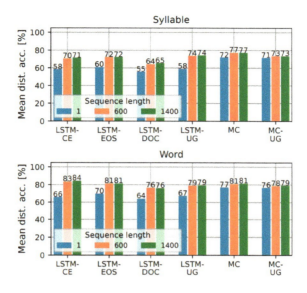

Figure 9.63 Mean distinction accuracies of the best configurations. (From [51].)

The MC outperforms all methods for syllables, especially with only one symbol. However, the LSTM-UG provides higher accuracies than the MC-UG, with longer sequences. For words, the MC is the most accurate with only one symbol but the MC-UG comes close. With longer sequences, the LSTM-CE gives the best performance on average. The LSTM-EOS and the MC achieve about the same results, just as the LSTM-UG and the MC-UG.

The cross entropy (CE) describes the differences between two probability distributions P and Q as

$$\mathbb{H}(P,Q) = -E_{x \sim P} \log Q(x) \qquad (9.46)$$

where E is the expected value. From information theory, the CE corresponds to the average number of bits required to encode an event of Q with a code that is optimal for P. The more similar P and Q are, the closer the required length is to the optimal code for P where P describes the true probability distribution that is to be learned and Q is the distribution estimated by the network. It is expected that during the training, the distributions will become more similar as the code length approaches the optimum [51].

The LSTM-DOC slightly falls behind the other methods however overall the distinction accuracies are higher for words since the unknown emitters have less words than syllables in common with the known emitters. This shows the benefit of the hierarchical emission model. To provide a more realistic scenario, the methods are also tested with corrupted data and are trained with ideals data only. As an example, the evaluation on sequences of length 1400 is presented with 20% missing or additional symbols. The symbols are removed or inserted randomly choosing additional symbols from the global dictionary Ω^l. Also symbols of other emitters might be inserted.

Figure 9.64 shows the results of the best configurations for the distinction accuracy with missing or additional syllables in comparison to the ideal results.

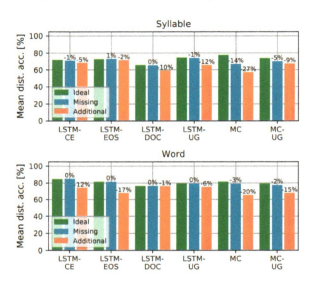

Figure 9.64 Mean distinction accuracies of the best configurations with 20% missing or additional symbols, resp., at a sequence length of 1,400 symbols. (©IEEE, reprinted with permission from [51].)

The labels of the bars are the accuracies relative to the ideal case calculated as

$$acc^{rel} = \frac{acc^{corrupt} - acc^{ideal}}{acc^{ideal}} \times 100\% \tag{9.47}$$

All LSTM-based methods are very robust with respect to missing syllables. The LSTM-CE and the LSTM-EOS only exhibit a small accuracy decrease with additional syllables.

A high rejection accuracy might be accompanied by a low acceptance accuracy and vice versa. Figure 9.65 depict these accuracies for each method and each test case with 1400 symbols using ideal data and 20% additional symbols. Figure 9.65(a) shows the results for Syllables and Figure 9.65(b) shows the results for Words.

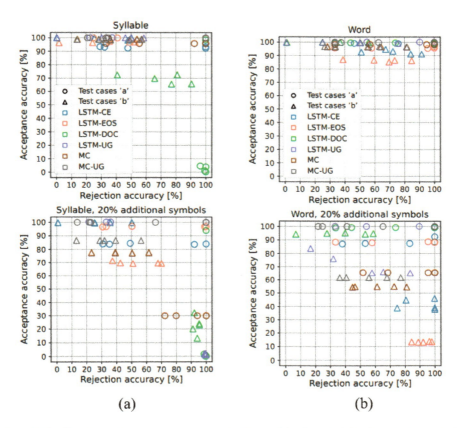

Figure 9.65 Rejection and acceptance accuracies with 1400 (a) *Syllables* and (b) *Words*. (©IEEE, reprinted with permission from [51].)

Most methods provide an acceptance accuracy $\geq 90\%$ for syllables and $\geq 80\%$ for words with ideal data, while the rejection accuracy varies greatly between the test cases. The LSTM-DOC, however, tends to reject nearly every input for the test cases "a" with syllables and also has the lowest acceptance accuracy in the test cases "b" however, it achieves much higher rejection accuracies for syllables.

As the LSTM-DOC classifies nearly every input as unknown in the cases "a," its rejection accuracy is high. As some unknown emitters are very similar to $\mathbb{K}$, correctly rejecting the unknown emitters also results in a false rejection of some of the known sequences. The best configuration of the LSTM-EOS includes a high threshold of 0.8 in the test cases "b" for words leading to a lower acceptance accuracy but with ideal data more than 80% of the input is accepted with a value ≥ 0.8. With additional symbols however, the high threshold leads to a low acceptance accuracy.

In summary, several variants of LSTM networks and Markov chains were tested which differ in the training and output classes. The LSTMs are employed with the cross-entropy (CE), entropic open-set (EOS), and deep open classification (DOC) loss. All methods are tested with different combinations of known and unknown emitters, as well as ideal and corrupted data [51].

An evaluation with corrupted data provides an estimate on the methods' accuracies under challenging conditions. Overall, the rejection results are better with words than with syllables. For most of the test cases and methods, the rejection accuracies are above 25% with words and ideal data, while the majority lies between 10% to 60% for syllables. With 20% additional symbols, the methods reject more sequences, leading to a higher rejection but a lower acceptance accuracy [51].

The investigation shows that unknown emitters that do not use known waveforms are reliably recognized even with corrupted data, while unknown emitters that are more similar to known ones are harder to detect. That is, the unknown emitters that are more similar to the known ones are less reliably recognized, while higher accuracies are achieved with words. For four of the five unknown unknown emitters, the best rejection rates are above 65% with words, which sufficient to tell that there are active unknown emitters.

9.9.3 InSAR Sentinel-1 Example

The LSTM has also been used for SAR imaging applications. Figure 9.66 shows the Sentinel-1 InSAR's several images that are used for a coherence evaluation.

9.9 Long-, Short-Term Memory

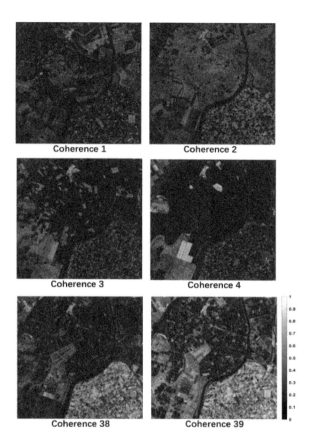

Figure 9.66 Several images from the InSAR-1 used to evaluate the complex interferometric coherence. (©IEEE, reprinted with permission from [47].)

Coherence between SAR scenes is important for two reasons. First its *phase* component is proportional to the vertical structure of the scene under observation. For example, relief topography can be obtained by means of two interferometric SAR scenes by extracting and utilizing the phase information. The second reason is that the *amplitude* contains information concerning the similarity between pairs of images, hence, it presents information about the phase quality (i.e., noise content).

The complex interferometric coherence between SAR scenes S_i and S_j is calculated as

$$\rho_{i,j} = |\rho|e^{j\phi} = \frac{E\{S_1 S_2^*\}}{\sqrt{E\{|S_1|^2\}E\{|S_2|^2\}}} \tag{9.48}$$

where $|\rho_{i,j}|$ denotes the magnitude and $e^{j\phi}$ the phase between the two images. The coherence is often used to assess the interferogram quality of the emitter.

In the coherence image series, changing seasonal texture of vegetation can be observed and can lead to temporal variation of features based on the gray-level appearance or gray-level co-occurrence matrix (GLCM) that calculates the gray scale relationship between a pixel and its adjacent neighbors. A flow chart summarizing the LSTM approach is shown in Figure 9.67. A summary of their classification performance is shown in Figure ??.

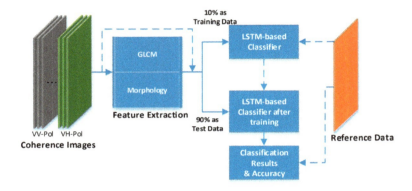

Figure 9.67 Flowchart showing the LSTM architecture used to evaluate the InSAR-1 coherence. (©IEEE, reprinted with permission (©IEEE, reprinted with permission from [47].)

LC classes	Test scale (pixels)	Accuracy (%)	
		UA	PA
Continuous urban fabric (111)	1635	56.68	90.28
Industrial or commercial units (121)	4517	90.00	86.07
Agricultural areas (200)	51548	81.25	60.49
Non-irrigated arable land (211)	9442	90.58	80.09
Permanently irrigated land (212)	181017	86.85	93.40
Rice fields (213)	491073	93.85	97.06
Pastures (231)	24234	90.90	92.01
Transitional woodland-shrub (324)	2242	78.76	67.66
Inland marshes (411)	55499	98.20	96.24
Salines (422)	8746	79.58	55.82
Inland waters-Water courses (511)	26996	79.23	93.03
Inland waters-Water bodies (512)	3925	95.98	85.15
Canals/Irrigation channels (0)	37384	54.36	29.48
Overall Accuracy (%)		90.26	
Kappa Coefficient		0.8476	

Figure 9.68 LC classes, test scale (pixels) and accuracy, overall accuracy and Kappa coefficient comparison. (©IEEE, reprinted with permission from [47].)

9.10 GATED RECURRENT UNITS

For hardware considerations, fast computation, and less memory consumption, the *gated recurrent units* (GRUs) are a choice and are similar to the LSTM as another recurrent unit that has hidden states. Just like LSTM, GRU uses gates to control the flow of information. They are relatively new as compared to LSTM. This is the reason they offer some improvement over LSTM and have a simpler architecture.

9.10.1 Original Concept

The GRU architecture was originally introduced in 2014 [53]. It learns to encode a variable-length sequence into a fixed-length vector representation and to decode a given fixed length vector representation back into a variable-length sequence. Figure 9.69 shows a schematic of the architecture. The encoder and decoder are jointly trained to maximize the conditional probability of a target sequence given a source sequence.

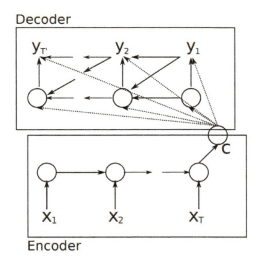

Figure 9.69 RNN Encoder–Decoder architecture that learns to encode a variable-length sequence into a fixed-length vector representation and to decode a given fixed-length vector representation into a variable length.

That is, the model presents a general method to learn the conditional distribution over a variable-length sequence conditioned on yet another another variable length sequence, or $p(y_1,\ldots,y_{T'}|x_1,\ldots,x_T)$ where the input and output sequence lengths T and T' can be different.

The encoder is an RNN that reads sequentially, each symbol of an *input* sequence **x**. As it reads each symbol, the hidden state of the RNN changes (see for example,

(9.37)). After the sequence is read entirely and marked by an end-of-sequence symbol, the hidden state of the RNN is a summary **c** of the entire input sequence [53].

The decoder is also an RNN that is trained to generate the *output* sequence by predicting the next symbol y_t given the hidden state $\mathbf{h}_t$. Unlike the RNN, both y_t and $\mathbf{h}_t$ are also conditioned on y_{t-1} and on the summary **c** of the input sequence. Thus, the hidden state of the decoder at time t is computed as [53]

$$\mathbf{h}_t = f(\mathbf{h}t-1, y_{t-1}, \mathbf{c}) \tag{9.49}$$

and likewise, the conditional probability of the next symbol is

$$P(y_t | y_{t-1}, y_{t-2}, \ldots, \mathbf{c}) = g(\mathbf{h}_t, y_{t-1}, \mathbf{c}) \tag{9.50}$$

for given activation functions f and g where the activation function g must produce valid probabilities by using, for example, a softmax.

Both the encoder and the decoder are jointly trained to maximize the conditional log-likelihood

$$\max_{\theta} \frac{1}{N} \sum_{n=1}^{N} \log p_\theta(\mathbf{y}_n | \mathbf{x}_n) \tag{9.51}$$

where θ is the set of model parameters and each $\mathbf{x}_n, \mathbf{y}_n$ is an input sequence, output sequence pair from the training set. Since the output of the decoder is differentiable, a gradient-based algorithm can be used to estimate the model parameters.

Once the RNN encoder–decoder is trained, the model can be used in two ways: (1) use the model to generate a target sequence given an input sequence, (2) use the model to score a given pair of input and output sequences, where the score is just the probability $p_\theta \mathbf{y} | \mathbf{x}$ calculated from

$$p(\mathbf{x}) = \prod_{t=1}^{T} p(x_t | x_{t-1}, \ldots, x_1) \tag{9.52}$$

and from (9.51).

Also used is a hidden activation unit that adaptively remembers and forgets. In addition to the new encoder-decoder a new type of hidden activation unit is presented in order to improve both the memory capacity and the ease of training, (for example, the tanh in (9.37)). The hidden activation function that adaptively remembers and forgets is shown in Figure 9.70.

9.10 Gated Recurrent Units

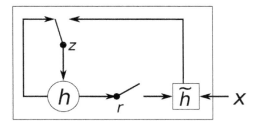

Figure 9.70 Hidden activation function that adaptively remembers and forgets.

Here the update gate z selects whether the hidden state is to be updated with a new hidden state $\tilde{h}$. The reset gate r decides whether the previous hidden state should be ignored.

The GRU is made up of an *update gate* (UG) and a *reset gate* (RG).[2] The UG controls the amount of information that must be passed forward as [53]

$$z_t = \sigma\left(W(z)X_t + U(z)h_{t-1}\right) \tag{9.53}$$

where $W(z)$ is the weight associated with X_t, $U(z)$ is the weight associated with input from the previous state that is h_{t-1}. and σ is the sigmoid activation function. The output z_t will be between 0 and 1 based on which information will be passed on [43].

The RG decides the amount of information to forget and is determined as

$$r_t = \sigma\left(W(r)X_t + U(r)h_{t-1}\right) \tag{9.54}$$

where $W(r)$ is the weight associated with X_t, and $U(r)$ is the weight associated with the input from the previous state that is h_{t-1}. The output r_t will be between 0 and 1 based on the information that will be forgotten.

The memory component (reset gate) now must be added into the network [43]. This reset gate pulls up the important information and assigns value = 1 and the remaining sentences/sequences will be assigned a value = 0. That is,

$$h'_t = \tanh(WX_t + r_t \odot Uh_{t-1}) \tag{9.55}$$

and now we have [43]

$$h_t = z_t \odot h_{t-1} + (1 - z_t) \odot h'_t \tag{9.56}$$

to calculate the current state h_t that will be passed onto the succeeding cells. The GRU was inspired by the LSTM unit but it is considered simpler to compute and implement.

Especially important in DRFM signaling is being able to process sequences. The idea behind *sequence to sequence learning* is that input data that is received in one

---

2   These are similar to the forget and input gates in the LSTM.

language is converted into another language. For example, the input data is a polyphase sequence where each n-tuple comes from a P4 + noise sequence. The input sequence is converter to an output sequence consisting of just the P4 sequence alone.

9.11 BOLTZMANN MACHINE

A Boltzmann machine is a network of symmetrically coupled stochastic binary units that contain a set of visible units $\mathbf{v} \in \{0,1\}^D$ and a set of hidden units $\mathbf{h} \in \{0,1\}^P$ as shown in Figure 9.71 [54].

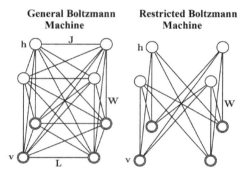

Figure 9.71 Left: A general Boltzmann machine. The top layer represents a vector of stochastic binary "hidden" features and the bottom layer represents a vector of stochastic binary visible variables. Right: A restricted Boltzmann machine with no hidden-to-hidden and visible-to-visible connections. (©IEEE, reprinted with permission from [54].)

The energy of the state $\{\mathbf{v}, \mathbf{h}\}$ is defined as

$$E(\mathbf{v},\mathbf{h};\theta) = -\frac{1}{2}\mathbf{v}^T\mathbf{L}\mathbf{v} - \frac{1}{2}\mathbf{h}^T\mathbf{J}\mathbf{h} - \mathbf{v}^T\mathbf{W}\mathbf{h} \tag{9.57}$$

where $\theta = \{\mathbf{W}, \mathbf{L}, \mathbf{J}\}$ are the model parameters with $\mathbf{W}, \mathbf{L}, \mathbf{J}$ representing the visible-to-hidden, visible-to-visible and the hidden-to-hidden symmetrical interaction terms. The diagonal elements of $\mathbf{L}$ and $\mathbf{J}$ are set to 0 [54].

The probability that the model assigns to a visible vector $\mathbf{v}$ is

$$p(\mathbf{v};\theta) = \frac{p^*(\mathbf{v};\theta)}{Z(\theta)} = \frac{1}{Z(\theta)} \sum_h \exp(-E(\mathbf{v},\mathbf{h};\theta)) \tag{9.58}$$

where

$$Z(\theta) = \sum_\mathbf{v} \sum_\mathbf{h} \exp(-E(\mathbf{v},\mathbf{h};\theta)) \tag{9.59}$$

where p^* represents the unnormalized probability and $Z(\theta)$ is the *partition function*. By setting both $\mathbf{J} = 0$ and $\mathbf{L} = 0$ the restricted Boltzmann machine is recovered. In contrast to the general Boltzmann machine, inference in the restricted Boltzmann machine is exact.

9.12 RESTRICTED AND DEEP BOLTZMANN MACHINES

The RBM is a two-layer generative stochastic artificial neural network, which was first proposed by Hinton in 2006 and later published as a technical report [55]. The schematic of a RBM is shown in Figure 9.72 [56].

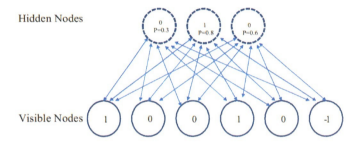

Figure 9.72 Schematic diagram of a restricted Boltzmann machine. (©IEEE, reprinted with permission from [56].)

The RBM estimates (generates) the probability distribution of the original input instead of associating a continuous/discrete value to an input example as in case of classification or regression. It is also useful for learning higher-level features of data in an unsupervised learning fashion.

The input visible layers in RBM are translated into higher feature space by means of an interlayer connection, where the translation has the objective to minimize the error between the input and the reconstructed version [56]. The pseudo-code for training a RBM is shown in Figure 9.73.

Algorithm 1 Training process of RBMs

Input: Input vector $v_1 = (v_{11}, \ldots, v_{1n})$ (size n), the size of hidden layer m, the learning rate γ, and the maximum epoch e.
Output: $H = (H_1, \ldots, H_m)$ (hidden vector), $V = (V_1, \ldots, V_n)$ (reconstruction of input vector v)
for $t=1, \ldots, e$ **do**
    for $j=1, \ldots, m$ **do**
        Compute (Gibbs Sampling):
$$P(h_j=1|v_1) = \sigma(b_j + \textstyle\sum_{i=1}^{n} W_{ij}\, v_{1i})$$
    end for
    Compute:
        positive gradient $= v_1 . h_1$
    for $i=1, \ldots, n$ **do**
        Compute (Gibbs Sampling):
$$P(v_{2i}=1|h_1) = \sigma(a_j + \textstyle\sum_{j=1}^{m} W_{ij}\, h_{1j})$$
    end for
    for $j=1, \ldots, m$ **do**
        Compute (Gibbs Sampling):
$$P(h_{2j}=1|v_2) = \sigma(b_j + \textstyle\sum_{i=1}^{n} W_{ij}\, v_{2i})$$
    end for
    Compute:
        Positive gradient $= v_2 . h_2$
    Update parameters (Contrastive Divergence):
$$W = W + \gamma(v_1 h_1 - v_2 h_2)$$
$$a = a + \gamma(v_1 - v_2),$$
$$b = b + \gamma(h_1 - h_2)$$
end for
$H_j = \sigma(b_j + \sum_{i=1}^{n} W_{ij}\, v_{1i})$, $V_i = \sigma(a_j + \sum_{j=1}^{m} W_{ij}\, h_{1j})$

Figure 9.73 Pseudo-code for training a RBM. (©IEEE, reprinted with permission from [56].)

The RBMs are variants of Boltzmann machines, with the restriction of interconnection between neurons in the same layer and fully connection between neurons in two layers as shown in Figure 9.72. The RBM training is accomplished in two passes, forward pass and backward pass using the Gibbs sampling algorithm, where the states of one layer are calculated based on the state of another layer until the distribution of the reconstructed input state is equilibrium with the initial input state and the contrastive divergence (CD) algorithm [55, 56]. The RBM network also works with *unlabeled data*, which is possible by its unsupervised learning process and is a major benefit. Similar to the RBM in these respects is the deep Boltzmann machine (DBM) network.

Deep Boltzmann machines are interesting for several reasons. First, like deep belief networks, DBMs have the potential of learning internal representations that become increasingly complex, which is considered to be a promising way of solving object and speech recognition problems. Second, high-level representations can be built from a large supply of unlabeled sensory inputs and very limited labeled data can then be used to only slightly fine-tune the model for a specific task at hand. Finally, unlike deep belief networks (discussed next), the approximate inference procedure, in addition to an initial bottom up pass, can incorporate top-down feedback, allowing deep Boltzmann machines to better propagate uncertainty about, and therefore deal more robustly with ambiguous inputs [57]. The DBM addresses the vanishing gradient problem, as evidenced in the RNN, by using a greedy layer-by-layer network; however, the high computational cost required in training is the major drawback [56].

9.13 DEEP BELIEF NETWORKS

The more precise deep belief network (DBN) can be built by using the stack of RBMs that is helpful in capturing the significance of higher order data in the input layer [56]. The DBM is probabilistic graphical model and combines graph theory with probability theory. The DBN uses unlabeled data to build unsupervised models consisting of multiple layers of hidden units with connections between the layers but not between units within each layer. That is, the nodes in each layer are connected to all the nodes in the previous and subsequent layer (restricted Boltzmann machines). They have use applications such as remote sensing [58] and the layers can be trained typically using an unsupervised learning algorithm (such as *contrastive divergence*).

Specifically, a DBN is a deep (large) directed acyclic graph (DAG). A deep Boltzmann machine (DBM), on the other hand, is a similar graph, but is undirected. Consequently, in a DBN, the data can flow both ways, and training can be bottom-up or top-down.

Hinton showed that a DBN can be viewed and trained (in a greedy manner) as a stack of simple unsupervised networks, namely restricted Boltzmann machines (RBMs), or generative AEs [59]. The CNNs have demonstrated better performance on various benchmark CV data sets. However, in theory DBNs are arguably superior since CNNs possess generally a lot more constraints [26].

Consider the DBM with D visible and P hidden units. The DBM energy function is

$$E(\mathbf{v}, \mathbf{h}; \theta) = -\frac{1}{2}\mathbf{v}^T \mathbf{L} \mathbf{v} - \frac{1}{2}\mathbf{h}^T \mathbf{J} \mathbf{h} - \mathbf{v}^T \mathbf{W} \mathbf{h} \qquad (9.60)$$

where the visible units are represented by a binary vector $\mathbf{v} \in (0.1)^D$, the hidden units are represented by a binary vector $\mathbf{h} \in (0, 1)^P$ and the DBM network parameters are in

$\theta = \{\mathbf{L}, \mathbf{J}, \mathbf{W}\}$ and $\mathbf{J}$ and $\mathbf{L}$ have zero diagonals. The conditional probabilities based on the visible and hidden input states can be evaluated as

$$p(\mathbf{v}, \mathbf{h}; \theta) = \frac{\sum_h \exp[-E(\mathbf{v}, \mathbf{h}; \theta)]}{\sum_v \sum_h \exp[-E(\mathbf{v}, \mathbf{h}; \theta)]} \qquad (9.61)$$

where the numerator is an unnormalized probability and the denominator is the partition function.

Figure 9.74 on the left is a deep belief network and on the right is a deep Boltzmann machine [57].

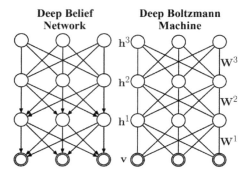

Figure 9.74 Left: A deep belief network and on the right: a deep Boltzmann machine. (From [54, 57].)

9.14 GENERATIVE ADVERSARIAL NETWORK

A GAN consists of a *generator* and a *discriminator*, and a pair of networks competing against each other. The generator G creates forgeries to fool the discriminator, whereas the adversarial discriminator D, receiving both real data and forgeries, attempts to determine the true identities of received data. Both D and G continuously fight each other and iteratively optimize their respective performances. When the process converges where the discriminator is no longer able to distinguish the forgeries from the true data, the generator is shown to produce a distribution matching the true data distribution. These networks, their derivatives, and some examples are given below.

The GAN is a class of neural networks that are used for unsupervised learning. It was developed and introduced by Ian J. Goodfellow in 2014 [60]. GANs are basically made up of a system of *two competing neural network models* that compete with each other and are able to analyze, capture, and copy the variations. The GAN framework is composed of two basic functions: a *generator* G to capture the feature distribution of real data and generate samples as real as possible, and a *discriminator* D to discriminate the authenticity of the generated samples. By the two-player minimax game between the generator and discriminator, the GAN is able to produce new images with high similarity to the original images. Figure 9.75 shows another view of the architecture showing the fine tuning as a result of a "no" decision on the discriminator output.

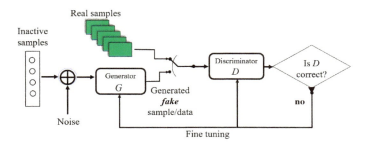

Figure 9.75 Generative adversarial network showing the fine tuning behavior output of the discriminator decision block at the end. (After [50].)

The *generative network* in GANs generates data (images) with a special kind of NN layer called a *deconvolutional* NN layer. A deconvolutional NN was developed by M. Seiler and R. Fergus at N.Y. U. as part of the development of ZF Net in the paper "Visualizing and Understanding Convolutional Neural Networks," [61].

During training, backpropagation is used on both networks to update the generating network's parameters to generate more realistic output images [49]. The objective here is to update the generating network's parameters to the point at which the discriminating network is sufficiently *fooled* by the generating network because the output is so realistic as compared to the training data's real images.

When modeling images, the *discriminator network* is typically a standard CNN. Using a secondary NN as the discriminator network allows the GAN to train both networks in parallel in an unsupervised fashion. These discriminator networks take images as input, and then output a classification.

9.14.1 SAR ATR Example

The design of a SAR ATR algorithm based on the combination of semisupervised CNN and dynamic multidiscriminator GAN is described in [62]. They describe that the involvement of a dynamically adjustable multidiscriminator architecture effectively improves the generating stability and image quality of the GAN. A block diagram of this approach is shown in Figure 9.76.

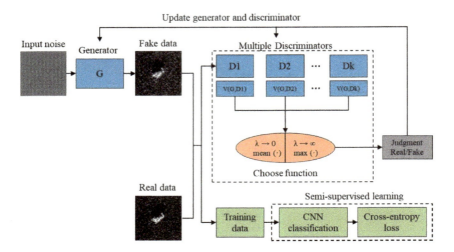

Figure 9.76 Architecture of the MGAN based semisupervised classification. The generator is trained with multiple discriminators that are dynamic adjusted by λ. If $\lambda \to 0$, G trains against the ensemble average of all discriminators; if $\lambda \to \infty$, G trains against the best discriminator. (©IEEE, reprinted with permission from [62].)

This example demonstrates that a label *smoothing regularization* algorithm will soften the classification model by reducing the confidence of label prediction and thus endues the CNN system higher robustness. Extensive experiments are carried out on the MSTAR dataset to validate the performance of the proposed method. They show that the method is able to achieve considerable improvement on the recognition ability of CNN compared with published SAR ATR methods, and it is more well-suited for the SAR recognition with a limited training dataset [62].

9.15 TRANSFER LEARNING

There is a stark difference between the traditional approach of building and training machine learning models, and using a methodology following transfer learning principles. Take, for instance, what we acquire as knowledge while learning about one task, we utilize in the same way to solve other related tasks. The more related the tasks, the easier it is for us to transfer, or cross-utilize our knowledge. Machine learning and deep learning algorithms, so far, have been traditionally designed to work in isolation. These algorithms are trained to solve specific tasks. The models have to be rebuilt from scratch once the feature-space distribution changes. Transfer learning is the idea of overcoming the isolated learning paradigm and utilizing knowledge acquired for one task to solve related ones.

A major assumption in many machine learning and data mining algorithms is that the training and any future data to be used must be from the same feature space and have the same distribution. However, in many real-world applications, this assumption may not hold [63].

Automatic modulation classification has attracted many important signal processing applications. Recently, deep learning models being adopted for modulation recognition and have been shown to outperform the traditional machine learning techniques that are based on feature extraction-neural network methods. However, automatic modulation classification is still challenging with DNN due to the deep learning methods being only applicable to data with the same distribution.

However, in practical scenarios, the data distribution varies with sampling frequency and, therefore, domains with different sampling rates are formed. That is, we sometimes have a classification task in one domain of interest, but we only have sufficient training data in another domain of interest, where the latter data may be in a different feature space or follow a different data distribution [63]. Accordingly, it is difficult to construct large-scale well-annotated datasets for all domains of interest [64]. In such cases, knowledge transfer, if done successfully, would greatly enhance the performance of learning by avoiding expensive data-labeling efforts.

To address these challenges, the work in [64] introduces an adversarial-based transfer learning for the task of automatic modulation classification. In recent years, *transfer learning* has emerged as a new learning framework to address this problem. Especially applicable for the DRFM, transfer learning relaxes the major assumption in many machine learning and data mining algorithms where the training and future data are required to be in the same feature space and have the same distribution [63].[3]

9.15.1 The Problem

For the problem, assume that in a communication system, received signal $x(n)$ can be given as

$$x(n) = F(s(n)) * h(n) + w(n), \quad n = 1, \ldots, N \tag{9.62}$$

where $s(n)$ denotes an initial baseband signal, and modulation function F transforms the baseband signal into a transmission signal that is sent through a communications channel $h(n)$ subject to additive white Gaussian noise $w(n)$. Here N represents the sampling rate of the receiver. In transfer learning, source domain S refers to the *existing knowledge*, while the target domain T requires to *learn*.

The problem considered is that the receivers possess different data domains S and T with sampling frequencies N_s and N_t. A well-annotated source dataset D_s

---

[3] The need for transfer learning arises when the data being used in the classification study can be easily outdated - such as the classification of emitter modulations from different configurations of a fire control ,radar system (AN/APG-65 vs. AN/APG-79). For this case, the labeled modulation data obtained from one system built earlier in one time period may not follow the same distribution from the system built in a later time period.

with sufficient signal samples X_s and labels Y_s also is available. Source model M_s is trained to predict source data distribution $p_s(x,y;N_s)$. For the signals with sampling frequency N_t, it is assumed that the samples X_t and labels Y_t are drawn from a target data distribution $p_t(x,y;N_t)$.

Insufficient data in the target dataset D_t results in a *weak* classification capability of model M_t. The knowledge in S however, is used to model the target domain T and improve the classification performance of M_t with overall learning of the target formulated as

$$\min_{S,D_t} H(M_t(\mathbf{X}_t), Y_t) \tag{9.63}$$

where H denotes cross entropy and is used to measure the difference between the two probability distributions p and q, with $H(p,q) = -\sum_i p(i)\log(q(i))$ being the entropy [65].

9.15.2 Framework for the ATLA Method

The adversarial transfer learning architecture (ATLA) method executes the adversarial transfer learning to minimize the domain shift and distribution difference between S and T. That is,

$$\min |p_s(x,y;N_s), f(p_t(x,y;N_t))|_D \tag{9.64}$$

where $|\cdot|_D$ denotes that the domain distance that is reflected in the discriminator D, and f represents the asymmetric mapping $f: p_t \to p_s$ [64]. Figure 9.77 shows an overview of the steps in the ATLA framework. The gray and green background represent the model being frozen and trainable, respectively.

9.15 Transfer Learning

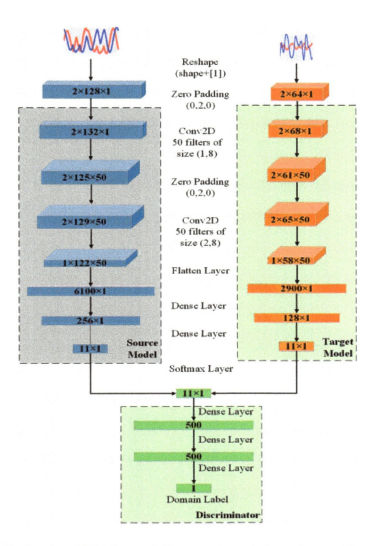

Figure 9.77 Overview of ATLA framework. The gray and green background represent the model being frozen and trainable, respectively. (©IEEE, reprinted with permission from [64].)

The deep learning models M_s and M_t are derived independently, not interfering with each other [64]. Recent efforts show that the residual network and inception modules obtain similar loss and accuracy to hyperparameter optimized CNN [65]. The CNN is chosen as the base model and the more applicable model structure is searched for in each domain M_s and M_t decreases the dimension of the first dense layer from 256 to 128. During the transfer learning, M_s is frozen and an adversarial training is conducted between M_t and the discriminator D. The output of the softmax layer in M_s or M_t is designated as the input of D. D contains two hidden layers that have 500 neurons that use a *leaky rectified linear unit* (ReLU) activation operation or (leakyReLu). The leaky ReLu performs a nonlinear threshold operation, where any input value less than zero is multiplied by a fixed scale factor. The output layer gives back the probability value whether the softmax representation belongs to the target domain or not. The objective of D is distinguishing the domain label of the input vector, while M_t attempts to confuse D and enhance the classification performance [64].

9.15.3 Losses for Target, Classification, and GAN

Two training epochs were conducted for D within each iteration. The loss of D is the standard choice used in GAN and expressed as [60]

$$L_D = -E_{x_s \sim X_s}[\log D(M_s(x_s))]$$

$$L_D = -E_{x_t \sim X_t}[\log(1 - D(M_t(x_t)))] \tag{9.65}$$

that aims at minimizing the error of domain label discrimination. To prevent mode collapse and maintain the category distinguishing ability of M_t, the loss of the target model L_T is the weighted sum of the GAN loss L_{GAN}, and the classification loss L_C given as

$$L_T = \alpha L_{GAN} + L_C$$

$$L_{GAN} = -E_{x_t \sim X_t}[\log D(M_t(x_t))]$$

$$L_C = \sum_{x_t \sim X_t, y_t \sim Y_t} y_t \log M_t(x_t) \tag{9.66}$$

where α is the weight coefficient of the GAN loss and the only hyperparameter. The L_{GAN} attempts to confuse the discriminator and L_C is consistent with the purpose of ATLA.

9.15.4 Experiments and Data

All experiments used the TensorFlow backend on a single NVIDIA Corporation GPU. Calculated by the number of multiply-accumulate operations (MACCs), the computational complexity of ATLA was slightly larger than a CNN and DNN, which was readily accepted considering the performance enhancement. The RadioML2016.10a dataset contained all communication modulation formats for the testing. These included 11 modulation formats: 8PSK, AM-DSB, AMSSB, BPSK, CPFSK, GFSK, PAM4, QAM16, QAM64, QPSK, and WBFM. Each signal sample is of size 2×128, including in-phase and quadrature (I,Q) components. The SNR ranged from -20 dB to $+18$ dB E_s/N_o, 2 dB apart. To verify the reliability of the transfer performance, a certain proportion of T is selected as D_T, and $T - D_T$ is used as the test set.

9.15.5 Results

The transfer learning performance is shown in Figure 9.78. This figure shows the top-1 classification accuracy comparison between ATLA, no transfer, parameter transfer and supervised learning. Train-test proportion denotes the proportion of training set and test set in the target domain, respectively. When it came to 100, supervised learning using the whole data was adopted [64].

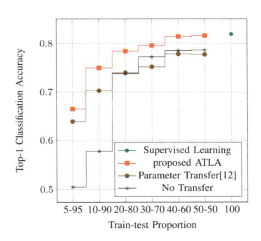

Figure 9.78 Top-1 classification accuracy comparison between ATLA, no transfer, parameter transfer, and supervised learning shown vs. the train-test proportion. Train-test proportion denotes the proportion of training set and test set in the target domain. When it comes to 100, supervised learning using the whole data set was adopted. (©IEEE, reprinted with permission from [64].)

A performance comparison of ATLA with the five autonomous classification methods is shown in Table 9.5.

Table 9.5
Performance Comparison Under the Eight Digital Modulation Types and One-Tenth Training Data

Method	KNN	SVM	DTree	DNN	CNN	ATLA
Top-1 accuracy	0.617	0.727	0.798	0.733	0.744	**0.801**

9.15.6 ATLA Summary

In summary, steps to construct an ATLA to significantly improve the performance of deep learning based automatic modulation classification with insufficient data was presented. This model utilizes adversarial training to reduce the difference between data distributions and realize transfer learning. The proposed ATLA outperforms the current parameter transfer approach and extends the tolerance of the distribution difference. Apart from the sampling rate (mentioned earlier), the proposed method works under various imperfections (frequency offset, etc.).

Transfer learning is especially useful for the classification and waveform recognition of known and unknown radar signals. Unknown radar signals are those that have not been observed or encountered previously. For the known modulations, they can be identified by the correlation between prior knowledge of the radar and the received signals by the noncooperative intercept receiver. For the unknown radar signals with a limited number of samples in a low SNR environment, even detecting the signals is a significant challenge.

9.15.7 Recognizing Unknown Radar Waveforms

The effort in [66] uses the ability of the deep features of the image from the CNN, the reconstructed features of the TFI of the known and unknown radar waveform signals have been mined. A decision fusion, unknown radar signal identification model based on transfer deep learning and linear weight decision fusion is designed to recognize the unknown. First, the CNN is trained using the known radar signals. Then, based on the transfer learning, the neurons obtained from the multiple underlying the CNN are used to represent the reconstruction feature [67].

9.16 ADDRESSING THE SECURITY IN DRFM NETWORKS

Consider the tactical battlefield where a network of DRFMs are all running a deep neural network machine learning model, to enable an EA. Training these machine learning models requires access to data. The training data must be aggregated in a distributed fashion as the DRFMs are not in the same location and some of them can also be owned by our *coalition partners*. An important data aggregation case arises when the training data is *privacy sensitive*. For example, a threat classification application on the DRFMs requires one or more machine learning models to enable multiple, simultaneous beamforming transmissions such as a high power microwave (HPM) to one threat, a decoy to another, and so on.

The most applicable training data for this type of model may be streamed from an intelligence source. As such the data may be highly sensitive and require a solution to preserve the privacy, for example, by ensuring that the raw training data never needs to the leave threat collection's database server.

9.16 Addressing the Security in DRFM Networks

Federated learning addresses this need and is a recent addition to distributed machine learning technology that is directed at training a machine learning or deep learning algorithm, across multiple local datasets. These datasets can be contained in decentralized DRFM sensors or DRFM servers holding local data samples, without exchanging their data. This addresses the critical issues such as data privacy, data security, and data access rights to heterogeneous data [25].

9.16.1 Federated Learning

Federated Learning is a decentralized form of machine learning. Federated machine learning was developed by McMahan et. al. from Google, who principally developed it for mobile devices, such as cell phones, to train a local model. These local models are aggregated centrally into a final central model [68]. As standard machine learning approaches require centralizing the training data on one machine or in a datacenter, Federated Learning enables these mobile phones to collaboratively learn a shared prediction model while keeping all the training data on the device, decoupling the ability to do machine learning from the need to store the data in the cloud where the privacy may not be preserved. Figure 9.79 shows how federated learning was originally conceived to allow smartphones, to have lower latency and less power consumption, all while ensuring privacy [68].

Introduced by Google AI in 2017 in a blog post "Federated Learning: Collaborative Machine Learning without Centralized Training Data" initiated a new approach called federated optimization [69]. Unlike the distributed learning algorithm, however, these different nodes (e.g., DRFMs, wearable devices, and so on) do not communicate among themselves due to *trust issues* and/or communications challenges (network jammed) and do not transfer raw data to the cloud due to *privacy concerns* - an important consideration in DoD DRFM networks.

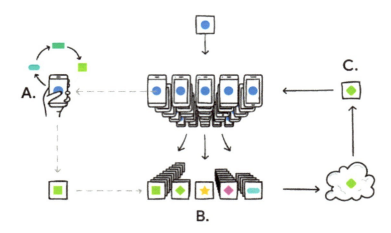

Figure 9.79 Your phone personalizes the model locally, based on your usage (A). Many users' updates are aggregated (B) to form a consensus change (C) to the shared model, after which the procedure is repeated. (From [68].)

9.16.2 How Federated Learning Works

The way federated learning works is that your device downloads the current model, improves it by, for example learning from data on your phone, and then summarizes the changes as a small focused update. Only this update to the model is sent to the cloud, using encrypted communication, where it is immediately averaged with other user updates, to improve the shared model. All the training data remains on the original server, and no individual updates are stored in the cloud [68].

Distributed optimization in machine learning is where the data defining the optimization are unevenly distributed over a large number of nodes. The goal to train a high-quality centralized model is called federated optimization. For this, communication efficiency is the key. Minimizing the number of rounds of communication is the principal goal [69]. Popular optimization algorithms use vanilla (stochastic) gradient descent for both local updates at clients and global updates at the aggregating server.

Recently, adaptive optimization methods such as AdaGrad have been studied for server updates [70]. Local adaptive methods can accelerate convergence, however, they can cause a non-vanishing solution bias. That is, the final converged solution for the model could be different from the stationary point of the global objective function. One correction technique to overcome these inconsistencies is described in [70].

The iterative learning is to ensure a good task performance of a machine learning model. The learning process is broken down into a set of client-server interactions known as a *federated learning round*. Each round consists of transmitting the current global model state to participating nodes and then training local models on these local nodes to produce a set of potential model updates at each node. Then the local updates are aggregated and processed into a single global update and applying it to the global model [71, 72].

As an example, consider a central C2 server is used for aggregation, while local DRFM nodes on unmanned air vehicles perform the local training depending on the C2's orders. Also, assume that the federated round is composed of one iteration of the learning process. The following steps are carried out asynchronously:

- *Initialization*: According to server inputs, machine learning model is chosen to be trained on local nodes and initialized. Nodes are activated and wait for C2 server to give calculation task;
- *Client selection*: Fraction of local nodes selected to start training on local data. Selected nodes acquire current statistical model while others wait for next federated round;
- *Configuration*: Central server orders selected nodes to undergo training of model on their local data in prespecified manner;
- *Reporting*: Each selected node sends local model to server for aggregation. Central server aggregates received models, and sends back model updates to nodes. Also handles failures for disconnected nodes or lost model updates. The next federated round is started returning to the client selection phase;
- *Termination*: Once pre-defined termination criterion met (e.g., maximum number of iterations reached or the model accuracy greater than threshold) central server aggregates updates and finalizes global model;

The asynchronous exchange of model updates is made as soon as the computations of a certain layer are available. These techniques are also commonly referred to as *split learning* and can be applied both at training and inference time regardless of centralized or decentralized federated learning settings.

The deep learning training mainly relies on variants of *stochastic gradient descent*, where gradients are computed on a random subset (of the total dataset) and then used to make one step of the gradient descent. *Federated stochastic gradient descent* is the direct transpose of this algorithm to the federated setting, however, is uses a random fraction of the nodes at each iteration for each node - using all the data on this node. The gradients are averaged by the server proportionally to the number of training samples on each node, and used to make a gradient descent step. One EW limitation for federated learning is that by hiding training data might allow attackers to inject backdoors into the global model.

9.16.3 Unmanned Vehicle Considerations

Federated learning is particularly useful in tactical surveillance operations where isolation among domains, for example, an intelligence database, is crucial. This is because of the security concerns related to the fact that a domain's possessing may have access to another domain's data. Federated learning techniques prevent this. However, with the enormous amount of data in classification and response management tasks, a very high bandwidth is required.

To help with this task, unmanned systems can use multi-access edge computing (MEC). MEC is a network architecture concept that moves the computing of traffic and services from a centralized cloud to the edge of the network and closer to the unmanned vehicle (customer). That is, instead of sending all data to a cloud for processing, via MEC, the network edge analyzes, processes, and stores the data. Collecting and processing data closer to the unmanned system reduces latency and brings real-time performance that can enable high-bandwidth applications such as (ES classification and EA response management).

MEC along with network function virtualization (NFV) are enablers to satisfy the demands of ultra low latency, extreme high throughput, and increased connectivity density. The NFV is a method to accelerate C2 operations deployment by decoupling networking functions like routers, switches, and firewalls from dedicated hardware and moving them to *virtual servers*. This collapses multiple applications into one physical server and reduces network power consumption – a critical advantage for unmanned jammer nodes.

NFV facilitates service provision and makes networks relatively more flexible. As a result, this enables C2 operations to tackle challenges linked to EA resource-changing demands. As well, NFV can make easier the capability to dynamically add, remove, or connect adhoc unmanned EA, ES systems just coming on-line. The networks can be enhanced to automatically configure, optimize, secure, and recover using the cognitive power of deep learning. The application of deep learning can also be used for network defense through anomaly detection and defending against attacks.

Deep learning models can be used for predictive virtual MEC application function (VMAF) autoscaling in a tactical, multi-domain setting that can better react to the changing DRFM service requirements. These deep learning models can optimize the network resource usage, and at the same time also comply with data protection policies imposed e.g., by intelligence sources for coalition warfighters.

The mobile edge nodes have a limited amount of physical and virtual resource capacity compared to the cloud data centers. Therefore, it is necessary to manage these resources efficiently. Consequently, the DRFM solution is to leverage on deep learning algorithms (e.g., time series forecasting) to develop *predictive autoscaling* solutions for effective resource utilization. Predictive autoscaling is far *more efficient* than reactive autoscaling [25]. Deep learning algorithms that are often used for predictive

9.16 Addressing the Security in DRFM Networks

VMAF autoscaling include feed forward neural networks, Long-, Short-Term Memory (LSTM) networks and Convolutional neural networks that are discussed below.

Developed principally for mobile devices such as cell phones to train a local model [71, 72]. The algorithm was developed for distributed optimization in machine learning, where the data defining the optimization are unevenly distributed over an extremely large number of nodes with the goal to train a high-quality centralized model. This framework or setting has been called *federated optimization* [69].

Federated learning can be also used in the field of machine learning called private and secure machine learning. The field of federated learning also allows the learning of a model without knowing details about the local data sources, as well as not having direct access to the data as in the case where the data is particularly sensitive or is protected by legal means [73].

In [69], distributed optimization in machine learning is introduced where the data defining the optimization are unevenly distributed over an extremely large number of nodes. The goal is to train a high-quality centralized model. This setting is referred to as federated optimization. In this setting, communication efficiency is of the utmost importance and minimizing the number of rounds of communication is the principal goal.

Instead, in federated learning, each entity locally updates the global model using its local data and then shares the updated model with a centralized entity, that intermittently passes that model to other entities for further updates and refinements of the global model [22].

A clear advantage of federated machine learning to electromagnetic warfare is that it has produced a number of tools that can be used as part of the DRFM networking strategy. For example, Tensorflow Federated is an open-source framework through which developers can simulate federated learning experiments. The API has a number of tutorials that allow the user to replicate the experiments of McMahan et al. [69].

Federated learning enables mobile DRFM network devices to collaboratively learn a shared inference model (e.g., threat IADS laydown) while keeping all the training data on a user's device, decoupling the ability to do machine learning from the need to store the data in the cloud [74]. The secure aggregation enables the server to learn an aggregate of at least a threshold number of device's model contributions without observing any individual device's contribution in unaggregated form.

9.16.4 Federated Transfer Learning

With the development of artificial intelligence and the Internet of Things (DRFMs), the use of the network in unmanned systems for library updating, model training, guidance and navigation is of high interest. With limited power available, it is hard for these DRFMs to perform will facing the tasks for training and updating models in federated learning. In order to ensure the security of data transmission, blockchain is introduced as the main algorithm of equipment authentication in the system. The blockchain network provides a "trustless" environment where users can conduct transactions without relying on a central trust agency. Currently, blockchain technology is used for applications to increase the efficiency and versatility of training model transfer learning is used to improve the system performance [75].

Transfer learning is discussed below in Section 9.7 as it has been used successfully in image processing, where large image classification networks such as GoogleLeNet have been successful upon hundreds of thousands of images. It has been adapted to classify images that GoogleLeNet has not been trained on. Transfer learning limits the amount of labeled data required for the new domain. Federated transfer learning as defined by Zang et al. applies where there the data on the local devices differs not only in their samples but also in feature spaces. This technique allows the algorithm to generate a solution for the sample and feature space of the complete dataset [73].

9.16.4.1 Federated Learning Summary

In summary, federated learning is similar to distributed learning as the data is still distributed across different, for example, DRFM platforms. However they do not communicate among themselves due to trust issues and/or poor network communications. As well, they do not transfer raw data to the cloud due to privacy concerns. Instead, in federated learning, each DRFM locally uploads a global model using its local data and then would share the uploaded model with a centralized entity such as a fusion center that intermittently passes the model to other DRFMs for further updates.

Federated learning is increasingly gaining popularity due to the increased emphasis on international cooperation and working with coalition warfighting partners. Federated learning enables multiple actors to build a common, robust machine learning model without sharing data, thus allowing to address critical issues such as data privacy, data security, data access rights, and access to heterogeneous data. Its applications are spread over a number of industries including defense, telecommunications, IoT, and pharmaceutics. That is, federated learning enables the DRFM devices to collaboratively learn a shared prediction model while keeping all the training data on device. This enables a decoupling of the ability to do machine learning from the need to store the data in the cloud.

Overall, ML is an exercise in optimization. With the networked DRFMs sending/receiving data sets over a network, *distributed optimization* is expected to play a major role in machine learning systems. Three particular distributed optimization techniques are (1) distributed-gradient methods for use in convex type learning problems, (2) stochastic first-order method for distributed machine learning, and (3) distributed stochastic optimization for machine learning considering the role of network topology. *Reinforcement learning* is another subdomain of ML that deals with the streaming data coming into the transceiver. Reinforcement learning is to take actions based on observed data that maximize some notion of a reward. The dynamics of the situation are completely unknown except for what the DRFM implicitly learns from the streaming data. Distributed reinforcement learning, where the streaming (incoming) observations are also distributed, is of high importance for a DRFM and takes high priority as a multiagent system.

As DRFMs will be using machine learning, their robustness and security against adversarial actions and malicious actors become paramount. While the initial focus in this direction within the open literature has mostly been on centralized problems, recent works have started to develop and analyze algorithms for distributed machine learning systems that can deal with unreliable data, malicious actors, and cyberattacks on individual entities in the network. A good survey of the recent developments pertaining to adversarial threats against distributed machine learning is given in [76].

9.17 NEUROMORPHIC COMPUTING FOR DRFMS

The question addressed here concerns the brain-inspired computing collectively called neuromorphic computing. Neuromorphic computing aims at uncovering novel computational frameworks that mimic the operation of the brain in a quest for orders-of-magnitude improvements in terms of energy efficiency and resource requirements. The recently developed deep and convolutional neural network algorithms can be trained to perform remarkably well in pattern-recognition talks, in some cases even outperforming humans. Running on conventional computing von Neumann architectures using large and power-hungry platforms, they are sometimes distributed over multiple server farms, the power required to run these algorithms and achieve impressive results is orders of magnitude larger than the power used by biological nervous systems that once served as inspiration for the artificial neural networks discussed above.

A new generation of ultralow-power and massively parallel computing technologies that are optimally suited to implement computing technologies are being developed by a host of researchers [77]. Also being developed are technologies that are optimally suited to implement neural network architectures that use the same principles of computation used by the brain. Examples of neuromorphic architectures that follow this mixed-signal approach include the Italian Istituto Superiore di Sanità (ISS) learning attractor neural network that has plasticity and a long-term memory chip (LANN-21); [78] ISS "final learning attractor neural network" (F-LANN) chip [79]; Georgia Tech learning-enabled neuron array integrated circuit (LENAIC) [80]; University of California San Diego integrate-and-fire array transceiver (IFAT) architecture [81]; Stanford University, California, Neurogrid system [82]; University of Zürich recurrent online learning spiking (ROLLS) neuromorphic processor [83]; and University of Zürich dynamic neuromorphic asynchronous processor-scalable (DYNAP-SE) chip [84].

In these devices, the analog synapse and neuron circuits have no active components as the circuits are driven directly by the input streaming data. Their synapses

Table 9.6
Quantitative Comparison of Mixed-Signal Neuromorphic Processor Specifications.

	LANN-21	F-LANN	LENA-IC	IFAT	NeuroGrid	ROLLS	DYNAP-SE
Technology	0.6 μm	0.6 μm	0.35 μm	90 nm	0.18 μm	0.18 μm	0.18 μm
Supply Voltage	3.3 V	3.3 V	2.4 V	1.2 V	3 V	1.8 V	1.8 V
Core Area	10 mm^2	68.9 mm^2	25 mm^2	139 μm^2	170 mm^2	51.4 mm^2	7.5 mm^2
Neurons/Core	21	128	100	2,000	65,636	256	256
Synapses/Core	129	16,384	30,000	N/A	4,096	128,000	16,000
Fan-In/Fan-Out	21/21	128/128	100/100	N/A	N/A	256/256	64/4,000
Synaptic Weight	Capacitor	Capacitor	>10 bits	8 bits	13-bit shared	Capacitor	1 + 1 bit
Online Learning	STRDP	STRDP	STDP	No	No	STRDP	No
Neurons/mm^2	174	N/A	27	7,142	360	1,089	812
Energy per SOP	N/A	N/A	10 pJ	22 pJ	31.2 pJ	77 fJ	17 pJ

fJ: femtojoule; pJ: picojoule.

(Adapted from [77].)

receive input spikes, and their neurons produce output spikes at the rate of the incoming data. If they are not processing data, there is no energy dissipated per synaptic operation (SOP) and no dynamic power consumption. Therefore, this approach is particularly attractive in the case of applications in which the signals have sparse activity in space and time. This approach is very power efficient. A quantitative comparison of the specifications of these devices is presented in Table 9.6. The spiking neural network device technology is described in a bit more detail below. An excellent review of the signal processing hardware and behavior modeling approaches are described in [77].

9.17.1 Spiking Neural Networks

There has recently been a lot of encouraging activity surrounding a new type of neural network called a *spiking neural network*. Spiking neural networks (SNNs) are distributed trainable systems whose computing elements, or neurons, are characterized by internal analog dynamics and by digital and sparse synaptic communications. In biological neural networks, neurons use spikes to communicate with each other. Incoming signals alter the voltage of a neuron and when the voltage exceeds a threshold value, the neuron sends out an action potential that is a short (1 ms) and sudden increase in voltage created in the cell body or soma. Due to the form and nature of this process, we refer to it as a spike or a pulse [50].

The sparsity of the synaptic spiking inputs and the corresponding event-driven nature of neural processing can be leveraged by energy-efficient hardware implementations, which can offer significant energy reductions as compared to conventional artificial neural networks [85]. Various new hardware solutions have recently emerged that attempt to improve the energy efficiency of artificial neural networks (ANNs) as inference machines by trading complexity for accuracy in the implementation of matrix operations. In [85], a different approach is taken that enables efficient online inference and low-power learning by taking inspiration from the working of the human brain. Figure 9.80 shows the model of a synaptic spiking neural network. SNNs have the unique capability to process information encoded in the timing of events, or spikes that are also used for synaptic communications, with synapses delaying and filtering the signals before they reach the postsynaptic neuron.

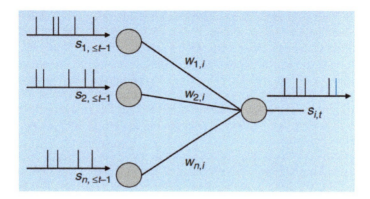

Figure 9.80 A synaptic SNN model.

Because of the presence of synaptic delays, neurons in an SNN can be naturally connected via arbitrary recurrent topologies, unlike the standard multilayer neural networks described above. When using rate encoding, they can approximate the performance of any ANN while also providing a graceful trade-off between accuracy, on the one hand, and energy consumption and delay, on the other. Most importantly, they have a unique capacity to process time-encoded information, yielding sparse, event-driven, and low-complexity inference and learning solutions. The accuracy rate of the signal recognition mainly relies on the feature extraction algorithm, but the artificial feature extraction relies mostly on the preprocessing and experience of the researchers as the extracted features are targeted to specific types of signals, and new features need to be extracted when identifying other signals. The feature vector has a number of constraints. It can't be too large to overwhelm the classifier and it can't be too small where there is not enough information to extract the details to separate the signals.

9.18 WIRELESS COMMUNICATIONS AND NETWORKING WITH UNMANNED AERIAL VEHICLES

Providing connectivity from the sky to ground wireless users is an emerging trend in wireless networking compared to terrestrial communications [50]. A wireless system with low-altitude UAVs is faster to deploy and more flexibly reconfigured, and likely to achieve better communications due to presence of line-of-sight links. This scenario and the DoD's use of UAVs as discussed in Chapter 1, brings on many challenges and opportunities for cognitive AI.

Within the wireless environment, most of the EMS data of interest collected with the DRFM transceivers will be time related. Data such as that pertaining to the adversary, UAV and troop movement, and human behavior, is of interest and is collected and processed along with the SoS as the first step (ES). An overview of the ANNs for the application of mining data results in several that are available. This effort is a perfect challenge for an embedded transfer learning within the UAV, and in particular, the adversarial transfer discussed in [64] to ensure the data is secure.

After transfer of the time-slice of data, SoS, and images to the decision maker (computer/human) over a secure network, they are able to understand the change in operation from the last time-slice. Here, as the human C2 analyzes the situation, a computer running deep learning techniques can aid in sorting out the spectrum white spaces spectrum gray spaces.

It has been shown that a CNN detector can sense the spectrum for energy and using only a small training set of a few hundred samples [86]. Here they compare the performance of a CNN detector to the more traditional energy detection as well as other published results of machine learning used for signal detection. The CNN detector is presented there and is shown to surpass other machine learning methods for signal detection. They also show that the CNN detector does not require a measurement of the noise floor, offering a significant improvement over the classic energy detector.

For detecting white spaces in the EMS, an auto encoder neural network is demonstrated as a spectrum sensing and monitoring tool with an automatic learning feature in [87]. The output from the auto encoder then becomes the input to a SVM classifier. The SVM then determines whether the spectrum of input samples can be concluded as being occupied by a primary user or secondary user.

We have studied in Section 9.9, how a LSTM network can sense the present and also retain the past history sensed. An LSTM based spectrum sensing (LSTM-SS) application is presented in [88] that learns the implicit features from the spectrum data for example, the correlation between the present and past time-stamps. By using this temporal correlation between the *present* intercepted sequence and a past *time-stamped* sequence, many features can be determined. The cognitive DRFM can also exploit the adversary user's spectrum activity statistics to improve the cognitive performance, that provides an essential tool for EA (and ES). By computing the adversary user activity statistics like on and off period duration, duty cycle, an LSTM-SS can be designed to quantify the primary user's activity statistics based spectrum sensing shown in Figure 9.81 [88].

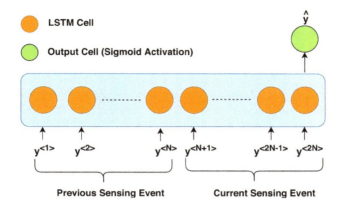

Figure 9.81 LSTM-SS architecture to identify the primary user's presence and their activity. (©IEEE, reprinted with permission from [88].)

The LSTM-SS architecture uses the current sensing event and the previous sensing event to drive an output cell that goes through a sigmoid activation unit. The sigmoid activation facilitates the conversion from continuous value output to binary output [88] LSTM-SS scheme proposed retains the previous sensing event and feeds it along with the present sensing event to give an increase in performance improvements, in terms of detection probability and classification accuracy, even at a low SNR situations. Figure 9.82 shows the training model used in the proposed scheme shown in Figure 9.81.

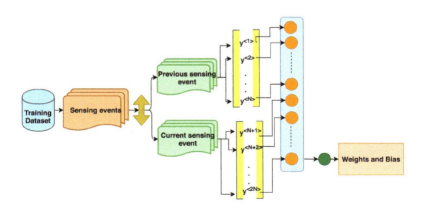

Figure 9.82 Training model considered in proposed LSTM-SS scheme. (©IEEE, reprinted with permission from [88].)

The training dataset which comprise of 60% of the total samples are fed in batches to different LSTM models, the error is backpropagated during the training procedure, the gradients are calculated and the parameters are updated. Figure 9.83 shows the pseudo-code for constructing the datasets for the LSTM-SS and the training algorithm for the proposed scheme. Figure 9.84 shows the pseudo-code for evaluating the LSTM-SS architecture.

```
Algorithm 1 Dataset Construction for Proposed LSTM-SS
 1: Procedure Create Dataset (Data, N, Label)
 2:   size ← length(Data)/N
 3:   PU_dataset ← zero matrix of dimensions size × N
 4:   for SNR ← −20 to 4 dB do
 5:     noisy_signal ← Data + AWGN {SNR is achieved}
 6:     for i ← 1 to size do
 7:       signal ← (i)ᵗʰ N samples from noisy_signal
 8:       PU_signal[i] ← signal {Row-wise assignment}
 9:     end for
10:   end for
11:   return PU_signal {The PU signal is returned}
```

```
Algorithm 2 Training of Proposed LSTM-SS Scheme
1: Procedure Train(Epochs, Batch_size, X, y, α)
2:   for i ← 1 to Epochs do
3:     s_event, label ← extract(Dataset, Batch_size)
       {Random training examples are extracted according to
       the batch size}
4:     Output ← Forward Propagate(LSTM_model, s_event)
5:     Error ← Backward Propagate(LSTM_model, label,
       output)
6:     Parameters ← Update(error,LSTM_model,α) {Parame-
       ters are updated according to the learning rate α}
7:   end for
```

Figure 9.83 Training model considered in proposed LSTM-SS scheme. (©IEEE, reprinted with permission from [88].)

```
Algorithm 3 Evaluation of the Proposed LSTM-SS Scheme
 1: Procedure Evaluate(LSTM_model, Dataset)
 2:   for i ← 1 to length(PU_signal)
 3:     s_event, label ← extract(Dataset, 1)
        {Test examples are extracted one by one}
 4:     $\mathcal{H}_0$_examples ← 0
 5:     $\mathcal{H}_0$_misclassified ← 0
 6:     $\mathcal{H}_1$_examples ← 0
 7:     $\mathcal{H}_1$_correct ← 0
 8:     Output ← Forward_Propagate(LSTM_model, s_event)
 9:     if Label is $\mathcal{H}_1$ then
10:       $\mathcal{H}_1$_examples ← $\mathcal{H}_1$_examples + 1
11:       if Output is $\mathcal{H}_1$
12:         $\mathcal{H}_1$_correct ← $\mathcal{H}_1$_correct + 1
13:       end if
14:     end if
15:     if Label is $\mathcal{H}_0$
16:       $\mathcal{H}_0$_examples ← $\mathcal{H}_0$_examples + 1
17:       if Output is $\mathcal{H}_1$
18:         $\mathcal{H}_0$_misclassified ← $\mathcal{H}_0$_misclassified + 1
19:       end if
20:     end if
21:     $P_d \leftarrow \frac{\mathcal{H}_1\_correct}{\mathcal{H}_1\_examples}$
22:     $P_f \leftarrow \frac{\mathcal{H}_0\_misclassified}{\mathcal{H}_0\_examples}$
23:   end for
```

Figure 9.84 Pseudo-code for the evaluation of the LSTM-SS scheme. (©IEEE, reprinted with permission from [88].)

Experimental results indicate that the proposed scheme has improved detection performance and classification accuracy at spectrum data of various technologies acquired at both low and high SNRs (Figure 9.85).

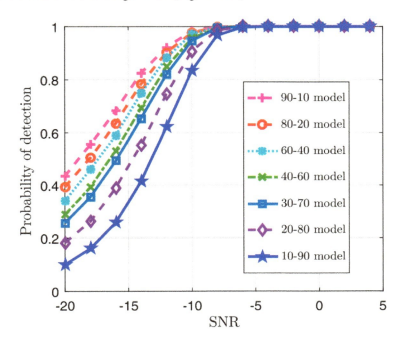

Figure 9.85 Probability of detection for the various model configurations ©IEEE, reprinted with permission from. (©IEEE, reprinted with permission from [88].)

Further wireless UAV applications are given in [50].

9.19 SUMMARY COMMENTS

It is clear that machines– more specifically, DRFM transceivers– can never do things humans can do and in return machines can do things no humans can do and hence we recall the Sheridan levels of human-computer performance trade-off:

> "And in any case, how sure are you for any clearly prespecified thing machines (programs) apparently cannot now do and in time still could not do it better than humans can?" (Perhaps "clearly specified" means you can write a program!) And in any case how relevant are these supposed differences to your career? [89].

> "The people are generally sure they are more than a machine, but usually can give no real argument as to why there is a difference unless they appeal to their

religion, and with foreign students of very different faiths around they are reluctant to do so-though obviously most (though not all) religions share the belief man is different, in one way or another, from the rest of life on Earth" [90].

REFERENCES

[1] J. R. Wilson, "Artificial intelligence and embedded computing for unmanned vehicles," *Military and Aerospace Electronics*, vol. 31, no. 4, pp. 12–19, 2020.

[2] ——, "Artificial intelligence AI in unmanned vehicles," *Military and Aerospace Electronics*, vol. 30, no. 4, pp. 12–22, 2019.

[3] R. W. Hamming, *The Art of Doing Science and Engineering, Learning to Learn*. Gordon and Breach Science Publishers, 1997.

[4] F. Parly, "The critical role of AI for operational security," *Defense News*, no. 12, p. 14, 2019.

[5] K. Z. Haigh and J. Andrusenko, *Cognitive Electronic Warfare An Artificial Intelligence Approach*. Norwood, MA: Artech House, 2021.

[6] J. Lynch, "US Army turns to AI to prevent cyberattacks," *Defense News*, no. 1, p. 23, 2019.

[7] Z. Parisa and M. Nova, "This AI can see the forest and the trees," *Spectrum*, no. 8, pp. 32–37, 2020.

[8] C. Christodoulou and M. Georgiopoulos, *Applications of Neural Networks in Electromagnetics*. Norwood, MA: Artech House, 2001.

[9] D. Schneider, "Deep learning at the speed of light: Lightmatter bets that optical computing can solve ai's efficiency problem," *IEEE Spectrum*, vol. 58, no. 1, pp. 28–29, 2021.

[10] N. Sadati and H. Berenji, "Coordination of large-scale systems using fuzzy optimal control strategies and neural networks," in *2016 IEEE International Conference on Fuzzy Systems (FUZZ-IEEE)*, 2016, pp. 2035–2042.

[11] P. E. Pace, *Detecting and Classifying Low Probability of Intercept Radar*. Norwood, MA: Artech House, 2009.

[12] M. F. L. de Vries, G. J. M. Koeners, F. D. Roefs, H. T. A. Van Ginkel, and E. Theunissen, "Operator support for time-critical situations: Design and evaluation," in *2006 IEEE/AIAA 25TH Digital Avionics Systems Conference*, 2006, pp. 1–14.

[13] R. Lippmann, "An introduction to computing with neural nets," *IEEE ASSP Magazine*, vol. 4, no. 2, pp. 4–22, 1987.

[14] A. G. Parlos, B. Fernandez, A. F. Atiya, J. Muthusami, and W. K. Tsai, "An accelerated learning algorithm for multilayer perceptron networks," *IEEE Transactions on Neural Networks*, vol. 5, no. 3, pp. 493–497, 1994.

[15] Y. Zhong and L. Zhang, "An adaptive artificial immune network for supervised classification of multi-/hyperspectral remote sensing imagery," *IEEE Transactions on Geoscience and Remote Sensing*, vol. 50, no. 3, pp. 894–909, 2012.

[16] Y. Jeon, M. Lee, and J. Y. Choi, "Differentiable forward and backward fixed-point iteration layers," *IEEE Access*, vol. 9, pp. 18 383–18 392, 2021.

[17] G. Nguyen, S. Dlugolinsky, and et al.,. M. Bobak, "Machine learning and deep learning frameworks and libraries for large-scale data mining: a survey," in *Artificial Intelligence Review*, vol. 52, no. 1, 2019, pp. 77–124. [Online]. Available: http://www.jstor.org/stable/3846811

[18] A. Nedic, "Distributed gradient methods for convex machine learning problems in networks: Distributed optimization," *IEEE Signal Processing Magazine*, vol. 37, no. 3, pp. 92–101, 2020.

[19] T.-H. Chang, M. Hong, H.-T. Wai, X. Zhang, and S. Lu, "Distributed learning in the nonconvex world: From batch data to streaming and beyond," *IEEE Signal Processing Magazine*, vol. 37, no. 3, pp. 26–38, 2020.

[20] S. Machiraju. How to train your deep learning models in a distributed fashion. [Online]. Available: https://towardsdatascience.com/how-to-train-your-deep-learning-models-in-a-distributed-fashion-43a6f53f0484

[21] W. U. Bajwa, V. Cevher, D. Papailiopoulos, and A. Scaglione, "Machine learning from distributed, streaming data [from the guest editors]," *IEEE Signal Processing Magazine*, vol. 37, no. 3, pp. 11–13, 2020.

[22] R. Nassif, S. Vlaski, C. Richard, J. Chen, and A. H. Sayed, "Multitask learning over graphs: An approach for distributed, streaming machine learning," *IEEE Signal Processing Magazine*, vol. 37, no. 3, pp. 14–25, 2020.

[23] R. Caruana. Multitask learning. autonomous agents and multi-agent systems. [Online]. Available: https://doi.org/10.1016/j.csl.2009.08.003

[24] S. Ruder. Natural language processing, machine learning, and deep learning. [Online]. Available: https://ruder.io

[25] T. Subramanya and R. Riggio, "Centralized and federated learning for predictive vnf autoscaling in multi-domain 5g networks and beyond," *IEEE Transactions on Network and Service Management*, vol. 18, no. 1, pp. 63–78, 2021.

[26] J. E. Ball, D. T. Anderson, and C. S. C. Sr., "Comprehensive survey of deep learning in remote sensing: theories, tools, and challenges for the community," *Journal of Applied Remote Sensing*, vol. 11, no. 4, pp. 1 – 54.

REFERENCES

[27] A. Saha, S. S. Rathore, S. Sharma, and D. Samanta, "Analyzing the difference between deep learning and machine learning features of eeg signal using clustering techniques," in *2019 IEEE Region 10 Symposium (TENSYMP)*, 2019, pp. 660–664.

[28] G. E. Hinton and R. R. Salakhutdinov, "Reducing the dimensionality of data with neural networks," *Science*, vol. 313, no. 5786, pp. 504–507, 2006. [Online]. Available: http://www.jstor.org/stable/3846811

[29] S. Russell and P. Norvig, *Artificial Intelligence, A Modern Approach, 4th Ed.* Hoboken, NJ: Pearson, 2021.

[30] A. Ali and F. Yangyu, "Automatic modulation classification using deep learning based on sparse autoencoders with nonnegativity constraints," *IEEE Signal Processing Letters*, vol. 24, no. 11, pp. 1626–1630, 2017.

[31] S. Kullback and R. A. Leibler, "On information and sufficiency," *The Annals of Mathematical Statistics*, vol. 22, no. 1, pp. 79–86, Mar. 1951. [Online]. Available: URL:https://www.jstor.org/stable/2236703

[32] H. Wang, M. F. Bugallo, and P. M. Djurić, "Adaptive importance sampling supported by a variational auto-encoder," in *2019 IEEE 8th International Workshop on Computational Advances in Multi-Sensor Adaptive Processing (CAMSAP)*, 2019, pp. 619–623.

[33] S. Kong, M. Kim, L. M. Hoang, and E. Kim, "Automatic LPI radar waveform recognition using CNN," *IEEE Access*, vol. 6, pp. 4207–4219, 2018.

[34] D. Bega, M. Gramaglia, M. Fiore, A. Banchs, and X. Costa-Perez, "Deepcog: Cognitive network management in sliced 5g networks with deep learning," in *IEEE INFOCOM 2019 - IEEE Conference on Computer Communications*, 2019, pp. 280–288.

[35] N. G. Shakev, S. A. Ahmed, V. L. Popov, and A. V. Topalov, "Recognition and following of dynamic targets by an omnidirectional mobile robot using a deep convolutional neural network," in *2018 International Conference on Intelligent Systems (IS)*, 2018, pp. 589–594.

[36] M. McGregor. (2/4/2021) What is a convolutional neural network? A beginner's tutorial for machine learning and deep learning. [Online]. Available: https://www.freecodecamp.org/news/convolutional-neural-network-tutorial-for-beginners/ ,OPTsubtitle={freeCodeCamp.org.},

[37] Z. Qu, W. Wang, C. Hou, and C. Hou, "Radar signal intra-pulse modulation recognition based on convolutional denoising autoencoder and deep convolutional neural network," *IEEE Access*, vol. 7, pp. 112 339–112 347, 2019.

[38] B. Wang, H. Wang, S. Wang, Y. Wang, D. Zeng, and M. Wang, "Recognition of the radar antenna scanning period based on convolutional neural network," in *2019 IEEE International Conference on Signal, Information and Data Processing (ICSIDP)*, 2019, pp. 1–5.

[39] X. Li, Z. Huang, F. Wang, X. Wang, and T. Liu, "Toward convolutional neural networks on pulse repetition interval modulation recognition," *IEEE Communications Letters*, vol. 22, no. 11, pp. 2286–2289, 2018.

[40] L. Cain, J. Clark, E. Pauls, B. Ausdenmoore, R. Clouse, and T. Josue, "Convolutional neural networks for radar emitter classification," in *2018 IEEE 8th Annual Computing and Communication Workshop and Conference (CCWC)*, 2018, pp. 79–83.

[41] Y. LeCun. (2015) Lenet-5, convolutional neural networks. [Online]. Available: http://yann.lecun.com/exdb/lenet/

[42] F. Liu, X. Tang, X. Zhang, L. Jiao, and J. Liu, "Large-scope polsar image change detection based on looking-around-and-into mode," *IEEE Transactions on Geoscience and Remote Sensing*, vol. 59, no. 1, pp. 363–378, 2021.

[43] T. A. Tejas, "Recurrent neural networks — complete and in-depth," 2020. [Online]. Available: https://medium.com/analytics-vidhya/what-is-rnn-a157d903a88

[44] Y. Bengio, P. Simard, and P. Frasconi, "Learning long-term dependencies with gradient descent is difficult," *IEEE Transactions on Neural Networks*, vol. 5, no. 2, pp. 157–166, 1994.

[45] W. Feixue, Y. Wenxian, and G. Guirong, "Recurrent neural network for high-resolution radar ship target recognition," in *Proceedings of International Radar Conference*, 1996, pp. 200–203.

[46] A. Sailaja, A. K. Sahoo, G. Panda, and V. Baghel, "A recurrent neural network approach to pulse radar detection," in *2009 Annual IEEE India Conference*, 2009, pp. 1–4.

[47] S. Ge, O. Antropov, W. Su, H. Gu, and J. Praks, "Deep recurrent neural networks for land-cover classification using sentinel-1 insar time series," in *IGARSS 2019 - 2019 IEEE International Geoscience and Remote Sensing Symposium*, 2019, pp. 473–476.

[48] S. Hochreiter and J. Schmidhuber, "Long short-term memory," *Neural Computation*, vol. 9, no. 8, pp. 1735–1780, 1997.

[49] J. Patterson and A. Gibson, *Deep Learning, A Practitioner's Approach*, 2017.

[50] M. Chen, U. Challita, W. Saad, C. Yin, and M. Debbah, "Artificial neural networks-based machine learning for wireless networks: A tutorial," *IEEE Communications Surveys Tutorials*, vol. 21, no. 4, pp. 3039–3071, 2019.

[51] S. Apfeld and A. Charlish, "Recognition of unknown radar emitters with machine learning," *IEEE Transactions on Aerospace and Electronic Systems*, pp. 1–1, 2021.

[52] S. Apfeld, A. Charlish, and G. Ascheid, "Identification of radar emitter type with recurrent neural networks," in *2020 Sensor Signal Processing for Defence Conference (SSPD)*, 2020, pp. 1–5.

[53] K. Cho, D. Bahdanau, F. Bougares, H. Schwenk, and Y. Bengio. (2014) Learning Phrase Representations using RNN Encoder–Decoder for Statistical Machine Translation. [Online]. Available: arXiv:1406.1078v3[cs.CL]

[54] R. Salakhutdinov and G. Hinton. (2009) Deep Boltzmann Machines. [Online]. Available: http://proceedings.mlr.press/v5/salakhutdinov09a/salakhutdinov09a.pdf

[55] G. Hinton. (2010) A Practical Guide to Training Restricted Boltzmann Machines. [Online]. Available: http://www.cs.toronto.edu/~hinton/absps/guideTR.pdf

[56] S. Ghimire, S. Ghimire, and S. Subedi, "A study on deep learning architecture and their applications," in *2019 International Conference on Power Electronics, Control and Automation (ICPECA)*, 2019, pp. 1–6.

[57] R. Saiakhutdinov and G. Hinton, "Deep boltzmann machines," in *Proceedings of the 12th International Confe-rence on Artificial Intelligence and Statistics (AISTATS)*, vol. 5, 2009, pp. 448–455.

[58] C. Qin, X. Song, and H. Chen, "Radar behavior classification based on dbn," in *2020 IEEE 9th Joint International Information Technology and Artificial Intelligence Conference (ITAIC)*, vol. 9, 2020, pp. 1169–1173.

[59] G. E. Hinton and R. R. Salakhutdinov, "Reducing the dimensionality of data with neural networks," *Science*, vol. 313, no. 5786, pp. 504–507, 2006. [Online]. Available: https://science.sciencemag.org/content/313/5786/504

[60] I. Goodfellow, J. Pouget-Abadie, M. Mirza, B. Xu, D. Warde-Farley, S. Ozair, A. Courville, and Y. Bengio, "Generative adversarial networks," *Commun. ACM*, vol. 63, no. 11, p. 139–144, Oct. 2020. [Online]. Available: https://doi.org/10.1145/3422622

[61] D. Mane and U. Kulkarni, "Visualizing and understanding customized convolutional neural network for recognition of handwritten marathi numerals," *Procedia Computer Science*, vol. 132, pp. 1123–1137, 2018, international Conference on Computational Intelligence and Data Science. [Online]. Available: https://www.sciencedirect.com/science/article/pii/S1877050918307592

[62] C. Zheng, X. Jiang, and X. Liu, "Semi-supervised sar atr via multi-discriminator generative adversarial network," *IEEE Sensors Journal*, vol. 19, no. 17, pp. 7525–7533, 2019.

[63] S. J. Pan and Q. Yang, "A survey on transfer learning," *IEEE Transactions on Knowledge and Data Engineering*, vol. 22, no. 10, pp. 1345–1359, 2010.

[64] K. Bu, Y. He, X. Jing, and J. Han, "Adversarial transfer learning for deep learning based automatic modulation classification," *IEEE Signal Processing Letters*, vol. 27, pp. 880–884, 2020.

[65] N. E. West and T. O'Shea, "Deep architectures for modulation recognition," in *2017 IEEE International Symposium on Dynamic Spectrum Access Networks (DySPAN)*, 2017, pp. 1–6.

[66] A. Lin, Z. Ma, Z. Huang, Y. Xia, and W. Yu, "Unknown radar waveform recognition based on transferred deep learning," *IEEE Access*, vol. 8, pp. 184 793–184 807, 2020.

[67] N. E. West and T. O'Shea, "Deep architectures for modulation recognition," in *2017 IEEE International Symposium on Dynamic Spectrum Access Networks (DySPAN)*, 2017, pp. 1–6.

[68] H. B. McMahan and D. Ramage. (2017) Federated learning: Collaborative machine learning without centralized training data. [Online]. Available: https://ai.googleblog.com/2017/04/federated-learning-collaborative.html

[69] J. Konecny, H. B. McMahan, D. Ramage, and P. Richtarik. (10/8/2016) Federated optimization, distributed machine learning for on-device intelligence. [Online]. Available: https://arxiv.org/pdf/1610.02527.pdf/

[70] J. Wang, Z. Xu, Z. Garrett, Z. Charles, L. Liu, and G. Joshi. (June 4, 2021) Local adaptivity in federated learning: Convergence and consistency. [Online]. Available: https://arXiv:2106.02305v1/

[71] J. Konečný, H. B. McMahan, F. X. Yu, P. Richtarik, A. T. Suresh, and D. Bacon, "Federated learning: Strategies for improving communication efficiency," in *NIPS Workshop on Private Multi-Party Machine Learning*, 2016. [Online]. Available: https://arxiv.org/abs/1610.05492

[72] K. Bonawitz, F. Salehi, J. Konečný, B. McMahan, and M. Gruteser, "Federated learning with autotuned communication-efficient secure aggregation," in *2019 53rd Asilomar Conference on Signals, Systems, and Computers*, 2019, pp. 1222–1226.

[73] B. Drury. Federated machine learning - collaborative machine learning without centralised training data. [Online]. Available: https://www.datasciencecentral.com/profiles/blogs/federated-machine-learning-collaborative-machine-learning-without

[74] K. Bonawitz, F. Salehi, J. Konečný, B. McMahan, and M. Gruteser, "Federated learning with autotuned communication-efficient secure aggregation," in *2019 53rd Asilomar Conference on Signals, Systems, and Computers*, 2019, pp. 1222–1226.

[75] P. Zhang, H. Sun, J. Situ, C. Jiang, and D. Xie, "Federated transfer learning for iiot devices with low computing power based on blockchain and edge computing," *IEEE Access*, vol. 9, pp. 98 630–98 638, 2021.

[76] Z. Yang, A. Gang, and W. U. Bajwa, "Adversary-resilient distributed and decentralized statistical inference and machine learning: An overview of recent advances under the byzantine threat model," *IEEE Signal Processing Magazine*, vol. 37, no. 3, pp. 146–159, 2020.

[77] G. Indiveri and Y. Sandamirskaya, "The importance of space and time for signal processing in neuromorphic agents: The challenge of developing low-power, autonomous agents that interact with the environment," *IEEE Signal Processing Magazine*, vol. 36, no. 6, pp. 16–28, 2019.

[78] E. Chicca, D. Badoni, V. Dante, M. D'Andreagiovanni, G. Salina, L. Carota, S. Fusi, and P. Del Giudice, "A vlsi recurrent network of integrate-and-fire neurons connected by plastic synapses with long-term memory," *IEEE Transactions on Neural Networks*, vol. 14, no. 5, pp. 1297–1307, 2003.

[79] M. Giulioni, P. Camilleri, V. Dante, D. Badoni, G. Indiveri, J. Braun, and P. Del Giudice, "A vlsi network of spiking neurons with plastic fully configurable "stop-learning" synapses," in *2008 15th IEEE International Conference on Electronics, Circuits and Systems*, 2008, pp. 678–681.

[80] S. Brink, S. Nease, P. Hasler, S. Ramakrishnan, R. Wunderlich, A. Basu, and B. Degnan, "A learning-enabled neuron array ic based upon transistor channel models of biological phenomena," *IEEE Transactions on Biomedical Circuits and Systems*, vol. 7, no. 1, pp. 71–81, 2013.

[81] J. Park, S. Ha, T. Yu, E. Neftci, and G. Cauwenberghs, "A 65k-neuron 73-mevents/s 22-pj/event asynchronous micro-pipelined integrate-and-fire array transceiver," in *2014 IEEE Biomedical Circuits and Systems Conference (BioCAS) Proceedings*, 2014, pp. 675–678.

[82] B. V. Benjamin, P. Gao, E. McQuinn, S. Choudhary, A. R. Chandrasekaran, J. Bussat, R. Alvarez-Icaza, J. V. Arthur, P. A. Merolla, and K. Boahen, "Neurogrid: A mixed-analog-digital multichip system for large-scale neural simulations," *Proceedings of the IEEE*, vol. 102, no. 5, pp. 699–716, 2014.

[83] J. Schemmel, D. Brüderle, A. Grübl, M. Hock, K. Meier, and S. Millner, "A wafer-scale neuromorphic hardware system for large-scale neural modeling," in *2010 IEEE International Symposium on Circuits and Systems (ISCAS)*, 2010, pp. 1947–1950.

[84] N. Qiao, H. Mostafa, and F. Corradi, "A re-configurable on-line learning spiking neuromorphic processor comprising 256 neurons and 128k synapses," in *Frontiers in Neuroscience*, vol. 9, no. 141, 2015, pp. 1–17.

[85] H. Jang, O. Simeone, B. Gardner, and A. Gruning, "An introduction to probabilistic spiking neural networks: Probabilistic models, learning rules, and applications," *IEEE Signal Processing Magazine*, vol. 36, no. 6, pp. 64–77, 2019.

[86] D. Chew and A. B. Cooper, "Spectrum sensing in interference and noise using deep learning," in *2020 54th Annual Conference on Information Sciences and Systems (CISS)*, 2020, pp. 1–6.

[87] A. Subekti, H. F. Pardede, R. Sustika, and Suyoto, "Spectrum sensing for cognitive radio using deep autoencoder neural network and svm," in *2018 International Conference on Radar, Antenna, Microwave, Electronics, and Telecommunications (ICRAMET)*, 2018, pp. 81–85.

[88] B. Soni, D. K. Patel, and M. López-Benítez, "Long short-term memory based spectrum sensing scheme for cognitive radio using primary activity statistics," *IEEE Access*, vol. 8, pp. 97 437–97 451, 2020.

[89] R. W. Hamming, "The art of doing science and engineering, learning to learn," p. 94, 1997.

[90] ——, *The Art of Doing Science and Engineering, Learning to Learn.* Gordon and Breach Science Publishers, 1997.

Chapter 10

Electronic Attack Using Deep Learning

In this chapter, the algorithms for EA are discussed. Emphasized are embedding these algorithms in a small DRFMs using GPUs and FPGAs, which can reside onboard unmanned airborne systems (UAS) and can interface with the communication network. As we recall, deep learning is a specific subfield of machine learning, where the learning is through successive layers of increasingly meaningful representations. The creation of obscuration electromagnetic attack and target electromagnetic decoys are discussed. As well, the creation of structured, multiple false target decoys are emphasized using UAS systems. Suppression of integrated air defense systems (IADS) and surface EA is also discussed. Creating embedded EA FPGAs and GPUs for EA are emphasized.

10.1 SPECTRUM DOMINANCE WITH UAVS AND DRFMS

Facing a complex EMS in the tactical environment, effectively identifying important RF signals by the DRFM and how to implement effective electromagnetic interference to the adversary's RF sensor and detection system under the condition of limited EA resources is a difficult problem that needs to be solved continuously and efficiently with persistent knowledge of the surrounding threats. This is one challenge for electromagnetic spectrum operations (EMSO). In EMSO, the function of EA resources is disruption to adversary RF sensor systems to distract, seduce, decoy, and even destroy RF sensor hardware from doing its job (e.g., image, track incoming attack aircraft).

An example illustrating an air-to-surface electromagnetic warfare engagement is shown in Figure 10.1. A distributed EMSO [1] is underway with two UAVs being flown in a geometrical position to synchronize with each other and launch an EA against an antiaircraft defense (AAD) radar system, the S-300V. The two UAVs are just a small part of a larger swarm, each with communication capabilities using a digital COSΦ (communication system via pheromone) [2].

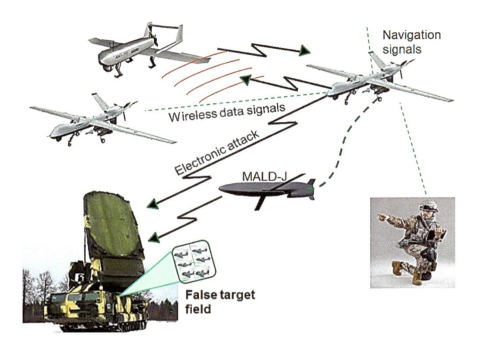

Figure 10.1 S-300V AAD radar system under EA from two UAVs with DRFMs, flying in escort formation. Note: wireless datalink UAVs and the launch of the miniature air-launched decoy (MALD). In addition, note that an adversary UAV is listening and attempting to apply deception to the friendly navigation signals (e.g., GPS) or the communications signals (cyberattack). In each instance, a DRFM transceiver is used.

The S-300V is designed for the defense of troops and important military and civilian objects from ballistic missiles and cruise missiles (e.g., an air-launched cruise missile) and strategic and tactical aircraft. An earlier version was arguably, the first universal mobile missile AAD which was developed and built by the Electromechanical Research Institute (NIEMI).

The EMS, SoS, and threat information for this situation is sent to the C2 over a network for their action. This attack scenario is then triggered by the decision maker on the ground (C2) transmitting a signal to one of the UAVs (e.g., a cluster-head) that sets in action the (a) launch of a coordinated attack from both UAVs, (b) launch of a Miniature Air Launched Decoy, (MALD) and (c) execution of software to launch a field of structured false targets to appear, overloading their track file formation capability. Within the next OODA loop, a reconnaissance on the S-300V AAD will confirm whether the actions taken were effective (or not), which will determine the next decisions by the C2.

10.1 Spectrum Dominance with UAVs and DRFMs

The MALD (ADM-160A/B) is equipped with a radio data link to expand an operator's situational awareness by sending situational awareness data to the electromagnetic warfare battle manager (EWBM) at the decision maker, or C2. The data link allows for in-flight targeting adjustments and coordinated EA with the other unmanned systems flying in escort, in order to confuse enemy air defenses. The MALD is also able to imitate radar signatures of manned fighter and rocket-powered vehicles. It sometimes attempts to duplicate the flight profile of the launch aircraft to trick enemy air defense batteries into firing SAMs at the decoy instead manned/high-value aircraft. The MALD also discloses the SAM's launcher location, broadcasting it on the *swarm's digital pheromone*.

One of the major difficulties that can arise is having the decoy, or any UAV, maneuver to the site autonomously (for example, at low altitude through difficult complex terrain). Deep learning is making a major impact in this area. Below, the development of a model is shown to demonstrate how this can be done efficiently.

10.1.1 Autonomous UAV Navigation Using Deep Reinforcement Learning

Long overlooked, as discussed briefly in Section 9.4, Google's DeepMind learned to play Go at the highest level [3] using *reinforcement learning* (RL). Reinforcement learning uses an agent to receive information about its environment and learns to choose actions that will maximize some reward and has become highly successful in nonlinear control.

10.1.1.1 Deep Learning Navigation Control Strategy

To dispatch the UAVs to the designated targets, one of the most efficient methods is to set up the autonomous navigation as a discrete-time, continuous control problem and use *deep* RL (DRL) as the solution. This method is most efficient as it applies deep learning to the traditional RL using a function approximation as a solution to a Markov decision process (MDP) and partially observable Markov decision process (POMDP) directly.

RL solves the model-free POMDPs by maintaining the memories of the observations and actions that occurred in the past, from which stochastic or deterministic control policies are derived. The *stochastic* policies map the memories of past observations and actions into distributions over the action space. The *deterministic* policies map memories into actions. Policy gradient methods such as the likelihood-ratio methods and actor critic methods derive parameterized stochastic policies for model-free POMDPs with continuous action spaces. This is done by performing gradient decent in the parameter space [4].

Here, the sensory inputs are projected onto the control signals in order to bridge the gap between an understanding of the natural world and the navigation control strategy [5]. In this manner, the UAV regards the (a) distances from itself to the obstacles in multiple directions, (b) its orientation angles, and (c) the distance and the angle between its present position and the destination (which can be derived from the GPS signals) as the sensory information. As a POMDP control solution, the UAV learns to navigate to the target position using the AI DRL. Note: as a POMDP, this method *does not* require learning a local or global map of the unknown complex environment or any path planning [5].

10.1.1.2 Reinforcement Learning and the Markov Decision Process

Recall from Section 9.4, RL is a type of DL that allows machines and software agents to automatically determine the ideal behavior within a specific context, in order to maximize their performance. That is, an agent has both a state S and a set of actions A. The action (or policy) the agent chooses to get to the next state can be either deterministic or stochastic where the next states have an initial state probability distribution $p(s_0)$ and a state transition probability distribution function $p(s_{t+1}|s_t, a_t)$. Using a simple (but required) reward-feedback reinforcement signal, the agent learns to correct its behavior over the series of changing states. This reinforcement learning with all its iterative solutions (reinforcement learning algorithms) are known as a Markov decision process (MDP). An important property of a MDP is the Markov property that the effects of an action taken in a state depend only on "that state" and not on the prior history. The goal of the DRL agents is to learn an optimal stochastic policy $\pi(s_t)$ or deterministic policy $\mu(s_t)$ that maximizes its long-term accumulated reward.

A POMDP is a generalization of a MDP in which it is assumed that the system dynamics are determined by an MDP, but the agent cannot directly observe the underlying state. Instead, it must maintain a probability distribution over the set of possible states based on a set of observations and observation probabilities and the underlying MDP [5].

10.1.1.3 Developing the Model

The POMDP framework is general enough to model a variety of real-world sequential decision processes based on an *actor-critic* architecture [6] with function approximation. A MDP becomes a POMDP when the agent cannot directly observe the state s_t but instead obtains an observation o_t where $o_t \sim p(o_t|s_t)$. That is, the policy instead of being a function of the state (as the MDP) it is now a function of the trajectory $h_t = (o_1, a_1, o_2, a_2, \ldots, a_{t-1}, o_t)$ automatically extracts state descriptions from raw sensory inputs using the deep network structures.

Environments such as those experienced by the stand-in jammer/DRFM platforms for the suppression of enemy air defense systems are dynamic and complex. To develop a model to autonomously navigate in a complex environment such as this, the complexity should be significant, such as that shown in Figure 10.2 [5].

10.1 Spectrum Dominance with UAVs and DRFMs

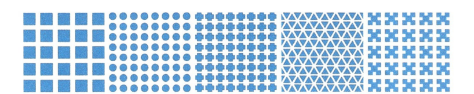

Figure 10.2 Five types of virtual environments, each of which covers one square kilometer [5].

The goal of the (virtual) UAV is to fly from an arbitrary starting position to arbitrary destinations using GPS signals and its perception of the local environment. Without loss of generality, we create a virtual UAV that flies at a fixed altitude with fixed speed, and its control profile only contains turning left or right.

Consider the UAV is mounted with five virtual range finders, illustrated in Figure 10.3(a), and we regard the distances from the UAV to obstacles in the five directions as the UAV's sensory information of the environment.

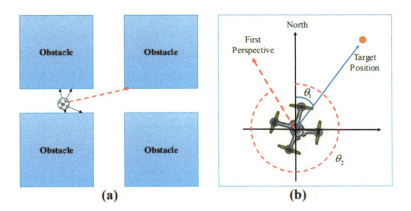

Figure 10.3 Geometry showing (a) range finders sensing obstacle in five different directions and (b) θ_1 : the angle between the present position and the target position, θ_2 : direction angle of the UAV. (©IEEE, reprinted with permission from [5].)

For the UAV navigation, the five distances with the UAV's orientation angle and distance and the angle between its present position and the destination, depicted in Figure ??(b), are used in the description of the final state. The reward specification is made up of four parts [5]:

- Environment Penalty: UAV is too close to any object during flying and should be penalized. If minimum distance (between UAV and obstacle decreases), penalty grows exponentially.

- Transition Reward: Distance from UAV to its target position decreases after time step and UAV rewarded proportional to reduced distance.
- Direction Reward: UAV gets a constant reward if its first-perspective direction is in accordance with the corresponding direction of the biggest distance among the five distances.
- Step Penalty: After every time step, UAV given a constant penalty so as to encourage it to reach the target position as soon as possible.

Using a *deterministic policy gradient* algorithm for the reinforcement learning is appealing as it is the expected gradient of the action-value function and can be estimated much more efficiently than the stochastic policy gradient by avoiding a problematic integral over the action space [7, 8]. However, as shown in Figure 10.4, the UAV can easily become trapped in a local area [5].

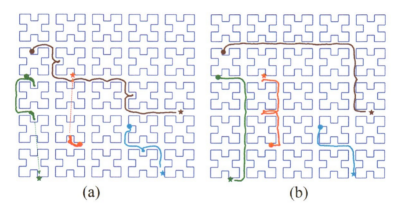

Figure 10.4 Illustration of trajectories, where circles represent starting positions and stars represent target positions. (a) Trajectories generated by a deterministic policy gradient algorithm and (b) trajectories generated by a fast-RDPG agent) (©IEEE, reprinted with permission from [5].)

In a different approach, a recurrent deterministic policy gradient (RDPG) algorithm solves the POMDP by using two recurrent neural networks Q^μ and μ^θ with the actor network updated using back propagation learning [9]. However, the updates to the actor are based on a sequence of highly correlated states.

To break up the correlation, a much faster algorithm is developed in [5] with the use of the actor-critic architecture with function approximation as in [6], which derives the policy update by directly maximizing the expected long-term accumulated discounted reward to obtain a new policy update for POMDP with continuous control [6].

RL involves estimating "value functions" of states and action-value functions of state-action pairs. The policy is explicitly updated with the history trajectory h_t rather

10.1 Spectrum Dominance with UAVs and DRFMs

than the entire trajectory τ. The stochastic action policy $a_t \sim \pi_\theta(a_t|h_t)$ and the state after observing the history h_t as s_t where $s_t \sim p(s_t|h_t)$ are used to derive the policy update. For stochastic policies, first define the *value function* as

$$V^{\pi_\theta}(h) = E_{p(s|h)}E_{\tau_1}\left[\sum_{t=1}^{\infty}\gamma^{t-1}r_t|s_1=s, h_t=h, \pi_\theta\right] \quad (10.1)$$

where γ is a discount factor ranging from 0 to 1 [4]. The *action value function* as

$$Q^{\pi_\theta}(h,a) = E_{p(s|h)}E_{p(o_{t+1}|s_t,a_t)p(\tau_{t+1})}\left[\sum_{k=1}^{\infty}\gamma^{k-1}r_{t+k}|s_t=s, h_t=h, a_t=a, \pi_\theta\right] \quad (10.2)$$

and the *expected reward* when observing history trajectory h as

$$R_h^a = E_{p(s|h)}E_{p(s_{t+1}|s_t=s,a_t=a)}\left[\sum_{k=1}^{\infty}r_{t+1}|s_t=s, h_t=h, a_t=a, \pi_\theta\right] \quad (10.3)$$

where

$$\tau_1 \sim \prod_{t=1}^{\infty}\pi(a_t|h_t)p(s_{t+1}|s_t,a_t)p(o_{t-1}|s_{t+1})$$

The *target function* can be defined as

$$J(\theta) = E_{p(h_0)}[V^{\pi_\theta}(h_0)] \quad (10.4)$$

where $p(h_0) = p(o_0)$ represents the initial observation distribution and

$$\frac{\partial V^{\pi_\theta}(h_0)}{\partial \theta} = \sum_a \left[\frac{\partial \pi_\theta(h_0,a)}{\partial \theta}Q^{\pi_\theta}(h_0,a) + \pi_\theta(h_0,a)\left(\gamma\sum_h p(h|h_0)\frac{\partial V^{\pi_\theta}(h)}{\partial \theta}\right)\right] \quad (10.5)$$

and can be written as

$$\frac{\partial V^{\pi_\theta}(h_0)}{\partial \theta} = \sum_h \rho^{\pi_\theta}(h|h_0)\frac{\partial \pi_\theta(h_0,a)}{\partial \theta}Q^{\pi_\theta}(h,a) \quad (10.6)$$

where $\rho^\pi(h|h_0)$ represents the historical distribution induced by policy π. The policy update can then be expressed as [7]

$$\frac{\partial J(\theta)}{\partial \theta}E_{\pi_\theta}E_{p(h_0)\rho^{\pi_\theta}(h|h_0)} = [\nabla_\theta \log(\pi_\theta(h,a))Q^{\pi_\theta}(h,a)] \quad (10.7)$$

with the expectation over the trajectory history h (rather than the entire trajectory τ). The new algorithm based on (10.7) and [10] is shown in Figure 10.5 and is called Fast-RDPG [5].

Fast-RDPG
Initialize critic network $Q^w(a_t, h_t)$ and actor $\mu^\theta(h_t)$ with parameters w and θ.
Initialize target networks $Q^{w'}$ and $\mu^{\theta'}$ with weights $w' \leftarrow w$, $\theta' \leftarrow \theta$.
Initialize replay buffer R.
for episodes=1, M, **do**
    Initialize a process N for action exploration
    Receive initial observation o_0 ($h_0 = o_0$)
    for t=1, T **do**
        Obtain action $a_t = \mu^\theta(h_t) + N_t$
        Execute action a_t, receive reward r_t, and obtain a new observation o_t
        Store transition (h_{t-1}, a_t, o_t, r_t) into R
        Update history trajectory $h_t = [h_{t-1}, a_t, o_t]$
        Sample a minibatch of L transitions (h_i, a_i, o_i, r_i)
        Set $y_i = r_i + \gamma Q^{w'}([h_i, a_i, o_i], \mu^{\theta'}([h_i, a_i, o_i]))$
        Compute critic update (using BPTT)

$$\Delta \omega = \frac{1}{L} \sum_i (y_i - Q^\omega(h_i, a_i)) \frac{\partial Q^\omega(h_i, a_i)}{\partial \omega}$$

        Compute actor update (using BPTT)

$$\Delta \theta = \frac{1}{L} \sum_i \frac{\partial Q^\omega(h_i, \mu^\theta(h_i))}{\partial a} \frac{\partial \mu^\theta(h_i)}{\partial \theta}$$

        Update actor and critic using Adam (Reference)
        Update the two target networks
            $w' \leftarrow \varepsilon w + (1-\varepsilon) w'$
            $\theta' \leftarrow \varepsilon \theta + (1-\varepsilon) \theta'$
    end for
end for

Figure 10.5 The Fast-RDPG algorithm. (©IEEE, reprinted with permission from [5].)

Before learning, all the sensory inputs are normalized between [0, 1] and the control signal is normalized between [−1, 1], where −1 means turning left 90 degrees and 1 means turning right 90 degrees. The hyperparameters in the learning algorithm, as in [10] but with uniformly distributed noise. To validate the efficiency of the Fast-RDPG against the RDPG, four pairs of starting positions and destinations were randomly generated in the most complex environment and let both our Fast-RDPG agent and a RDPG agent fly from these starting positions to target positions. Figure 10.4 (a) shows the trajectories of the RDPG and Figure 10.4(b) shows the trajectories of the Fast-RDPG trajectories [5].

Simulation results for the deep deterministic policy gradient algorithm (DDPG), the RDPG and the Fast-RDPG are shown in Figure 10.6. For the DDPG, the action of the deterministic policy is certain, so the agent cannot visit other states, thus making learning of these states impossible [11].

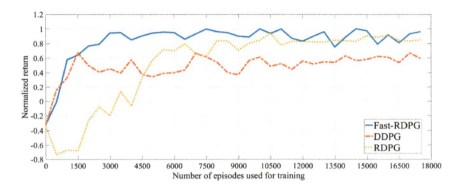

Figure 10.6 Convergence curves for DDPG, RDPG, and Fast-RDPG algorithms. (©IEEE, reprinted with permission from [4].)

Due to the break up of high correlation present in observation sequences, the Fast-RDPG converges much faster than the DDPG. This is because the Fast-RDPG agent knows more local information than the RDPG agent and its learned navigation policy is less greedy than that of the RDPG agent and it possesses the ability to avoid or escape from traps while the DDPG agent is easily trapped.

10.1.2 Autonomous UAV Navigation Using Deep Transfer Learning

A generic model of a CNN for transfer learning is shown in Figure 10.7 [12]:

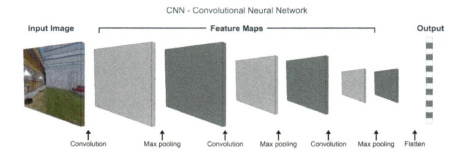

Figure 10.7 Convolutional neural network model for transfer learning [12].

Here a sequence of filters is applied to the input image (convolutional layer) and the results are resized and flattened into a 1D feature vector. The transfer learning concept is used by the CNN as a feature extractor. The attributes returned by CNN are used by the classifiers for the construction of the computer vision system. Figure 10.8 presents the methodology used to implement the proposed approach. Other methods of precise UAV navigation in complex areas have been explored, for example, with the use of topological maps. Classifiers that were considered include the k-nearest neighbor (kNN), multilayer perceptron (MLP), optimum-path forest (OPF), SVM, the radial basis function (RBF) kernels, and the Bayesian classifier.

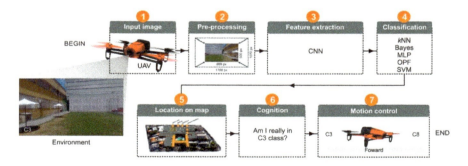

Figure 10.8 Methodology of the proposed approach for the UAVs navigation. (©IEEE, reprinted with permission from [12].)

Figure 10.9 shows the topographical map of the environment and the path steps of the UAV [12].

10.2 Joint Electromagnetic Spectrum Operations

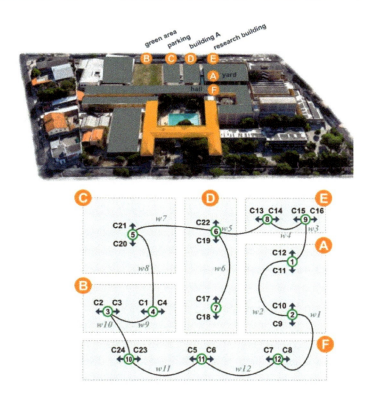

Figure 10.9 Topographical map of the environment showing the UAVs navigation steps. (©IEEE, reprinted with permission from [12].)

According to the results shown in [12], the CNN confirms itself is a convincing alternative for the task of locating and navigating the UAVs, reaching 99.97% in accuracy.

10.2 JOINT ELECTROMAGNETIC SPECTRUM OPERATIONS

Each adversary is well aware of the importance of the EMS and have taken steps to improve their capabilities to threaten the other's ability to use and control the EMS. What can be done? The pervasiveness of the EMS across all warfighting domains means that ensuring superiority against an adversary is critical to battlefield success [13].

Fortunately, after nearly 12 years in the making, the DoD has published the first-ever doctrine on *joint electromagnetic spectrum operations* (JEMSO) with the release of Joint Publication (JP) 3-85 [14, 15]. So why is this important for today's DRFM EA capabilities? The reason is that the JEMSO is now the major focal point for all joint force exploitation, attack, and protect missions using the electromagnetic environment. That is, the JEMSO is the center of gravity for all integration, deconfliction, and dynamic control of high-energy military operations in the EMS to achieve EMS superiority.

The DoD defines EMS superiority as:

> "EMS superiority is that control in the electromagnetic spectrum that permits the conduct of operations at a given time and place without prohibitive interference, while affecting an adversary's ability to do the same" [15].

The result is a fully integrated scheme of maneuver in the EMOE to achieve EMS superiority and joint force commander (JFC) objectives.

The JEMSO promotes a common perspective from which to plan, train, and conduct military operations. Most of these efforts were slow to develop or stalled due to myriad issues until November 2017, when USSTRATCOM J3E proposal at the Joint Doctrine Planning Conference to develop the first joint publication on JEMSO and received unanimous approval to proceed, beginning of a new chapter in institutionalizing JEMSO across the joint force[14].

One superior strategic vision is to first dominate the networks while using information operations to control the conflict in its early stages. It is also recognized that this winning strategy is to emphasize EW dominance by suppressing, degrading, disrupting, or deceiving enemy electronic equipment. This includes targeting adversary systems that operate in radio, radar, microwave, infrared, and optical frequency ranges, as well as computers and information systems [13]. The use of AI to monitor these contiguous bands continuously is an incredible companion to the joint force information operations officer [16]. Consider "Russia trails China and the U.S. in all metrics of AI" [17], "AI Looms Large in Race for Global Superiority [18]," and "Russia trails China and the U.S. in all metrics of AI [19]." Note, a DRFM now can enter an environment with no knowledge of the adversary and quickly gain an understanding of the signals present.

10.3 TYPES OF ELECTRONIC ATTACK

Electronic attack is the use of EM energy, including directed energy or antiradiation weapons, to attack personnel, facilities, or equipment with the intent of degrading, neutralizing, or destroying enemy combat capability. Outlined in Chapter 1 was the importance of the EA capabilities and how they can directly produce effects in the EMOE. The capabilities contained within a DRFM can be used to deny (i.e., disrupt, degrade, destroy) and/or deceive an enemy's military EMS activities.

There are three types of EA:

- *Obscuration Electromagnetics*: Create RF waveforms that change the electrical, magnetic characteristics of environment in an effort to reduce radar/thermal detectability of the platforms;

- *Target Electromagnetics*: Create RF waveforms that are similar to the signals scattered by the target;

- *Destruction/Non-destruction Electromagnetics*: Create RF waveforms for EM suppression. Note that this category includes directed energy weapons and antiradiation missiles. Note also that obscuration and target jamming are nondestructive.

Typical EA capabilities include EM-EA and EM-intrusion. EM-EA is the deliberate radiation, re-radiation, or reflection of EM energy for the purpose of preventing or reducing an enemy's effective use of the EMS to degrade or neutralize the enemy's combat capability [15]. EM-intrusion involves the intentional insertion of EM energy into transmission paths to deceive or confuse enemy forces and can be either active (i.e., radiating) or passive (i.e., nonradiating/reradiating).

Examples of active EA systems (to include lethal and nonlethal directed energy) include lasers, electro-optical, IR, and RF weapons such as high-power microwaves or those employing an electromagnetic pulse. Examples of passive EA include chaff and corner reflectors. EA can also be divided into offensive and defensive categories and can be applied from the air or the surface.

10.3.1 Airborne EA: Suppression Techniques

The classification of airborne operational EA techniques for suppression of surface-to-air-missile (SAM) and rocket sites is one of the most heavily used EA assets.[1] The U.S. Navy involved in suppression operations use the EA-18G (*Growler*), the upgrade of the *Prowler*. The U.S. Air Force involved in suppression operations use the F-16CJ.

Airborne operational EA techniques for suppression engagement that support an EA attack are categorized by their location with respect to the target and the SAM's lethal range. Figure 10.10 shows the *offensive* (support an attack) and *defensive* (save your skin) breakdown of the type classifications. These techniques are broken down into three separate categories. The offensive techniques that support an attack include the

---

[1] For instance, the Electronic Attack Wing, Naval Air Weapons Stations, Whidbey Island, home of the EA-18G *Growler*, is busy all the time constantly training–day in and day out.

standoff jammer (SOJ), escort jammer (EJ) modified escort jammer (MEJ), mutual support jammer (MSJ), and stand-in jammer mission scenarios.

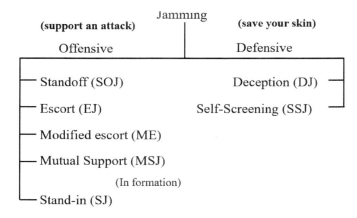

Figure 10.10 Offensive and defensive breakdown of EA operational techniques.

Critical issues concerning the DRFM time-line - during these operational tactical engagement scenarios, including manned and unmanned EA platform placement must be optimized. Most engagements are contiguous and are connected in time. These tactical engagements are highly nonlinear and the AI soft/firm/hardware techniques chosen must be capable of high-speed computations.

10.3.2 Surface EA: Defeating Anti-Ship Missiles

Surface electronic warfare relies on a combination of passive EW (e.g., low RCS values, tethered ES sensors for beyond LOS) and various active EA methods. The most important, high-priority item to a ship's captain is an incoming anti-ship missile (ASM) aimed straight for their ship coming in at, for example, Mach 2. That is, nothing else takes precedent! For the U.S. Navy, active EA on surface vessels against an incoming ASM is conducted in layers. An example is shown in Figure 10.11.

10.3 Types of Electronic Attack   695

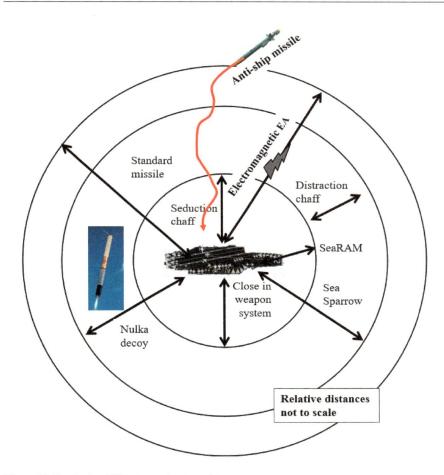

Figure 10.11 Surface EW in layers (not to scale).

Active EA using electromagnetic energy for the attack on the seeker can be engaged as soon as the missile seeker is detected beyond the LOS. The seeker can also remain silent until ≈0.5 nmi and then light up. The EA programs continue up until the missile misses the ship. Due to the onboard missile altimeters, they cannot be driven into the water [20]. The Extended Range Active Missile (ERAM) or Standard Missile Block-6 (SM-6) is used at the outermost range ring as it has a good deal of range and speed [21]. Next for example, the Sea Sparrow, a short-range point defense weapon is used along with the Nulka off-board repeater rocket. Also included in this ring is the use of distraction chaff Mk-216. Active EA if the missile gets close-in consists of seduction chaff (e.g, Mk-214 combined with the Gatling gun of the Phalanx, *close in weapon system*, or CIWS).

10.3.2.1 Surface EW Improvement Program

The AI coordination of the surface EW is accomplished through the upgraded AN/SLQ-32 or the Surface EW Improvement Program (SEWIP) [22]. Developed in several increments, SEWIP Block 1 focused on rapid development and fielding of low-risk upgrades by refreshing the display console and display/pulse processing; adding stand-alone *specific emitter identification* (SEI) capability and the display of combat system tracks to the operator to improve threat correlation and situational awareness; integration of SEI functionality, net-centric, and mission planning and introduction of a high gain high sensitivity (HGHS) capability to SEI and allowing the operator to launch both the Nulka active offboard decoys and passive EA on combat system tracks [22].

Block 2 is the ES interferometer-based antenna array upgrade for fine resolution direction finding (DF) delivering improvements in being able to see spread spectrum signals and LPI radars.

Block 3 is the integrated EA subsystem to protect against RF-guided threats using AESA technology based on gallium nitride transmit/receive modules based on the Office of Naval Research's (ONR's) Integrated Topside (InTop) Innovative Naval Prototype (INP) program. The system has four AESAs, and there are four quadrants on each ship that equals 16 AESAs working coherently to provide total ship self-protection from RF threats from any angle. NRL's Tactical EW Division (TEWD) has also supplied a transportable EW module (TEWM) to address separate urgent operational needs statements (UONS) raised by the US Sixth Fleet (Europe) and Seventh Fleet (Pacific) [22]. Conceived as a modular, portable, and platform-agnostic testbed incorporating an ES receiver integrated with a wideband DRFM-based EA capability. TEWM has previously demonstrated noise jamming and high-resolution false targets with realistic amplitude and Doppler modulation, providing a capability to engage multiple threats simultaneously and generate multicomponent waveforms that combine false targets with obscuration jamming [22].

10.3.3 Offensive EA

Offensive EA is the use for power projection to support operations within the time and tempo of the scheme of maneuver [15]. The JEMSO, along with joint force planning, coordinates the EA effects in order to minimize both the risk and the collateral damage. In some cases, maybe the suppression of a threat for only a limited period of time is necessary. Offensive EA examples include [15]:

- Employing self-propelled decoys;
- EA radar emitters or C2 systems;
- Using antiradiation missiles to suppress air defenses;

- Using EM deception techniques to confuse intelligence, surveillance, and reconnaissance (ISR) systems;
- Using directed energy weapons to disable personnel, facilities, or equipment;
- Using directed energy weapons to destroy satellites in orbit, airborne optical sensors, or massed land forces.

The DRFM using an AI operating system can be highly efficient in the management of time and tempo for the EA. This on/off time of the various offensive EA techniques is critical to their effectiveness within the scheme-of-maneuver [23].

10.3.4 Defensive EA

Defensive EA is the use of EA to protect against threats by denying enemy use of the EMS to target, guide, and/or trigger weapons. EA used for defensive purposes in support of force protection or self-protection is often mistaken as EP [15]. Although defensive EA actions and EP protect personnel, facilities, capabilities, and equipment, EP protects from the effects of EA or EMI, and missile defensive EA is primarily used to protect against lethal attacks by denying enemy use of the EMS to target, guide, and/or trigger weapons. Both self-screening jamming and deception jamming are two examples of defensive EA in order to save your skin. In the next section, a range-Doppler imaging summary tutorial is presented to provide the reader with the proper context for the remaining sections in this chapter and the next.

10.4 RANGE-DOPPLER IMAGING EMITTER: SIGNAL PROCESSING

In this section, pulse Doppler synthetic aperture and inverse synthetic aperture radar signal processing is reviewed. This is for the benefit of the reader for the section on inserting multiple, structured false targets as a deception measure.

10.4.1 Pulse Doppler Signal Processing

Typical waveforms used in modern pulsed seekers and radar systems are constructed to enable the extraction of the Doppler profiles from targets. The pulse Doppler signal starts with continuous complex sinusoid, which has a frequency spectrum of a delta functions located at the transmit frequency. Then, the continuous complex sinusoid is modulated by a pulsed waveform, yielding a spectrum consisting of weighted discrete spectral lines at multiples of the PRF centered at the transmit frequency (i.e., the complex sinusoid spectrum is replicated at multiples of the PRF with weighting determined by the pulsewidth).

These pulse Doppler radar waveforms are coherent and produced by modulating a complex sinusoid with a pulsed waveform. Considering an infinite time-on-target, the pulsed waveform can be written as

$$p(t) = p_\tau(t) * \sum_{n=-\infty}^{\infty} \delta(t - nT_r) \quad (10.8)$$

where $*$ indicates convolution and τ is the pulsewidth, T_r is the pulse repetition interval and

$$p_\tau(t) = \begin{cases} 1, & -\tau/2 < t \leq \tau/2 \\ 0, & \text{otherwise} \end{cases} \quad (10.9)$$

The spectrum of the pulsed waveform can be shown to be [24]

$$P(f) = \frac{\tau}{T_r} \sum_{m=-\infty}^{\infty} \frac{\sin(\pi\tau m f_r)}{\pi\tau m f_r} \delta(f - mf_r) \quad (10.10)$$

and consists of discrete spectral lines (delta functions δ) spaced at the pulse repetition frequency $f_r = 1/T_r$ that are weighted by a sinc function, where $\text{sinc}(x) = \sin(\pi x)/\pi x$. The sinc weighting is determined by the pulsewidth. Thus, the pulsewidth only determines the overall shaping of the spectrum and does not affect the location of the spectral lines [24].

10.4.2 Synthetic Aperture Signal Processing

Consider the SAR transmitted signal $s_t(t)$ given by

$$s_t(t) = \text{rect}\left[\frac{t}{T}\right] \exp j \left[-2\pi f_0 t + \pi \gamma_r t^2\right] \quad (10.11)$$

where t is the fast time dimension, T is the SAR pulse duration, γ_r is the rate of chirp in range dimension, f_0 is the carrier frequency of the SAR, and

$$\text{rect} = \begin{cases} 1; & 0 \leq X \leq 1 \\ 0; & \text{otherwise} \end{cases} \quad (10.12)$$

Recall the SAR as an airborne, spaceborne, coherent radar that provides a high resolution, range-Doppler image by traversing a path across the scene while radiating. Commonly used frequency bands and wavelength ranges used are shown in Table 10.4.2.

Table 10.4.2. Commonly Used Frequency Bands for SAR Systems and the Corresponding Frequency and Wavelength Ranges

Freq. Bnd	Ka	Ku	X	C	S	L	P
Freq.(GHz)	40-25	17.6-12	12-7.5	7.5-3.75	3.75-2	2-1	0.5-0.25
Wave.(cm)	0.75-1.2	1.7-2.5	2.5-4	4-8	8-15	15-30	60-120

Using the motion of the radar to create the Doppler, the objects being illuminated can be reconstructed with fine resolution [25]. The transmitted energy interacts with the Earth's surface and only a portion of it is backscattered to the receiving antenna, which can be the same as the transmit antenna (for a monostatic radar) or a different one (for a bi- or multistatic radar). The amplitude and phase of the backscattered signal depends on the physical (i.e., geometry, roughness) and electrical properties (i.e., permittivity) of the imaged object [25].

A SAR map is generated from reflectivity data collected as the radar platform moves past a stationary target area to be mapped, whereas ISAR target imagery is generated from reflectivity data collected as the target rotates within a stationary beam. It consists of coherently processing the large bandwidth returns received at different target viewing angles. The processed image consists of the magnitude and position of scatterers in the slant range and cross range. The *cross-range resolution* is dependent on the resolvable difference in the Doppler frequencies from two scatterers in the same slant-range cell. The *range resolution* depends upon the bandwidth of the transmitted waveform. Typically the expression is

$$\Delta R = \frac{c}{2B} = \frac{c\tau}{2} \qquad (10.13)$$

where B is the bandwidth of the transmitted waveform. This expression holds for most of the RF emitters except the CW frequency shift keying radar systems. In this case the range resolution depends on the duration of the frequency [20].

10.4.3 Inverse Synthetic Aperture Imaging

The ISAR is a version of SAR that can be used operationally to image targets such as ships, aircraft, and space objects. The technique also has application to instrumentation radar for evaluating radar cross section of targets and target models. The amplitude of the transmitted waveform is constant during the pulse time, while the instantaneous frequency is varied in some fashion over time and over a particular bandwidth. The target's range, bearing, and geodetic positional data is collected for display and recording.

However, today's trend is to remain as quiet as possible. These LPI radar systems employ measures to avoid detection by passive, noncooperative radar detection receivers (such as a radar warning receiver (RWR), or electronic support receivers) while it is searching for a target or engaged in target tracking [20]. For example a SAR using Costas frequency shift keying (FSK) modulation was reported in [26] and a polyphase P3 modulation was reported in [27].

ISAR-AI signal processing is often used to identify a ship uniquely by its hull structure and movement. Identification of airborne platforms and aircraft are also being identified uniquely not only by their ISAR images and air frame, but also by their jet engine modulation (JEM) signature within the frequency domain [28]. The JEM has a unique frequency structure that contains the *body line*, the *blade*, and the *shaft* [29]. These frequencies are illustrated in Figure 10.12.

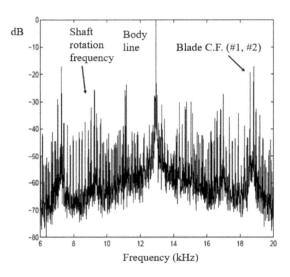

Figure 10.12 Jet engine modulation illustrated in the frequency domain showing the presence of a shaft rotation frequency, the blade chopping frequency (C. F.), and the body line. (©IEEE, reprinted with permission from [29].)

Only small and simple databases are required with short ToT and with AI. In addition, a large SNR is required and obtaining a good frequency spectrum of the JEM depends on the *aspect angle*. Without the use of AI and significantly detailed database, the classification results sometimes have questionable reliability.

ISAR (and SAR) use either chirp, or LFM, or a stepped frequency waveform SFW where the carrier frequency changes linearly from pulse to pulse. As expected, the *range resolution* ΔR is inversely proportional to the modulation bandwidth as $\Delta R = c/2\Delta F$ and the *unambiguous range* R_u is inversely proportional to the processing bandwidth Δf as $R_u = c/2\Delta f$ and should be greater than twice the target length [29].

The first step is to understand what a ship looks like under the illumination of an ISAR as the ship rolls back and forth, creating the Doppler signature necessary to form the image within the ISAR. The motion of the target (relative to emitter) imposes Doppler frequencies on the reflected signal as shown in Figure 10.13.

10.4 Range-Doppler Imaging Emitter: Signal Processing

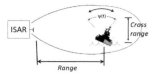

Figure 10.13 The motion of the target relative to the emitter over the frame time, T, imposes Doppler frequencies on the reflected signal.

These imposed Doppler frequencies are contained within the returned pulses as a pulse-to-pulse phase rotation of the scatterers within each range bin.

As an example, Figure 10.14 shows the series of six images of a rolling ship, the U.S.S. *Crockett*, as it rolls within the ISAR illumination, here the AN/APS-137(V)5. In Figure 10.14 a sequence of six range-Doppler images are shown of the U.S.S *Crockett*, rolling under the illumination of the AN/APS-137(V)5 ISAR.

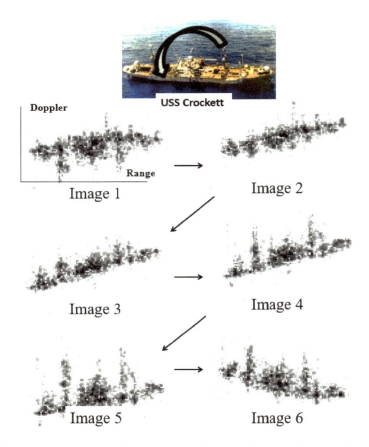

Figure 10.14 Sequence of six range-Doppler images of the U.S.S *Crockett* rolling under the illumination of the AN/APS-137(V)5. Note the image flips upside down as the ship rolls to and from the ISAR (Adapted from [30]).

Note the image flips upside down as the ship rolls *toward* the ISAR emitter and *away* from the ISAR emitter. The ISAR processor shown in Figure 10.15, converts the return pulses into a range-Doppler (cross-range) image. The compression for both the range (using the complex conjugate of the transmitted waveform) and the Doppler are shown including their resolution in both dimensions.

10.5 Obscuration-EA and Deep Learning

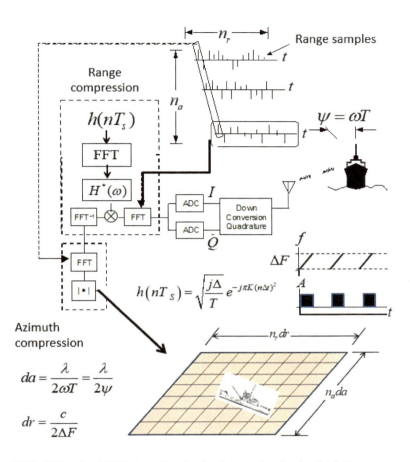

Figure 10.15 Schematic of ISAR range-Doppler signal processing showing the fast range compression of range n_r samples followed by the Doppler compression for the n_a range bin samples over a frame time T. Also shown is the range-Doppler map resulting from the signal processing and the range resolution $dr = c/2\Delta F$ and azimuth resolution $da = \lambda/2\psi = \lambda/2\omega T$, where ω is the rotation rate in rad/s.

10.5 OBSCURATION-EA AND DEEP LEARNING

Obscuration EA is still quite important and can be created actively (with a DRFM) or can be created passively (e.g., with chaff.) Chaff, consisting of aluminized glass fiber with the proper core diameter (bandwidth) cut at the proper length (frequency), can be quite devastating to an RF sensor. Still at the forefront of EW techniques, it can be used by itself or used in conjunction with the other DRFM programs. For *surface ships* it should include surface coattack programs for distraction chaff

(MK-216) and seduction chaff (MK-214) launched from the MK-36 Super Rapid Blooming Off-Board Chaff (SRBOC). For *airborne platforms*, the DRFM program should include the interface with components within the IDECM (Integrated Defensive Electronic Countermeasures) system. For example, the AN/ALE-47 is the Airborne Countermeasures Dispenser System used to protect military aircraft from incoming radar and infrared homing missiles. It works by dispensing flares or chaff. It is used on a variety of U.S. Air Force, Navy, and Army aircraft, as well as in other military systems. The EA operation techniques outlined in Figure 10.10 are all important program techniques for a DRFM to have in memory during a suppression engagement as shown in Figure 10.16. Note the three zones that distinguish the suppression engagement areas. For the EA systems in a reactive operation, the information being used within the control loop is highly time-critical. The resource manager must interleave these techniques in a time-sensitive fashion depending on whether the platform is outside enemy airspace, within enemy airspace-outside SAM lethal range, or inside enemy airspace and also enemy's lethal range.

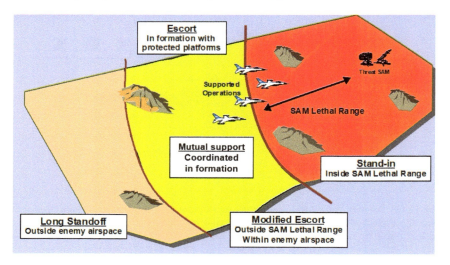

Figure 10.16 Operational classifications of EA showing outside enemy airspace (long standoff jammers), within enemy airspace-outside SAM lethal range (escort jammer, modified escort jammer and mutual support jammer configuration), and inside enemy airspace and also enemy's lethal range (stand-in jammer).

10.5.1 Obscuration Analysis

Noise is still a highly effective tool and the DRFM can use it correctly for the offensive engagement at hand. These engagements first are analyzed below before details of the

10.5 Obscuration-EA and Deep Learning

algorithm are presented. The obscuration algorithm objective is to hide the target and ensure actions in an effort to prevent radar lock-on to the target and, hopefully, prevent the firing of a weapon (missile). The four main algorithm categories the DRFM EA has to choose from are spot, barrage, swept spot, and multispot, shown in Figure 10.17.

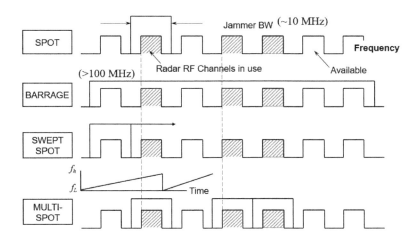

Figure 10.17 Obscuration noise EA categories.

Note the power level shown in the spot noise is higher within the bandwidth of the radar's frequency channel than that of the barrage jammer. This is because there is only so more power available within the jammer and spreading it out over multiple channels will bring the power density down considerably. The swept spot noise is particularly effective showing up in each channel for only a brief period depending on the serrodyne rate. The other noise modulations are described in the Appendix A.

10.5.2 The Beacon Equation

Consider the radar and the DRFM in Figure 10.18. Also consider that the bandwidth of the DRFM output noise filter B_j is matched to the radar's bandwidth B_r.

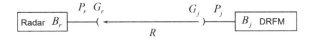

Figure 10.18 Radar and EA jammer active configuration.

The output of the DRFM is P_j W (continuous power) and the DRFM antenna gain is G_j. The radar, at a range R has a receive antenna gain of G_r. The power at the radar from the DRFM (jammer) is

$$P_{rj} = \underbrace{\left(\frac{P_j G_j}{4\pi R^2}\right)}_{\text{Power density at Radar}} \underbrace{\left(\frac{G_r \lambda^2}{4\pi}\right)}_{\substack{A_e \\ \text{Receiving Area}}} \quad \text{W} \quad (10.14)$$

This is known as the *beacon equation*. Note the order of the subscripts.
Now, consider the free-space spreading loss (in dB) as [31]

$$L_s = 32 + 20\log_{10}(f_c^{MHz}) + 20\log_{10}(R^{km}) \quad \text{dB} \quad (10.15)$$

where f_c is the carrier frequency in MHz and R is the range in km. However, if you are considering the situation close to the ground as shown in Figure 10.19, then the loss is a bit different. Here it shows the two heights of, for example, a radar and a target or a communication transmitter and a receiver.

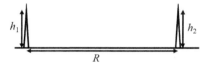

Figure 10.19 Two towers on the ground at separation distance of R to determine if the range is greater than the Fresnel distance ($R > FZ = 4\pi h_1 h_2/\lambda$).

In Figure 10.19 consider for example, h_1, h_2 are the heights of a DRFM and the adversary radar (in m) and separated by a range R. If $R > FZ$ (Fresnel zone) where $FZ = 4\pi h_1 h_2/\lambda$, then in this case, the spreading loss is better estimated as [31]

$$L_{SF} = 120 + 40\log_{10}(R^{km}) - 20\log_{10}(h_1) - 20\log_{10}(h_2) \quad \text{dB} \quad (10.16)$$

Then the beacon equation in dB can be written as

$$P_{rj} = P_j + G_j - L_{SF} + G_r \quad \text{dB} \quad (10.17)$$

Note that the bandwidth of the radar and DRFM output noise filter cancel out as they are equal.

Consider the airborne DRFM engagement shown in Figure 10.20 where the DRFM resides on a UAV platform with RCS of σ_T.

10.5 Obscuration-EA and Deep Learning

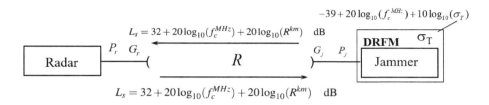

Figure 10.20 Engagement for computing jam-to-signal ratio with DRFM on a platform with RCS σ_T m².

Using linear analysis, the power at the radar from the DRFM platform RCS (σ_T) is

$$P_{rT} = \underbrace{\left(\frac{P_r G_r}{4\pi R^2}\right)}_{\substack{\text{Poynting}\\\text{Vector}}} \underbrace{\left(\frac{\sigma_T}{4\pi R^2}\right)}_{\substack{\text{Portion}\\\text{Reflected}\\\text{Back @ Radar}}} \underbrace{\left(\frac{G_r \lambda^2}{4\pi}\right)}_{\substack{A_e\\\text{of Radar}}} \quad (10.18)$$

where we have shown the Poynting vector transmitted, the portion reflected back at the radar, and the effective aperture of the radar. It can also be shown from Figure 10.20 that (in dB)

$$P_{rT} = P_r + 2G_r - 103 - 20\log_{10}(f_c^{MHz}) - 40\log_{10}(R^{km}) + 10\log_{10}(\sigma_T) \quad (10.19)$$

The jam-to-signal ratio (JSR) is then $JSR = J/S = (10.14)/(10.18)$ in linear units or $JSR = J - S = (10.17) - (10.19)$ in dB.

It can also be shown that the power received at the DRFM receiver from the radar can be expressed as

$$P_{jr} = \underbrace{\left(\frac{P_r G_r}{4\pi R^2}\right)}_{\substack{\text{Poynting}\\\text{Vector}}} \underbrace{\left(\frac{G_j \lambda^2}{4\pi}\right)}_{\substack{A_e\\\text{of Jammer}}} \quad (10.20)$$

This can also be expressed (in dB) as a

$$P_{jr} = P_r + G_r - 32 - 20\log_{10}(f_c^{MHz}) - 20\log_{10}(R^{km}) + G_j \quad (10.21)$$

and will be used in a later section when we discuss target jammers and repeater analysis.

10.5.3 Offensive Engagement Algorithms:

10.5.3.1 Self-Screening Jammer Engagement

The self-screening jammer engagement scenario is shown in Figure 10.21.

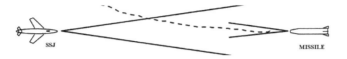

Figure 10.21 Self-screening operational EA.

Here the DRFM jammer is using a noise jammer within the beamwidth of the missile seeker to prevent any range or Doppler detection of its platform. All the seeker will get within the receiver (continuously) is noise. However, the noise waveform does enable the platform to be efficiently angle tracked in space if it continues [23].

Considering the self-protection engagement in Figure 10.21, to determine the per-pulse target-to-jammer ratio using linear units for a self-screening jammer (SSJ) where the jammer is the dominate noise source over the thermal noise ($P_{rj} \gg N_T$), the per-pulse

$$\frac{P_{rT}}{P_{rj}} = \left(\frac{P_r G_r^2 \sigma_T \lambda^2}{(4\pi)^3 R^4}\right) / \left(\frac{P_j G_j \lambda^2 G_r B_r}{(4\pi)^2 R^2 B_j}\right) \quad (10.22)$$

or

$$\frac{P_{rT}}{P_{rj}} = \frac{P_r G_r \sigma_T B_j}{4\pi R^2 P_j G_j B_r} = \left(\frac{ERP_r}{ERP_j}\right)\left(\frac{B_j}{B_r}\right)\left(\frac{\sigma_T}{4\pi R^2}\right) \quad (10.23)$$

The JSR is the inverse of $(P_{rT}/P_{rj})_p$ or

$$JSR_p = \left(\frac{J}{S}\right)_p = \left(\frac{P_{rj}}{P_{rT}}\right)_p = \left(\frac{ERP_j}{ERP_r}\right)\left(\frac{B_r}{B_j}\right)\left(\frac{4\pi R^2}{\sigma_T}\right)\rho \quad (10.24)$$

where we have included ρ to account for the mismatch (offset angle ϕ) between radar E-field E_r and DRFM jammer's E-field E_j. The antenna polarization mismatch, ϕ is shown in Figure 10.22.

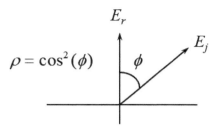

Figure 10.22 Radar electric field E_r and DRFM jammer electric field E_j showing a mismatch (offset angle ϕ) between them with a polarization mismatch factor of $\rho = \cos^2(\phi)$ offset.

Rule: To achieve greater resistance to the radar protection measures in place, the DRFM should initially randomize its radiation polarization until the cognitive deep learning can receive information on its effectiveness.

To show why initial randomization is truly an easy calculus problem, we let the angle ψ in the plane of rotation of the EA radiation electric field vector be a random value with a *uniform pdf*

$$p(\psi) = \frac{1}{2\pi}$$

in the range of angles 0 to 2π. Then taking the expected value of $\gamma_j = \cos^2(\psi)$ or

$$M(\gamma_j) = \frac{1}{2\pi} \int_0^{2\pi} \cos^2(\psi) d\psi = \frac{1}{2} \tag{10.25}$$

10.5.3.2 Stand-Off Jammer Engagement

For the stand-off jamming (SOJ) engagement, the DRFM receiver determines the frequency of the SAM radar. The SOJ engagement shown in Figure 10.23 shows the DRFM protecting a friendly penetrator by obscuration EA. The surface-to-air missile (SAM) radar attempts to hit the SOJ at its maximum weapons range. However, the penetrator jams right straight into their target since they know they are protected by a high-power SOJ, that makes sure their skin return is masked.

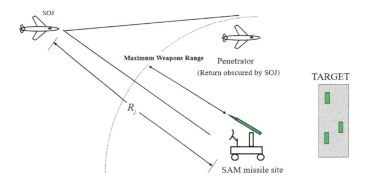

Figure 10.23 SOJ operational EA.

In this stand-off EA DRFM situation, the SOJ engagement requires the DRFM to take out all the threat bands (both SAM and air intercept). The DRFM jammer has to have an output amplifier that can project a high ERP with wideband coverage required for multiple threat emitters. A couple of definitions are in order here:

Burnthrough Range R_{BT}: Range at which the target can be *detected above* the EA noise power;
Crossover Range R_{CO}: Range at which the target signal power equals the EA noise power or $S = J$.

As an example, calculate the burnthrough range of an SSJ ($\sigma_T = 100$ ft^2) if the required JSR_p for single-pulse detection is -5 dB.
  For the jammer, consider it has an $ERP_j = 30$ dBW, $B_j = 70$ dBHz (10 MHz) and a platform with a $\sigma_T = 20$m^2. The radar has an $ERP_r = 67$ dBW and $B_r = 70$ dBHz (10 MHz). To determine R_{BT}, we solve the JSR_p for the range R and substituting $JSR_p = -5$ dB,

$$R_{BT} = \left(\frac{JSR_p ERP_r B_j \sigma_T}{ERP_j B_r 4\pi} \right)^{1/2} \quad \text{(SSJ)}$$

Or we can solve for the burnthrough range R_{BT} using dB (often easier to do in our heads) as

$$R_{BT} = (-5 + 67 + 70 + 20 - 30 - 70 - 11)/2 = 20.5$$
$$R_{BT} = 112 \, \text{ft}!$$

This of course is an extremely short range, and denies useful information until it is too late! Note when the DRFM jammer is remote from the target and not on the platform being detected, the EA energy will come in through the sidelobes of the radar antenna. In this case care has to be used when formulating the dB linear equations above and use the correct distances R and the correctly predicted sidelobe levels at the radar antenna.

In addition, so far we have considered 1 DRFM jammer platform vs. 1 radar. Consider, for example, two DRFM noise jammers in the main beam of the radar as shown in Figure 10.24. Here the two DRFMs put out power densities P_1 and P_2 at

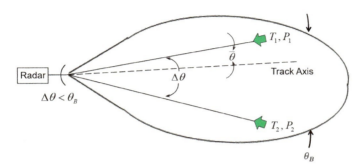

Figure 10.24 Two DRFM noise jammers in main beam showing tracking error as measured as an error in tracking T_1.

positions T_1 and T_2 within the main beam of the emitter within beamwidth θ_B. Note that the angle between the two jammer sources $\Delta\theta < \theta_B$. The track axis of the radar $\bar{\theta}$ is the *tracking error* measured as an error in tracking T_1 and is due to the power centroid computation due to the powers within the beam as

$$\bar{\theta} = \frac{P_2}{P_1 + P_2}\Delta\theta \qquad (10.26)$$

Note the power centroid above can be extended to any number of unresolved noise sources.

10.5.3.3 Escort Engagements

An escort situation is shown in Figure 10.25.

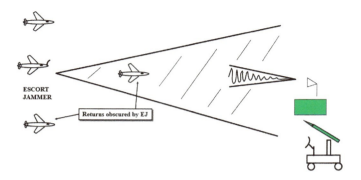

Figure 10.25 Escort operational EA.

One particularly useful technique that can be flown in a mutual support formation when their main beam is the coordinated blinking attack. This is shown in Figure 10.26.

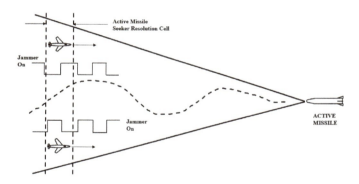

Figure 10.26 Mutual support jammer demonstrating coordinated blinking.

Note this requires multiple DRFMs in formation and the jammer mode programmed to do an on/off blinking or swept spot noise. This mode will increase the track error and ultimately create a larger miss distance. Although this technique is very effective it does require coherent time-synchronization between platforms.

10.5.4 Cognitive Deep Learning Control

The DRFM's AI control algorithm not only can optimally make the required EA assignments above but during a network-enabled operation, such as a suppression, the AI can maintain selective control over several *autonomous subnets* that are involved in the operation [23]. As shown in Figure 10.27, several types of subnets can be formed in support of the operation. This includes the stand-in jammers and stand-in sensors

10.5 Obscuration-EA and Deep Learning

that may be miniature air-launched vehicles. Separate *EA nets* and *sensor nets* can also be separately controlled. Particularly challenging is the resource management and its power efficiency behind this type of selective control. Using the *detection reports* from the sensor nets, the stand-in sensors, the *engagement reports* from the EA nets, and any lethal SEAD, the AI (along with the C2 in the EA-18G) selectively controls the autonomous subnets as shown graphically in Figure 10.27.

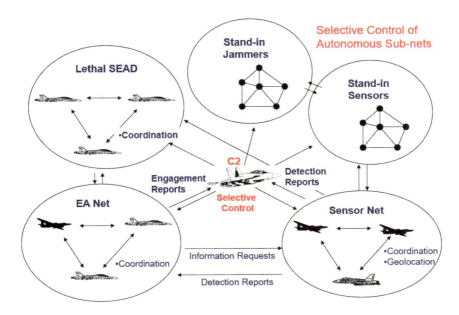

Figure 10.27 Selective control of the autonomous subnets for DRFM ES/EA during lethal SEAD, stand-in EA, and ES.

To handle the interleave of suppression-type signaling waveforms efficiently without a dispersion of resources, an effective control system must be used. AI, deep learning control techniques discussed in the last chapter have been applied to beamforming [32] and for generating smart interference waveforms [33]. Interleaving these EA modes in high-tempo situations requires a deep learning control system.

Figure 10.28 shows a nonlinear, deep learning system receives the sensor inputs and the *information damage* that represents the quality of information lost by the threat radar system. The deep learning control system is embedded within a *cognitive feedback loop* around the jammer-sensor pair. Note that the sensor inputs, knowledge database inputs, and any dual-ported memory resident information must be streamed to a cognitive data fusion process together with the information coming from the threat in order to (1) determine the effectiveness of the EA, (2) deceptively adjust/tune the EA waveform generation, and (3) continue transmitting/monitoring [23].

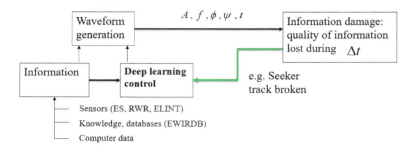

Figure 10.28 Cognitive deep learning control system for managing autonomous subnets.

AI deep learning can be used to manage this control efficiently. In [32], a DL neural network is demonstrated to provide a joint optimization approach controlling the beamforming in a wireless communication system. Because the offensive operations can overwhelm an EA-18G pilot or back-seater, efficient control is a requirement. Within an unmanned system or miniature air vehicle, it goes without question that a deep learning control system is the optimum approach. On the contested EA battlefield, intelligent (AI) control of unmanned combat systems is the key to dominating the air-combat engagements. These engagements have high dynamics and large uncertainties. In order to form an air-combat system with highly efficient offensive and defensive capabilities, it is necessary to obtain the OODA advantage in intelligence, detection, EA, comms, and firepower strike capability [34]. Therefore the integration of AI technology with traditional unmanned autonomous combat systems has important scientific research value and is also of practical significance.

10.5.5 Practical Considerations

Traditional noise jamming techniques with a DRFM typically have included continuous and intermittent spot noise, noise swept amplitude modulation, continuous and intermittent barrage noise, barrage noise swept amplitude modulation, and swept noise.

You happen to face an air defense system with early warning (EW), acquisition, and ground control intercept (GCI) radar is to set up attacks on you and the surface to air SAM, air intercept (A) SAM, AI, and AAA radar to prosecute the attacks. *Who do you jam?* Since you can't jam all of the emitters due to SWAP considerations, go after the lethal systems. You have to take out the SAM, AI, and AAA radar system– *they can kill you.* Of special consideration in noise EA in particular is how the energy and power is put into the adversary's receiver; that is, does it have the right key? To consider how far down into the receiver you have to go to be effective, consider the receiver frequency-function diagram shown in Figure 10.29.

10.5 Obscuration-EA and Deep Learning

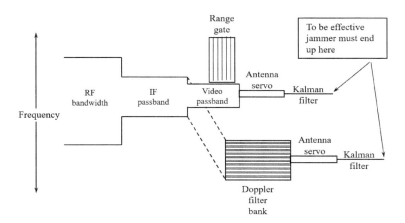

Figure 10.29 Schematic diagram of a RF receiver showing the RF band, IF passband, video band, range gate with (range bins), and Doppler filter bank. Also shown is the antenna servo and Kalman filter that typically drives a closed-loop homing seeker or track formation.

The input RF bandwidth of the adversary emitter is typically governed by their transmit waveforms and their antenna and unless the DRFM is only attempting just to stop there, it is important to consider how to move forward. Getting through the IF and video passbands into where the EA is digitized is not always easy. To be truly effective, the DRFM must end up at the end-state where the final emitter action is taking place–the Kalman filter, for example, in a track loop. Modulation of a noise waveform in amplitude, frequency, time, and polarization to attack the detection, angle tracking, range/Doppler tracking, AGC setting, and SNR estimation can all be attempted depending on the EP the emitter has. Below we describe some newer techniques that, for example, use AI, deep learning against SAR.

10.5.6 Deep Reinforcement Learning for Unmanned Combat Systems

In 2016, US Air Force combat simulation *Alpha AI* and DARPA's *Deep Green* for operational decision-making, using deep reinforcement learning technology to develop combat intelligence can solve decision problems in complex battlefield environments. Combat intelligence based on deep reinforcement learning combines the perception of deep learning with decision making capabilities of reinforcement learning. This enables combat agents to have *anthropomorphic intelligence* when dealing with highly complex and flexible tasks. A deep neural network is used to represent the Q function taking the state and decision as input and the corresponding Q value as the output. At the same time, consider for a multitask feature of resource management and control, the multiagent DQN network can be established as shown in Figure 10.30.

716 ELECTRONIC ATTACK USING DEEP LEARNING

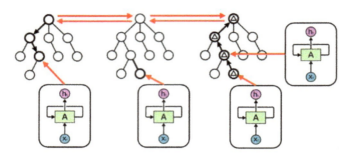

Figure 10.30 Information spreading diagram of multi-DQN agents. (©IEEE, reprinted with permission from [34].)

For every agent, the tree search structure can be used to realize the policy update and the parallel computing and iteration of DQN network parameters and uncertainty set constraints can then be realized between agents. A deep neural network is combined with reinforcement learning. A separate deduction platform can be used to comprehensively analyze the EMS and SoS.

The training model is shown in Figure 10.31. Feature extraction is performed on the current state and the feature is integrated to state $S_t - 1$. By deep learning network, the assessment of the current state can be completed. Then the output return, R_t, of the action a_t is the reward for the deep network through R. For the entire game countermeasure combat process, the system needs to evaluate each node to find the optimal behavior sequence. Fig. 10.31 shows the game countermeasure model that shows the integration of an intelligent assistant decision-making system for driving a C2.

10.5 Obscuration-EA and Deep Learning

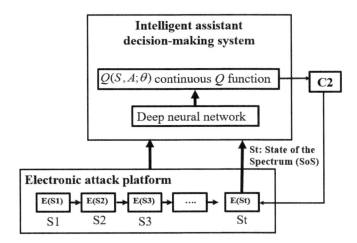

Figure 10.31 Game countermeasure model. (©IEEE, reprinted with permission from [34].)

The authors note in [34] that it is not advisable for the machine to adopt a zero-learning strategy. This is especially true for decision-making in complex environment, where training in the initial phase of the model should be based on human experience. After the model can handle some simple tasks on its own, the reinforcement learning of self-simulation countermeasure will be more effective. Due to the complex nature and diverse objectives of complex information and constantly changing goals of unmanned operations, the continuity of information perception, the reliability of target recognition, and real-time requirements of operational decision making are in urgent demand. The following technologies can provide help with these requirements:

- Autonomous sensing;
- Self-identification;
- Autonomous mission planning and decision;
- Autonomous launch;
- Independent monitoring and security.

The reinforcement learning process only needs to obtain the evaluation of a feedback signal and take the max reward as the learning goal. The characteristic is the initiative temptation to the environment; the environment generates feedback to the temptation action that is evaluated, and the agent adjusts the future behavior according to the feedback.

10.5.7 Selective-Reactive EA

The time between a selected emitter being jammed until the time the emitter's operating (AI) system (or human) initiates a frequency change leading to the emitter being unjammed must be maximized. This is what is known as selective-reactive EA. The selective-reactive timeline is shown in Figure 10.32. As jammers come up and emitters shut down, what becomes important is the minimization of δt_j or the difference between when the emitter was on and when the emitter was jammed (just before the operator or computer changes frequency).

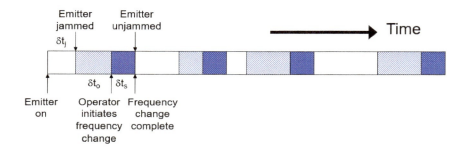

Figure 10.32 Information timelines for reactive EA operation emphasizing DRFM throughput and speed as a key driver.

If a single platform is considered.

$$\delta t_j = L_s + L_p \qquad (10.27)$$

where L_s is the ES sensor's latency and L_p is the processing latency. For a distributed system as shown in Figure 10.33

10.5 Obscuration-EA and Deep Learning

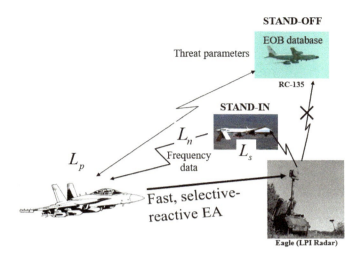

Figure 10.33 Selective-reactive solution in a distributed system. Stand-in sensor senses LPI radar (L_s), cues on-board DRFM receiver over network (L_n), and tunes DRFM jammer (L_p).

$$\delta t_j = L_s + L_p + L_n \qquad (10.28)$$

where L_n is the network latency. Note that the stand-off EA asset cannot receive the LPI radar emission from the stand-off distance. However, they can relay the threat parameters and the electronic order of battle down to the EA-18G. Also note accordingly, this data link may impose no significant delay to reactive assignment!

In summary, selective-reactive EA allows the jammer to focus more power effectively on specific radar frequencies and to counter modern frequency-hopping radars designed to defeat older jamming systems. Selective-reactive technology enables the DRFM to rapidly sense and locate threats with a significantly higher degree of accuracy than was previously possible, which improves the accuracy thus enabling a greater concentration of energy against threats.

10.5.8 DRFM Subsumptive AI Rules

During an ingress, the DRFM senses any EMS emitters, detects the modulations, and stores and classifies the modulation. However, there are other rules that are must be included in a subsumptive manner as a component element. For example, what does the DRFM platform AI do when facing an adversary integrated air defense (IADS)? That is, when an early warning radar, an acquisition radar, and a ground control intercept radar are to set up attacks against you? A SAM, an air intercept (AI) radar, and a anti-aircraft artillery (AAA) radar are set to prosecute the attacks? What does the DRFM AI do? The need for re-forming the attack algorithms in order to deal with the complexity is possible if it makes feasible "subsumptive transformations".

Consequently, since you can't jam *all* the emitters, you go after the lethal systems or *the ones that can kill you!* So you jam the SAM, AI, and the AAA since they can take you out. The AI commands the DRFM to put all of the power into the victim's receiver modulating in amplitude, frequency, time, and polarization, to attack target detection capability, angle tracking, range/Doppler tracking, AGC setting, and SNR estimation.

10.5.9 Smart Interference Using Generative Adversarial Nets

As discussed in Chapter 8, GANs can be used for modifying and reconstructing image information. The DRFM intercepts the signal and based on the analysis and preprocessing of the intercepted signals, a GAN can be suitably constructed to generate interference or jammer signals against the radar and emitter signals that have been identified using the DRFM detection and classification network.

In [33], a detection and classification network followed by a GAN is set up to detect a specific radar emitter modulation for EA against 11 test signals. They use 11 types of radiation signals including four linear frequency modulation (LFM) signals (two with negative slopes, two with positive slopes). Their parameters were specifically designed relevant to communication and navigation radar signals, with their amplitudes and initial phases randomly chosen within the specified range. Also used were white Gaussian noise, colored noise, two CW signals, and three complex modulations.

For the signal processing, the first step after digitization is a T-F detection using the Wigner-Ville Distribution (see Chapter 8). The T-F plane for all the signals is preprocessed to truncate the signal length of each file to the same number of samples. It is then used as a fingerprint-input to an arrangement of three CNNs in parallel, each having a full connection layer output. Figure 10.34 shows a diagram of this CNN architecture.

10.5 Obscuration-EA and Deep Learning

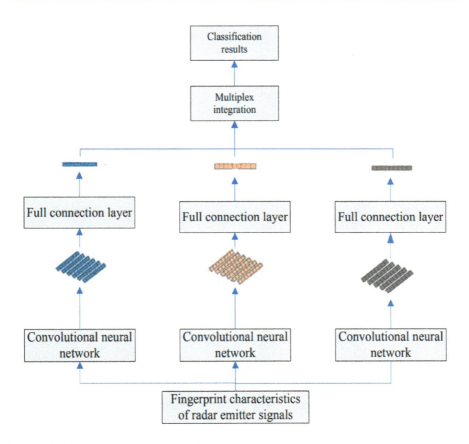

Figure 10.34 Multichannel integrated multitag identification and classification network for radar emitters. (©IEEE, reprinted with permission from [33].)

The three classification results are then multiplexed into a single, multilabeled result (see Chapter 8). The final classification step is the efficient integration required to combine the results from these multiple channels and output the final classification result (specific emitter identified). The confusion matrix for the 11 signals is shown in Figure 10.35 showing the probability of correct classification P_{cc}. The classification output is then sent to the GAN interference generator.

722 ELECTRONIC ATTACK USING DEEP LEARNING

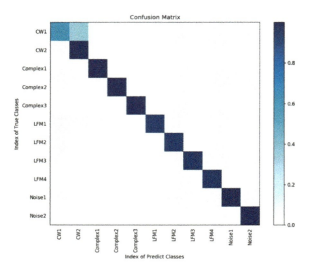

Figure 10.35 Confusion matrix showing the probability of correct classification for the 11 input signals. ©IEEE, reprinted with permission from [33]

After the analysis and preprocessing of the intercepted signals, a GAN suitable to those expected signals is called up (or constructed) for interference generation. Construction of the GAN shown in Figure 10.36 includes the relevant learning algorithms, such as structure of the generator, structure of the discriminator, connecting style of the networks, and the loss function [35, 36]. Note that for the data distributions, the use of transfer learning can also be of great assistance in this type of situation [37].

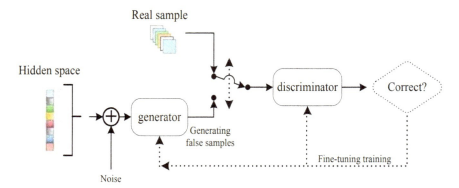

Figure 10.36 Network model of GAN reported in [33] to preprocess the intercepted radar signals using proper generator, discriminator algorithms, and loss function to maximize probability of a high-fidelity interference.

10.5 Obscuration-EA and Deep Learning

The discriminator for these signals runs on an optimization algorithm as discussed in Chapter 9. The overall objective of the optimization is to maximize the probability of correct discrimination by discriminating what comes over the network into the discriminator. The discriminating network receives the real processed radar data and the fake processed radar data (generated by the GAN generator). It maximizes the probability of correct discrimination by minimizing the probability that the radar signal samples generated by the GAN generator are identified. After a number of iterations, parameter tuning is performed based on the output. If the discriminating network gives a *correct judgment*; that is, the radar data is fake, the parameters of the GAN generator need to be adjusted to make the generated fake radar signal data more realistic. If the discriminating network reaches a *wrong judgment*; that is, the radar data is true, the GAN generator parameters need to be adjusted to avoid making a similar mistake next iteration.

During the training process, the network keeps approaching the true value through fitting, gradually evolving from the initial unstable output to correct judgment on whether the data distribution is from a real radar signal or from a fake one generated by the GAN. Ideally, after the training is complete, the probability distribution of the data generated by the GAN is completely the same as that of the real data, and the discriminating network can no longer tell which is which. The result of the training is a high-quality automatic generator and a highly capable discriminator.

After the radar signal GAN is constructed, feature learning is performed on the radar signals in various environments. Then, a noise input is introduced, and the learned GAN is employed to generate the fake radar data. When implementing the interference, the jammer achieves a perfect match with the radar echo signal by transmitting the high-fidelity target radar signal, thus achieving smart interference. The transmitting of a incomplete radar signal in the traditional transponder interference methodology is avoided. This relaxes the requirements on the radiated energy of the interference signal and improves the performance of the EA, which is especially important in the case of limited resources.

In the test of the GAN's interference signal generation, the GAN was employed to generate fake radar data. Test results are shown in Figure 10.37. Figure 10.37(a) shows the true output sample image of one of the radars intercepted. Figure 10.37(b) shows the GAN output sample image for one of the radars.

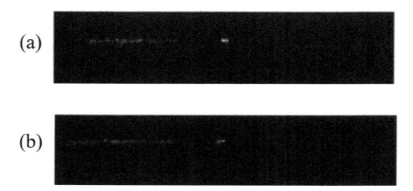

Figure 10.37 For one of the radars intercepted; (a) image data for radar 1, and (b) high-fidelity radar GAN interference data generated by radar 1 using GAN in Figure 10.36. (©IEEE, reprinted with permission from [33].)

Results show the radar EA obtained from the GAN network can generate waveforms nearly identical to the original signal with all its characteristics. The high-fidelity interference waveform can support not only interference noise but also deception and smart interference patterns. This interference implementation process, effectively reduces the interference power dependence of the interference generation mechanism and improves the EA capability of the GAN machine, especially under limited resources. High-fidelity signal imitation and simulation can also be achieved for certain communication signal sets as well.

10.6 DECEPTION-EA AGAINST SYNTHETIC IMAGING APERTURES

High-resolution imaging of air, land, and sea targets in clutter are some of the main tasks of SAR and inverse SAR. They are used by scientists and the military for applications such as object detection, recognition, and classification. Deception architectures of these SAR and ISAR imaging capabilities using a deception-EA DRFM architecture is the subject of this section.

10.6.1 Generating Noise Patches in SAR Images

One critical capability of the DRFM is to prevent the SAR surveillance activities from acquiring high-resolution images especially where important areas on the ground are concerned. These areas might hide targets, civilians, hospitals, and so forth. The DRFM's effectiveness depends on both noise (obscuration) EA and deception (targeting) EA techniques and how well they can hide the backscatter power from the image within the SAR transmit power.

10.6 Deception-EA Against Synthetic Imaging Apertures   725

10.6.1.1 SAR vs. Multiplicative Noise

For example, *multiplicative noise* can be generated against an LFM SAR waveform and used to generate narrow strips or small extended areas and targets. For the DRFM to generate multiplicative noise, the SAR signal is intercepted then multiplied by a pulsed-noise signal and varying the *noise to SAR bandwidth ratio* (NSBR). The block diagram of the 1D and 2D multiplicative noise EA signal generation techniques are illustrated in Figure 10.38 [38]. Figure 10.38(a) shows the 1D multiplicative noise jamming the spectrum of the pulsed noise after being low-pass filtered in the fast time dimension, and Figure 10.38(b) shows the 2D EA signal and the spectrum of a 2D generated random noise is filtered in both fast and slow time dimensions. In order to properly generate the two-dimensional random noise in both fast and slow time dimensions it is required either to accurately estimate or to get a prior knowledge of the SAR system parameters necessary to generate the SAR Doppler or azimuth bandwidth [38].

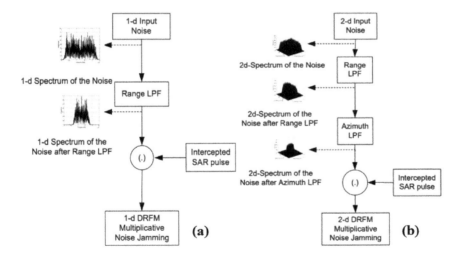

Figure 10.38 (a) Shows 1D multiplicative noise jamming the spectrum of the pulsed noise after being low-pass filtered in the fast time dimension, and (b) the 2D jamming signal and the spectrum of a 2D generated random noise is filtered in both fast and slow time dimensions. (©IEEE, reprinted with permission from [38].)

In Figure 10.39 the output real image is shown in the presence of 1D multiplicative noise EA at a JSR or 20 dB with different values of NSBR.

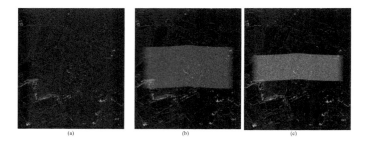

Figure 10.39 The output real image in presence of 1D multiplicative noise jamming signal at JSR= 20 dB at different values of NSBR (a) NSBR=1 (b) NSBR=0.5 (c) NSBR=0.25. (©IEEE, reprinted with permission from [38].)

The NSBR of the multiplicative noise EA has a direct impact on the effectiveness of this technique. Reduction of the NSBR causes the EA to disrupt the SAR system impulse response hence degrading the SAR image quality at lower values of JSR. However, this comes at the cost of a reduced affected covered swath on the SAR image.

10.6.1.2 SAR vs. Multichannel Convolution Noise

With limited available power within an unattended ground sensor, UAV or MALD, it is important that the EA technique be power-efficient. With AI, the DRFM is able to produce and interleave smart EA techniques efficiently by condensing its power on specific areas on the SAR image to cover certain sensitive targets. Convolution noise jamming has proven to allow a high level of control. Control signals developed in [39] include the *noise-to-SAR pulse width ratio* (NSPR) and the *noise-to-synthetic time ratio* (NSTR). These controls are used to reduce the JSR required to achieve an effective EA on the SAR image by condensing the jamming power in a limited extent. Using multiple parallel channels of convolution noise, this technique has the ability to control each channel to produce a certain noise patch with certain position and size. Considering the SAR relative motion in the azimuth dimension, the input *to* the jammer *from* the SAR is

$$X(t,s) = \text{rect}\left[\frac{t - R(s)/c}{T}\right] \exp j \left[-2\pi f_0 \left(t - \frac{R(s)}{c}\right) + \pi \gamma_r \left(t - \frac{R(s)}{c}\right)^2\right] \quad (10.29)$$

where s is the slow time dimension, c represents the speed of EM wave propagation (free space), $R(s)$ is the range from the SAR antenna phase center to the scene center, and $R(s)/c$ is the one-way delay time from the SAR sensor to the jammer position.

The DRFM is used to convolve the pulsed noise $n(t)$ with the intercepted SAR pulses. In fast time dimension, $NSPR = N_t/T$ where N_t is the noise duration. In the slow time dimension $NSTR = N_s/T_s$ where N_s is the noise duration in the

10.6 Deception-EA Against Synthetic Imaging Apertures

slow time dimension and T_s is the synthetic duration corresponding to the synthetic aperture length [39]. The DRFM controls the 1D convolutional noise EA by controlling NSPR for range dimension or the NSTR for the cross-range dimension, while the 2D convolutional noise EA is generated by the controlling both the NSPR and NSTR simultaneously in both receive and transmit mode, respectively. The DRFM core where the EA signal processing takes place and the noise waveform is created is shown in Figure 10.40.

The core consists of a noise generator and N number of parallel channels inside the DRFM with each channel associated with two sequential variable delay lines. These delay lines control the delay time of the pulsed noise in fast time, t_i, and slow time, d_i, dimensions. These two time parameters control the noise position in range and azimuth dimensions in the SAR image, respectively.

The intercepted SAR signal from the receiver is convolved in each channel with the pulsed noise waveform. The use of the FFT block in each channel transforms both the radar pulses and the pulsed noise waveforms to the frequency domain. The

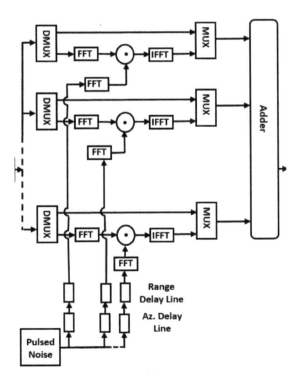

Figure 10.40 SAR noise jammer for creating clutter patches within the SAR image. (©IEEE, reprinted with permission from [39].)

convolution process is executed by a multiplying process in the frequency domain. The waveforms are then converted back to the time domain. The use of an IFFT block in each channel is used to convert the signals back to the time domain.

The convolutional EA waveform development process from N number of channels can be written as

$$j_{MCNJ}(t,s) = \sum_{i=1}^{N} J_i * X_i \tag{10.30}$$

where J_i is the pulsed noise in channel i, δ_i is the delay time given to the intercepted SAR pulses in the slow time dimension in channel i, and

$$J_i = \{n(t - \tau_i - R(s - \delta_i)/c)\}, \quad \text{for } i = 1, 2, \ldots, N$$

and X_i is the intercepted SAR signal in channel i as

$$X_i = (\gamma_a T_s \ A)(\gamma_r T \ B) \exp(-j4\pi f_0 R_0/c) \tag{10.31}$$

where

$$A = \text{sinc}[\gamma_a T_s (s - \gamma_i)] \tag{10.32}$$

and

$$B = \text{sinc}[\gamma_r T (t - \tau_i - 2R(s - \gamma_i)/c] \tag{10.33}$$

Considering the DRFM is in the center of the scene, the position of each noise strip within the SAR image in range $P_R(i)$ and azimuth $P_{Az}(i)$ dimensions is a function of the amount of delay time τ_i and δ_i assigned to each channel given as [39]

$$P_R(i) = R_{near} + \frac{\tau_i c}{2} \tag{10.34}$$

and

$$P_{Az}(i) = Y_{near} + v \delta_i \tag{10.35}$$

where v is the sensor velocity component in the azimuth dimension, Y_{near} is the near range in the azimuth dimension and R_{near} represents the short range in the range dimension and given as

$$R_{near} = H \tan(\theta_i - \theta_r) \tag{10.36}$$

and

$$Y_{near} = R_s / \cos(\theta_{sq} - \theta_{Az}) \tag{10.37}$$

where θ_r is the antenna beamwidth in the range dimension, θ_i is the beam incident angle calculated to the scene center, H is the height of the SAR sensor, R_s is the closest distance, θ_{sq} is the squint angle, and θ_{Az} is the azimuth beamwidth. Figure 10.41 shows a configuration diagram of the sensor flight path and SAR squint map image formation.

10.6 Deception-EA Against Synthetic Imaging Apertures

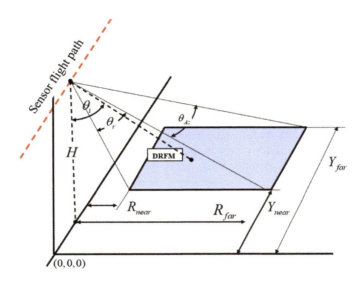

Figure 10.41 Configuration diagram of the sensor flight path and SAR squint map image formation. Note the position of the DRFM.

The range of values include:

$$0 \leq \tau_i \leq (T_{far} - T_{near})$$

and

$$0 \leq \delta_i \leq (S_{far} - S_{near})$$

SAR simulation results against the convolutional noise EA described above gathered with the SAR parameters as given in Figure 10.42.

Parameter	Value
Frequency	10 [GHz]
Pulse width	3 [u sec]
Bandwidth	200 [MHz]
Pulse repetition frequency (PRF)	1500 [Hz]
Platform velocity	250 [m/s]
Height	1400 [m]

Figure 10.42 SAR sensor parameters going against the DRFM convolutional noise EA. (From [39].)

Simulation results of the SAR system against the multichannel convolutional noise jammer (MC-CNJ) are shown in Figure 10.43 and demonstrate the ability of the EA technique to generate noise strips and/or batches to protect targets distributed over the SAR scene. Figure 10.43 (a) shows the selected targets to be protected with the jammer. Figure 10.43(b) shows the resulting SAR image when applying a conventional noise EA technique with a $JSR = 20$ dB, which isn't enough to cover the entire scene. Figure 10.43(c) shows the results when applying the MC-CNJ with JSR=20 dB utilizing only two channels. The results show that smart use of the MC-CNJ can produce an effective EA against the SAR through power management of the technique (compared to the same JSR shown in the conventional EA test results).

10.6 Deception-EA Against Synthetic Imaging Apertures

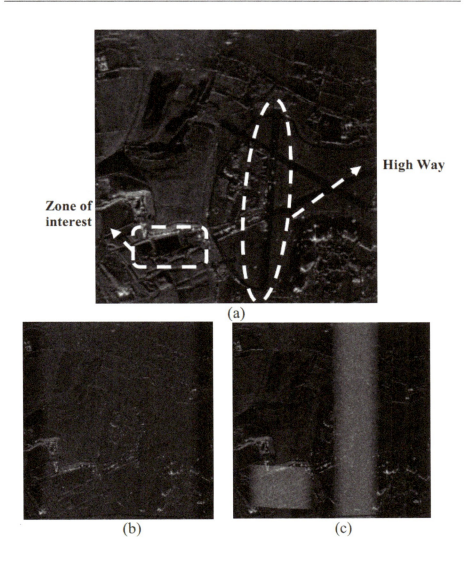

Figure 10.43 (a) The output image from SAR sensor without EA, (b) the output image resulting when a conventional noise EA is applied, and (c) the output image resulting when the MC-CNJ with two channels is applied to cover two different targets selected from the entire scene. (©IEEE, reprinted with permission from [39].)

In summary, two types of noise EA, multiplicative and multichannel convolutional, were discussed above with simulation results shown for the later case. For this case the convolution noise EA technique allows a high level of control and a closed form of the multichannel, MC-CNJ EA technique that governs the effect on the SAR receiver was also introduced. Parameters introduced to control each EA channel independently are able to produce noise patches with determined size and position in range and azimuth dimensions. The significance of this technique is that high intensity noise patches can be formed to mask limited areas instead of scattering the jammer power over the entire SAR image.

Noise jamming techniques suffer from the SAR processing gain (PG). The PG it reaches to 65 dB for 2D Fourier transform, which is applied in SAR receiver. So, the jammer must manage its power to affect the SAR image quality. For the previous reason a lot of recent work [40–42] investigates smart noise jamming techniques to overcome the SAR processing gain. Others study the SAR deceptive jamming techniques as a more effective solution. Actually all the deceptive jamming techniques are more complex than the noise jamming techniques.

10.7 TARGET EA AND DEEP LEARNING ALGORITHMS

10.7.1 Characteristics of Target Jamming

Sophisticated receivers have been developed to process radar target return signals to distinguish between characteristics that are expected from the true target signal of an aircraft, such as jet engine modulation, and unexpected characteristics, such as spurious signals, signal jitter, and phase noise. The latter are not normally expected in target return signals. Such spurs and phase noise are generated during the process of frequency downconversion and upconversion in EA systems and frequency downconversion in EW systems.

10.7.2 Transponder False Targeting

Transponders are devices for receiving a RF signal and automatically transmitting a different signal. They are characterized by, for example, their data rate and the maximum distance their transmitted signal can travel. Transponders have been used for RFID tagging [43], data tagging, and logging [44]. They have also been employed for many years to give air traffic control the identification and height of the aircraft in response to an interrogating signal from an airport's secondary surveillance radar [45]. These systems are similar to those used for identifying friend or foe (IFF), such as the XPC-TR-50 Mode C IFF transponder from Sagetech (for use on UAS)[2].

---

2   To reduce the possibility of civilian air crashes, every civilian aircraft is equipped with a special transmitter, the ADS-B (Automatic Dependent Surveillance - Broadcast) transponder, which receives requests at $f_{req} = 1,030$ MHz, and sends a special data package at $f_c = 1,090$ MHz, and contains the flight parameters, aircraft type, its coordinates, destination, etc. One of the drawbacks of using this system is the inaccuracy of the transmitted location and speed of flight data [46].

10.7.2.1 Constant Power False Target Operation

When the DRFM goes into the transponder mode for false target generation, it serves as an excellent surface-to-air EA technique for deception against tactical fire control airborne radar systems, especially their air-to-ground modes. These include modes such as real beam ground map (RBGM), radar navigation ground map (RNGM), and Doppler beam sharpening (DBS). The transponder can also be used against SAR and ISAR and makes a superior unattended false target generator and intruder alarm system.

The transponder false target generator works by the use of a *trigger threshold*. When the DRFM receives a signal at the receive antenna G_{jrcv}, it is processed by a wideband receiver, magnitude detected, and saved coherently in the dual-port memory. That is, if the magnitude level of the received pulse is above the threshold, it triggers a sequence of pulses – each one a replica, coherently derived from the received pulse that triggered the sequence and was stored coherently in memory. When the level of the input signal is above the threshold, the transmitter coherently generates a sequence of identical (amplified) pulses and transmits them out the output antenna G_j back to the radar.

The pulse sequence that is sent out is made up of a sequence of coherent duplicates (derived from the input pulse that exceeded the threshold, or trigger pulse). Note that isolation is also not a problem as the receiver is gated off during transmit. As the sequence sent out is an amplified version of the stored trigger pulse, the transponder is considered a *constant power* false target generator.

10.7.2.2 Transponder Model & JSR

The input and output characteristics of the constant power transponder DRFM mode can be modeled generally as shown in Figure 10.44. To compute the JSR we consider the model to be installed on a UAS (e.g., in the air), and have a platform RCS σ_T. To compute the transponder JSR, the transmitted power from the radar

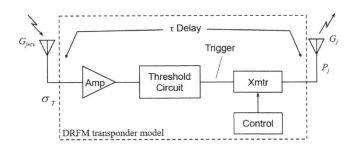

Figure 10.44 Block diagram of a DRFM operating as a transponder.

reflecting monostatically off the target UAS's σ_T can be found in linear units as

$$P_{rT} = \frac{P_r G_r^2 \sigma_T \lambda^2}{(4\pi)^3 R^4} \tag{10.38}$$

The power at the radar from the DRFM transponder can be calculated as

$$P_{rj} = \frac{P_j G_j G_r \lambda^2}{(4\pi R)^2} \tag{10.39}$$

Now a first-order approximation to the JSR in linear units can be calculated as

$$\frac{J}{S} = \frac{P_j G_j 4\pi R^2}{P_r G_r \sigma_T} \tag{10.40}$$

Note that the JSR contains no bandwidth ratio because the transmitted signal is coherently derived from the received signal.

10.7.2.3 Transponder Example

Below we consider an example to illustrate the effect the transponder has as a surface-to-air EA technique. The transponder and radar (real beam ground map mode) parameters are shown in Figure 10.45. The transponder surface-to-air false target parameters are given in the left column and the radar system's real beam ground map mode (RBGM) parameters are given in the right column. The RBGM is an important mode used in all airborne radar systems to achieve situational awareness out of the front-end, by displaying the real ground return formations about the velocity vector. No Doppler filters are formed since the integration time is too long.

The transponder's pulsed deception sequence that is streamed out is intended to be coherently integrated up in the radar receiver to create a dense field of false targets. The question then arises, "What does the display look like to the pilot?"

10.7 Target EA and Deep Learning Algorithms

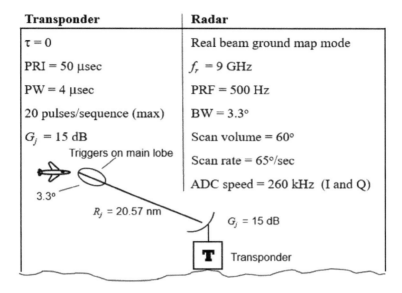

Figure 10.45 Transponder example showing the transponder parameters on the left and the radar real beam ground mapping parameters on the right for the example shown in Figure 10.48 and Figure 10.49.

A block diagram of the display signal processing for the RBGM map display is shown in Figure 10.46. The main objective of the display processing is to convert the raw I/Q to a magnitude and to increase the SNR with minimal change to the characteristics of the signal [47]. The display of ground backscatter, after being coherently received, begins with automatic gain control (AGC) processing. The AGC attempts to even out the display appearance across the map. It first determines the peak value being received over a set of pulses. This peak value is then used to set the sensitivity time control (STC) values used to control the attenuation of the RF in the receiver and the displayed intensity across the range bins on the display. STC effectively takes out the R^4 round-trip range effects, and gives the image on the map a more balanced look.

736 ELECTRONIC ATTACK USING DEEP LEARNING

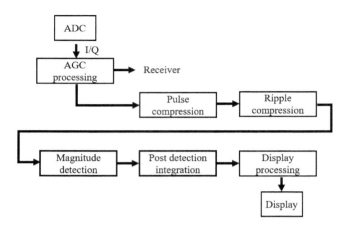

Figure 10.46 Example of real beam ground map RBGM signal processing.

Pulse compression and ripple suppression are used to increase the SNR while minimizing the sidelobe structure. Then the complex signal is magnitude detected. After the magnitude detection a postdetection integration (PDI) is run again to increase the SNR before display.

Figure 10.47 shows an example of a RBGM display map from a tactical fire control radar, multipurpose display. The map display shows the ground profile and the range ring markers at 20, 60, 100, and 120 nmi. Also shown is the position of the radar system creating the map and the position of the transponder (at 20 nmi), currently turned off. The transponder triggers on the 3-dB beamwidth of the radar's antenna beamwidth and the question then arises, "What does the display look like to the pilot when a transponder located at 20 nmi directly ahead is turned on?"

10.7 Target EA and Deep Learning Algorithms

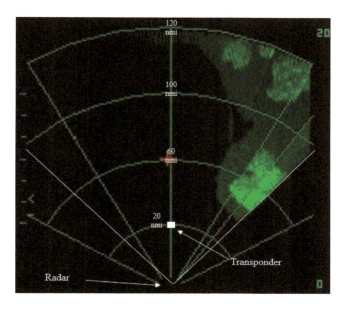

Figure 10.47 Example of a RBGM display map from a tactical fire control radar multipurpose display.

To see what the map looks like under the effects of EA from a transponder, we first have to do our homework and compute the total time on target (*tot*) as

$$tot = \frac{\theta_B}{\omega_s} \quad (10.41)$$

where ω_s is the scan rate of the antenna during map formation. Considering $\omega_s = 65$ deg/s and $\theta_B = 3.3$ deg., $tot = 51$ ms. From Figure 10.45 the number of pulses N_p during tot is $N_p = PRF(tot) = 25$. Also, the range bin size is

$$R_b = [(260 \text{ kHz}) * 2 \text{ ns/ft}]^{-1} = 1923 \text{ ft} = 586.2 \text{m}$$

This is a pulsewidth of $PW = 4\mu s$. Note also the $PRI = 2$ ms and the unambiguous range is $R_u = c(PRI)/2 \approx 160$ nmi.

The first step in computing the display properties is constructing a *pulse density timing diagram* (PDTD) for when the transponder is turned on. As shown in Figure 10.45, the transponder is $R_j = 20.57$ nmi away from the radar when the antenna θ_{3dB} scans by the transponder. As the antenna scans by the transponder, it triggers the 20 pulse sequence for every pulse intercepted. Using the approximation of 12.4 μs/nmi round-trip, radiation travel time, the radar will receive its own pulse back (second line down) from the transponder's RCS and the 20- pulse transponder sequence (one-way travel) at $t = 250$ μs. This allows us to construct, at least part, of the PD timing diagram as shown in Figure 10.48.

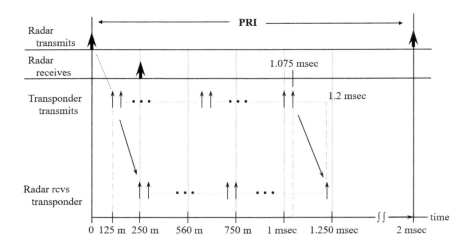

Figure 10.48 Transponder and radar pulse density timing diagram (PDTD) showing the transponder sequence for the example in the text.

As the transponder is halfway between a radar pulse's round trip, the sequence starts transmitting at 125 μs transmitting the 20-pulse sequence at the radar. Considering this begins at the halfway mark and extends all the way until the end of the 20 pulses, this is a highly deceptive move.

Figure 10.49 is a simulated display of what it should look like. As the antenna passes over the transponder, the trigger is set off by the pulse starting at the antenna's 3-dB beamwidth and for each pulse that is intercepted. Within tot, the radar sends 25 pulses consequently, and 25 sets of sequences are strobed out of the transponder. Can you predict from a PDTD, what the display looks like to the pilot if the transponder's pulse sequence (shown in Figure 10.48), extends *beyond* the radar's PRI (or unambiguous range)?

10.7 Target EA and Deep Learning Algorithms 739

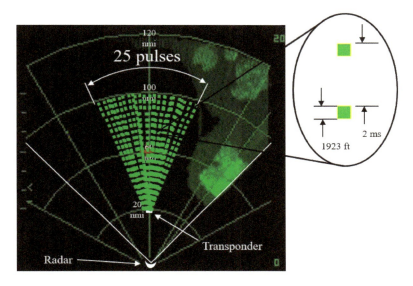

Figure 10.49 Transponder and radar, pulse density timing diagram (PDTD) showing the transponder sequence for the example in the text.

10.7.3 Repeater False Targeting

The application of a DRFM as an EA repeater is discussed in the section. A repeater intercepts the radar or communication signal, modulates it in time, frequency, amplitude, and so on, and repeats or transmits it back to the radar or communication system in order to change or reduce the information it is receiving.

10.7.3.1 Constant Gain False Targeting

The DRFM repeater is most accurately modeled as a *constant gain* system. Figure 10.50 shows a DRFM modeled as an EA repeater against an incoming missile seeker at a range R. Note that the repeater is applying a constant gain G to the intercepted signal. In addition, the missile seeker has a peak transmit pulse power of P_r and an antenna gain of G_r in the direction of the DRFM.

740 ELECTRONIC ATTACK USING DEEP LEARNING

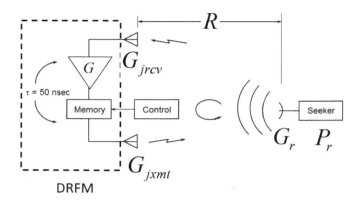

Figure 10.50 DRFM coherent repeater model against a seeker with pulsed peak power P_r and antenna gain in direction of DRFM G_r.

In this situation the DRFM's transmitted signal is coherently derived from the intercepted signal as discussed in earlier chapters. To evaluate the jam-to-signal ratio, the power at the seeker's *radar* from the DRFM *jammer* P_{rj} can be shown to be (note the position of the subscripts in P_{rj})

$$P_{rj} = \underbrace{\left(\frac{P_r G_r}{4\pi R^2}\right)}_{\substack{\text{Poynting}\\\text{vector}}} \underbrace{\left(\frac{G_r \lambda^2}{4\pi}\right)}_{\substack{A_e\\\text{radar}}} \underbrace{\left(\frac{G_{jrcv} G G_{jxmt} \lambda^2}{4\pi R^2}\right)}_{\substack{\text{sent back}\\\text{to radar}}} . \qquad (10.42)$$

The DRFM receive antenna gain is G_{jrcv} and the transmit antenna gain is G_{jxmt}. Also note that the P_{rj} is formed by multiplying the Poynting vector × the antenna effective aperture A_e × the EA jammer signal sent back to the seeker's radar.

The power at the seeker's radar from the jammer's *radar cross-section* (RCS) σ_T is (again note the subscripts)

$$P_{rT} = \underbrace{\left(\frac{P_r G_r}{4\pi R^2}\right)}_{\substack{\text{Poynting}\\\text{vector}}} \underbrace{\left(\frac{G_r \lambda^2}{4\pi}\right)}_{\substack{A_e\\\text{radar}}} \underbrace{\left(\frac{\sigma_T}{4\pi R^2}\right)}_{\substack{\text{portion reflected}\\\text{back @ radar}}} . \qquad (10.43)$$

The jam-to-signal ratio can be calculated by dividing (10.42) by (10.43) or

$$\frac{P_{rj}}{P_{rT}} = \frac{J}{S} = \left(\frac{\lambda^2}{4\pi}\right) \frac{G_{jrcv} G G_{jxmt}}{\sigma_T^2} . \qquad (10.44)$$

10.7 Target EA and Deep Learning Algorithms

This is an important result. Note that the range R between the DRFM repeater and the seeker's radar does not enter into the calculation.

Consider the pulse Doppler radar that does range-Doppler imaging. The target signal received by the pulse Doppler radar is proportional to

$$y_P(t) = p(t)\exp[j\phi(t)] \tag{10.45}$$

where $\phi(t) = 2\pi f_0 t$ and the frequency f_0 takes into account the downconversion performed by the DRFM and the Doppler frequency of the target. That is, $f_0 = f_{Idrfm} + f_D$ where f_{Idrfm} is the intermediate frequency of the DRFM and f_D is the Doppler frequency.

As the information content of the intercepted radar signal is mainly carried in the phase of the signal, the amplitude information is extracted and reserved for the later proper reconstruction of the return signal and only the phase is quantized. Typically, a CORDIC processor is utilized for I/Q to magnitude and ϕ translation. The phase is quantized using $N = 2^M$ levels where M is the number of bits.

The EA signal can be written as [48]

$$x(t) = e^{j\hat{\phi}(t)} = \sum_{m=-\infty}^{+\infty} \text{sinc}\,(m+1/N)\, e^{j(Nm+1)\phi(t)} \tag{10.46}$$

where $\hat{\phi}(t)$ is the quantized phase. The spectrum of the quantized signal (Fourier transform of $x(t)$) is

$$\hat{X}(f) = \sum_{m=-\infty}^{+\infty} \text{sinc}\,(m+1/N)\, \delta[f - (Nm+1)f_0] \tag{10.47}$$

where $\delta(f - f_x)$ is the Dirac delta function centered around $f = f_x$ and demonstrates that the spectrum of the deception EA consists of the superposition of several terms. The contribution for $m = 0$ is called the *primary image* term and located at f_0 [48]. The other terms $m \neq 0$ are attenuated by the factor $\text{sinc}(m+1/N)$. See Chapter 5 for further discussion on DRFM quantization and spurious signals.

10.7.4 EA: Range Gate Pull-Off

One of the most important functions of a DRFM is instrumenting a range gate pull-off (RGPO). Using time-delay/advance modulations of the form

$$v_j(t) = KA(t-t_j)\cos\left(2\pi f_r(t-t_j)\right) \tag{10.48}$$

the jammer can walk the range gate of the radar completely off the target. This is done by delaying the jammer pulse using the DRFM. As shown in Figure 10.51, the jammer pulse amplitude is slightly higher (3 dB) than the seeker's radar return from the jammer skin.

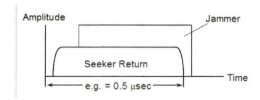

Figure 10.51 Seeker's radar return pulse from jammer skin with DRFM repeater jammer pulse. Note the EA is 3 dB stronger. Also, it is slightly delayed.

As the DRFM implements the delay over a number of pulses, the range gate of the seeker is pulled off the jammer. As shown in Figure 10.52, when the seeker range gate is pulled off the target, the $JSR \to \infty$.

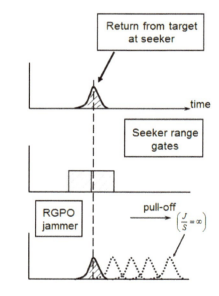

Figure 10.52 Showing seeker's early-late tracking gate with return from target and the RGPO jammer pull-off sequence in time, range resulting in a $J/S = \infty$.

Once the split gate range tracker is centered on the target using search information, the control system attempts to keep the tracker centered on the target by balancing the signal energy between the early and late range gates [24]. The range gate pull-off technique is one common method used by EA systems to defeat or degrade the performance of split range gate trackers.

10.7.4.1 RGPO Algorithm

First we consider the input signal to the DRFM is a sinusoid. The RGPO output can then be expressed as

$$y(t) = \exp[j2\pi f_0 t] \exp\left[2\pi f_0 \beta \sum_n u(t - nT_c)\right] \quad (10.49)$$

where the first term is the emitter carrier frequency f_0 and the second term is the phase where T_c is the step length as shown in Figure 10.53. Shown are the notional time delay functions for an ideal and real DRFM. For an ideal DRFM $c(t) = \alpha t$ where α is the delay time (pull-off) rate. For a nonideal (real) DRFM, $c(t) = r(t)$ where $r(t)$ is a piece-wise linear continuous function of time that approximates αt with a stair-step function.

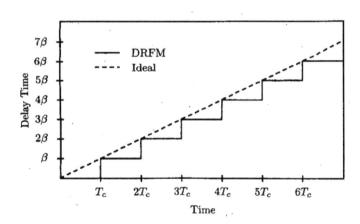

Figure 10.53 Time delay functions for ideal and DRFM linear RGPO.

The discrete time delay updates of the DRFM's linear RGPO signal represent discrete changes in phase of the output signal relative to input signal. The phase change per update cycle can be expressed as

$$\Delta \phi = -2\pi (f_0 \beta - \lfloor 2f_0\beta \rfloor + \lfloor f_0\beta \rfloor) \quad (10.50)$$

The output signal is a sinusoid with time-varying phase that translates into a frequency shift of output signal relative to input signal. The frequency shift of the output signal relative to input signal is

$$f_s = \frac{\Delta\phi}{2\pi\Delta t} = \frac{-(f_0\beta - \lfloor 2f_0\beta \rfloor + \lfloor f_0\beta \rfloor)}{T_c} \tag{10.51}$$

The output signal can then be written as

$$y(t) = \sum_n p_n(t)\delta(t - nT_c) \tag{10.52}$$

and the spectrum $Y(f)$ will have spectral lines at $f = f_0 + f_s \pm nf_c$.

10.7.4.2 Spectrum of an RGPO

Using the infinite time assumption, it is clear from Figure 10.53 that the real DRFM linear RGPO time delay function can be written as [49]

$$r(t) = \beta \sum_{n=-\infty}^{\infty} np_{T_c}(t - nT_c - T_c/2) \tag{10.53}$$

where the step size β and step length T_c are determined by the time delay resolution and update rate of the DRFM and the linear GPO parameter α. β and T_c can be determined by AI and deep learning in a cognitive feedback approach. The magnitude frequency spectrum of the DRFM linear RGPO signal

$$b(t) = \exp[-2\pi f_0(\beta \sum_{n=-\infty}^{\infty} np_{T_c}(t - nT_c - T_c/2))] \tag{10.54}$$

is given as [49]

$$|B(f)|^2 = \frac{\sin^2(\pi fT_c)}{(\pi fT_c)^2} \left(\sum_{n=-\infty}^{n=\infty} \delta(f + \beta f_0 f_c - nf_c) \right) \tag{10.55}$$

Figure 10.54 shows a sample of a DRFM's RGPO output magnitude spectrum for three delays. The walk time is 10 s with an update rate of 1.5 ms and a time delay resolution of 4.44 ns with a carrier frequency $f_0 = 9.655$ GHz.

10.7 Target EA and Deep Learning Algorithms 745

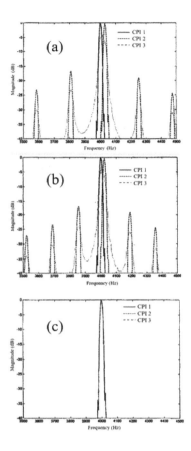

Figure 10.54 Output magnitude spectrum showing a walk time of 10 s, an update rate of 1.5 ms, a time delay resolution of 4.44 ns, and carrier frequency, f_0 =9.655 GHz showing (a) delay time of 10 μs, and (b) delay time of 10 μs, and (c) delay time of 10 μs[49].

10.7.4.3 Practical Considerations

The DRFM rate of pull-off must be within the design capabilities of the seeker tracking loop. If the rate of pull is too great then the tracking loop is *not* able to follow the DRFM pulse and the tracking gate will fall back to the real target. Consequently, the acceleration already being pulled by the DRFM must be considered in the algorithm. The range pull-off delay Δr is determined by the time delay τ of the RGPO program that is, $\Delta r = c\tau/2$. In general, there are two basic models to generate the time delay

$$\tau = v_0(t_k - t_0)/c \quad \text{Linear} \quad (10.56)$$

$$\tau = a_0(t_k - t_0)^2/2c \quad \text{Parabolic} \quad (10.57)$$

where t_0 is the start time of the walk-off program, v_0 is the velocity, and a_0 is the acceleration.

For aviation jet pilots, the typical acceleration rates can reach 10-11g's [50]. However, for example, the situation such that the maximum acceleration was limited (by the platform) to $a = 3g = 3(32\text{ft}/\text{sec}) = 96\text{ft}/\text{sec}^2$.

A typical RGPO DRFM cycle is shown in Figure 10.55. Note the DRFM throughput delay τ (≤ 2 ns) that is always present. This, however, can fluctuate depending on the sensing architecture, the signal processor, and its technology.

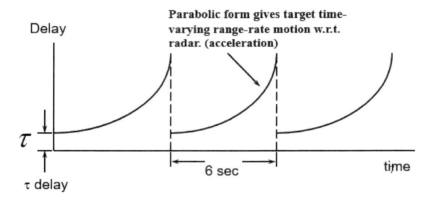

Figure 10.55 Missile seeker's pull-off delay as a function of pull-off time for the example above.

The delay is how fast the signal can be intercepted, detected, and classified and a response called up (or produced) to send back out the DRFM. In this example, the return pulse is pulled off over a period of 6 s. The pull-off range is then

$$R_{PO} = \frac{1}{2}at^2 \quad (10.58)$$

where a is the acceleration and t is the period of pull-off. After six seconds,

$$R_{PO} = \frac{1}{2}\left(96\frac{\text{ft}}{\text{sec}^2}\right)(36 \text{ sec}^2) = 1728 \text{ ft}$$
$$= 3.5\mu\text{sec}.$$

10.7 Target EA and Deep Learning Algorithms

Considering the example shown in Figure 10.51, the pull-off is

$$R_{PO} = 7 \text{ pulsewidths} \tag{10.59}$$

Consider, for example, a seeker PRF = 1000 Hz. In this case the pull-off is accomplished in 6,000 pulses. Note also the *parabolic form* of the target time varying range-rate delay motion with respect to the radar acceleration. This parabolic form of the delay as a function of time must be formed to match a realistic maneuver.

The DRFM algorithms for this type of application must be adaptive to counter most seeker tracking loops. The seeker tracking loops are smart and have limited acceleration. That is, if the target is already accelerating, then the higher acceleration will be neglected. Statistical processing techniques have been used quite successfully to detect the DRFM repeated signal and are discussed in the next chapter. Since the recycle may happen at any inopportune time, this DRFM mode is most often used with other techniques such as RGPO-to-chaff cloud, or RGPO-to-decoy such as an AN/ALE-55 towed decoy (jammer).

A bidirectional RGPO algorithm is examined in [51] and briefly investigates its effectiveness against the early-late gate tracking type algorithms. Simulation results are shown that show the algorithm defeats these types of narrow gate target trackers often used in airborne tactical multimode radar systems in their air-to-air tracking modes.

10.7.4.4 RGPO Delay Kinematics

For the DRFM's RGPO mode, the delay kinematics are given as

$$x = x_{in} + v_{in}t + \frac{1}{2}a_{in}t^2 \tag{10.60}$$

where x is the position of the resulting position of the seeker's range gate, x_{in} is the target's initial position, v_{in} is the target's velocity, and a_{in} is the target's acceleration. The pull-off range is given by the last term in the equation as

$$R_{PO} = \frac{1}{2}a_{in}t^2 \tag{10.61}$$

Following the example above, consider after 6 seconds,

$$R_{PO} = \frac{1}{2}\left(96\frac{\text{ft}}{\text{sec}^2}\right)(36\text{sec}^2) = 1728\,\text{ft}$$
$$R_{PO} = 3.5\,\mu\,\text{sec}$$
$$R_{PO} = 7\,\text{pulse widths}$$

If the seeker's radar PRF was low (e.g., 1000 Hz), the pull-off would be over 6000 pulses!

It is also important to mention that any angle EA techniques against the seeker's radar antenna will certainly not do well and at the very least require a very large *JSR* if the tracking gate in range is still centered on the target. That is, the return signal from the RCS underneath the track gate is very large. Consequently, any EA technique trying to pull the antenna off the target will be contending with this range gate success.

Therefore, the RGPO is typically *used first* within a sequence of other techniques. That is, RGPO is used first as a means to an end. For example, or RGPO→ distraction chaff cloud. This cloud should have a higher RCS and be located at a specific range away from the platform target. Techniques that can also be used include, for example, RGPO→Cross Polarization and RGPO→Swept Square Wave (angle techniques). Chaff is also an alternate method for pull-off in range. Depending on the wind characteristics (how fast and from which direction), the pull-off in range can be attempted with the launch of *seduction chaff*. Seduction chaff, such as the Mk-214 shipboard SRBOC, can be launched such that it blooms immediately then begins to move off in range, moving the track gate off the target (ship).

10.7.4.5 Hook Concepts

The RGPO objective is to have the range tracker follow the jamming pulse denying the radar system accurate range information. If the range tracker follows the jamming pulse, the EA system turns off (or recycles) the jamming sequence of pulses, leaving the range tracker without any energy to track ($JSR = \infty$). Without any energy in the range gate, the radar system breaks range lock and initiates a search, or reacquisition, of the target [24]. Since the radar antenna starts at a position that's pointed at the target, a reacquire can be nearly immediate.

To keep the radar's track-gate off the target at the end of a cycle (peak delay) a *hook* can be applied. Figure 10.56 shows the hook concept as a secondary pulse that is sent directly after the first pulse. The *first pulse* does the job of pulling the track-gate off the target in time. The first pulse delay can be considered as one sample of a sampled version of Figure 10.55. The hook is the *second pulse* and is transmitted at a fixed delay offset in time (or range). The offset, for example, could be 4-7 pulsewidths from the original target position.

10.7 Target EA and Deep Learning Algorithms

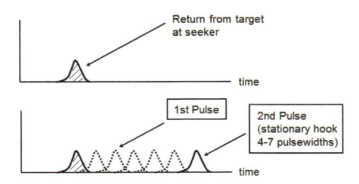

Figure 10.56 The RGPO hook concept.

Figure 10.57 shows an antiship missile seeker ship self-protection EA RGPO engagement showing the emitted pulse and the echo back at the seeker. The RGPO begins and starts to pull the range gate off the target. The hook is generated by using distraction chaff at the end of a delay cycle. The distraction chaff is a cloud with a higher RCS that blooms at a specific range away from the target.

750   ELECTRONIC ATTACK USING DEEP LEARNING

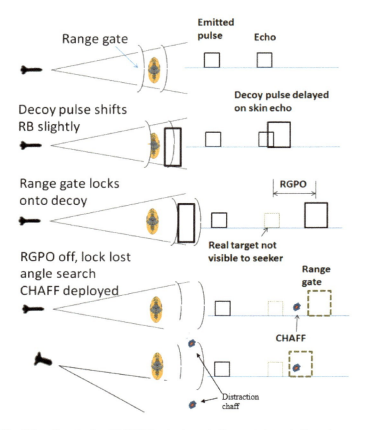

Figure 10.57 Ship self-protection EA RGPO to hook-to-chaff concept in a missile seeker engagement.

In summary, RGPO is always used as a means to an end. That is, it is typically followed by other techniques; for example, modulation in antenna angle modulation, antenna polarization modulation, or frequency (or velocity) modulation. Also, RGPO not only consists of repeating a single pulse but should use AI deep learning techniques to configure two or more hook pulses as seduction and distraction decoys. An ultrahigh wideband range-velocity false target technique for coordinated range and velocity gate pull-off is given in [52].

10.7.5 EA: Velocity Gate Pull-Off

Tracking radar systems can provide the range, velocity, and angle measured values of the target continuously, smoothly, and accurately. Velocity gate pull off (VGPO) EA is an active deception technique focused on modulation of the frequency or velocity tracking system (e.g., pulse Doppler).

10.7 Target EA and Deep Learning Algorithms

The velocity is directly related to frequency as range is related to time. The VGPO modulation is

$$v_j(t) = GA\cos(2\pi(f_r - f_j)t) \quad (10.62)$$

where G is the (constant) gain applied to the intercepted signal with amplitude A and f_j is the jammer frequency. The frequency modulation EA is effective against any RF sensor that uses narrowband filtering such as Doppler radar, FMCW, and LPI FSK to name a few. The two-way Doppler shift is

$$f_d = \frac{2\dot{R}_T}{\lambda_r} \quad (10.63)$$

It is much easier to modulate the target Doppler frequency up or down in order to capture the Doppler frequency tracking gate. That is, for the range gate to be moved in, special emitter parameters must be fixed. For example, the EA is predicting the PRF to move the target further in. Therefore, the emitter's PRF cannot change.

$$f_d = f_0 + (f_{v0})t + \frac{1}{2}(f_{a0})t^2 \,(\text{Hz}) \quad (10.64)$$

where f_0 is the initial Doppler shift (Hz) f_{v0} is the initial Doppler velocity – initial Doppler shift rate of change in time (Hz/s) and f_{a0} is the initial Doppler acceleration – initial rate of change in time of f_{v0} (Hz/s^2) We know the Doppler velocity is related to the closing velocity V_C as

$$f_d = -\frac{2cV_C}{f_c} \quad (10.65)$$

and emphasized above, the closing velocity is related to the change in range

$$V_C = -\frac{f_d \lambda}{2} = \frac{dR}{dt} = \frac{d}{dt}\left(\frac{1}{2}at^2\right) = at \quad (10.66)$$

Continuing the antiship missile seeker example above, Figure 10.58 shows the missile's RCS Doppler profile. The Doppler frequency within the seeker is shown as a function of frequency f_d. The VGPO (first repeated pulse) is shown on both sides of the target's initial Doppler position for illustration-middle plot. Also shown is the hold out hook (second pulse). The third plot on the bottom shows the offset frequency (kHz) as a function of pull-off time. The pull-off time in our example is 4s. Every 4s, the VGPO recycles (with a new acceleration profile). Unless you know that your EA was successful (i.e., you are not being tracked), then the recycle continues. AI plays a large part in the pull-off profile as it represents an acceleration or the slope ($\dot{f}_d$). That is, anywhere along the curve cannot exceed the maximum gs (for example, we used 3 gs). Typically 9 gs is usually considered a maximum limitation or three times the acceleration due to gravity [50]. Note that if the platform is already accelerating, this acceleration also has to be taken into account. The vector sum of these accelerations must be less than the limits for the situation at hand. Embedded AI is ideal for this as the maneuver time constants can be much faster than a manned aircraft due to their smaller size.

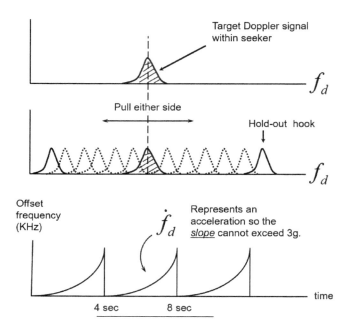

Figure 10.58 Target RCS profile as a function of Doppler frequency f_d and the VGPO, first pulse (shown both sides) pulling the tracking gate off the target and the hold out hook, second pulse. Also shown is the offset frequency (kHz) as a function of the pull-off time.

Figure 10.59 shows the engagement where the EA pulls the seeker's tracking gate into the clutter. This is especially effective since the seeker now has to detect the radar's platform within the clutter, which can be a very difficult task. Note also the track gate can either be pulled up or down into the clutter.

10.7 Target EA and Deep Learning Algorithms

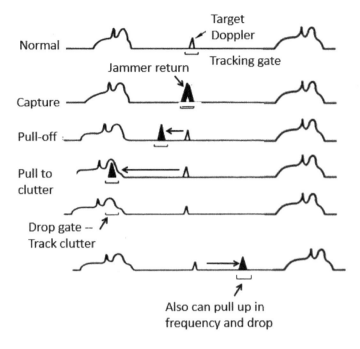

Figure 10.59 Airborne target (white triangle) against missile seeker tracking gate (black triangle). DRFM EA applies a VGPO pulling the seeker's track gate into the clutter.

10.7.6 Coordinated RGPO-VGPO

The RGPO and VGPO techniques above are useful and are most often combined with other techniques. The *coordinated range velocity gate pull-off* (RVGPO) technique is a highly effective EA modulation on its own since in range-Doppler space, it acts like a *true target*. Figure 10.60 shows the range bin number - filter number at two contiguous frame times T_1 and T_2. Shown are the target's, movement both in range and velocity at the same time.

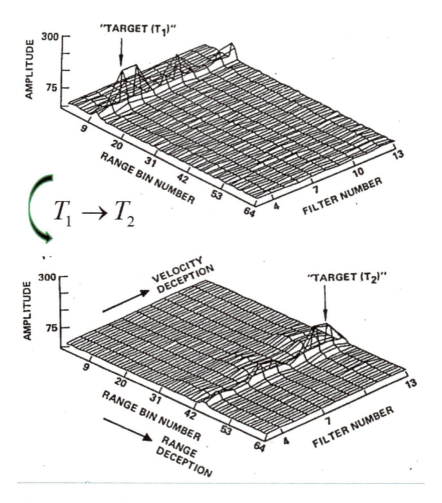

Figure 10.60 Coordinated RGPO and VGPO (RVGPO) showing targets movement from range bin 9 filter number 5 (at T_1) to range bin 42 filter number 11 (at T_2).

The advantage this technique can be emphasized by examining how the radar system typically determines if it is being deceived–an electronic protection measure (EP) against RGPO and VGPO. This method is to compare the target's range rate with the target's Doppler frequency. If these do not agree, then deception modulation has been detected. This is more clearly illustrated in Figure 10.61. Both the range tracker and the Doppler tracker activity markers are plotted out as a function of time. For the range tracker, the range rate and range acceleration, $\ddot{R}$ are plotted. The same quantities are also plotted out except they are derived from the Doppler tracker.

10.7 Target EA and Deep Learning Algorithms

If for example, an RGPO (only) EA waveform attempts deception of the seeker, comparison of the range and Doppler tracker clearly shows a difference. The same is true for the VGPO. If the VGPO is the only deception, again comparison clearly shows a difference between the range tracker and the range-rate tracker. If however, a *coordinated RVGPO* EA is present, note that it looks like a true target maneuver over time and the comparison of the range and Doppler tracker reveal no deception present.

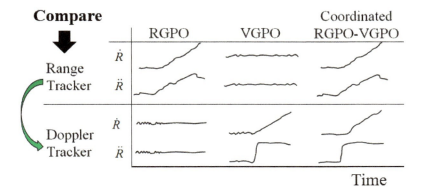

Figure 10.61 Comparison of range tracker and Doppler tracker to identify RGPO or VGPO.

The accuracy of the RVGPO in range and Doppler is another DRFM AI control area as it can get quite complicated since it depends on both the seeker and the DRFM. For example, the range accuracy depends on the seeker's range resolution, ΔR, and therefore directly on the seeker's ADC sampling frequency. Consequently, for our EA technique of RVGPO deception, we want

$$\text{DRFM } f_s^{ADC} > f_s^{ADC} \text{ Missile}$$

The Doppler 3-dB filter width depends on the seeker's integration time. Unfortunately, the EA technique has no control over this.

10.7.6.1 Example: Linear Frequency Modulation with RVGPO

Consider for example, a LFM signal being intercepted by the DRFM. That is, the radar transmits a pulsed signal of the form

$$S(t) = \exp(\phi(t) + j\phi_0) \quad \text{for} \quad t \in [0, \tau] \quad (10.67)$$

where $\phi(t) = j\pi(2f_0 t + kt^2)$, f_0 is the intermediate frequency, k is the frequency modulation "slope" or $k = \Delta/\tau$, τ is the pulsewidth, and ϕ_0 is the initial phase of the transmitted signal. This type of waveform is typically wideband and used with high-resolution SAR, inverse SAR (ISAR) imaging sensors, and has a range resolution $\Delta R = c/2\Delta$ where Δ is the modulation bandwidth.

With a DRFM target at a position R_0 away from radar system, the signal received back at the radar from DRFM can be expressed as

$$S_{re}(t) = A_{re} \exp\left[\phi\left(t - \frac{2R_0}{c}\right) + j\phi_0\right] \tag{10.68}$$

where A_{re} is the amplitude of the target signal.

The time domain signal received by the radar when the DRFM switches on a *coherent RGPO* is

$$J_{RGPO}(t) = A_{re} \exp\left[\phi\left(t - \frac{2R_0}{c}\right) + j\phi_0\right]$$

$$+ A_R \exp\left[\phi\left(t - \frac{2R_0}{c} - \Delta t_J - \Delta t_J(t)\right) + j\phi_J\right] \tag{10.69}$$

where A_R is the amplitude of the RGPO EA, Δt_J is the inherent delay required by the DRFM to receive, store, process, and forward the radar signals back, $\Delta t_J(t)$ is the modulation time delay of RGPO jamming, and ϕ_J is the initial phase of the EA signal.

When DRFM jammer implements a *coherent VGPO*, the jamming signal and the target echo are coherently synthesized, the synthetic signal received by the radar is

$$J_{VGPO}(t) = A_{re} \exp\left[\phi\left(t - \frac{2R_0}{c}\right) + j\phi_0\right] + A_V \exp\left[\phi\left(t - \frac{2R_0}{c} - \Delta t_J\right) + j\phi_J\right]$$

$$\times \quad \exp j2\pi\Delta f_{dJ}(t)\ t \tag{10.70}$$

where A_V is the amplitude of VGPO EA and $\Delta f_{EA}(t)$ is the Doppler shift of the VGPO EA signal. When DRFM jammer implements a *coherent RVGPO*, the synthetic signal received by the radar is

$$J_{RVGPO}(t) = A_{re} \exp\left[\phi\left(t - \frac{2R_0}{c}\right) + j\phi_0\right]$$

$$+ A_{RV} \exp\left[\phi\left(t - \frac{2R_0}{c} - \Delta t_J - \Delta t_J(t)\right) + j\phi_J\right] \exp(j2\pi\Delta f_{dJ}(t)t) \tag{10.71}$$

The linear frequency modulation will be revisited later when digital image synthesizer EA techniques are examined against ISAR and SAR.

10.7.7 SAR Active Decoy EA Technique

Because SAR has such a high processing gain based on 2D correlation processing that is coded with, many different pulse compression codes in addition to LFM, EA techniques typically do not resort to straight noise due to its being decorrelated or spread out at the correlator. This makes noise EA even less effective.

An effective EA technique against SAR for the stripmap geometry shown in Fig. 10.62 shows only one EA source described in [53]. Figure 10.62 shows the aircraft flying along the y axis at the point $(0, u)$ at some moment moving with constant velociy v_a. The EA source is located at (x, y).

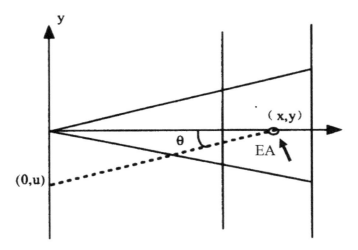

Figure 10.62 Synthetic aperture array electronic attack scenario.

In [53], a *range deception modulator* is used to produce a decoy target at $(x + \Delta x, 0)$ and is configured as [53]

$$H_n(\omega) = \sigma_n \exp(j2k\Delta x_n) \tag{10.72}$$

where $k = \omega/c$ is the wavenumber. Note that the characteristic of range deception is only determined by the offset in the range direction. It is independent of the jammer's location relative to the radar that is convenient to its realization. This means that we can preload the parameters of the range deception modulator.

Considering azimuth deception is to also produce a decoy target at $(x, \Delta y)$, the representative *deception modulator in azimuth* can be represented as [53]

$$H_n(\omega, u) = \sigma_n \exp\left(jk((\Delta y)^2 - 2u\Delta y)/R\right) \tag{10.73}$$

Note that in contrast to range deception, the deception modulator in azimuth deception is more complicated and more difficult to realize. It is determined not only by the azimuth offset but also by the EA's platform location relative to the radar. This means that for the sake of accurate azimuth deception the jammer must know the radar location at every moment. This makes accurate azimuth deception very difficult because measuring and estimation of the radar location will increase the complexity of the EA.

However, the azimuth deception modulator can be decomposed into two portions according to the phase in the equation above and the azimuth deception modulator can be simplified as

$$H_n(\omega, \theta) = \sigma_n \exp(-j2k\Delta y \sin \theta) \qquad (10.74)$$

and only needs the estimation of radar azimuth angle relative to the EA platform. Figure 10.63 shows the azimuth deception modulator and range deception modulator processing to create the EA waveform discussed above [53]. Figure 10.64 shows the creation of a lattice within the SAR image, which is similar to the mosaic seen in military photographs and is especially useful for the protection of important targets in military applications.

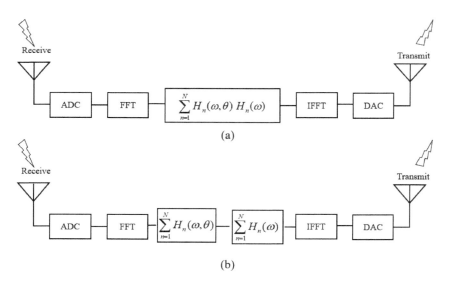

Figure 10.63 Block diagram of a SAR deception EA showing (a) azimuth deception modulator and (b) range deception modulator. (©IEEE, reprinted with permission from [53].)

10.7 Target EA and Deep Learning Algorithms   759

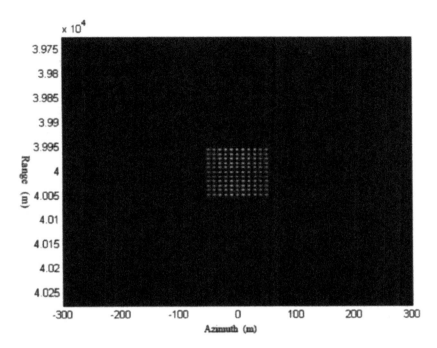

Figure 10.64 Creation of lattice within the SAR image useful for protection of important targets. (©IEEE, reprinted with permission from [53]).

In summary, based on the analysis above, an active-decoy can be realized at any location through two steps: (1) Make azimuth deceptive modulation using $H_n(\omega, \theta)$, and (2) make range deceptive modulation using $H_n(\omega)$. For many decoys, the accuracy deception modulator should be as follows:

$$H_n(\omega, \theta) = \sum_{n=1}^{N} H_n(\omega, \theta) H_n(\omega) \tag{10.75}$$

However, if the goal is to only jam the SAR, the two-step deception modulator below is strongly recommended.

$$H_n(\omega, \theta) = \sum_{n=1}^{N} H_n(\omega, \theta) \sum_{n=1}^{N} H_n(\omega) \tag{10.76}$$

10.8 COUNTERTARGETING THE IMAGING SENSOR

Chapter 1 discussed force maneuver and sustainment over long distances being the key characteristics of a successful organization. This success comes in part from the use of *precision engagement* against the adversary in an effort toward a decisive effect on their forward campaigns and major operations. The entire precision engagement process is often called *targeting*. The process involves targeting the adversary and then allocating the selected targets to the most appropriate hard-kill weapon systems. Recall that the time between the surveillance activities, location, acquisition, and identification of the correct targets, and the weapon arrival time, is the *maneuver time*, $t_{Maneuver}$ s (10.77) (1.14).

Some of the insights revealed in Chapter 1 were that disruption *anywhere* in the engagement timeline can cause changes in the targeting results. As expressed by $t_{Maneuver}$,

$$t_{Maneuver} = \frac{1}{Tempo_{OODA}} - (t_{Sense} - t_{Process} - t_{Analyze}) \qquad (10.77)$$

the time required to sense, process, and analyze the operational targeting activities or spectrum activities (or both) directly affects the hard-kill maneuver or spectral dominance, respectively. The frequency (tempo) of the EM maneuver OODA is

$$Tempo_{OODA} \leq 1/t_{OODA} =$$

$$\frac{1}{t_{Sense} + t_{Process} + t_{Analyze} + t_{Configure} + t_{Transmit}} \text{ (Hz)} \qquad (10.78)$$

and is the speed at which a weapon can be delivered to the correct target and is of great significance [54]. The hard-kill targeting process is summarized in Figure 10.65 and is called the hard-kill timeline. The timeline lays out the steps involved in the identification of a target, the assurance that it is correct, and the selection of the best weapon as well as its launch and guidance.

Consider an adversary's hard-kill timeline as shown in the figure. Their process begins with locating, acquiring, and identifying their target. During this ISR phase they use high-resolution imaging sensors such as SAR and ISAR sensors.

10.8 Countertargeting the Imaging Sensor

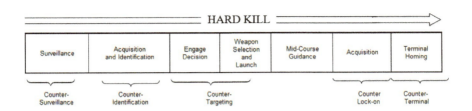

Figure 10.65 Hard-kill chain starting from surveillance through the selection of a weapon to the terminal homing kill stage [55].

Figure 10.66 shows an example of a high-value unit (the aircraft carrier) being targeted by a antiship capable missile (ASCM) whose guidance commands while in flight, are networked to it from the, Transport, Erector, Launch vehicle (TEL), the SAR in space, ISAR on shore, and the Over the Horizon Radar (OTHR) that uses skywave propagation to image the HVU.

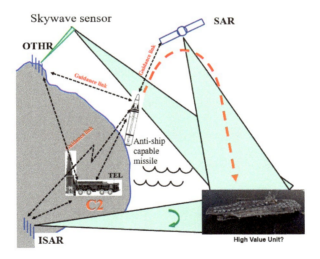

Figure 10.66 ASCM targeting a high-value unit (aircraft carrier) being guided mid-course by a SAR, ISAR, its TEL, (C2) and OTHR and in its terminal phase by its onboard seeker.

Depending on the collective target identification at the C2, the decision to engage the target and which weapon to launch must be made and only the ability to quickly confuse this targeting process could have prevented the weapon from being launched [55]. After the launch of the weapon, the effort to confuse the targeting sensors, including the ASCM seeker, which in some cases can also be an ISAR in the terminal phase. In the next sections we begin the discussion about how to deceive these imaging sensors and seekers.

The use of ISAR for classification purposes also has interesting surface-to-air applications for obtaining information about the shape, class, size, and trajectory orientation of a target in flight within a surveillance or no-fly-zone. It can also be useful for navigation purposes, especially in risky situations in which a clear description of the ships or objects present in the scenario is extremely important. High spatial measurement of the RCS of the target reflecting center scan be used to design the shape of an object to decrease the strength of its RCS hot spots to reduce its radar visibility [56]. Another challenging issue for ISAR imaging is to acquire knowledge of noncooperative target rotational motion typically characterized by the rotation rate of target around a center point. Lack of knowledge of target rotation rate can lead to an unrecognizable and defocused image because the cross-range scale is unknown [57].

10.8.1 Confusing the Adversary's Targeting Process

Actions taken to confuse or deceive the adversary's hard-kill time line are discussed in this section. Prelaunch weapons designation and targeting efforts are known as countertargeting techniques. This encompasses the techniques used to counter surveillance, acquisition, and identification phase of the target. The engage decisions and the weapon selections and launch decisions are also part of the focused counter-targeting efforts as well as any of the seeker acquisition and terminal homing EA measures that are carried out. Depending on the ISAR target identification, the decision to engage the target and launch the weapon is made and only the ability to quickly confuse this targeting process can prevent the weapon from being launched. Actions taken to confuse or deceive the adversary's prelaunch weapons designation and targeting efforts are known as *countertargeting techniques*. The use of noise jamming must be done continuously, interfering with normal radar and communication operations. Few key targets are stationary and "deception" is a major part of the entire countertargeting process.

Surveillance, acquisition, and ID actions within the hard-kill chain often use a wideband imaging sensor like an ISAR or SAR. To prevent this, countersurveillance and counter-ID techniques use low-RCS materials and stealth techniques to prevent the skin echo from being detected causing the platforms not to be acquired and identified. However, these are largely ineffective against the wideband imaging sensors such as an ISAR.

To ensure the adversary's engagement decision is correct and the proper weapon is selected and launched, they often use a wideband RF imaging sensor. To prevent this, a countertargeting effort is put forth to prevent this engagement decision, target and weapon selection from being made. Countertargeting methods are generated using deception devices of various types but for the same reasons above, these methods are mostly not effective against a wideband imaging radar.

If the weapon is launched, and acquires the target out of mid-course guidance, and acquisition and it goes into a terminal phase, a targeting seeker such as an ISAR is selected. The ISAR seekers are especially useful for ASCMs as they provide:

- High range-resolution profiling;
- Use of coherent processing and possibly LPI modulation;
- Forms an image of the target;
- Allows decoy rejection;
- Good aimpoint accuracy;
- Greater probability of kill!

Counter-lock-on includes the use of low radar cross-section, paints, materials, stealth, and deception devices in order to disrupt the weapons targeting prior to valid lock-on, distraction chaff, thus preventing the enemy from obtaining an accurate fire control solution. Unfortunately, these actions are largely ineffectual against wideband imaging RF seekers such as the ISAR. Once locked on, even seduction chaff or decoys are ineffectual against an ISAR as it can easily distinguish between chaff, especially when trying to hide a large, high-value target. DRFMs can generate false point targets against search radar, fire control radar, weather radar, and Doppler radar, to name just a few. However, generation of false, structured targets against the imaging radar system is quite a different matter. The need for coherent countering of these imaging sensors and seekers remains a high priority for EW systems and DRFMs.

10.9 DIGITAL SYNTHESIS FOR STRUCTURED FALSE TARGETS

In this section we address the subject of digitally synthesizing multiple, structured false targets against RF, range-Doppler imaging radar systems. The ISAR discussed previously represents a unique example and a system that is now used most successfully for the air-to-surface imaging of ships and the imaging of aircraft and satellites. For the air-to-ground targeting, the separation of the features requires higher resolution and the use of AI because the magnitude of the rotation is considerably less in most cases.

10.9.1 Early Analog Technology

About the year 2000, a number of techniques were attempted to create false targets with realistic signatures that took advantage of acoustic charge transport (ACT) devices. An overview of the functional capabilities, operational performance along with a description of the basic ACT device architectures and operation of the device technology is presented in [58]. A ship rendering within an emitter using the ACT technology is shown in Figure 10.67, and to the author's knowledge, is the first attempt to generate this type of false image of a ship. The ACT devices provide delay times on the order of μs [59]. Figure 10.67 shows the ship image using acoustic charge transport technology to generate the image of a ship within the emitter.

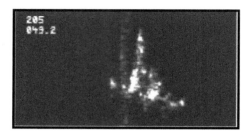

Figure 10.67 Ship image using ACT devices that are no longer available.

The ACT bandwidth, however, presented a severe limitation when trying to counter wideband signals. In fact, it was a technology that couldn't be improved upon past a certain point. Hence, like the dinosaurs, it too has close to disappeared off the face of the Earth. Fiber-optics has also been tried. Fiber-optic tapped delay-lines have a flat spectral response in a broad range of wavelengths and have been configured as tunable filters for optical add/drop multiplexers for selectively dropping/inserting wavelength channels from/into wavelength division multiplexing networks [60]. Four 3-dB couplers, a piezoelectric transducer, and four metal-coated fiber-optic tapped delay-lines with a flat spectral response were used.

However, cleaving taps into a fiber-optic line for a series of range bins proved to be cumbersome. In addition, it was a complex interface and was not programmable. Consequently, for a false target generator transverse filter application, fiber optics were not useful.

10.9.2 Embedded Digital Integration

With embedded, digital integration, for example, sub-0.1-μm CMOS and GaAs, the objective finally comes down to constructing a digital, programmable scatterer that requires no knowledge of the interrogating waveform, and is capable of synthesizing multiple, large false targets.

This at first seems like a tall order but as demonstrated in Figure 10.68, a closer look at the scattering physics provides a solution that seems to present itself. Notice, for example, that a scatter first reflects the scattering centers within the *first range bin* closest to the emitter at time t. That is, the scattering centers all combine within ΔR_1 to reflect the waveform back toward the emitter. As the interrogating waveform then continues to travel forward, it is reflected in the *second range bin* ΔR_2 at $t + \Delta t$.

10.9 Digital Synthesis for Structured False Targets

This reflection then combines with the reflection from scatterers in the first range bin noncoherently, and propagates back toward the emitter, and so on.

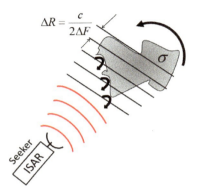

Figure 10.68 Scattering concept showing that the composite scatters within the nearest range bin combine and are reflected back first. The second range bin scatterers combine and are then reflected next and combine at a later time $\Delta \tau = 2\Delta R/c$ with the scattering from the first range bin.

In Figure 10.69, the composite scattering model of a ship structure showing the range bin slices (dashed lines running vertically) laid across perpendicular to the LOS to the emitter. Each range bin i that is shown, combines all of the scatterers together to generate the complex backscattering coefficient magnitude α_i and the backscattering coefficient's phase, α_i. The magnitude is typically given in units of dB below 1 m² per m². The phase of the complex return, ϕ_i, for $i \in \{1, \ldots, n\}$ can also be used together with the magnitude to begin the development of a false target-imaging architecture. This then marks the beginning point of a programmable digital image synthesizer, the concept developed in the next section.

Figure 10.69 Composite scattering model of a ship structure that can be used in target imaging showing range bin slices (dashed lines running vertically), with each having the combined scatterers into a complex backscattering coefficient with magnitude α_i, in dB below 1 m²/m² and phase ϕ_i, of the complex return for $i \in \{1, \ldots, n\}$.

10.10 PROGRAMMABLE DIGITAL IMAGE SYNTHESIZER

Figure 10.70 shows the original false target image synthesis concept where the digital image synthesizer (DIS) on the center ship is being imaged by a seeker; however, a confusion factor is injected into the seeker by the creation of several false targets about the true target.

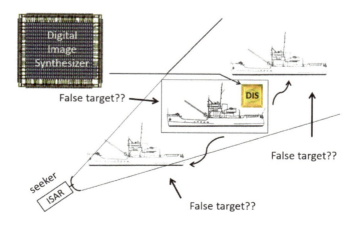

Figure 10.70 Original concept showing the implementation of a digital image synthesizer, first prototype. This was the first work to ever produce a synthetic structured image within an RF imaging ISAR [61, 62].

10.10 Programmable Digital Image Synthesizer

The digital, programmable imaging architecture to generate multiple, structured false target signatures that were realistic against a high-resolution ISAR was developed in a highly integrated, large array of generic complex range bin processors [61, 62]. Enabled by new VLSI sub-0.01 micron bulk CMOS fabrication processes, these technologies are especially timely due to recent advances in FMCW imaging radars and LPI seeker technology.

Working in conjunction with the DoD laboratories, the objective was to design, fabricate, and test an all-digital target imaging device capable of generating multiple large false targets from an intercepted wideband chirp signals of any duration, and any PRF.

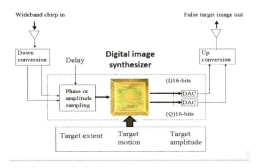

Figure 10.71 Architecture block diagram showing the DIS along with the downconversion, the sampling by either an amplitude analyzing ADC or a phase sampling arrangement of comparators, a DAC arrangement for I and Q, and the upconversion of the output target before transmission.

In Figure 10.71 an application specific integrated circuit (ASIC) design (DIS-512) to be integrated into a KOR (now known as Mercury Defense Systems) DRFM architecture. It contained 512 fully programmable complex range bin processors. The ASIC clock speed was (adaptable) and depended on the range resolution desired. It was constructed in a scalable/generic mask design and compatible with a sub-0.1-μm CMOS process.

10.10.1 Warfighter Payoffs

To be able to provide a countertargeting capability against the many high-range resolution RF sensors in the hard-kill chain provides a significant payoff to a warfighter. Deception using miniature air vehicles, ships, or manned aircraft platforms to generate large, multiple false targets against high-resolution imaging radars (ISARs, SARs) provides a superior RF decoy capability.

A special note, however, is that providing this coherent, digital false-targeting and deception capability requires a special DRFM augmentation. That is, countering of wideband imaging seekers and profiling radars requires a special architecture as a DRFM alone cannot provide the detailed amplitude Doppler modulation of the RF waveform to produce the realistic target image that moves as a ship moves over time.

The DIS is shown in Figure 10.72 and consists of a parallel arrangement of complex range bin processors that is configured depending on distance from the interrogating radar to each part of the false target. That is, a separate processor is allocated for each range bin. The pipelined, all-digital image synthesizer is capable of generating multiple false-target images from a series of intercepted ISAR chirp pulses to provide a novel RF imaging decoy capability. The DRFM image synthesizer is able to modulate the phase samples from a phase-sampling front-end that strobes the phase of the intercepted ISAR pulses into each complex range bin processor. The image synthesizer

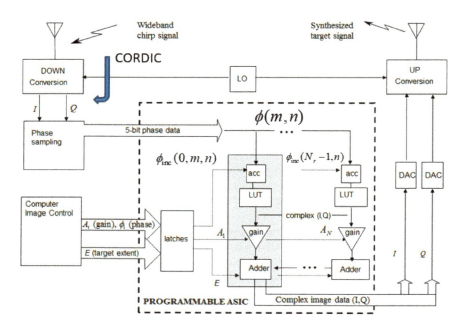

Figure 10.72 Block diagram of a DIS consisting of a series of tapped delay lines with each tap, a complex range bin processor with an accumulator (Doppler), a LUT, a gain block (RCS), and an adder/delay.

must synthesize the temporal lengthening and amplitude modulation caused by the many recessed and reflective surfaces of the target and must generate a realistic Doppler profile for each surface. Consequently, the synthesizer contains a parallel array of N_r (identical) complex digital modulators with one modulator for each false target range bin [63].

The k_p-bit binary phase samples from each intercepted ISAR pulse are applied one at a time to the modulator array. To synthesize the image, each modulator requires a set of *phase and gain coefficients* that are derived from the range-Doppler description of the false-target to be synthesized. To generate the false target Doppler spectrum within a range bin, each DRFM phase sample within a pulse is added to a k_p-bit phase coefficient that increments the phase on a pulse-to-pulse basis [63]. This phase rotation is accomplished with a binary adder and generates the desired motion profile of the range bin. To change the range bin's radar cross-section (RCS) characteristics, the incremented phase value is then converted to a normalized complex signal in-phase (I) and quadrature (Q) using a lookup table (LUT). The RCS is modulated by a gain block that multiplies the complex signal by a gain coefficient. Multiplication is accomplished with a left-shift of the binary I and Q signals using a parallel array of multiplexers that are controlled by a k_g-bit binary word.

10.10.2 DIS Complex Range Bin Processing

A block diagram of the bit-level complex range bin processing is shown in Figure 10.73.

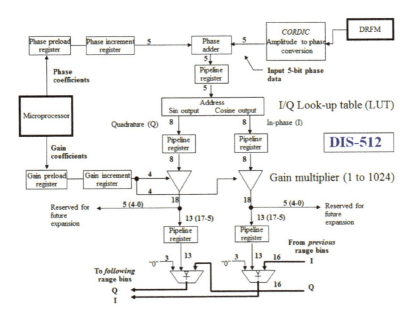

Figure 10.73 Example of a bit-level complex range bin processor showing the DRFM input (through the CORDIC) and the phase increment and gain coefficient inputs (double buffered).

The programmable DIS processes the phase samples from a phase sampling DRFM and uses a high-performance microprocessor for image control. Each k_p-bit DRFM phase sample $\phi(m,n)$ is applied in parallel to all N_r range bin modulators. Each range bin modulator contains a phase adder, an LUT, a gain block, and a summation adder. Note that to maintain synchronization as the bits propagate through the parallel range bin processors, the architecture must be pipelined and use registers between each of the DSP components. This helps increase the execution throughput by using the clock resources in the most efficient manner.

10.10.2.1 CORDIC Processing

The CORDIC processor was an algorithm first described in 1959 by Jack Volder, and first used in navigation processing for elementary function evaluation such as $\sin(\theta)$, $\tan^{-1}(\theta)$. It is used today in many diverse fields such as communications, signal processing, and electronic warfare. For the application at hand, it is used to take an I/Q value in and put out a magnitude and phase. The phase is easier to work with when trying to implement the Doppler on the intercepted signal samples.

For example, consider the complex number $z_0 = 2 + 7j$ or $I = 2$ and $Q = 7$. The rotations start as shown in Figure 10.74.

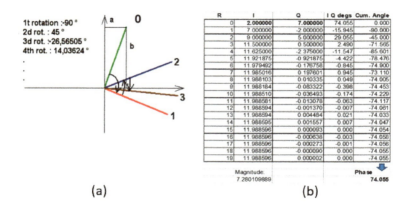

Figure 10.74 CORDIC algorithm showing (a) four initial rotations of the CORDIC algorithm for the complex number example $z_0 = 2 + 7j$ and (b) the series of rotating the phase of the complex number by multiplying (shifts and adds) by succession of constant values to determine the phase (and magnitude).

Note that many FPGA and ASIC vendors (e.g., Xilinx) have CORDIC cores that can easily be downloaded but that ties the developer to the vendor for any products that they may sell. For most DRFM developers today, this can easily be avoided with the CORDIC as this is a two-step algorithm as shown below and outlined in Figure 10.75. Outlined in Figure 10.76 are the constructs involved with Step 1 and Step 2.

The iterations made by the CORDIC and MATLAB will tally until decimal numbers begin to appear in the iteration. Amplitude errors appear in the amplitude computation after Verilog begins to round off the decimal number. The error is significant if (1) decimal numbers appear in the early iteration stage, and (2) the amplitude of I and Q are small. The phase error begins to appear when the sign of Q/I is wrong at any stage of the iteration.

10.10.2.2 Accuracy

The erroneous sign is attributed to the current CORDIC's limitation in handling small numbers. Accuracy of the CORDIC can still meet the requirement of DIS since the input resolution is just 5 bits. These design considerations have to be revisited if further accuracy in the phase rotations are required.

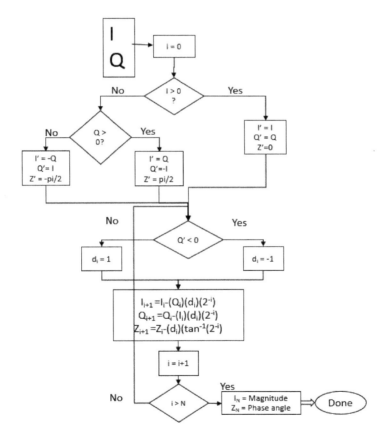

Figure 10.75 Flow diagram of the CORDIC algorithm showing the two-step iterative process for computing the magnitude and phase from a complex number.

Cordic Algorithm

Step 1: Check if vector is in $-\pi/2$ to $\pi/2$ regions by checking if I >0. If it is, no pre-rotation required.	**Step 2:** Check if Q is positive or negative and determine the direction of subsequent rotation.
If not, 2 possible cases:	Update I_{i+1} Q_{i+1} Z_{i+1}
Case 1: Q> 0 => vector is in $\pi/2$ to π quadrant, need to rotate vector clockwise by $\pi/2$	Repeat step 2 until N iterations are completed
Case 2: Q<0 => vector is in $-\pi/2$ to $-\pi$ quadrant, rotate vector counter clockwise by $\pi/2$	I_N – represents the magnitude of the vector
	Z_N – represents the phase of the vector

Figure 10.76 CORDIC algorithm showing the 2-step iterative algorithm for computing magnitude and phase from a complex number.

10.10.3 Bit-Level Simulation of the DIS

The last stage in the range bin modulator is a summation adder that sums the gain block results with the adjacent (delayed) summation adder output and sends the results forward to the next summation adder. The results are passed forward until they reach the summation adder in the first range bin modulator. The synthesizer's complex output samples containing the image modulation are taken from the summation adder in the first range bin. These samples are then sent to the DAC for conversion to an analog signal for amplification and coherent retransmission.

In order to evaluate the quality of the test image, an ISAR compression routine is constructed. To reconstruct the test image using the output pulses from the synthesizer, an ISAR fast correlation image compression algorithm is presented below. As an example, the range-Doppler description of a ship with 32 range bins is presented with a configuration that closely matches an actual AN/APS-137 ISAR image of the USS *Crockett*. Simulation results are shown using the ship as a synthesizer input to demonstrate the feasibility of the architecture to generate false target returns.

10.10.4 LFM Chirp Pulse Model

Consider an ISAR that emits a linear frequency modulated pulse where the frequency is swept upward across the pulse (up-chirp). The complex envelope of the transmitted signal can be expressed as

$$s(t) = \text{rect}\left(\frac{t}{\tau}\right) e^{j2\pi\left(f_c t + f_d t + \Delta t^2/2\tau\right)} \tag{10.79}$$

where

$$\text{rect}\left(\frac{t}{\tau}\right) = \begin{cases} 1 & \text{for } \left|\frac{t}{\tau}\right| < \frac{1}{2} \\ 0 & \text{for } \left|\frac{t}{\tau}\right| < \frac{1}{2} \end{cases}$$

The magnitude spectrum of the chirp signal showing the baseband modulation bandwidth Δ of the waveform is as shown in Figure 10.77.

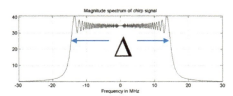

Figure 10.77 Baseband magnitude spectrum of the chirp signal showing the modulation bandwidth Δ.

Here f_d is the Doppler frequency due to the closing rate between the ISAR and the DRFM platform intercepting the pulse. If the Doppler shift is only tens of Hertz compared with the MHz chirp bandwidth, a constant phase change within a chirp pulse can be assumed. The phase-sampling DRFM intercepts and stores N_p pulses from the ISAR. To generate the present and future sample values of the intercepted pulses we let

$$t = (m/fs) + nPRI \tag{10.80}$$

where m represents the sample number within the chirp pulse of width τ, n is the pulse number index with $n \in \{0, 1, \ldots, N_p - 1\}$, PRI is the pulse repetition interval, and f_s is the sampling frequency ($f_s > \Delta$ if using a superherterodyne DRFM approach).

In this work we assume that the PRI remains a constant for all N_p pulses. Substituting (10.80) into (10.79) and using the notation $|a|_m$ for the reduced residue of a modulo m; that is, $a \equiv |a|_m \pmod{m}$ and $(0 \le |a|_m < m)$, the *phase samples* from the chirp pulses can be expressed as

$$\phi_0(m,n) = \left| 2\pi f_d \left(\frac{m}{f_s} + nPRI\right) + \frac{\pi\Delta}{\tau}\left(\frac{m}{f_s} + nPRI\right)^2 \right|_{2\pi} \tag{10.81}$$

where N_p is the number of ISAR pulses and PRI is the pulse repetition interval.

The integer value of the intercepted chirp phase samples with k_p-bit resolution is then expressed as

$$\phi(m,n) = \left\lfloor \frac{\phi_0(m,n)}{2\pi/2^{k_p}} \right\rfloor \tag{10.82}$$

The chirp phase samples are then strobed into the parallel arrangement of complex range bin processors and processed by each one to generate the false target image.

In the next section, a test-target is laid out in range-Doppler space and the *phase and gain coefficients* are extracted from the image. The test target is shown in Figure 10.78 and was constructed to represent a similar structure as that of the U.S.S. *Crockett* as imaged by an AN/APS-137 ISAR.

10.10 Programmable Digital Image Synthesizer 775

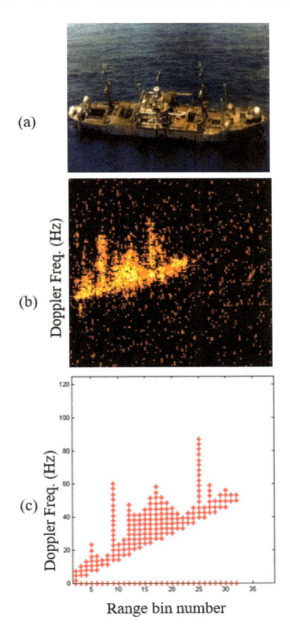

Figure 10.78 Images in the process of creating a false test target range-Doppler image showing (a) the U.S.S. *Crockett*, (b) the AN/APS-137 ISAR range-Doppler image of the U.S.S. *Crockett* and (c) the range-Doppler matrix of scatters composed to serve as a likeness to the example image in (b).

10.10.5 Deriving Digital Image Synthesizer Coefficients for the Test Target

The ISAR forms an image using the digital image synthesizer output pulses that contain modulations derived from a false target configuration placed on the intercepted chirp. To derive the coefficients for testing the digital synthesizer architecture, the test target configuration to be synthesized is laid out in a 2D range-Doppler grid with each range bin (indexed by r) being composed of $N_d(r)$ scatterers (indexed by d).

10.10.5.1 Test-Target: Complex Image

The target return is a function of the PRI and can be expressed as a complex exponential:

$$T(r,d,n) = A(r,d) e^{-j2\pi f(r,d)nPRI} \tag{10.83}$$

where $f(r,d)$ is the target's range-Doppler frequency configuration and $A(r,d)$ is the corresponding scattering amplitudes. The combined returns for range bin r are the sum of the individual returns due to each scatterer. *Superposition* holds and the overall complex return signal is

$$T'(r,n) = \sum_{d=1}^{N_d(r)} T(r,d,n) \tag{10.84}$$

The return signal $T'(r,n)$ can then be used to extract the required phase and gain coefficients for each range bin modulator.

10.10.5.2 Test-Target: Phase Increment Coefficients

To extract the phase coefficients, the phase angles are obtained and quantized with k_p-bit resolution as

$$\phi_T(r,n) = \angle\left\{\frac{Im(T'(r,n))}{Re(T'(r,n))}\right\} \bigg/ \frac{2\pi}{2^{k_p}} \tag{10.85}$$

To create the proper Doppler frequency, the image synthesizer rotates the DRFM phase samples $\phi(m,n)$ on a pulse-to-pulse basis. Consequently, instead of the extracted phase value being applied directly, *a phase increment* is required. That is, for the first pulse ($n = 0$)

$$\phi'_{inc}(r,0) = \phi_T(r,0) \tag{10.86}$$

For every pulse after this $(1 \leq n \leq N_p - 1)$

$$\phi'_{inc}(r,n) = \phi'_{inc}(r,n-1) + (\phi_T(r,n-1) - \phi_T(r,n)) \tag{10.87}$$

Integer values of the phase increment coefficients are obtained as

$$\phi_{\text{inc}}(r,n) = \lfloor \phi'_{\text{inc}}(r,n) \rfloor \tag{10.88}$$

and are applied to the phase adders in the N_r range bin modulators.

10.10.5.3 Test Target: Gain Coefficients

The output of the phase adder is converted into a complex signal by the LUT. The gain block multiplies the complex signal by $2^{g(r,n)}$ where the gain coefficient $g(r,n)$ represents the number of left shifts applied. The first step in extracting the gain coefficients is to normalize the return signal as

$$T'_N(r,n) = \frac{|T'(r,n)|}{\max|T'(r,n)|} \tag{10.89}$$

The normalized return $T'_N(r,n)$ is then amplitude analyzed into 2^{k_g} levels where k_g represents the number of bits controlling the left shifts. Many types of thresholding schemes can be used to produce $g(r,n)$ such as uniform, exponential, and log thresholds. An exponential thresholding scheme is used for the results in this book. From the target's range-Doppler description above, $T'(r,n)$ is formed in order to generate the N_r-by-N_p, phase increment and gain coefficient matrices. From (10.82) and (10.88), $\phi_{\text{inc}}(r,n)$ is generated with $k_p = 5$ bits. For range bins with many scatterers, $\phi_{\text{inc}}(r,n)$ will contain many frequencies. To compare, Figure 10.79 shows the $\phi_{\text{inc}}(r,n)$ for range bin 1 (two scatterers) and range bin 9 (the ship mast) for all 32 pulses. The entire $\phi_{\text{inc}}(r,n)$ matrix is shown in Figure 10.80.

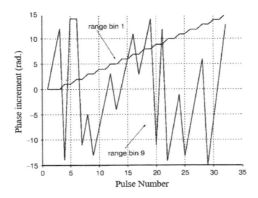

Figure 10.79 Phase increment values for range bins one and nine.

778 ELECTRONIC ATTACK USING DEEP LEARNING

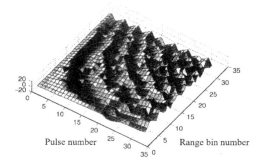

Figure 10.80 Phase increment matrix for ship target with number of pulses $N_p = 32$.

To derive $g(r,n)$ an *exponential* set of thresholds is used to amplitude analyze $T'_N(r,n)$. That is, if

$$0.8 < T'_N(r,n) \leq 1.0 \tag{10.90}$$

then $g(r,n) = 3$. If

$$0.4 < T'_N(r,n) \leq 0.8 \tag{10.91}$$

then $g(r,n) = 2$. If

$$0.2 < T'_N(r,n) \leq 0.4 \tag{10.92}$$

then $g(r,n) = 1$. If

$$0.0 < T'_N(r,n) \leq 0.2 \tag{10.93}$$

then $g(r,n) = 0$. The $g(r,n)$ matrix for $N_p = 32$ pulses is shown in Figure 10.81.

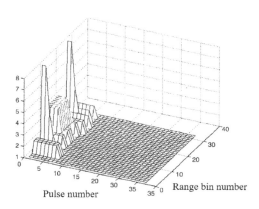

Figure 10.81 Gain matrix for ship test target with number of pulses $N_p = 32$.

10.10.5.4 Test Target: Summary

In summary, the phase increment $\phi_{inc}(r,n)$ and gain $g(r.n)$ coefficients are applied to each complex range bin modulator r where $\hat{r} \in \{0,1,2\ldots,N_r-1\}$ for each pulse n.

10.10.5.5 Doppler Processing: Phase Rotation

The Doppler is created in each range bin by a pulse-to-pulse phase rotation. The first adder is used to create the Doppler modulation within each range bin by a phase rotation of each received chirp pulse using $\phi_{inc}(r,n)$ as shown in Figure 10.82.

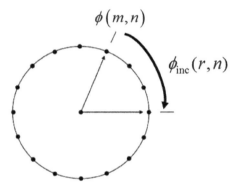

Figure 10.82 Phase increment commanded $\phi_{inc}(r,n)$ to rotate the phase at a certain rate to create the desired Doppler frequency content within a range bin.

The phase samples at the output of the phase adder are

$$\hat{\phi}(r,m,n) = \phi(m,n) + \phi_{inc}(r,n) \tag{10.94}$$

In the adder, $\phi(m,n)$ and $\phi_{inc}(r,n)$ are summed using 2's complement. The phase increment values are double buffered as shown in Figure 10.73. Here $\phi_{inc}(r,n-1)$ is stored in the second register while $\phi_{inc}(r,n)$ is loaded into the first register. The double buffer consists of two k_p-bit registers to enhance the speed in going from one set of phase coefficients to the next. A different $\phi_{inc}(r,n)$ is applied to *each* range bin modulator *every PRI*, providing the flexibility to create multiple Doppler frequencies per range bin and also the capability to synthesize multiple targets.

10.10.5.6 Complex Signal Construction

The k_p-bit phase values at adder output $\hat{\phi}(r,m,n)$ are used by the LUT to index a sine and cosine value to construct the normalized I and Q amplitude components.

Although $\cos(\hat{\phi}(r,m,n))$ and $\sin(\hat{\phi}(r,m,n))$ could readily be calculated, doing so is less computationally efficient than the use of an LUT. The LUT output is

$$L(r,m,n) = e^{j(\phi(m,n)+\phi_{\text{inc}}(r,n))} \quad (10.95)$$

The complex signal from the LUT is applied to the gain block to modulate the RCS.

10.10.5.7 RCS Modulation

To modulate the RCS, the gain block multiplies the complex signal and the gain coefficient as

$$S(r,m,n) = 2^{g(r,n)} e^{j(\phi(m,n)+\phi_{\text{inc}}(r,n))} \quad (10.96)$$

Multiplication of the complex signal in both I and Q channels is performed by a $g(r,n)$-bit shift to the left using a parallel array of multiplexers. The shift operation is controlled by k_g-bits that are also double buffered as shown in Figure 10.73 and discussed next.

10.10.5.8 Gain Shift Operation

To perform the gain shift operation, Figure 10.83 shows two multiplexer arrays that allow a maximum shift of 3 bits. To control the gain, $k_g = 2$-bits are used (gain0, gain1).

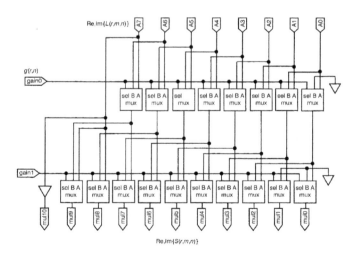

Figure 10.83 Fast convolution to compress the DIS output $I(m,n)$ with range compression first and cross-range compression for the output.

10.10 Programmable Digital Image Synthesizer

The first multiplexer row connects its "select" input to gain0 and the second row connects its "select" to gain 1. Table 10.1 shows the multiplication factor for each binary input. With a maximum shift to the left of 3 bits, the gain block output resolution is 11 bits. Note the gain output signal mul10 contains a single buffer since no multiplexer is used with this output. The size of the last adder determines the maximum number of modulators within the parallel array that can be summed without overload, assuming all range bins are generating a maximum amplitude signal. A 16-bit adder is used for 32 taps without the possibility of arithmetic overload. The dynamic range of the image is $DR = 20\log_{10}(2^{16}) = 96$ dB and is greater than or equal to that for most modern radar receivers. The 16-bit adder also allows separate image synthesizers to be daisy chained. Every multiple of 32 modulators requires an additional bit in the adder to guarantee against overflow. However, it would be highly unusual for all range bins to generate a maximum amplitude signal at the same time. Thus, when cascading multiple groups of 32 range bins, it might not actually be necessary to increase the number of bits in the adder. (More on the overflow analysis below.)

Table 10.1

Gain Multiplication Table

gain0	gain1	$g(r,n)$ (shift)	Multiplication factor
0	0	0	1
0	1	1	2
1	0	2	4
1	1	3	8

10.10.5.9 Output Pulse Summation Adder

The summation adder within each range bin modulator sums its results with the adjacent (delayed) summation adder output and sends the results forward to the summation adder in the next modulator. The complex output pulse is taken from the summation adder in the first range bin modulator. Thus, each output pulse is the superposition of N_r copies of the pulse, each delayed with respect to one another by the adder delay, scaled differently by the gains $2^{g(r,n)}$ and phase rotated by $\phi_{inc}(r,n)$ as

$$\begin{aligned} I(m,n) = & 2^{g(0,n)} e^{j\phi(m,n)+\phi_{inc}(0,n)} \\ & + z^{-1} \left[2^{g(1,n)} e^{j\phi(m,n)+\phi_{inc}(1,n)} \right] \\ & + z^{-2} \left[2^{g(2,n)} e^{j\phi(m,n)+\phi_{inc}(2,n)} \right] \\ & + \cdots + z^{-(N_r-1)} \left[2^{g(N_r-1,n)} e^{j\phi(m,n)+\phi_{inc}(N_r-1,n)} \right] \end{aligned} \quad (10.97)$$

where z^{-1} represents the (same) delay inherent in each adder. Rewriting, the sampled output pulses from the digital image synthesizer are

$$I(m,n) = \sum_{r=0}^{N_r-1} 2^{g(r,n)} e^{j(\phi(m-r,n) + \phi_{\text{inc}}(r,n))} \tag{10.98}$$

(indexing the first range bin at 0). In practice, the digital output pulses are then converted into an analog signal using a DAC. The analog signal is then upconverted onto a carrier frequency for retransmission back to the ISAR or other multi-static receiver.

10.10.6 ISAR Image Compression for Testing DIS Architecture

An ISAR that performs an image compression on the N_p return pulses from the synthesizer will interpret them as having come from a target with N_r range bins with spectral characteristics determined by the applied phase and gain coefficients. To evaluate the imaging capability of the synthesizer, an image compression algorithm based on fast correlation is presented below.

The compression algorithm correlates $I(m,n)$ with a reference function $s(m,n)$ and is convenient since it uses FFTs. Other techniques for image compression include complex spectral estimation algorithms, the Capon estimation algorithm, and 3D algorithms based on the use of a near-field focusing function. The ISAR range compression process using fast correlation is shown in Figure 10.84.

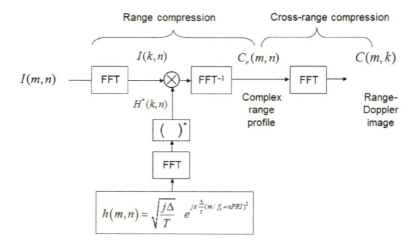

Figure 10.84 Fast convolution to compress the DIS output $I(m,n)$ with range compression first and cross-range compression for the output.

10.10 Programmable Digital Image Synthesizer

From (10.79) and (10.80), the reference function is

$$s(m,n) = \text{rect}\left(\frac{t}{\tau}\right) e^{j(\pi\Delta/\tau)(m/f_s + nPRI)^2} \tag{10.99}$$

and is a periodic discrete time signal with transfer function

$$S(k,n) = \frac{1}{N} \sum_{m=0}^{N-1} s(m,n) e^{-j\omega T_s km} \tag{10.100}$$

where N is the total number of samples within the pulse (including zero padding) and $T_s = 1/f_s$. Each pulse from the image synthesizer is also transformed into the frequency domain as

$$I(k,n) = F\{I^*(m,n)\} = \frac{1}{N} \sum_{m=0}^{N-1} I^*(m,n) e^{-j\omega T_s km} \tag{10.101}$$

After multiplying $I(k,n)$ by $S^*(k,n)$, the discrete complex range profile is computed as

$$\begin{aligned} C_r(m,n) &= F^{-1}\{I(k,n) H^*(k,n)\} \\ &= \frac{1}{N} \sum_{k=0}^{N-1} I(k,n) H^*(k,n) e^{j\omega T_s km} \end{aligned} \tag{10.102}$$

The ISAR range bin size is $dr = c/2\Delta$. Note that the ISAR range resolution should be larger than the image synthesizer range resolution in order that the image should not be detected as a false target.[3] The image synthesizer range resolution depends on the DRFM sampling frequency f_s and the delay between each range bin modulator.

The azimuth compression process is also shown in Figure 10.84. An azimuth profile is computed for each range bin m using the range bin samples from the N_p pulses. The range-Doppler image is formed by Fourier transforming the complex range samples and taking the absolute value as

$$C(m,k) = |F\{C_r(m,n)\}| = \left| \frac{1}{N} \sum_{n=0}^{N_p-1} C_r(m,n) e^{-j\omega T_s km} \right| \tag{10.103}$$

The azimuth resolution of the ISAR $da = \lambda/2\omega_r T_i$ where ω_r is the rotation rate of the object and T_i is the integration (frame) time $T_i = N_p PRI$. The azimuth resolution of the synthesizer depends on the bit resolution of the DRFM phase samples and $\phi_{inc}(r,n)$.

---

[3] In addition, although not addressed here, the coefficients for the correct *clutter distribution* should be added to any false target decoy.

10.10.7 Pipeline Architecture: Simulation Results

The pipeline architecture shown in Fig. 10.73 shows a range bin modulator r, a phase adder, LUT, gain shifter, and summation adder. Consider the phase adder $k_p = 5$ bits to match the phase values coming from the phase-sampling DRFM. The LUT is a composition of two building blocks, a 5-to-32-bit decoder and a read-only memory (ROM) array. The 5-to-32-bit decoder takes the 5-bit binary input from the phase adder and converts it into an address space for the ROM. Although the 2's complement addition of two 5-bit binary words could result in a 6-bit word, this fact can be ignored since the sine and cosine waveform repeat over a period of 2π. The output resolution of the LUT is 8 bits providing a reduced quantization error in the sin and cosine waveforms.

10.10.7.1 Constructing the Test Target

A ship is used as a synthesizer input example. A photo of the U.S.S. *Crockett* is shown in Figure 10.85(a) and an AN/APS-137 ISAR range Doppler image of the ship is shown in Figure 10.85(b) [64]. The presence of noise spikes is characteristic of the background sea clutter in these types of images. The naturally occurring relative movement between the target and the radar also generates in the image that is also known as *glint*. In terms of the physics, these spikes are due to the difference between individual measured distance (or intensity) and "actual" target range or error, and is known as *range glint*. Range glint is caused mostly by the constant changes in the transmitted signal and in case of the imaging radar, also caused by the varying intensities within each range bin. Also evident in the image is the ship superstructure and the two masts (on each side).

10.10 Programmable Digital Image Synthesizer

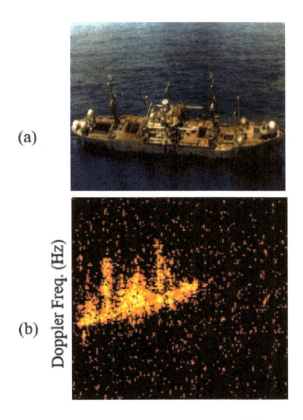

Figure 10.85 (a) Photo of the U.S.S. *Crockett* and (b) the AN/APS-137 ISAR range Doppler image of the ship [64].

For the ISAR to automatically determine the class of the target, the target must be extracted from the rest of the image using segmentation techniques to leave the image with only the target area illuminated. AI and deep learning have a significant role to play in this mission and many results have been reported.

An example test target is constructed with 32 range bins to closely match the AN/APS-137 ISAR range-Doppler image of the U.S.S. *Crockett* as shown in Figure 10.85(b) The file is shown in Figure 10.86. The test target shown in Figure 10.86 has 32 range bins with each range bin having its own unique set of scatterers giving it a separate Doppler profile. (The reader can imagine this as a ship in Doppler-range space.)

786 ELECTRONIC ATTACK USING DEEP LEARNING

```
freq=zeros(32,dp_pts);
freq(1,1:2)=    [15 49];
freq(2,1:3)=    [17 47 72];
freq(3,1:3)=    [   49 74 100];
freq(4,1:4)=    [   51 77 103 130];
freq(5,1:6)=    [       79 105 132 160 195 231];
freq(6,1:4)=    [       80 108 134 162];
freq(7,1:2)=    [       111 137];
freq(8,1:3)=    [       113 139 166];
freq(9,1:16)=   [           141 168 199 230 260 291 321 351 379 409 440 471 505 536 571 601];
freq(10,1:4)=   [           142 169 201 232];
freq(11,1:4)=   [           171 204 234 262];
freq(12,1:11)=  [           173 206 236 264 294 323 353 381 411 442 475];
freq(13,1:8)=   [               208 237 265 295 324 354 382 412];
freq(14,1:8)=   [               212 238 266 296 325 355 383 413];
freq(15,1:8)=   [                   241 267 297 326 356 384 414 445];
freq(16,1:10)=  [                   242 268 298 327 357 385 415 446 481 515];
freq(17,1:11)=  [                       269 299 328 358 386 416 447 482 516 546 582];
freq(18,1:9)=   [                       270 300 329 359 387 417 448 483 518];
freq(19,1:7)=   [                           301 329 360 389 418 449 484];
freq(20,1:6)=   [                           302 330 361 390 419 450];
freq(21,1:4)=   [                               331 362 391 420];
freq(22,1:3)=   [                               332 363 392];
freq(23,1:3)=   [                                   364 393 422];
freq(24,1:4)=   [                                   365 394 423 454];
freq(25,1:16)=  [                                       395 425 455 491 526 556 590 620 651 681 714 741 776 808 840 869];
freq(26,1:3)=   [                                       396 426 456];
freq(27,1:6)=   [                                           427 457 493 528 558 592];
freq(28,1:3)=   [                                           428 458 494];
freq(29,1:3)=   [                                               459 495 530];
freq(30,1:4)=   [                                               460 496 531 560];
freq(31,1:2)=   [                                                   497 532];
freq(32,1:2)=   [                                                   498 533];
```

Figure 10.86 Creating the test target with 32 range bins each having a separate Doppler profile.

In Figure 10.87, the simulation input file is shown in Figure 10.86 in range Doppler space.

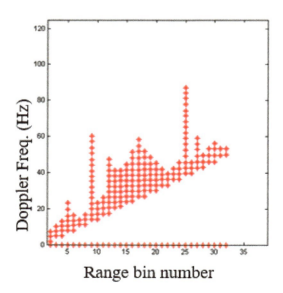

Figure 10.87 Simulation input file shown in Figure 10.86 in range Doppler space.

The image in Figure 10.87 was constructed to represent the ship in Figure 10.85(a) (the U.S.S. *Crockett*). The amplitude of the scatterers $A(r,d) = 1$ for all r,d. The presence of noise spikes is characteristic of the background clutter in these types of images.

Also evident in the image is the ship superstructure and the two masts on each side as in the original image.

Note that we are only concerned here with the synthesis of the test target. That is, no clutter is being considered. However, to be an actual false target, the presence of the correct clutter is necessary and those clutter coefficients can be added.

10.10.8 False Target in ISAR

To reconstruct the image, the ISAR compression was used with parameters that closely match those given in [61]. Figure 10.88(a) shows the reconstructed image of the ship when $N_p = 32$ pulses are integrated. The scatterers are correctly resolved with the Doppler frequencies corresponding to those specified above. Figure 10.88(b) shows the reconstructed image of the ship when $N_p = 64$ pulses are integrated. An increase in the number of pulses integrated provides a better focused image. In practice, a larger number of range bin modulators and more scatterers per range bin in the input target file can make the presence of a false target much harder to discern.

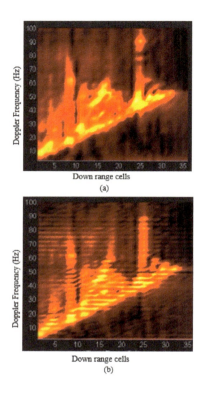

Figure 10.88 ISAR image with input as the test target (no clutter) using (a) 32 pulses and (b) 64 pulses integrated coherently by the ISAR.

10.10.9 Overflow is the Biggest Problem

Simulations indicate that the DIS can sustain quantization errors with very low image degradation. The gain modulation and adder overflow, however, have been identified as the main source of image degradation; overflow can be reduced by managing the gain quantization scheme. That is, most of image information is contained in the upper 8 MSBs.

10.10.9.1 16-Bit Summation Adder Overflow

The complex signal for each bin is added and the adder can sum in the range of −8192.00 to 8191.75 (two fraction bits and 14 integer bits). The overflows typically

10.10 Programmable Digital Image Synthesizer

will *change the sign* of the result! For a 512 complex range bin processor, the peak sum is 9936. For the 512 range bins, the number of adder overflows was 25,774!

Adder overflow management can be done in one of three ways:

(1) Set flag and ignore overflow;
(2) Sum may be capped at maximum (minimum) value in 16-bit summation adder;
(3) Set flag and let system controller reduce number of gain levels used.

Figure 10.89 investigates comparing the ideal image in (a) with lowering the number of quantization bits. Note the holes developing in the image as the number of quantization bits representing the image gets smaller. Figure 10.90 summarizes the results for setting the flag and then letting the system controller reduce the number of gain levels. This is a cognitive approach that also uses AI. Note the holes developing in the image as the number of quantization bits representing the image gets smaller.

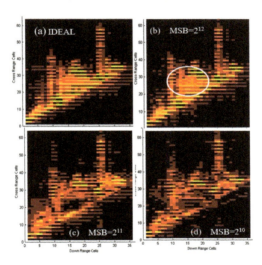

Figure 10.89 ISAR image of test target using (a) ideal (no quantization) (b) with 2^{12} as MSB, (c) with 2^{11} as MSB, and (d) with 2^{10} as MSB. Note the holes developing in the image as the number of quantization bits representing the image gets smaller.

Setting the flag and letting the AI system controller reduce the number of gain levels to 10 and 9 reduced the number of overflows as summarized in Figure 10.90. Note that when the gain levels were set to 10, 51 overflows occurred, reducing the correlation coefficient. However, when the number of gain levels were set to 9, overflows were zero, resulting in the highest correlation coefficient. As a final note, the final output signal level is determined by the DAC, the single side-band modulator, and the high-power amplifier in the transmit chain.

Gain normalized to 10 levels (2^0 to 2^{-9} : 1 to 512)		Gain normalized to 9 levels (2^0 to 2^{-8} : 1 to 256)	
• Overflows	51	• Overflows	0
• Peak	8373	• Peak	4332
• Noise Avg	3.4432e-004	• Noise Avg	3.4315e-004
• Corel Coeff	0.9898	• Corel Coeff	0.9898

Figure 10.90 Set the flag and let the system controller reduce number of gain levels.

The floor plan of a single complex range bin modulator is shown in Figure 10.91(a) and a micrograph of the entire 8-range bin prototype is shown in Figure 10.91(b) [62] indicating that it is working correctly.

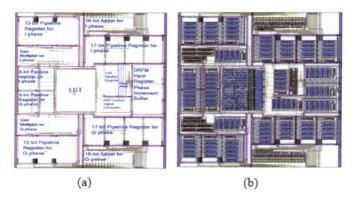

(a) (b)

Figure 10.91 DIS showing (a) the floor plan and (b) the micrograph of a single complex range bin processor for the DIS. (©IEEE, reprinted with permission from [62].)

Figure 10.92 shows the fabricated CMOS ASIC test results for the 8-range bin DIS [62].

10.11 AI and DRFM DIS CONOPS 791

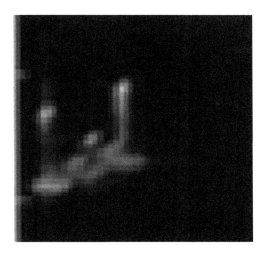

Figure 10.92 Fabricated CMOS ASIC test results for the 8-range bin DIS. (©IEEE, reprinted with permission from [62].)

10.11 AI AND DRFM DIS CONOPS

Use of the ASIC (or FPGA) for false targeting requires not only the embedded technology but also AI within the CONOPS. Also, for use against a synthetic aperture, where the movement is being created by the platform, to create the correct orientation and kinetics, the DRFM must know that position of the synthetic platform for each intercepted pulse as well as the geodetic position of its own platform.

10.11.1 Experimental Determination of False Target: Phase and Gain Coefficients

To generate realistic coefficients (not the test target), a series of high-fidelity EM simulations can be used, such as CST or FEKO. One of the major expenditures for governments around the world is to fund their national defense laboratories to build highly accurate simulations of adversary systems as well as their own for modeling and simulation purposes. For example, to generate the coefficients for the DIS ASIC, a high-fidelity simulation of one of their ISAR systems is used to illuminate a highly accurate model of an adversary ship as shown in Figure 10.93. As the waveform is reflected off the surface of the ship, the complex waveform is collected (either bistatically or monostatically) and sampled in series and the gain and phase increment coefficients can then be derived for any movement of the ship target at all (Figure 10.93).

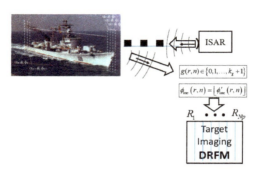

Figure 10.93 Experimentally deriving the gain $g(r, n)$ and phase $\phi_{inc}(r, n)$ for each range bin to image a ship using either a high-fidelity simulation model of a ship or an actual ship.

10.12 DERIVING THE SEA CLUTTER COEFFICIENTS

To consider the required clutter return along with the false target, the geometry in Figure 10.94 shows the geometry for an airborne ISAR with a height of h. The clutter model presented below was developed initially at the Naval Research Laboratory.

10.12 Deriving the Sea Clutter Coefficients

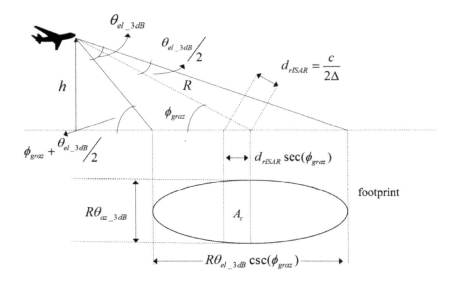

Figure 10.94 Geometry of airborne platform with ISAR at a height h m. (adapted from [65].)

From an airborne radar at height h above the ground, the grazing angle ϕ_{graz} at a given range R can be expressed as [66]

$$\phi_{graz} = \sin^{-1}\left(\frac{h}{R} + \frac{h^2}{2r_eR} - \frac{R}{2r_e}\right) \quad (10.104)$$

where r_e is the effective Earth radius. The grazing angle ϕ_{graz} as a function of the range R at a constant h is shown in Figure 10.95 The area of the illuminated patch A_c can be approximated as a function of the grazing angle as

$$A_c = R\theta_{az\_3dB}d_{rISAR}\sec(\phi_{graz}) \quad (10.105)$$

where d_{rISAR} is the slant range resolution and θ_{az_3dB} is the ISAR azimuth 3-dB beamwidth.

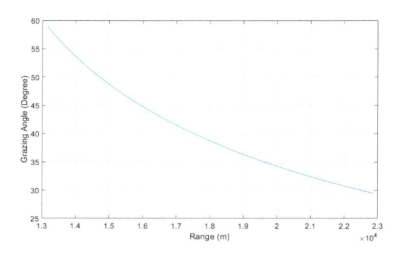

Figure 10.95 Grazing angle as a function of range.

10.12.1 NRL Normalized RCS Model

The Naval Research Laboratory RCS model describes an empirical model of the sea clutter that is in close agreement with the measurements taken by Nathanson [67]. The radar range equation is used to calculate the mean clutter return power P_{RC}^0 and is expressed as

$$P_{RC}^0 = \frac{P_T G_t G_r \lambda^2 \sigma^0 A_c}{(4\pi)^3 R^4 L} \quad (10.106)$$

where σ^0 is the normalized mean sea clutter backscatter coefficient, which can be obtained using the NRL model. As well, G_t and G_r are the transmit and receive gain, respectively, and P_T is the peak power of the pulsed-ISAR waveform transmitted, L represents the losses, and λ is the carrier wavelength.

The NRL *empirical relationship* describes a model that is in close agreement with the behavior of sea clutter during various sea states and is expressed as a function of five free parameters as

$$\sigma_{HH,VV}^0 = c_1 + c_2 \log_{10} \sin(\phi_{graz}) + \frac{(2.75 + c_3 \phi_{graz}) \log_{10} f_{GHz}}{(1 + 0.95 \phi_{graz})} \\ + c_4(1+SS)^{\frac{1}{2+0.085\phi_{graz}+0.0335S}} + c_5 \phi_{graz}^2 \quad (10.107)$$

where $\sigma_{HH,VV}^0$ is the normalized copolarized RCS and SS is the Douglas sea state. The carrier frequency is f_{GHz} in GHz, and c_1, c_2, c_3, c_4, and c_5 are free parameters that are

10.12 Deriving the Sea Clutter Coefficients

adjusted to minimize the average absolute deviation between the empirical equation and a set of mean sea backscatter data collected by Nathanson [67]. The normalized RCS (dBsm) as a function of range is shown in Figure 10.96 [68].

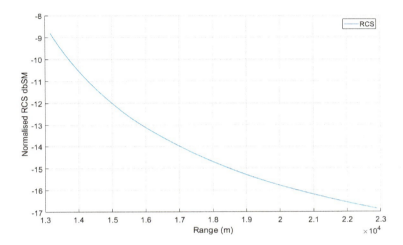

Figure 10.96 Normalized RCS (dBsm) as a function of range.

The Douglas sea state describes the roughness of the sea wave surfaces and is shown in Table 10.2. The free parameter values for the NRL model are shown in Table 10.3.

Table 10.2
Summary of Sea State Conditions

Sea State	Description	Wave height (ft)	Wind speed (kn)	Fetch (nmi)	Duration (h)
1	Smooth	0 - 1	0 - 6	–	–
2	Slight	1 - 3	6 - 12	50	5
3	Moderate	3 - 5	12 - 15	120	20
4	Rough	5 - 8	15 - 20	150	23
5	Very rough	8 - 12	20 - 25	200	25
6	High	12 - 20	25 - 30	300	27
7	Very high	20 - 40	30 - 50	500	30

From: [67, 69].

Table 10.3
Summary of NRL Model Free Parameters

Constants	Horizontal pol.	Vertical pol.
c_1	-73.00	-50.79
c_2	20.78	25.93
c_3	7.351	0.7093
c_4	25.65	21.58
c_5	0.00540	0.00211

From: [69].

10.12.2 Clutter Parameter Comparison

The inputs for the NRL RCS model are grazing angle, RF polarization, carrier frequency, and sea state condition. For this text, the RF polarization and the carrier frequency are fixed to horizontal polarization and 10 GHz, which are normally the specifications for an ISAR radar such as the APS-137 ISAR. In this section, the effect of the grazing angle and sea state on the normalized RCS are examined [68].

10.12.2.1 Grazing Angle

Assuming that the height of the aircraft h remains constant, we have shown that the grazing angle ϕ_{graz} decreases as the slant range R increases (see Figures 10.95 and 10.96). The RCS of a clutter patch σ_c is the product of A_c and σ^0 or $\sigma_c = A_c \sigma^0$. The relationship between σ_c and R, assuming h is constant, is displayed in Figure 10.97. The illuminated area increases significantly less and results in an overall decrease in the clutter patch σ_c.

10.12 Deriving the Sea Clutter Coefficients

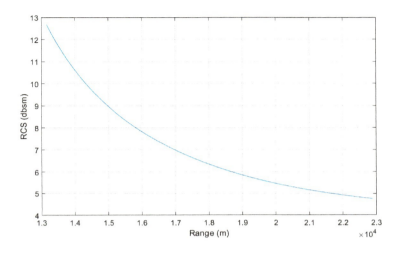

Figure 10.97 RCS σ_c as a function of range for $SS = 2$.

10.12.2.2 Effect of Sea State on σ^0

The effect the sea state SS has on σ^0 is displayed in Figure 10.98, which shows the normalized RCS as a function of R for sea state 2 through 7. An increase in SS leads to an increase in σ^0 as well. Based on (10.106), the increase in σ^0 consequently increases the return P_{RC}^0 as well and is shown in Figure 10.99 as a function of range.

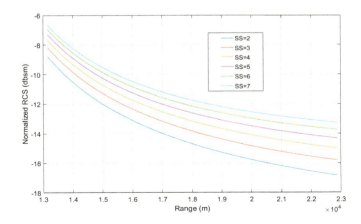

Figure 10.98 RCS σ_c as a function of range.

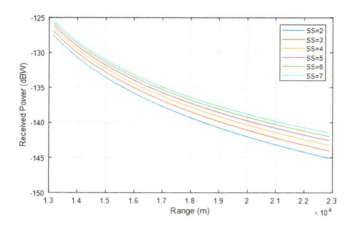

Figure 10.99 Received power P_{RC}^0 function of range.

10.12.2.3 Power Fluctuation via Compound KA Distribution

The sea clutter simulation model developed by Brooks is comprised of a slow-moving component called texture that represents the local sea structures such as Bragg scatterers and distributed whitecaps [70, 71]. It also includes a fast-moving component called speckle that represents structures such as sea spray, discrete whitecaps, and bursts [66]. The texture is modeled using a gamma distribution, and the speckle follows an exponential distribution. The final expression for the fluctuating sea clutter power is expressed as

$$P(z|x) = \left(\sum_{n=0}^{\infty} \frac{1}{x(1+n\rho)} \exp\left(\frac{-z}{x(1+n\rho)} \right) \right) P_{poiss}(n) \qquad (10.108)$$

where x is the texture, z is the clutter power, n is the number of instantaneous bursts found in a range cell [71, 72], and ρ is the ratio of the burst power to the Bragg power. A Poisson distribution is used to describe the occurrence of n, which is approximated by

$$P_{poiss}(n) = \left\{ \begin{array}{ll} 1-\bar{N}, & n=0 \\ \bar{N}, & n=1 \\ 0, & n \geq 2 \end{array} \right\} \qquad (10.109)$$

where $\bar{N}$ is the probability of a spike.

10.12.2.4 Modeling of Fluctuating Sea Clutter Spectra

The model for the Doppler spectrum for a single range cell is [70]

$$G(v,x,s) = \frac{x}{\sqrt{2\pi}s} \exp\left(\frac{-(v-m_f(x))^2}{2s^2}\right) \qquad (10.110)$$

where v is the Doppler frequency, $m_f(x)$ is the mean Doppler frequency as a function of texture x, and s is the standard deviation. The mean frequency can be expressed as

$$m_f(x) = \left(\alpha + \beta \frac{x}{P_{RC}^0}\right)\cos\theta_w + f_D \qquad (10.111)$$

where α and β are expressed as [70, 73, 74]

$$11\beta = \alpha + \beta = \frac{2}{\lambda}(0.25 + kU) \qquad (10.112)$$

where $k = 0.25$ for horizontal polarization and $k = 0.18$ for vertical polarization [70, 73]. The Doppler shift introduced by the motion of the transmitting platform is f_D and is set to zero for the relative motion to the clutter as is the case for a stationary ISAR radar. The wind velocity U is approximated as [70]

$$U = 3.16\left(SS^{0.8}\right) \qquad (10.113)$$

The standard deviation of the clutter spectrum s follows a Gaussian distribution as

$$p(s) = \frac{\alpha}{\sqrt{2\pi}s} \exp\left(\frac{-(s-m_s)^2}{2\sigma_s^2}\right) \qquad (10.114)$$

where $\sigma_s \approx 20$ Hz, and the mean m_s is given by

$$m_s = 0.2\frac{U\cos\theta_w}{\lambda} \qquad (10.115)$$

where θ_w is the headwind direction.

10.12.2.5 Generating Random Sea Clutter Power and Doppler Spectrum

The fluctuating sea clutter power spectral density is generated when the mean clutter power (10.106) is used in distribution (10.108), (10.110). The mean clutter power and the fluctuating power for each range bin within the radar main beam is shown in Figure

10.100. The power spectral density for $SS = 2$ is shown in Figure 10.101, and together they provide the information on phase and amplitude that can be used to create the phase coefficient and gain coefficient for the sea clutter profile to augment the false target coefficients.

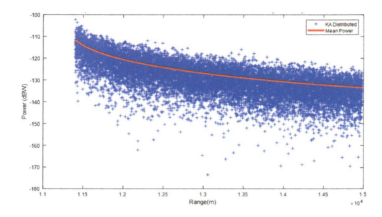

Figure 10.100 Mean sea clutter power and fluctuation.

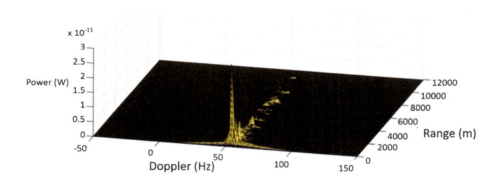

Figure 10.101 Power spectral density for sea state 2.

10.13 PHASE AND GAIN COEFFICIENTS FOR SEA CLUTTER

The simulation of sea clutter returns to an ISAR radar are discussed in this section. The first step is the extraction of the phase and the amplitude information from the sea clutter spectrum. Depending on the level of the sea state, these values can change dramatically. After the complex information is extracted, the phase and gain coefficients can be derived. Table 10.4 shows the radar parameters used to simulate the sea clutter return. Information on the operating environment is also included. The radar parameters used are similar to that of the AN/APS-137 ISAR radar [64].

Table 10.4
ISAR Operating Parameters and Environment

Parameter	Values	Parameter	Values
Frequency	10 GHz	Elevation Beamwidth	4.5°
Range Resolution	0.3 m	Azimuth Beamwidth	1.05°
Transmit Gain	32 dB	PRF	200 Hz
Receive Gain	32 dB	Bandwidth	500 MHz
Loss	3 dB	Pulse integration	128
Power	500 W	Polarization	Horizontal
Operating Environment			
Range	3,000 m	Radar height	8,000 ft
Grazing Angle	54.4°	Head wind direction	20°
Sea state	2		

From: [68]

The main beam power spectrum is shown in Figure 10.102. The target is constructed to span 32 range bins and is located in the center of the radar beam. The spectra at these range bins are extracted and enlarged in Figure 10.103.

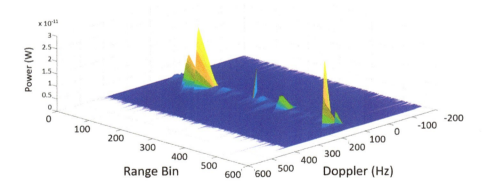

Figure 10.102 Main beam power density spectrum, clutter only [68].

802 ELECTRONIC ATTACK USING DEEP LEARNING

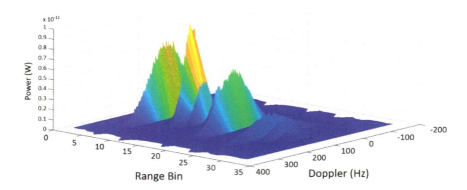

Figure 10.103 Main beam power density spectrum, target, and clutter [68].

The power spectral density diagrams for range bins 7, 13, 22, and 32 are shown in Figure 10.104. The power spectra have different magnitudes and different bandwidths. The spikiness is also notable and reflects the complex probabilistic nature of the NRL sea clutter model. Note that the false target can be placed at any point along the line-of-sight.

10.13 Phase and Gain Coefficients for Sea Clutter

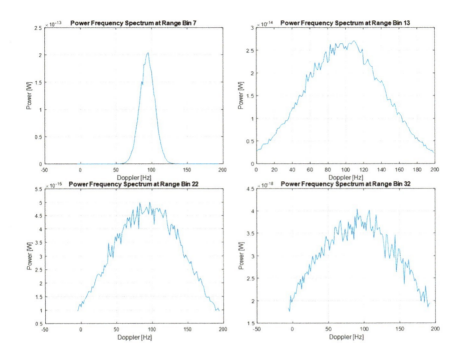

Figure 10.104 Power spectral density diagrams for range bins 7, 13, 22, and 32 [68].

The power and Doppler components of the spectrum for each range bin are saved and exported to disc1.m where they are used to reconstruct the sea clutter return. The clutter return is complex and can be represented as a function of the transmitted pulse index and coordinate location on the range-Doppler map. It can also be consolidated at each range bin for each pulse by

$$ClutterSum = \sum_{d=1}^{N_d} Clutter(r,d,n)$$

The clutter return is added to the target return to form a new complex signal $T''(r,n)$, which is then used to generate the phase coefficient and gain coefficient. The new complex signal is expressed as

$$T''(r,n) = T'(r,n) + ClutterSum(r,n) \qquad (10.116)$$

That is, 10.116 now contains the target and the clutter for this particular frame time set of coefficients. Note that usually the target movement will be streamed from the dual-ported memory so the use of AI for dynamic handling of the clutter distribution model is necessary.

This is, due to the constantly changing weather and sea conditions that necessitates the use of measurement sensors to determine the *refractive index* of the atmosphere (modified index of refraction) as a function of altitude. These inversions as shown in Figure 10.105 can set up different trapping layers affecting the RF propagation. In addition, sonobouys are used to determine the correct sea state. All of these variables are then used by the AI deep learning to correctly choose the proper coefficients for the clutter model.

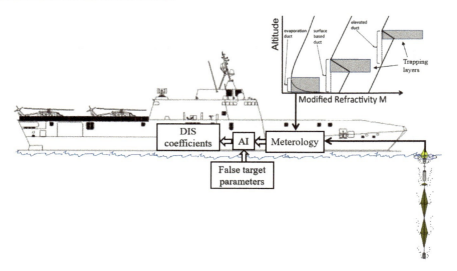

Figure 10.105 Drawing of an littoral combat ship with meteorology sensors to determine the modified refractivity profile (*M*) for setting up RF trapping layers in the evaporation duct, surface-based duct, and an elevated duct and sonobouys to determine the current sea state.

10.13.1 Phase Coefficient for Sea Clutter

The phase angle of $T''(r,n)$ is derived using the real and imaginary components as

$$\phi'_T(r,n) = \angle \left(\frac{imag\{T''(r,n)\}}{real\{T''(r,n)\}} \right) = \angle \left(\frac{imag\{T'(r,n) + ClutterSum(r,n)\}}{real\{T'(r,n) + ClutterSum(r,n)\}} \right)$$
(10.117)

In a similar manner, the phase increment coefficient is generated as

$$\phi''_{inc}(r,n) = \phi''_{inc}(r,n-1) + \phi'_T(r,n-1) - \phi'_T(r,n)$$
(10.118)

If a bit-level simulation is being developed, then the phase increment is then quantized (e.g., 5 bits).

10.13.2 Gain Coefficient for Sea Clutter

The gain coefficient for generating the sea clutter is derived using

$$T_N''(r,n) = \frac{|T''(r,n)|}{\max(|T''(r,n)|)} = \frac{|T'(r,n) + ClutterSum(r,n)|}{\max(|T'(r,n) + ClutterSum(r,n)|)} \quad (10.119)$$

Using (10.118) and (10.119) the phase and gain coefficients are used to modulate the phase samples of the ISAR LFM signal. The output from the DIS is returned to the DRFM and subsequently transmitted back to the ISAR. The resultant range-Doppler images are discussed in the following section.

To demonstrate the addition of clutter, Figure 10.106(a) shows the test target without any additional clutter added. In Figure 10.106(b), the addition of sea clutter $SS = 3$ is added to the coefficients of the test target. In addition, the Doppler coefficients are inverted. That is, the ship is rolling away from the emitter.

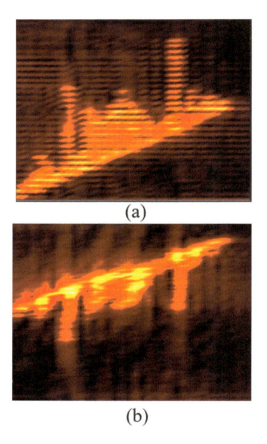

Figure 10.106 Demonstration of the addition of clutter coefficients to the test target with (a) no clutter and (b) clutter $SS = 2$ added to the test target. In addition, the ship is rolling away from the emitter (negative Doppler frequencies).

It is noticeable that the artificial blind effect is filled in, giving more of an appearance of a true target.

10.13.3 Summary

The creation of a sea clutter target profile for the DIS was also studied in this thesis. An existing sea clutter simulation model for the surface platform was adapted and modified to generate sea clutter for an airborne ISAR platform. The modified sea clutter model retained the ability in generating sea clutter at different sea states, waveform polarizations, and wind heading angle while using the normalized RCS model developed by NRL for a high grazing angle up to 60 degrees.

Using a random probabilistic model, we generated the sea clutter Doppler power spectrum to create the phase and gain coefficients for the sea clutter in the DIS. Since the sea clutter had a high Doppler resolution, the sea clutter maintained a convincing appearance as the ISAR increased its Doppler resolution. In addition to the phase angle, the amplitude of the complex vector is another output from the phase converter, but it was not used in the DIS design. This piece of information is required to reconstruct the proper DRFM output power. Since the amplitude is also scaled by a factor of 1.647, the CORDIC algorithm shown in Figure 10.75 must be modified in order to have this scale factor taken into account.

We evaluated the fidelity and image creation process of the DIS using a test pattern image. Further, insight into the bit-resolution influence on image quality was examined. Suggested future effort is to investigate a realistic image formation process. Going forward, validation of the sea clutter model by collecting data in the Monterey Bay using a range-Doppler radar would aid in the creation of a realistic false target. The banding gaps described in this research were due to the finite resolution of the false target used to extract the imaging coefficients. A realistic false target can be generated by illuminating and collecting the ISAR waveform from a detailed target model using EM simulation software like CST Microwave Studio.

10.14 CONCLUDING REMARKS

Since the first publications detailing the architecture for synthesizing multiple structured false targets against range-Doppler imaging RF sensors [61, 62], a flurry of activity followed. For example, to improve the ISAR countertargeting capability, a kinetic model is used to synthesize a series of real-time images of a ballistic target is presented in [75]. Then in [76], an approach is proposed to evaluate the efficiency of the false target decoys quantitatively according to aiming error, EA bandwidth, EA pulse duration error, and PRF error. In [77], to improve the efficiency of the jamming signal generation, a jamming signal generation method based on the ISAR echo model is proposed. According to the jammer's position as well as the false targets' expected position, the method calculates the false target's minimum resolution cell's approximate delay time and its accurate phase match and then forms the false target pattern. While receiving the ISAR signal, the DRFM can then generate the EA signal based on the jamming pattern and the ISAR signal in a much faster manner.

As an alternate approach, an *invchirp* method is proposed in [78] for false target generation. The algorithm only requires one complex multiplier in the FPGA to generate the EA signal when the ISAR signal is intercepted, which can be obtained by multiplication of radar signal samples and the equivalent dechirped target echo in the time domain. As the complex synthesis of the dechirped target echo can be realized by a DSP within the interpulse time, the overall cost is reduced and the false-target size and number of false targets is not limited by the size of the FPGA. Other published results are given more recently in [79, 80].

APPENDIX 10A: OBSCURATION WAVEFORMS

10A.1 CONTINUOUS SPOT NOISE (SPT)

SPT is generated by frequency modulating a CW carrier to create a Gaussian noise signal. SPT must be preset to the exact RF prior to operating. SPT denies range and range-rate information to the radar being exercised.

10A.2 INTERMITTENT SPOT NOISE (NCDB)

Noise count down blink (NCDB) is generated in the same way as SPT except that a pin switch is used to modulate the RF output. NCDB forces the victim radar to switch back and forth between passive angle track and velocity or range track.

10A.3 NOISE SWEPT AMPLITUDE MODULATION (NSAM)

NSAM is generated by amplitude modulating the spot noise at a frequency that is linearly varied in a sawtooth fashion between preset frequency limits while the duty factor is held constant. The frequency limits of NSAM are set to cover the expected lobing rate of the victim radar. Each time the NSAM frequency corresponds to the lobing frequency or angle processing rate of the radar, errors are generated in the radar's angle-tracking circuitry. NSAM, therefore, degrades or breaks the angle track of radars, which have specific lobing frequencies or angle-processing rates. It will not degrade angle track monopulse-type radars.

10A.4 CONTINUOUS BARRAGE NOISE (BAR)

BAR is generated by noise modulating a voltage controlled oscillator (VCO) to provide a broadband RF noise signal. Because of its broadband RF spectrum, it is not necessary to know the exact frequency of the victim radar, and it is possible to jam more than one radar simultaneously. As with SPT, BAR denies range and range-rate information to the radar.

10A.5 INTERMITTENT BARRAGE NOISE (BCDB)

Barrage count down blink (BCDB) is generated in the same way as BAR except that a pin switch is used to modulate the barrage noise off and on. As with NCDB, BCDB forces the victim radar to switch back and forth between passive angle track and velocity or range track.

10A.6 BARRAGE NOISE SWEPT AMPLITUDE MODULATION (BSAM)

BSAM is generated by amplitude modulating the barrage noise signal in exactly the same manner as NSAM. As with NSAM, BSAM degrades or breaks the angle track of nonmonopulse-type radars.

10A.7 SWEPT NOISE (SWPT)

SWPT is generated by sweeping the RF signal from a VCO across preset limits in a sawtooth fashion. SWPT can affect a radar's automatic gain control (AGC) circuit and thus reduce information to its tracking circuits.

APPENDIX 10B: TARGET DECEPTION REPEATER ALGORITHMS

10B.1 MULTIPLE FREQUENCY REPEATER (MFR)

MFR employs an amplitude-modulated, coherent repeater that produces a number of equally spaced signal frequencies each with greater amplitude than that of the target return. MFR introduces errors in the range and range-rate computations of pulse Doppler (PD) radars or introduces false targets into a PD radar while it is in the search mode. MFR can also affect a radar's AGC operation and thus degrade its track.

10B.2 REPEATER SWEPT AMPLITUDE MODULATION (RSAM)

RSAM is generated in exactly the same manner as NSAM except that the repeated radar signal is modulated instead of the spot or barrage noise signal. As with NSAM and BSAM, RSAM degrades or breaks the angle track of nonmonopulse radars.

10B.3 VELOCITY GATE STEALER (VGS)

VGS is generated by serrodyning a CW, traveling-wave-tube-amplifier (TWTA) repeater to produce slow-changing, false Doppler frequencies. VGS deception is employed against Doppler systems that use a speed gate or velocity tracker. VGS pulls the velocity tracker off the target return and drops it. The radar may then lock onto clutter or be forced into a reacquisition sequence.

10B.4 COMBINATION OF VGS AND RSAM (VSG RSAM)

VGS RSAM is generated by combining the VGS and RSAM programs discussed above. First, VGS is produced by itself and then, during the latter portion of the VGS program, RSAM is applied. VGS degrades or breaks angle track of nonmonopulse type radars. With the velocity tracker pulled off, the angle jamming does not have to compete with the target return. Therefore, VGS is typically more effective than RSAM alone.

10B.5 CHIRP GATE STEALER (CGS)

CGS is generated in exactly the same manner as VGS except that the maximum deviation of CGS is approximately 20 times that of VGS. The CGS false signal interacts with the intrapulse frequency modulation of radars using a pulse compression mode. The interaction is then translated from a frequency change into a time change by the dispersive delay line in the radar. Thus the radar interprets the periodic frequency changes of the CGS as periodic range changes. Therefore, CGS pulls the range tracker off the true target, and then drops it.

10B.6 NARROWBAND REPEATER NOISE (NBRN)

NBRN is generated by serrodyning a CW, TWTA repeater with a rapidly swept sawtooth waveform. The sawtooth frequency is swept rapidly from about 50 Hz to a maximum that is adjustable from 1 to 30 kHz. While the frequency is being swept, the slope of the sawtooth is alternately switched between positive and negative. This technique cause false signals to appear both above and below that of the the target return frequency. Amplitude modulation is also used with the serrodyning to add more lines to the RF spectrum. The result is a relatively even distribution of noise-like power over the selected bandwidth that masks the target return. In a Doppler radar NBRN can severely degrade the target tracking and may even force the radar into passive angle track.

10B.7 RANDOM DOPPLER (RD)

RD is generated by serrodyning a CW, TWTA repeater with a sawtooth waveform that is randomly varied in frequency. Each of the sawtooth frequencies is held for a period of 20 milliseconds. The sawtooth frequency typically deviates from a minimum of 50 Hz to a maximum that is adjustable from 1 to 50 kHz. The output of the TWTA then contains, in addition to the true target return signal, false signals that are greater in amplitude than that of the target return and are changing in frequency randomly about the target return. This technique introduces false Doppler targets and can cause confusion during the search and acquisition sequence of Doppler radars.

REFERENCES

[1] I. Burton and J. Straub, "Autonomous distributed EW system of systems," in *2019 14th Annual Conference System of Systems Engineering (SoSE)*, 2019, pp. 96–101.

[2] F. Arvin, T. Krajník, A. E. Turgut, and S. Yue, "Cosϕ: Artificial pheromone system for robotic swarms research," in *2015 IEEE/RSJ International Conference on Intelligent Robots and Systems (IROS)*, 2015, pp. 407–412.

REFERENCES

[3] F. Chollet, *Deep Learning with Python*. Shelter Island, NY: Manning Publications Co., 2018.

[4] C. Wang, J. Wang, Y. Shen, and X. Zhang, "Autonomous navigation of uavs in large-scale complex environments: A deep reinforcement learning approach," *IEEE Transactions on Vehicular Technology*, vol. 68, no. 3, pp. 2124–2136, 2019.

[5] C. Wang, J. Wang, X. Zhang, and X. Zhang, "Autonomous navigation of UAV in large-scale unknown complex environment with deep reinforcement learning," in *2017 IEEE Global Conference on Signal and Information Processing (GlobalSIP)*, 2017, pp. 858–862.

[6] R. Sutton, D. McAllester, S. Singh, Y. Mansour, "Policy gradient methods for reinforcement learning with function approximation," in *Neural Information Processing Systems (NIPS)*, 1999, pp. 1057–1063.

[7] D. Silver, G. Lever, N. Heess, T. Degris, D. Wierstra, and M. Riedmiller, "Deterministic policy gradient algorithms," in *Proceedings of the 31st International Conference on Machine Learning*, ser. Proceedings of Machine Learning Research, E. P. Xing and T. Jebara, Eds., vol. 32, no. 1. Bejing, China: PMLR, 22–24 Jun 2014, pp. 387–395. [Online]. Available: http://proceedings.mlr.press/v32/silver14.html

[8] O. Bouhamed, H. Ghazzai, H. Besbes, and Y. Massoud, "Autonomous UAV navigation: A ddpg-based deep reinforcement learning approach," in *2020 IEEE International Symposium on Circuits and Systems (ISCAS)*, 2020, pp. 1–5.

[9] N. Heess, J. J. Hunt, T. P. Lillicrap, and D. Silver, "Memory-based control with recurrent neural networks," 2015.

[10] T. P. Lillicrap, J. J. Hunt, A. Pritzel, N. Heess, T. Erez, Y. Tassa, D. Silver, and D. Wierstra, "Continuous control with deep reinforcement learning," 2019.

[11] Z. Gongsheng, P. Chunmei, D. Jiang, and S. Junfeng, "Deep deterministic policy gradient algorithm based lateral and longitudinal control for autonomous driving," in *2020 5th International Conference on Mechanical, Control and Computer Engineering (ICMCCE)*, 2020, pp. 740–745.

[12] S. P. P. da Silva, P. Honório Filho, L. B. Marinho, J. S. Almeida, N. M. M. Nascimento, A. W. d. O. Rodrigues, and P. P. R. Filho, "A new approach to navigation of unmanned aerial vehicle using deep transfer learning," in *2019 8th Brazilian Conference on Intelligent Systems (BRACIS)*, 2019, pp. 222–227.

[13] United States Government Accountability Office, "Electromagnetic Spectrum Operations, GAO-21-64," Dec 2020.

[14] D. Rocha and M. H. Houchin, "The development of the JEMSO doctrine," *The Journal of Electromagnetic Dominance*, vol. 43, no. 11, pp. 44 – 50, 2020.

[15] Joint Chiefs of Staff, "Joint Electromagnetic Spectrum Operations, JP 3-85," May 2020.

[16] D. L. Adamy, *EW 103: Tactical Battlefield Communications Electronic Warfare 1st Edition*. Norwood, MA: Artech House, 2008.

[17] N. Friedrich, "AI and machine learning redefine the EW landscape," *Microwave Journal*, vol. 295, no. 3, pp. 344 – 353, 1990.

[18] R. K. Ackerman, "AI Looms Large in Race for Global Superiority," in *Signal*, no. 9, 2020.

[19] N. Friedrich, "AI and Machine Learning Redefine the EW Landscape," in *Microwave Journal*, no. 12, 2020.

[20] P. E. Pace, *Detecting and Classifying Low Probability of Intercept Radar*. Norwood, MA: Artech House, 2009.

[21] J. Keller, "Raytheon to build RIM-174 air-defense missiles to protect shipboard forces from planes and missiles," *Military and Aerospace Electronics*, vol. 29, no. 10, 2018.

[22] R. Scott, "Block by block – surface navy EW evolves under SEWIP," *Journal of Electromagnetic Dominance*, vol. 43, no. 1, pp. 12–19, 2020.

[23] P. Sharma, K. K. Sarma, and N. E. Mastorakis, "Artificial intelligence aided EW systems- recent trends and evolving applications," *IEEE Access*, vol. 8, pp. 224 761–224 780, 2020.

[24] S. D. Berger, "Digital radio frequency memory linear range gate stealer spectrum," *IEEE Transactions on Aerospace and Electronic Systems*, vol. 39, no. 2, pp. 725–735, 2003.

[25] A. Moreira, P. Prats-Iraola, M. Younis, G. Krieger, I. Hajnsek, and K. P. Papathanassiou, "A tutorial on synthetic aperture radar," *IEEE Geoscience and Remote Sensing Magazine*, vol. 1, no. 1, pp. 6–43, 2013.

[26] Z. A. Wagner, D. A. Garren, and P. E. Pace, "Sar imagery via frequency shift keying costas coding," in *2017 IEEE Radar Conference (RadarConf)*, 2017, pp. 1789–1792.

[27] D. A. Garren, P. E. Pace, and R. A. Romero, "Use of p3-coded transmission waveforms to generate synthetic aperture radar images," in *2014 IEEE Radar Conference*, 2014, pp. 0765–0768.

[28] N. Ricardi, A. Aprile, and F. Dell'Acqua, "A novel technique for feature-based aircraft identification from high resolution airborne isar images," in *2012 IEEE International Geoscience and Remote Sensing Symposium*, 2012, pp. 2082–2085.

[29] A. Karakassiliotis, G. Boultadakis, G. Kalognomos, B. Massinas, and P. Frangos, "Advanced signal processing techniques for inverse synthetic aperture radar (ISAR) imaging," in *International Technical Laser Workshop on SLR Tracking of GNSS Constellations*, 2009.

[30] D. Fouts, P. Pace, C. Karow, and S. Ekestorm, "A single-chip false target radar image generator for countering wideband imaging radars," *IEEE Journal of Solid-State Circuits*, vol. 37, no. 6, pp. 751–759, 2002.

[31] D. L. Adamy, *EW 104: Electronic Warfare Against a New Generation of Threats*. Norwood, MA: Artech House, 2015.

[32] G. Du, L. Wang, Q. Liao, and H. Hu, "Deep neural network based cell sleeping control and beamforming optimization in cloud-ran," in *2019 IEEE 90th Vehicular Technology Conference (VTC2019-Fall)*, 2019, pp. 1–5.

[33] J. Yuan, Y. Bu, S. Yang, and Q. Chi, "Refined recognition and intelligent smart interference of radar signal," in *2019 IEEE International Conference on Unmanned Systems (ICUS)*, 2019, pp. 616–622.

[34] R. Tang, Z. Zhuo, C. Zhang, and L. Li, "The applications of artificial intelligence in situation assessment and game countermeasure during unmanned air combat," in *2019 IEEE International Conference on Unmanned Systems (ICUS)*, 2019, pp. 909–913.

[35] I. Goodfellow, J. Pouget-Abadie, M. Mirza, B. Xu, D. Warde-Farley, S. Ozair, A. Courville, and Y. Bengio, "Generative adversarial networks," *Commun. ACM*, vol. 63, no. 11, p. 139–144, Oct. 2020. [Online]. Available: https://doi.org/10.1145/3422622

[36] C. Zheng, X. Jiang, and X. Liu, "Semi-supervised SAR ATR via multi-discriminator generative adversarial network," *IEEE Sensors Journal*, vol. 19, no. 17, pp. 7525–7533, 2019.

[37] K. Bu, Y. He, X. Jing, and J. Han, "Adversarial transfer learning for deep learning based automatic modulation classification," *IEEE Signal Processing Letters*, vol. 27, pp. 880–884, 2020.

[38] M. A. Ammar, H. A. Hassan, M. S. Abdel-Latif, and S. A. Elgamel, "Performance evaluation of SAR in presence of multiplicative noise jamming," in *2017 34th National Radio Science Conference (NRSC)*, 2017, pp. 213–220.

[39] M. A. Ammar and M. S. Abdel-Latif, "A novel technique for generation of multiple noise patches on SAR image," in *2020 12th International Conference on Electrical Engineering (ICEENG)*, 2020, pp. 277–280.

[40] B. Lv, "Simulation study of noise convolution jamming countering to sar," in *2010 International Conference On Computer Design and Applications*, vol. 4, 2010, pp. V4–130–V4–133.

[41] M. A. Ammar and M. S. Abdel-Latif, "A novel technique for generation of multiple noise patches on sar image," in *2020 12th International Conference on Electrical Engineering (ICEENG)*, 2020, pp. 277–280.

[42] M. A. Ammar, M. S. Abdel-Latif, S. A. Elgamel, and A. Azouz, "Performance enhancement of convolution noise jamming against sar," in *2019 36th National Radio Science Conference (NRSC)*, 2019, pp. 126–134.

[43] S. Preradovic, N. C. Karmakar, and I. Balbin, "RFID transponders," *IEEE Microwave Magazine*, vol. 9, no. 5, pp. 90–103, 2008.

[44] D. A. Evans, "Applications of transponders in the electrical industries-keeping a tab on components," in *IEE Colloquium on Use of Electronic Transponders in Automation*, 1989, pp. 3/1–3/2.

[45] S. P. Curtis, "Transponder technologies, applications and benefits," in *IEE Colloquium on Use of Electronic Transponders in Automation*, 1989, pp. 2/1–2/8.

[46] V. Y. Meltsov, A. A. Lapitsky, V. A. Lesnikov, and A. S. Kuvaev, "Features of decoding transponder signal of an aircraft using FPGA," in *2019 IEEE Conference of Russian Young Researchers in Electrical and Electronic Engineering (EIConRus)*, 2019, pp. 120–124.

[47] L. Zhang, Y. Yu, T. Xu, and J. Zhang, "Simulation of airborne radar real beam ground map based on digital terrain," in *2013 International Conference on Computational and Information Sciences*, 2013, pp. 26–29.

[48] M. Greco, F. Gini, and A. Farina, "Combined effect of phase and RGPO delay quantization on jamming signal spectrum," pp. 37–42, 2005.

[49] S. D. Berger, "Digital radio frequency memory linear range gate stealer spectrum," *IEEE Transactions on Aerospace and Electronic Systems*, vol. 39, no. 2, pp. 725–735, 2003.

[50] J. D. Townsend, M. A. Saville, S. M. Hongy, and R. K. Martin, "Simulator for velocity gate pull-off electronic countermeasure techniques," in *2008 IEEE Radar Conference*, 2008, pp. 1–6.

[51] M. Xie, L. Liu, C. Zhang, and X. Fu, "Bidirectional false targets RGPO jamming," in *2018 13th IEEE Conference on Industrial Electronics and Applications (ICIEA)*, 2018, pp. 2345–2348.

[52] M. Attygalle, D. Dissanayake, P. Hall, K. Hue, and K. Brown, "Analysis of an ultra high resolution wideband false target technique," in *2018 International Conference on Radar (RADAR)*, 2018, pp. 1–5.

[53] D. Dai, X. F. Wu, X. Wang, and S. Xiao, "Sar active-decoys jamming based on drfm," in *2007 IET International Conference on Radar Systems*, 2007, pp. 1–4.

[54] Rear Admiral James Stavridis and Captain Frank Pandolfe, "From sword to shield: Naval forces in the war on terror," *Proceedings, U.S. Naval Institute*, vol. 130, no. 8, 2004.

[55] P. E. Pace, D. J. Fouts, and D. P. Zulaica, "Digital image synthesizers: are enemy sensors really seeing what's there?" *IEEE Aerospace and Electronic Systems Magazine*, vol. 21, no. 2, pp. 3–7, 2006.

[56] F. Berizzi, "ISAR imaging of targets at low elevation angles," *IEEE Transactions on Aerospace and Electronic Systems*, vol. 37, no. 2, pp. 419–435, 2001.

[57] Y. J. Huang, X. Wang, X. Li, and B. Moran, "Inverse synthetic aperture radar imaging using frame theory," *IEEE Transactions on Signal Processing*, vol. 60, no. 10, pp. 5191–5200, 2012.

[58] M. Hoskins and B. Hunsinger, "Recent developments in acoustic charge transport devices," in *IEEE 1986 Ultrasonics Symposium*, 1986, pp. 439–450.

[59] R. J. Davisson and W. M. Dougherty and H. J. Lubatti and R. J. Wilkes and D. Fleisch and R. Kansy, G. Peters, "Evaluation of acoustic charge transport delay lines for ssc/lhc applications," *Nuclear Instruments and Methods in Physics Research*, vol. 295, no. 3, pp. 344 – 353, 1990.

[60] Chan-Ho Yoon, Hyuk Kim, and Jong-Dug Shin, "A tunable optical add/drop multiplexer using a fiber-optic tapped delay-line transversal filter," in *Technical Digest. CLEO/Pacific Rim '99. Pacific Rim Conference on Lasers and Electro-Optics (Cat. No.99TH8464)*, vol. 3, 1999, pp. 785–786 vol.3.

[61] P. E. Pace, D. J. Fouts, S. T. Ekestorm, and C. Karow, "Digital false-target image synthesiser for countering ISAR," *IET Proceedings - Radar, Sonar and Navigation*, vol. 149, no. 5, pp. 248–257, 2002.

[62] D. J. Fouts, P. E. Pace, C. Karow, and S. T. Ekestorm, "A single-chip false target radar image generator for countering wideband imaging radars," *IEEE Journal of Solid-State Circuits*, vol. 37, no. 6, pp. 751–759, 2002.

[63] P. E. Pace, R. E. Surratt, "Signal synthesizer and method therefor," Patent US Patent 6,721,358, 2004.

[64] A. W. Doerry, "Ship Dynamics for Maritime ISAR Imaging," no. SAND2008-1020, Technical Report, Sandia National Laboratories.

[65] B. R. Mahafza, *Radar Systems Analysis and Design Using MATLAB*, 3rd ed. Boca Raton, FL: CRC Taylor and Francis, 2013.

[66] K. D. Ward, S. Watts, and R. A. Tough, *Sea Clutter: Scattering the K-Distribution and Radar Performance.* "London": Institution of Engineering and Technology, 2006.

[67] F. E. Nathanson, *Radar Design Principles: Signal Processing and the Environment*, 2nd ed. London: SciTech Publishing, 1999.

[68] S. A. Pak, "DRFM CORDIC Processor and Sea Clutter Modeling for Enhancing Structured False Target Synthesis."

[69] V. Gregers-Hansen and R. Mital, "An improved empirical model for radar sea clutter reflectivity," *IEEE Transactions on Aerospace and Electronic Systems*, vol. 48, no. 4, pp. 3512–3524, 2012.

[70] P. E. Pace, S. Teich, O. E. Brooks, D. C. Jenn, and R. A. Romero, "Extended detection range using a polyphase cw modulation with an efficient number theoretic correlation process," in *2017 IEEE Radar Conference (RadarConf)*, 2017, pp. 1669–1674.

[71] S. Watts, K. D. Ward, and R. A. Tough, "The physics and modelling of discrete spikes in radar sea clutter," in *IEEE International Radar Conference*, 2005, pp. 72–77.

[72] K. D. Ward and R. J. A. Tough, "Radar detection performance in sea clutter with discrete spikes," in *RADAR 2002*, 2002, pp. 253–257.

[73] S. Kemkemian, L. Lupinski, V. Corretja, R. Cottron, and S. Watts, "Performance assessment of multi-channel radars using simulated sea clutter," in *2015 IEEE Radar Conference (RadarCon)*, 2015, pp. 1015–1020.

[74] S. Watts, "Modeling and simulation of coherent sea clutter," *IEEE Transactions on Aerospace and Electronic Systems*, vol. 48, no. 4, pp. 3303–3317, 2012.

[75] F. An and X. Xu, "Real-time ISAR image synthesis of ballistic targets," in *2005 IEEE International Symposium on Microwave, Antenna, Propagation and EMC Technologies for Wireless Communications*, vol. 1, 2005, pp. 326–329 Vol. 1.

[76] P. Shi-rui, L. Xin, P. Yi-chun, and Y. Chun-lai, "Study on quantitative efficiency evaluation for deception jamming to ISAR," in *2009 2nd Asian-Pacific Conference on Synthetic Aperture Radar*, 2009, pp. 544–547.

[77] G. Zheng, L. Wang, and Y. Zeng, "An effective method of active deception jamming for countering ISAR," in *2013 IEEE Third International Conference on Information Science and Technology (ICIST)*, 2013, pp. 1346–1349.

[78] X. Wei, S. Xu, B. Peng, and Z. Liu, "False-target image synthesizer for countering ISAR via inverse dechirping," *Journal of Systems Engineering and Electronics*, vol. 27, no. 1, pp. 99–110, 2016.

[79] Q. Shi, J. Huang, T. Xie, C. Wang, and N. Yuan, "An active jamming method against ISAR based on periodic binary phase modulation," *IEEE Sensors Journal*, vol. 19, no. 18, pp. 7950–7960, 2019.

[80] J. Wang, D. Feng, R. Zhang, L. Xu, and W. Hu, "An inverse synthetic aperture radar image modulation method based on coding phase-switched screen," *IEEE Sensors Journal*, vol. 19, no. 18, pp. 7915–7922, 2019.

Chapter 11

Counter-DRFM Methods

This chapter presents techniques that can be used for the protection of coherent range-Doppler imaging radar systems from the DRFM's structured false target deception methods. From its initial development [1] and demonstration [2] the concept of digitally generating multiple, structured false-targets and clutter (e.g., for deception) has attracted increased attention and the architectures have gotten more sophisticated. For example, in [3], the addition of a polyphase code to the intercepted waveform for additional phase control and the use of sub-Nyquist sampling for movement of the false-target in ISAR the range and cross-range dimensions. With the modern DRFM transceiver, the jammer can intercept, modulate, and retransmit the radar signal with minimal delay. As a result, the true target signature is highly correlated with the jammer's signal. Consequently, using ISAR and SAR we now have to stop and ask ourselves: Are we really seeing what's there? [4]

In this chapter, we have taken off our DRFM design and development hat and are now wearing our range-Doppler, synthetic imaging EP protection hat! Methods are presented for the protection and identification of false targets. These techniques include the use of *pulse diversity* techniques and signal processing techniques.

11.1 PULSE DIVERSITY TECHNIQUES

Consider that the radar system transmits the conventional pulse LFM chirp signal. The DRFM jammer can intercept, sample, store, and retransmit the radar signal. As a result, the replica from the DRFM creates a false target with an apparent negative range offset by delaying the stored pulse until a fixed time before the expected arrival of the next incoming radar pulse. However, creating false targets in both negative and positive range offsets is also possible to confuse the radar system.

A method used for suppressing the effect of the DRFM EA is to measure the SAR signature of a target scene using a radar that varies its transmitted pulse in the slow-time domain (from one PRI to another). The radar maintains the same carrier frequency and bandwidth; however, pulse diversity is used such that each pulse is orthogonal to the next [5]. This radar is less susceptible to a DRFM adaptive repeater EA since the DRFM cannot adapt easily as the radar signal varies each PRI (slow-time) and/or the signal transmitted by the DRFM repeat jammer at a given PRI is the radar signal used by the SAR system during the *previous PRI* and is approximately orthogonal to the signal that the SAR system is utilizing during the current PRI.

A comprehensive model of the received signal at the radar that includes both the presence of the DRFM's false target and the true target's echo is given as

$$s(t,m) = \underbrace{\sigma p_m(t-\Delta t)}_{\text{True Target Signal}} + \underbrace{\rho p_{m-1}(t-\Delta t_j)}_{\text{Jamming Signal}} \qquad (11.1)$$

where the pulse $p_m(t)$ is transmitted at slow-time interval m and since the DRFM needs a certain amount of time to analyze and construct the false target waveform, it is acknowledged that the pulse is retransmitted by the DRFM *one pulse* behind the radar. That is, the DRFM retransmits $p_{m-1}(t)$ in slow-time m where t is in fast time. As done previously, let σ_T represent the true target reflectivity and let Δt be the round trip delay for the true target. Also let Δt_j be the round trip delay for the false-target, which is generated by the DRFM, and let ρ be the false target's reflectivity.

11.1.1 Perturbing the Phase of the LFM Chirp

Consider the expression for the LFM chirp signal given in Section 10.4. To combat the DRFM jammer one very effective strategy is varying the slope of the frequency modulation waveform [6]. In this case our radar is transmitting an LFM chirp signal with a varying slope at each PRI. For the slope varying (SV) chirp signal, let $a(t)$ be the amplitude of the SAR transmitted signal $s_t(t)$ given by

$$s_t(t) = a(t)\text{rect}\left[\frac{t}{T}\right]\exp j\left[-2\pi f_0 t + \pi \gamma_r t^2\right] \qquad (11.2)$$

where t is the fast time dimension, T is the SAR pulse duration, and γ_r is the chirp slope or rate of chirp in range dimension,

$$\gamma_r = \Delta/T_p \qquad (11.3)$$

where T_p is the chirp pulse duration. and $a(t) = \exp[j2\pi\phi(t)]$ and $\phi(t)$ at the mth PRI is defined as

$$\phi_m(t) = \gamma_m t^2 \qquad (11.4)$$

where γ_m is the chirp slope perturbation chosen by a random number generator known to the radar. In order to keep the bandwidth of amplitude perturbation signal $a(t)$ within

11.1 Pulse Diversity Techniques

the original LFM chirp signal, γ_m should be chosen within a certain percentage of chirp slope γ_r.

After perturbation, the chirp slope is $\gamma_r + \gamma_m$. In order to keep all the transmitted signals occupying the same bandwidth, the pulse duration T_p is modified at the m th PRI as

$$T'_p = \frac{k}{\gamma_r + \gamma_m} t_p \quad (11.5)$$

The radar transmits the LFM chirp signal with the varying slope at each PRI. Therefore the signal retransmitted by the DRFM jammer is approximately orthogonal to the signal transmitted by radar at current PRI. Matched filtering can be used to enhance the SJR. As shown in Figure 11.1(a) when two different SV LFM signals

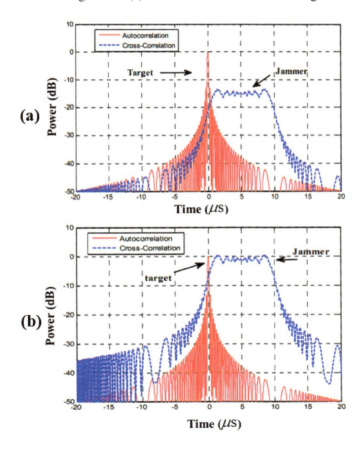

Figure 11.1 Slope varying LFM signals (a) with the same power level and (b) with different power levels. (©IEEE, reprinted with permission from [6].)

have the same power, the peak power of autocorrelation is higher than that peak power of cross-correlation with another SV LFM (DRFM EA signal) [6]. However, the effects of the EA signal still remain for the reason that two different SV LFM signals are *not strictly orthogonal*. Consequently, when the power of DRFM EA signal is higher than the target as shown in Figure 11.1(b), it is still difficult to detect the target.

The first step to cancel the jammer is to pass the received signal at slow-time interval m, through a *matched filter*, which is matched to the current pulse transmitted by the radar [6]. That is,

$$s_1(t, m) = s(t, m) * p_m^*(-t) \tag{11.6}$$

or

$$s_1(t, m) = \sigma_T p_m(t - \Delta t) * p_m^*(-t) + \rho p_{m-1}(t - \Delta t_j) * p_m^*(-t) \tag{11.7}$$

A flow diagram of the cancellation algorithm is shown in Figure 11.2.

To demonstrate the algorithm, a simulation of the flow diagram is shown using two LFM chirp signals that occupies the same 5-MHz bandwidth; however, they have different chirp slopes and pulse durations: 20 μs and 10 μs. A target is considered to be 5 km away from the radar. The DRFM generates the false targets and is 3.5 km away from the radar, and has a reflectivity that is 5 times the true target.

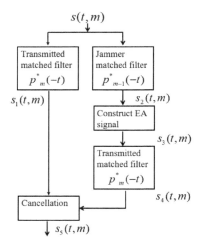

Figure 11.2 Flow diagram of the slope varying LFM chirp method for cancellation of DRFM EA. (©IEEE, reprinted with permission from [6].)

The range false-target $\rho p_{m-1}(t - \Delta t_j)$ will not range-focus in the fast time. Since the previous radar pulse p_{m-1} is, however, not strictly orthogonal with the current radar pulse, the effects of the EA can still remain in the filter output $s_1(t, m)$ as shown in Figure 11.3.

11.1 Pulse Diversity Techniques

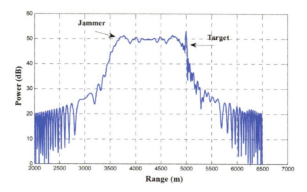

Figure 11.3 Matched filter output $s_1(t,m)$ showing the effects of the EA signal still remaining (©IEEE, reprinted with permission from [6].)

To further reduce the effects of the EA $s(t,m)$ is passed through the *DRFM matched filter*. That is, $s(t,m)$ is match filtered with the previously returned radar pulse $p^*_{m-1}(-t)$ from the DRFM as

$$s_2(t,m) = s(t,m) * p^*_{m-1}(-t) \tag{11.8}$$

or

$$s_2(t,m) = \sigma_T p_m(t - \Delta t) * p^*_{m-1}(-t) + \rho p_{m-1}(t - \Delta t_j) * p^*_{m-1}(-t) \tag{11.9}$$

The matched filter output $s_2(t,m)$ in Figure 11.4 that the false-target gets range-focused in fast time while the true target gets smeared in range. Of course this does nothing to aid detection of the target; however, the false target can be isolated

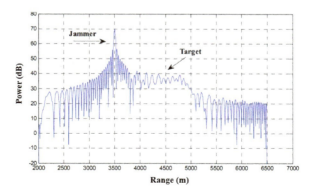

Figure 11.4 Matched filter output $s_1(t,m)$ showing the effects of the EA signal still remaining (©IEEE, reprinted with permission from [6].)

easily and a threshold set at a level that typically reduces the energy of the EA without affecting the true target's return. However, without prior knowledge of the true target this is difficult.

Alternately, to detect the false target in $s_2(t,m)$, first the round-trip $\Delta t'_j$ and the peak power P_{peak} of the false target are estimated in order to calculate the reflectivity of ρ', estimated as

$$\rho' = \frac{P_{peak}}{P} \qquad (11.10)$$

where P denotes the power of the previous transmitted radar pulse $p_{m-1}(t)$. In this manner the DRFM EA signal can be uniquely constructed by itself as

$$s_3(t,m) = \rho' p_{m-1}(t - \Delta t'_j) \qquad (11.11)$$

The reconstructed DRFM EA signal is then match filtered with the transmitted matched filter $p_m^*(-t)$ to get $s_4(t,m)$ as

$$s_4(t,m) = \rho' p_{m-1}(t - \Delta t'_j) * p_m^*(-t) \qquad (11.12)$$

Finally, the false target will be filtered out, effectively subtracting the range profile of $s_1(t,m)$ from $s_4(t,m)$ as shown in Figure 11.5.

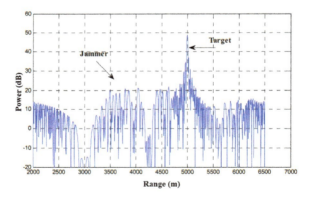

Figure 11.5 Cancellation block output showing $s_4(t,m) - s_1(t,m)$, suppression of the DRFM EA, and the recovery of the true target. (©IEEE, reprinted with permission from [6].)

In summary, for detection at the receiver, the method used by [6] and shown in Figure 11.2 first reconstructs the EA signal with parameters the of false target, estimated from a matched-filter output. The received signal is first matched-filtered with the previous pulse transmitted by the radar focuses the false target signal. The false target is then subtracted from the received signal to produce the remaining target signal. The two signals are discernable because they are based on different chirp frequencies known only to the transmitting radar. According to [7] where they also investigate the use of compressed sensing, the slope-varying LFM signal has more merit than a multi tone phase modulated signal or a pulse repetition interval varying signal.

A variation of this method was used by [5] and generates the diverse pulses by adding a slow-fluctuating phase modulated LFM chirp in the slow-time domain. The received signal is first matched filtered with the pulse used by the DRFM (i.e., the time-reversed and conjugated $p_{m-1}(t)$) to range focus the false targets. The dominating targets are then removed using a thresholding procedure similar to the above procedure before being inverted and matched filtered. Another matched filtering with respect to the current pulse $p_m(t)$ then reveals the true targets, where the EA effects have been minimized.

Drawbacks of the scheme in [5] are that the several matched filtering operations and the clipping operation can possibly eliminate the true targets if the pulses being reflected off the true targets and the replicated pulses coming from the DRFM have similar time delays (close in range). In this case the clipping will incorrectly eliminate the real targets especially if they show very strong reflecting properties in the range profile [8].

This scheme requires several matched filtering operations and if the pulses being reflected off the true targets and the replicated pulses coming from the jammer happen to have similar time delays (close in range), then clipping will incorrectly also eliminate the real targets. There is also the possibility that real targets may mistakenly be clipped off if they end up showing very strong reflecting properties in the range profile. This can have fatal results in a practical setting if the objects of interest are erroneously removed.

11.1.2 Orthogonal Coding

The counter-DRFM EA method outlined in [5] involves coding the radar signals so that the waveform at a given PRI is orthogonal to the previous PRI. The technological foundation of this technique is similar to methods used in orthogonal frequency division multiplexing (OFDM) communications. Because the signals are orthogonal to one another, the reflected pulse from the target from a given PRI will have a waveform that is orthogonal to the waveform generated by the DRFM jammer from characteristics of the prior pulse. Similar to the previous pulse-diversity technique discussed using an LFM chirp signal, the DRFM jammer on the target will stay at a minimum one pulse behind the transmitting radar. This will limit the effectiveness of the jamming signal at the radar during the correlation process when the received signals are compared to the proper known signal.

Beside transmitting the pulses as they are, we also make use of the conjugate pulses, both for transmission, and more obviously, matched filtering operations. Notationally speaking, the Fourier transform of a pulse $p_i(t)$ is $F\{p_i(t)\} = P_i(\omega)$. Also, a term often used in photonic detectors, the *point spread function* (PSF) is defined as

$$PSF_{p_i(t)} = p_i(t) * p_i(-t) = F^{-1}\{|P_i(\omega)|^2\}$$

where * denotes the time-domain convolution operator.

The first of the techniques, denoted method I, manages to completely separate the false and true reflectors assuming a certain amount of integration time while the second scheme, method II, does not fully separate the reflecting signal in two independent components but gives a clear indication of the false targets and poses no requirements on the integration time [8]. Both of the schemes operate with a simple one-step, low-complexity matched filtering operation at the receiver.

The following two concepts are utilized: (1) the potential for continuous transmission of new/perturbed pulses as long as a range compression can be achieved with a matched filtering operation, and (2) an orthogonal design is emulated that manages to separate the pulses reflected off the true targets and those emitted from the jammer while posing no orthogonality requirements on the pulses [8]. The received signal at the radar can therefore be modeled similar to (11.1), as the sum of two reflective blocks with two different pulses as [8]

$$s(t, u_m) = \sum_n \sigma_n p_m(t - \Delta_n(u_m)) + \sum_l \rho_l p_{m-1}(t - \Delta_l(u_m)) \qquad (11.13)$$

where u_m denotes the slow-time domain at placement m and t is the fast-time domain as before. Also we let σ_n be the reflectivity levels of real targets while $\Delta_n(u_m)$ is the signal delay associated with each reflector. Similarly, ρ_l are the false target reflectors emulated by the jammer, with imitated delays given by $\Delta_l(u_m)$. Note that this above expression is restricted to the case where the antenna and the targets are relatively stationary to each other but it has been shown that this also be applied for slow-moving, start-and-stop approximation platforms often used in SAR/ISAR systems.

11.1.2.1 Method I

For this algorithm in Method I, the reflectors and their reflectivity levels can be considered stationary over T slow-time intervals. Starting at $u = 0$, the following sequence of equal length pulses is emitted from the radar at each slow-time interval: $p_0(t)$, $p_1(t)$, $-p_1^*(-t)$, and $p_0^*(-t)$. The repeat jammer, trying to keep up with the radar, emits the previously sent pulse from the radar. This results in the following pulse order from the DRFM for this block period: $p_{-1}(t)$, $p_0(t)$, $p_1(t)$, and $-p_1^*(-t)$. An summary of the pulse transmissions is shown in Table 11.1. No restrictions are imposed on the pulses $p_0(t)$ and $p_1(t)$ and they can be chosen by the radar to be the most suitable for confusing the DRFM.

Table 11.1
Summary of Pulse Transmissions Method I.

u	Radar	DRFM Repeater
0	$p_0(t)$	$p_{-1}(t)$
1	$p_1(t)$	$p_0(t)$
2	$-p_1^*(-t)$	$p_1(t)$
3	$p_0^*(-t)$	$-p_1(-t)$

From: [8].

Detection of the received signals can most conveniently be expressed in the Fourier domain as

$$S(\omega,1) = \sum_n \sigma_n P_1(\omega)e^{-j\omega\Delta_n} + \sum_l \rho_l P_0(\omega)e^{-j\omega\Delta_l} \quad (11.14)$$

$$S(\omega,3) = \sum_n \sigma_n P_0(\omega)e^{-j\omega\Delta_n+2\delta} + \sum_l \rho_l P_1^*(\omega)e^{-j\omega\Delta_l+2\delta} \quad (11.15)$$

The range measurement estimates for the reflectors can be directly obtained from $S(\omega,1)$ and $S(\omega,3)$ by matched filtering over both of the received signals with regard to either the true or false targets and summing together the results [8]. For the true targets

$$r_\sigma(t) = \sum_n \sigma_n (PSF_{p_0}(t-\Delta_n) + PSF_{p_1}(t-\Delta_n)) \quad (11.16)$$

and separates the true reflectors and the false targets emulated by the jammer with simple matched filtering. The false targets may not always be of much interest; however, due to orthogonality, a symmetrical matched filtering operation can be performed in an equivalent manner to range-focus the jamming targets while discarding the real ones. In the time domain,

$$r_\rho(t) = \sum_l \rho_l (PSF_{p_0}(t-\Delta_l) + PSF_{p_1}(t-\Delta_l)) \quad (11.17)$$

Note that the above two PSF equations are the sum of two independent PSFs implying pulse diversity and this can be put to use for various advantages [8]. This method also is very efficient in separating the two signals being reflected off targets and those emitted by a DRFM repeater using simple matched filtering. It does however, require integration over several pulses in order to separate the incoming signals.

11.1.2.2 Simulation Results Method I

Simulation results using standard chirp signals with identical frequencies are employed. That is, $p_0(t) = \exp[j(\omega_0 t + \Delta t^2)]$ but the chirp rate for $p_1(t) = \exp[j(\omega_0 t + 2\Delta t^2)]$ is set for twice that of the first pulse. Also, $\omega_0 = 1$ and $\Delta = 1$ and both have the same modulation pulse width in time.

Fig.11.6(a) shows the simulation results for (11.16) and Figure 11.6(b) shows the results for (11.17).

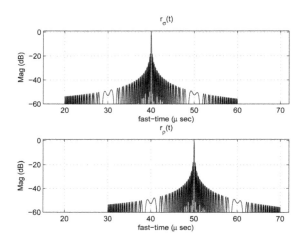

Figure 11.6 Simulation results for (a) (11.16) to recover actual targets and (11.17) to recover false targets. (©IEEE, reprinted with permission from [8].)

11.1.2.3 Method II

In the second method a strategy is used where the pulses are no longer spread over several slow-time intervals but rather stacked together to combine joint pulses. This scheme also has the potential of making the DRFM jammer perceive that a few long pulses have been emitted rather than several short free-standing pulses.

At slow-time step $u = 0$ the pulse emitted from the radar is defined as transmitting *two pulses*

$$p_0(t) = [q_0(t) \ \ q_1(t)] \tag{11.18}$$

The pulse $p_1(t)$ transmitted at slow-time step $u = 1$ is

$$p_1(t) = [q_1^*(t) \ -q_0^*(t)] \tag{11.19}$$

Table 11.2

Summary of Pulse Transmissions Method II.

u	Radar	DRFM Repeater
0	$p_0(t) = [q_0(t) \; q_1(t)]$	$p_{-1}(t)$
1	$p_1(t) = [q_1^*(t) \; -q_0^*(t)]$	$p_0(t) = [q_0(t) \; q_1(t)]$

From: [8].

and are the conjugated pulses emitted at $u = 0$ and can be chosen at will. For example, these could be designed such that they are observed as one continuous waveform. A summary of pulse transmissions for Method II from the radar and the DRFM respectively, is provided in Table 11.2. Note that $p_1(t)$ is based on the conjugated pulses emitted at $u = 0$, with the pulse order being reversed and with $q_{0,1}(t)$ chosen at will. As well, the signal received at $u = 0$ is also not applied directly in the matched filtering process, although the radar is free to use it for other signal processing tasks [8].

The range-profile of the true reflectors are detected with traditional matched filtering over $s(t, 1)$ in the time domain as

$$r_{\sigma_n}(t) = p_1^*(t) * s(t, 1) = \sum_n \sigma_n PSF_{p_1}(t - \Delta_n) + \sum_n \rho_l p_1^*(-t) * p_0(t - \Delta_l) \quad (11.20)$$

Here the true targets are now focused through the PSF while the DRFM pulses sometimes introduce significant interference. Due to the orthogonal properties of the pulses and the fact that a convolution is equivalent with a time-reversed correlation, the second term in (11.20) typically shows up as a deep fade in a dB plot, making it easy to distinguish the false targets from the true targets.

A matched filtering operation with $p_0^*(-t)$ on $s(t, 1)$ can focus the false targets and dampen the true targets. Due to corresponding matched filtering operation then is

$$r_{\rho_l}(t) = p_0^*(-t) * s(t, 1) = \sum_n \sigma_n p_0^*(-t) * p_1(t - \Delta_n) + \sum_l \rho_l PSF_{p_0}(t - \Delta_l) \quad (11.21)$$

The ability to distinguish between the true and false reflectors by placing one of them in a fade provides an easy method to identify targets while discarding the falsely emulated reflector [8].

11.1.2.4 Simulation Results Method II

Figure 11.7 displays the real part of the observed signal over the second slow-time interval, $s(t, 1)$. The DRFM is repeating the signal $p_0(t)$ emitted at $u = 0$ (left side, lesser delay). In Figure 11.8 the outcome of the matched filtering (11.20) with respect to true reflector are shown.

830   COUNTER-DRFM METHODS

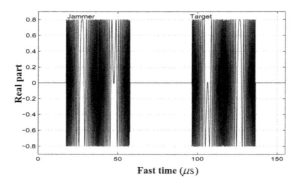

Figure 11.7 Real part of observed signal over second slow-time interval $s(t,1)$ (©IEEE, reprinted with permission from [8]).

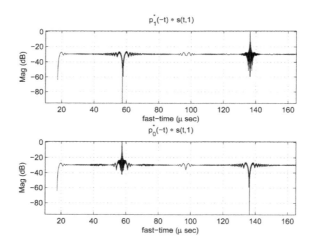

Figure 11.8 Simulation results for the upper plot showing the outcomes of matched filtering (11.20) with respect to true reflector. For the lower plot, matched filtering (11.21) with respect to the false target. (©IEEE, reprinted with permission from [8].)

The occurrence of a deep fade can be seen at the instance of maximum correlation that an AI deep learning technique can be trained to easily recognize. The lower plot demonstrates the alternative use of matched filtering (11.21) with respect to the false target putting the true reflector into a fade.

In summary, the techniques above allow for continuous transmission of new and various types of waveforms in which their parameters can be chosen randomly in a general fashion. The orthogonal properties of the detection methods allow the focusing of the true and false (DRFM) targets with matched filtering operations whereby AI techniques can be easily used.

11.2 SIGNAL PROCESSING TECHNIQUES

11.2.1 Polarization Discrimination Methods

A reflected target and a DRFM jammer signal will generally have different polarization characteristics that can be measured and compared using either a polarization *scattering matrix* or a polarization *discrimination algorithm*. As we have previously diagnosed, the successive layers of understanding available to deep learning algorithms can solve these types of discrimination algorithms quickly and with minimal resources.

Generally, jammer antennas are circularly polarized as this maximizes their chances against a number of linear polarizations as discussed earlier. In addition, the polarization characteristics of the jamming signal will not change with the *attitude* of the jammer antenna relative to the transmitting and receiving radar. Changes in the amplitude and phase of the false target signal will not change the polarization state. Even when a jammer is operating in a linear polarization mode, the polarization will be highly correlated between all of the false targets generated.

The actual target return polarization, by contrast, is dependent on the transmitter radar polarization and target scattering characteristics and will almost certainly be less correlated with any of the false targets than the false targets are with each other. Where these polarization differences can be discriminated, the true target can be picked out of a field of false images. However, such polarization detection is generally thought to require specific and more sophisticated electronic protection hardware that limits practical application.

Consider a 4-element Tx/Rx monopulse antenna configuration, typically used in airborne active radar. The sum and difference pattern for a monopulse antenna is shown in Figure 11.9 as a function of the gain (dB) for both H and V polarizations. With this configuration, if the two beams are added together, an on-axis beam is formed that is the sum beam Σ. If the two beams are subtracted, the resulting pattern will have a positive lobe on one side of the axis and a negative lobe out of phase on the other and a null that occurs, the difference beam Δ. The beams can be compared and subtracted for developing and error signal and tracking. Also the direction of arrival (DOA) can be calculated easily. The monopulse antenna configuration is much more difficult to attack electromagnetically. In addition, with both polarizations, copolarization processing is feasible, which can increase the range resolution beyond the $\Delta R = c\tau/2$.

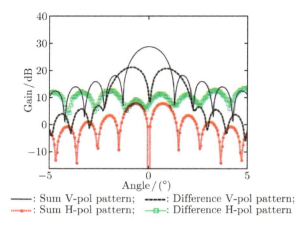

—— : Sum V-pol pattern;   ----- : Difference V-pol pattern;
······ : Sum H-pol pattern;   —□— : Difference H-pol pattern

Figure 11.9 The monopulse sum and difference pattern as a function of gain (dB) for both polarizations.

A new polarization-measurement technique shown in Figure 11.10 allows a single polarized monopulse radar (which has the sum and difference beam differential antenna property) to make polarization measurements and discriminate the false targets by discriminating the received polarization [9].

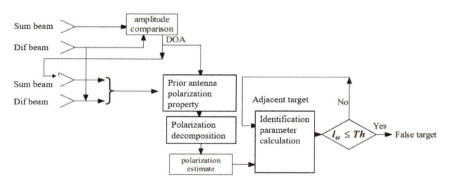

Figure 11.10 Block diagram of polarization measurement technique [9].

That is, based on the *generation mechanism differences* between the target scattering and the DRFM false-targets, the actual targets can be determined from the measurement of the measurement's polarization characteristics. Note that the DOA is also used to sort out the true/false targets. In addition, the matched filtering identification step prior to thresholding is robust in sorting out the real and false targets as shown in the example in Figure 11.11. Then, the amplitude is compared to the threshold T_h to determine the final output.

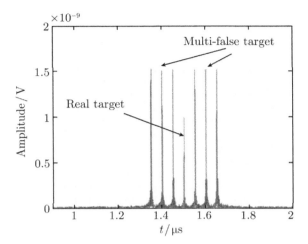

Figure 11.11 Matched filter output showing the real and false targets (from [9]).

The goal is to train our deep learning discriminator network to recognize a false target using the discrimination measurement output. To do this, we first apply a structured false target. When the discrimination algorithm runs and the output has been identified as a false target, the measurement of the polarization vector reveals that the DRFM jammer signal only modulates the polarization characteristics of the DRFM's jammer antenna. That is, the target's polarization scattering characteristics and the radar's transmitting antenna's polarization pattern are not disturbed.

Next, a true target is used for training. When the discrimination output is a true target, the polarization measurement reveals the target signal modulates *both* the target's polarization scattering characteristics and the radar transmitting antenna's polarization pattern [9]. The polarization of each received pulse (real target or false target) can be compared to adjacent pulses using an angle cosine measurement. For example, adjacent pulses with identical polarizations will have a value of one.

11.2.1.1 Polarization Measurement Simulation Results

Figure 11.12(a) shows a sample target polarization amplitude estimates for both H and V polarizations. Figure 11.12(b) shows a sample interference polarization amplitude estimates for both H and V polarizations.

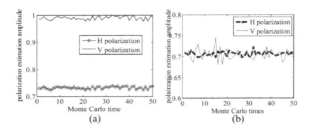

Figure 11.12 Plots showing the sample amplitude estimates for both H and V polarizations with (a) sample target and (b) sample interference amplitude. (©IEEE, reprinted with permission from [9].

The polarization Euclidean distance distribution of the 50 intensive targets is shown in Figure 11.13. The 49 calculated Euclidean distances are almost near to 0 while the value of the 14th and 15th calculation is about 1.6. The polarization estimation of the 15th target is quite different from those of the other targets.

By using the polarization identification algorithm and process, Monte Carlo simulation experiments are carried out. The right identification probability curve for the false target about SNR/INR is shown in Figure 11.14.

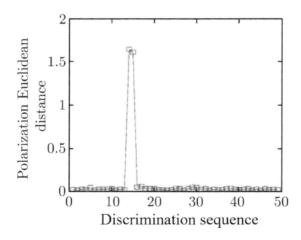

Figure 11.13 Polarization Euclidean distance of adjacent target (©IEEE, reprinted with permission from [9].)

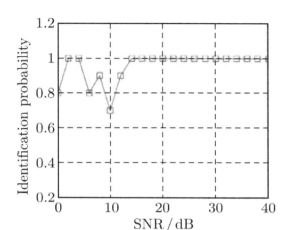

Figure 11.14 False target identification performance curve (©IEEE, reprinted with permission from [9].)

The identification threshold $T_h = 0.2$ and only when the interference is relatively high ($INR \approx 10$ dB) is the false target detected as a real target (false detection).

In summary, this technique is robust in that there is no need to measure the full polarization scattering matrix (PSM) completely. Also this method easily realizes the identification of the real and false targets by estimating the relative orthogonal polarization components, thus reducing the difficulties in the calculating and processing of the full PSM. Studies have shown the identification probability can reach up to 90% or better, even when the DRFM false target number is large and the SNR is below 20 dB. Finally, the method only requires knowledge of the full polarization characteristics of the antenna and direction of signal arrival (which sometimes is difficult to get). However, it is expected that this method can help improve the signal processing and protection capability of today's modern, multi-mode radar system.

11.2.2 Statistical Counter-DRFM Theory Using Estimation and Neyman-Pearson Receivers

The problem of sampling and testing statistical hypotheses is an old one whose origin is usually connected with the name of Thomas Bayes, who gave the well-known theorem on the probabilities a posteriori of the possible causes of a given event [10]. Introduced by Jerzy Neyman and Egon Pearson in a paper in 1933 [11], the Neyman-Pearson Lemma provides a basis for choosing criteria for testing any set of statistical hypotheses, H_0 (e.g., adversary targets detected) with regard to an alternative H_1 (DRFM false targets present). Below we begin we first begin with a discussion on estimation theory followed by hypothesis testing.

11.2.2.1 Estimation of a Serious Threat

The presence of an adversary DRFM generating false targets in an AOR is a significant problem to all RF emitters attempting to detect or image targets and image areas on the ground. Many books have been written on the theory of statistical estimation and detection that use radar as an example [12], and these can form the basis for uncovering the presence of false targets and false clutter produced by these very capable synthetic generators.

For radar signal processing, *estimation* is the data analysis framework that uses a combination of sample sizes and confidence intervals to analyze and interpret the received hits to place a numerical value on the presence of a DRFM false target and false clutter. The main objective of the estimation methods are to report an effect size (a point estimate) along with its confidence interval, the latter of which is related to the precision of the estimate. The confidence interval summarizes a range of likely values of the underlying population where we let θ be the unknown DRFM population parameter; that is, $\theta = \mathcal{E}\{x\}$ with the samples $x_1, x_2, \ldots, x_n$ are the observed values of the random values of the DRFM random variable. The point estimator is then $\hat{\theta} = \hat{\theta}(x_1, x_2, \ldots, x_n)$.

Hypothesis testing is a form of inferential statistics that allows the radar to draw conclusions about an entire DRFM engagement based on a representative sample. You gain tremendous benefits by working with a sample. In most cases, it is simply impossible to observe the entire engagement to understand its properties. The only alternative is to collect a random sample and then use statistics to analyze it.

11.2.2.2 What Makes a Good Estimator?

When the radar is detecting structured targets among false targets, clutter, and noise, and trying to determine which are false (if any), it is important to set up the best estimator possible so that the false targets can be identified. For a good estimator, four conditions should hold:

First,

$$P(|\hat{\theta}(x_1 x_2 \cdots x_n) - \theta| < \varepsilon) \underset{n \to \infty}{\to} 1$$

That is, as the number of samples n goes to ∞, the probability approaches one that the estimator is true.

Second, the estimator should be centered within the confidence region as shown in Figure 11.15.

11.2 Signal Processing Techniques

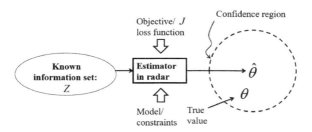

Figure 11.15 Estimation of $\hat{\theta}$ from a time series of data Z using the objective function J.

This ensures an equal distance to the outermost edge of the confidence region. That way, subsequent minimization of the distance between the estimator $\hat{\theta}$ and each and every value within the confidence region will be $\leq \theta$.

Third, the estimator should be an *unbiased* estimator. That is,

$$\mathcal{E}\{\hat{\theta}(x_1 x_2 \cdots x_n)\} = \theta$$

Fourth, the estimator should have a *minimal variance*. In other words, for each θ, there is a $\hat{\theta}$. For the many different situations there are a lot more estimators $\hat{\theta}$.

The known information set Z is an information set, shown on the left in Figure 11.15, and is the collection of data that the radar signal processor has received concerning the presence of false targets, such as the communications that relay the presence of DRFMs, any UAS in the area, or the presence of DRFMs revealed within the time series data, etc. To connect the known information set Z, with the space of unknowns (confidence region), a model is developed.

But how does the model know to reach the true value within the unknowns? The model must establish a bridge (or map) between the *knowns* and the *unknowns*. To establish this connection, an *objective function J* that specifies the goals that have to be achieved by the estimator is developed. The objective function typically uses the minimization of some loss or risk function or a distance measurement, where the actual form depends on the problem at hand. Some commonly used objective functions include negative log-likelihood and squared error approximations.

As we know, the estimator is a mathematical model that computes the estimate using the information set Z, the objective function J, and the model to compute an *estimate*. That is, it takes the information time series and applies a filter to derive the true solution. Some examples you may be familiar with are the Kalman filter and the Wiener filter which estimate states and signals.

11.2.2.3 Hypothesis Testing for Counter-DRFM Detection

This section starts with the two types of definitions of what a hypothesis actually is:
Definition 1:

If a hypothesis specifies the values of all of the parameters of a density function, it is called a *simple* hypothesis; otherwise it is called a *composite* hypothesis.

For example, consider the density function

$$p(x;\theta_1,\theta_2) = \exp\left\{\left(-\frac{1}{2}\right)\left(\frac{x-\theta_1}{\theta_2}\right)\right\}^2 / \sqrt{2\pi\theta_2^2} \tag{11.22}$$

For example:
$H_0 : \theta_1 = 10, \theta_2 = 2$ (simple hypothesis)
$H_0 : \theta_1 = 10, \theta_2 < 2$ (composite hypothesis)

Definition 2:
Hypothesis is an assumption about the density function of a random variable.
For example, $p(x : \theta) = \theta e^{-\theta x}$ where θ is an unknown parameter such as the presence of a DRFM false target. A hypothesis test is then the procedure for deciding whether to accept or reject the hypothesis. As an example, from our frame of data tested, we have determined that θ in the above is either 1 or 2 and with other observations we believe 2.

We consider that:
$$p(x : \theta) = \theta e^{-\theta x} \quad x \geq 0$$

H_0: (Null hypothesis): $\theta = 2$
H_1: (Alternate hypothesis): $\theta = 1$
Based on a single observation of the random variable x, we wish to establish a precedure for either accepting H_0 or rejecting H_0 (accepting H_1)? That is, what value of x will correspond to the rejection of the hypothesis?

Definition 3:
The *Critical Region* is that part of the sample space which corresponds to the rejection of the hypothesis.

11.2.2.4 Types of Error

Suppose we arbitrarily assign $x > 1$ as the critical region in the above experiment. Then there are two types of errors in the hypothesis testing of the data. There is the Type I error where H_0 is true, $x > 1$ but you rejected H_0. The Type II error is where H_1 is true but you accepted H_0 and $x \leq 1$. These are summarized in Table 11.3

Table 11.3
Summary of Type I and Type II Errors in Hypothesis Testing

	H_0 True	H_1 True
$x > 1$; Reject H_0	Type I Error	Correct
$x \leq 1$; Accept H_0	Correct	Type II Error

The rate of occurrence (or the size) of the Type I error or α is also called the level of *significance* and is the probability that the sample will fall into the critical region when H_0 is true. These errors are caused by sampling error. The significance is an evidentiary standard that you set to determine whether your sample data are strong enough to reject the null hypothesis. The null hypothesis tests define that standard using the probability of rejecting a null hypothesis that is actually true. You set this value based on your willingness to risk a false positive. The size of the Type II error rate or β cannot be set but it can be estimated by approximating the alternative hypothesis being studied. This type of estimation is called *power analysis*.

11.2.2.5 Neyman-Pearson Hypothesis Testing Procedures: Best Practices

In this framework, we define the two hypothesis that we would like to choose between. Because there are two possible choices, there are two possible errors: Type I and Type II. The idea is that one error is more important than the other and this determines how we choose the hypothesis. The *null hypothesis* typically represents a "no effect" type choice while the *alternative hypothesis* represents the main choice of interest, which could be good or bad (like the presence of a false target from a DRFM).

The best procedure, considering the Type I and II errors, is to fix the (significance) Type I error and chose a procedure that minimizes the Type II error (Neyman-Pearson). Usually, as the Type I α significance decreases, the Type II β increases.

From Figure 11.16 and the example above, the critical region is $x > 1$ with:

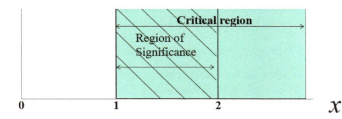

Figure 11.16 Critical region for the example shown.

$H_0 : \theta = 2$;
$H_1 : \theta = 1$

and integration over the critical region then takes the form

$$\alpha = \Pr[x > 1 | H_0 \text{ true}] = \int_1^2 2e^{-2x} dx = 0.135$$

and for the Type II error or β the integration takes the form

$$\beta = \Pr[x \leq 1 | H_1 \text{ true}] = \int_0^1 e^{-x} dx = 0.632$$

However, suppose as an alternate test, the critical region was $x < x_0$ and

$$\alpha = \int_0^{x_0} 2e^{-2x} dx = 0.135$$

and we *fix* this Type I error. Then $x_0 = 0.07$ and

$$\beta = \int_{0.07}^{\infty} e^{-x} dx = 0.932$$

This error is much larger. Consequently, we see that the first test is quite superior to the second test.

11.2.2.6 Power Functions

Sometimes a simple alternative is not always obvious and easily recognized. For example, suppose we wish to test the hypothesis:
$H_0 : \theta = 2$;
$H_1 : \theta < 2$
Now β is a function of θ. That is,

$$\beta(\theta) = \Pr[\text{Sample point falls in non-critical region} | \theta \text{ True}]$$

and

$$P(\theta) = 1 - \beta(\theta) = \Pr[\text{Sample point falls in Critical region} | \theta \text{ True}]$$

The function $P(\theta)$ is what we now define as a *power function*. It is important to note that minimizing $\beta(\theta)$ is equivalent to maximizing $P(\theta)$. So consider in the example above

$$p(x : \theta) = \theta e^{-\theta x} \quad \text{for } x \geq 0$$

and
$H_0 : \theta = 2$;
$H_1 : \theta \geq 2$

So the possible critical regions for $\alpha = 0.135$ are $x > 1$ and $x < 0.07$.

$$P_1(\theta) = \int_1^\infty \theta e^{-\theta x} dx = e^{-\theta}$$

and

$$P_2(\theta) = \int_0^{0.07} \theta e^{-\theta x} dx = 1 - e^{-0.07\theta}$$

The power functions $P_1(\theta)$ and $P_2(\theta)$ are shown in Figure 11.17.

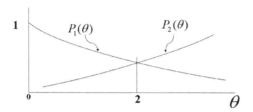

Figure 11.17 Critical region for the example shown.

Note that they intersect at $\theta = 2$. Also, since $P_1(\theta) > P_2(\theta)$ for all $\theta < 2$, the first test is superior.

11.2.2.7 Neyman-Pearson Lemma

Consider that a critical region A of size α and a constant κ exist such that

$$\frac{\prod_{i=1}^N p(x_i : \theta_0)}{\prod_{i=1}^N p(x_i : \theta_1)} \leq \kappa \quad inside \ A$$

and

$$\frac{\prod_{i=1}^N p(x_i : \theta_0)}{\prod_{i=1}^N p(x_i : \theta_1)} \geq \kappa \quad outside \ A$$

then A is the *best* critical region. The proof of the Neyman-Pearson lemma is given in Appendix II A at the end of the chapter.

Example

Consider the pdf given by $p(x : \theta) = \theta e^{-\theta x}$ for $x \geq 0$, $H_0 : \theta = \theta_0$ and $H_1 : \theta = \theta_1 < \theta_0$. Also,

$$L_0 = \prod_{i=1}^N p(x_i : \theta_0) = \theta_0^n e^{-\theta_0 \sum_{i=1}^N x_i}$$

$$L_1 = \prod_{i=1}^{N} p(x_i : \theta_1) = \theta_1^n e^{-\theta_1 \sum_{i=1}^{N} x_i}$$

A is defined by

$$\frac{\theta_0^n e^{-\theta_0 \sum x_i}}{\theta_1^n e^{-\theta_1 \sum x_i}} \leq \kappa$$

Dividing the left side and multiplying both sides through by the ratio of the random variables gives

$$e^{(\theta_1 - \theta_0)\sum_i x_i} \leq \kappa \left(\frac{\theta_1}{\theta_0}\right)^n$$

Then by taking the natural logarithm of both sides

$$(\theta_1 - \theta_0)\sum x_i \leq \ln \kappa \left(\frac{\theta_1}{\theta_0}\right)^n$$

For $\theta_1 - \theta_0 \leq 0$ so $\theta_1 \leq \theta_0$ and

$$\sum x_i \geq \frac{\ln \kappa (\theta_1/\theta_0)^n}{\theta_1 - \theta_0}$$

In this example $\theta_1 = 1, \theta_0 = 2, n = 1$ where κ is chosen to make $\alpha = 0.135$ and $x_0 = -\ln \kappa / 2$ (or 1). That is,

$$0.135 = \int_{x_0}^{\infty} 2e^{-2x} dx = \alpha$$

or

$$0.135 = e^{-2x_0}$$

or $x_0 = 1$. Note that the critical region does not depend on the value of θ_1 because we know that $\theta_1 < \theta_0$.

11.2.3 Detecting DRFM Decoys Using Neyman-Pearson Receivers

In the examples above, the concept of *glint* is important in target signature scattering and is covered in almost every introductory radar textbook (e.g., [13, 14]). Glint can be defined as the apparent radial direction from where a single-scatter electromagnetic field originates, and is perpendicular to the phase front of the wave. However, complex targets have multiple, independent scattering centers. In the angular regions where there is destructive interference between the scattered fields, there can be warps or significant distortion of this phase front. This distortion in the RF phase front is the *angular glint*, or scintillation. It can also be observed as an angular deviation of the direction of power-flow tilt from the *true* radial direction and gives rise to RF pointing errors.[1]

Interestingly, there was some argument about the equivalence of the distortion in the phase front explanation for angular glint and the power-flow tilt explanation [15]. However, when polarization was taken into account and geometrical optics used, the results were identical.

A target with multiple scatterers distributed in range can cause tracking errors because of glint similar to the angle glint error experienced in angle tracking. The phase front distortion due to scatterers distributed in range is known as *range glint*. The pointing errors in range and angle due to glint can be reduced with frequency agility, however, quantifying this improvement is difficult [14].

If the target can be represented by a line of scatterers perpendicular to the LOS, there will be a *time-correlated* angle error signal for a single-frequency radar with the spectral energy of the error signal concentrated near zero frequency (neglecting receiver noise) [14]. Since the time constant of angle tracking loops is generally quite long, the angle error signals will follow the slowly varying phase front of the complex target echo. The spectrum of the angle error signals for a target of linearly distributed scatterers can be written as [14]

$$W_u(f) = \frac{\bar{U}^2}{2f_m} \left(2 - \frac{3f}{f_m} + \frac{3f^2}{f_m^2} - \frac{f^3}{f_m^3} \right) \quad (11.23)$$

for $f \leq 2f_m$ and zero elsewhere. Here $\bar{U}^2$ is the mean square value of the perturbation signal $U(t)$, $f_m = \omega_r(L\cos\theta)/c$ where we have previously defined ω_r as the rotation rate of the (line) target about its center (small angles assumed) and $L\cos(\theta) = L_x$ is the projection of the target perpendicular to the LOS.

The variance of $U(t)$ is independent of ω_r and the spectral width of (11.23) is proportional to ω_r. In addition, a radar with frequency agility then spreads the glint's spectral energy out, allowing the angle of arrival to be averaged over a number of independent samples within its passband. Range error fluctuations due to glint can also be reduced by frequency agility [14].

11.2.3.1 Hypothesis Testing Concepts

In radar detection, the returned signal is observed and a decision is made as to whether a target is present or absent. At the receiver, the received signals representing the zeros and ones are corrupted in the medium by additive noise, by the receiver, noise and the possibility of EA from (one or more) jammers always exists. The receiver does not know which signal represents a zero and which signal represents a one but it must make a decision as to whether the received signal is a zero or a one. This process that

---

[1] The mechanism that gives rise to these errors is also responsible for the angular errors observed in radar multipath conditions.

the receiver undertakes in selecting a decision rule falls under the theory of *statistical decision theory* and the outputs are referred to as *hypotheses*. The *null hypothesis* H_0 represents a zero (target not present) and the *alternate hypothesis* H_1 represents a one (target present) [12]. The range of values the random variable takes is the observation space.

At the near distance, the angular error caused by the target's angular glint is an important noise source. For longer range targets ($R > 100$ km) the above method is not applicable because the measurement error caused by the angular glint is small within the resolution cell and can therefore be neglected. Compared with the angular glint error, range glint error is independent of the absolute distance, and it will not change with the increase of the target's distance.

Consequently range glint can be utilized to discriminate between the repeating decoy and the actual target [16] and the method presented below has the advantage that it can be used with sequential sampling since it is a hypothesis technique. The method is based on the assumption that the actual target's reflection is a multifeed radiation source. That occurs, because an actual target will scatter an incident pulse from many range points or range bins. On the other hand, a false target decoy is a single-feed radiation source or a single DRFM antenna.[2] The total received echo is a sum of all scattered waves that combine in a single sampling range bin, based on the sampling rate of the radar receiver.

The repeating decoy is single-feed radiation and there is no range glint effect, so the distance measuring values can be expressed as [16]

$$d(n) = d_i(n) + d_n(n) \tag{11.24}$$

While the actual target is multifeed radiation and range glint exists, so the distance measuring values can be expressed as

$$d(n) = d_t(n) + d_n(n) + d_g(n) \tag{11.25}$$

where $d_t(n)$ is the target's actual distance and $d_n(n)$ is the ranging random error caused by receiver's systems noise. The range glint of actual target is $d_g(n)$ and provides a basis to discriminate between the repeating decoy and the actual target.

Because the range glint noise and other observation noise are independent, the range observation value can be modeled as a Gaussian distribution expressed as

$$Y = [y_1, y_2, \ldots, y_N]^T \tag{11.26}$$

---

[2] Note that this assumption does not apply for a field of distributed, synchronized DRFMs that generate a coordinated EA false target decoy.

The joint PDF of the distance observation is

$$f(y_1, y_2, \ldots, y_N) = \prod_{i=1}^{N} \frac{1}{\sqrt{2\pi\sigma_N^2}} \exp\left[-\frac{(y_i - \bar{y})^2}{2\sigma_N^2}\right] \quad (11.27)$$

Here we let H_0 represent the repeating decoy where there is only one scatterer point in range and H_1 represents the actual target with multiple scatterers in the range dimension.

For these two cases the likelihood functions are given by

$$H_0: \quad f(y_1, y_2, \ldots, y_N | H_0) = \prod_{i=1}^{N} \frac{1}{\sqrt{2\pi\sigma_0^2}} \exp\left[-\frac{(y_i - \bar{y})^2}{2\sigma_0^2}\right] \quad (11.28)$$

and

$$H_1: \quad f(y_1, y_2, \ldots, y_N | H_1) = \prod_{i=1}^{N} \frac{1}{\sqrt{2\pi\sigma_1^2}} \exp\left[-\frac{(y_i - \bar{y})^2}{2\sigma_1^2}\right] \quad (11.29)$$

From the above, the likelihood ratio is then

$$L(y_1, y_2, \ldots, y_N) = \frac{(y_1, y_2, \ldots, y_N) | H_1}{(y_1, y_2, \ldots, y_N) | H_0} \quad (11.30)$$

and the likelihood ratio after dividing (11.28) by (11.29) is

$$L(y_1, y_2, \ldots, y_N) = \frac{\sigma_0^N}{\sigma_1^N} \exp\left[\sum_{i=1}^{N}(y_i - \bar{y})^2 \left(\frac{1}{2\sigma_0^2} - \frac{1}{2\sigma_1^2}\right)\right] \quad (11.31)$$

The maximum likelihood estimator (MLE) for the mean value of the range $\bar{y}$ that also maximizes the likelihood function is

$$\hat{\bar{y}} = \frac{1}{N} \sum_{i=1}^{N} y_i \quad (11.32)$$

so referring to (11.24) and (11.25) $\sigma_0^2 < \sigma_1^2$ so $1/2\sigma_0^2 - 1/2\sigma_1^2 > 0$ and

$$T = \sum_{i=1}^{N}(y_i - \bar{y})^2 \quad (11.33)$$

then the detection thresholding algorithm can be expressed as [16]

$$T \underset{H_0}{\overset{H_1}{\gtrless}} \lambda \quad (11.34)$$

Here T is the detection statistic discussed above and λ is the detection threshold value obtained by the Neyman-Pearson theorem also discussed in this section and obtained by the false alarm probability P_{fa} and the detection probability P_d.

For the null hypothesis H_0, $(y_i - \bar{y})/\sigma_0 \sim N(0,1)$ and so then $T/\sigma_0^2 \sim \chi^2(N-1)$. Consequently, the detection threshold $\lambda_1 = \lambda/\sigma_0^2$ and the P_{fa} can be expressed as [16]

$$P_{fa} = \Pr(T/\sigma_0^2 > \lambda_1 | H_0) = \frac{1}{2^{(N-1)/2}\Gamma((N-1)/2)} \int_{\lambda_1}^{+\infty} y^{((N-1)/2)-1} e^{-y/2} dy \quad (11.35)$$

So now the threshold value λ can be expressed as

$$\lambda = \lambda_1(P_{fa})\sigma_0^2 \quad (11.36)$$

and from the SNR equation

$$\sigma_0^2 = \frac{c^2 \tau^2}{8 SNR} \quad (11.37)$$

resulting in the expression for the probability of detection as [16]

$$P_D = P(T > \lambda | H_1) = \frac{1}{2^{(N-1)/2}\Gamma((N-1)/2)} \int_{\lambda}^{+\infty} y^{((N-1)/2)-1} e^{-y/2} dy \quad (11.38)$$

Considering the above, the final detection algorithm is given by

$$T \underset{H_0}{\overset{H_1}{\gtrless}} \lambda_1(P_{fa})\sigma_0^2 = \lambda_1(P_{fa}) \frac{c^2 \tau^2}{8 SNR} \quad (11.39)$$

The DRFM jammer signal will be generated by a single antenna. Consequently the false targets will appear from a single source. Either the naturally occurring relative movement between the target and radar or constant changes in the transmitted signal will cause slight differences in the measured target range between successive pulses, or in the case of an imaging radar, varying intensities within each range bin. This is because of the slightly varying intensities of the received signal that will combine in each range bin with slight changes in the range to the target between pulses.

Simulation results using a linear FM signal with a bandwidth of $B = 30$ MHz and a pulsewidth of $\tau = 160\mu s$ is used with a peak power of 200 kW at 2 GHz. A matched filter with a Hamming window is used upon reception sampling at 120 MHz

[16]. One thousand Monte Carlo runs were run with a $P_{fa} = 1 \times 10^{-2}$. The P_{fa} vs. SNR are shown in Figure 11.18(a).

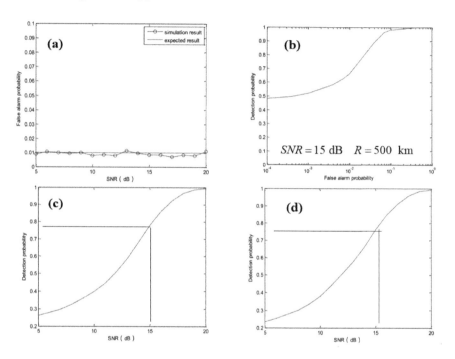

Figure 11.18 Counter-DRFM results using glint showing (a) P_{fa} results vs. SNR and (b) ROC: probability of detection P_d vs. probability of false alarm P_{fa} with $SNR = 15$ dB and (c) ROC: P_d vs. P_{fa} and (d) ROC: P_d vs. P_{fa}. (©IEEE, reprinted with permission from [16].)

In summary, the range glint is completely independent from the target's absolute distance and can be utilized at any range. This is the reason why discriminating differences based on range glint is superior to related methods based in a measure of angular glint. Angular glint, by contrast, only works at relatively short distances where azimuthal error is more pronounced.

11.2.4 Frequency Diversity Characteristics

Similar to the counter-DRFM techniques using range glint methods, the frequency diversity techniques present a method designed for countering RGPO-based DRFM jammers but has applicability to countering false-image synthesizing DRFMs and are discussed in [17]. A DRFM repeated signal can be countered by a receiver able to use frequency diversity to discriminate the amplitudes of the target echoes. The amplitudes of the target echoes are modeled as being Rayleigh distributed while the amplitudes of the RGPO echoes will be fixed across the frequencies.

The specific detection algorithm used in [17] compares each echo received to a generalized maximum likelihood (GML) detection algorithm and a variance test to determine if the specific return echo is likely Rayleigh or a fixed-amplitude target. The GML ratio test for discriminating with N subpulses is given by [17]

$$T\left(\{R_{oi}\}_{i=1}^{N}\right) = \prod_{i=1}^{N} \frac{f\left(R_{oi}|H_1, R_F = \hat{R}_F\right)}{f\left(R_{oi}|H_0, R_R = \hat{R}_R\right)} \tag{11.40}$$

where R_R denotes the SNR of the Rayleigh target, R_{oi} denotes the observed SNR for subpulse i, and R_F is the SNR of a fixed-amplitude target. Note that the conditioning on the measured amplitude of the signal associated with target detection has been omitted from (11.40) so the presence of a target is assumed.

$$\delta\left(\{R_{oi}\}_{i=1}^{N}\right) = \begin{cases} H_0, & T\left(\{R_{oi}\}_{i=1}^{N}\right) \leq \lambda_1 \\ H_1, & T\left(\{R_{oi}\}_{i=1}^{N}\right) > \lambda_1 \end{cases} \tag{11.41}$$

and the dash lines correspond to the discrimination with the *variance test* of

$$\delta\left(\{R_{oi}\}_{i=1}^{N}\right) = \begin{cases} H_0, & \tilde{T}\left(\{R_{oi}\}_{i=1}^{N}\right) \leq \lambda_2 \\ H_1, & \tilde{T}\left(\{R_{oi}\}_{i=1}^{N}\right) > \lambda_2 \end{cases} \tag{11.42}$$

where we have skipped the derivation given in [17] and

$$\tilde{T}\left(\{R_{oi}\}_{i=1}^{N}\right) = \frac{1}{2\hat{R}_F + 1}\left[-\hat{R}_R + \sum_{i=1}^{N} R_{oi}^2\right] \tag{11.43}$$

where $\hat{R}_F$ can be found by

$$\hat{R}_F^{(0)} = \frac{1}{N}\left[\sum_{1}^{N} \sqrt{R_{oi}}\right]^2 - \frac{1}{2} \tag{11.44}$$

with the maximum likelihood estimate $\hat{R}_F^{(0)}$ obtained through the use of

$$I_1|0 \approx 1 - \frac{1}{2x} \quad x > 0.5$$

and are the first two terms in a binomial series.

11.2 Signal Processing Techniques   849

The results of Monte Carlo simulations with 100,000 experiments are summarized in Figure 11.19 where the solid lines correspond to the GML discrimination of (11.41) and the dash lines correspond to the discrimination with the variance test of (11.42). In Figure 11.19(a) and Figure 11.19(b) show the detection threshold as a function of the probability of false detection for fixed amplitude targets where the solid lines correspond to the *GML discrimination* of

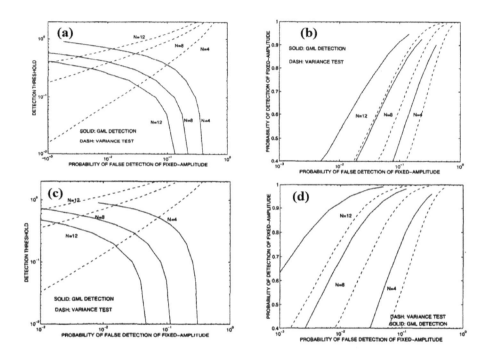

Figure 11.19 Monte Carlo simulations with 100,000 experiments are summarized in (a) showing detection thresholds and (b) ROC curves for $R = 7$ dB detection threshold as a function of the probability of false detection for fixed amplitude targets where the solid lines correspond to the *GML discrimination* of 11.41 and the dashed lines correspond to the discrimination with the *variance test* 11.42; (c) and (d) are for $R = 13$ dB; (©IEEE, reprinted with permission from [17].)

Multiple pulses are required to make an assessment with as few as four pulses (at *SNR* of 13 dB) being necessary to achieve the necessary balance between (1) the probability of detection of fixed amplitude and (2) the probability of false detection of fixed-amplitude. More pulses are required at lower *SNRs* but detection is still possible. Using GML detection with 4 subpulses at SNRs greater than 10 dB provides reliable discrimination between echoes from Rayleigh targets and echoes from an RGPO or a fixed-amplitude target.

850 COUNTER-DRFM METHODS

11.2.5 Counter-DRFM Pull-Off Using LSTM

In Chapter 9 we saw the kinematics of the RGPO, VGPO, and coordinated RVGPO along with their consequences in the range-Doppler domain. Since their development by Hochreiter and Schmidhuber in 1997 (while doing research on the *vanishing gradient problem* [18]), the LSTM AI networks have been particularly capable of both extracting features from sequential, time series input signals automatically and learning their long-range dependencies.[3] In the traditional RNN the recurrent hidden layer has one state h_t at step t while the LSTM adds a memory cell c_t to encode the sequence in memory [19].

Using fast range-Doppler signal processing to encode the separation process between the true and false targets within the tracking gate, a DRFM pull-off attack in either *time or frequency* can be recognized using the recurrent LSTM method. Since the method applies to either the range pull-off (RGPO) or Doppler pull-off (VGPO) the example below demonstrates the method on a VGPO using the Doppler domain [19]. Figure 11.20 shows the VGPO timeline broken down into three-stages.

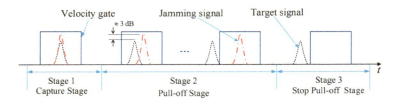

Figure 11.20 VGPO timeline broken down into three-stages.

The three stages are summarized below.

- Capture stage: After receiving radar pulses, jammer retransmits these pulses with the same Doppler frequency but increased power (≈ 3 dB). Velocity gate of radar can capture false target pulses at this stage.

- Pull-off stage: Jammer EA signal increases (or decreases) Doppler frequency of attack pulses to guarantee separation of true, false targets.

- Stop pull-off stage: When velocity gate loses true target and tracks false target, jammer stops transmitting false signals. Radar does not see any pulse energy in velocity gate, all velocity tracking is lost.

Recall the Doppler frequency of the VGPO can be expressed as:

3 Artificial neural networks (such as the MLP) and convolutional neural networks only process fixed-length inputs and any inherent dependencies present that depend on the inputs before and inputs after are not considered.

11.2 Signal Processing Techniques

$$f_{dj} = \begin{cases} f_d & 0 \leq t \leq t_1 \\ f_d + v_f(t - t_1) & t_1 \leq t \leq t_2 \\ 0 & t_2 \leq t \leq t_3 \end{cases} \quad (11.45)$$

where v_f is the pull-off velocity with its sign depending on the direction of pull and T_j is the time value of the attack period. The attack signal received by the radar is then

$$J(t) = A_j \exp\left[j2\pi(f_c + f_{dj})\left(t - \frac{2R_0}{c}\right) \right] \quad (11.46)$$

A fast recognition method of a track gate pull-off is shown in Figure 11.21. This model uses the LSTM network to implement the automatic feature extraction and sequence dependency learning [19].

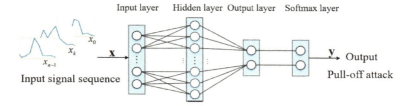

Figure 11.21 Fast network recognition of track gate pull-off.

The network demonstrated here consists of an input layer, one hidden layer with 32 units, a softmax layer, and an output layer.

Without a loss of generality, we assume the radar signal processing for tracking a target includes the coherent interception, pulse compression (PC), clutter filtering (or moving target detection) (MTD) and constant false alarm rate (CFAR) detection. This leads to an output target information sequentially, frame by frame as $x = [x_0, x_1, \ldots, x_k, \ldots, x_{n-1}]$ where x_k is the signal in the tracking gate of each frame.

As discussed in Section Chapter 9.5, the LSTM as a RNN, maintains a hidden state. As it processes the sequential inputs it activates the hidden state at each input depending on the previous one. In this matter, the LSTM has the potential to represent and learn sequential dependencies contained in the data.

As Hochreiter and Schmidhuber first discovered using simple RNNs, as one goes deeper into the network, the gradient gets very small, nearly zero. The LSTM from Chapter 8 is redrawn in Figure 11.22 to show the weights in each of the forget gate, the input gate, and the output gate as well as the three inputs X_t at time step t and h_{t-1} and c_{t-1} to denote the output and the cell state information at time step $t-1$, respectively.

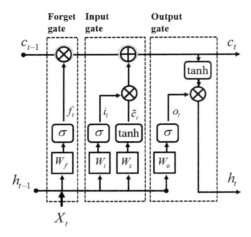

Figure 11.22 LSTM redrawn to show the weights in the forget gate W_f, the input gate W_i, W_c, and the output gate, W_o.

The forget gate decides which contents of the existing memory cell should be forgotten (0) or retained (1). The input gate modulates the input information that should be added to the memory cell (2-steps). The output gate determines which memory contents should be exposed in the current time step. Summarizing, the equations that calculate h_t and c_t are

$$f_t = \sigma(W_f[h_{t-1}, X_t] + b_f])$$
$$i_t = \sigma(W_i[h_{t-1}, X_t] + b_i])$$
$$o_t = \sigma(W_o[h_{t-1}, X_t] + b_o])$$
$$\tilde{c}_t = \tanh(W_c[h_{t-1}, X_t] + b_c])$$
$$c_t = c_{t-1} \odot f_t + \tilde{c}_t \odot i_t$$
$$h_t = \tanh(c_t) \odot o_t$$
(11.47)

where $[h_{t-1}, X_t]$ indicates that h_{t-1} and $X(t)$ are concatenated into a new feature vector and f_t, i_t, and o_t denote the activation vectors of the forget gate, the input gate and the output gate, respectively. Also, c_t represents the new contents of the memory cell, with $W_{f,i,o,c}$ and $b_{f,i,o,c}$ representing the weight matrices and bias values.

In the LSTM network shown in Figure 11.21, the input layer receives the signal from the tracking gate each coherent processing frame. The hidden layer calculates the hidden state information for the current signal and also transmits this information to the next cell. The Softmax layer transforms the features extracted by the hidden layer into the form of a probability distribution [19]. Finally, the output layer determines the label from the binary decision made by the softmax layer (pull-off or not). The statistical average of this binary label can then reveal insight into the recognition probability.

11.2.5.1 Pull-Off Recognition Simulation Results

The radar has a 5-ms coherent processing interval (cpi). The example data set consists of 4000 sequences with 40 frames per sequence. Of the 4000 sequences, 2000 of them are pull-off sequences and 2000 are no pull-off sequences. The pull-off time was 200 ms resulting in complete separation between the true and false target within the velocity gate.

The range of selected pull-off velocities $v_1, v_2, \ldots, v_{10}$ with each velocity was simulated 100 times (both 3 dB and 7 dB JSR).

The non-pull-off attack sequences are generated in two different ways: (1) signal sequence without EA, only the target echo in velocity gate, and (2) single false target deception EA mixed with the true target. That is, the true and false targets are colocated in the velocity gate (without separation). The simulation parameters are shown in Table 11.4.

Table 11.4
Pull-Off Simulation Parameters

Category	Parameter	Value
Radar Parameter	Waveform	Chirp
	BW(MHz)	10
	PW(us)	5
	Accumulation time(ms)	5
	Sampling Frequency(MHz)	20
Jamming Parameter	JSR(dB)	[3,7]
	Pull-off Velocity (v_1 v_{10}, KHz/s)	[3,3.4,3.8,4.2,4.6,5, 5.4,5.8,6.2,6.6]
	Pull-off Time(ms)	200

In order to select a suitable hidden layer structure, simulations were run to compare a different number of hidden layer units $L = 2, 4, 8, 16, 32, 64, 128$. The 4000 sample sequences were divided into two data sets. Data set 1 consists of 2800 sequences selected randomly (range of velocities is v_1 to v_{10}. Data set 2 then consists of the remaining 1200 sequences. For each sequence sample in data set 1, a new sample of length N is intercepted from the beginning of this sequence for training, and N is randomly chosen from 1 to 40. Then each sequence in data set 2 is tested to obtain the recognition rate at the 4th frame, 10th frame, and 20th frame separately, corresponding to 20 ms, 50 ms, and 100 ms.

To compare the recognition performance with different pull-off velocities, the training data are selected from data set 1 with the range of velocities v_1 to v_{10}. The testing data are selected from data set 2 with the velocities are v_1, v_6, v_{10}. Figure 11.23 shows the plot of the recognition rate for the three groups of signals sequences with different velocities in Figure 11.23.

854 COUNTER-DRFM METHODS

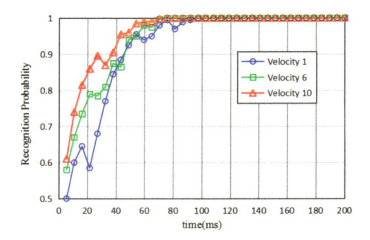

Figure 11.23 Recognition performance with different pulling velocities. (©IEEE, reprinted with permission from [19].)

In comparison, the V_{10} is the fastest followed by V_6 and v_1 is the slowest. Also noted is that their recognition rates all approach 100% at 90 ms and the faster the velocity, the faster the separation of the true and false targets in the tracking gate and the faster the recognition of the pull off attack.

With the LSTM currently the latest branch of recurrent neural networks different statistical distributions, which implies a change occurring in the measurement distribution after the RGPO onset or termination. Further, the subsequent tests after detection of the RGPO are unclear [20].

Consequently, the sequential detection problem formulation is segmented into two parts: (1) RGPO onset detection and (2) RGPO termination detection [20]. They are both simple hypothesis testing problems since all the parameters are attainable. In addition, sequential hypothesis testing methods are chosen for the following good reasons:
(a) Measurements in the tracking systems are usually available sequentially;
(b) A sequential test usually makes a decision faster than a nonsequential test (on average) under the same decision error rates;
(c) A sequential test does not need to determine the sample size in advance;
(d) The thresholds of (SPRT)-based sequential tests can be approximately determined without knowing the distribution of the test statistics.

The CUSUM test and Shiryayev's sequential probability ratio test (SSPRT) are two of the most popular approaches for sequential change detection. CUSUM is non-Bayesian while SSPRT is Bayesian. CUSUM minimizes the average detection delay given the decision error probabilities, and SSPRT minimizes a risk function at each time assuming that the change point has a geometric distribution a priori.

The statistical modeling of the measurements consists of considering a phased array monopulse radar for single target tracking obtaining multiple measurements in range, bearing, and elevation providing clutter, true target measurement, and false target (RGPO) at time k. Each individual measurement must be a return from a true target, clutter, or false target. Here, the term "clutter" includes clutter returns (scattering off of objects in the environment such as land, water, and buildings) and false alarms resulting from thermal noise in the radar receiver.

For a RGPO measurement, the bearing and elevation are nearly the same as the true target, but the range is progressively walking away from the true target return, denoted as $r_k^f = r_k^t + \Delta r$. The range pull-off Δr is determined by the time delay τ of the RGPO program; that is, $\Delta r = c\tau/2$. In general, there are two basic models to generate the time delay

$$\tau = v_0(t_k - t_0)/c \quad \text{Linear} \tag{11.48}$$

$$\tau = a_0(t_k - t_0)^2/2c \quad \text{Parabolic} \tag{11.49}$$

where t_0 is the start time of the walk-off program, v_0 is the velocity, and a_0 is the acceleration.

For a single target environment there are five hypothesis possibilities for any given pair of measurements with close angles referred to as the common-angle measurements:

(a) Pure clutter pair (hypothesis H_{cc});

(b) Clutter and target measurement pair (H_{ct});

(c) False-target measurement pair (including the cover pulse) or target and false-target measurement pair (H_{tf}), both referred as T-FT measurement pair;

(d) Multiple false-target measurement pair, excluding the cover pulse (H_{ff});

(e) Clutter and false-target measurement pair (H_{cf}).

Hypothesis H_{cc} actually consists of three subhypotheses H'_{cc} (a clutter return pair), H''_{cc} (a clutter return and a false-alarm), and H'''_{cc} (a false alarm pair).

Hypothesis H_{ct} actually consists of two subhypotheses H'_{ct} (a clutter return and a target), H''_{ct} (a false alarm and a target).

Hypothesis H_{tf} actually consists of two subhypotheses H'_{tf} (a target and false-target measurement pair) and H''_{tf} (a false-target pair, one being the cover pulse).

H_{cf} is neglected due to its extremely small probability, which occurs when neither the target measurement nor the cover pulse is detected and a clutter measurement happens to have a common angle with the other false-target measurement [21].

11.2.5.2 Sequential Detection Techniques for RGPO Onset and Termination

Sequential change detection methods for binary hypothesis testing problems have been successfully applied to spectrum sensing. Walds, sequential probability ratio test (SPRT) is the basis of such tests as discussed; before however, it *does not fit change detection*.Therefore, extended versions of SPRT for change detection have been proposed, the two most well known of which are the CUSUM test and the Shiryayev's SSPRT. They are both optimal for simple binary hypothesis testing under different criteria.

11.2.5.3 Cumulative Sum Log-Likelihood Ratio Test

The CUSUM test is in a non-Bayesian framework, which assumes the unknown change time n is deterministic. One of its derivations is based upon a repeated use of SPRT with the lower threshold $\log A = 0$ and the upper threshold $\log A = B$, depending on decision error rates. The key difference is to restart the SPRT whenever H_0 is declared, which makes it fit to change detection problems[21].

The CUSUM of the log-likelihood ratio is

$$L^k = \max\left\{ L^{k-1} + \log \frac{f(\Delta\gamma^k|H_1,\gamma^{k-1})}{f(\Delta\gamma^k|H_0,\gamma^{k-1})}, 0 \right\}, \quad L^0 = 0 \qquad (11.50)$$

* If $L^k \geq \lambda$, declare H_1 and then stopping time $\hat{n} = \min\{k : L^k \geq \lambda\}$ is taken as the RGPO onset or termination time.
* Else continue the test $(k \to k+1) if L^k < \lambda$.

11.2.5.4 Shiryayev's Sequential Probability Ratio Test

The SSPRT minimizes an expected cost at each time. It is the quickest detection of a change in a sequence of conditionally independent measurements under the given decision cost. SSPRT is a Bayesian approach, which assumes the change time is random. It needs to know the prior probability p_0^1 of the change at $n = 0$ and the transition probability π from H_0 to H_1.

The posterior probability, defined as $p_k^1 = P\{n \leq k|\Delta\gamma^k\}$, stands for RGPO onset or termination at an unknown time N no later than tiime k given the available measurements $\Delta\gamma^k = \{\gamma_1,\ldots,\gamma_k\}$. Then $p_k^0 = 1 - p_k^1$ is the posterior probability that no change occurred up to time k.

The test statistic of the SSPRT is obtained recursively by

$$P_k = \frac{p_k^1}{p_k^0} = \frac{f(\Delta\gamma_k|H_1,\Delta\gamma^{k-1})}{f(\Delta\gamma_k|H_0,\Delta\gamma^{k-1})} \frac{P_{k-1} + \pi}{1 - \pi}, P_0 = \frac{p_0^1}{p_0^0} \qquad (11.51)$$

and the decision rule is:[20].
* Declare H_1, if $P_k \geq P_T$.
* Else, continue the test $k \to k+1$ where P_T is an appropriate threshold.

Note as pointed out in [21] it is difficult to obtain the prior information for the RGPO detection problems. Fortunately, the performance of the tracking filter using the SSPRT is not sensitive to the prior information [20].

Simulation results are run to test out the decisions rules with only one T-FT pair but multiple clutter returns and false alarms during target tracking. The radar revisit time to constant at 0.1s. The entire track lasts 100s. The RGPO is active from time $k = 201$ to 400 and from $k = 501$ to 700. For the SSPRT detector for RGPO onset, the prior probability of H_0 is 0.9 and transition probability from H_0 to H_1 is 0.02. The transition probability from H_0 to H_1 is set to 0.1 for RGPO termination detection. Monte Carlo simulation of 100 runs are used unless otherwise specified.

The ROC curves (P_d vs. P_{fa}) are shown in Figure 11.24 for $k = 202, 203$ and 204. The miss detections for RGPO onset at k=202, 303, and 204 with 1000 runs is shown in Figure 11.25.

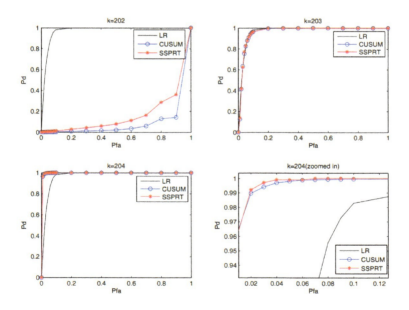

Figure 11.24 ROC curves of RGPO onset detectors at time $k = 202, 203, 204$ [20].

Detector	Type I error rate P_{fa}		
	1%	5%	10%
LR test	(606, 469, 403)	(73, 74, 59)	(2, 10, 8)
CUSUM test	(976, 884, 48)	(909, 173, 2)	(873, 28, 0)
SSPRT test	(976, 874, 51)	(907, 161, 1)	(869, 27, 0)

Figure 11.25 ROC curves of RGPO onset detectors at time $k = 202, 203, 204$. (©IEEE, reprinted with permission from [20].)

Simulation results show that CUSUM and SSPRT have comparable performance, but SSPRT is slightly better than CUSUM in terms of average detection delay and ROC curve, and they both substantially outperform the likelihood ratio test. Further results for countering RGPO using frequency diversity are shown using pulse diversity methods in [22]. Here a joint slow/fast time method implements a transmission pulse block in half of the PRI, which is designed based on a orthogonal block structure. Then a one-step matched-filtering operation can be used to efficiently suppress the RGPO. Since the cancellation process can be completed within one PRI, the stationary hypothesis is no longer needed.

11.2.6 Countering the Coordinated RGPO-VGPO

Synchronous range-velocity gate pull-off EA is an important technique because it will cause a tracking seeker/radar to break track (with a high enough JSR) due to the fact that it appears as an actual target maneuver. That is, correlation of a range-rate $\dot{R}$ tracker and a Doppler frequency tracker would lead to no differences indicating no presence of EA. The protection against this DRFM EA mode has been studied with most relying on a degree of nonsynchronization or phase noise.

A more sophisticated approach was taken in [23] where the EA interference is modeled as a complex correlated Gaussian process with the signal belonging to a cone whose axis is the true target signal. Note that after baseband conversion the EA signal spectrum and the target signal spectrum are centered around the target Doppler frequency f_D. Also consider that both are filtered by an antialiasing filter and sampled by the ADC and that the filter suppresses all of the DRFM components outside the filter's bandwidth.

To quantify the errors introduced by the DRFM on the EA signal with respect to the target signal, a new measure of effectiveness is introduced: the *jammer signal error angle* (JSEA) μ as [23]

$$\rho = \cos(\mathbf{x}, \mathbf{y}) = \cos(\mu) = \frac{|\mathbf{x}^H \mathbf{y}|}{||\mathbf{x}|| \cdot ||\mathbf{y}||} \quad (11.52)$$

and is the angle between the two complex vectors **x** and **y**.[4] The angle μ depends on the number of quantization levels $N = 2^M$, the normalized frequency f_0 and the number of samples K. The results for $M = 1, 2, 3$, and 4 are shown as a function of f_0 in Figure 11.26 for $K = 128$ samples.

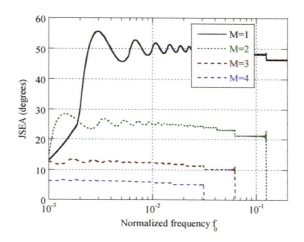

Figure 11.26 JSEA as a function f_0, $K = 128$. (©IEEE, reprinted with permission from [23].)

In Figure 11.27, the target and EA signal normalized spectrum are shown centered around f_0. The differences between the two spectra increase with decreasing number of bits. Numerical results show that when the number of quantization bits is low, the angle μ between the target steering vector and the EA steering vector is significant [23]. Consequently, this measure of effectiveness can be used to recognize the EA signal and avoid a false detection. If M is large, the angle is almost zero, and therefore, it is not useful.

---

[4] Note that this is analogous to the angle ρ to account for the mismatch (offset angle ϕ) between radar E-field E_r and DRFM jammer E-field E_j. The antenna polarization mismatch, ϕ, is shown in Figure 10.22 in Chapter 10 for EA as an offset angle between the true return and the EA – a mismatch factor of $\rho = \cos^2(\varphi)$ offset.

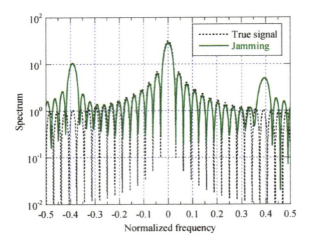

Figure 11.27 Jammer EA and true signal spectra signal centered around f_0. (©IEEE, reprinted with permission from [23].)

The problem of detecting and classifying the signal as coming from the actual target or deciding if the received signal is a coordinated RGPO-VGPO from a DRFM EA (RVGPO) can also be formulated as a two-step multiple hypothesis testing (MHT) problem starting with the complex measured vector

$$\mathbf{z} = [z(0) z(1) \cdots z(K-1)]^T$$

recorded by the radar during the time on target where K is the number of samples collected by the radar during a single pulse [23]. Under the null hypothesis (H_0) it is assumed that the data consist of only clutter plus thermal noise. Under the alternative hypothesis, it is instead assumed that the data consists of the sum of noise and signal backscattered by the radar target (H_1) or by the DRFM (H_2). Consequently, we formulate our multiple hypotheses problem as follows:

$$H_0 : \mathbf{z} = \mathbf{d}$$

$$H_k : \mathbf{z} = \mathbf{d} + \mathbf{s}_k, \quad \text{for } k = 1, 2 \tag{11.53}$$

where $\mathbf{s}_1 = \alpha \mathbf{p}$ is the target vector, α is a complex Gaussian random variable $\alpha \sim CN(0, \sigma_\alpha^2)$ (zero mean and variance σ_α^2), $\mathbf{p}$ is the target steering vector, and known a priori to be of the form $p(n) = \exp(j2\pi f_D n)$ where the target Doppler frequency is normalized to the radar's PRF [23].

Let the DRFM deception EA be given by $\mathbf{s}_2 = \beta \mathbf{p}_j$ where $\beta \sim CN(0, \sigma_\beta^2)$. Note that the vector $\mathbf{p}_j$ is unknown to the receiver since it is generated by the DRFM. Also, let the disturbance be formed by the superposition of thermal noise and correlated clutter and be modeled as complex Gaussian distributed random vector $\mathbf{d}$ with zero-mean and correlation matrix $\mathbf{M} = E\{\mathbf{dd}^H\}$.

Now assign the hypothesis probabilities $P_k = Pr\{H_k\}$ the a priori[5] probability of the hypothesis H_k for $k = 0, 1, 2$. The detection/classification has the goal to detect the presence of a target or DRFM EA signal and correctly classify it. If P_0, P_1, P_2, and $\mathbf{p}_j$ were known we could use the maximum a posteriori (MAP) criterion and decide in favor of one of the three hypothesis according to the following rule[23].
Decide in favor of $H_{\bar{k}}$, where $\bar{k}$ is the value of k for which the a posteriori probability[6] of H_k, $\Pr\{H_k|\mathbf{z}\}$, is maximal. That is,

$$\hat{H}_{\bar{k}} : \bar{k} = \arg\max_k \Pr\{H_k|\mathbf{z}\} = \arg\max_k f_{\mathbf{z}|H_k}(\mathbf{z}|H_k)P_k \quad k = 0, 1, 2 \tag{11.54}$$

where $f_{\mathbf{z}|H_k}(\mathbf{z}|H_k)$ is the data pdf conditioned on hypothesis H_k. However, this is practically not feasible since the a priori probabilities are usually unknown. But fortunately, we can assume to know the steering vectors $\mathbf{p}$ and $\mathbf{p}_j$ and can use the Neyman-Pearson hypothesis criterion with two matched filters – one for the target signal, one for the DRFM EA, testing H_1 against H_0 *and* H_2 against H_0. However, $\mathbf{p}_j$ is actually unknown also and so the second matched filter cannot be constructed.

Consequently, based on these considerations, a two-block detection/classification scheme can however be constructed as shown in Figure 11.28[23].

[5] A priori probabilities proceed from theoretical deduction rather than from observation or experiments.
[6] A posteriori probabilities relate to those probabilities that are calculated from reasoning or knowledge that proceeds from actual observations or upon calculations from experimental data.

862 COUNTER-DRFM METHODS

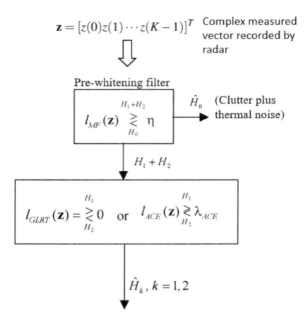

Figure 11.28 Summary of the two-step detection and classification scheme to decide if the received signal is a coordinated RGPO-VGPO (RVGPO) from a DRFM or an actual target. (Adapted from [23].)

The first block is devoted to the *detection task*, testing the cumulative hypothesis $H_{12} = H_1 + H_2$ against hypothesis H_0. It is made up by a whitening matched filter (WMF) (matched to **p**) (typically found in a coherent radar system) and given as

$$l_{\text{WMF}}(\mathbf{z}) = \frac{|\mathbf{p}^H \mathbf{M}^{-1} \mathbf{z}|^2}{\mathbf{p}^H \mathbf{M}^{-1} \mathbf{p}} \underset{H_0}{\overset{H_1+H_2}{\gtrless}} \eta \qquad (11.55)$$

If the threshold η is exceeded, a detection is declared. The threshold is fixed based on the probability of false alarm P_{fa} that the radar system can tolerate.

The second block is a clairvoyant detector, and in this case, performs the actual task of classifying the signal as an actual target return or a DRFM EA. In the second decision block, either the *generalized likelihood ratio test* (GLRT) classifier is passed or the *adaptive coherent estimator* (ACE) classifier is passed.

The *GLRT hypothesis detector* is:

$$l_{\text{GLRT}}(\mathbf{z}) = \frac{|\mathbf{p}^H \mathbf{M}^{-1} \mathbf{z}|^2}{\mathbf{p}^H \mathbf{M}^{-1} \mathbf{p}} - 2\Re\{\mathbf{z}^H \mathbf{M}^{-1} \hat{\mathbf{s}}_{2\text{ML}}\} + \{\hat{\mathbf{s}}_{2\text{ML}}^{\mathbf{H}} \mathbf{M}^{-1} \hat{\mathbf{s}}_{2\text{ML}}\} \underset{H_2}{\overset{H_1}{\gtrless}} 0 \qquad (11.56)$$

Here, one might notice that the first term on the right side of (11.56) is (11.55).

The *ACE hypothesis detector* is:

$$l_{ACE}(\mathbf{z}) = \frac{|\mathbf{p}^H \mathbf{M}^{-1} \mathbf{z}|^2}{(\mathbf{z}\mathbf{M}^{-1}\mathbf{p})(\mathbf{z}^H \mathbf{M}^{-1}\mathbf{z})} \underset{H_2}{\overset{H_1}{\gtrless}} \lambda_{ACE} \tag{11.57}$$

The performance of the classifier was analyzed by means of correctly estimating the true hypothesis. However, it was not possible since the a priori probabilities were not known. Therefore, the performance of the scheme was computed by calculating the

* *Probability of false alarm* (P_{fa}):

$$P_{fa} = Pr\{l_{WMF}(\mathbf{x} > \eta | H_0)\} \tag{11.58}$$

or

$$P_{fa} = \exp(-\eta).$$

*Probability of detection (P_d):

$$P_d = \frac{P_1}{P_1 + P_2} P_{d|H_1} + \frac{P_2}{P_1 + P_2} P_{d|H_2} \tag{11.59}$$

where

$$P_{d|H_1} = Pr\{\hat{H}_{12}|H_1\} = Pr\{l_{WMF}(\mathbf{x}) > \eta | H_1\}$$

and

$$P_{d|H_2} = Pr\{\hat{H}_{12}|H_2\} = Pr\{l_{WMF}(\mathbf{x}) > \eta | H_2\}$$

and since $P_1 = P_2$ and

$$P_{d|H_i} = \exp\left(-\frac{\lambda}{2\sigma_i^2}\right), \quad i = 1, 2 \tag{11.60}$$

where

$$\sigma_1^2 = \frac{1 + \sigma_\alpha^2 \mathbf{p}^H \mathbf{M}^{-1} \mathbf{p}}{2}$$

and

$$\sigma_2^2 = \frac{1}{2}\left[1 + \frac{\sigma_\alpha^2 (\mathbf{p}^H \mathbf{M}^{-1} \mathbf{p}_j)}{\mathbf{p}^H \mathbf{M}^{-1} \mathbf{p}}\right]$$

and we get

$$P_d = \frac{1}{2}\exp\left(-\frac{\lambda}{2\sigma_1^2}\right) + \frac{1}{2}\exp\left(-\frac{\lambda}{2\sigma_2^2}\right) \tag{11.61}$$

Finally, the probability of correct classification (P_c) is

$$P_c = \frac{P_1}{P_1+P_2}P_{c|H_1} + \frac{P_2}{P_1+P_2}P_{c|H_2} \tag{11.62}$$

where we define $P_{c|H_k} = Pr\{\hat{k}|H_k\}$ for $k=1,2$. Then for the WMF and GLRT system we have

$$P_{c|H_1} = \Pr\{l_{WMF}(\mathbf{x}) > \eta, l_{GLRT}(\mathbf{x}) > 0 | H_1\} \tag{11.63}$$

and

$$P_{c|H_2} = \Pr\{l_{WMF}(\mathbf{x}) > \eta, l_{GLRT}(\mathbf{x}) < 0 | H_2\} \tag{11.64}$$

For the WMF and ACE system, we have

$$P_{c|H_1} = \Pr\{l_{WMF}(\mathbf{x}) > \eta, l_{ACE}(\mathbf{x}) > \lambda | H_1\} \tag{11.65}$$

and

$$P_{c|H_2} = \Pr\{l_{WMF}(\mathbf{x}) > \eta, l_{ACE}(\mathbf{x}) < \lambda | H_2\} \tag{11.66}$$

and in this case $P_{c|H_2} = (1/2)P_{c|H_1} + (1/2)P_{c|H_2}$.

Monte Carlo results were run by defining the target SNR as $SNR_t = \sigma_\alpha^2/\sigma_d^2$ and the DRFM EA-to-interference (clutter plus noise) SNR as $SNR_j = \sigma_\beta^2/\sigma_d^2$. Also $M=2$ (recall M is the number of bits of the DRFM and the number of levels is $N = 2^M$. Also, $K = 32$, the target Doppler frequency $f_d = 100$ kHz. The intermediate frequency of the DRFM $F_{IFdrfm} = 100$ MHz and the sampling frequency is $f_s = 1.024$ GHz.

The EA steering vector $\mathbf{p}_j$ has been normalized to the same norm as $\mathbf{p}$. Concerning the clutter-plus-noise matrix, it was set to $\{\mathbf{M}_{i,j} = \sigma_d^2 \rho^{|i-j|}\}$ and two cases were analyzed; that for $\rho = 0$ (white Gaussian noise) and $\rho = 0.9$ (Gaussian colored noise to simulate the presence of clutter).

In Figure 11.29(a) shows the probability of correct classification P_c as a function of the signal-to-disturbance power ratio SNR where the $SNR_t = SNR_j = SNR$ for the WMF and GLRT system using 1000 Monte Carlo trials with the decision threshold set to achieve $P_{fa} = 10^{-4}$. with $\rho = 0$. Figure 11.29(b) shows the same probability of correct classification P_c but with $\rho = 0.9$.

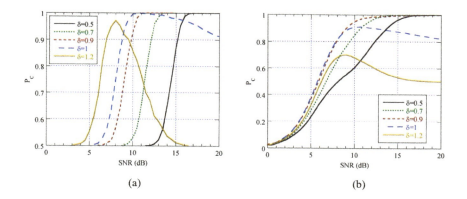

Figure 11.29 Probability of correct classification for the WMF and the GLRT showing (a) $\rho = 0.0$ and (b) $\rho = 0.9$. (©IEEE, reprinted with permission from [23].)

The curves in these figures correspond to different values of θ, the angle characterizing a Lagrange multiplier cone of the GLRT formulation [23]. We set $\theta = \delta\mu$ where $\delta = 0.5, 0.7, 0.9, 1.0$ and 1.2 and μ are as defined before, the JSEA. The performance of the GLRT depends on the value of δ. For $\delta < 1$, the EA steering vector $\mathbf{p}_j \notin C$, then there is a mismatch between the cone in which the ML estimator looks for the best estimate of $\mathbf{s}_2 = \beta \mathbf{p}_j$ and the true position of $\mathbf{s}_2$. This mismatch decreases with increasing values of δ and the performance of the GLRT improves.

11.2.7 Statistical Signal Processors

Below we describe a DRFM RF quantization detection concept that does not depend directly on a spectral estimation technique but uses an enhanced statistical processing of the received signals and ignores the DRFM limitations such as DC offsets, differential gains, and quadrature errors. The basis for the statistical detection relies on a linear discriminant analysis (LDA) and ANN processing to discriminate to 3-bit DRFM resolution depending on jammer-to-noise ratio.

The basis of a higher-order moment structure of the quantized signal detection algorithm relies on the ADC quantization step size process that occurs when sampling the radar signal at the jammer's receiver. This ADC quantization will produce inevitable harmonic spurs that persist through processing and rebroadcast of the jamming signal. When the combined signal is received back at the initial transmitting radar, the victim radar can detect the harmonic quantization spurs through a statistical analysis.

The statistical process seeks to detect the mean square error power of the RF pulse after channel noise contamination. It uses a three-way multivariate analysis of variance (MANOVA) that assumes there are distinct causes of variance in the data as well as an *interaction* between the distinct causes. This can be described with the equation adapted from [24]

$$T = H_a + H_b + H_c + H_{ab} + H_{ac} + H_{bc} + H_{abc} + E$$

where the overall sum of squares and products matrix (SSPM) T is comprised of the SSPMs of factors A, B, and C as well as the interactions between any two factors, all three factors, and a term E, which represents the residual unexplained variance.

Next, a neural network is applied for pattern recognition and trained to discriminate from among the factors. The specific model presented in [24] summarized in Figure 11.30 specifically seeks to classify the received signal based on (1) the state of quantization, (2) the number of jammer bits/sample, and (3) the SNR or JNR. Figure 11.30 shows the calculated signal characteristics that are first calculated before being split off to the effects.

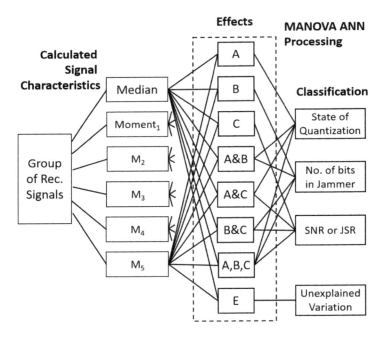

Figure 11.30 Diagram of the statistical signal processing method.

The process was able to discriminate with successful results for SNRs in the range of 20-40 dB and assumes present DRFMs are low-bit devices because of high ADC sampling rate demands. Study results showed that both the MANOVA and a multilayer neural network statistical processing approach are promising solutions for implementing a DRFM detector in radar systems.

11.2.8 Use of Phase Noise Measurements

Since the transmission of the DRFM signals have to be on a carrier signal a counter-DRFM technique is based on the measurement of the phase noise *difference of oscillators* in the received signals. In this technique, a transmitted LPI signal is CW generated by a master oscillator (MO) with a certain amount of phase noise. That is, the transmitted signal is marked with slightly nonlinear frequency modulation produced by this oscillator Since the base station radars can use many kinds of subsystems and devices in their systems. Consequently, they have low phase noise oscillators.

The DRFM block does not have this advantage and the LO of the DRFM system has for example, a high phase noise level. Furthermore, a simulated signal made through complex circuits including mixer, analog filters, SSB modulator DSP, and/or FPGA have different PSD and higher phase noise levels in comparison to a real backscatter one.

That is, the phase noise level is increased when a signal is made through imperfect electronic devices. It can be considered very low phase noise MO and LO for carrier generation and downconversion in radar systems. A system is provided to measure the phase noise of oscillators to investigate the phase noise spectrum of any kind of oscillators, which have their own patterns and power levels. Subsequently, P_D and P_{fa} are obtained for evaluating the performance of distinguishing real targets from false ones.

We consider different SNRs and SJRs. In predefined P_{fa} and analyze P_D to verify the proposed method performance.

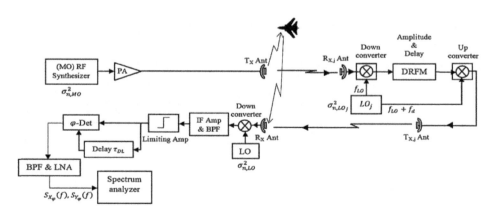

Figure 11.31 Block diagram of proposed method to measure the phase noise of target signals. (©IEEE, reprinted with permission from [25].)

The phase noise powers are measured through the same sets of circuits and coherent time periods. Then, the phase noise spectrum shape and its power can be used to recognize the real target backscatter from the DRFM signal.

Sometimes, control of the amplitude fluctuation of the received signal is very difficult. Therefore, to quantify the phase noise of DRFM received signal with respect to the target signal, a different metric can be used to compare phase noise levels that are normalized by signal power in predefined bandwidths. The likelihood ratio test (LRT) is used for target discrimination. In this case, a threshold level is achieved by minimum P_{fa} to discriminate the targets.

The proposed method has a simple structure based on the inherent characteristics of oscillators. It has no additional complexities and can easily be adapted to common radar systems. All presented results are investigated using real DRFM systems including subblocks in their structure. In [25] they consider both L-band and X-band radar sensors with time delay and Doppler frequency in the presence of additive white Gaussian noise (AWGN). Results are also given with additional details on the methodology.

11.2.9 Converse Beam Cross Sliding Spotlight SAR

Modern SAR is able to produce high-quality pictures with high resolution and for this reason remains a critical technology in the world of remote sensing. Consequently, many active EA techniques to degrade these types of sensors have been produced recently, some of which have been discussed in the previous chapter.

The image patch of a high-resolution *spotlight SAR* is very small, which limits its usefulness in many wide-swath, remote sensing applications. To address this and other limitations, a *sliding spotlight SAR* has been described [26]. For the sliding spotlight SAR, a the beam velocity V_f is smaller (or larger) than its platform velocity V_a. Note that a classic stripmap SAR results if the two velocities are equal ($V_f = V_a$) and a pure spotlight if the beam velocity is zero ($V_f = 0$). The classical inverse sliding spotlight mode is characterized by an antenna beam velocity, $V_f > V_a$, implying that the azimuth scene extension is larger than that of a classic stripmap mode [27].

The CBC sliding spotlight mode has the moving direction of V_f and V_a opposing each other. During the data acquisition time, the CBC sliding spotlight SAR mode image processing algorithm is based on a data blocking fast back projection for imaging.

During data acquisition, the SAR antenna is steered about the virtual rotation point as in a classical standard sliding spotlight mode and inverse sliding spotlight mode and a sliding *spotlight factor* α can be defined to describe the *grade of the slide* as

$$\alpha = \frac{r_{rot} - r_0}{r_{rot}} = 1 - \frac{r_0}{r_{rot}} = 1 - \frac{V_f}{V_a} \qquad (11.67)$$

for $V_f \leq V_a$ where $X_f = V_f \Delta T$ depicted the distance covered by the antenna beam and $X_a = V_a \Delta T$ is the distance covered by the platform during the data acquisition time

ΔT. This is shown in Figure 11.32 and shows the concept of the conventional converse beam cross (CBC) sliding spotlight SAR geometry [28].

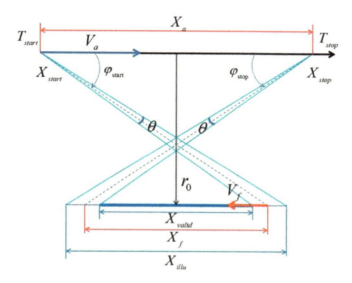

Figure 11.32 Geometry of the converse beam cross sliding spotlight synthetic aperture radar. (©IEEE, reprinted with permission from [28].)

Figure 11.34 shows the geometry for the cross beam sliding spotlight SAR mode and aircraft flight path.

870 COUNTER-DRFM METHODS

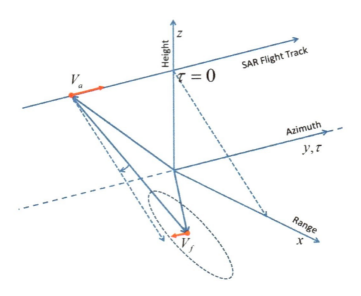

Figure 11.33 Geometry of CBC sliding spotlight SAR mode. (From [27].)

Here the x-component corresponds to the range direction, the y-component corresponds to the azimuth direction and direction of the velocity vector, and the z-component completes the right-handed Cartesian frame of reference. The SAR platform's velocity vector is parallel to the y-axis; the SAR's x and y position is 0 at $\tau = 0$ in azimuth. Figure 11.34 shows a block diagram of the data blocking, fast-back projection algorithm that can be executed in four steps as described below [27].

11.2 Signal Processing Techniques   871

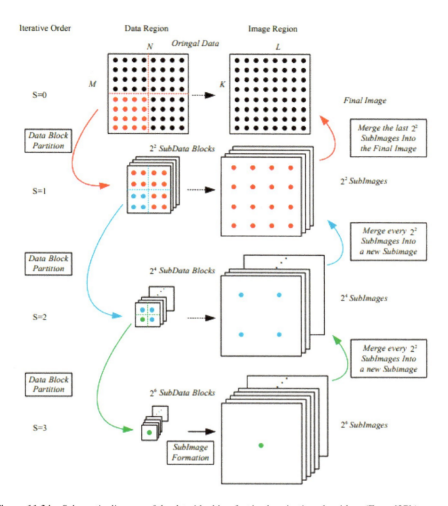

Figure 11.34 Schematic diagram of the data blocking fast backprojection algorithm. (From [27].)

The return echo data matrix is $s_r(m,n)$ and the data blocking based fast back-projection algorithm (DB-FBPA) can be executed following four steps. The algorithm below is taken from [27]:

Step 1:
Data matrix $s_r(m,n)$ is demodulated (dechirped) and deskewed. First determine the data block partition factor q and iterative order s according to the specific imaging requirements. Then calculate the subdata matrix $\hat{s}_{block}(m',n')$ and the SAR position

vector $\mathbf{p}'_t$ corresponding to each subdata matrix.

Step 2:
Single-point FFT operation of the subdata matrix in range direction. After data block partition operations, we can get q^{2s} subdata matrices $\{\hat{\mathbf{s}}_{block}\}_{q^{2s}}$ and corresponding new transmitter position vector $\mathbf{p}'_t$. In oder to do the single-point FFT operations of $\{\hat{\mathbf{s}}_{block}\}_{q^{2s}}$ in the range direction, the distance R'_{ref} and R'_{kT}, which between new transmitter position vector $\mathbf{p}'_t$ and the central point of the scene O and subimage grids $I_{block}(k', l')$, respectively, must be calculated. The expressions for R'_{kT} and R'_{ref} are

$$R'_{kT} = 2|\mathbf{p}'_{kT} - \mathbf{p}'_t| \tag{11.68}$$

$$R'_{ref} = 2|\mathbf{p}'_O - \mathbf{p}'_t| \tag{11.69}$$

Let $R'_{\Delta kT}$ be the value of the difference between R'_{kT} and R'_{ref}. According to dechirp theory, the corresponding frequency value of dechirped signal with a distance of $R'_{\Delta kT}$ is $f'_i = -K_r R'_{\Delta kT}/c$. Doing single-point FFT operations on every image grids according to their $R'_{\Delta kT}$ value as

$$\mathbf{S}'_{block}(f'_i, n') = \sum_{m'=0}^{\frac{M}{q^s}-1} \hat{\mathbf{s}}_{block}(m', n') \exp\left(-j2\pi f'_i \frac{m'}{f_s}\right) \tag{11.70}$$

Step 3:
Subimage formation. After processing every subdata matrix $\{\hat{\mathbf{s}}_{block}\}_{q^{2s}}$ with backprojecting operation, we get a set of subimages $\{I_{block}(k', l')\}_{q^{2s}}$ as

$$\{I_{block}(k', l')\}_{q^{2s}} = \left\{\sum_{n'=1}^{N/q^s} \mathbf{S}'(f'_i, n') \exp\left[j\frac{2\pi f_c}{c} R'_{\Delta kT}\right]\right\}_{q^{2s}} \tag{11.71}$$

which includes a data block partitioning operation. The compensated phase function is

$$\{\Theta_{block}\}_{q^{2s}} = \left\{\exp\left[-j\frac{2\pi}{c}(f'_i + K_r R'_{\Delta p}/c)\left(\frac{aM}{q^s}+1\right)\right]\right\}_{q^{2s}} \tag{11.72}$$

Multiplying (11.71) with the conjugate of $\{\Theta_{block}\}_{q^{2s}}$ after compensating the initial phase can be expressed as $\{\hat{I}_{block}(k'l')\}_{q^{2s}}$. The details can be found in reference [27].

Step 4:
Subimage combination and output the final imaging results. During the process of

subimage combination, the data blocking, fast backprojection algorithm merges adjacent q^2 subimages to generate a new subimage, and then do the same operation to newly generated subimages that are next to each other and so forth, until there is only one image finally left. The grid points contained in the new subimage (obtained by the combination of the former subimages) are twice the size of the former subimage grid points in both range direction and azimuth direction. This is due to the fact that SAR image is a complex image, the interpolation of the subimage should be operated on both in phase and amplitude.

In [27] the interpolation methods of phase and amplitude are selected as "nearest" and "spline" Figure 11.35 compares simulations for a point spread function for the classical sliding spotlight SAR and CBC sliding spotlight SAR sliding spotlight SAR and the fast backprojection algorithm. The simulation parameters and numerical results are given in Figure 11.36.

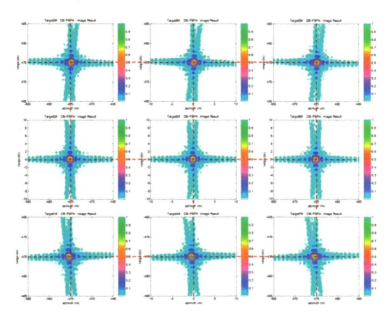

Figure 11.35 Geometry of CBC sliding spotlight SAR mode. (©IEEE, reprinted with permission from [27].)

SAR Mode	Parameters	Value		PSF
Classical Sliding spotlight	Wavelength	0.1874m	R	PSLL -13.69dB
	Beam width	5.376⁰		
	Speed	100m/s		ISLL -10.40dB
	Middle Range	7.072km		
	Steer Angle	5⁰	A	PSLL -13.21dB
	Bandwidth	300MHz		
	PRF	45.23Hz		ISLL -10.57dB
	Pulse Width	1us		
CBC sliding spotlight	Wavelength	0.1874m	R	PSLL -13.88dB
	Beam Width	5.376⁰		
	Speed	100m/s		ISLL -12.37dB
	Middle Range	7.072km		
	Steer Angle	15.58⁰	A	PSLL -13.89dB
	Bandwidth	300MHz		
	PRF	45.23Hz		ISLL -12.42dB
	Pulse Width	1us		

Figure 11.36 Simulation Parameters and PSF Properties [27].

11.3 CONCLUDING REMARKS

The miniaturization of microelectronics for electromagnetic warfare presents a new opportunity for DRFMs and jammers alike. By embedding this advanced EA functionality within UAV, UAS stand-in DRFMs, a new generation of smart jammers are emerging. This new class of DRFMs with embedded AI and deep machine learning features a self-protection, DRFM EA capability to mitigate adversarial sensors to a degree never seen before.

On the other hand, the inclusion of deep learning, machine learning embedded in the RF sensors to counter this advance DRFM threat, a disrupting, advance sensor capability is now forcing DRFMs to have a shorter turn-around time and turn the EA towards the guidance and control loops. This chapter gives an introduction to some of the counter-DRFM techniques and solutions that take advantage of the DRFM drawbacks including the limited turn-around time.

In this regard, the cat-n-mouse game continues driven only by the technological solutions offered and the deep, machine learning intelligence embedded!

APPENDIX 11: PROOF OF THE NEYMAN-PEARSON LEMMA

To start the proof, let B be any other critical region of size α and let

$$L_0 = \prod_{i=1}^{N} p(x_i : \theta_0)$$

and
$$L_1 = \prod_{i=1}^{N} p(x_i : \theta_1)$$

Next consider the regions A and B and the regions a, c and the region of intersection, b in Figure 11.37.

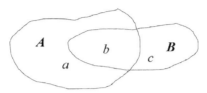

Figure 11.37 Regions A, B, a, c and the region of intersection b for the proof of the Neyman-Pearson lemma.

Then over region A we have,

$$\int \cdots \int_A \prod_{i=1}^{N} p(x_i : \theta_i) dx_1 \cdots dx_N = \int_A L_0 dx$$

and also

$$\int_A L_0 dx = \int_B L_0 dx$$

and since the α's are the same, the integrals over the region b will cancel and

$$\int_a L_0 dx = \int_c L_0 dx$$

Let us now calculate the size of a Type II error. That is, let us calculate the probability that the sample will fall outside the critical region when H_1 is true. The first step is to calculate

$$\beta_B = 1 - \int_B L_1 dx \quad \text{and} \quad \beta_A = 1 - \int_A L_1 dx$$

Then subtracting

$$\beta_B - \beta_A = \int_A L_1 dx - \int_B L_1 dx$$

while the integral over the common region cancels out

$$= \int_a L_1 dx - \int_c L_1 dx$$

Since a lives inside A, it then follows from the definition of A that every point a satisfies $L_0 \leq \kappa L_1$ and

$$\int_a L_1 dx \geq \frac{1}{\kappa} \int_a L_0 dx$$

As well, since c lies outside A, every point c satisfies $L_0 \geq \kappa L_1$ and

$$\int_c L_1 dx \leq \frac{1}{\kappa} \int_c L_0 dx$$

and also

$$\beta_B - \beta_A \geq \frac{1}{\kappa} \int_a L_0 dx - \frac{1}{\kappa} \int_c L_0 dx$$

However, we also know that

$$\int_a L_0 dx = \int_c L_0 dx$$

Therefore, $\beta_B \geq \beta_B$. The constant κ should be chosen to make A a critical region of size α. Q.E.D.

REFERENCES

[1] P. E. Pace, D. J. Fouts, S. T. Ekestorm, and C. Karow, "Digital false-target image synthesiser for countering ISAR," *IET Proceedings - Radar, Sonar and Navigation*, vol. 149, no. 5, pp. 248–257, 2002.

[2] D. J. Fouts, P. E. Pace, C. Karow, and S. T. Ekestorm, "A single-chip false target radar image generator for countering wideband imaging radars," *IEEE Journal of Solid-State Circuits*, vol. 37, no. 6, pp. 751–759, 2002.

[3] G. Li, Q. Zhang, L. Su, Y. Luo, and J. Liang, "A digital false-target image synthesizer method against isar based on polyphase code and sub-nyquist sampling," *IEEE Geoscience and Remote Sensing Letters*, vol. 17, no. 3, pp. 372–376, 2020.

[4] P. E. Pace, D. J. Fouts, and D. P. Zulaica, "Digital image synthesizers: are enemy sensors really seeing what's there?" *IEEE Aerospace and Electronic Systems Magazine*, vol. 21, no. 2, pp. 3–7, 2006.

[5] M. Soumekh, "SAR-ECCM using phase-perturbed LFM chirp signals and DRFM repeat jammer penalization," *IEEE Transactions on Aerospace and Electronic Systems*, vol. 42, no. 1, pp. 191–205, 2006.

[6] L. Wei, Y. Xiaopeng, Z. Chao, L. Shuai, and N. Liyue, "Drfm range false target cancellation method based on slope-varying lfm chirp signal," in *2016 IEEE 13th International Conference on Signal Processing (ICSP)*, 2016, pp. 1629–1632.

REFERENCES

[7] Hongyan Yang, Zhen Liu, Xizhang Wei, Dongping Liao, and Bo Peng, "Anti-jamming range imaging using slope-varying lfm signal based on compressed sensing," in *2015 3rd International Workshop on Compressed Sensing Theory and its Applications to Radar, Sonar and Remote Sensing (CoSeRa)*, 2015, pp. 199–203.

[8] J. Akhtar, "Orthogonal block coded ECCM schemes against repeat radar jammers," *IEEE Transactions on Aerospace and Electronic Systems*, vol. 45, no. 3, pp. 1218–1226, 2009.

[9] H. Dai, X. Wang, and Y. Li, "Novel discrimination method of digital deceptive jamming in mono-pulse radar," *Journal of Systems Engineering and Electronics*, vol. 22, no. 6, pp. 910–916, 2011.

[10] T. Bayes, "An essay towards solving a problem in the doctrine of chances," *Philosophical Transactions of the Royal Society; Series A, Containing Papers of a Mathematical or Physical Character*, vol. 53, no. 1, pp. 370–418, 1763.

[11] J. Neyman and E. S. Pearson, "On the problem of the most efficient tests of statistical hypotheses," *Philosophical Transactions of the Royal Society; Series A, Containing Papers of a Mathematical or Physical Character*, vol. 231, no. 1, pp. 289–337, 1933.

[12] M. Barkat, *Signal Detection and Estimation*. Norwood, MA: Artech House, 2nd Edition, 2005.

[13] M. S. Skolnik, *Introduction to Radar Systems*. New York, NY: McGraw Hill Education, 3rd Edition, 2002.

[14] F. E. Nathanson, *Radar Design Principles*. New York, NY: McGraw Hill Education, 2nd Edition, 1991.

[15] H. C. Yin and P. K. Huang, "Further comparison between two concepts of radar target angular glint," *IEEE Transactions on Aerospace and Electronic Systems*, vol. 44, no. 1, pp. 372–380, 2008.

[16] W. Jun-Jie, Z. Feng, L. Jin, A. Xiao-Feng, and W. Lian-Dong, "Repeating decoys discrimination based on range glint," in *2016 Sixth International Conference on Instrumentation Measurement, Computer, Communication and Control (IMCCC)*, 2016, pp. 964–967.

[17] W. Blair and M. Brandt-Pearce, "Discrimination of target and rgpo echoes using frequency diversity," in *Proceedings The Twenty-Ninth Southeastern Symposium on System Theory*, 1997, pp. 509–513.

[18] S. Hochreiter and J. Schmidhuber, "Long short-term memory," in *Neural Computation (1997)*, vol. 9, no. 8, 1997, p. 1735–1780.

REFERENCES

[19] Y. Qin, J. Yang, M. Zhu, and Y. Li, "Fast recognition of pull-off jamming using lstm," in *2019 IEEE International Conference on Signal, Information and Data Processing (ICSIDP)*, 2019, pp. 1–5.

[20] J. Hou, X. R. Li, V. P. Jilkov, and Z. Jing, "Sequential detection of RGPO in target tracking by decomposition and fusion approach," in *2012 15th International Conference on Information Fusion*, 2012, pp. 1800–1807.

[21] J. D. Townsend, M. A. Saville, S. M. Hongy, and R. K. Martin, "Simulator for velocity gate pull-off electronic countermeasure techniques," in *2008 IEEE Radar Conference*, 2008, pp. 1–6.

[22] W. Li, W. Liu, L. Guo, and X. Yang, "Joint slow/fast-time pulse diversity method for countering range gate pull off jamming," in *IET International Radar Conference 2015*, 2015, pp. 1–4.

[23] M. Greco, F. Gini, and A. Farina, "Radar detection and classification of jamming signals belonging to a cone class," *IEEE Transactions on Signal Processing*, vol. 56, no. 5, pp. 1984–1993, 2008.

[24] P. Hill and V. Truffert, "Statistical processing techniques for detecting drfm repeat-jam radar signals," in *IEE Colloquium on Signal Processing Techniques for Electronic Warfare*, 1992, pp. 1/1–1/6.

[25] M. Nouri, M. Mivehchy, and M. F. Sabahi, "Novel anti-deception jamming method by measuring phase noise of oscillators in lfmcw tracking radar sensor networks," *IEEE Access*, vol. 5, pp. 11 455–11 467, 2017.

[26] J. Mittermayer, R. Lord, and E. Borner, "Sliding spotlight sar processing for terrasar-x using a new formulation of the extended chirp scaling algorithm," in *IGARSS 2003. 2003 IEEE International Geoscience and Remote Sensing Symposium. Proceedings (IEEE Cat. No.03CH37477)*, vol. 3, 2003, pp. 1462–1464.

[27] C.-b. Yin and D. Ran, "Converse beam cross sliding spotlight sar imaging processing with data-blocking based fast back projection," in *2016 IEEE International Geoscience and Remote Sensing Symposium (IGARSS)*, 2016, pp. 1070–1073.

[28] J. Xin, Y. C. Bin, Q. W. Dong, and G. Z. Hai, "Anti-jamming property of converse beam cross sliding spotlight SAR," in *2009 2nd Asian-Pacific Conference on Synthetic Aperture Radar*, 2009, pp. 548–551.

About the Author

Phillip Pace is a distinguished professor (emeritus) in the Department of Electrical and Computer Engineering at the Naval Postgraduate School (NPS) and has also taught as an adjunct professor, at Southern Methodist University (SMU), Dallas, TX. He was a senior scientist at L3Harris Technologies, Advanced Systems & Technologies, Plano, Texas from Jan. 2020 to March 2022.

He received B.S. and M.S. degrees from the Ohio University in 1983 and 1986, respectively, and a Ph.D. from the University of Cincinnati in 1990 — all in electrical and computer engineering. Prior to joining NPS, he spent 2 years at General Dynamics Corporation, Air Defense Systems Division, as a design specialist in the Radar Systems Research Engineering department. Before that, he was a member of the technical staff at Hughes Aircraft Company, Radar Systems Group, El Segundo, CA for 5+ years.

Dr. Pace was the Chairman of the N2/N6 Threat Missile Simulator Validation Working Group for 21 years (from Oct. 1996 to Sept. 2017) and was one of four members on the Navy's NULKA Blue Ribbon Panel in Jan. 1999. He is the founding director of the NPS Center for Joint Services Electronic Warfare, 1993 and the author of the books, *Advanced Techniques for Digital Receivers*, 2000, and *Detecting and Classifying Low Probability of Intercept Radar*, 2004, and the 2nd Edition in 2009 all through Artech House. He has 80+ refereed conference papers, 11 patents, and 36 refereed transactions, journal, and magazine publications.

Dr. Pace was a technical editor, *IEEE Trans. on Aerospace and Electronic Systems,* from 2013 to 2020. and has been a principal investigator on research projects in the areas of microwave-photonics, digital and photonics/IR transceiver design and signal processing, electromagnetic warfare, and weapon systems analysis. He was the recipient of the Department of Defense *Superior Civilian Service Medal* and is a life member of the AOC and is a Fellow of the IEEE.

Index

A
A-to-I (analog-to-information) defined, 142
activation functions, 574
active decoy EA
    against SAR, 757
Agent
    reward-feedback, 684
ADRV9009,
    Analog Devices example transceiver, 180
ADC
    analog LPF, 211
    analog LPF, stopband attenuation, 211
    CMOS, FinFET, 329
    digital sampling clock jitter, 202
    dynamic range summary, 237
    ENOB, 236
    figures of merit (FOM), 323–328
    flash, 222–229
    flash, calibration, 227
    flash, fat tree encoder, 223
    flash, ROM encoder, 223
    flash, Wallace tree encoder, 227
    gain and offset errors, 215
    in early DRFM development, 6
    jitter, 199
    least significant bit, 208
    mathematical model, 212
    maximum slew rate, 200
    noise floor, 228
    noise floor, magnitude squared, 232
    noise floor, magitnude squared example, 233
    quantization error, 234
    quantization noise, 213
    real-life model, 196
    sample time uncertainty, 199
    sampling clock uncertainty, 202
    sample time uncertainty, compared to fundamental limit, 203
    SFDR, 294, 296, 304, 309
    SiGe BiCMOS, 289
    SINAD, 239
    SNR, 239
    THD, 240
    time interleave pipeline, 310
    transfer function, 212
    transfer function, characteristic, 214
ADC, DAC
    CMOS, FinFET, 330
ADC, DAC accuracy
    absolute, 210
    relative, 210
ADC, DAC linearity
    differential, 211
    gain error, 211
    integral, 211
    operational amplifiers, 218
AESA (Active Electronically Scanned Array)
    advantages of, 87
    analog beamforming, 82–83

AESA (Active Electronically Scanned Array) (continued)
  array factor, 88
  beamforming concepts, 81–87
  beamwidth, 89
  digital beamforming, 83–84
  fully digital array, 127
  future materials, architectures, 127
  gain, 90, 91
  hybrid beamforming, 84–86
  hybrid beamforming, use in MIMO, 87
  hybrid beamforming, use in MIMO OFDM, 85
  integrated sidelobe level, 91
  interface circuits,receiver, 138
  linear array theory, 88
  noise figure, 123
  performance comparison, 107
  transmit/receive module, 107
  use in UAV's, requirements, 92
  T/R module, analog beamforming, 82
  typical operating frequencies, 81
AI
  generative adversarial nets, 720
AI control
  autonomous subnets, 713
  cognitive subnets, 713
  air-to-surface EMW engagement, example of, 681
Airborne EMW
  different directions, 9–10
AN/APS-137 ISAR
  U.S.S. Crockett, 785
analysis/synthesis kernel, 52
antenna aperture
  effective area, 707
anthropomorphic intelligence
  combat agents, 715
Artificial intelligence, 595
  in commercial products, 15
  tactics, techniques, and procedures, 563
ASCM ISAR seeker advantages, 763
asymmetrical CMOS switch, 140
atomic decomposition, 558
  atoms, 559
autonomous modulation, 568
autonomy, 563
autonomous mission planning AI, 717
autonomous navigation

deep reinforcement learning, 683

B
bandit problems
  Gittins index, 507
Beamwidth
  azimuth, 90
  elevation, 90
blind compressive sensing w/o a priori basis matrix, 153
Boltzmann machine, 648
Boyd, John (Col.)
  theory of energy-maneuverability, 11
broadband techniques
  polarization agility, 102
  spectrum management, 102

C
C4ISR, 3
  open system standards, part of, 13
CASCADE, 16
centralized EMS sensing, 500
chaff
  aluminized glass, 703
  distraction (MK-216), 704
  seduction (MK-214), 704
chaff, distraction (MK-216), 695
chaff, seduction (MK-214), 695
channelized kernel, 51
channelizer
  bandpass sampling, 154
  improving frequency resolution, options for, 165
  nonmaximally decimated, 164
  perfect reconstruction, use of nonmaximally decimated, 164
  polyphase analysis synthesis, 159
  split N−point FFT, 170
  split N−point FFT, use in FPGA, 172
  using noble identity, 167
channelizers
  use of resampling filters, 161
Chris Haynes
  DRFM patent, early developer, 5
circulator, 138
classification results
  FMCW, 592
  Frank code, 593
  P4, 594
close-in-weapon system (CIWS), 695

INDEX

CMOS
    inverter, nMOS, pMOS, 139
CMOS FinFET transistors, 329
CNN
    11-input confusion matrix, 721
cognitive AI
    control unmanned combat systems, 714
    waveform control, 714
cognitive EMW
    described, 11–13
    SDRs, use of, 12
Cohen's time-frequency
    Choi-Williams, 511
    continuous Wigner-Ville, 515
    kernel function, 513
    Wigner-Ville distribution, 514
comparator circuits, 219-220
    charge steering, 304
    comparison, 222
    regenerative latch, 304
    time interleave pipeline, 316
    ultra-low power, 304
comparator design, 279–285
    double-tail, clock gating, 281
    double-tail, conventional, 281
    latched, 282
    two-stage open-loop, 278
combat agents
    anthropomorphic intelligence, 715
commercial technologies, 14-15
    compare trackers (range, velocity), 755
    bit error probability, 503
    test statistic, 502
compressive sensing
    l_1 minimization, 146
    examples, 146
    interpolation filter, in NYFR, 146
    interpolation filter, output, 183
    Nyquist folding A2I, 145
    Nyquist zones, 147
    photonic techniques, 438–469
CONCERTO, 3
    meeting SWaP requirements, 16
convolutional EA
    N-channels, 728
convolutional neural network (CNN)
    classification, 720
convolutional noise EA
    against SAR, 730

cooperative sensing coordinated on/off blinking, 712
coordinated RGPO-VGPO, 753, 755
    compare trackers (range, velocity), 755
CUSUM
    in quickest detection, 504
cyclostationary analysis
    spectral correlation density, 545
cyclostationary spectral analysis, 544
cyclostationary algorithms
    channel pair regions, 551
    direct frequency smoothing, 553
    discrete time, 550
    FFT accumulation method, 550
    test signals, 556
    cyclic autocorrelation, 544
    spectral correlation density, 547

D

DAC
    in early DRFM development, 6
    inverted R−2R, 246
    lowpass filter, 240
    mathematical model, 239
    R−2R, 246
    switched capacitor, 346
    time-interleaved $\Sigma\Delta$, 339
    transfer function, 241
    weighted resistor, 245
DARPA program
    Alpha AI, Deep Green, 715
    CONCERTO, 15
    Mosaic, 15
DDPG, 689
DDS CORDIC vs. ROM
    SFDR comparison, 359
deep Boltzmann, 651
deep learning
    antenna modulation, 614
    auto encoder, 602
    autoencoder, 603
    CNN antenna scan, 617
    CNN for radar scan, 615
    CNN LPI modulation, 621
    convolutional NN, 602, 608
    deep belief nets, 601
    EMW apps, 601
    generative adversarial, 601
    LPI recognition, 619
    recurrent NN, 602

deep learning (continued)
    regularization, 606
    variational auto encoder, 607
decentralized EMS sensing, 500
deception EA
    2D noise patches, SAR, 726
    against SAR, 726
    multi-channel convolution noise, 726
    multiplicative noise EA, 725
deception jammer
    defensive EA, 697
deception repeater
    chirp gate stealer (CGS), 810
    combo VGS RSAM (VGS RSAM), 810
    multiple frequency repeater (MFR), 810
    narrowband repeater noise (NBRN), 811
    random Doppler (RD), 811
    repeater swept amp modulation (RSAM), 810
    velocity gate stealer (VGS), 810
Deep Learning
    Navigation Control Strategy, 683–691
deep reinforcement learning
    autonomous navigation, 683
    combat agents, 715
    unmanned combat systems, 715
defeating anti-ship missiles surface EA, 694
defensive EA, 697
detecting DRFM decoys
    Neyman-Pearson, 842
detector
    cutoff frequency, 394
    detectivity (D°), 393
    figures of merit (FOM), 392
    noise equivalent power (NEP), 392
    photodiode schematic, 392
    responsivity, 393
detection methods, 509
    Choi-Williams, 511
    classification, 511
    Cohen's time-frequency distributions, 511
    continuous Wigner-Ville, 515
    for efficient jammer, 511
    kernel function, 513
    time-frequency distribution, 511
    transceiver, 510
    Wigner-Ville distribution, 514
digital image synthesizer (DIS), 767
    architecture, 768
    phase, gain coefficients, 769
    structured false target, 767
    warfighter payoffs, 767
digital receiver/exciter, 179
diplexer, 138
direct digital synthesis (DDS), 353
    elementary version, 353
    phase wheel, 355
    using CORDIC, 359
DIS
    bit-level simulation, 772
    complex range bin processor, 769
    complex signal reconstruction, 779
    CORDIC, 770
    CORDIC algorithm, 772
    derivation, test target, 776
    gain coefficients, 777
    ISAR with LFM, 772
    overflow problem solutions, 788
    RCS gain shift, 781
    summation adder, 781
    test target, 776
    U.S.S. Crockett, 775
DIS false target
    CMOS ASIC, 791
    realistic coefficients, 791
DIS sea clutter
    coefficients, airborne, 792
    coefficients, modified index of refractivity, 804
    coefficients, Nathanson, 794
    coefficients, phase, gain, 801, 804
    KA distribution., gamma, Poisson, 798
distributed sensing, 499
    spatial diversity order, 499
    double sideband kernel, 50
    coefficients, airborne, 792
    coefficients, modified index of refractivity, 804
    coefficients, Nathanson, 794
    coefficients, phase, gain, 801, 804
    KA distribution, gamma, Poisson, 798
DRFM
    compressive sensing, 143, 50
    phase sampling architecture, 263–264
    technical issues, 248–264

INDEX

early development, 5
in electromagnetic warfare, 3
requirements, 47
wideband software-defined, 143
DRFM EA
adversary attack, four steps, 20
DRFM architectures
analysis/synthesis kernel, 52
channelized kernel, 51
double sideband kernel, 50
phase-sampling kernel, 52
received power, 69
requirements, 47
robust design with AI and ML, 71
single sideband kernel, 48
types, 47–61
DRFM EMW
diagram of EA, EP and ES activities, 9
spectrum warfare, 11
DRFM oscillators phase noise, 360
DRFM receivers
architectures, 142, 179
DRFM problems
clock jitter, 251
image frequencies, 250, 262
in frequency conversion, 260
intermodulation distortion, 261
LO leakage, 249
mixer isolation, 262
reducing spurious signals, 258
spurious frequencies, 249
spurious signals, 252
spurious signals, 1-bit, 252
spurious signals, 2–5 bits, 253
spurious signals, summary, 256
third-order intercept, 262
tuning errors, 2
DRFM security
federated learning, 661
DRFM transceiver
designer challenges, 270
direct sampling, SNR for, 267
electronic support, 270
figures of merit, 264, 268
flash ADC, requirements, 269
requirements, 266
spurious signal comparison, quant.
scheme, 89
Douglas sea state, 795
DRL agent
goal, 684

duplexer, 138
DSP & FPGA, use of, 61–63
dual-port memory, 330
FinFET for SRAM, 331
SRAM configurations, 332
SRAM MOSFET, 333
SRAM NEMFET, 334
duplexer, 138
dynamic programming
Bellman equation, 506
spectrum sensing, 506
DyNAMO, 16

E
EA-18 (Growler)
AN/ALQ-99 jammer pods, 10
Boeing EA platform, 10
EA-18G (Growler), 10
EA
AI, 697
antiradiation missiles, 696
deception (DJ), 694
directed energy weapons, 697
EMSO, MALD, 681–691
escort (EJ), 694
F-16CJ, 693
Growler, 693
modified escort (ME), 694
mutual support, formation, 694
Prowler, 693
self-screening (SSJ), 694
stand-in (SJ), 694
standoff (SOJ), 694
UAVs and DRFMs, 681
EA deception
deception modulator, 758
effective isotropic radiated power
(EIRP), 89
effective isotropic radiated power density
(EIRPD), 89
Electromagnetic Maneuver Warfare
(EMMW)
described, 10–11
electromagnetic maneuver warfare
(EMW)
as an add-on activity, 5
electromagnetic warfare
role of the DRFM, 3
electro-optical sampling
amplitude jitter, 405

electro-optical sampling (continued)
   laser pulse limitations, 407
   MZM bias controller, 405
   sampling error, 404
   temporal jitter, 406
electronic attack (EA)
   as part of EMW, 8
electronic protection (EP)
   as part of EMW, 8
electronic support (ES)
   as part of EMW, 8
   high power microwaves, 270
   high temperature superconductors, 272
   receiver considerations, 270
EMS, 4–6
   diagram showing different bands, 6
EMS frequency bands
   models for, 12
EMW
   applications, 601
   unique role, 8
   airborne assets, 9
   cognitive objective, 12
   defined, 7
   new categories of, 8
   offensive (EA) and defensive (EP) activities, 9
epochs, 591
ES
   unknown sensor emissions, 17

F
false target
   using transponder, 760
false-targets
   DRFM technique, 763
fast-RDPG, 688, 689
feature extraction, 577
federated learning
   DRFM security, 661
fiber laser
   erbium-doped fiber amplifier (EDFA), 383
   Fabry-Perot, 383
   mode locking, 382
   NPS sigma laser, 391
   NRL sigma laser, 389
   polarization maintaining (PM) fiber, 386
   ring laser, 383

   sigma MLL, 384
figures of merit (FOM)
   Walden/Murmann, 266
flash ADC, 285
   24 GS/s CMOS, 284
   24 GS/s FDSOI CMOS, 285
   4-bit thermometer code, 294
   4-bit-to-Gray encoder, 294
   40-Gb/s optical receiver, 291
   interleave techniques, 296
fractal antenna
   antenna shapes, 106
   multiband planar, 105
fractal antennas
   fractal, meaning of word, 104
Fresnel zone
   spreading loss, 706
   wavelength dependence, 706
frequency diversity
   counter-DRFM, 855

G
gated recurrent units, 645
generative adversarial nets (GAN), 652
   detection and classification, 720
   EA, smart interference, 724
   interference EA, 720
   interference generation, 723, 724
   preprocessing radar signals, 722
gradient decent, 683
   SAR ATR, 653
Global positioning system (GPS)
   antennas, 64–67
   codes used, 68–69
Global positioning system receivers, 63–76
   graphical processing unit (GPU)
   circular shift ×2 oversampling, 175
   DDR4 SDRAM, resources for, 15
   filter kernel operation, 176
   multi-instance GPU (MIG), 14
   overlapping data across kernel launches, 176
   part of open systems technology, 13
   polyphase channelizer, 173, 174
Greenert, Jonathan ADM
   30th CNO (2011–2015), 11

H
Hamming, Richard, 564

hard-kill chain
    steps involved, 761
hidden units, 573
high speed anti-radiation missile (HARM)
    electronic attack, 10
high-gain high sense (HGHS), 696
hook concept
    RGPO, 748
HOST
    open system standards, part of, 13
hypothesis testing
    counter-DRFM, 845

I
IDECM, 704
    countermeasure dispenser
        (AN/ALE-47), 704
index policy
    sample mean term, part of cognitive
        processing, 13
integrated defensive electronic
        countermeasures (IDECM), 704
intelligence, surveillance, and
        reconnaissance (ISR), 697
InTop, integrated topside, 696
inverse SAR
    signal processing, 698
Intrepid Tiger II
    AN/ASQ-239 payload, 10
ISAR
    AI, 699
    false test target, 788
    image compression, 782
    JEM line detection, 704
    signal processing, 702
ISAR JEM
    lines, 700
ISAR, 827
ISAR seekers
    for ASCMs, 763
ISR, 16

J
jam-to-signal ratio (JSR)
    repeater, 740
    SSJ, 708
JEMSO, 696
    DRFM importance, 691
jet engine modulation (JEM), 700
jitter

fundamental limit, 202
joint electromagnetic spectrum operations
        (JEMSO)
    (JP) 3-85, 691
JSR, repeater
    Poynting vector, 740

L
layered defense, 695
laser
    fiber mode locked laser (MLL), 381
    pulse train, 381
    pump actions, 380
    transitions, 380
loss function, 625
LSTM, 631
    InSAR Sentinel-1, 642
LSTM AI
    counter-DRFM, 850
    RVGPO, 850
LSTM forget gate
    counter-DRFM, 852
LSTM input gate
    counter-DRFM, 852

M
Mach-Zehnder modulators (MZM)
    extinction ratio, 400
    half-wave voltage $V\pi$), 400
    number of folds, 400
    schematic, push-pull, 377
    thin film, 401
    transfer function, 399
    velocity matching, 401
machine learning (ML), 595
    TIDAC calibration, 349
machines think, 564
MALD
    SAM's launcher location, disclosed,
        683
MALD (ADM-160A/B), 683
MAYO clinic
    early DRFM, 6
marginal frequency, 581
Markov decision process (MDP), 683
Markov process
    positively correlated, 509
mars chip, 566
Max Planck
    constant, 4

Max Planck (continued)
  law of thermal radiation, 4
maximum slew rate, 200
Mercury Defense Electronics
  Recent 4-bit DRFM, 7
microwave-photonic transceiver
  crystal technologies, 376
  detector dark current, 390
  detectors, 389
  digital antenna system, 376
  Henry Taylor, ADC, 402
  LiNbO3 crystal, 377
  Mach-Zehnder modulator (MZM), 377
  MLL master equation, 385
  optical link, 394
  oscillator, 396
  phase noise, 396
  photonic ADC, 425
  photonic antennas, 412
  photonic memory, 407
mixer design
  balanced bridge, 260
model parallelism, 597
multiplicative noise
  control signals, 725
multi-task learning, 599
multilayer perceptron, 573
mode-locked laser (MLL)
  master equation model, 385
modulated wideband converter
  described, 149
  reconstruction techniques, 152
  related to spread spectrum, 150
Mosaic, 3
  tile approach, 16
Mosaic program
  description, 15–17
multi-DRFM
  game theory, 506
  spectrum sensing, 506
multifunction AESA
  conformal system, metamaterials, 103
  fire control radar, T/R module counts, 95
  fire control radar, used on, 93
  phase comparison monopulse theory, 96
multifunction DRFM
  AESA interface, 137
  MIMO, 137

multifunction phased array (MPAR) panel
  assembly, 97

N
Naval Postgraduate School (NPS)
  fiber sigma laser, 389
network latency
  selective-reactive EA, 718
Neural Networks
  for use in EMW, 565
neuromorphic computing, 667
Neyman-Pearson detection, 841
  in cognitive processing, 12
  likelihood ratio, 501
nonlinear classification
  networks for, 569
Nulka, 695
NVIDIA corporation
  CUDA parallel computing, 172
  commercial DRFM technologies, 14
  part of open systems technology, 13
  GPU cores, 172
  MAC operations, increased performance, 177
NYFR, 182
  Nyquist folding receiver, 145

O
objective function, 837
obscuration
  appendix A, 705
  barrage noise, 705
  barrage swept amplitude modulation (BSAM), 809
  beacon equation, 705
  burnthrough range, 710
  chaff, 703
  continuous barrage noise (BAR), 808
  continuous spot noise (SPT), 808
  cross-over range, 710
  escort EA, 712
  intermittent barrage (BCDB), 809
  intermittent SPT, 808
  multi-spot noise, 705
  mutual support EA, 712
  noise, 705
  noise modulation, 715
  noise swept amplitude modulation, 808
  spot noise, 705
  stand-off jammer (SOJ), 709

INDEX 889

swept noise (SWPT), 809
swept spot noise, 705
obscuration EA, 703–712
  multi-channel convolutional, 728, 731
  multiplicative, 731
offensive engagement
  self-protection, 708
  self-screening, 708
offensive jamming
  support an attack, 694
  save your skin, 693
OODA advantage
  intelligent strike, 714
OODA loop
  S-300V AAD reconnaissance, 682
Open System Standards
  described, 13–14
open systems
  FPGA, DSP boards, 15
open-systems electronics standards, 13
Optisys Ka-band tracking array, 128
orthogonal coding, 825
oscillator choices, 361
  Allan variance, 396
  photonic LO, 396

P

partially observable Markov decision process (POMDP), 683
periodic structures
  creating EM response, 129
persistent sensing, 498
  classification in, 498
  network-enabled, 498
phase noise
  definition, 362
  DRFM oscillators, 360
  spectral density, 363, 366
phase noise measurements
phase noise measurements
  counter-DRFM, 874
phase-sampling kernel, 52
  comparator comparison with amplitude analyzing kernel, 54
  phase quantization levels vs. angular resolution and no. amplitude states, 56
photonic ADC
  2-bit Taylor demo, 428
  4-bit Taylor scheme, 428
  aperture jitter, early estimates, 426
  high resolution scheme, 432
  optimum symmetrical number system (OSNS), 432
  robust symmetrical number system (RSNS), 433
  Taylor scheme, 427
  Taylor scheme limitations, 430
photonic antenna
  flared monopole, 415
  phased array beamforming, 420
  photo-diode μ–strip patch, 413
  quasi-Yagi, 416
  SSBFC, 422
photonic channelizer
  fiber Bragg grating, 469
photonic CS
  detection latency, 468
  double MZM, extending bandwidth, 465
  double parallel MZM, 461
  orthonormal basis matrix, 459
  sparsity, 459
photonic CS techniques
  Otsu's noise reduction, 454
  pNYFR, 442
  pNYFR, hardware results, 452
  pNYFR, implications, 448
  pNYFR, two MZMs, 444
photonic fiber Bragg transmission intensity, 472
photonic memory
  Doppler frequency control, 409
  extended storage time, 408
  FS-RDL, 408
  optical BPF, 410
photonic multiplier, 427
photonic sampling
  phase-encoded, 432
photonic sensing
  direct conversion, 474
photonic spectrum sensing, 467
planar-fed folded notch (PFFN)
  comparison, conventional notch, 99
  metamaterials, used for cooling, 100
polarization discrimination, 831
POMDP, 683–686
  generalization, MDP, 684
  recurrent deterministic policy gradient (RDPG), 686

power functions
  hypothesis testing, 846
power centroid tracking error, 711
Poynting vector
  pull-off recognition
    counter-DRFM, 858
  dB analysis, 707
principal components, 586

Q

quadrature conversion process, 190–193
quadrature mirror filter bank
  QMFB tree receiver, 533
quadrature mirror filtering
  complex signal example, 534, 537
  modified sinc filter, 531
  tree receiver, 542
  wavelet filters, 530
  wavelet transform, 528
quickest detection
  Bayesian, non-Bayesian, 504
  cumulative sum CUSUM, 504
  Markov decision process, 505

R

range-Doppler imaging
  signal processing, 697
RDPG, 689
  recurrent neural networks, 686
receive process, 196–208
recurrent deterministic policy gradient
  (RDPG), 686, 689
reinforcement learning, 717
  autonomous UAV navigation, 683
reinforcement learning (RL)
  stochastic action policy, 684
reinforcement learning (RL)
  deep RL (DRL), 683
reinforcement signal
  reward-feedback, 684
repeater constant gain, 739
  EA, 739
  range gate pull-off, 741
repeater JSR
  range independence, 742
restricted Boltzmann, 649
RGPO
  algorithm description, 743–750
  applying the hook, 748
  delay kinematics, 747

magnitude spectrum, expression for, 744
pull-off time delay, characteristics of, 746
  time delay functions, 743
  to chaff EA, 748
ROC
  counter-DRFM, 857
RVGPO
  LFM, expression with, 755
Rosenblatt, 569
Rosenblatt perceptron, 570

S

sample time uncertainty, 199
sampling circuit
  closed-loop S/H, 208
  InP HBT and Si CMOS closed-loop S/H, 206
  open-loop differential CMOS S/H, 205
  open-loop S/H, 206
  sample and hold (S/H), 205
sampling operations, 196
SAR, 826
SAR squint map, 729
sea sparrow, 695
selective-reactive EA, 718
  network latency, 719
self-propelled decoys, 696
self-screening jammer
  defensive EA, 708
self-screening jammer (SSJ)
  self-protection, 708
sequential detection
  in cognitive processing, 12
  bandit problems, 507
  composite hypotheses, 504
sequential sensing
  quickest detection, 503
  sequential probability test, 504
Sheridan level, 567
Sheridan levels, 565
Shiryayev's sequential probability ratio
  counter-DRFM, 856
slope varying LFM, 825
single DRFM
  spectrum sensing, 505
single sideband kernel, 48
sample time uncertainty, 200
sampling circuit

closed-loop S/H, 208
 InP HBT and Si CMOS closed-loop
  S/H, 208
 open-loop differential CMOS S/H, 205
 open-loop S/H, 206
 sample and hold (S/H), 205
sampling operations, 196
single perceptron, 569
 logic units, 572
 sigmoid, 570
superheterodyne kernel, 48
software defined radios (SDRs), 11
SOSA
 open system standard, 13
 open system standards, part of, 13
 openVPX, 13
spatial-spectral sensing
 crystal photonics, S2, 478
 cueing receiver, 483
specific emitter ID (SEI), 696
spectrum dominance
 UAVs, DRFMs, 681
spreading loss, 706
SRBOC
 MK-36, 704
SSBFC
 four components, 422
stand-in jammer
 model, 684
stand-off jammer
 obscuration, 709
standard missile block, 695
statistical decision theory
 hypothesis testing, 844
statistical estimation
 counter-DRFM, 839
stochastic policies
 policy update, 683
 target function, 687
 value function, 687
structured false target
 ACT, use of, 763
 composite scattering model, 765
 fiber optics, use of, 764
structured false targets
 against range-Doppler imaging, 763
 early technology, 763
super rapid blooming offboard chaff
  (SRBOC), 704
suppression EA
 airborne EA, 693

suppression of enemy of air defense
  (SEAD), 684
surface EA
 defeating anti-ship missiles, 694
surface electronic warfare improvement
  program (SEWIP), 696
swarm
 digital pheromone, MALD, 683
synthetic aperture radar (SAR)
 signal processing, 698
spectrum dominance
 key to quantifying, 3
spectrum sensing
 detection methods, 509
 DRFM fusion center, 500
 game theory, 506
 multi-DRFM, 506
 Neyman-Pearson, 501
 single DRFM, 505
state of the spectrum, 498, 503
STITCHES, 16
SWaP-C, 3

T
target EA with deep learning, 732–759
Taylor weighting, 91
spiking NN, 668
time-frequency methods
 Wigner-Ville distribution, 720
time interleave
 successive approximation (SAR), 316
time interleave ADC, 310
time interleave pipeline
 comparator, 316
 decision stage, 319
 resolution, 312
time interleave SAR
 advantages, 320
 SFDR, 323
time-interleaved DAC
 machine learning calibration, 347
 SFDR, 347
topside, integrated (InTop), 696
tracking error
 two or more DRFMs within beam, 711
transponder
 against RBGM, 735
 constant power, 733
 effects on MPD, 737
 JSR, 758

transponder (continued)
 pulse density timing diagram, 738
 types of EA, 692
  quadrature mirror filtering, 529
  wavelet decomposition, 528
transceiver process
 mathematical model of, 191–198
transmit/receive module
 alternating T/R, 116
 alternating transmit simultaneous
  receive, 118
 early development, 107
 fully digital array, 127
 GaN, 113
 receive power, 110
 schematic diagram, 109
 SiGe BiCMOS, 110
 transmit power, 110
 zero IF example, 119
transmit/receive tech band technologies, 108
types of error
 hypothesis testing, 843

U

unmanned platforms, 563
unmanned systems
 in spectrum warfare, 3

UAS
 control signals, 683
 range finders, 685
 requirements, 16
 spectrum operations, part of, 40
UAV control signals
 sensory inputs, mapped, 683

V

velocity gate pull-off, 750–753
VGPO
 AI, 751

W

Wald
 sequential probability ratio test SPRT, 503
Wiener filtering, 583
Wigner-Ville
 complex signal example, 521
 real signal example, 518
 two-tone results, 525

Z

Zadoff-Chu sequence
 modulated wideband converter, use in, 152

Artech House Electronic Warfare Library

Dr. Joseph R. Guerci, Series Editor

Active Radar Electronic Countermeasures, Edward J. Chrzanowski

Antenna Systems and Electronic Warfare Applications, Richard A. Poisel

Battlespace Technologies: Network-Enabled Information Dominance, Richard S. Deakin

Cognitive Electronic Warfare: An Artificial Intelligence Approach, Karen Zita Haigh and Julia Andrusenko

Developing Digital RF Memories and Transceiver Technologies for Electromagnetic Warfare, Phillip E. Pace

Digital Techniques for Wideband Receivers, Second Edition, James Tsui

Electronic Intelligence, Richard G. Wiley

Electronic Intelligence: The Analysis of Radar Signals, Second Edition, Richard G. Wiley

Electronic Warfare for the Digitized Battlefield, Michael R. Frater and Michael Ryan

Electronic Warfare in the Information Age, D. Curtis Schleher

Electronic Warfare Receivers and Receiving Systems, Richard A. Poisel

Electronic Warfare Receiving Systems, Dennis Vaccaro

Electronic Warfare Signal Processing, James Genova

Electronic Warfare Target Location Methods, Richard A. Poisel

ELINT: The Interception and Analysis of Radar Signals, Richard G. Wiley

Emitter Detection and Geolocation, for Electronic Warfare, Nicholas A. O'Donoughue

EW 101: A First Course in Electronic Warfare, David L. Adamy

EW 102: A Second Course in Electronic Warfare, David L. Adamy

EW 103: Tactical Battlefield Communications Electronic Warfare, David L. Adamy

EW 104: EW Against a New Generation of Threats, David L. Adamy

EW 105: Space Electronic Warfare, David L. Adamy

Foundations of Communications Electronic Warfare, Richard A. Poisel

Fundamentals of Electronic Warfare, Sergei A. Vakin, Lev N. Shustov, and Robert H. Dunwell

Information Warfare and Electronic Warfare Systems, Richard A. Poisel

Introduction to Communication Electronic Warfare Systems, Richard A. Poisel

Introduction to Electronic Defense Systems, 3rd Edition, Filippo Neri

Introduction to Electronic Warfare, D. Curtis Schleher

Introduction to Electronic Warfare Modeling and Simulation, David L. Adamy

Introduction to Modern EW Systems, 2nd Edition, Andrea De Martino

Military Communications in the Future Battlefield, Marko Suojanen

Modern Communications Jamming Principles and Techniques, Richard A. Poisel

Modern Communications Receiver Design and Technology, Cornell Drentea

Practical ESM Analysis, Sue Robertson

Radar Electronic Countermeasures, Stephen L. Johnston

RF Electronics for Electronic Warfare, Richard A. Poisel

Synthetic Aperture Radar and Electronic Warfare, Gisela Goj

Tactical Communications for the Digitized Battlefield, Michael Ryan and Michael R. Frater

Target Acquisition in Communication Electronic Warfare Systems, Richard A. Poisel

For further information on these and other Artech House titles, including previously considered out-of-print books now available through our In-Print-Forever® (IPF®) program, contact:

Artech House
685 Canton Street
Norwood, MA 02062
Phone: 781-769-9750
Fax: 781-769-6334
e-mail: artech@artechhouse.com

Artech House
16 Sussex Street
London SW1V 4RW UK
Phone: +44 (0)20-7596-8750
Fax: +44 (0)20-7630-0166
e-mail: artech-uk@artechhouse.com

Find us on the World Wide Web at: www.artechhouse.com